HANDBOOK OF

Biological Effects of Electromagnetic Fields

SECOND EDITION

HANDBOOK OF

Biological Effects of Electromagnetic Fields

SECOND EDITION

Edited by
Charles Polk
Elliot Postow

CRC Press
Boca Raton New York London Tokyo

Library of Congress Cataloging-in-Publication Data

Handbook of biological effects of electromagnetic fields / edited by
Charles Polk, Elliot Postow. -- 2nd ed.
p. cm.
Includes bibliographical references and index.
ISBN 0-8493-0641-8 (alk. paper)
1. Electromagnetic fields--Physiological effect--Handbooks,
manuals, etc. I. Polk, Charles. II. Postow, Elliot.
QP82.2.E43C73 1995
574.19'17—dc20 95-32947
CIP

International Standard Book Number 0-8493–0641-8
Library of Congress Card Number 95-32947
Printed in the United States of America 1 2 3 4 5 6 7 8 9 0
Printed on acid-free paper

FOREWORD TO THE FIRST EDITION

The objective of this book is to present in a concise manner what is actually known at the present time about the biological effects of time invariant, low frequency, and radio frequency (including microwave) electric and magnetic fields. In reviewing the vast amount of experimental data which have been obtained in recent years, the authors tried to select those results that are, in their opinion, of major importance and of lasting value. In discussing mechanisms of interaction of electromagnetic fields with living matter, they have tried to differentiate between what is clearly established, what is suggested by available evidence without being convincingly proven, and what is conjecture at the present time.

No quantitative discussion of the interaction of electric and magnetic fields with living matter, and in fact no reasonable qualitative discussion, is possible without knowledge of the macroscopic electric and magnetic properties of living tissue. Therefore, in Part I* of the book, theoretical and experimental aspects of dielectric permittivity and electrical conductivity of biological materials are treated in some detail. Description and analysis of the variation of these properties with frequency lead also to a discussion of microscopic level phenomena, although some additional molecular level processes are discussed in subsequent chapters.

Part II of the book is then concerned with phenomena that do not involve radiation, while Part III deals with electromagnetic energy which mostly enters living tissue in the form of radiated electromagnetic waves. The introduction is concerned primarily with two topics: (1) fundamental considerations important for evaluating how much of the extremely low frequency (ELF) to radio frequency (RF) electromagnetic energy present in the environment of an organism can reach its interior, and (2) the difference between "ionizing radiation" and "nonionizing radiation."

Parts I, II, and III of this book are relatively independent of each other, so they can be consulted directly by the reader whose interest is restricted to a particular, specialized topic. On the other hand, the sequence of chapters is such that the book can also be used as a text for a one-semester or one-year course on biological effects of electromagnetic fields. One of the editors (Charles Polk) has used the manuscript in this manner with a group of graduate students at the University of Wisconsin-Madison drawn from Electrical Engineering, Medical Physics, Medicine, and Biology. Prerequisites for enrollment in the course were a good intermediate-level undergraduate course in Electricity and Magnetism and at least an introductory course in Biology. The reader with such background should find this book useful as an introduction to the topics listed in the Table of Contents, as a review of recent research, and as a guide to the specialized literature.

* Consecutive chapter numbers are employed in the Second edition although the division into Parts I, II and III is maintained.

FOREWORD TO THE SECOND EDITION

Since publication of the first edition of this book in 1986, many new research reports on the biological effects of electric and magnetic fields have appeared, some old topics have provoked new interest, some new topics have become important, and new, useful computational methods have become available. We have therefore found it desirable to prepare a new, updated edition of a book that was intended to provide a comprehensive overview of "Bioelectromagnetics" from DC through the radio frequency range.

New chapters now include topics that were treated only briefly within more general reviews in the first edition. Thus, the use of relatively low frequency and pulsed electric and magnetic fields for bone and soft tissue repair is described in chapter 5, because this clinical application of electromagnetics has become more widely employed. Numerous epidemiological studies suggesting possibly adverse health effects of power frequency fields have been completed in recent years; a description and evaluation of these studies, such as now given in chapter 7, must therefore be part of any comprehensive review of what is known about biological effects of extremely low frequency fields. Much progress has been made recently in understanding the interaction of short duration, high intensity electric field pulses with biological cells. This topic is now discussed in chapter 6; "electroporation" has become important not only for understanding the nature of electric injury, but also for local drug delivery to tissue and as a tool in biotechnology for transferring genetic material between different cells.

Several chapters that were already part of the first edition had to be changed substantially to incorporate the description of new developments since 1986. These are chapter 3 on the effects of static magnetic fields, chapter 4 on the effects of extremely low frequency (ELF) magnetic fields, chapter 9 on numerical methods for field computation, chapter 11 on non-modulated microwave fields, and chapter 12 on "window" effects. The untimely death of Professor Sol Michaelson also made it necessary to find a new author for the revision of chapter 11.

Also revised were the chapters that deal with the most fundamental aspects of bioelectromagnetics, which have changed relatively little since 1986. While some new material was included, the emphasis here was more on rewriting to provide more concise discussions that can be followed more easily by the non-specialist. In this category are the following: chapter 1 on dielectric permittivity and electrical conductivity of biological materials; chapter 2 on what is known about physical mechanisms involved in the interaction of DC and ELF electric fields with living matter; chapter 8 on experimental radiowave and microwave dosimetry; and chapter 10 on thermoregulation in the presence of microwave fields.

Appendix 2 on "Safety Standards" needed substantial changes in view of the many new "standards" and "guidelines" published since 1986 by different agencies to insure protection of individuals exposed to electric and magnetic fields from DC through the millimeter wave region.

Some topics dealing with the biological effects of electric and magnetic fields are controversial—perhaps even more so than in 1986. This applies particularly to potential health effects of very low intensity, "non-thermal" fields. In this book, we have tried above all to present known science. As in the first edition, all authors have tried "to differentiate between what is clearly established, what is suggested by available evidence without being convincingly proven, and what is conjecture at the present time." But as editors, we have not acted as censors, and some controversial topics are treated by different authors from different points of view. We believe that this is as it should be in a "handbook" that is designed to provide a comprehensive overview of a science that is still rapidly developing, and where many observations still lack an adequate explanation.

Charles Polk
Elliot Postow

THE EDITORS

Dr. Charles Polk is Professor Emeritus of Electrical Engineering at the University of Rhode Island. He was Chairman of the Electrical Engineering Department at the University of Rhode Island from 1959 to 1979; during leaves, he was Visiting Professor at Stanford University, Stanford, Calif. (1968/1969) and the University of Wisconsin/Madison (1983/1984). From 1975 to 1977 he was Head of Electrical Sciences and Analysis and Acting Director of the Engineering Division of the National Science Foundation in Washington, D.C. A native of Vienna, Austria, Dr. Polk has studied at the University of Paris (Sorbonne) and is a graduate of Washington University in St. Louis (B.S.E.E.) and the University of Pennsylvania in Philadelphia (M.S., Physics; Ph.D., Electrical Engineering). He worked on electrical devices, antennas, and radio propagation at R.C.A. in Camden and Princeton, N.J. He contributed to several books and wrote papers on electromagnetic wave propagation, antennas, electromagnetic noise of natural origin, and interaction of electromagnetic fields with living systems. The Institute of Electrical and Electronic Engineers elected him "Fellow" "for contributions to understanding earth-ionosphere cavity resonances, and for leadership in engineering education." Currently Chairman of the Power Frequency Subcommittee of the IEEE Committee on Man and Radiation, he is also a member of the American Society of Engineering Education, the New York Academy of Sciences, and AAAS, and is a former President (1988/89) of the Bioelectromagnetics Society.

Dr. Elliot Postow is currently Chief of the Technology and Applied Sciences Review Section, Division of Research Grants, National Institutes of Health. Dr. Postow received training in physics at the City College of New York. He earned his B.S. in 1961. He earned his Ph.D. in biophysics at Michigan State University in 1968. He served as Scientific Coordinator with the Office of Naval Research from 1968 to 1972. From 1972 to 1978, he was the Assistant Director of the Electromagnetic Radiation Project Office of the Navy Bureau of Medicine and Surgery. From 1979 to 1988, he was at the Naval Medical Research and Development Command, first as the Program Manager for Research on the Biological Effects of Electromagnetic Radiation and later as the Command's Associate Director for Program Implementation and Review. In 1988, he joined the National Institutes of Health as a Scientific Review Administrator in the Division of Research Grants, and in 1990 he was appointed Chief of the Special Review Section.

Dr. Postow is a member of the American Association for the Advancement of Science and the Bioelectromagnetics Society. He served as president of the Society during 1985–1986 and was the founding editor of its journal, *Bioelectromagnetics*—a position he held for five years. In 1984, he was co-organizer of an international symposium on the Interaction of Electric Fields with Biological Systems, which was sponsored by the International Union of Radio Sciences (URSI).

CONTRIBUTORS

Eleanor R. Adair
Associate Fellow
John B. Pierce Foundation
and
Senior Research Associate in Psychology
Yale University
New Haven, Connecticut

Frank S. Barnes
Professor
Department of Electrical and Computer Engineering
University of Colorado
Boulder, Colorado

Kenneth R. Foster
Associate Professor
Department of Bioengineering
University of Pennsylvania
Philadelphia, Pennsylvania

Richard B. Frankel
Senior Research Scientist
F. Bitter National Magnet Laboratory
Massachusetts Institute of Technology
Cambridge, Massachusetts

James C. Lin
Professor and Head
Department of Bioengineering
University of Illinois
Chicago, Illinois

Sol M. Michaelson
Professor
Department of Radiation Biology and Biophysics
University of Rochester
Rochester, New York

Morton W. Miller
Associate Professor
Radiation Biology and Biophysics Department
University of Rochester
Rochester, New York

Charles Polk
Professor
Department of Electrical Engineering
University of Rhode Island
Kingston, Rhode Island

Elliot Postow
Electromagnetic Radiation Program Manager
Naval Medical Research and Development Command
National Naval Medical Center
Bethesda, Maryland

Herman P. Schwan
Alfred Fitler Moore Professor Emeritus
Department of Bioengineering
University of Pennsylvania
Philadelphia, Pennsylvania

Richard G. Stevens
Cancer Epidemiologist
Life Sciences Center
Pacific Northwest Laboratory
Richland, Washington

Maria A. Stuchly
Research Scientist
Radiation Protection Bureau
Health and Welfare Canada
Ottawa, Ontario, Canada

Stanislaw S. Stuchly
Professor Electrical Engineering
Department Electrical Engineering
University of Ottawa
Ottawa, Ontario, Canada

Mays L. Swicord
Center for Devices and Radiological Health
Food and Drug Administration
Rockville, Maryland

T. S. Tenforde
Senior Scientist
Biology and Medicine Division
Lawrence Berkeley Laboratory
University of California
Berkeley, California

James C. Weaver
Senior Scientist
Harvard University–Massachusetts
Institute of Technology
Division of Health Sciences
and Technology
Cambridge, Massachusetts

CONTENTS

INTRODUCTION

Charles Polk

CONTENTS

I. NEAR FIELDS AND RADIATION FIELDS

In recent years it has become, unfortunately, a fairly common practice—particularly in nontechnical literature—to refer to the entire subject of interaction of electric (E) and magnetic (H) fields with organic matter as biological effects of nonionizing radiation; although fields that do not vary with time and, for most practical purposes, also slowly time-varying fields do not involve radiation at all. The terminology had its origin in an effort to differentiate between relatively low-energy microwave radiation and high-energy radiation, such as UV and X-rays, capable of imparting enough energy to a molecule or atom to disrupt its structure by removing one or more electrons. However, when applied to direct current (DC) or extremely low frequency (ELF), the term "nonionizing radiation" is inappropriate and misleading.

A structure is capable of efficiently radiating electromagnetic waves only when its dimensions are significant in comparison with the wavelength λ. But in free space $\lambda = c/f$, where c = velocity of light in vacuum ($3 \cdot 10^8$ m/sec) and f = frequency in hertz (c/sec); therefore the wavelength at the power distribution frequency of 60 Hz, e.g., is 5000 km, guaranteeing that most available man-made structures are much smaller than one wavelength.

The poor radiation efficiency of electrically small structures (i.e., structures whose largest linear dimension $L \ll \lambda$) can be illustrated easily for linear antennas. In free space the radiation resistance. R_r of a current element, i.e., an electrically short wire of length ℓ carrying uniform current along its length,[1] is

0-8493-0641-8/96/$0.00+$.50

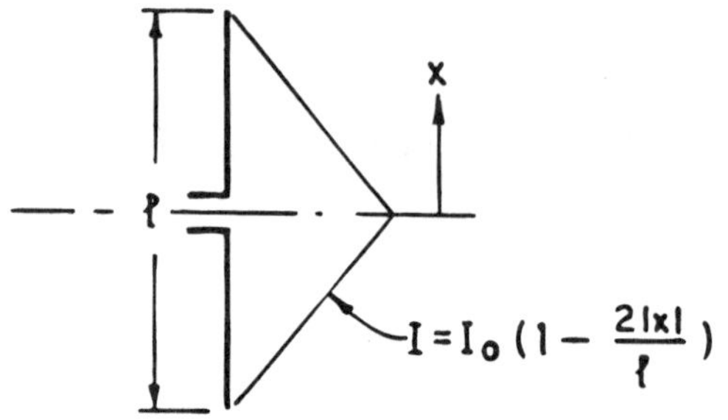

FIGURE 1. Current distribution on short, thin, center-fed antenna.

$$R_r = 80\pi^2\left(\frac{\ell}{\lambda}\right)^2 \quad (1)$$

while the R_r of an actual center-fed radiator of total length ℓ with current going to zero at its ends, as illustrated in Figure 1 is

$$R_r = 20\pi^2\left(\frac{\ell}{\lambda}\right)^2 \quad (2)$$

Thus, the R_r of a 0.01-λ antenna—50 km long at 60 Hz—would be 0.0197 Ω. Since the radiated power $P_r = I^2R_r$ where I is the antenna terminal current, while the power dissipated as heat in the antenna wire is I^2R_d when I is uniform,* the P_r will be very much less than the power used to heat the antenna, given that the ohmic resistance R_d of any practical wire at room temperature will be very much larger than R_r (e.g., the resistance of a 50-km long, $^1/_2$-in. diameter solid copper wire would be 6.65 Ω). At DC, of course, no radiation of any sort takes place, since acceleration of charges is a condition for radiation of electromagnetic waves.

The second set of circumstances which guarantees that any object subjected to low frequency E and H fields usually does not experience effects of radiation is that any configuration that carries electric currents sets up E- and H-field components which store energy without contributing to radiation. A short, linear antenna in free space (short electric dipole) generates in addition to the radiation field E_r, an electrostatic field E_s, and an induction field E_i. Neither E_S nor E_i contribute to the P_r.[2,3]. While E_r varies as $1/r$, where r is the distance from the antenna, E_i varies as $1/r^2$, and E_s as $1/r^3$. At a distance from the antenna of approximately one sixth wavelength ($r = \frac{\lambda}{2\pi}$) the E_i is equal to the E_r, and when $r \ll \lambda/6$ the E_r quickly becomes negligible in comparison with E_i and E_s. Similar results are obtained for other antenna configurations.[4] At 60 Hz the distance $\lambda/2\pi$ corresponds to about 800 km and objects at distances of a few kilometers or less from a 60-Hz system are exposed to nonradiating field components which are orders of magnitude larger than the part of the field that contributes to radiation.

A living organism exposed to a static (DC) field or to a nonradiating near field may extract energy from it, but the quantitative description of the mechanism by which this extraction takes place is very different than at higher frequencies, where energy is transferred by radiation:

1. In the near field the relative magnitudes of E and H are a function of the current or

* For nonuniform current $I = I_0 - \frac{2|x|}{l}$ the dissipated power becomes $I_0^2R_d/3$.

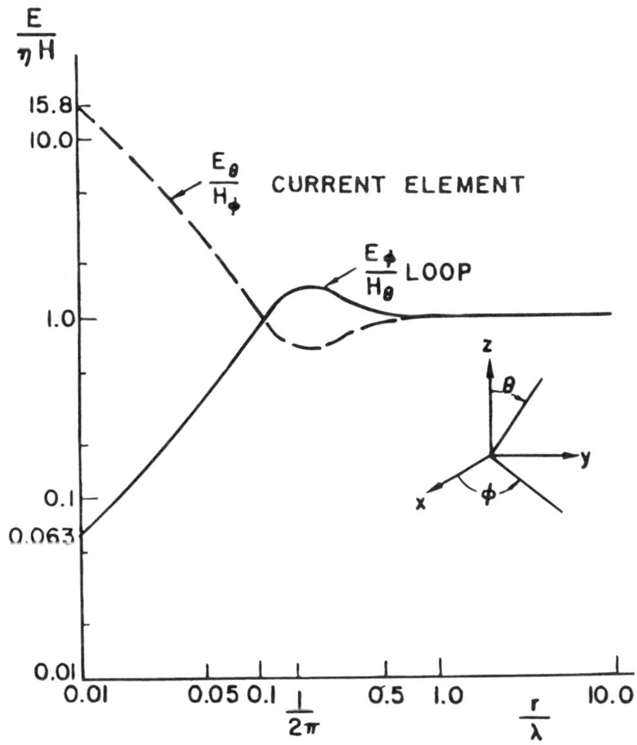

FIGURE 2. Ratio of E to H field (divided by wave impedance of free space $\eta = 377\Omega$) at $\theta = 90°$: for electric current element at origin along z-axis and for electrically small loop centered at the origin in the x-y plane.

charge configuration and the distance from the electric system. The E field may be much larger than the H field or vice versa (see Figure 2).

2. In the radiation field the ratio of the E to H is fixed and equal to 377 in free space (when E is given in volt per meter and H in ampere per meter).
3. In the vicinity of most presently available man-made devices or systems carrying static electric charges, DC, or low frequency (< 1000 Hz) currents, the E and H fields will only under very exceptional circumstances be large enough to produce heating effects inside a living object, as illustrated by Figure 3. (This statement assumes that the living object does not form part of a conducting path that permits direct entrance of current from a wire or conducting ground.) However, nonthermal effects are possible; thus an E field of sufficient magnitude may orient dipoles, or translate ions or polarizable neutral particles (see Chapters 2 to 5).
4. With radiated power it is relatively easy to produce heating effects in living objects with presently available man-made devices (see Chapters 8 to 11). This does not imply, of course, that all biological effects of radiated radio frequency (RF) power necessarily arise from temperature changes.

Results of experiments involving exposure of organic materials and entire living organisms to static E and ELF E fields are described in Chapter 2. There, various mechanisms for the interaction of such fields with living tissue are also discussed. In the present Introduction we shall only point out that one salient feature of static (DC) and ELF E-field interaction with living organisms is that the external or applied E field is always larger by several orders of magnitude than the resultant average internal E field.[5,6] This is a direct consequence of boundary conditions derived from Maxwell's equations.[1–3]

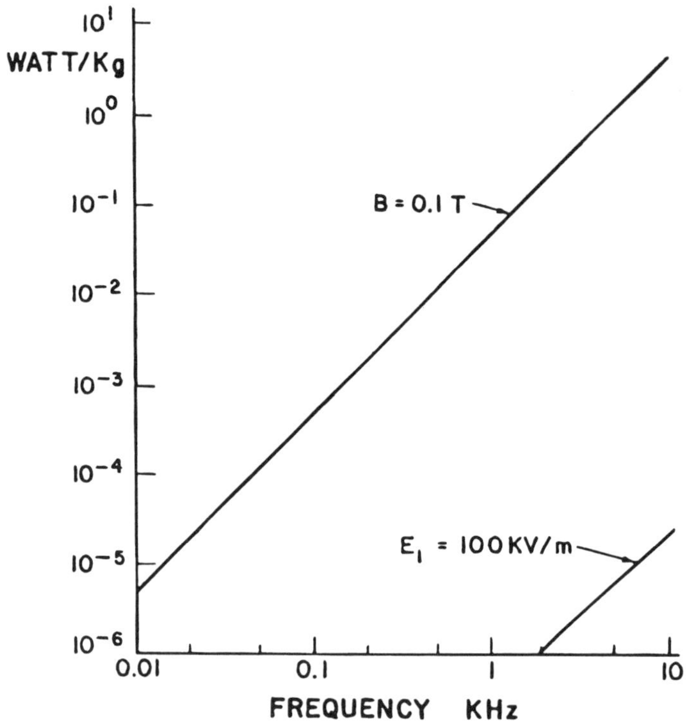

FIGURE 3. (a) Eddy current loss produced in cylinder by sinusoidally time-varying axial H field. Cylinder parameters are conductivity $\sigma = 0.1$ S/m, radius 0.1 m, density D = 1100 kg/m^3, RMS magnetic flux density 0.1 T = 1000 G. Watt per kilogram = $\sigma B^2 r^2 \omega^2/8D$; see Equation 15 and use power per volume = J^2/σ; (b) loss produced by 60 Hz E field in Watt per kilogram = $\sigma E_{int}^2/D$ where external field E_1 is related to E_{int} by Equation 9 with $\epsilon_2 = \epsilon_0 \cdot 10^5$ at 1 kHz and $\epsilon_0 8 \cdot 10^4$ at 10 kHz.

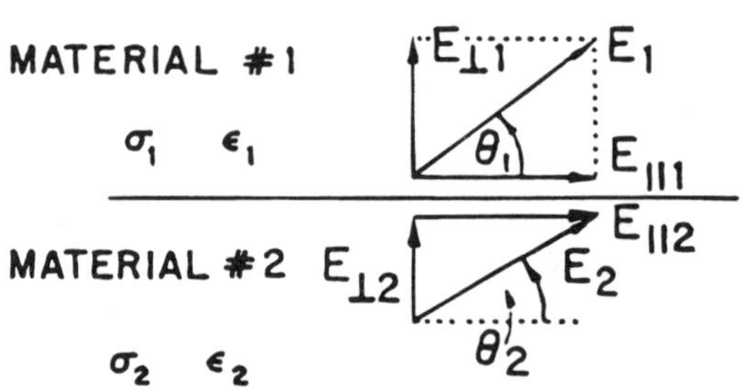

FIGURE 4. Symbols used in description of boundary conditions for E-field components.

II. PENETRATION OF DC AND LOW FREQUENCY ELECTRIC FIELDS INTO TISSUE

Assuming that the two materials illustrated schematically in Figure 4 are characterized, respectively, by conductivities σ_1 and σ_2 and dielectric permittivities ϵ_1 and ϵ_2, we write E field components parallel to the boundary as $E_{\parallel}$ and components perpendicular to the boundary as $E_{\perp}$. For both static and time-varying fields

$$E_{\parallel 1} = E_{\parallel 2} \tag{3}$$

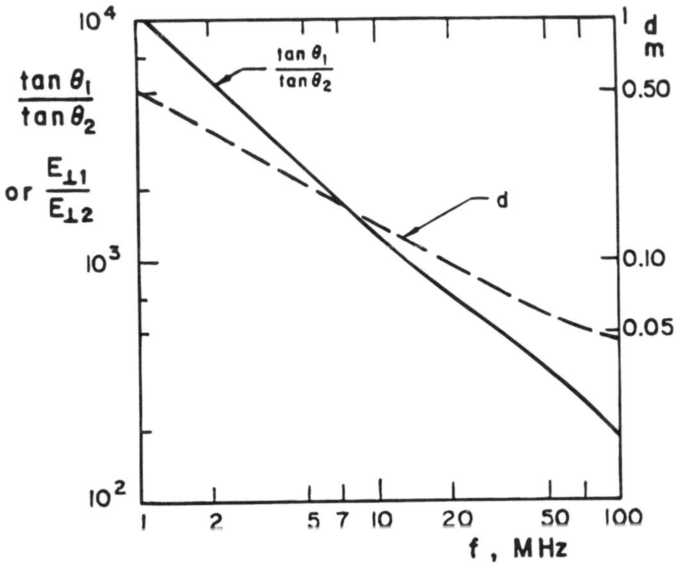

FIGURE 5. Orientation of E-field components at air-muscle boundary (or ratio of fields perpendicular to boundary); depth (d) at which field component parallel to boundary surface decreases by approximately 50% (d = 0.693δ).

and for static (DC) fields

$$\sigma_1 E_{\perp 1} = \sigma_2 E_{\perp 2} \tag{4}$$

as a consequence of the continuity of current (or conservation of charge). The orientations of the total E fields in Media 1 and 2 can be represented by the tangents of the angles between the total fields and the boundary line

$$\tan \theta_1 = \frac{E_{\perp 1}}{E_{\parallel 1}}, \qquad \tan \theta_2 = \frac{E_{\perp 2}}{E_{\parallel 2}} \tag{5}$$

From these equations it follows that

$$\tan \theta_1 = \frac{\sigma_2}{\sigma_1}\frac{E_{\perp 2}}{E_{\parallel 1}} = \frac{\sigma_2}{\sigma_1}\frac{E_{\perp 2}}{E_{\parallel 2}} = \frac{\sigma_2}{\sigma_1}\tan \theta_2 \tag{6}$$

If Material 1 is air with conductivity[7] $\sigma_1 = 10^{-13}$ S/m and Material 2 a typical living tissue with $\sigma_2 \approx 10^{-1}$ S/m (compare Chapter 1), $\tan \theta_1 = 10^{12} \tan \theta_2$, and therefore even if the field in Material 2 (the inside field) is almost parallel to the boundary so that $\theta_2 \approx 0.5^\circ$ or $\tan \theta_2 \approx (1/100)$, $\tan \theta_1 = 10^{10}$ or $\theta_1 = \frac{\pi}{2} - 10^{-10}$ radians. Thus an electrostatic field in air, at the boundary between air and living tissue, must be practically perpendicular to the boundary. The situation is virtually the same at ELF although Equation 4 must be replaced by

$$\sigma_1 E_{\perp 1} - \sigma_2 E_{\perp 2} = -j\omega\rho_s \tag{7}$$

and

$$\epsilon_1 E_{\perp 1} - \epsilon_2 E_{\perp 2} = \rho_s \tag{8}$$

where $j = \sqrt{-1}$, ω is the radian frequency ($= 2\pi \times$ frequency), and ρ_s is the surface charge density. In Chapter 1 it is shown that at ELF the relative dielectric permittivity of living tissue may be as high as 10^6 so that $\epsilon_2 = 10^6\epsilon_0$, where ϵ_0 is the dielectric permittivity of free space $(1/36\pi)10^{-9}$ F/m; however, it is still valid to assume that $\epsilon_2 \leq 10^{-5}$. Then from Equations 7 and 8

$$E_{\perp 1} = \frac{\sigma_2 + j\omega\epsilon_2}{\sigma_1 + j\omega\epsilon_1} E_{\perp 2} \tag{9}$$

which gives at 60 Hz with $\sigma_2 = 10^{-1}$ S/m, $\sigma_1 = 10^{-13}$ S/m, $\epsilon_2 \approx 10^{-5}$ F/m, and $\epsilon_1 \approx 10^{-11}$ F/m

$$E_{\perp 1} = \frac{10^{-1} + j4(10^{-3})}{10^{-13} + j4(10^{-9})} E_{\perp 2} \approx \frac{\sigma_2}{j\omega\epsilon_1} E_{\perp 2} = -j(2.5)10^7 E_{\perp 2} \tag{10}$$

This result, together with Equations 3 and 5, shows that for the given material properties, the field in air must still be practically perpendicular to the boundary of a living organism: $\tan\theta_1 \sim 2.5(10^7)\tan\theta_2$.

Knowing now that the living organism will distort the E field in its vicinity in such a way that the external field will be nearly perpendicular to the boundary surface, we can calculate the internal field by substituting the total field for the perpendicular field in Equation 4 (DC) and Equation 9 (ELF). For the assumed typical material parameters we find that in the static (DC) case

$$\frac{\text{E internal}}{\text{E external}} \approx 10^{-12} \tag{11}$$

$$\rho_f = \frac{3(\sigma_2\epsilon_1 - \sigma_1\epsilon_2)E_0}{2\sigma_1 + \sigma_2} \cos\theta \quad \text{C/m}^2$$

and for 60 Hz

$$\frac{\text{E internal}}{\text{E external}} \approx 4(10^{-8}) \tag{12}$$

Thus, a 60-Hz external field of 100 kV/m will produce an average internal E field of the order of 4 mV/m.

If the boundary between air and the organic material consists of curved surfaces instead of infinite planes, the results will be modified only slightly. Thus, for a finite sphere (with ϵ and σ as assumed here) embedded in air, the ratios of the internal field to the undisturbed external field will vary with the angle θ and distance r indicated in Figure 6, but will not deviate from the results indicated by Equations 7 and 8 by more than a factor of three.[3,8]

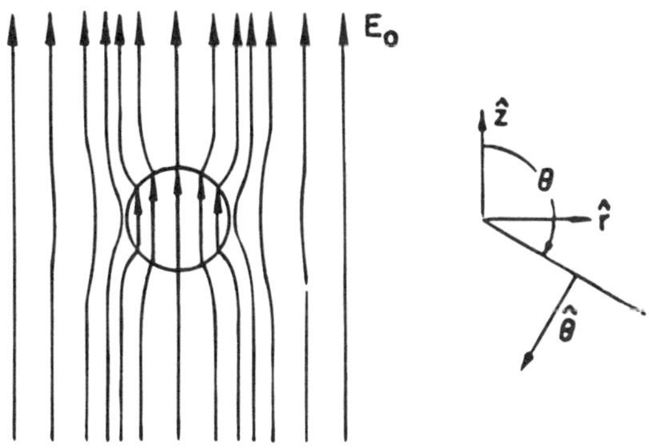

$$r<R \quad \bar{E} = \frac{3\sigma_1 E_0}{2\sigma_1+\sigma_2}\hat{z} \qquad \epsilon_2,\ \sigma_2$$

$$r>R \quad \bar{E} = E_0\cos\theta\left[1+\frac{2R^3(\sigma_2-\sigma_1)}{r^3(2\sigma_1+\sigma_2)}\right]\hat{r} \qquad \epsilon_1,\ \sigma_1$$

$$-E\sin\theta\left[1-\frac{R^3(\sigma_2-\sigma_1)}{r^3(2\sigma_1+\sigma_2)}\right]\hat{\theta}$$

FIGURE 6. E field when sphere of radius R, conductivity σ_2, and dielectric permittivity ϵ_2 is placed into an initially uniform static field ($E = \hat{z}\, E_0$) within a medium with conductivity σ_1 and permittivity ϵ_1. The surface charge density is $\rho_f = \dfrac{3(\sigma_2\epsilon_1 - \sigma_1\epsilon_2)E_0}{2\sigma_1 + \sigma_2} \cos\theta \quad C/m^2$.

III. DC AND LOW FREQUENCY MAGNETIC FIELDS

DC H fields are considered in Part II, Chapter 3 and 6, and ELF H fields in Part II, Chapter 4 and 5. Since the magnetic permeability μ of most biological materials is practically equal to the magnetic permeability μ_0 of free space, $4\pi(10^{-7})$ H/m, the DC or ELF H field “inside” will be practically equal to the H field “outside.” The only exception is organisms such as the magnetotactic bacteria described in Chapter 3, which synthesize ferromagnetic material. The known and suggested mechanisms of interaction of DC H fields with living matter discussed in, Chapters 3 and 4 are

1. Orientation of ferromagnetic particles, including biologically synthesized particles of magnetite.
2. Orientation of diamagnetically or paramagnetically anisotropic molecules and/or cellular elements.[9]
3. Generation of potential differences at right angles to a stream of moving ions (Hall effect) as a result of the magnetic force $F_m = qvB\sin\theta$, where q = the electric charge, v = velocity of the charge, B = magnetic flux density, and $\sin\theta$ = sine of the angle θ between the directions of v and B. One well documented result of this mechanism is a “spike” in the electrocardiogram of vertebrates subjected to large DC H fields.[10,11]
4. Changes in intermediate products or structural arrangements in the course of light-induced chemical (electron transfer) reactions, brought about by Zeeman splitting of molecular energy levels or effects upon hyperfine structure. (The Zeeman effect is the splitting of spectral lines, characteristic of electronic transitions, under the influence of an external H field; hyperfine splitting of electronic transition lines in the absence of an external H field is due to the magnetic moment of the nucleus; such hyperfine splitting can be modified by an externally applied H field.) The magnetic flux densities

involved depend upon the particular system and can be as high as 2000 G but also < 100 G. Bacterial photosynthesis and effects upon the visual system are prime candidates for this mechanism.[12,13]

5. Induction of E fields with resulting electrical potential differences and currents within an organism by rapid motion through a large static H field. Some magnetic phosphenes are due to such motion.[14]

Time-varying H fields, which are discussed in, Chapter 4, may interact with living organisms through the same mechanisms that can be triggered by static H fields, provided the variation with time is slow enough to allow particles of finite size and mass, located in a viscous medium, to change orientation or position where required (Mechanisms 1 and 2), and provided the field intensity is sufficient to produce the particular effect. However, time-varying H fields, including ELF H fields, can also *induce* electric currents into stationary conducting objects. Thus, all modes of interaction of time-varying E fields with living matter may be triggered by time-varying, but not by static, H fields.

In view of Faraday's law, a time-varying magnetic flux will induce E fields with resulting electrical potential differences and "eddy" currents through available conducting paths. Since very large external ELF E fields are required (as indicated by Equations 9 to 12) to generate even small internal *E* fields, many man-made devices and systems generating both ELF E and H fields are more likely to produce physiologically significant internal E fields through the mechanism of *magnetic* induction.

The induced voltage V around some closed path is given by

$$V = \oint \mathbf{E} \cdot d\ell = - \iint \frac{\partial \mathbf{B}}{\partial t} \cdot ds \qquad (13)$$

where $\mathbf{E}$ is the induced E field. The integration $\oint \mathbf{E} \cdot d\ell$ is over the appropriate conducting path, $\partial\mathbf{B}/\partial t$ is the time derivative of the magnetic flux density, and the "dot" product with the surface element, ds, indicates that only the component of $\partial\mathbf{B}/\partial t$ perpendicular to the surface (i.e., parallel to the direction of the vector **ds**) enclosed by the conducting path, induces an E field. To obtain an order-of-magnitude indication of the induced current that can be expected as a result of an ELF H field, we consider the circular path of radius r, illustrated by Figure 7. Equation 13 then gives the magnitude of the E field as

$$E = \frac{\omega B r}{2} \qquad (14)$$

where $\omega = 2\pi f$ and f = frequency. The magnitude of the resulting electric current density J in ampere per square meter is*

$$J = \sigma E = \frac{\sigma \omega B r}{2} \qquad (15)$$

where σ is the conductivity along the path in Siemans per meter. In SI (Systeme Internationale)

* Equation 15 neglects the H field generated by the induced eddy currents. If this field is taken into account, it can be shown that the induced current density in a cylindrical shell of radius r and thickness Δ is given by $[-jrH_0/\delta^2]/[1 + j\Delta r/\delta^2]$, where $H_0 = B_0/\mu_0$ and δ is the skin depth defined by Equation 17 in this Introduction. However, for conductivities of biological materials ($\sigma < 5$ S/m) one obtains at audio frequencies $\delta > 1$ m, and since for most dimensions of interest $\Delta r < 0.01$ m^2, the term $j\Delta r/\delta^2$ becomes negligible. The result—jrH_0/δ^2—is then identical with Equation 15.

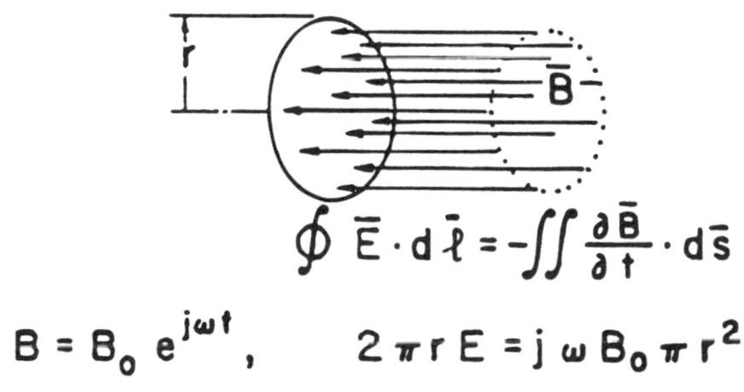

FIGURE 7. Circular path (loop) of radius r enclosing uniform magnetic flux density perpendicular to the plane of the loop. For sinusoidal time variation $B = B_0e_{j\omega t}$.

units used throughout this book, B is measured in tesla (1 T = 10^4 G) and r in meters. Choosing for illustration a circular path of 0.1-m radius, a frequency of 60 Hz, and a conductivity of 0.1 S/m; Equations 14 and 15 give E = 18.85 B and J = 1.885 B. The magnetic flux density required to obtain a current density of 1 mA/m^2 is 0.53 mT or about 5 G. The E field induced by that flux density along the circular path is 10 mV/m. To produce this same 10-mV/m internal E field by an external 60-Hz **E** field would require, by Equation 12, a field intensity of 250 kV/m!

Since the induced voltage is proportional to the time rate of change of the H field (Equation 13), implying a linear increase with frequency (Equation 14), one would expect that the ability of a time-varying H field to induce currents deep inside a conducting object would increase indefinitely as the frequency increases; or conversely, that the magnetic flux density required to induce a specified E field would decrease linearly with frequency, as indicated in Figure 8. This is not true however, because the displacement current density $\partial \mathbf{D}/\partial t$, where $\mathbf{D} = \epsilon \mathbf{E}$, must also be considered as the frequency increases. This leads to the wave behavior discussed in Part III, implying that at sufficiently high frequencies the effects of both external E *and* H fields are limited by reflection losses (Figures 9 and 10) as well as by skin effect[15] (i.e., limited depth of penetration; d in Figure 5).

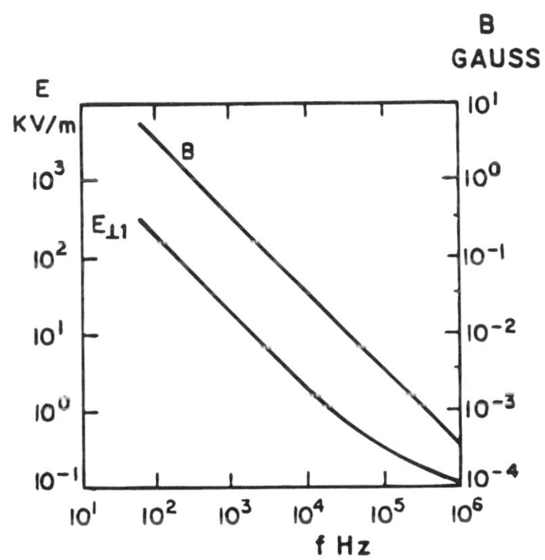

FIGURE 8. External E and H field required to obtain an internal E field of 10 mV/m (conductivity and dielectric permittivity for skeletal muscle from Chapt. 1, Figure 10—H-field calculation assumes a circular path of 0.1-m radius perpendicular to magnetic flux).

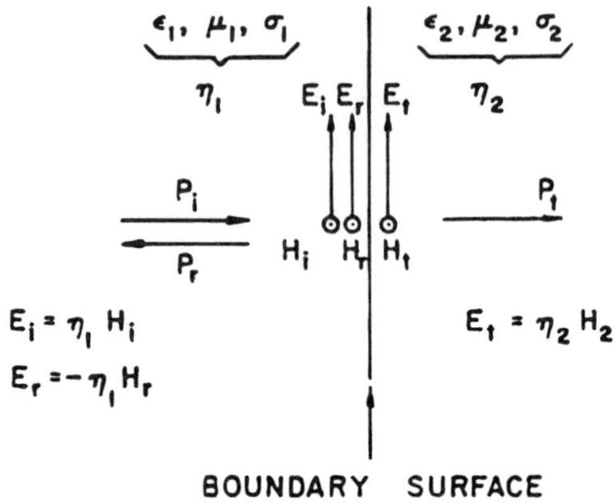

FIGURE 9. Reflection and transmission of an electro-magnetic wave at the boundary between two different media: perpendicular incidence (P_i = incident power, P_r = reflected power, P_t = transmitted power).

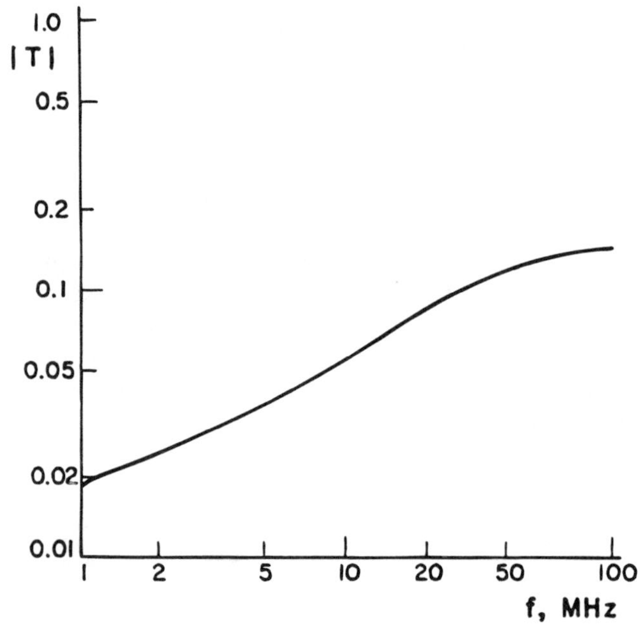

FIGURE 10. Magnitude of transmission coefficient T for incident E field parallel to boundary surface. $T = E_t/E_i$: reflection coefficient $\Gamma = E_r/E_i = T - 1$, Γ and T are complex numbers (ϵ_r and σ for skeletal muscle from Chapt. 1).

IV. RF FIELDS

At frequencies well below those where most animals and many field-generating systems have dimensions of the order of one free space wavelength (e.g., at 10 MHz where λ = 30 m) the skin effect limits penetration of the external field. This phenomenon is fundamentally different from the small ratio of internal to external E fields described in Equation 4 (applicable to DC) and Equation 9.

Equation 9 expresses a "boundary condition" applicable at all frequencies, but as the radian frequency ω increases (and in view of the rapid decrease with frequency of the dielectric permittivity ϵ_2 in biological materials—see Chapt. 1), the ratio of the normal component of the external to the internal E field at the boundary decreases with increasing frequency. This is illustrated by Figure 5 where $\tan \theta_1/\tan \theta_2$ is also equal to $E_{\perp 1}/E_{\perp 2}$ in view of Equations 3, 5, and 9. However, at low frequencies the total field inside the boundary can be somewhat larger than the perpendicular field at the boundary and any field variation with distance from the boundary is not primarily due to energy dissipation, but in a homogeneous body is a consequence of shape. At RFs, on the other hand, the E *and* H fields of the incoming electromagnetic wave, after reflection at the boundary, are further decreased due to energy dissipation. Both E and H fields decrease exponentially with distance from the boundary

$$g(z) = Ae^{-\frac{z}{\delta}} \tag{16}$$

where g(z) is the field at the distance z and A the magnitude of the field just *inside* the boundary.

As defined by Equation 16 the skin depth δ is the distance over which the field decreases to 1/e (= 0.368) of its value just *inside* the boundary. (Due to reflection, the field A just inside the boundary can already be very much smaller than the incident external field; see Figures 9 and 10.)

Expressions for δ given below were derived[2,3,15,16] for plane boundaries between infinite media. They are reasonably accurate for cylindrical structures if the radius of curvature to skin depth (r_0/δ) is larger than about five.[15] For a good conductor

$$\delta = \frac{1}{\sqrt{\pi f \mu \sigma}} \tag{17}$$

where a good conductor is one for which the ratio p of conduction current, $J = \sigma E$, to displacement current, $\partial D/\partial t = \epsilon(\partial E/\partial t) = j\omega\epsilon E$ is large:

$$p = \frac{\sigma}{\omega\epsilon} \gg 1 \tag{18}$$

Since for most biological materials p is of the order of one ($0.1 < p < 10$) over a very wide frequency range (see Chapt. 1), it is frequently necessary to use the more general expression[15]

$$\delta = \frac{1}{\omega\left[\frac{\mu\epsilon}{2}\left(\sqrt{1 + p^2} - 1\right)\right]^{1/2}} \tag{19}$$

The decrease of field intensity with distance from the boundary surface indicated by Equation 16 becomes significant for many biological objects at frequencies where $r_0/\delta \geq 5$ is not satisfied. However, the error resulting from the use of Equations 16 and 17 or 19 with curved objects is less when $z < \delta$; thus at $z = 0.693\ \delta$, where $g(z) = 0.5\ A$ from Equations 16 and 17, the correct values of g(z), obtained by solving the wave equation in cylindrical coordinates, differs only by 20% (it is 0.6 A) even when r_0/δ is as small as 2.39.[16] Therefore, on Figure 5 the distance $d = 0.693\ \delta$, at which the field decreases to half of its value just inside the boundary surface, rather than δ was plotted for the frequency range 1 to 100 MHz, using Equation 19 with typical values for σ and ε for muscle from Figure 11 of Chapt. 1. It is apparent that the skin effect becomes significant for humans and larger vertebrates at frequencies > 10 MHz.

Directly related to skin depth, which is defined for fields varying sinusoidally with time, is the fact that a rapid transient variation of an applied magnetic flux density constitutes an exception to the statement that the DC H field inside the boundary is equal to the H field outside. Thus, from one viewpoint one may consider the rapid application or removal of a DC H field as equivalent to applying a high frequency field during the switching period, with the highest frequencies present being of the order of $1/\tau$, where τ is the rise time of the applied step function. Thus, if $\tau < 10^{-8}$ sec, the skin effect will be important during the transient period, since d in Figure 5 is < 5 cm above 100 MHz. It is also possible to calculate directly the magnetic flux density inside a conducting cylinder as a function of radial position r and time t when a magnetic pulse is applied in the axial direction.[17,18] Assuming zero rise time of the applied field B_0, i.e., a true step function, one finds that the field inside a cylinder of radius a is

$$B = B_0\left[1 - \sum_{k=1}^{\infty} J_0\left(r\frac{\nu_k}{a}\right)e^{-t/T_k}\right] \tag{20}$$

where $J_0(r\,\nu_k/a)$ is the zero order Bessel function of argument $r\,\nu_k/a$ and the summation is over the nulls of J_0 designated ν_k (the first four values of ν_k are 2.405, 5.520, 8.654, and 11.792).* T_k is the rise time of the k'th term in the series and is given by

$$T_k = \frac{\mu_0\sigma a^2}{\nu_k} \tag{21}$$

As ν_k increases, the rise time decreases and therefore the longest delay is due to the first term in the summation with $k = 1$:

$$T_1 = \frac{\mu_0\sigma a^2}{2.405} \tag{22}$$

For a cylinder with 0.1-m radius and a conductivity $\sigma \approx 1$ S/m, which is a typical value for muscle between 100 and 1000 MHz, Equation 22 gives $T_1 = 2.6(10^{-8})$ sec. This finite rise time (or decay time in case of field removal) of the internal H field may be of some importance when pulsed H fields are used therapeutically.[19] It might also be used to measure noninvasively the conductivity of biological substances in vivo through determination of the final decay rate of the voltage induced into a probe coil by the slowly decaying internal field after the applied field is removed.[18]

Properties of biological substances in the intermediate frequency range, above ELF (> 300 Hz), and below the higher RFs, where wave behavior and skin effect begin to be important (~ 20 MHz), are discussed in Part I of this book. However, while Part II is concerned with biological effects at DC and frequencies below a few kilohertz, Part III deals primarily with the higher RFs, > 50 MHz. One reason for this limited treatment of the intermediate frequency range is that very little animal data are available for this spectral region in comparison with the large number of experiments performed at ELF and microwave frequencies in recent years. Another reason is that most electrical processes known to occur naturally in biological systems—action potentials, EKG, EEG, ERG, etc.—are limited to DC and ELF frequencies. Thus one might expect some physiological effects from external fields of appropriate intensity in the same frequency range, even if the magnitude of such fields is not large enough to produce thermal effects. As illustrated

* This result is based upon the solution of $\partial\mathbf{B}/\partial t = (1/\mu\sigma)\nabla^2\mathbf{B}$ which is a consequence of Ampere's and Faraday's laws when the displacement current is disregarded. Equations 20 to 22 are therefore only correct when $p \gg 1$.

by Figures 3 and 8, most E fields below 100 kHz set up by presently used man-made devices, and most except the very largest H fields below 10 kHz, are incapable of producing thermal effects in living organisms. (This excludes, of course, fields accompanying currents directly introduced into the organism via electrodes.) Thus, the frequencies between about 10 and 100 kHz have been of relatively little interest because they are not very likely to produce thermal or other biological effects. On the other hand, the higher RFs are frequently generated at power levels where enough energy may be introduced into living organisms to produce local or general heating. In addition, despite skin effect and the reflection loss to be discussed in more detail below, microwaves modulated at an ELF rate may serve as a vehicle for introducing ELF fields into a living organism of at least the same order of magnitude as would be introduced by direct exposure to ELF. Any effect of such ELF-modulated microwaves would, of course, require the existence of some amplitude-dependent demodulation mechanism to extract the ELF from the microwave carrier.

Part III is divided into five parts. Chapters 8 and 9 give the information necessary for establishing the magnitude of the fields present in biological objects: (1) experimental techniques and (2) analytical methods for predicting field intensities without construction of physical models made with "phantom" materials (i.e., dielectric materials with properties similar to those of living objects which are to be exposed). Since thermal effects at microwave frequencies are certainly important—although they may not be the only biological effects of this part of the spectrum—and since some (but not all) thermal effects occur at levels where the thermoregulatory system of animals is activated, thermoregulation in the presence of microwave fields is discussed in Part III, Chapter 10. Not only are therapeutic applications of microwaves based upon their thermal effects, but experimental establishment of possible nonthermal effects at the threshold of large scale tissue heating in particular living systems also requires thorough understanding of thermoregulatory mechanisms:

The vast amount of experimental data obtained on animal systems exposed to microwaves is discussed in Part III, Chapters 11 and 12. Part III, Chapter 11 emphasizes primarily application of nonmodulated fields or of modulated fields, where the type of modulation had no apparent effect other than modification of the average power level, while Part III, Chapter 12 considers in some detail modulated fields and experimental evidence for nonlinear effects characterized by "amplitude windows" or "frequency windows."

At the higher radio frequencies, the external E field is not necessarily perpendicular to the boundary of biological materials (see Figures 4 and 5), and the ratio of the *total* external E field to the *total* internal field is not given by Equation 9. However, the skin effect (Equations 16 to 19) and reflection losses still reduce the E field within any biological object below the value of the external field. As pointed out in Chapt. 1, dielectric permittivity and electrical conductivity of organic substances both vary with frequency. At RFs, most biological substances are neither very good electrical conductors nor very good insulators—with the exception of cell membranes, which are good dielectrics at RF (at ELF cell membranes can act as intermittent conductors or as dielectrics, but are ion selective[20–22]). The ratio p (Equation 18) is neither much smaller, nor very much larger than one as shown for typical muscle tissue[23,24] in Table 1.

TABLE 1

f (MHz)	σ	ϵ_r	$\frac{\sigma}{\omega\epsilon_0\epsilon_r}$
1	0.40	2000	3.6
10	0.63	160	7.1
100	0.89	72	2.2
10^3	1.65	50	0.59
10^4	10.3	40	0.46
10^5	80	6	2.4

Reflection loss at the surface of an organism is a consequence of the difference between its electrical properties and those of air. Whenever an electromagnetic wave travels from one material into another with different electrical properties, the boundary conditions (Equations 3 and 8) and similar relations for the H field require the existence of a reflected wave. The expressions for the reflection coefficient

$$\Gamma = \frac{E_r}{E_i} \tag{23}$$

and the transmission coefficient

$$T = \frac{E_t}{E_i} \tag{24}$$

become rather simple for loss-free dielectrics ($p \ll 1$) and for good conductors ($p \gg 1$). Since biological substances are neither, the most general expressions for Γ and T, applicable at plane boundaries, are needed.[3,15] For perpendicular incidence, illustrated by Figure 9,

$$\Gamma = \frac{\eta_2 - \eta_1}{\eta_2 + \eta_1} \tag{25}$$

$$T = \frac{2\eta_2}{\eta_2 + \eta_1} = 1 + \Gamma \tag{26}$$

where η_1 and η_2 are the wave impedances, respectively, of Medium 1 and 2. The wave impedance of a medium is the ratio of the E to the H field in a plane wave traveling through that medium; it is given by[15]

$$\eta = \left(\frac{j\omega\mu}{\sigma + j\omega\epsilon}\right)^{1/2} \tag{27}$$

Clearly Γ and T are in general complex numbers, even when Medium 1 is air for which Equation 27 reduces to the real quantity $\eta_0 = \sqrt{\mu_0/\epsilon_0}$, because Medium 2, which here is living matter, usually has a complex wave impedance at RFs.

The incident, reflected, and transmitted powers are given by[15]

$$P_i = \text{Rl}|E_i|^2 \frac{1}{\eta_1^*} = \frac{|E_i|^2}{|\eta_1|^2} R_1 \tag{28}$$

$$P_r = \text{Rl}|E_r|^2 \frac{1}{\eta_1^*} = \frac{|E_r|^2}{|\eta_1|^2} R_1 \tag{29}$$

$$P_t = \text{Rl}|E_t|^2 \frac{1}{\eta_2^*} = \frac{|E_t|^2}{|\eta_2|^2} R_2 \tag{30}$$

where the E fields are effective values ($E_{eff} = E_{peak}/\sqrt{2}$) of sinusoidal quantities, Rl signifies "real part of," η^* is the complex conjugate of η, and R_1 and R_2 are the real parts of η_1 and η_2. If Medium 1 is air, $\eta_1 = R_1 = 377\ \Omega$. It follows from Equations 23, 24, and 28 to 30

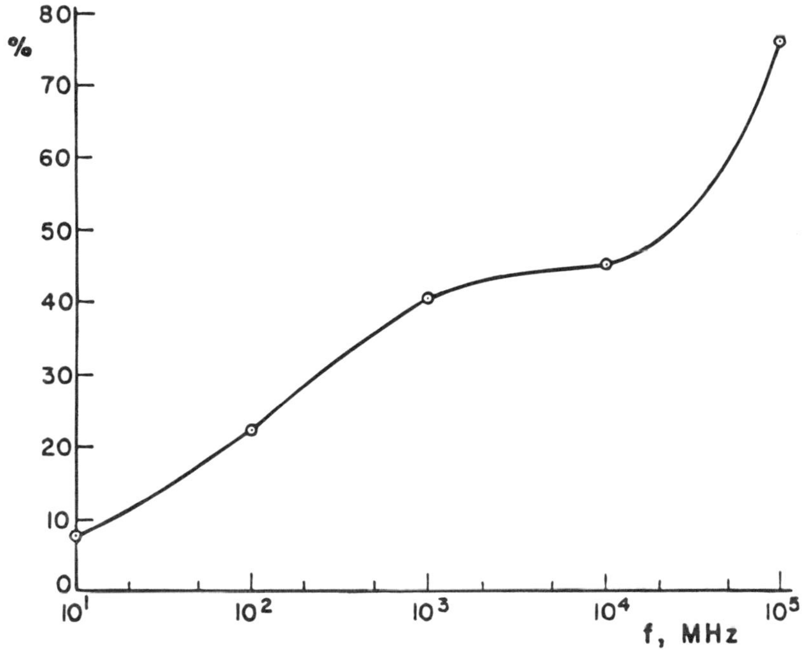

FIGURE 11. Ratio of transmitted to incident power expressed as percent of incident power. Air-muscle interface, perpendicular incidence (Equation 31, Table 1).

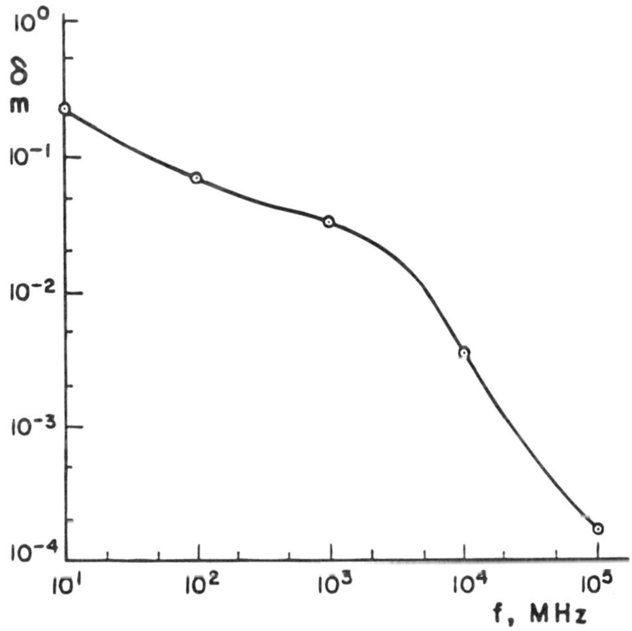

FIGURE 12. Electromagnetic skin depth in muscle tissue from plane wave expression (Equation 19, Table 1).

and conservation of energy that the ratio of the transmitted to the incident real power is given by

$$\frac{P}{P_i} = |T|^2 \frac{\eta_1\eta_2^* + \eta_1^*\eta_2}{2|\eta_2|^2} = 1 - \frac{P_r}{P_i} = 1 - |\Gamma|^2 \tag{31}$$

The magnitude of the transmission coefficient T for the air-muscle interface over the 1- to 100-MHz frequency range is plotted in Figure 10, which shows that the magnitude of the transmitted E field in muscle tissue is considerably smaller than the E field in air. The fraction of the total incident power that is transmitted (Equation 31) is shown in Figure 11, indicating clearly that reflection loss at the interface decreases with frequency. However, for deeper lying tissue this effect is offset by the fact that the skin depth δ (Equation 19) also decreases with frequency (Figure 12) so that the total power penetrating beyond the surface decreases rapidly.

In addition to reflection at the air-tissue boundary, further reflections take place at each boundary between dissimilar materials. For example, the magnitude of the reflection coefficient at the boundary surface between muscle and organic materials with low water content, such as fat or bone, is shown in Table 2.

TABLE 2

f (MHz)	Fat or bone σ (S/m)	Fat or bone ϵ_r	Muscle[a]—fat Γ
10^2	0.048	7.5	0.65
10^3	0.101	5.6	0.52
10^4	0.437	4.5	0.52

[a] σ and ϵ_r for muscle from Table 1.

The situation is actually more complicated than indicated by Figures 10 and 11, because the wavefront of the incident electromagentic wave may not be parallel to the air-tissue boundary. Two situations are possible: the incident E field may be polarized perpendicular to the plane of incidence defined on Figure 13 (perpendicular polarization—Figure 13A) or parallel to the plane of incidence (parallel polarization—Figure 13B). The transmission and reflection coefficients[8] are different for the two types of polarization and also become functions of the angle of incidence α_1:

Perpendicular polarization

$$T_\perp = \frac{2\,\eta_2 \cos\alpha_1}{\eta_2\cos\alpha_1 + \eta_1\cos\alpha_2} \tag{32}$$

$$\Gamma_\perp = \frac{\eta_2\cos\alpha_1 - \eta_1\cos\alpha_2}{\eta_2\cos\alpha_1 + \eta_1\cos\alpha_2} \tag{33}$$

Parallel polarization

$$T_\parallel = \frac{2\,\eta_2 \cos\alpha_1}{\eta_2\cos\alpha_2 + \eta_1\cos\alpha_1} \tag{34}$$

$$\Gamma_\parallel = \frac{\eta_1\cos\alpha_1 - \eta_2\cos\alpha_2}{\eta_2\cos\alpha_2 + \eta_1\cos\alpha_1} \tag{35}$$

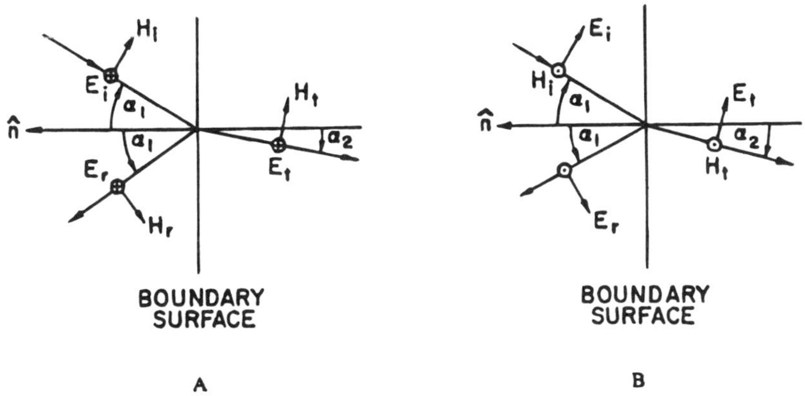

FIGURE 13. Oblique incidence of an electromagnetic wave at the boundary between two different media. (A) Perpendicular polarization (E vector perpendicular to plane of incidence); (B) parallel polarization (E vector parallel to plane of incidence). The plane of incidence is the plane formed by the surface normal (unit vector $\hat{n}$ and the direction of the incident wave); ⊗ indicates a vector into the plane of the paper; ⊙ indicates a vector out of the plane of the paper. The orientation of the field vectors in the transmitted field is shown for loss-free dielectrics. For illustration of the transmitted wave into a medium with finite conductivity, where the wave impedance η_2 becomes a complex number, see Reference 3.

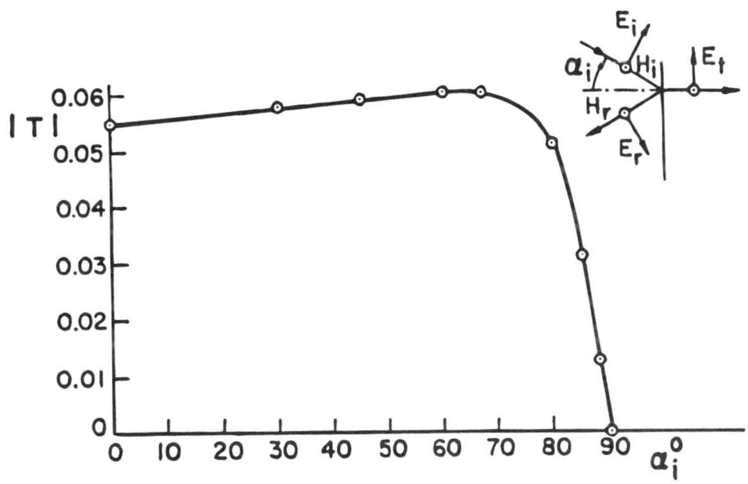

FIGURE 14. Magnitude of complex transmission coefficient for parallel polarization vs. angle of incidence α_1 at 10 MHz (E field in plane of incidence, H field parallel to boundary plane; $\sigma_2 = 0.7$ S/m, $\epsilon_{r2} = 150$, $T = E_t/E_r$).

where α_2 is given by the generalized Snell's law (when both media have the magnetic permeability of free space) by

$$\sin \alpha_2 = \frac{\sqrt{\epsilon_1}}{\sqrt{\epsilon_2 - j\dfrac{\sigma_2}{\omega}}} \tag{36}$$

so that $\cos \alpha_2 = \sqrt{1 - \sin^2 \alpha_2}$ is a complex number unless $p_2 = (\sigma_2/\omega\epsilon_2) \ll 1$.

As illustration, the variation with angle of incidence of the transmission coefficient for parallel polarization at the air-muscle interface at 10 MHz, is shown in Figure 14. It is apparent that the transmitted field is not necessarily maximized by perpendicular incidence in the case of parallel polarization. Furthermore, whenever $p \approx 1$ or $p > 1$ (see Table 1 above) α_2 is complex, which causes the waves entering the tissue to be inhomogeneous—they are not simple plane waves, but waves where surfaces of constant phase and constant amplitude do not coincide;[3,25] only the planes of constant amplitude are parallel to the boundary surface.

Analytical solutions for nonplanar structures taking into account size and shape of entire animals have been given[26] and are also described in Chapter 9.

V. IONIZATION, IONIZING RADIATION, CHEMICAL BONDS, AND EXCITATION

Since RF fields can be characterized as nonionizing radiation, it is desirable to review the differences between ionizing and nonionizing radiation, to explain ionization phenomena and also to discuss related excitation phenomena which require less energy than ionization. We note first that the energy of electromagnetic waves is quantized with the quantum of energy (in joules) being equal to Planck's constant (see Appendix I) times the frequency. This energy can also be expressed in electronvolts, i.e., in multiples of the kinetic energy acquired by an electron accelerated through a potential difference of 1 V: 1 eV $\approx$ 1.6 (10^{-19}) J. Energy quanta for a few frequencies are listed in Table 3.

Quantized energy can "excite" molecules; appropriate frequenceis can couple to vibrational and rotational oscillations, and if the incident energy quantum has sufficient magnitude it can tear an electron away from one of the constituent atoms. The energy required to remove one electron from the highest energy orbit of a particular chemical element is called its "ionization potential." Typical ionization potentials are of the order 10 eV—e.g., for the hydrogen atom it is 13.6 eV and for gaseous sodium it is 5.1 eV. Since chemical binding forces are essentially electrostatic, ionization implies profound chemical changes. Therefore ionization by any outside agent of the complex compounds that make up a living system leads to profound and often irreversible changes in the operation of that system.

Table 3 shows that even the highest RFs (millimeter waves) have quantum energies well below the ionization potential of any known substance; thus one speaks of nonionizing radiation when referring to electromagnetic waves below UV light frequencies. UV and higher frequency waves (X-rays, γ-rays) constitute ionizing radiation.

This explanation of the difference between ionizing and nonionizing radiation should not imply that nonionizing electromagnetic radiation cannot have profound effects upon inorganic and organic substances. Since excitation of coherent vibrational and rotational modes requires considerably less energy than ionization, it could occur at RFs; this will be discussed further on. In addition, many other possible biological effects require energies well below the level of ionizing potentials. Examples are tissue heating, dielectrophoresis, depolarization of cell membranes, mechanical stress due to piezoelectric transduction, or dielectric saturation

TABLE 3

Name of radiation or application	Frequency (Hz)	Wavelength (m)	Energy of 1 quantum of radiation (eV)
UHF TV	7 (10^8)	0.43	2.88 (10^{-6})
Microwave radar	10^{10}	3 (10^{-2})	4.12 (10^{-5})
Millimeter wave	3 (10^{11})	1 (10^{-3})	1.24 (10^{-3})
Visible light	6 (10^{14})	5 (10^{-7})	2.47
Ionizing UV	10^{16}	3 (10^{-4})	41.2
Soft X-ray	10^{18}	3 (10^{-10})	4120
Penetrating X-ray	10^{20}	3 (10^{-12})	4.12 (10^5)

resulting in the orientation of the polar sidechains of macromolecules leading to the breaking of hydrogen bonds. These and other mechanisms will be discussed by the authors of several chapters, who will also give estimates of rates at which energy must be delivered to produce particular effects.

Returning to the discussion of ionization, it is important to note that ionization of a chemical element can be brought about not only by absorption of electromagnetic energy, but also by collision either with foreign (injected) elementary particles of the requisite energy, or by sufficiently violent collision among its own atoms. The latter process constitutes ionization by heating, or thermal breakdown, of a substance, which will occur when the kinetic energy of the colliding particles exceeds the ionization potential. Since the kinetic energy of elementary particles is related to temperature[27] by $W = kT$ where k is Boltzman's constant (see Appendix I), we find that the required temperature is

$$1.38\,(10^{-23})T \approx 5\text{ eV} \approx (5)\,1.6\,(10^{-19})\text{ J}$$

$$T \approx 5\,(10^{4})\text{ K}$$

which is about twice the temperature inside a lightning stroke[28] and orders of magnitude higher than any temperature obtainable from electromagnetic waves traveling through air. Actually, initiation of lightning strokes is an example of ionization by collision with injected energetic elementary particles. A few free electrons and ions always present in the air due to ionization by cosmic rays are accelerated by the E fields generated within clouds to velocities corresponding to the required ionization energy. Only when the field is large enough to impart this energy over distances shorter than the mean free path of the free electrons or ions at atmospheric pressure can an avalanche process take place: each accelerated electron separates a low-energy electron from the molecule with which it collides, and in the process loses most of its own energy; thus, one high-energy free electron is exchanged for two free low-energy electrons and one positive ion. Both electrons are in turn accelerated again by the field, giving them high kinetic energy before they collide with neutral molecules; their collision produces four free electrons, etc. The breakdown field strength for air at atmospheric pressure is approximately $3(10^6)$ V/m, implying a mean free path of electrons

$$\Delta\ell \approx [5\text{ eV}/3(10^{6})\text{ V/m}] \approx 10^{-6}\text{ m}$$

However, this model is not entirely accurate since the actual mean free path corresponds to energies of the order of 0.1 eV, which is only sufficient to excite vibrational modes in the target molecule. Apparently such excitation is sufficient to cause ionization if the collision process lasts long enough.[29]

Except for some laboratory conditions where a sufficiently high potential difference can be applied directly across a biological membrane to bring about its destruction, collisional ionization is generally not a factor in the interaction of electromagnetic waves with tissue: the potential difference required for membrane destruction[30] is between 100 and 300 mV, corresponding to a field strength of the order of $2(10^7)$ V/m (assuming a membrane thickness $d = 100$ Å; $E = V/d$). However, there is a third mechanism of ionization that is particularly important in biological systems. When a chemical compound of the type wherein positive and negative ions are held together by their electrostatic attraction, such as the ionic crystal NaCl, is placed into a suitable solvent, such as H_2O, it is separated into its ionic components. The resulting solution becomes an electrolyte, i.e., an electrically conducting medium in which the only charge carriers are ions.

In this process of chemical ionization the Na^+ cations and Cl^- anions are separated from the original NaCl crystal lattice and individually surrounded by a sheet of solvent molecules,

TABLE 4

Type of bond	Change in free energy (binding energy) kcal/mol	eV/molecule
Covalent	50–100	2.2–4.8
Van der Waals	1–2	0.04–0.08
Hydrogen	3–7	0.13–0.30
Ionic[a]	5	0.2
Av thermal energy at 310 K	0.62	0.027

Note: For conversion from kcal/mol to eV/molecule see Appendix I.

[a] For ionic groups of organic molecules such as COO^-, $NH3_3^-$ in aqueous solution.

the "hydration sheath." If the solvent is H_2O, this process is called "hydration," or more generally, for any solvent, "solvation."

A dilute solution of NaCl crystals in H_2O is slightly cooler than the original constituents were before the solvation process, indicating that internal energy of the system was consumed. Actually energy is consumed in breaking up the original NaCl bonds and some, but less, energy is liberated in the formation of the solvation sheath. The process is due to the interaction between the dipole moment of the solvent molecule (H_2O in our example) and the electric charges on the ions. Thus, solvents with higher relative dielectric constant ϵ_r indicating higher inherent electric dipole moment per unit volume (P), solvate ions more strongly ($\epsilon_r = 1 + P/[\epsilon_0 E]$ where E is the electric field applied during the measurement of ϵ_r). For example, H_2O with $\epsilon_r \approx 80$ solvates more strongly than methanol with $\epsilon_r \approx 33$. For biological applications it is worth noting that solvation may affect not only ionic substances, but also polar groups (i.e., molecular components which have an inherent dipole moment) such as $-C{=}O$, $-NH$, or $-NO_2$. Details of the process are discussed in texts on electrochemistry.[31,32]

In biological processes not only chemical ionization and solvation of ionic compounds but, of course, all kinds of chemical reactions, take place. One of the central questions in the study of biological effects of E and H fields is therefore not only whether they can cause or influence ionization, but whether they can affect—speed up, slow down or modify—any naturally occurring biologically important chemical reaction.

In Table 4, typical energies for various types of chemical bonds are listed. For comparison the thermal energy per elementary particle at 310 K is also shown.

Complementing the numbers in Table 4 one should also point out that:

1. The large spread in the energies of thermal motion guarantees that at physiological temperatures some molecules always have sufficient energy to break the strongest weak bonds.[33]
2. The average lifetime of a weak bond is only a fraction of a second.
3. The weak binding forces are effective only between surfaces in close proximity and usually require complementary structures such as a (microscopic) plug and hole (as are thought to exist between antigen and antibody).[33]
4. Most molecules in aqueous solution form secondary bonds.
5. The metabolism of biological systems continuously transforms molecules and therefore also changes the secondary bonds that are formed.

Comparison of the last columns in Tables 3 and 4 shows that millimeter waves have quantum energies which are only about one order of magnitude below typical Van der Waals energies (waves at a frequency of 10^{12} Hz with a quantum energy of 0.004 eV have a wavelength of 0.3 mm and can still be classified as millimeter waves). One might expect therefore that such waves could initiate chemically important events (i.e., configurational changes), e.g., by multiple transitions between closely spaced vibrational states at successively higher energy levels.[47]

Energies associated with transition from one to another mode of rotation of a diatomic molecule are given by $W = \ell(\ell + 1)A$,[27,34] where $\ell = 0,1,2,3 \ldots$ and $A = 6(10^{-5})$ eV; thus already an electromagnetic wave with a frequency of 29 GHz—still in the microwave region—can exicte a rotational mode. *Vibrational* modes of diatomic molecules[27,34] correspond to energies of the order of 0.04 eV, requiring excitation in the IR region. Vibrational frequencies in a typical H-bonded system[35] are of the order of 3000 GHz; however, attenuation at this frequency by omnipresent free H_2O may prevent any substantial effect.[35]

Kohli et al.[36] predict that longitudinal and torsional modes of double helical DNA should not be critically damped at frequencies > 1 GHz (although relaxation times are of the order of picoseconds) and Kondepudi[37] suggests the possibility of an influence of millimeter waves at approximately $5(10^{11})$ Hz upon oxygen affinity of hemoglobin due to resonant excitation of heme plane oscillations. Although Furia et al.[38] did not find resonance absorption at millimeter waves in yeast, such was reported by Grundler et al.[39,48] The latter experiment has been interpreted[40] as supporting Fröhlich's theory of cooperative phenomena in biological systems. That theory postulates "electric polarization waves" in biological membranes which are polarized by strong biologically generated[20] fields ($\sim 10^7$ V/m). Fröhlich[41–43] suggests that metabolically supplied energy initiates mechanical vibrations of cell membranes. The frequency of such vibrations is determined by the dimensions and the elastic constants of the membranes: based on an estimate of the sound velocity in the membrane of 10^3 m/sec and a membrane thickness of 100 Å (equal to one half wavelength) one obtains a frequency of $5(10^{10})$ Hz. Individual molecules within and outside the membrane may also oscillate, and frequency estimates vary between 10^9 Hz for helical RNA[44] and $5(10^{13})$ Hz for hydrogen-bonded amide structures.[45] Since the membranes and molecules involved are strongly polarized, the mechanically oscillating dipoles produce oscillating electromagnetic fields that are then able to transmit energy, at least in some situations, over distances much larger than the distance to the next adjacent molecule.

Electromagnetic coupling of this type may produce long range cooperative phenomena. In particular, Fröhlich[46] has shown that two molecular systems may exert strong forces upon each other when their respective oscillation frequencies are nearly equal, provided the dielectric permittivity of the medium between them is strongly dispersive or excitation is supplied by pumping (i.e., by excitation at the correct frequency from an external source). The mechanism is nonlinear in the sense that it displays a step-like dependence on excitation intensity. Possible long range effects may be, for example, attraction between enzyme and substrate.[43] These and related topics have been discussed in detail by Illinger[35] and are reviewed in the present volume in Chapter 12.

ACKNOWLEDGMENTS

Preparation of this Introduction and my part of the editorial work for this book were done mostly while I was a visitor at the University of Wisconsin in Madison. I wish to express my thanks to the University of Rhode Island for granting the sabbatical leave which made this visit possible and to the University of Wisconsin for providing the opportunity for stimulating discussions with faculty and students on the subject matter of this book. I also want to sincerely thank Ms. Alyce Hegge and Ms. Kay Ewers for first-rate secretarial assistance.

REFERENCES

1. **Jordan, E. C.,** *Electromagnetic Waves and Radiating Systems.* Prentice-Hall, Englewood Cliffs, N.J., 1950.
2. **Schelkunoff, S. A.,** *Electromagnetic Waves,* D Van Nostrand, New York, 1943, 133.
3. **Stratton, J. A.,** *Electromagnetic Theory,* McGraw-Hill, New York, 1941, 435.
4. **Van Bladel, J.,** *Electromagnetic Fields,* McGraw-Hill, New York, 1964, 274.
5. **Kaune, W. T. and Gillis, M. F.,** General properties of the interaction between animals and ELF electric fields, *Bioelectromagnetics,* 2, 1, 1981.
6. **Bridges, J. E. and Preache, M.,** Biological influences of power frequency electric fields—a tutorial review from a physical and experimental viewpoint, *Proc. IEEE,* 69, 1092, 1981.
7. **Iribarne, J. V. and Cho, H. R.,** *Atmospheric Physics,* D. Reidel, Boston 1980, 134.
8. **Zahn, M.,** *Electromagnetic Field Theory, A Problem Solving Approach,* John Wiley & Sons, New York, 1979.
9. **Raybourn, M. S.,** The effects of direct-current magnetic fields on turtle retina in vitro, *Science,* 220, 715, 1983.
10. **Gaffey, C. T. and Tenforde, T. S.,** Alterations in the rat electrocardiogram induced by stationary magnetic fields, *Bioelectromagnetics,* 2, 357, 1981.
11. **Tenforde, T. S., Gaffey, C. T., Moyer, B. R., and Budinger, T. F.,** Cardiovascular alterations in *Macaca* monkeys exposed to stationary magnetic fields: experimental observations and theoretical analysis, *Bioelectromagnetics,* 4, 1, 1983.
12. **Schulten, K.,** Magnetic field effects in chemistry and biology, *Festkörperprobleme/Advances in Solid State Physics,* Vol. 22, Heyden, Philadelphia, 1982, 61.
13. **Blankenship, R. E., Schaafsma, T. J., and Parson, W. W.,** Magnetic field effects on radical pair intermediates in bacterial photosynthesis, *Biochim. Biophys. Acta,* 461, 297, 1977.
14. **Sheppard, A. R.,** Magnetic field interactions in man and other mammals: an overview, in *Magnetic Field Effect on Biological Systems,* Tenforde, T. S., Ed., Plenum Press, New York, 1979, 33.
15. **Jordan, E. C.,** *Electromagnetic Waves and Radiating Systems,* Prentice-Hall, Englewood Cliffs, N.J., 1950, 132.
16. **Ramo, S., Whinnery, J. R., and Van Duzer, T.,** *Fields and Waves in Communication Electronics,* John Wiley & Sons, New York, 1965, 293.
17. **Smyth, C. P.,** *Static and Dynamic Electricity,* McGraw-Hill, New York, 1939.
18. **Bean, C. P., DeBlois, R. W., and Nesbitt, L. B.,** Eddy-current method for measuring the resistivity of metals, *J. Appl. Phys.,* 30(12), 1959, 1976.
19. **Bassett, C. A. L., Pawluk, R. J., and Pilla, A. A.,** Augmentation of bone repair by inductively coupled electromagnetic fields, *Science,* 184, 575, 1974.
20. **Plonsey, R. and Fleming, D.,** *Bioelectic Phenomena,* McGraw-Hill, New York, 1969, 115.
21. **Houslay, M. D. and Stanley, K. K.,** *Dynamics of Biological Membranes,* John Wiley & Sons, New York, 1982, 296.
22. **Wilson, D. F.,** Energy transduction in biological membranes, in *Membrane Structure and Function,* Bittar, E. D., Ed., John Wiley & Sons, New York, 1980, 182.
23. **Johnson, C. C. and Guy, A. W.,** Nonionizing electromagnetic wave effects in biological materials and sytems, *Proc. IEEE,* 60, 692, 1972.
24. **Schwan, H. P.,** Field interaction with biological matter, *Ann. N.Y. Acad. Sci.,* 303, 198, 1977.
25. **Kraichman, M. B.,** *Handbook of Electromagnetic Propagation in Conducting Media,* NAVMAT P-2302, U.S. Superintendent of Documents, U.S. Government Printing Office, Washington, D.C., 1970.
26. **Massoudi, H., Durney, C. H., Barber, P. W.,and Iskander, M. F.,** Postresonance electromagnetic absorption by man and animals, *Bioelectromagnetics,* 3, 333, 1982.
27. **Sears, F. W., Zemansky, M. W., and Young, H. D.,** *University Physics,* 5th ed., Addison-Wesley, Reading, Mass., 1976, 360.
28. **Uman, M. A.,** *Lighting,* McGraw-Hill, New York, 1969, 162.
29. **Coelho, R.,** *Physics of Dielectrics for the Engineer,* Elsevier, Amsterdam, 1979, 155.
30. **Schwan, H. P.,** Dielectric properties of biological tissue and biophysical mechanisms of electromagnetic field interaction, in *Biological Effects of Nonionizing Radiation,* Illinger, K. H., Ed., ACS Symposium Series 157, American Chemical Society, Washington, D.C., 1981, 121.
31. **Koryta, J.,** *Ions, Electrodes and Membranes,* John Wiley & Sons, New York, 1982.
32. **Rosenbaum, E. J.,** *Physical Chemistry,* Appleton-Century-Crofts, Educ. Division, Meredith Corp., New York, 1970, 595.
33. **Watson, J. D.,** *Molecular Biology of the Gene,* W. A. Benjamin, Menlo Park, Calif.,, 1976, 91.
34. **Rosenbaum, E. J.,** *Physical Chemistry,* Appleton-Century-Crofts, Educ. Division, Meredith Corp., New York, 1970, 595.
35. **Illinger, K. H.,** Electromagnetic-field interaction with biological systems in the microwave and far-infrared region, in *Biological Effects of Nonionizing Radiation,* Illinger, K. H., ACS Symposium Series 157, American Chemical Society, Washington, D.C., 1981, 1.

36. **Kohli, M., Mei, W. N., Van Zandt, L. L., and Prohofsky, E. W.,** Calculated microwave absorption by double-helical DNA, in *Biological Effects of Nonionizing Radiation,* Illinger, K. H., Ed., ACS Symposium Series 157, American Chemical Society, Washington, D.C., 1981, 101.
37. **Kondepudi, D. K.,** Possible effects of 10^{11} Hz radiation on the oxygen affinity of hemoglobin, *Bioelectromagnetics,* 3, 349, 1982.
38. **Furia, L., Gandhi, O. P., and Hill, D. W.,** Further investigations on resonant effects of mm-waves on yeast, Abstr. 5th Annu. Sci. Session, Bioelectromagnetics Society, University of Colorado, Boulder, June 12 to 17, 1983, 13.
39. **Grundler, W., Keilman, F., and Fröhlich, H.,** Resonant growth rate response of yeast cells irradiated by weak microwaves, *Phys. Lett.,* 62A, 463, 1977.
40. **Fröhlich, H.,** Coherent processes in biological systems, in *Biological Effects of Nonionizing Radiation,* Illinger, K. H., Ed., ACS Symposium Series 157, American Chemical Society, Washington, D.C., 1981, 213.
41. **Fröhlich, H.,** What are non-thermal electric biological effects?, *Bioelectromagnetics,* 3, 45, 1982.
42. **Fröhlich, H.,** Coherent electric vibrations in biological systems and the cancer problem, *IEEE Trans. Microwave Theory Tech.,* 26, 613, 1978.
43. **Fröhlich, H.,** The biological effects of microwaves and related questions, in *Advances in Electronics and Electron Physics,* Marton, L. and Marton, C., Eds., Academic Press, New York, 1980, 85.
44. **Prohofsky, E. W. and Eyster, J. M.,** Prediction of giant breathing and rocking modes in double helical RNA, *Phys. Lett.,* 50A, 329, 1974.
45. **Careri, J.,** Search for cooperative phenomena in hydrogen-bonded amide stuctures, in *Cooperative Phenomena,* Haken, H. and Wagner, W., Eds., Springer-Verlag, Basel, 1973, 391.
46. **Fröhlich, H.,** Selective long range dispersion forces between large systems, *Phys. Lett.,* 39A, 153.
47. **Barnes, F. S. and Hu, C.-L. J.,** Nonlinear Interactions of electromagnetic waves with biological materials, in *Nonlinear Electromagnetics,* Uslenghi, P. L. E., Ed., Academic Press, New York, 1980, 391.
48. **Grundler, W., Keilmann, F., Putterlik, V., Santo, L., Strube, D., and Zimmermann, I.,** Nonthermal resonant effects of 42 GHz microwaves on the growth of yeast cultures, in *Coherent Excitations of Biological Systems,* Frölich, H., and Kremer, F., Eds., Springer-Verlag, Basel, 1983, 21.

Chapter 1

DIELECTRIC PROPERTIES OF TISSUES

Kenneth R. Foster and Herman P. Schwan

CONTENTS

0-8493-0641-8/96/$0.00+$.50

I. INTRODUCTION

The bulk electrical properties of tissues and cell suspensions have been of interest for many reasons for over a century. These properties determine the pathways of current flow through the body and are therefore important in the measurement of physiological parameters using impedance techniques, studies of biological effects of electromagnetic fields, electrocardiography, muscle contraction and nerve transmission. Indeed, biological impedance studies have long been important in electrophysiology and biophysics; one of the first demonstrations of the existence of the cell membrane was based on dielectric studies on cell suspensions.

It has been nearly 40 years since the publication of a comprehensive review of the bulk electrical properties of tissues and cell suspensions by Schwan,[1] from 10 Hz to almost 10 GHz. The subsequent years have seen the development of computer-controlled instrumentation for precise and rapid measurements of dielectric properties, and a greater understanding of dielectric phenomena in tissues and other complex materials.

This critical review emphasizes the mechanisms that are responsible for the observed properties. We believe that sufficient dielectric data are presently available for most engineering applications, and that the mechanisms for the dielectric properties of tissues at hertz through low gigahertz frequencies are largely understood even though much remains to be done.

The specific aims of this chapter are:

1. To summarize some of the very early work on this subject and establish a historical perspective.

2. To review the basic concepts of dielectric phenomena. The most important relaxation mechanisms observed in tissues will be summarized.
3. To review the dielectric properties of several tissues at frequencies from extra-low-frequency through microwave ranges. We discuss in some detail more recent data and the underlying mechanisms.
4. To discuss effects of cellular structure on the dielectric properties of tissues. We describe the coupling between externally applied electromagnetic (EM) fields and intracellular compartments, including membranes, as a logical extension of this subject.
5. To examine some questions of interpretation of dielectric relaxation spectroscopy, in particular, the limits to which the distribution of relaxation times can be inferred from dielectric data. The discussion will extend the earlier comments on this subject by Schwan[1] using a more precise analysis that has recently become possible. We hope that this will help resolve several disputes involving different interpretations of dielectric data from tissues and other biological materials.
6. To review the dielectric relaxation properties of tissue water and their relation to other transport properties of biological systems.

This review is a revised and updated version of the chapter published in the first edition of this book.[2] We have rewritten the material in places, and updated it to reflect developments that occurred after its preparation in 1984–5. We have extended the discussion on the dielectric properties of some tissues (in particular, neoplastic and adipose tissues), and on the coupling of electromagnetic fields with cells and tissues; important recent developments in the theory of counterion polarization have been summarized. Finally, we have corrected some errors in the earlier review.

Several other major reviews of the dielectric properties of tissues and cell suspensions are available. Cole,[3] Pethig,[4] Grant et al.,[5] and Schanne and P.-Ceretti[6] have published noteworthy books on dielectric properties of biological materials; see also a chapter in the recent book by Duck.[7] Pethig and Kell[8] have reviewed the passive electrical properties of biological systems with a perspective that is quite different from this review. Stuchly and Stuchly[9] have published an exhaustive tabulation of data from tissue, other biological materials, and "phantom" tissue materials used to mimic real tissues for engineering applications. Schwan and Foster[10] and Stuchly[11] have reviewed the mechanisms of interaction of EM fields with biological materials. The Stuchlys' papers complement our own and we highly recommend them.

A. HISTORICAL INTRODUCTION

Interest in the electrical properties of biological materials began about 100 years ago, as instruments for the measurement of electrical resistance, and eventually capacitance, became available. This interest gradually developed into an important branch of biophysics and yielded, even from the first, important insights into electrophysiology.

Several authors have summarized this early work. Gildemeister[12] reviewed the extensive literature already in existence in 1928. Rajewsky, in his 1938 monograph[13] *Ultrashort Waves and Their Medical and Biological Applications*, offered a particularly complete survey of the field up to that time. Schanne and P.-Ceretti[6] and Cole[3] have provided many references to the older literature; Cole's historical survey covering the 1920s through 1940s is particularly interesting in view of the many important contributions he made during that period.

Over a century ago investigators became interested in the bulk electrical properties of tissues and blood. They quickly discovered that the electrical properties of such materials differ greatly from those of simple electrolyte solutions. For example, they observed that upon the application of a direct current (DC) potential step, the current changes with time, beginning with a charging spike; they also observed hysteresis phenomena. These effects were poorly understood for some time, and in retrospect the data might be suspected of being influenced by electrode polarization artifacts. This early work led to the concept of a tissue "polarization." Du Bois-Reymond, the eminent neurophysiologist, summarized this early work in his book *Investigations about Animal Electricity* that was published in Berlin in 1849.[14]

Some early discoveries can be summarized as follows. Peltier[15] discovered in 1834 that "animal bodies" behave as "polarization cells" if exposed to DC currents, i.e., they are able to store electricity and release it after the termination of the current. Hermann[16] reported in 1872 that the DC resistance of muscle varies by a factor of 4 to 9 with orientation in freshly excised tissue, but this anisotropy slowly disappears after death. He correctly interpreted this as arising from the microscopic structure of the tissue, but incorrectly proposed that it arose from different ionic permeabilities in directions perpendicular and parallel to the fibers. Hermann probably was the first to measure the resistance of tissues to "alternating" currents (AC), using an apparatus of his own construction that alternately applied current pulses of opposite sign to the tissue.

As electrical technology improved, investigators began to perform measurements on tissue using sinusoidally alternating currents. Several authors reported a complex frequency dependence of the electrical properties of tissues. Du Bois[17] found in 1898 that skin behaves like a capacitor if subjected to DC currents. Galler[18] reported in 1913 the first true AC measurements on a tissue (frog skin) and found that resistance at 1 kHz is appreciably smaller than at DC. Gildemeister[19] in 1919 and Einthoven[20] in 1923 further examined the frequency dependence of tissue resistance. Gildemeister and others observed the current peak that occurs when skin is exposed to a voltage step and interpreted this as arising from a resistance that increases with time; they considered the initial small resistance the "true" resistance, and the larger final resistance as "pretended." The time constants they observed for transition from "true" to "pretended" resistances were of the order of milliseconds. Other investigators interpreted this apparent increase in resistance as arising from induced counter potentials of a "polarization" kind. In retrospect, this early "time domain spectroscopy" might have been troubled by electrode polarization effects.

Several investigators studied the impedance properties of blood, and gained important insights that could be applied to more complex tissues. For example, Stewart[21] measured the low frequency resistance of blood while developing indicator dilution measurements of blood flow. These early workers recognized that erythrocytes appear at low frequencies as nonconducting particles in a conducting medium. Attempts to apply dielectric mixture theories (such as that developed by Maxwell in his famous treatise[22]) to blood met with limited success. In retrospect, this difficulty might have arisen in part because of the failure of Maxwell's theory for concentrated suspensions, and in part because erythrocytes are not spherical as the theory demanded.

The work of Höber[23,24] helped clarify the underlying reason for the strong frequency dependence of the resistance of blood. Because no continuous-wave, high-frequency generators were available, Höber inferred the resistance of erythrocyte suspensions at high frequencies by observing the damping of oscillations that were excited in a resonant circuit by a spark discharge generator. By substituting various salt solutions, he measured the apparent ionic strength of blood, and showed that it increased significantly as the frequency of measurement was raised. These studies were conducted only a few years after Bernstein had proposed the final version (1902) of his hypothesis of the cell membrane. Höber speculated that the membrane excluded low frequency currents, and passed high frequency currents. In short, he discovered and correctly identified the mechanism of what is now called the "beta dispersion." Höber's interpretation was hotly debated for many years, but eventually prevailed.

The physiological effects of electric currents excited great popular and scientific interest in the 19th Century.[25] "Electrotherapy" came to be widely practiced both by charlatans and reputable physicians; Licht[26] cites 923 dissertations and books on the subject, most written in the 19th century! Shortly after technology was developed for producing high frequency currents, investigators began to study its biological effects, frequently using themselves as subjects. D'Arsonval was the first in 1892 to conduct experiments using high frequency currents of uncertain waveshape. Tesla followed using alternating currents of an unspecified frequency, probably in the range of 10 to 100 kHz. One remarkable finding of these investigators was that high frequency currents do not produce shocks and muscle contractions, and can be passed through the body with little effect.

This finding was surprising to contemporary physiologists. Tesla proposed that tissues are "condensors," i.e., they offer low resistance at high frequencies.[27] Nernst, the famous physiologist, developed a theory in 1908 for the threshold for stimulation of tissue by alternating currents, that predicted the threshold should increase in proportion to the square root of the frequency.[28] Later studies demonstrated an increase in the threshold,[29] although with a frequency dependence that was different from Nernst's prediction. These problems interested some of the most prominent physicists, physiologists, and engineers of the time: Wien, Nernst, and Tesla.

The understanding gained before World War I can be summarized as follows. Scientists of the day recognized that tissue conducts electricity, that its resistance varies with frequency, and that the mechanism for conduction is the movement of ions. They recognized that cellular structure and cell membranes determined the electrical properties of tissues, and they recognized—if not fully understood—the anisotropy in the electrical properties of muscle and nerve tissues. The concept of tissue "polarization" was established, which reflects the capacitive properties of tissues as we know them today. Clearly, this was an extremely active period and much insight was gained. However, the origin of the tissue "polarization" remained unknown and the complex events that occur when tissue is electrically excited could not be clearly separated from passive linear processes such as the capacitive charging of the cell membrane.

After World War I, dielectric studies encompassed a steadily increasing range of frequencies and variety of materials, and more sophisticated attempts were made to understand the underlying mechanisms. Philippson[30] conducted measurements on many tissues over a wide frequency range, and demonstrated that the electrical properties of all tissues are frequency dependent. McClendon[31] reported the resistance and capacitance of blood from 266 Hz to 2 MHz, and was perhaps the first to measure both components of the complex admittance of tissues over this extended range. Investigators proposed all sorts of equivalent circuits to describe the frequency dependence of the properties; they also recognized that cellular structure and cell membranes were largely responsible for these effects.

During the 1920s and 1930s various cellular systems were successfully characterized. Osterhout[32] performed an influential study of the conductance of the marine kelp, and interpreted the changes in tissue conductance after death as arising from changes in the membrane permeability. Fricke, Curtis, and Cole entered the field and developed advanced instrumentation that permitted the accurate measurement of both conductivity and permittivity over a wide frequency range. In a famous series of experiments, they studied the dielectric properties of various cell suspensions and applied the complex extension of the Maxwell mixture theory to analyze their data. Dänzer[33] developed a rigorous theory for the beta dispersion in cell suspensions. Rajewsky and colleagues (Dänzer, Osken, Schaefer, Wachter, and Schwan) measured the conductivity of many materials, including tissues and biological solutions from a few kilohertz to 100 MHz, using bridges at audio and radio frequencies, and transmission lines at very high frequencies.

These efforts and others not mentioned led to extensive conductivity and permittivity data from many systems, and a quantitative understanding of the mechanisms responsible for the beta dispersion for cell suspensions. The work of Fricke, Curtis, and Cole, in particular, is noteworthy for its application of sophisticated physical concepts and advanced measurement techniques to the solution of important problems in biology.

The Second World War stimulated rapid developments in microwave technology, including the development of transmission lines, waveguides and oscillators operating at frequencies up to 10 GHz. Investigators soon adapted the technology for dielectric studies on biological materials. Wachter was probably the first to measure the absorption of water, blood, milk, ascites, and other biological fluids at 1 to 10 GHz using damped oscillation.[34] He obtained values 50 years ago that are close to those accepted today. In 1948, Rajewsky and Schwan[35] were the first to report complex permittivity data on blood at frequencies up to nearly 1 GHz, and their data were soon confirmed and extended to still higher frequencies by Cook[36] in England. Dielectric properties of several canine tissues were obtained in the late 1940s from 1 to 8.5 GHz by Herrick (published in[37]) as part of an early effort to develop microwave diathermy. In the late 1940's, several British investigators—including Cook, Buchanan, Grant, and England—began to study the dielectric properties of biological systems at microwave frequencies.

Experimental and theoretical advances continued. The early theories by Cole, Fricke, and Dänzer for the dielectric properties of cell suspensions were further simplified and extended by Schwan and coworkers. Schwan discovered the existence of additional dispersions, first at microwave frequencies then at audio frequencies and below. He recognized that counterion displacement was one of several mechanisms responsible for this extraordinary low-frequency dispersion. He and his coworkers were the first to demonstrate the phenomena in suspensions of materials such as cells, fat and latex particles. Thus, since World War II, the frequency range of measurements has been extended both upward and downward, additional dispersions have been discovered and explained, and previously known dispersions in cellular suspensions have been better studied.

Studies of the dielectric properties of biological materials remain an active area of research. The laboratory that Schwan established at the University of Pennsylvania in 1950 remains active today. Other groups have carried out excellent dielectric studies on biological materials in recent years; among the more active have been those led by Grant in London, Carstensen and colleagues at Rochester, Hanai and Irimajiri in Japan, Pethig in Wales, and Stuchly in Ottawa. Theoretical advances have been made by Hanai, Irimajiri, Carstensen, Chew, Grant, Grosse, and others.

With the introduction of modern network analyzer and time domain measurement systems, the determination of the dielectric properties of biological samples is easier which has encouraged increased efforts in this field. Many opportunities still exist, and fundamental problems remain to be solved. Moreover, it is now possible to measure *electrical* parameters of cells by observing their *mechanical* response to an applied electric field. This has potential application to cell sorting and electric field-induced cell fusion, both important to biotechnology. To understand and optimize these techniques, the electrical properties of the cells and their mechanical response to an applied field must be better understood.

B. DEFINITIONS AND BASIC CONCEPTS

The dielectric permittivity ϵ and conductivity σ of a material are, respectively, the charge and current densities induced in response to an applied electric field of unit amplitude. We illustrate these definitions by the example of an idealized parallel plate capacitor, of plate area A and separation d, which contains the material under investigation (Figure 1). Initially let the region between the plates be a vacuum. A constant voltage difference V between the plates will induce a charge density D given by

$$D = \epsilon_0 V/D \qquad (1)$$

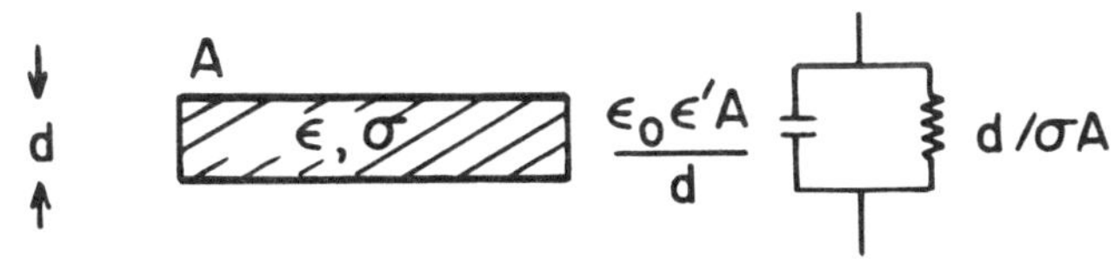

FIGURE 1 Equivalent circuit of an idealized parallel plate capacitor filled with material of relative permittivity ϵ and conductivity σ. The plate area is A and the distance between plates is d. In a real circuit, a small additional shunt capacitance would occur from the stray fields in the air outside of the material.

where ϵ_0 is the permittivity of free space (8.85×10^{-12} F/m). The capacitance C is the ratio of the total induced charge on the plates to the applied voltage. Ignoring the effects from the fringing fields outside of the region between the plates it is:

$$C = \epsilon_0 A/d \tag{2}$$

If a material is now introduced between the plates, an additional charge density P_s will be induced on the plates from the polarization of charges within the material. For sufficiently low voltages and in the limit of long times, this charge density is proportional to E and can be written

$$\begin{aligned} D &= \epsilon_0 E + P_s \\ &= \epsilon_s \epsilon_0 E \end{aligned} \tag{3}$$

and the capacitance is now

$$C = \epsilon_s \epsilon_0 A/d \tag{4}$$

where ϵ_s is the static permittivity of the material. The circuit might also include a conductance G in parallel with the capacitance

$$G = \sigma_s A/d \tag{5}$$

where σ_s is the static conductivity of the material. If relaxation processes are present, the response of the material to a voltage step will be a function of time, and the "static" permittivity and conductivity will be found from Equations 3 and 5 only in the limit of infinite time. In that case, the corresponding response to a sinusoidal field will be characterized by dielectric properties (ϵ, σ) which vary with frequency.

For sinusoidally oscillating fields, the electrical properties of the circuit of Figure 1 can be specified in several equivalent ways:

1. At circular frequency ω the complex admittance Y^* can be written

$$\begin{aligned} Y^* &= G + j\omega C \\ &= (A/d)\,(\sigma + j\omega\epsilon_0\epsilon) \end{aligned} \tag{6a}$$

from which is defined the complex conductivity σ^* (sometimes called the admittivity or specific admittance) of the material

$$\sigma^* = \sigma + j\omega\epsilon\epsilon_0 \tag{6b}$$

2. Equivalently, the complex capacitance C* can be defined as

$$C^* = Y^*/j\omega \tag{7}$$

which leads to the definition of the complex relative permittivity ϵ^*

$$\epsilon^* = \epsilon' - j\epsilon'' \tag{7b}$$
$$= \epsilon' - \frac{j\sigma}{\omega\epsilon_0}$$

The complex conductivity and permittivity are related by

$$\sigma^* = j\omega\epsilon^*\epsilon_0 \tag{8}$$

3. Finally, the series equivalent impedance Z* can be written

$$Z^* = 1/Y^*$$
$$= R + jX \tag{9a}$$
$$= \frac{\sigma - j\omega\epsilon_0\epsilon'}{\sigma^2 + (\omega\epsilon_0\epsilon)^2}$$

from which the complex specific impedance of the material z* is defined as

$$z^* = 1/\sigma^* \tag{9b}$$
$$= \frac{\sigma - j\omega\epsilon'\epsilon_0}{\sigma^2 + (\omega\epsilon'\epsilon_0)^2}$$

There is much confusion in the literature about nomenclature. The conductivity σ usually has a frequency-independent part (due to ionic conduction) and a frequency-dependent part (due to dielectric relaxation), and ϵ'' (called the "loss") is sometimes used referencing only the frequency-dependent part of the conductivity. In this review, we always use ϵ'' as equivalent to $\sigma/\omega\epsilon'\epsilon_0$, where σ is the measured conductivity. The chemical literature contains many references to the "dielectric constant" of materials, which means the real part of the complex permittivity (that is, ϵ'), typically at low frequencies at which ϵ' is essentially independent of frequency. The "resistivity" ρ indicates the inverse of the conductance, or $1/\sigma$, which is quite different from the real part of the tissue impedance, z*. We (and many other writers) frequently use the terms "permittivity" or "relative permittivity" to indicate the real part of the complex permittivity relative to that of free space, ϵ_0. Thus, the "permittivity" of water below the microwave range is about 78, although in most theoretical expressions the SI system of units (used here) would require also the constant ϵ_0.

Dielectric data can be represented in several equivalent ways to best suit the purposes at hand. In the older physiologically oriented literature, authors frequently presented the dielectric properties of tissues in the series equivalent form (as tissue impedances) rather than in the parallel equivalent (complex permittivity) form. However, the fundamental processes, buildup of charge density and electrical conduction, occur in parallel, and the dielectric properties of a material are more logically presented as a parallel combination of a capacitive (permittivity) and a conductive (conductivity) element. For example, the permittivity and conductivity of most aqueous electrolyte solutions are practically independent of frequency up to the microwave range; the corresponding specific impedance values would be highly frequency dependent as seen from Equation 9b. For this reason, data are best represented in the parallel equivalent form, in terms of permittivities and conductivities.

The illustration of an ideal parallel plate capacitor is only meaningful at frequencies at which the free space wavelength is much larger than the dimensions of the capacitor. Even at relatively low frequencies, real circuits exhibit a variety of parasitic effects that can be modeled as series or shunt impedance elements; any practical measurement of the dielectric properties of a material must involve correction for them. An extensive review of practical measurement techniques is given by Schwan.[36]

C. GENERAL RELAXATION THEORY

The response of a material to a voltage step involves the physical displacement of charge, and the kinetics of this displacement determines the frequency dependence of the bulk properties. In principle, the response could be either relaxational or resonance in nature. A few investigators have claimed to observe resonance-type effects (for example in DNA solutions at low-Gigahertz frequencies), but these claims could not be confirmed by subsequent work and are probably in error. Relaxational effects may be described by first-order differential equations which lead to single time constant responses. However, some dielectric processes are known that are not single time constant in nature, for example those associated with diffusion of counterions (Section II C). Moreover, in real physical and biological materials, several relaxation processes might occur in parallel and the total electrical response of the material might be characterized by several time constants.

In this section, we review the dielectric response of a material with a single relaxation process; in Section II we will describe several physical mechanisms leading to relaxation in biological systems, and in Section IV we discuss the dielectric properties of several tissues and their relation to these mechanisms.

In the simplest case, the polarization of a sample will relax towards the steady state as a first order process characterized by a relaxation time τ:

$$D = D_\infty + (D_0 - D_\infty)(1 - e^{-t/\tau}) \tag{10a}$$

In this equation, D_∞ is the "instantaneous" response, that might arise in a real system from polarization of electronic clouds (a very fast process), while D_0 is the response long after the application of the step. The time constant τ depends on the physical process involved: it can be as short as picoseconds (for reorientation of molecular dipoles) or as long as seconds (for counterion effects). The dielectric response of a first order system in the frequency domain can be obtained by Laplace transformation:

$$\epsilon = \epsilon_\infty + \frac{\epsilon_s - \epsilon_\infty}{1 + j\omega\tau} \tag{10b}$$

where ϵ_∞ now refers to the permittivity at infinite frequencies due to the electronic polarizability. Equation 10a does not include the possible existence of currents at infinite time, such as would arise from movement of ions in a constant field. We expand the model by including a static conductivity σ_s:

$$\epsilon^* = \epsilon_\infty + \frac{\epsilon_s - \epsilon_\infty}{1 + j\omega\tau} - \frac{\sigma_s}{j\omega} \tag{10c}$$

Separating Equation 10c into real and imaginary parts yields:

$$\epsilon^* = \epsilon' - j\epsilon'' \tag{11a}$$

$$\epsilon' = \epsilon_\infty + \frac{\epsilon_s - \epsilon_\infty}{1 + (\omega\tau)^2} \tag{11b}$$

$$\epsilon'' = \frac{\sigma_s}{\omega\epsilon_0} + \frac{(\epsilon_s - \epsilon_\infty)\omega\tau}{1 + (\omega\tau)^2} \tag{11c}$$

$$\sigma = \sigma_s + \frac{(\sigma_\infty - \sigma_s)(\omega\tau)^2}{1 + (\omega\tau)^2} \tag{11d}$$

in which the limit values ϵ_∞, ϵ_s, σ_∞, σ_s are interrelated by

$$\frac{(\epsilon_s - \epsilon_\infty)\epsilon_0}{\tau} = \sigma_\infty - \sigma_s \tag{12}$$

Thus, the permittivity and conductivity cannot vary independently, and the frequency dependence of ϵ' will determine the frequency dependence of σ and conversely (see Section I. E).

One might just as well look at the time dependence of the current density J:

$$J/E = \sigma_\infty + (\sigma_s - \sigma_\infty)(1 - e^{-t/\tau}) \tag{13}$$

This transforms into the admittance equivalent of Equation 10b:

$$\sigma^* = \sigma_\infty + \omega\epsilon_\infty\epsilon_0 + \frac{(\sigma_s - \sigma_\infty)}{1 + j\omega\tau} \tag{14}$$

where we have included $\omega\epsilon_\infty\epsilon_0$ to account for a frequency independent addition to the relaxation effect not included in Equation 13.

Figure 2 shows the variation in the permittivity and conductivity with frequency for a singe time constant relaxation. This dispersion extends over roughly a decade in frequency

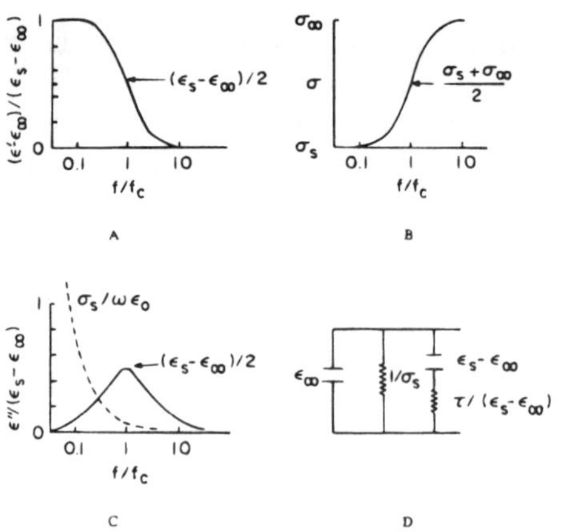

FIGURE 2 Single time constant relaxation. (A) Relative permittivity vs. frequency; (B) conductivity vs. frequency; (C) loss vs. frequency showing (broken line) a contribution from a DC conductivity; (D) an equivalent circuit that models the Debye equations. In this single time constant situation, the relaxation frequencies defined from (A) and (B) would be identical; in the presence of a distribution of time constants, the relaxation frequency from the conductivity plot (B) would exceed that from the permittivity plot (A).

from one third to three times f_c. At the relaxation (or characteristic) frequency $f_c = (2\pi\tau)^{-1}$ the permittivity is halfway between its low and high frequency values. Alternately, the relaxation frequency could be taken as that at which the conductivity is halfway between its two limiting values. For the single time constant relaxation considered here, these two frequencies are the same; they are different if there is a distribution of relaxation times (cf. I D).

Dielectric relaxation behavior can be conveniently plotted on the complex plane. Different plots can be used to emphasize different aspects of the relaxation behavior:

In the normalized complex admittance plane Y*, or equivalently, the complex conductivity plane, (see Equation 6) the imaginary part of Y* or σ^* are plotted against their real parts. Equation 14 can be transformed into

$$[\sigma - (\sigma_s - \sigma_\infty)/2]^2 + [\omega\epsilon_0(\epsilon' - \epsilon_\infty)]^2 = [(\epsilon_s - \epsilon_\infty)\epsilon_0/2\tau]^2 \tag{15}$$

Thus, a plot of $\omega\epsilon_0(\epsilon' - \epsilon_\infty)$ vs. σ is a circle centered at $\sigma' = (\sigma_s + \sigma_\infty)/2$, of which only the half corresponding to positive values of $\omega\epsilon_0(\epsilon' - \epsilon_\infty)$ has physical meaning. At the relaxation frequency when $\omega = 2\pi f_c = 1/\tau$, the ordinate becomes $(\sigma_\infty - \sigma_s)/2$, which corresponds to the apex of the semicircle.

In the complex permittivity plane

$$[\epsilon' - (\epsilon_s + \epsilon_\infty)/2]^2 + [\epsilon'' - \sigma_s/\omega\epsilon_0]^2 = [(\epsilon_s - \epsilon_0)/2]^2 \tag{16}$$

Thus a plot of $(\epsilon'' - \sigma_s/\omega\epsilon_0)$ vs. ϵ' will produce a semicircle for single time constant relaxation. The center of the semicircle is on the real axis at the point $(\epsilon_s + \epsilon_\infty)/2$, and its apex occurs at $\omega = 2\pi f_c$.

If a distribution of relaxation times is present, the corresponding plots in the complex permittivity and complex admittance planes no longer are semicircles centered on the real axis, but might resemble semicircular arcs with centers below the real axis, or skewed arcs. In that case, the relaxation frequencies obtained from the peaks of the plots in the complex permittivity or admittance planes are different, the larger frequency being the one obtained from the complex admittance plane.

Plots in the complex impedance plane are generated from those in the complex admittance (or permittivity) plane by the transformation Z = 1/Y, which corresponds to a conformal mapping. Since this mapping will transform a circle into a circle, a circle in the admittance plane will yield a circle in the complex impedance plane. However, the frequencies corresponding to the apices of the arcs (i.e. the relaxation frequencies) will be different, by the ratio σ_∞/σ_s. Thus the peak of the circle in the impedance plane does not directly give the relaxation frequency of the data expressed in the complex permittivity or complex conductivity form.

A special case occurs when ϵ and σ are independent of frequency. The plot in the admittance plane is then a straight line intersecting the real axis at σ. In the impedance plane this corresponds to a circle which approaches the origin as ω approaches ∞. Thus the data, if plotted on the impedance plane, will exhibit a frequency dependence even though the conductivity and permittivity are independent of frequency. An analogous situation is found in a parallel combination of a resistor and capacitor: the series equivalent impedance of the combination varies with frequency even though the values of the components themselves do not.

D. DISTRIBUTION OF RELAXATION TIMES

Few substances, in fact, show single time constant relaxation. For some materials (e.g. water) the dielectric properties are very closely fitted by the simple Debye equations, while for others (e.g. tissues) the departure is very great.

There are several reasons why a substance might depart from single time constant behavior. One might be the presence of multiple relaxation processes, each with a different relaxation time; another might be the existence of dielectric processes whose kinetics are not of first order. Moreover, in concentrated suspensions, electrical interactions between the suspended particles will usually lead to a distribution of relaxation times in the bulk properties even though the individual particles themselves might show single time constant relaxation. To distinguish reliably among these possibilities can be a difficult problem.

The simplest case is when the dielectric response arises from the presence of independent first order processes. In that case, the dielectric response to a step change E in field strength can be described as:

$$D = D_\infty + D_1\,(1 - e^{-t/\tau_1}) + D_2\,(1 - e^{-t/\tau_2}) + \dots \tag{17}$$

or

$$\epsilon^*(\omega) = \epsilon_\infty + \frac{\Delta\epsilon_1}{1 + j\omega\tau_1} + \frac{\Delta\epsilon_2}{1 + j\omega\tau_2} + \cdots \tag{18}$$

If the relaxation times are well separated, i.e. if

$$\tau_1 \ll \tau_2 \ll \tau_3 \dots$$

a plot of the permittivity and conductivity vs. frequency will exhibit clearly resolved dispersions, separated by plateaus at intermediate frequencies. A plot of $\epsilon'' - \sigma_s/\omega\epsilon_0$ vs. ϵ' in the complex permittivity plane will then appear as a series of semicircles, each with a center on the real (permittivity) axis, and they will intersect the real axis at points

$$\epsilon_\infty,\ \epsilon_\infty + \Delta\epsilon_1,\ \epsilon_\infty + \Delta\epsilon_1 + \Delta\epsilon_2,\ \dots$$

If the relaxation times are not well separated, the material will exhibit a broad, perhaps featureless dispersion extending from roughly one third the lowest to three times the highest relaxation frequency that is present. In the limit of a continuous distribution of relaxation times, Equation 18 would be replaced by an integral

$$\epsilon^* = \epsilon_\infty + (\epsilon_s - \epsilon_\infty)\int_0^\infty \frac{\rho(\tau)d\tau}{1 + j\omega\tau} - j\sigma_s/\omega\epsilon_0 \tag{19}$$

where

$$\int_0^\infty \rho(\tau)\,d\tau = 1 \tag{20}$$

In the absence of resonance phenomena, Equation 19 can represent all dielectric relaxation data, provided one chooses an appropriate distribution function $\rho(\tau)$.

Macdonald has described some of the problems in the interpretation of dielectric data that are characterized by a distribution of relaxation times.[38] In principle—and sometimes in practice—one can invert dielectric relaxation spectra to determine $\rho(\tau)$ directly.[39] However, this process is numerically unstable unless there are only a few components that are widely separated in frequency. More commonly, an investigator will assume a distribution function (or equivalently, a function that describes the frequency-dependence of the dielectric properties) and fit the data to the function. This introduces a model-dependence in the interpretation of the data that the investigators themselves do not always acknowledge.

Many investigators have proposed different functions for use in fitting dielectric data. These fall broadly into one of two classes: those with some theoretical justification, and simple empirical functions. For example, it is possible to derive distribution functions from the physically plausible assumption of a Gaussian distribution of activation energies. But, the resulting distribution functions may be too complicated to be very helpful in interpreting data. Empirical distribution functions, that lack theoretical justification, serve chiefly to parameterize the data without clarifying the underlying mechanisms involved.

Cole and Cole[40,41] proposed a relaxation function has been widely used:

$$\epsilon^* = \epsilon_\infty + \frac{\epsilon_s + \epsilon_\infty}{1 + (jf/f_c)^{1-\alpha}} - j\sigma_s/\omega\epsilon_0 \tag{21}$$

This separates into real and imaginary parts

$$\frac{\epsilon' - \epsilon_\infty}{\epsilon_s - \epsilon_\infty} = \frac{1 - (f/f_c)^{1-\alpha}\sin(\alpha\pi/2)}{1 + (f/f_c)^{2(1-\alpha)} + 2\,(f/f_c)^{1-\alpha}\sin(\alpha\pi/2)} \tag{22}$$

$$\frac{\epsilon'' - \sigma_s/\omega\epsilon_0}{\epsilon_s - \epsilon_\infty} = \frac{(f/f_c)^{1-\alpha}\cos(\alpha\pi/2)}{1 + (f/f_c)^{2(1-\alpha)} + 2\,(f/f_c)^{1-\alpha}\sin(\alpha\pi/2)} \tag{23}$$

At frequencies $f \gg f_c$, the dielectric relaxation approaches constant power-law behavior. If $\alpha > 0$, in the limit of high frequencies

$$\begin{aligned} \epsilon' &\propto (f/f_c)^{\alpha-1}\sin(\alpha\pi/2) \\ \epsilon'' &\propto (f/f_c)^{\alpha-1}\cos(\alpha\pi/2) \\ \sigma &\propto (f/f_c)^{\alpha}\sin(\alpha\pi/2) \end{aligned} \tag{24}$$

Note that the conductivity diverges as $f \rightarrow \infty$ which is physically unreasonable.

The Cole-Cole function leads to distinctive semicircular plots in the complex permittivity plane. As the frequency is varied, a plot of $(\sigma - \sigma_s)/\omega\epsilon_0$ vs. ϵ' describes a circular arc given by

$$(x - x_1)^2 + [y + (r - y_1)]^2 = r^2 \tag{25}$$

where

$$x = (\epsilon' - \epsilon_\infty)/(\epsilon_s - \epsilon_\infty)$$

$$y = (\epsilon'' - \sigma_s/\omega\epsilon_0)/(\epsilon_s - \epsilon_\infty)$$

$$x_1 = 1/2$$

$$y_1 = \frac{\cos(\alpha\pi/2)}{2[1 + \sin(\alpha\pi/2)]}$$

$$r = [2\cos(\alpha\pi/2)]^{-1}$$

The apex of the arc corresponds to the mean relaxation frequency f_c. A radial line from the

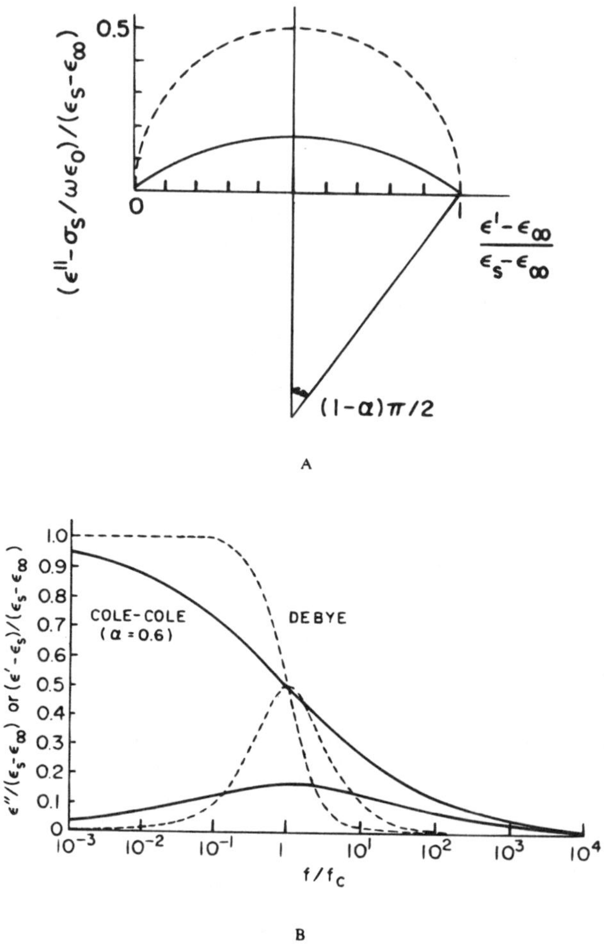

FIGURE 3 (A) The complex permittivity plot of the Cole-Cole equation (Equation 21), showing the depressed semicircular behavior that is characteristic of many biological materials. The parameters in this figure refer to Equation 21; (B) the corresponding permittivity and conductivity plot. The corresponding Debye relaxation is shown by the broken lines. The distribution parameter α is 0.6, corresponding to a broad range of relaxation times.

center of the arc to its intersection with the horizontal axis forms an angle of $(1 - \alpha)\pi/2$ radians with the vertical (Figure 3).

The distribution function $p(\ln(\tau))$ that corresponds to the Cole-Cole function is

$$p[\ln(\tau)] = \frac{1}{2\pi} \frac{\sin(\alpha\pi)}{\cosh[(1 - \alpha)\ln(\tau/\tau_0)] - \cos(\alpha\pi)} \tag{26}$$

where $\tau_0 = (2\pi f_c)^{-1}$ is the mean relaxation time. This distribution is logarithmically symmetrical about τ_0. We note that Equation 26 is written in a form that is normalized for the variable $\ln(\tau)$, i.e.

$$\int_{-\infty}^{\infty} p[\ln(\tau)]d[\ln(\tau)] = 1 \tag{26a}$$

A comparable function can be written for the dielectric properties expressed in the complex conductivity form:

$$\sigma^* = \sigma_\infty + \frac{\sigma_s - \sigma_\infty}{1 + (-jf/f_c)^{1-\alpha}} + j\omega\epsilon_0\epsilon_\infty \tag{27}$$

and for the complex impedance:

$$z^* = R_\infty + \frac{R_s - R_\infty}{1 + (jf/f_c')^{1-\alpha}} \tag{28}$$

These equations are formally similar but are not equivalent; any of them could be used to summarize empirically the dielectric properties of tissues over broad frequency ranges. All behave in physically unreasonable ways in the limit of low or high frequency. Thus, in Equation 21 the conductivity diverges as $f \rightarrow \infty$; in Equation 27 the permittivity diverges as $f \rightarrow 0$; and in Equation 28 the series-equivalent capacitance vanishes as $f \rightarrow \infty$.

Circles in the admittance and impedance planes can be related as follows. The circle in the admittance plane can be written

$$y^* = \sigma_\infty + \frac{\sigma_s - \sigma_\infty}{1 + (jf/f_c)^{1-\alpha}} \tag{29}$$

which can be changed to

$$\frac{\sigma_s - y^*}{y^* - \sigma_\infty} = (j\omega\tau)^{1-\alpha} \tag{30}$$

Equation 28 may be written

$$\frac{z^* - \rho_s}{\rho_\infty - z^*} = \frac{\sigma_\infty}{\sigma_s}(j\omega\tau)^{1-\alpha} \tag{31}$$

and compared with the circle in the impedance plane

$$\frac{z^* - \rho_s}{\rho_\infty - z^*} = (j\omega\tau')^{1-\alpha} \tag{32}$$

where ρ_s and ρ_∞ are the limiting values of the resistivity at low and high frequencies. The time constants τ and τ' of the relaxations in the admittance and impedance planes are different; in fact

$$(\tau/\tau')^{1-\alpha} = \sigma_s/\sigma_\infty \tag{33a}$$

or

$$\tau^{1-\alpha}\frac{\sigma_\infty}{\sigma_s} = (\tau')^{1-\alpha} \tag{33b}$$

Thus, the peak frequency of the impedance circle will always be lower than that for the admittance circle. The difference can be very pronounced. For example, the beta dispersion of muscle (see Section IV.A.2.) can be described by a Cole-Cole equation with conductivity ratio (σ_∞/σ_s) of 4 to 5, with a distribution parameter α of 0.3. Thus, $\tau'/\tau \approx 7$ to 9.

The Cole-Cole function implies a fractional power law behavior, or (equivalently) a constant phase angle behavior of the tissue impedance. In fact, the dielectric properties of most tissues, if plotted suitably on a log-log scale, appear as straight lines over broad frequency ranges, and can be adequately fitted to Equation 21, 27, or 28, with α values between 0.3–0.5. Indeed, many investigators customarily fit tissue data to the Cole-Cole function.

This "fractional power law" behavior has elicited much discussion. However, its biophysical significance is unclear. Cole and Fricke suggested that cell membranes have an impedance that is characterized by a phase angle that is independent of frequency. There is little independent evidence for such behavior, however. A distribution of relaxation times (which might mimic constant-phase angle behavior) will arise from many factors, including a broad range of sizes of cells and membrane-bound organelles in a tissue, tight packing of cells, the existence of cellular junctions, or irregular sizes and shapes of the cells.

Recently, Markx et al. offered interesting comments on this question.[42] In a review of data from 18 cell suspensions, they noted that the distribution parameter α for the beta dispersion is in the range of 0.1–0.4 in bacterial or animal cell suspensions, close to 0.1 in red cell suspensions, and nearly 0 for liposome suspensions.

For example, when fitted to the Cole-Cole equation, the permittivity of suspensions of the yeast *S. cerevisiae* yields α values close to 0.3 (using data from 0.1–10 MHz). This value is far larger than expected on the basis of the (quite narrow) distribution in cell sizes. However, the data (Equation 3 in their paper) show systematic deviations from the best-fit Cole-Cole function. An additional dispersion is clearly visible at the lower end of the frequency range these investigators covered. The broad dispersion may reflect the presence of multiple relaxation processes in the frequency range covered in that study. Yeast and bacteria (organisms that show the largest values of α) have thick cell walls, which themselves might be a source of dispersion.

"Constant power law" behavior can be related formally to fractal geometries, i.e. to self-similar structures. Dissado proposed a fractal model to account for the broad dielectric relaxation of tissues in the beta dispersion range.[43] This theory, while mathematically attractive, lacks a clear physical basis. Multiple relaxation mechanisms have been definitely established in cells and tissues, and a satisfactory theory must take them into account.

Many nonbiological materials also show a fractional power law relaxation; indeed such behavior is the rule rather than the exception in disordered solids. Hill and Jonscher[44] have proposed that this behavior in many solids is due to cooperative relaxation effects. Macdonald[38] reviewed several theoretical models giving rise to "fractional power law" behavior, including an exponential distribution of activation energies, fractal geometry, and equivalent electrical circuits arranged in ladder or hierarchial networks.

Thus, there is no shortage of reasons why something resembling "fractional power law" behavior should occur, over some frequency range, in complex solids including tissues. In tissues such behavior may be suggested by the common practice of plotting dielectric data using log-log scales. In the words of one investigator, "anything is straight line in a log-log plot."[45] It remains unclear to what extent the "constant power law" relaxation in tissues arises from a single mechanism, or is merely the consequence of the overlapping of multiple poorly resolved relaxations. Sorting out these questions is a very difficult task.

E. KRAMERS-KRONIG RELATIONS

The complex permittivity ϵ characterizes the linear, causal response of a material to an applied stimulus (the electric field). As such, the Kramers-Kronig relations provide a necessary connection between the conductivity and permittivity[46]. Thus

$$\epsilon''(f) - \sigma_s/\omega\epsilon_0 = \frac{-2f}{\pi}\int_0^{\infty} \frac{\epsilon'(x) - \epsilon_\infty}{x^2 - f^2}\,dx \tag{34}$$

$$\epsilon'(f) - \epsilon_\infty = \frac{2}{\pi}\int_0^{\infty} \frac{x\epsilon''(x)}{x^2 - f^2}\,dx \tag{35a}$$

where x is a variable of integration. This implies that the permittivity and conductivity of a material cannot vary independently with frequency; any decrease with frequency in the permittivity must be accompanied by an increase in the conductivity. This yields a consistency test for data (which some published data fail, strongly suggesting error). The Kramers-Kronig relations apply to dielectric materials with any arbitrary distribution of relaxation times, and hold even in the presence of resonance-type behavior as is found at optical frequencies. In Equation 35, $\epsilon''(\omega)$ is the loss excluding the ionic conductivity.

One interesting consequence of the Kramers-Kronig relations is that the area under a loss peak must be independent of the distribution of relaxation times, when $(\epsilon'' - \sigma_s/\omega\epsilon_0)$ is plotted against ln(f). This can be easily shown by letting $f \to 0$ in equation 35:

$$\epsilon_s - \epsilon_\omega = \frac{2}{\pi}\int_0^{\infty} \frac{\epsilon''(x)}{x}\,dx \tag{35b}$$

$$= \frac{2}{\pi}\int_{-\infty}^{\infty} \epsilon''(\ln(f))d(\ln(f))$$

where ϵ'' refers to the part of ϵ'' that remains after subtracting the DC conductivity.

The Kramers-Kronig relations are based on the assumptions that the response is linear and causal or that the applied field must be of sufficiently low amplitude to ensure linear response. This condition is practically always met in dielectric measurements on biological materials. In Section II.D., we briefly consider situations that might produce nonlinear responses.

II. DIELECTRIC RELAXATION MECHANISMS

The discussion above made no reference to specific dielectric polarization mechanisms. In this section, we briefly review three relaxation processes that are chiefly responsible for the dielectric properties of tissues: interfacial polarization, dipolar orientation, and ionic diffusion.

A. INTERFACIAL POLARIZATION (MAXWELL-WAGNER EFFECT)

If a material is electrically heterogeneous, a dispersion occurs in the bulk properties from the charging of the interfaces within the material. This phenomenon does not arise from dielectric relaxation in the bulk phases of the material (which might also be present), but is a consequence of the boundary conditions on the fields at the interfaces between phases. Interfacial effects typically dominate the dielectric properties of colloids and emulsions.

The theory of the bulk electrical properties of suspensions is one aspect of the general theory of the transport properties of mixtures. Many investigators have contributed to this theory, working within several engineering and scientific traditions. Grosse and Greffe,[47] Hanai,[48] and Dukhin[49,50] (among others) have published excellent reviews of mixture theory applied to dielectric phenomena. We consider here the dielectric properties of a few simple model systems and later discuss the application of these results to biological systems.

1. Two Slabs in Series

Consider two slabs of thickness d_1 and d_2, in contact with their interface oriented perpendicular to the external field. If the permittivity and conductivity of the two materials (ϵ_1, ϵ_2, σ_1, σ_2) are independent of frequency, the boundary condition on the electric field component normal to the interface is

$$\epsilon_1 E_1 = \epsilon_2 E_2 \tag{36}$$

if the interface is free of charge. The current densities at the interface are therefore related by

$$j_1/j_2 = \sigma_1 E_1/\sigma_2 E_2 = \sigma_1 \epsilon_2/\sigma_2 \epsilon_1 \tag{37}$$

Thus, if $\sigma_1\epsilon_2, \neq \sigma_2\epsilon_1$, the interface will be charged at a rate that is proportional to the difference between j_1 and j_2.

The bulk permittivity ϵ and conductivity σ can be found by regarding the composite as two capacitors in series and using the well known rule for series addition of capacitances. Thus

$$\frac{d_1 + d_2}{\epsilon + \sigma/j\omega} = \frac{d_1}{\epsilon_1 + \sigma_1/j\omega} + \frac{d_2}{\epsilon_2 + \sigma_2/j\omega} \tag{38}$$

This can be written as a dispersion with a relaxation time of

$$\tau = \epsilon_0 \frac{\epsilon_1 d_2 + \epsilon_2 d_1}{\sigma_1 d_2 + \sigma_2 d_1} \tag{39}$$

and limiting values at low and high frequencies:

$$\epsilon_s = \frac{(\epsilon_2\sigma_1 - \epsilon_1\sigma_2)^2(d_1 + d_2)(d_1 d_2)}{(\epsilon_1 d_2 + \epsilon_2 d_1)(\sigma_1 d_2 + \sigma_2 d_1)^2} + \epsilon_\infty \tag{40}$$

$$\sigma_s = (d_1 + d_2)\frac{\sigma_1\sigma_2}{\sigma_1 d_2 + \sigma_2 d_1} \tag{41}$$

$$\epsilon_\infty = \frac{(d_1 + d_2)\epsilon_1\epsilon_2}{(\epsilon_1 d_2 + \epsilon_2 d_1)} \tag{42}$$

$$\sigma_\infty = \frac{(\epsilon_2\sigma_1 - \epsilon_1\sigma_2)^2(d_1 + d_2)(d_1 d_2)}{(\epsilon_1 d_2 + \epsilon_2 d_1)^2(\sigma_1 d_2 + \sigma_2 d_1)} + \sigma_s \tag{43}$$

where

$$\sigma_\infty = \lim_{\omega \to \infty} (\epsilon''\omega\epsilon_0)$$

Equation 38 applies even when the bulk material properties of the constituent phases vary with frequency. In that case, however, the frequency dependence of the composite can no longer be characterized by a single relaxation time.

2. Dilute Suspension of Spherical Particles

Maxwell[22] was the first to calculate the conductivity of a dilute suspension of spherical particles. His result (which applies for static fields) relates the conductivity of the continuous and suspended phases (σ_a and σ_i, respectively) to that of the suspension (σ) by

$$\frac{\sigma - \sigma_a}{\sigma + 2\sigma_a} = p\frac{\sigma_i - \sigma_a}{\sigma_i + 2\sigma_a} \tag{44}$$

where ρ is the volume fraction of the suspended particles. Maxwell's derivation consisted of two steps: First, he solved Laplace's equation to determine the electric field around a single particle. Second, he calculated an equivalent conductivity for a larger sphere that contained many individual particles. The second of these steps is based on two assumptions: (1) the concentration of particles inside the larger sphere is small so that electrical interactions are negligible among the particles; (2) the larger sphere is viewed from a great distance to permit the easy calculation of the contributions from the individual particles.

Wagner extended this analysis to the case of alternating currents and transformed Equation 44 into a set of dispersion equations.[51] This can be done by inserting the complex conductivities $\{\sigma^*, \sigma_a^*, \sigma_i^*\}$ for the real conductivities $\{\sigma, \sigma_a, \sigma_i\}$ of the suspension, medium, and particle in Equation 44. After some algebra (which is cumbersome but easily done with the help of a computer algebra program), the result can be rearranged into a dispersion equation of the form of Equation 21, with

$$\epsilon_\infty = \epsilon_a \frac{2\epsilon_a + \epsilon_i - 2p(\epsilon_a - \epsilon_i)}{2\epsilon_a + \epsilon_i + p(\epsilon_a - \epsilon_i)} \tag{45}$$

$$\epsilon_s - \epsilon_\infty = \frac{9(\epsilon_a\sigma_i - \epsilon_i\sigma_a)^2 p(1 - p)}{[2\epsilon_a + \epsilon_i + p(\epsilon_a - \epsilon_i)][2\sigma_a + \sigma_i + p(\sigma_a - \sigma_i)]^2} \tag{46}$$

$$\sigma_s = \sigma_a \frac{2\sigma_s + \sigma_i - 2p(\sigma_a - \sigma_i)}{2\sigma_a + \sigma_i + p(\sigma_a - \sigma_i)} \tag{47}$$

$$\sigma_\infty - \sigma_s = \frac{9(\sigma_a\epsilon_i - \sigma_i\epsilon_a)^2 p(1 - p)}{[2\sigma_a + \sigma_i + p(\sigma_a - \sigma_i)][2\epsilon_a + \epsilon_i + p(\epsilon_a - \epsilon_i)]^2} \tag{48}$$

$$\tau = \epsilon_0 \frac{2\epsilon_a + \epsilon_i + p(\epsilon_a - \epsilon_i)}{2\sigma_a + \sigma_i + p(\sigma_a - \sigma_i)} \tag{49}$$

where ρ is the volume fraction of the suspension, and (ϵ, σ), (ϵ_i, σ_i), and (ϵ_a, σ_a) are the dielectric properties of the suspension, particle, and medium, respectively.

According to this analysis, a dispersion will occur in the dielectric properties of the suspension whenever $\sigma_a\epsilon_i \neq \sigma_i\epsilon_a$, which is practically always. But the dispersion is not necessarily large. If $\epsilon_i \ll \epsilon_a$ and $\sigma_i \ll \sigma_a$, Equation 44 can be expanded to first order in ϵ_i/ϵ_a and σ_i/σ_a to yield:

$$\epsilon \simeq \epsilon_a \frac{1 - p}{1 + p/2} + \epsilon_i \frac{9p}{(2 + p)^2} \tag{50}$$

$$\sigma \simeq \sigma_a \frac{1 - p}{1 + p/2} + \sigma_i \frac{9p}{(2 + p)^2} \tag{51}$$

In this approximation, the properties of the mixture are just the weighted averages of those of the constituent phases. In the opposite case ($\epsilon_i \gg \epsilon_a$, $\sigma_i \gg \sigma_a$) a large dispersion can occur.

Fricke[52,53] extended the Maxwell-Wagner theory to the case of prolate or oblate spheroids; other investigators considered additional geometries. The results can be expressed in a form similar to (the complex extension of) Equation 44:

$$\frac{\sigma^* - \sigma_a^*}{\sigma^* + \gamma\sigma_a^*} = p\,\frac{\sigma_i^* - \sigma_a^*}{\sigma_i^* + \gamma\sigma_a^*} \tag{52}$$

In this equation the shape factor γ has been introduced, which takes on the value 2 for spheres, and 1 for cylindrical particles oriented normal to the field. Velick and Gorin[54] solved the still more complex problem of suspended ellipsoids with three different principal axes. We will refer to Equation 52 as the "Maxwell-Fricke equation" in the discussion below.

Equation 44 can be rewritten for the case of a spherical particle of radius R′ contained within a larger sphere of radius R. Then

$$\frac{\sigma^* - \sigma_a^*}{\sigma^* + 2\sigma_a^*} = (R'/R)^3\,\frac{\sigma_i^* - \sigma_a^*}{\sigma_i^* + 2\sigma_a^*} \tag{44a}$$

where R′ and R are the radii of the small and larger spheres, and $p = (R'/R)^3$ is the volume fraction of the larger sphere that is occupied by the smaller particle. Equation 44a is rigorous for a single composite particle located in an otherwise uniform space. However, it does not apply strictly when other particles are nearby. A more complete analysis would entail a multipole expansion for the field experienced by the particle, with higher order terms representing contributions from nearby particles or groups of particles.

3. Dilute Suspension of Membrane-Covered Spheres

The Maxwell-Wagner theory can be easily extended to a suspension of spheres of radii R and dielectric properties ϵ_i and σ_i surrounded by a thin shell of thickness d, of properties ϵ_{sh} and σ_{sh}.[1,55] One first calculates the equivalent conductance of the composite membrane covered sphere (i.e. the complex conductivity of a uniform sphere whose electrical properties appear from the outside to be the same as that of the composite), and then the complex conductivity of the suspension by two consecutive applications of Equation 44a. The algebra in this calculation is complex. The final expression can be simplified by keeping only terms to order (d/R) in both the numerator and denominator (by forming the (1,1) Padé approximate) to yield:†

$$\sigma^* \approx \frac{\sigma_i^* + (2d/R)(\sigma_{sh}^*)}{1 + (d/R)(\sigma_i^* - \sigma_{sh}^*)/\sigma_{sh}^*} \tag{53}$$

Further substitutions are appropriate for tissues or cell suspensions.[1,55] The membrane capacitance C_m and conductance per unit area of membrane surface G_m are defined as:

$$C_m = \epsilon_{sh}\epsilon_o/d \ (F/m^2)$$

$$G_m = \sigma_{sh}/d \ (S/m^2)$$

$$\sigma_{sh} = G_m d + j\omega C_m d$$

For cells suspended in typical physiological fluids, at radiofrequencies and below,

† This is slightly different from similar expressions given in the first edition of this Handbook or in Reference 1. The present expression, calculated as a Padé approximate, is correct to $O(d/R)^2$ whereas the previous expressions were calculated in a slightly different approximation and are correct to O(d/R).

$$\sigma_i \gg \omega\epsilon_o\epsilon_i$$

$$\sigma_a \gg \omega\epsilon_o\epsilon_a$$

Using these approximations, expanding the resulting equations to first order in p, and rearranging, dispersion equations are obtained[1] with:

$$\epsilon_s - \epsilon_\infty \simeq \frac{9pRC_m}{4\epsilon_o[1 + RG_m(1/\sigma_i + 1/2\sigma_a)]^2} \rightarrow \frac{9pRC_m}{4\epsilon_o} \quad (54a)$$

$$\sigma_s \simeq \sigma_a\left[1 - \frac{3p}{2}\frac{1 + RG_m(1/\sigma_i - 1/\sigma_a)}{1 + RG_m(1/\sigma_i + 1/2\sigma_a)}\right] \rightarrow \sigma_a(1 - 3p/2) \quad (55a)$$

$$\tau \simeq RC_m\frac{\sigma_i + 2\sigma_a}{2\sigma_i\sigma_a + RG_m(\sigma_i + 2\sigma_a)} \rightarrow RC_m(1/2\sigma_a + 1/\sigma_i) \quad (56a)$$

$$\epsilon_\infty \simeq \left[1 - 3p\frac{\epsilon_a - \epsilon_i}{2\epsilon_a + \epsilon_i}\right]\epsilon_a \quad (57a)$$

where p is the volume fraction of cells in the suspension. The limits above correspond to the situation where the membrane conductance is small, which nearly always applies.

This approximate analysis shows that the dielectric properties of the suspensions are insensitive to the membrane conductance. Since $\sigma \approx \sigma_a \approx 1$ S/m, while G_m is typically in the range 1 to 10 S/m^2 for biological membranes, the product RG_m is negligible compared to σ_i or σ_a for spheres of typical cellular dimensions, and the dispersion equations take on the simple limiting forms given above. These approximations may fail if σ_i is low, which is often the case in studies involving electrical manipulation or electrofusion of cells.

Equations 54a–57a incorporate numerous approximations; in particular, terms quadratic and above in d/R were dropped, and the permittivity of the cytoplasm and extracellular fluid is assumed to be negligible. These assumptions are reasonable for dilute cell suspensions, but might not be valid for small cells, concentrated cell suspensions, or at high frequencies.

Analogous equations can be derived readily using a computer algebra program (we used *Maple,* Waterloo Software,) to rederive these results without these simplifications, but they are quite involved.[56] A set with fewer approximations (which are frequently called the Pauly-Schwan equations) apply to suspensions of small cells and vesicles, or when the conductivity of the medium is very low. For example, the equations corresponding to Equations 54a–57a that are valid (under the Maxwell theory) for all values of p are:

$$\epsilon_s - \epsilon_\infty = \frac{9pRC_m}{4\epsilon_o\left[1 + \frac{p}{2} + \frac{RG_m}{\sigma_i\sigma_a}(2\sigma_a + \sigma_i + p(\sigma_a - \sigma_i))\right]^2} \quad (54b)$$

$$\sigma_s = \sigma_a\left[\frac{1 - p + \frac{RG_m}{\sigma_i\sigma_a}\left(\sigma_a + \frac{\sigma_i}{2} + p(\sigma_i - \sigma_a)\right)}{1 + \frac{p}{2} + \frac{G_m}{\sigma_i\sigma_a}\left(\sigma_a + \frac{\sigma_i}{2} - \frac{p}{2}(\sigma_i - \sigma_a)\right)}\right] \quad (55b)$$

$$\tau = RC_m \frac{\sigma_i + 2\sigma_a - p(\sigma_i - \sigma_a)}{2\sigma_a\sigma_i(1 + p) + RG_m[(\sigma_i + 2\sigma_a) - p(\sigma_i - \sigma_a)]} \tag{56b}$$

$$\epsilon_\infty = \epsilon_a \frac{2\epsilon_a + \epsilon_i - 2p(\epsilon_a - \epsilon_i)}{2\epsilon_a + \epsilon_i + p(\epsilon_a - \epsilon_i)} \tag{57b}$$

In the derivation of these equations using *Maple,* numerous simplifications were made, including $\sigma_i \gg \omega\epsilon_0\epsilon_i$ and $\sigma_a \gg \omega\epsilon_0\epsilon_a$, and dropping terms in d subject to the constraints of fixed membrane conductance and capacitance. This last assumption is equivalent to ignoring the volume of the membrane in computing the permittivity of the suspension. The above equations differ somewhat from those published earlier (e.g. Ref. 1) because of different simplifications taken in deriving the final results.

In the full treatment, a second dispersion occurs, generally at frequencies in the range of 100–1000 MHz. This is simply the Maxwell-Wagner dispersion for the cytoplasm interacting with the medium as given by the equations for a suspension of homogeneous particles (cf. eqs. 45–49). For cells in normal physiological media, this second dispersion is much smaller than that associated with charging of the membrane and may be too small to observe experimentally.

Unfortunately, the full equations that result from the Maxwell mixture theory (which can be obtained easily using a computer algebra program) are complex and difficult to obtain physical insight from, and cannot be manipulated readily to allow easy extraction of cellular parameters from data. The simpler forms given above are excellent approximations under nearly all circumstances. But in deriving them many terms were dropped that, in unusual circumstances, might be significant.

The circuit shown in Figure 4 suggests a physical interpretation of these results. At low frequencies the current flows around the cells; in the limit of high frequencies the current flows without restriction through the cells. This gives rise to a dispersion whose relaxation frequency $f_c = 1/2\pi\tau$ corresponds to the midpoint of the transition between these two limiting situations, and is determined by the time constant for charging the cell membrane capacitance through the extracellular and intracellular fluids.

4. Dilute Suspensions of Spheres Surrounded by Multiple Membranes

Fricke[56] and, more recently, Irimajiri[57] have extended the Maxwell-Wagner theory to the case of spheres surrounded by multiple membranes. Their procedure was to calculate repeatedly the equivalent electrical properties of the sphere, including successive shells in turn. The resulting equations yield multiple dielectric dispersions, corresponding to the number of interfaces lying between successive shells. This theory was used to analyze the dielectric properties of cultured lymphoma cells, in which the cell nucleus contributes significantly to the permittivity of the suspension at high frequencies.[58] We will apply this analysis to the coupling of EM fields with intracellular membranes (Section V.C).

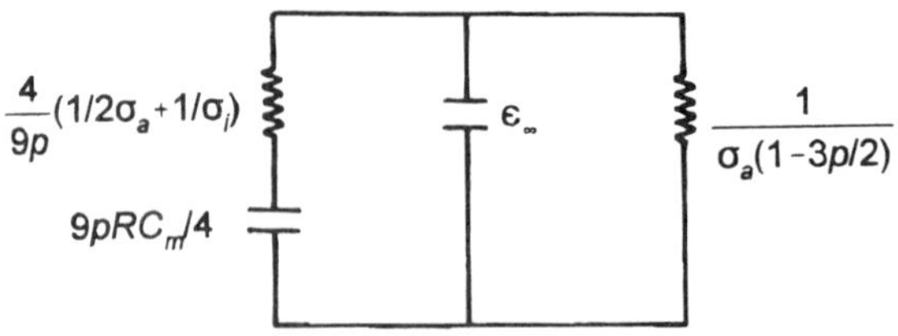

FIGURE 4 Equivalent circuit for the Maxwell-Wagner dispersion due to RC charging of the cell membranes. The parameters σ_i = conductivity of the cytoplasm (S/m), σ_a = conductivity of the extracellular fluid, C_m = specific membrane capacitance (F/m^2), R = cell radius (m), p = volume fraction of cells in suspension.

5. More Concentrated Suspensions

Calculating the properties of concentrated suspensions is a difficult problem, and no rigorous closed-form solution exists even for "simple" suspensions of identical spheres. Dukhin[49] nicely describes the difficulties involved as well as the approaches that different investigators have taken to extend the theory.

One approach has been to replace the particle and its surrounding medium by a (fictional) homogeneous medium with an "effective" permittivity, and then calculate the incremental change in the effective permittivity with an incremental change in the volume fraction of particles. This was first developed in 1935 by Bruggeman[59] and later by Hanai.[48] If ϵ_i^* and ϵ_a^* are the complex permittivities of the suspended and continuous phases, respectively, and p is the volume fraction of the suspension, the complex permittivity ϵ_s^* of the suspension is found to be:

$$\left[\frac{\epsilon^* - \epsilon_i^*}{\epsilon_a^* - \epsilon_i^*}\right]\left[\frac{\epsilon_a^*}{\epsilon_s^*}\right]^{1/3} = 1 - p \tag{58}$$

Boned and Peyrelasse[60] recently extended the theory to include the case of ellipsoidal particles.

Not surprisingly, for dilute suspensions the results of these theories are numerically similar to those of the Maxwell-Wagner theory. However, the equations are far more complicated. The following limiting expressions can be obtained from Equation 58:

$$\left[\frac{\epsilon_\infty - \epsilon_i}{\epsilon_a - \epsilon_i}\right]\left[\frac{\epsilon_a}{\epsilon_s}\right]^{1/3} = 1 - p \tag{59}$$

$$\epsilon_s\left[\frac{3}{\epsilon_s - \sigma_i} - \frac{1}{\sigma_s}\right] = 3\left[\frac{\epsilon_a - \epsilon_i}{\sigma_a - \sigma_i} + \frac{\epsilon_i}{\sigma_s - \sigma_i}\right] - \frac{\epsilon_i}{\sigma_i} \tag{60}$$

$$\left[\frac{\sigma_s - \sigma_i}{\sigma_a - \sigma_i}\right]\left[\frac{\sigma_a}{\sigma_s}\right]^{1/3} = 1 - p \tag{61}$$

$$\sigma_\infty\left[\frac{3}{\epsilon_\infty - \epsilon_i} - \frac{1}{\epsilon_\infty}\right] = 3\left[\frac{\sigma_a - \sigma_i}{\epsilon_a - \epsilon_i} + \frac{\sigma_i}{\epsilon_\infty - \epsilon_i}\right] - \frac{\sigma_a}{\epsilon_a} \tag{62}$$

In these expressions, σ_s, σ_∞, and ϵ_∞ are the values of the conductivity or permittivity of the suspension in the limits of zero or infinite frequency.

The dispersion predicted from these equations is broader than single time constant (Debye) relaxation, because of electrical interactions among the particles. Hanai and colleagues measured the dielectric properties of water-in-oil suspensions (which exhibit large interfacial polarization effects), and showed that the data were in substantially better agreement with Hanai's theory than the simpler Maxwell-Wagner result.[61]

Equations 59 to 62 were derived on the assumption that the dielectric properties of the suspended and continuous phases are independent of frequency. If this is not the case, Equation 58 must be solved numerically to obtain the bulk dielectric properties of the suspension.

The Hanai-Bruggemann theory can be extended to the case of membrane-covered spheres by calculating an "effective" complex permittivity of the sphere using Equation 53, and then solving Equation 58 numerically to obtain the dielectric properties of the suspension. Hanai et al. used this approach to calculate the dielectric properties of erythrocyte suspensions.[62] The result illustrates the effects of electrical interactions among the cells on the dielectric

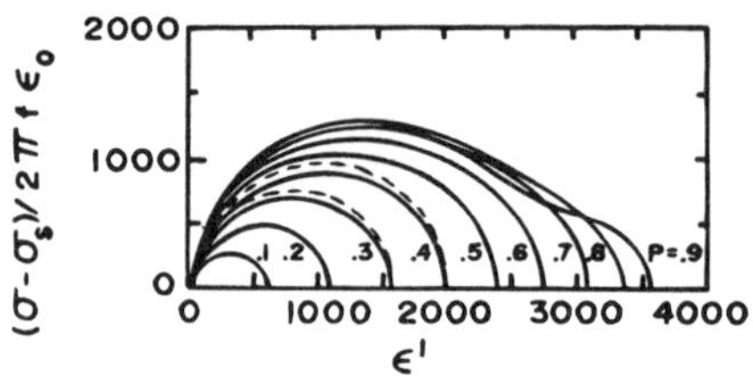

FIGURE 5 Complex permittivity plots of the dielectric properties of suspensions of shell-spheres calculated from Equations 53 and 58 (solid curves). The broken lines are semicircles representing single, time-constant relaxation. The parameters that were used are ϵ_a = 80, σ_a = σ_a = 0.25 S/m, ϵ_i = 50, ϵ_s = 6.5, σ_s = S/m, d = 5 nm, D = 3.8 mm. This calculation assumes that the cells are of identical size; thus the results at the higher volume fractions are unphysical. (From Hanai, T., Asami, K., and Koizumi, N., Bull. Inst. Chem. Res. Kyoto Univ., 57, 297, 1979. With permission.)

dispersion properties of the suspension (Figure 5). Each cell in isolation would show a single time constant relaxation. At low concentrations of cells ($p < 0.3$) the dielectric properties of the suspension closely approach the single time constant behavior expected from the Maxwell-Wagner theory, while at higher concentrations ($p \approx 0.3$–0.5), the complex permittivity plots resemble the depressed circular arcs characteristic of the Cole-Cole equation. For highly concentrated suspensions ($p > 0.8$) the permittivity calculated using the Hanai-Bruggemann theory shows a complex frequency dependence that deviates significantly from the Cole-Cole equation.

The Hanai-Bruggemann theory is a reasonable approximation for cell suspensions, but is not rigorous. In particular, the combination of Equations 53 and 58, which underlies this analysis, is questionable. Equation 58 was developed to take into account interparticle interactions, to overcome the limitations of the Maxwell theory. However, Equation 53 yields the correct result for the "equivalent homogeneous permittivity) of the shell-surrounded particle only if it is placed in an initially uniform field. The equation neglects the contribution of higher order terms in the solution to Laplace's equation for the field that result from the presence of nearby objects. Thus, Hanai's theory, as extended to cells, only partly accounts for interparticle interactions.

A more rigorous approach would be to calculate the dielectric properties as a power series in p with terms of order (p^n) representing n-wise interactions among the particles. Maxwell's results can be shown to be exact only to first order in p for random suspensions.[63] In fact, experimental evidence cited above shows that it is better than would be expected from this consideration. The reason might be associated with the factor $(1 - p) / (1 + p/2)$ that appears in the theory. This ensures that the properties of the mixture equal those of the appropriate component phase when $p = 0$ or 1. Hanai's theory includes some correction for interparticle interactions, but the order of its approximation is difficult to judge.

Despite its approximations, the Maxwell theory is remarkably successful in some applications to concentrated suspensions. For example, Cole found that the Maxwell theory could accurately predict the low frequency conductivity of suspensions of sea urchin eggs with volume fractions as high as 80%[3] and of electrical analogues of tissues with volume fractions as high as 90%.[64] At low frequencies a cell is essentially nonconducting (compared to the surrounding electrolyte) and electrical interactions among cells have little effect on the bulk conductivity of the suspension. Davey measured the permittivity of various bacterial and yeast suspensions at volume fractions up to approximately 50%, and found that the data could be fitted accurately to the Maxwell theory.[65]

We conclude that the Maxwell theory is useful in spite of its limitations. It has the important advantage that its results are simple, and it offers insights into pertinent mechanisms even in situations in which interparticle interactions might be important. Modest improvements in the theory lead to more complex mathematical expressions that require numerical solution in most cases—and are still not rigorously correct. However, the Maxwell theory is certainly not exact and must be used with caution.

6. Very Small Cell Size

Equation 53 applies when the membrane thickness d is small compared with the cell radius R. This is usually the case, since d is about .01 μm and typical cell diameters are in the micrometer range. Schwan and Morowitz[66] derived an extension of this theory for particles of radii as low as 0.1 μm; Pauly and Schwan[55] gave a general treatment for any size particle. The Pauly-Schwan theory yields two dispersions: a large dispersion due to membrane charging effects, and a small dispersion at higher frequencies due to the electrical dissimilarity of the cytoplasm and extracellular fluids. For micron-sized cells in typical biological electrolytes, the lower (larger) dispersion typically occurs near 1 MHz, while the higher (smaller) dispersion is in the 100 MHz range. If the cell diameter is smaller than about 0.1 μm, these two dispersion overlap in frequency, and numerical techniques must be employed to separate them.[67,68] This higher frequency dispersion does not appear in the theory presented above because of the simplifications employed.

B. DIPOLAR RELAXATION MECHANISMS

A second polarization mechanism is the partial orientation of permanent dipoles. Excellent reviews of dipolar relaxation theory and its applications to protein solutions are available.[69,70,71] The simple Debye theory, described below, suggests the size of the effects that are involved.

Consider an ensemble of independent dipoles, each with dipole moment μ. An externally imposed electric field of intensity F at each dipole will exert a torque μF sin(θ), where θ is the angle between the dipole moment and the field. This torque will act to orient each dipole, but will be opposed by random thermal agitation, and the ensemble will approach an equilibrium orientation $<\cos\theta>$. Evaluation of the thermodynamic partition function yields[69]

$$<\cos\theta> = \coth(\mu F/kT) - kT/\mu F \tag{63a}$$

where k is Boltzmann's constant and T the absolute temperature. For reasonable field strengths, the maximum orientational energy of a dipole (μF) is much less than its mean kinetic energy (which is of the order of kT and Equation 63a reduces to

$$<\cos\theta> \approx \mu F/kT \tag{63b}$$

The net polarization of the sample, p, is given by

$$p = N\mu<\cos\theta> \tag{64}$$

where N is the number of molecules per unit volume, or approximately

$$p \approx N\frac{\mu^2 F}{3kT} \tag{65}$$

The local field intensity F at each molecule can be related to the external field strength E by the Lorentz expression

$$F = \frac{2 + \epsilon_s}{3} E \tag{66}$$

which leads to an expression for the relative static permittivity

$$(\epsilon_s - 1)/(\epsilon_s + 2) = N_o\mu^2/9kTV\epsilon_o \tag{67}$$

where N_o is Avogadro's number and V is the molar volume equaling N_o/N. This result holds only for a lossless medium with relative permittivity 1, and in which the concentration of dipoles is vanishingly small.

After application of an electric field, the ensemble will approach thermal equilibrium with first order kinetics leading to dielectric relaxation behavior of the Debye type. The relaxation time τ, which characterizes rotational diffusion, can be estimated from Stokes' law[69]

$$\tau = \frac{4\pi\eta a^3}{kT} \tag{68}$$

where a is the radius of the dipole and η the viscosity of the medium. (The macroscopic dielectric relaxation time of the liquid will be larger by a small factor because of local field effects.) The time constant for dipolar relaxation ranges from picoseconds for small dipolar molecules such as water, to microseconds for large globular proteins. The corresponding dielectric dispersion will have center frequencies in the gigahertz to megahertz range.

This theory is entirely macroscopic and does not strictly apply to molecular systems. Moreover, the Lorentz expression (Equation 66) is a poor approximation for liquids; Equation 67 is significantly in error for water. (The Stokes' law calculation, Equation 68, is more satisfactory.)

In tissues, several dipolar relaxation effects can be anticipated. Globular proteins typically exhibit total dielectric increments (defined as the maximum increase in permittivity above that of the water divided by the concentration) in the range of 1 to 10/(g of protein per 100 g of solution) with relaxation frequencies in the range 1–10 MHz[70,71] (Table 1). Polar side-chains on protein molecules relax at somewhat higher frequencies, which has been suggested as contributing to the small dispersion observed in protein solutions at 0.1 to 1 GHz.[1,72] However, these are likely to be relatively small effects in tissues at radiofrequencies and

TABLE 1
Dielectric Properties of Various Proteins at 25°C

Protein	Mol wt ($\times 10^3$)	$\Delta\epsilon$/g/l	μ (D)	$\tau \times 10^8$ (s)	a/b
Ovalbumin	44	0.10	250	18; 4.7	5
Horse serum albumin	70	0.17	380	36; 7.5	6
Horse carboxyhemoglobin	67	0.33	480	8.4	1.6
Horse serum pseudoglobulin	142	1.08	1100	250; 28	9
β-Lactoglobulin (in 0.25 *M* glycine)	40	1.51	730	15; 5.1	4
Myoglobin	17	0.15	170	2.9	

Note: mol wt, molecular weight (Daltons); μ, dipole moment (Debye units); τ, rotational relaxation times; a/b, axial ratio; $\delta\epsilon$/g/l, total increase in permittivity per gram of protein per gram of solution, measured at low frequencies above the permittivity of pure water.

From Oncley, J.L., *Proteins, Amino Acids, and Peptides,* Cohn, E.J. and Edsall, J.T., Eds., Reinhold, New York, 1943, 543.

below for which membrane charging phenomena or counterion effects dominate the dielectric properties.

Dipolar relaxation of water is a dominant effect in tissues at microwave frequencies. Pure water undergoes nearly single time constant relaxation centered at 20 GHz at room temperature and 25 GHz at 37°C.[73,74] Water associated with protein surfaces has a lower relaxation frequency than that of the bulk liquid,[75,76,77] and this water fraction will contribute noticeably to the dielectric dispersion at frequencies near 1 GHz. This is discussed further in Section V below.

C. COUNTERION POLARIZATION EFFECTS

A third major class of dielectric phenomena arise from ionic diffusion in the electrical double layers adjacent to charged surfaces. Characteristically, the time constants for such effects are of the form L^2/D, where L is a length over which diffusion occurs and D is a diffusion coefficient. (In contrast, the time constant for the Maxwell-Wagner effect is of the form of RC or $\epsilon\epsilon_o/\sigma$, where R (or $1/\sigma$) is a resistance and C (or $\epsilon\epsilon_o$) represents a capacitance.) Counterion effects are typically dominant in biological systems at kilohertz frequencies and below; Maxwell-Wagner effects typically are dominant at radio frequencies. Counterion polarization effects can be large at low frequencies: suspensions of submicron-sized polystyrene spheres have permittivity values approaching 10^4 with relaxation frequencies in the kilohertz range (Table 2).

Many investigators have reported counterion polarization effects in systems that contain charged surfaces: emulsions,[48,49,50] suspensions of charged polystyrene spheres,[78] microorganisms,[79] and long chain macromolecular polyions such as DNA.[80,81] Counterion effects contribute to the alpha dispersion that is found in tissues at low frequencies.

Counterion phenomena are difficult to analyze rigorously, since they involve coupled electrical and hydrodynamic phenomena in the double layers surrounding the particle that are governed by nonlinear equations. Schwarz[82,83] proposed the first theory that successfully accounted for the amplitude of the dispersion in latex particles. In spite of its severe approximations and questionable assumptions about the nature of the double layer, this model remains the only one that is widely used by experimentalists in interpreting their results.

Schwarz considered the case of a macroscopic sphere of radius a with counterion surface charge density δ_o (in units of m^{-2}), in which the thickness of the electrical double layer is much less than the particle diameter. The frequency dependent surface conductivity was found by solving the electrodiffusion equation to obtain the effective permittivity $\epsilon_p{}^*$ of the particle:

TABLE 2
Electrical Properties of Suspensions of Polystyrene Spheres, of Volume Fraction Close to 30%

Particle radius (μm)	ϵ_s	f_c (kHz)
0.59	10,000	0.6
0.28	3,000	1.8
0.094	2,450	15
0.044	540	80

Adapted from Schwan, H. P., Schwarz, G., Maczuk, J., Pauly, H., *J. Phys. Chem.*, 66, 2626, 1962.

$$\epsilon_p^* = \epsilon_b + \frac{1}{1 + j\omega\tau}\frac{e_0^2\delta_0 a}{\epsilon_0 kT} \tag{69}$$

where ϵ_b is the permittivity of the bulk material of the particle, and e_o is the charge of the counterion. The relaxation time τ is proportional to the square of the radius:

$$\tau = e_o a^2/2ukT \tag{70}$$

where μ is the surface mobility of the counterions (m^2/Vsec). (The bulk permittivity of the suspension is then calculated from ϵ_p* using the Maxwell mixture theory.) Einolf and Carstensen[79] extended this model and applied it to dielectric data from bacteria, taking into account the diffusion of ions in the surface layers both inside and outside the cell wall.

In recent years, quite a different model for counterion polarization has emerged that places more emphasis on the diffusion of ions in the bulk electrolyte surrounding the particle and the exchance of ions between the counterion layer and the bulk electrolyte; an excellent review is by Mandel and Odijk.[84] Dukhin,[49,50] Fixman,[85,86] and Chew and Sen[87,88,89] have made important contributions to this problem. The theories are complex and we refer the reader to the original papers for more detailed discussion.

Recently, Grosse proposed a simplified model that preserves important features of these latter theories yet can be exactly solved.[90,91] The model assumes that the counterion layer is thin, and the diameter of the particle is much larger than the Debye screening length of the electrolyte. In addition, the model assumes the counterion layer contains only ions that are opposite in sign to that of the fixed charges on the particles, and the counterions can exchange freely only with ions of the same sign in the bulk electrolyte. The surface conductivity G_s (whose units are Siemens) was assumed to be high.

In this model, when an electric field is applied, the ions in the system will redistribute under the influence both of the field and diffusion. Grosse obtained a set of coupled differential equations for the ion concentrations and current densities. The solution of these equations yields a broad, asymmetrical, low-frequency dispersion in the permittivity of the particle. The time constant τ of this dispersion is similar to that derived by Schwarz:

$$\tau = a^2/D \tag{71}$$

except that D is the diffusion coefficient of ions in the *bulk* electrolyte (not a surface diffusion constant as in Schwarz's theory). The parameter that determines the low frequency limit of permittivity is χa, where χ is the inverse of the Debye screening length of the electrolyte:

$$\chi^2 \approx 2\sigma/(\epsilon\epsilon_o D) \tag{72}$$

where σ and ϵ are the conductivity and permittivity of the bulk electrolyte. (The Debye length $1/\chi$ for tissue electrolytes is of the order of 1 nm). For particles with high surface conductivity, the amplitude of the dispersion is approximately,

$$(\epsilon_s - \epsilon_\infty)/p\epsilon_m = 9(\chi a)^2/16 \tag{73}$$

This dispersion in ϵ is quite broad, and can be fitted approximately to a Cole-Cole relaxation function with $\alpha = 0.4$. (This is one example of a physical process that yields a relaxation that is far broader than the Debye result by virtue of its kinetics.) Remarkably, Equation 73 yields the permittivity of the suspension with no adjustable parameters whatever.

By now several investigators have measured the dielectric properties of suspensions of polystyrene particles; the results are in reasonable agreement with Equation 73.[90,91]

Grosse's model lends itself to a simple physical interpretation of counterion polarization: the large dielectric increments of the polystyrene spheres are direct consequences of the asymmetric charge distribution in and near the surface of the particles. The motion of an ion in the bulk electrolyte near the particle depends greatly on whether its sign is the same or opposite that of the ions in the counterion layer. Ions of the same sign can enter the counterion layer and their charge is quickly conducted to the opposite side of the particle. Those of opposite sign are excluded from the layer and must travel around the particle in the bulk electrolyte surrounding the particle. Thus, for an ion in the bulk electrolyte, the particle acts either as a good conductor or an insulator, depending on whether its charge is the same or opposite that of the counterions. A cloud of charge accumulates in the electrolyte near (within a Debye length) the particle, giving rise to an induced dipole moment of the system and hence a large permittivity of the suspension.

So far, the theories of Dukhin, Fixman, Grosse, and Chew and Sen have not been applied in detail to biological materials. Given the complexity of the situation (both biological and physical), it is difficult to see how they could be applied in any compelling way to tissues. But clearly the surfaces of membranes and proteins are highly charged, and counterion polarization effects are prominent. Remarkably, these theories lead to large permittivities and far broader than Debye relaxation behavior, even for "simple" charged dielectric spheres in an electrolyte.

D. SURFACE CONDUCTIVITY EFFECTS

An additional mechanism for dielectric relaxation in cells arises from a frequency-dependent surface conductivity. This problem is discussed at length by Schwan[92] and (for frequency-independent surface conductance) by Chizmadzhey.[93] Equation 53 then reduces to

$$\sigma^*_{\text{cell}} = \sigma^* + 2\sigma^*_{\text{shell}}\frac{d}{R} = \sigma^* + 2\frac{Y^*_{\text{shell}}}{R} \tag{53a}$$

where $Y^*_{\text{shell}} = G_{\text{shell}} + j\omega C_{\text{shell}}$ is the surface admittance, and G_{shell} and C_{shell} are the surface conductance and capacitance. Any dispersion in Y^*_{shell} will give rise to a dispersion in the effective permittivity of the cell, and of the suspension.

Typically, the surface charge density of cell membranes is quite high, which leads to a large and typically frequency-dependent surface admittance. Thus, two dispersions are associated with the cell membrane: one (the beta dispersion, which typically occurs in the low MHz range) from the charging of the membrane capacitance, the other from a frequency-dependent surface admittance. The contribution from surface conductance will clearly be most pronounced for small cells.

E. NONLINEAR EFFECTS

The dipolar and counterion relaxation theories summarized above apply to linear responses to weak fields. As the field intensity is increased, at some level the response will no longer be linear. The threshold at which nonlinearities become noticeable depends on the system, and the dielectric effect in question. We summarize the expected thresholds for producing nonlinear responses for the effects considered above.

1. Molecular Phenomena

The response of an ensemble of dipoles (Equation 63a) remains linear as long as $\mu F << kT$. At higher fields, the mean orientation $<\cos(\theta)>$ is given by

$$\langle \cos(\theta) \rangle \approx \frac{\mu F}{3kT} - \frac{1}{45}\left(\frac{\mu F}{3kT}\right)^3 + O\left(\frac{\mu F}{kT}\right)^5 \qquad (63c)$$

This implies that the polarization induced by an alternating field will have components at odd harmonics of the frequency of the applied field, whose amplitude will depend strongly on the field strength. At very high fields, when the interaction energy of a molecular dipole (μF) exceeds the mean thermal energy kT, the molecules are substantially all aligned with the field and dielectric saturation occurs. For typical globular protein molecules, dielectric saturation typically requires field strengths of the order of 10^6 to 10^7 V/m.

2. Cellular Phenomena

Many different nonlinear responses might occur at high field strengths. We summarize a few:

a. Due To Active Response of Cell Membranes

Nonlinear effects can obviously arise from the physiological response of the membrane to the applied field. This might include action potentials or, at lower levels of stimulation, changes in conductance due to gating phenomena. Since the bulk dielectric properties of tissues and cell suspensions are normally insensitive to the membrane conductance, we would not expect such responses to markedly affect the dielectric properties of a suspension of cells.

b. Membrane Breakdown

Another nonlinear effect is the breakdown of cell membranes. This typically requires potential differences of 0.1–1 V across the membrane. The effect can be used to fuse cells to form hybridomas, a fundamental process in biotechnology.[94]

The thresholds for membrane breakdown and active membrane phenomena depend on E_m, the maximum voltage difference across the membrane. For spherical cells of radius R, in an originally uniform field E, the induced potential is

$$E_m = 1.5ER \qquad (74)$$

at frequencies below the relaxation frequency for the suspension (cf. Section V C). Thus the minimum field strength required to rupture a 10 μm cell is in the range of 10–30 kV/m. Fields of this magnitude will certainly elicit nonlinear responses in a tissue or cell suspension.

c. Electrically-induced Distortion or Aggregation of Cells

Electric fields exert mechanical forces on cells[95,96] and can distort cells, orient nonspherical cells, or cause cells to aggregate and line up along the field lines (the so called "pearl chain" effect).[97] The formation of pearl chains typically requires high field strengths (1–10 kV/m for micron-sized cells). Somewhat lower fields might produce smaller mechanical effects leading to detectable nonlinear responses if the sensitivity of the measurement were high enough.

3. Counterion Polarization

No rigorous calculation of thresholds for nonlinear polarization of counterions has apparently been published. Nevertheless, a simple argument based on energy principles can estimate the field strengths needed to produce a nonlinear dielectric response. The energy gained by an ion of charge e_o in diffusing across a sphere of radius R in field E is e_oER. This is comparable to kT for a 1 μm sphere in a field of about 25 kV/m. This suggests that sufficiently high field strengths might lead to saturation of counterion polarization or other nonlinear responses.[98]

All of these mechanisms require quite high field strengths to produce dielectric saturation or other strongly nonlinear effect. However, with sensitive techniques, harmonics may be observable in the dielectric response at field strengths that are substantially lower than those needed to produce complete saturation. For example, Woodward and Kell reported nonlinear dielectric responses of bacterial suspensions with fields of 2 V/m at 20 Hz.[99,100] The responses included the presence of even and odd harmonics of the frequency of the applied field, whose amplitude depended on respiratory activity and vanished when the cells were killed. Woodward and Kell attributed these responses to biologically generated signals resulting from energy that is transduced by physiological processes in the cells.

If correct, these results suggest that studies of nonlinear dielectric responses of cells are a promising new way to study their physiological responses to electric fields. We note, however, that at low field strengths the harmonics have very small amplitude (it is difficult to judge their magnitude from the papers of Woodward and Kell) and great care is needed to exclude artifacts.

The distinction between linear and nonlinear effects can be related to the question of "signal to noise" in producing measurable changes in biological systems by electromagnetic fields. Most dielectric phenomena discussed in this chapter are *linear* effects, which is to say that the perturbation of the system by the applied field is small. Thus, the forces exerted on an ensemble of molecular dipoles, or the counterion layer surrounding a particle, by fields used in the measurement (which are typically of the order of a few volts per meter) are sufficient to cause only a small net polarization in the system; the motions of individual ions or dipoles are hardly affected at all. Only because of the high sensitivity of the measurement can the net effect be observed at all. By contrast, nonlinear effects imply that the system has undergone large changes, for example near-complete alignment of molecular dipoles or significant distortion or aggregation of cells. The thresholds for such effects depend on the frequency of the field and the response time of the system, but are generally quite high. One needs to be very careful in extrapolating from dielectric relaxation mechanisms in tissue to prediction of biological effects of low-level electromagnetic fields.

III. DIELECTRIC DISPERSION IN TISSUES

A. CONDUCTIVITY

It is useful to consider the conductivity of tissues separately from their permittivity, even though these two properties are inextricably linked through the Kramers-Kronig relations. In tissues at low frequencies, $\epsilon'' >> \epsilon'$ and for most practical purposes only their conductivity must be taken into account.

We consider several frequency regions; in each, the Maxwell-Wagner theory can be applied to data from tissues with a different volume fraction of the tissue being excluded from the current flow. In a later section (Section V.B.) we will describe the use of conductivity data from tissues to study the properties of "biological" water.

1. Low frequencies ($<$ 0.1 MHz)

For direct currents at times longer than the charging time τ of the cell membranes (Equation 56), or for alternating currents below the corresponding relaxation frequency $1/2\pi\tau$ (i.e. the beta relaxation frequency), a cell is poorly conducting compared to the surrounding electrolyte. Thus, in the limit of low frequencies only the extracellular fluid is available to current flow.

This simple idea is useful for interpreting the low frequency conductivity of blood. Early investigators applied this model quite literally, using sand as an electrical model for blood at low frequencies.[52] The conductivity of blood can be fitted to the Maxwell-Fricke mixture theory. This, however, requires the use of empirical values of the form factor. This form factor varies among different studies, presumably due to species differences or aggregation effects (Section IV. A.1.).

Solid tissues are still more complex; however, a simple analysis based on the Maxwell-Fricke theory is informative. Assuming a typical fraction of the extracellular fluid in tissue to be 0.1, the conductivity of extracellular fluids σ_a to be 2 S/m, and the cells themselves to be nonconductive, Equation 51 predicts a DC conductivity of 0.14 S/m. This is a typical conductivity of soft, high water content tissues at low frequencies. The conductivity at low frequencies depends strongly on the volume fraction of extracellular fluid, which can be expected to vary with physiological changes in the cells.

2. Frequencies between 1–100 MHz

In this frequency range, the cell membranes are largely shorted out and do not offer significant barrier to current flow. Thus, the tissues can be considered to be electrically equivalent to suspensions of nonconducting protein (and other solids) in electrolyte. The conductivity of most tissues approaches a plateau between 10–100 MHz.

3. UHF and Microwave Frequencies (> 100 MHz)

Three effects add significantly to the conductivity at these frequencies:

1. A Maxwell-Wagner effect due to interfacial polarization of tissue solids through the tissue electrolyte (cf. Equations 44 to 49). The Maxwell theory predicts a small dispersion in tissues at several hundreds of MHz because of the dissimilar electrical properties of tissue protein and electrolyte. Assuming that the conductivity of electrolyte in tissues is about 1 S/m and the protein has a permittivity of around 10 but negligible conductivity, Equations 44 to 49 predict increases in conductivity of a few hundredths of a S/m with a relaxation frequency of about 300 MHz. This effect may be a minor contribution to the dielectric relaxation in muscle.[101]
2. Dielectric loss of small polar molecules and polar side-chains on proteins. The time constant for reorientation (and thus for dipolar relaxation) of globular proteins varies as the cube of the protein radius (cf. Section II B). Globular protein molecules undergo dipolar relaxation at low megahertz frequencies; because of their smaller size, polar side-chains on protein molecules relax at much higher frequencies. Since the maximum conductivity increase associated with a relaxation process is proportional to the mean relaxation frequency (Equation 12), even a small component with high relaxation frequency can contribute noticeably to the total conductivity at these high frequencies. Thus, dipolar relaxation of proteins or polar sidechains on proteins might contribute a small amount to the conductivity of tissues and protein solutions at UHF frequencies.[72]
3. Dielectric relaxation of water. Pure water exhibits a dielectric relaxation centered at 20 GHz at 25°C that is characterized by nearly single time constant (Debye) behavior. Thus, its conductivity will rise nearly quadratically with frequency below ca. 10 GHz (cf. Equation 11d). For typical high water content tissues, the conductivity at 3 to 5 GHz is due in roughly equal proportions to ionic conductivity and to dipolar loss of water.

B. PERMITTIVITY

The increase in conductivity with frequency is associated with a decrease in permittivity. The permittivities of tissues at low frequencies can be astonishingly high, 10^5 to 10^6 (or higher) at subaudio frequencies. Three major dispersion regions can be identified (Figure

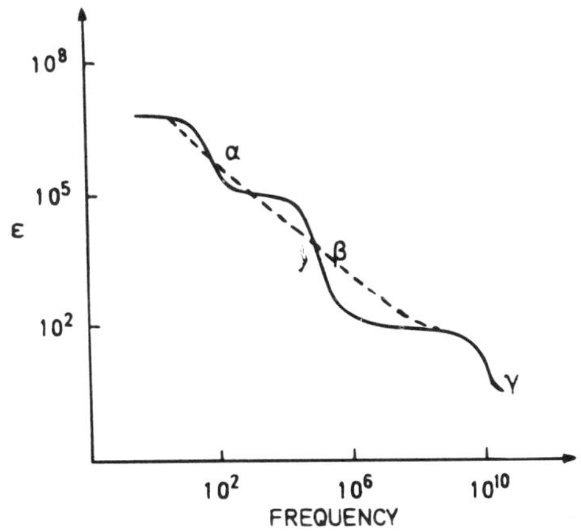

FIGURE 6 A schematic illustration of three major dispersion regions found in typical tissues, with permittivity and conductivity values typical of soft tissues. The dotted line represents an extreme case in which the dispersion regions overlap, which frequently occurs in soft tissues.

6): the alpha, beta, and gamma dispersions, with dielectric increments, $\delta\epsilon_\alpha$, $\delta\epsilon_\beta$, and $\delta\epsilon_c$, respectively. The corresponding relaxation frequencies are in the ranges of kHz, hundreds of kHz, and GHz. In tissues, these dispersion regions frequently overlap and the permittivity undergoes a continuous monotonic decrease over the entire frequency range.

The alpha (low frequency) dispersion is manifested by the very large increase in permittivity at audio frequencies. These effects can be ascribed, at least in part, to counterion diffusion effects (cf. Section II.C.). We can crudely estimate the dielectric increment that such effects are likely to produce in tissue. Suspensions of polystyrene particles of radius 0.6 μm and volume fraction 0.3 show dielectric increments of about 10^4 in media of conductivity typical of biological fluids[78]. The theories of Schwarz and Grosse, when applied to suspensions of cells of radius 30 μm and volume fraction 0.9, (Equations 69 and 73) predict a dielectric increment of roughly 10^6 which is not greatly different from the measured permittivities of tissues at audio frequencies (which are usually in the range 10^6–10^7).

Other tissue-specific effects might contribute to the alpha dispersion. These include active membrane conductance phenomena, the charging of intracellular membrane-bound organelles that connect with the outer cell membrane, and perhaps a frequency dependence in the membrane impedance itself. Schwan reviewed the present understanding and unsolved questions about the alpha dispersion.[102] An important difficulty is the lack of adequate data at low frequencies, in particular a lack of data for the low-frequency plateau in permittivity.

The alpha dispersion is very striking in the permittivity data, but hardly appears at all in the conductivity. Assuming a dielectric increment $\delta\epsilon$ of 10^6 and relaxation frequency of 100 Hz, the Kramers-Kronig relations suggest that the total increase in conductivity associated with the alpha dispersion is about 0.005 S/m. For comparison, the ionic conductivity of typical tissues is about 200 times higher. Thus, at low frequencies, tissues are overwhelmingly resistive in spite of their tremendous permittivity values. Thus, for engineering applications, the alpha dispersion has little significance. Its biological significance remains to be established.

The beta dispersion occurs at radiofrequencies, and arises principally from the capacitive charging of cellular membranes in tissues (Section II.A.). In addition, a smaller component might arise at high radiofrequencies from dipolar orientation of tissue protein.

To illustrate the magnitude of the effect, blood exhibits a total dielectric increment $\Delta\epsilon$ of 2000 and a beta relaxation frequency of 3 MHz. The associated increase in conductivity is about 0.4 S/m. For other tissues, the total change in permittivity associated with the beta dispersion is 10^3–10^4 relative to free space, with a relaxation frequency of about 500 kHz.

The static permittivity values and relaxation times of this dispersion are typically larger in tissues than in blood, a consequences of the typically larger cell sizes in tissues. We discuss the implications of the beta dispersion for bioeffects studies in Section V.C.

The gamma dispersion occurs with a center frequency near 25 GHz at body temperature, due to the dipolar relaxation of the water that accounts for 80% of the volume of most soft tissues. For a total dielectric increment of 50 (typical of soft tissues containing 80% water) and relaxation frequency of 25 GHz, the total increase in conductivity is about 70 S/m.

Tissues and other biological materials typically exhibit a small dispersion between 0.1 and 3 GHz, which several authors have termed the delta dispersion. In tissues between 0.1 and 1 GHz, the permittivity typically decreases by 10–20 relative to free space, and the conductivity increases by 0.4 to 0.5 S/m. Mechanisms that can be suggested include dipolar relaxation of water (including water associated with tissue surfaces), rotational relaxation of polar sidechains (Section V.B.), Maxwell-Wagner effects, and counterion diffusion along small regions of charged surfaces. The lack of a single dominant mechanism makes the interpretation of this dispersion region difficult.

IV. DIELECTRIC PROPERTIES OF SELECTED TISSUES

A. PROPERTIES OF EXCISED TISSUES

1. Blood

The dielectric properties of blood were studied by early investigators; the subject continues to be important for practical applications. Cole[3] presented an interesting review of early work by Höber, Fricke, and others; Schanne and P.-Ceretti[6] summarized the results of the many dielectric studies these and more recent investigators have been conducted on this tissue.

Trautman and Newbower[103] have prepared an excellent review of conductivity data from blood, including many references to early work. The data can be fitted to the Maxwell-Fricke mixture theory (Equation 52), although the shape factor γ that is needed to fit the data takes on values ranging from 0.65 to 1.39, depending on the species of animal and measurement conditions. (Trautman and Newbower actually used empirical functions to fit the data.) In most species the red cell is nonspherical and is subject to changes in aggregation and orientation with flow; these variations in the form factor are understandable and important.

The dielectric relaxation properties of whole blood and hemoglobin have been reviewed by Schwan,[104] with emphasis on the mechanisms responsible for them. Whole blood exhibits beta, gamma, and delta relaxations but (curiously) no alpha dispersion. For blood with hematocrit of 40%, the beta dispersion has a total permittivity increment of about 2000 and is centered at about 3 MHz. The gamma and delta dispersions are similar to those found in hemoglobin solutions of similar concentration.

The dielectric properties of more concentrated erythrocyte suspensions have been studied by several investigators. Pfutzner reported the permittivity of blood from several species at frequencies at 50 kHz, as a function of the hematocrit.[105] The permittivity reached a peak at a hematocrit of roughly 60% and declined at higher volume fractions, which he attributed to changes in the current pathways around regions of packed cells. Such effects would be difficult to explain using a mixture theory, and presumably arise from aggregation of the erythrocytes.

The dielectric properties of blood at microwave frequencies have been reported by Pethig (0.1–4 GHz)[106] and Alison and Sheppard (29–90 MHz).[107] At these high frequencies, the dielectric properties of blood (as with other high water content tissues) are dominated by the dipolar relaxation of water.

2. Skeletal Muscle

Geddes and Baker have published a comprehensive summary of the low frequency conductivity of various muscle tissues.[108] Figure 7 shows recent data from canine skeletal muscle.

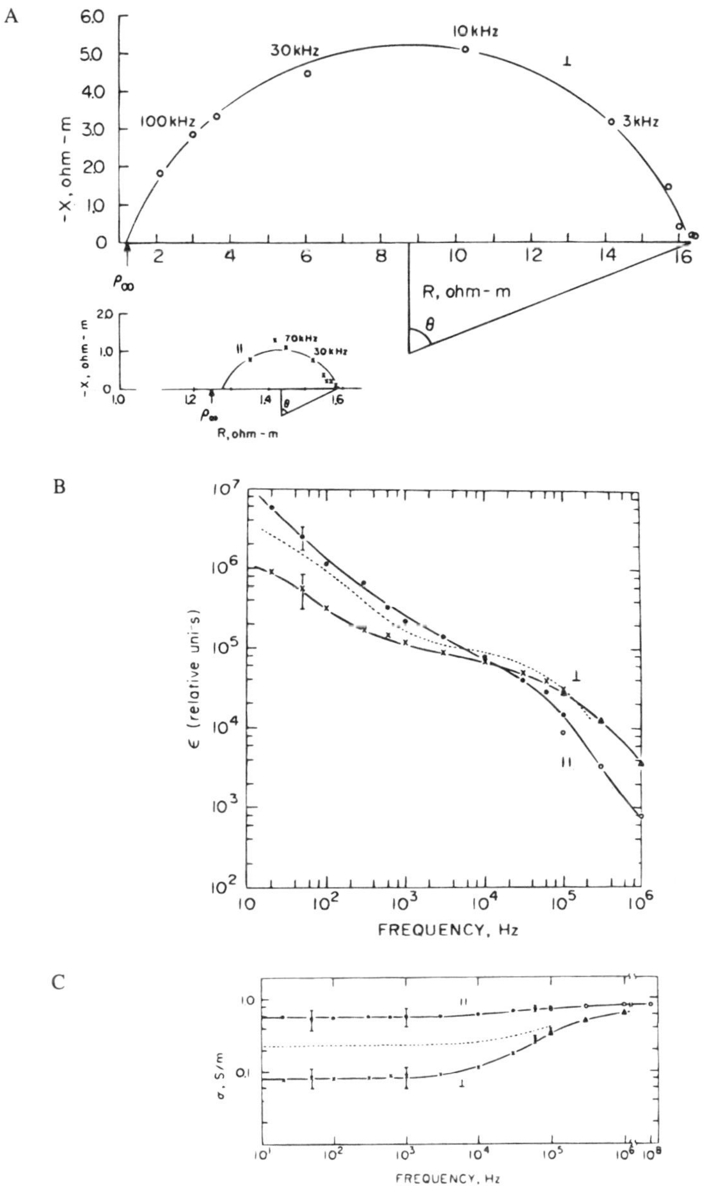

FIGURE 7 The electrical properties of dog skeletal muscle at 37°C in the perpendicular and parallel orientations. (A) Specific impedance plots. The arrow indicates the specific resistance measured at 100 MHz in nonoriented tissue, representing the limiting resistivity of the beta dispersion; (B) permittivity; (C) conductivity vs. frequency. Also shown (dotted lines) are the dielectric properties of nonoriented frog skeletal muscle at 25°C. The symbols v and h refer to tissues in the transverse and the parallel orientations, respectively. (From Epstein, B. R. and Foster, D. R., *Med. Biol. Eng. Comput.*, 21, 51, 1983. With permission.)

Muscle exhibits an extreme anisotropy in its electrical properties, with a seven to tenfold variation in conductivity and permittivity (of dog skeletal muscle) at low frequencies (Figure 7). Similar variations have been reported in skeletal and heart muscle tissue from many other species.[109,110,111] Moreover, from Figure 7, it appears that some degree of anisotropy persists at all frequencies below the microwave range.

This anisotropy has interesting consequences. For example, when measuring the impedance of muscle tissue with a four electrode arrangement, the conductivity that is obtained depends on the spacing between the electrodes, compared to the space constant of the muscle fibers.[112] As current flows parallel to the fiber axes, it redistributes itself between intracellular and extracellular spaces over distances comparable to the space constant of the fiber. When the interelectrode distance is smaller than the space constant, current flows only through the extracellular space; otherwise it flows through the intracellular space as well. This effect has significant consequences for volume conductor theories for muscle electrophysiology.

We summarize briefly the dielectric properties at the three major dispersion regions described above.

a. Alpha Dispersion

Muscle exhibits a large alpha dispersion, which is more pronounced in the longitudinal than transverse orientation. The effect is clearly seen in Figure 7.

Schwan had earlier suggested that the alpha dispersion in muscle arises principally from polarization of counterions near the membrane surface,[1]however, Falk and Fatt[113,114] argued that this would require unrealistically high fixed charge densities on the membrane. (A simple calculation based on Grosse's or Schwartz's theory for counterion polarization, presented above, does not support this point, however.)

Fatt and Falk proposed instead that the alpha dispersion arises from polarization of the sarcotubular system, which is a complex membrane system that permeates the interior of the muscle fibers and connects to the exterior of the cell. The sarcotubular system was modeled electrically as a series of tubes, filled with electrolyte, that pass into the body of the cell. At low frequencies, current penetrates deeply into the sarcotubular system, while at higher frequencies it is shunted instead through the cell membranes. The frequency corresponding to the transition between these two limits would be determined by a combination of the sarcotubular capacitance and resistance of the fluid within these tubules. As a consequence, the total capacitance between the sarcoplasm and the extracellular medium will vary with frequency, from the capacitance of the outer call membrane plus the sarcotubular system, to that of the outer cell membrane only. The capacitance of the sarcotubular system exceeds that of the outer cell membrane by a factor of ten or more, leading to a tenfold increase in the permittivity of the tissue. There will be a corresponding variation in the permittivity of the tissue (when measured in the perpendicular direction), as shown by Equation 54.

Experimental evidence is roughly in agreement with this expectation. Schwan[110] first reported the alpha dispersion in a tissue (nonoriented frog skeletal muscle), which appeared as a 30-fold increase in permittivity with a center relaxation frequency of about 100 Hz. In measurements on single fibers of the frog skeletal muscle using a vaseline gap technique, Takashima observed a dispersion in the apparent capacitance of the outer membrane centered at 100 Hz.[115] The total increase in apparent membrane capacitance, however, was only a factor of 5. The difference between these two results might in part arise from different experimental techniques, i.e., the use of single fiber preparations vs. dielectric measurements on bulk muscle tissue.

In short, the underlying mechanism for the alpha dispersion in muscle tissue is likely to reflect a combination of two effects—polarization of the sarcotubular system and counterion polarization—which cannot be clearly separated at present.

b. Beta Dispersion

For muscle tissue oriented perpendicular to the external field, the characteristics of the beta dispersion can be calculated using the theory presented in Section II. The radii and volume fraction of muscle cells in the tissue depend on the muscle type and condition, but typical values for dog skeletal muscle are 50 μm (average cell radius) and 0.85 (average cell volume fraction). Equations 54 to 57, modified using the shape factor for a cylinder oriented normal to the field, then yield

$$\begin{aligned} \epsilon_s - \epsilon_0 &\approx aC_m/\epsilon_0 \approx 5 \times 10^4 \\ \sigma_s &\approx \sigma_a(1 - p)/(1 + p) \approx 0.06 \text{ S/m} \\ f_c &\approx \sigma_i/(2\pi a C_m) \approx 250 \text{ kHz} \end{aligned} \tag{75}$$

These estimates assume that σ_a and σ_i are both equal to the measured conductivity of canine skeletal muscle at 100 MHz, and the membrane capacitance C_m is 10^{-2} F/m^2. The data in Figure 7 exhibit, as expected, a plateau at about 10^5 relative to free space, and a relaxation frequency (at which the conductivity is halfway between its high and low frequency limits) around 250 kHz. In the longitudinal orientation, the conductivity at low frequencies is higher and varies much less with frequency.

c. Gamma and Delta Dispersions

Figure 8 shows the dielectric properties of nonoriented dog skeletal muscle and barnacle muscle fibers (perpendicular orientation). While the data suggest a plateau in the permittivity near 0.1 GHz, there is no distinct separation between the beta and higher frequency dispersions. Figure 9 shows the dielectric properties of barnacle muscle fibers (perpendicular orientation) at 1°C on the complex permittivity plane. The barnacle muscle at 1°C shows a gamma relaxation centered at 8 to 9 GHz, the same frequency as shown by pure water at that temperature.

The magnitude $\Delta\epsilon_\gamma$ of the gamma dispersion is found in Figure 9 from the intercept of the semicircle with the real axis. Thus $\delta\epsilon_\gamma$ is about 57 relative to free space. The total non-water fraction of the tissue, calculated with the mixture theory,. is about 0.2, which is 25% higher than the experimentally deduced value (obtained by drying the tissue to constant weight and using the known density of dry protein). We attribute this to the presence of motionally hindered water that does not contribute significantly to the dielectric properties at these high frequencies.

The delta dispersion is seen in the variation in permittivity with frequency below 3 GHz, at which the permittivity of pure liquid water has essentially reached its low frequency limiting value. It can be more clearly seen in a plot of the conductivity vs. the quantity $f^2/[1 + (f/f_c)^2]$ (Figure 10) where f_c is the relaxation frequency of pure liquid water. For pure water, this should yield a linear plot; however at low gigahertz frequencies pronounced deviations appear in the data from the tissue.

In summary, the dielectric relaxation properties of muscle are complex. The various dispersion regimes—alpha, beta, gamma, and delta—are easily recognized but overlap somewhat in frequency. It is easy to appreciate the difficulties that similar data must have presented to the early investigators.

3. Other Soft Tissues

Figures 11 and 12 show the dielectric properties of several soft, high water content normal and tumor tissues.[116,117] The dielectric properties of all of these tissues are quite similar and we discuss them as a group.

a. Beta Dispersion

The dielectric relaxation shown in Figure 11 reflects a broad distribution of relaxation times which arises principally from the presence of membrane-bound structures of widely varying dimension. Stoy et al.[116] established this for liver tissue, using a simple structural

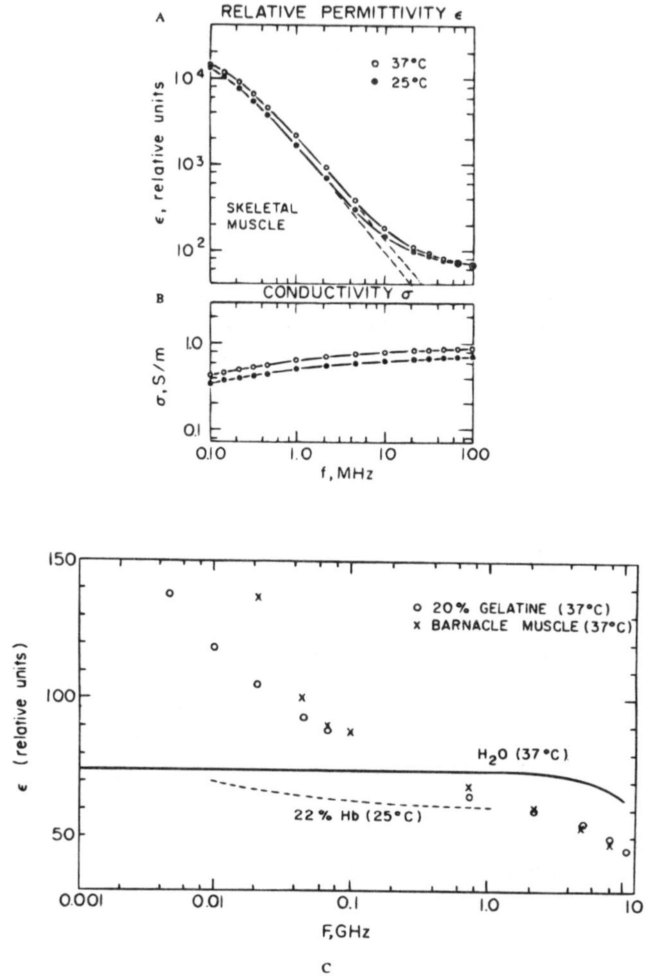

FIGURE 8 Dielectric properties of nonoriented muscle tissue at higher frequencies. (A) Permittivity of dog skeletal muscle (from Ref. 116); (B) conductivity (Ref. 116); (C) permittivity of barnacle muscle fibers (perpendicular orientation) at UHF and microwave frequencies, with the permittivity of pure water and two protein solutions of the same volume fraction shown for comparison (Ref. 101). (From *Biophysical Journal and Physics in Medicine and Biology.* With permission).

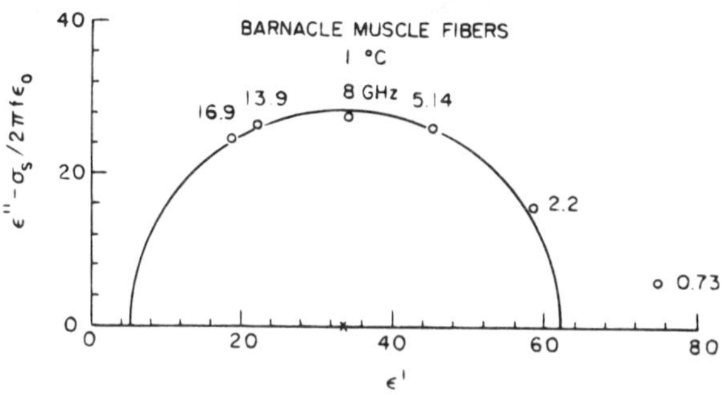

FIGURE 9 Complex permittivity plot of the dielectric properties of barnacle muscle fibers (perpendicular orientation) at 1°C, showing a single time constant relaxation with center frequency near 8 to 9 GHz with an additional delta dispersion evident at low gigahertz frequencies. (From Foster, K. R., Schepps, J. L., and Schwan, H. P., *Biophys. J.*, 29, 271, 1980. With permission.)

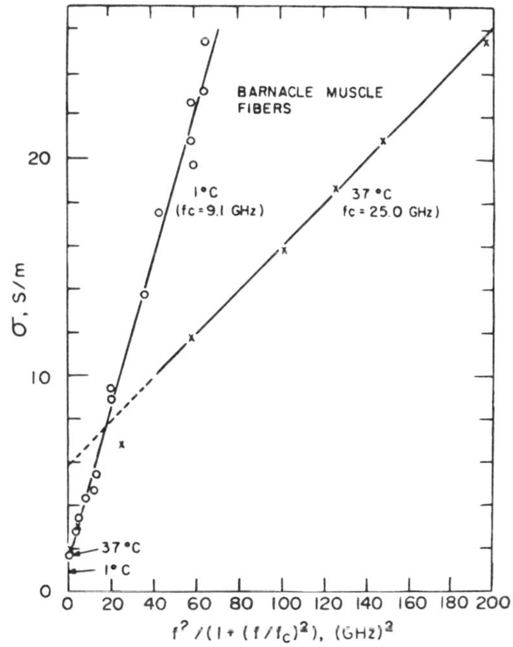

FIGURE 10 The conductivity of barnacle muscle fibers (perpendicular orientation) plotted vs. the quantity $f^2/[1 + (f/f_c)^2]$, with f_c assumed equal to the relaxation frequency of pure water at the indicated temperatures. The conductivity of the tissue at 100 MHz at these two temperatures is indicated by the arrows at left. The data show the delta dispersion that adds significantly to the conductivity of the tissue above 0.1 GHz, evident by the 4.5 S/m difference at 37°C between the zero frequency intercept of the data (6 S/m) and the conductivity at 100 MHz (1.5 S/m). (From Foster, K. R., Schepps, J. L., and Schwan H. P., *Biophys. J.*, 29, 271, 1980. With permission.)

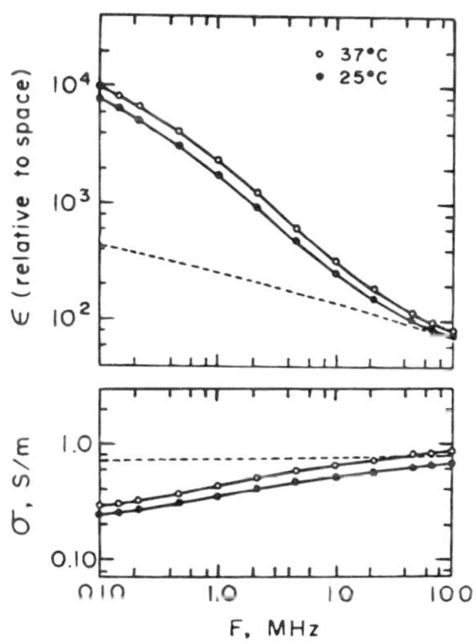

FIGURE 11 Dielectric permittivity and conductivity of dog pancreas at 25 and 37°C, together with the fitted Cole-Cole function (Equation 21). The parameters are ϵ_s = 10,480 (12,960) relative to free space, ϵ_∞ = 41.3 (36.7) relative to free space, σ_s = 0.226 (0.268) S/m, f_c = 230 (254) kHz, α = 0.232 (0.228) for the data. The figures in parentheses refer to results at 37°C; otherwise at 25°C. Also shown (dotted line) are the dielectric properties (25°C) after homogenization of the tissue, showing the elimination of most of the beta dispersion by disruption of the cells. (From Stoy, R. D., Foster, K. R., and Schwan, H. P., *Phys. Med. Biol.*, 27, 501, 1982. With permission).

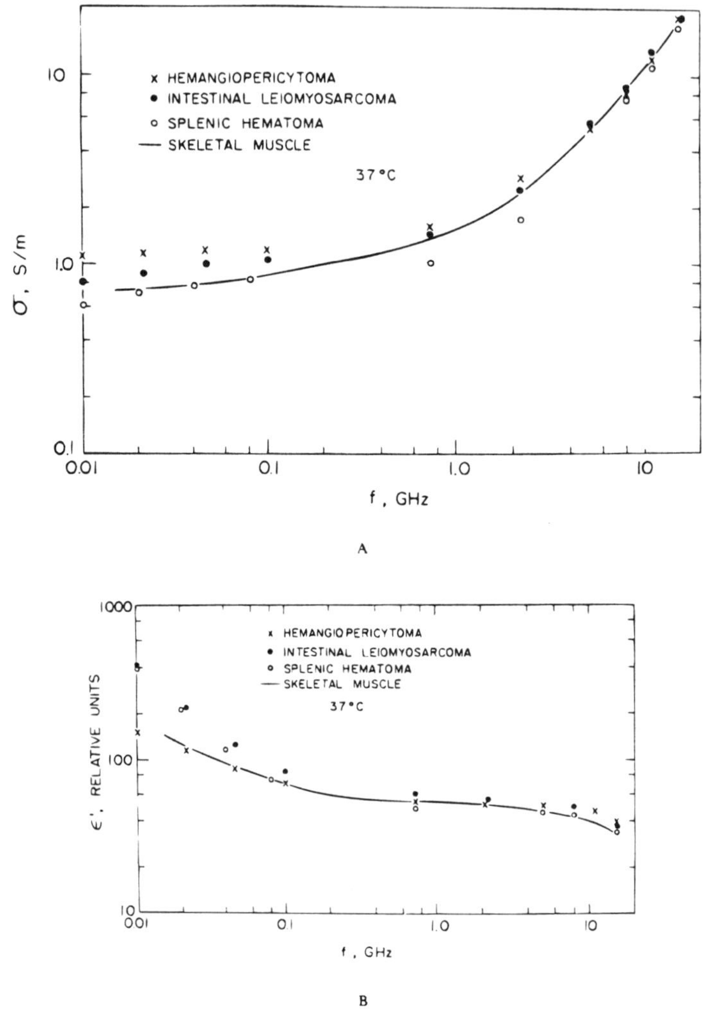

FIGURE 12 Dielectric properties at RFs of three representative tumor tissues and dog skeletal muscle at 37°C. (A) Conductivity; (B) permittivity. (From Schepps, J. L. and Foster, K. R., *Phys. Med. Biol.*, 25, 1149, 1980. With permission.)

model for the tissue and assuming that the contributions to the permittivity from each major tissue structure are additive. Thus

$$\epsilon' = \Sigma_i \frac{\Delta\epsilon_i}{1 + (f/f_{ci})^2} + \epsilon_\infty \tag{76}$$

where ϵ_∞ is the contribution of tissue water and protein, and $\delta\epsilon_i$ and f_{ci} are, respectively, the total increase in permittivity and relaxation frequency of the ith structure.

Table 3 shows the results of this analysis for liver tissue. Stoy et al. obtained the dimensions of major membrane-bound structures from a previous ultrastructural analysis, and calculated the values $\delta\epsilon_i$ and f_{ci} using Equations 54a to 57a. The permittivity ϵ_∞ (corresponding to that of the tissue electrolyte and solid content excluding membrane charging effects) was taken as the measured permittivity at 100 MHz; the dielectric increment and relaxation frequency of tissue protein were assumed to be similar to those of globular proteins as given in Table

TABLE 3
Morphology of Rat Liver: Contributions to Permittivity

Structure	Average radius (m)	Volume fraction	$\Delta\epsilon$	f_c (MHz)
Hepatocyte	8.9×10^{-6}	0.83	9400	0.72
Total nuclei	3.9×10^{-6}	0.05	470	1.6
Mitochondria	5.0×10^{-7}[a]	0.22	277	13
Endoplasmic reticulum	2.5×10^{-8}[b]	0.15	8	250
Protein	c	0.16	100	3

[a] The mitochondria are cylindrical in shape, with diameters of 2.4×10^{-6}m by 0.66×10^{-6}m. The "average radius" is taken as the radius of a sphere with the same total volume as the cylinder.

[b] The "average radius" of the endoplasmic reticulum is taken as half of the "mean cisternal width".

[c] The total tissue protein was assumed to have a static permittivity and relaxation frequency typical of most globular proteins. While these values are uncertain for the mostly fibrous tissue proteins, the contribution of tissue protein to the total dielectric increment is probably rather small at most frequencies, compared with the contribution from membrane polarization effects.

Adapted from Stoy, R. D., Foster, K. R., and Schwan, H. P., *Phys. Med. Biol.,* 27, 501, 1982.

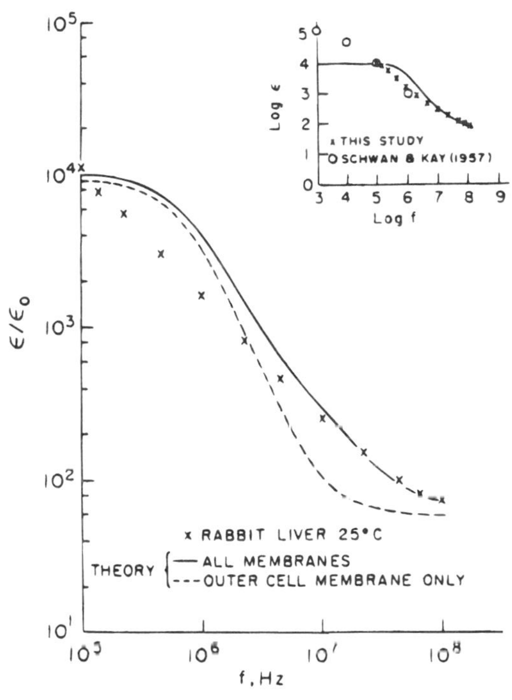

FIGURE 13 Comparison between the measured permittivity of rabbit liver tissue (crosses and open circles) and calculated values obtained from Equation 76 (solid line). Also shown (broken line) is the calculated contribution of the outer cell membranes. (From Stoy, R. D., Foster, K. R., and Schwan, H. P., *Phys. Med. Biol.,* 27, 501, 1982. With permission.)

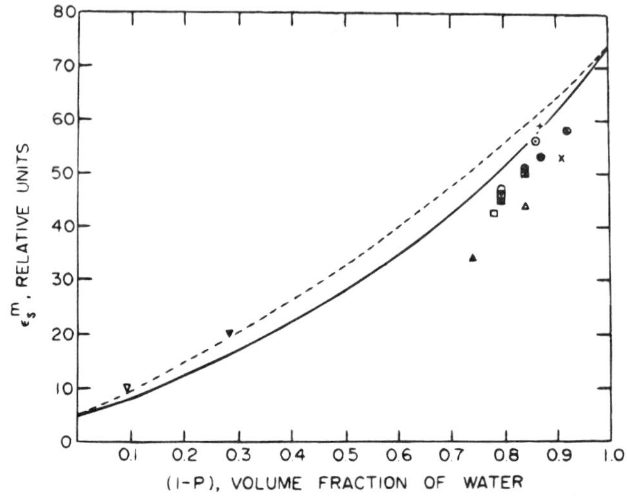

FIGURE 14 Extrapolated permittivity is obtained by fitting data from several normal and tumor tissues above 3 GHz to the Debye equation with the same relaxation time as pure water vs. (1-p) the volume fraction of solid in the tissues. The solid and broken lines are the predictions of the Maxwell-Fricke theory for suspended spheres (broken curve) and suspended ellipsoids of revolution of axial ratio 5:1 (solid curve) for comparison. Identification of the symbols is in Table 4. The parameter ϵ_s is obtained by fitting the data above 3 GHz to single time constant relaxation (Equations 11a to d) with center frequency equal to that of water at the same temperature ($\epsilon_\infty = 4$). (From Schepps, J. L. and Foster, K. R., *Phys. Med. Biol.* 25, 1149, 1980. With permission.)

1. The results of this simple model agree remarkably well with the measured properties of the tissue above 1 MHz[118](Figure 13), but the model underestimates the permittivity at lower frequencies. The little "hump" seen in Figure 13 near 10 MHz might arise from mitochondria, whose dispersion is centered at this frequency.[119]

A few words are needed about this approach. Equation 76 is only an approximate solution for the permittivity of a series of concentric membrane-bound structures. The permittivity should be calculated using the approach described in Section II.A.4. or by solving Laplace's equation directly. At low frequencies, the interior of the cells is shielded from the externally applied electric fields and intracellular organelles will not contribute to the bulk permittivity. Thus, Equation 76 will overestimate the permittivity of the tissue at low frequencies, compared to a more rigorous calculation based on the mixture theory. In fact, it significantly underestimates the actual properties because it neglects the alpha dispersion.

In spite of its limitations, Stoy's analysis shows that:

1. The dielectric dispersion in liver tissue in the range of 1 to 100 MHz represents the high frequency end of the beta dispersion of the cells together with the organelles.
2. The large range in dimensions of major membrane-bound structures within the tissue contributes to the broad spread of relaxation frequencies observed at radiofrequencies.
3. The permittivity values observed below 100 kHz greatly exceed those expected due to the polarization of the outer cell membranes.
4. Tissue protein is a small, and probably negligible, contribution to the permittivity of the tissue. The dielectric properties of the tissue are principally due to membrane-bound structures.

b. Gamma Dispersion

The contribution of the tissue water to the permittivity at microwave frequencies can be estimated by fitting the complex permittivity data above 5 GHz to the Debye equations

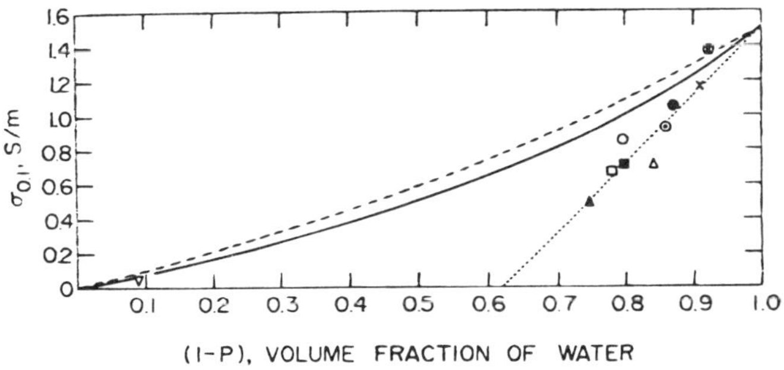

FIGURE 15 The conductivity of several tumor and normal tissues at 100 MHz vs. the volume fraction of solids in the tissue (1-p). Identification of the other symbols is in Table 4. The solid and broken lines are the predictions of the Maxwell-Fricke mixture theory for suspended, nonconductive spheres (broken curve) and prolate ellipsoids of revolution of axial ratio 5 (full curve). The dotted line is the locus of data from various proteins at 1 kHz. (From Schepps, J. L. and Foster, K. R., *Phys. Med. Biol.* 25, 1149, 1980. With permission.)

TABLE 4
Summary of Data

	Symbol	Volume fraction of water, (1 − p)	Extrapolated microwave permittivity, ϵ_s^m	Conductivity at 0.1 GHz, $\sigma_{0.1}$ (S/m)	Extrapolated microwave conductivity $\sigma_{0.1}$ (S/m)
Brain (gray matter)	Δ	0.84	44	0.70	1.13
Brain (white matter)	▲	0.74	34	0.48	0.75
Splenic hematoma	○	0.86	56	0.92	1.60
Splenic hematoma (different sample)	○	0.795	47	0.85	1.30
Skeletal muscle	■	0.795	47	0.70	2.40
Hemangiopericytoma	×	0.91	53	1.16	2.90
Hemangiopericytoma (different sample)	⊗	0.92	58	1.37	3.60
Fat	∇	0.09	10	0.005	0.10
Intestinal leiomyosarcoma	•	0.87	53	1.05	2.30
Liver	□	0.795	43	0.67	2.30
Vaginal fibroleiomyona	+	0.87	59	—[a]	1.80
Pulmonary papillary adenosarcoma	⊕	0.84	51	—	1.40
Lipoma	▼	0.28	20	—	0.40
Renal tubular adenosarcoma	⊞	0.84	50	—	240
Spleen	□	0.795	46	—	1.90

[a] Dash indicates that no value was measured.

From Schepps, J. L. and Foster, K. R., *Phys. Med. Biol.*, 25, 1149, 1980. With permission.

(Equations 11b and 11c) with the same relaxation frequency as pure liquid water.[120] In fitting the data, we have chosen the parameter ϵ_∞ to be 4, and ϵ_s^m (corresponding to ϵ_s in Equations 11b and 11c) allowed to vary. Figure 14 shows the results, for several tumor and normal tissues, as a function of the water content of the tissue. Figure 15 shows the conductivity values of these tissues at 0.1 GHz ($\sigma_{0.1}$) which we assume approximates that of the cytoplasm. The fitted parameters are summarized in Table 4.

c. Delta Dispersion

The tissues exhibit a broader relaxation near 1 GHz can arise from dipolar relaxation of bulk water alone. After subtracting the contribution of the free tissue water, the delta dispersion

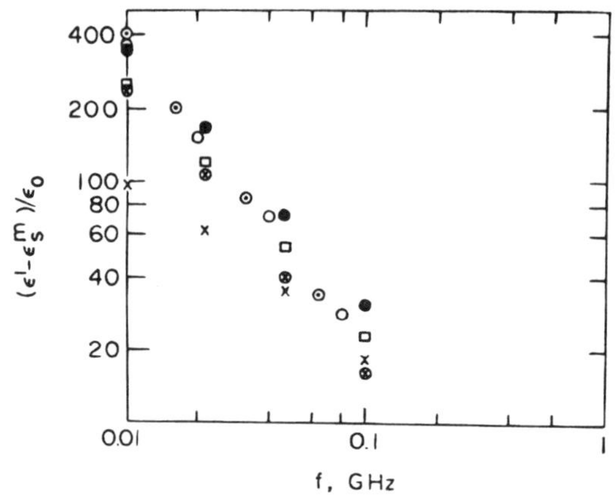

FIGURE 16 Delta dispersion of several tumor and normal tissues found by subtracting the parameter es from the measured permittivity (see text). The symbols are identified in Table 4. (From Schepps, J. L. and Foster, K. R., Phys. Med. Biol., 25, 1149, 1980. With permission.)

is revealed extending down into the radiofrequency region. This dispersion is remarkably similar in several tissues (Figure 16).

d. Parameterization of Dielectric Properties at UHF-Microwave Frequencies

The data from the tissues represented in Figures 14 and 15 can be summarized by empirical equations[120]

$$\epsilon'/\epsilon = 1.71\ f^{-1.13} + \frac{\epsilon_s^m - 4}{1 + (f/25)^2} + 4 \tag{77}$$

$$\sigma\ 1.35\ f^{0.13}\ \sigma_{0.1} + [0.00222(\epsilon_s^m - 4)f^2]/[1 + (f/25)^2]$$

(f in GHz, σ in S/m). In these equations, the Debye terms represent the contribution of the bulk tissue water to the dielectric properties; the power terms represent the delta dispersion. The parameters $\epsilon_s{}^m$ and $\sigma_{0.1}$ would be estimated from Figures 14 and 15.

These equations were obtained from data over the frequency range 0.01–18 GHz. They might be used to estimate the dielectric properties at somewhat higher frequencies; they do not include the effects of the beta dispersion and are not valid at frequencies below this range.

e. Tumor vs. Normal Tissues

A variety of factors will lead to pronounced differences in the dielectric properties of normal compared with neoplastic tissues. We consider several that have been explored in recent studies.

Differences in water content. Tumor tissues have significantly higher water content than homologous normal tissues.[121,122] We quote from the text by Homburger and Fishman:[121]

> Of the gross constituents . . . , the most constant difference between normal and tumor tissues is the increased water content of the latter. Most authors who have studied the chemistry of tumors have made the observation that the dry weight of tumor tissue is less than that of the corresponding normal tissue. The high water content of tumor tissue is shared by embryonic tissue, and therefore cannot be considered as uniquely characteristic of neoplasia. The significance of the increased water content is not known,

but the suggestion has been made that it represents a tendency toward increased protein hydration.

For example, the water content of normal epidermis by weight is 60.9%, while that of carcinoma of the skin is 81.7%; normal rat liver has a water content of 71.4% while that of hepatoma tissue is 81.9%.[122]

Associated with these differences, we expect that at UHF and microwave frequencies neoplastic tissues will exhibit somewhat larger permittivity and conductivity values than homologous normal tissues. Schepps et al.[120] noted a consistent trend in this direction in measurements on samples of spontaneous canine tumor tissue.

Presence of necrosis. At audio- and radiofrequencies, substantial differences are expected between normal and neoplastic tissues, in particular those associated with necrosis in tumor nodules. Smith et al. observed that the conductivity at audio frequencies of an implanted liver tumor in the rabbit is 7 to 10 times higher than that of the surrounding normal tissue; the permittivity of the tumor was generally lower than that of the normal tissue.[123] These differences apparently arise from necrosis in the tumor tissue, which leads to breakdown of cell membranes and thus a larger fraction of the tissue that can carry current at low frequencies. Inflammation and necrosis are commonly found in tumor tissues, and we should not be surprised if similar differences were observed in many other tissues.

However, noticeable differences are not always present. Astbury et al.[124] observed that the dielectric properties of neoplastic and normal canine splenic tissues were indistinguishable from 1 kHz–10 MHz. These tissues are highly perfused with blood and necrosis was absent in the tumor specimens, as confirmed by histological analysis. Moreover, the conductivity of the splenic tissues was relatively high (compared with other tissues such as liver), presumably due to their large blood volumes. Both factors would tend to minimize any differences in the dielectric properties between the normal and neoplastic splenic tissues.

Breast tissues. The normal breast largely consists of adipose tissue, and the presence of infiltrating neoplastic tissues of high water content will certainly lead to pronounced changes in the dielectric properties of breast tissue. Nearly 60 years ago, Fricke and Morse reported that the permittivity of neoplastic breast tissues (at 20 kHz) was higher than that of nonmalignant tumor and normal breast tissues, and suggested the feasibility of diagnosis based on this difference.[125]

Recently, Surowiec et. al.[126] reported that the permittivity and conductivity of breast carcinoma tissues (20 kHz–100 MHz) were both higher than those of normal breast tissues. The differences were striking—orders of magnitude differences were observed in both the permittivity and conductivity. The properties varied with the location (within or surrounding the tumor nodule) from which the sample had been taken. It appears that these differences arise, in part, from the replacement of low-permittivity adipose tissue by high water content neoplastic tissue. By contrast, Campbell and Land, in a dielectric study on excised normal human breast tissue at 3.2 GHz, observed no pronounced difference in the dielectric properties of normal and benign or malignant tumor tissues;[127] however, the data showed high variability.

Thus, the differences in dielectric properties of tumor and normal tissues may often be large. However, the practical significance of any such differences is unclear. Some investigators have proposed the use of impedance methods for diagnostic or screening purposes, but their practicality remains to be shown. For diagnosis of symptomatic patients, few doctors would choose impedance methods over more definitive techniques. On the other hand, the screening of asymptomatic individuals demands far higher sensitivity and specificity than is likely to be possible with impedance methods.

The differences in bulk electrical properties between neoplastic and normal tissues are likely to be more significant in hyperthermia treatments employing radiofrequency or microwave energy. Certainly, when the differences are large (as is the case with a carcinoma of the breast surrounded by fatty normal breast tissue) their consequences for heating are likely to be large. Joines[128] has suggested that such differences be considered in choosing an optimal frequency for radiofrequency hyperthermia treatment.

4. Bone

Kosterich et. al. recently investigated the dielectric properties of bone under near-normal physiological conditions.[129,130] The conductivity of bone at low frequencies is associated with fluid-filled channels that permeate the tissue, and is proportional to the conductivity of the medium surrounding the tissue.[130] The DC conductivity varies by a factor of three with orientation in the several mammalian haversian bones that have been studied,[129] arising from the orientation and dimensions of the major channels through the tissue. The corresponding variation in permittivity with orientation is largely unknown.

Fluid-saturated bone exhibits a broad, featureless dispersion with permittivities of about 1000 at audio frequencies (Figure 17), decreasing to 10 to 20 at 100 MHz; the conductivity increases by about 0.05 S/m over this same frequency range. This dispersion can be closely fitted by a Cole-Cole function with distribution parameter (α) of 0.5 and center relaxation frequency of a few kilohertz for bone in physiologically normal solutions.

Curiously, the relaxation shifts in frequency in proportion to the conductivity of the electrolyte in which the bone is immersed. The reasons for this are unclear. The Maxwell-Wagner effect shows relaxation frequencies that characteristically have the form σ/ϵ, where σ is a conductivity and ϵ a permittivity (cf. Equation 39), while counterion effects characteristically have relaxation times of the form D/a^2, where D is a diffusion coefficient and a the

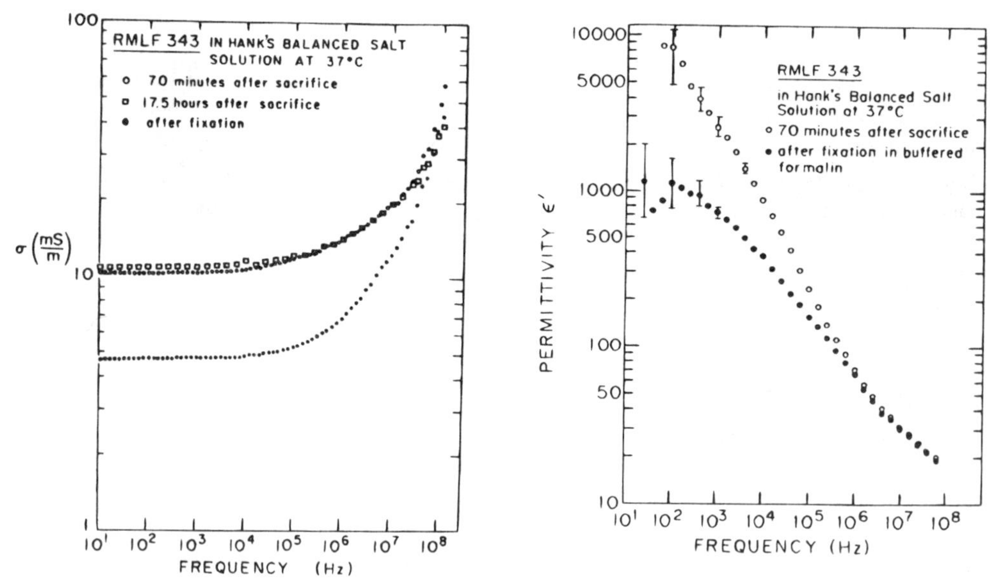

A B

FIGURE 17 Dielectric properties of fluid-saturated rat femur (radial direction) in freshly excised samples and after formalin fixation. (A) Conductivity; (B) relative permittivity. (From Kosterich, J. D., Foster, K. R., and Pollack, S. R., IEEE Trans. Biomed. Eng., 30, 81, 1983. With permission.)

distance over which diffusion occurs (cf. Equation 71). The conductivity-dependence of the relaxation in bone would suggest a Maxwell-Wagner mechanism; however, the magnitude of the dispersion is larger than can be readily explained by such an effect. Similar results are observed in other disordered solids (eg. porous rock,[131]), which other investigators have interpreted as Maxwell-Wagner effects.

5. Adipose tissues (fat and bone marrow)

The dielectric properties of fat and bone marrow were investigated by Schwan and Li[132] and more recently by Smith.[133] These tissues are distinguished by their low water content, and by cells largely filled with lipid. (By contrast, the chief nonwater component of other soft tissues is protein.) Figure 18 shows the permittivity and conductivity of bone marrow and fat, in comparison to that of bone and liver. Fat and bone marrow show a large alpha dispersion at audio low radiofrequencies; the beta dispersion is either absent or small in comparison to that of liver. Curiously, the conductivity of many of the fat samples was higher than that of liver at low frequencies, presumably due to its larger extracellular fluid fraction.

Figure 19 shows the permittivity and conductivity of many samples of adipose tissue 1000 MHz vs. the volume fraction of water (as inferred from the water content measured

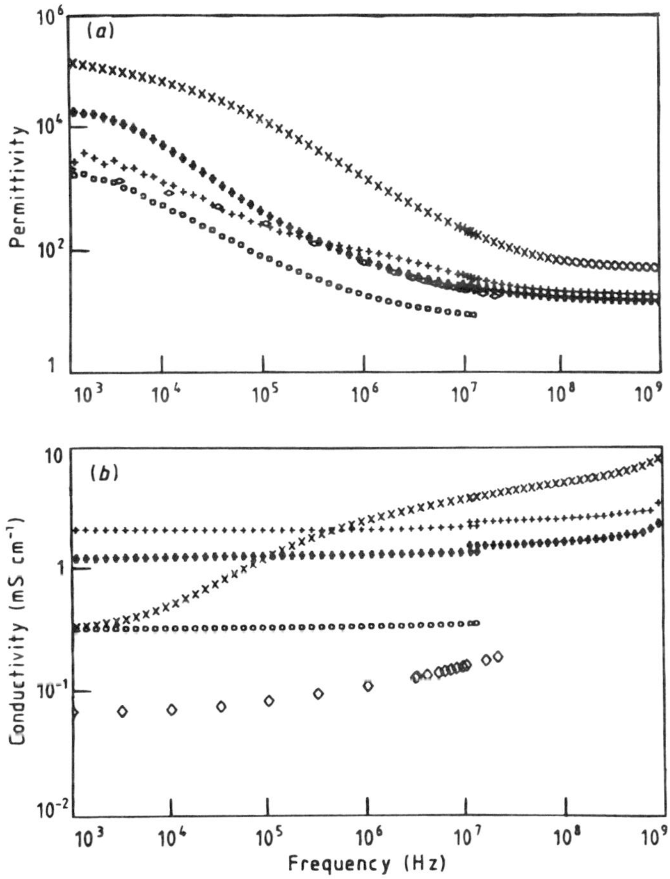

FIGURE 18 (a) Dielectric permittivity relative to free space and (b) electrical conductivity of bone marrow (+), adipose tissues (○, +), liver (×) and bone (◇) as a function of frequency at 25°C, for all tissues except bone (37°C). Each symbol represents measurements on two tissue samples, one for high- and one for low-frequency ranges. The discontinuities in the electrical conductivity data from fat and bone marrow at 10 MHz arise from the use of different samples. (From S. R. Smith and K. R. Foster, *Phys. Med. Biol.* 30, 965, 1985. With permission).

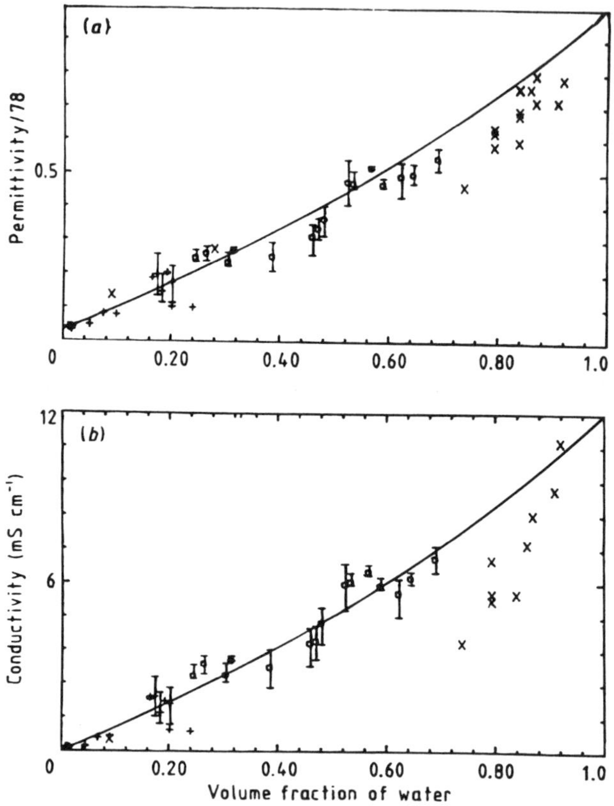

FIGURE 19 (a) Tissue permittivity at 1000 MHz, normalized by the permittivity of water at 25°C (b) conductivity at 100 MHz. The data represent single measurements, or averages of repeated measurements on different areas from the same sample, and the bars indicate the range of the values obtained. Tissues are bone marrow (○) and fat (+). The full curve represents values predicted by Eq. (50–51), assuming that the permittivity of the suspended and continuous media are 2.5 and 78, respectively, and that the conductivity of the non-water fraction is negligible and that of the tissue electrolyte is 1.2 S/m. The data from tumor and normal high water content tissues is the same as that shown in Figures 14–15. Their conductivity values were multiplied by 0.8 to take into account the different measurement temperatures (20°C compared with 37°C). (From S. R. Smith and K. R. Foster, *Phys. Med. Biol.* 30, 965, 1985. With permission).

on a weight basis by drying the specimens). In contrast to analogous data from high water content tissues, the dielectric properties of adipose tissues at these high frequencies agree reasonably well with the Maxwell-Fricke theory and the assumption that the excluded volume consists of the lipid fraction. Perhaps these tissues contain less motionally altered water per unit dry volume than the high water content tissues; no doubt much of their lipid is shielded from contact with water.

B. *IN VIVO* VS. *IN VITRO* PROPERTIES

The dielectric properties of the tissues described above were for the most part measured in excised tissues that are far removed from a physiologically normal state. The question arises whether these properties are likely to be representative of tissues *in situ.* A related question is whether the impedance of tissues *in situ* reflects physiological changes in a way that can be exploited for physiological measurements.

These questions have long been of interest, and much can be said about them. Back in the 1930's, Rajewsky and his colleagues studied the changes in electrical properties of tissues after excision from the animal.[13] In the late 1950's, Schwan and colleagues measured the dielectric properties of tissues in living animals using invasive methods.[134,135] More recently (during the 1970's) several groups began developing noninvasive methods for measuring the dielectric properties of tissues at radio and microwave frequencies *in situ* in living animals.[136,137,138,139,140] Based on this work, we offer the following comments.

1. Changes after Death

At low-frequencies (below the beta dispersion), electrical current largely passes through the extracellular space. Any changes in the fluid distribution in the tissue can obviously lead to a pronounced change in the low-frequency conductivity.

Löfgren[141] clearly demonstrated this effect in 1951, in a careful study of the conductivity of the kidney in a living rat between 2 kHz and 15 MHz. Immediately after the rat was killed, the conductivity of the tissue began to decrease. The conductivity fell to 50% of its baseline value within an hour, and fell even further within the next few hours. In other experiments, Löfgren induced edema in the kidney by forcing electrolyte into the organ, which resulted in a pronounced increase in resistivity. All of these changes were most pronounced at low-kHz frequencies. By contrast, over longer times (a day or more) Löfgren noted pronounced increases in conductivity compared to baseline values. Löfgren interpreted these changes as arising, at short times, from swelling of cells in the dying tissue and, at long times, from disruption of the cells with tissue putrefaction.

These observations are consistent with the report by Rajewsky and colleagues back in the 1930s[13]of a progressive decline in the beta dispersion in several days after the excision of tissue. These long-term changes obviously result from gross disruption of the tissue, and were associated with the gradual decay of metabolic activity in the tissue.

Remarkably similar results were reported recently by Fallert et al. from the myocardium of living sheep, either in a normal physiological state or at various times after production of infarction by tying off a coronary artery.[142] The measurements were performed using a four-electrode technique at low frequencies (1–10 kHz). Within a few minutes after the infarction, the investigators noted a twofold *decrease* in the conductivity of the myocardium; over longer times (weeks) there was a twofold *increase* in the conductivity compared to the baseline state. Fallert et al. attributed the increase at longer times to death of myocites and their replacement by fibrotic scar tissue, and the rapid decrease in conductivity immediately after infarction to swelling of cells due to ischemia. This effect has potential clinical application for the localization of infarcted tissue by impedance during catheter ablation procedures.

Other changes have been reported in tissues over shorter periods, or with less disruption of the samples. Surowiec et al. measured the dielectric properties of bovine brain[143] and bovine liver, kidney, and spleen,[144]frequencies from 20 kHz to 100 MHz. The organs had been removed at the death of the animal and stored at room temperature; periodically samples were removed for measurements. Substantial changes were observed, depending on the type of tissue and length of storage. For example, the conductivity at low frequencies of spleen increased from 0.08 to 0.2 S/m over a 31 hour period; that of brain (white matter) increased from 0.07 to 0.11 S/m over 25 hours. The permittivity at low frequencies decreased by comparable factors over the same period. In their study of the dielectric properties of the canine spleen, Astbury et al.[124] also observed a small increase in the low-frequency conductivity within ten hours or so after the death of the animal, if the tissues were stored at room temperatures. These changes were accompanied by breakdown of cell membranes (confirmed histologically as autolysis).

By contrast, Stuchly et al. reported the dielectric properties of various feline tissues from 10 kHz to 100 MHz, both *in vivo* and *in vitro* (after euthanasia of the animal by overdose of pentobarbital).[145] The changes that occurred in the dielectric properties of the tissues immediately after death were small, comparable to the animal-to-animal variability.

Thus, much (but not all) evidence indicates that the electrical properties of tissue, in particular the conductivity at low frequencies, change greatly a few minutes or hours after the death of the animal. These changes are almost certainly a result of changes in fluid distribution in the tissue (at short times) and membrane breakdown (at long times). This might be one reason for the considerable variability in the reported dielectric data from tissues, particularly at frequencies below about 1 MHz. Many of the data in the Tables are from excised tissues, and the reader needs to be cautious in using them.

2. Other Changes

A few investigators have sought to correlate changes in dielectric properties of tissues at radio- or microwave frequencies with physiological changes. Stuchly et al. observed no changes in the permittivity of cat brain tissue *in situ* between 0.1 and 1 GHz within the first minute after sacrifice of the animal by overdose of pentobarbital.[146]

By contrast, Burdette et al. reported pronounced changes in the permittivity at 2.45 GHz of isolated kidneys with changes in blood flow into the renal artery, with less pronounced changes in the conductivity.[147] Since the kidney has an unusually high blood volume, amounting to 25% or so of its total volume,[148] some changes in electrical properties with blood perfusion are to be expected. However, the size of the changes (a decrease in permittivity from 63 to 45, with an increase in total organ perfusion from 0 to 60 ml/min) is too large and in the wrong direction to arise from a simple increase in the blood volume, and the effect at present is difficult to explain. These measurements were performed using a probe that effectively sensed the dielectric properties of the tissues within a millimeter or so of the surface, and the observed changes might not reflect those of the bulk tissue. Burdette et al. also reported measurable changes in the dielectric properties of kidney after perfusion with 6% DMSO (a cryoprotectant).[149] DMSO has a lower static permittivity than water (47 vs. 78), and these changes are understandable and perhaps of practical use.

Other changes are easier to understand. Schwan and Kay,[135] and more recently Surowiec et al.,[150] measured the resistivity of lung tissue as a function of air content, comparing normal with collapsed and inflated lung tissues. Deflating the lung greatly increases its conductivity (from 0.06 to 0.1 S/m for inflated and deflated feline lung at low frequencies, according to Surowiec et al.). Surowiec et al. observed a broad dispersion at radiofrequencies, which they attributed to the charging of capillary erythrocytes and cellular structures forming the alveoli.

C. TEMPERATURE COEFFICIENTS

Information about the variation with temperature of the dielectric properties of tissues is needed for various applications. For example, several investigators have proposed noninvasive measurement of tissue temperature using impedance imaging techniques.[151,152] The temperature coefficients may reflect reversible changes (due, for example to changes in ionic conductivity) or irreversible changes due to tissue damage.

1. Low Frequencies

The electrical conductivity of tissues at low frequencies (below the beta dispersion) should reflect the volume fraction of extracellular space (since most of the current passes through that tissue compartment), as well as, the conductivity of tissue electrolyte.

McRae and Esrick recently reported extensive measurements of the variation with temperature of the conductivity properties of tumor and normal tissues, both exercised and *in vivo*.[153,154,155] For small temperature increases, McRae and Esrick reported that the conductivity of the tissues (at 44 kHz or 1 MHz) varied reversibly, with a temperature coefficient of about 2% per degree C. These changes evidently reflect thermally induced changes in the conductivity of tissue electrolyte.

However, for large temperature increases (above about 44.5°C), the dielectric properties exhibited abrupt and irreversible changes, reflecting thermal damage to the tissue. These irreversible changes depended on the tissue type, duration of heating, and other factors. In excised tissues maintained at 44°C, the low-frequency conductivity decreased initially by approximately 10% during the first hour, then gradually increased by approximately 50% during the next 8 hours; tenfold changes were reported in the permittivity over similar periods. These changes are an obvious source of uncertainty when attempting to monitor tissue temperature by impedance as proposed by several investigators.

2. High Frequencies

At high frequencies, the conductivity of tissues should reflect that of the tissue electrolyte. The thermal coefficients of conductivity will reflect two (sometimes opposing) effects. Increasing the temperature will increase the ionic conductivity, and also increase the relaxation frequency f_c. At frequencies below f_c this will result in a *decrease* in the loss, and hence in a decrease in the dipolar contribution to the conductivity. At low frequencies this contribution varies inversely as f_c, and thus will show a temperature coefficient close to −2%/°C. In tissues, the contributions of dipolar loss and ionic conduction to the total conductivity are comparable at about 2 GHz. Thus, the temperature coefficient of the conductivity should change sign close to this frequency.

Figures 20 and 21 show the variation in dielectric properties of muscle and another tissue (brain) with temperature. As expected, the conductivity below 1 GHz has the same temperature coefficient as that of simple electrolytes, i.e., about 2%/°C. There is a sign change in the temperature coefficient near 2 GHz where the contributions from ionic and dipolar effects are equal. Above this frequency, the temperature coefficient is close to −2%/°C. At microwave frequencies the permittivity of tissues reflects that of water; the static permittivity of water varies little with temperature.

For tissues at radiofrequencies, the situation is more complicated. At these frequencies, the dielectric properties are determined by Maxwell-Wagner effects associated with the charging of cell membranes. The low-frequency permittivity increase due to this effect depends on the cell size and membrane capacitance and is insensitive to temperature (cf. Equation 54a); but the relaxation frequency should scale as the conductivity of tissue electrolyte and exhibit temperatuire coefficients of about 2% per degree (Equation 56a). We thus expect complex temperature coefficients, that are in the range of 0 to 2%/°C, depending on frequency.

For many practical applications, these temperature coefficients are negligible. However, they must be considered when applying literature data from tissues at 20°C to predict absorption of electromagnetic energy at 37°C.

V. OTHER TOPICS

A. DISTRIBUTION OF RELAXATION TIMES

We earlier touched (Section I.D.) on the problem of determining the distribution times from dielectric spectra. Dielectric data that form symmetric curves on the complex permittivity plane can be represented by a wide variety of distributions of relaxation times that are logarithmically symmetric.

A simple example (based on an earlier treatment by Schwan[1]) illustrates the difficulty. Figure 22 compares the dielectric relaxation of a system calculated using Equation 19, assuming three different logarithmically symmetric distribution functions. These functions are (Figure 23)

Cole-Cole Distribution:

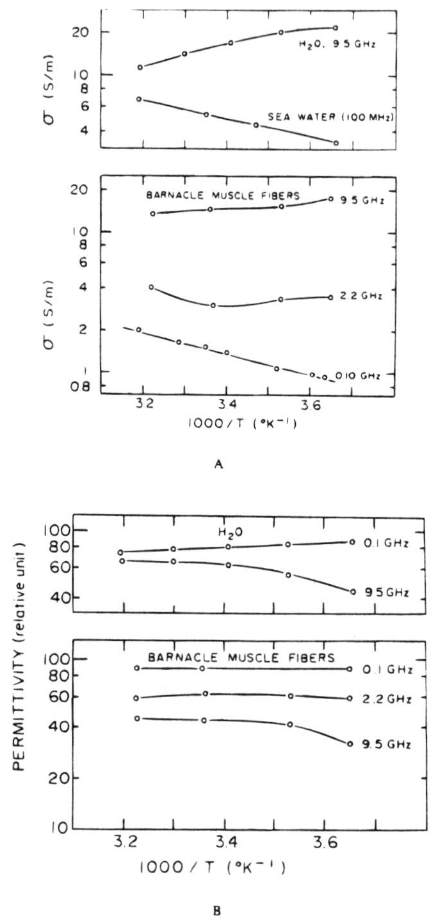

FIGURE 20 Variation in the dielectric properties of barnacle muscle fibers with reciprocal temperature. The corresponding properties of pure water or seawater are shown for comparison. At frequencies near 1 to 2 GHz, two opposite changes (increase with temperature in the ionic conductivity of tissue electrolyte and decrease in the dipolar loss of water) combine so that the temperature coefficient of the tissue is very small. (A) conductivity; (B) permittivity. (From Foster, K. R. and Schepps, J. L., *J. Microwave Power,* 16, 107, 1981. With permission.)

$$p\left[\ln\left(\frac{\tau}{\tau_0}\right)\right] = \frac{1}{2\pi}\frac{\sin(\alpha\pi)}{\cosh[(1-\alpha)\ln(\tau/\tau_0)] - \cos(\alpha\pi)} \tag{78}$$

Log-Rectangular Distribution

$$\begin{aligned} p[\ln(\tau/\tau_0)] &= 0.3, \qquad |\ln(\tau/\tau_0)| \leq 1.67 \\ &= 0, \qquad \ln(\tau/\tau_0)| > 1.67 \end{aligned} \tag{79}$$

Log-Normal Distribution

$$p[\ln(\tau/\tau_0)] = \frac{k}{\sqrt{\pi}}\exp[-k^2(\ln(\tau/\tau_0))^2] \qquad k = 0.60 \tag{80}$$

(The points shown in Figure 22 were calculated for the Cole-Cole distribution with

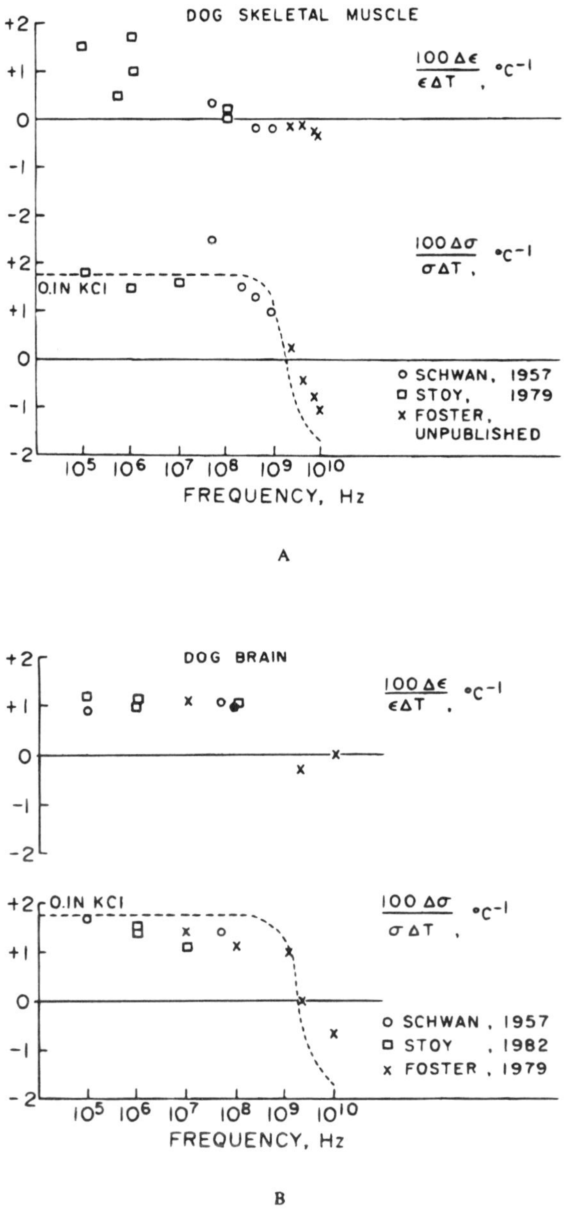

FIGURE 21 Fractional change with reciprocal temperature of dielectric properties of (A) dog muscle and (B) brain near 37°C. The sources of the data are Schwan (Ref. 1), Stoy (Ref. 116), and Foster (Ref. 117). In most cases, the temperature coefficients have been calculated from two measurements at 25 to 28 and 37°C.

$\alpha = 0.23$). The dielectric properties of these systems are close, and would probably be experimentally indistinguishable.

Thus, dielectric data that describe a depressed semicircle in the complex impedance or admittance plane could be well fitted by a variety of functions, possibly supporting quite different interpretations. The question arises: what is the range of distribution functions that is consistent with a given set of data, within specified maximum errors?

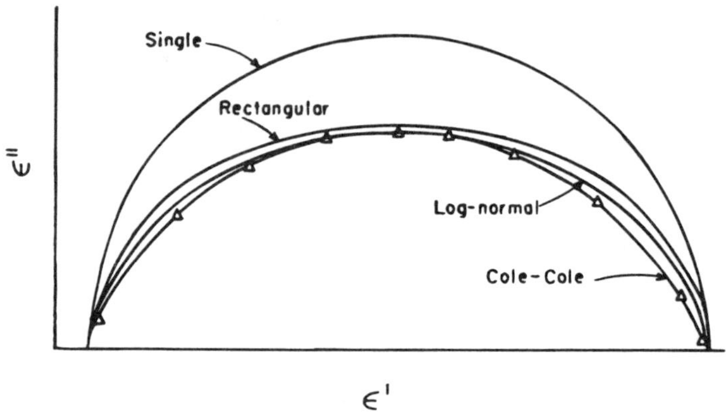

FIGURE 22 Complex permittivity plot obtained from the Cole-Cole distribution (Equation 78), log-rectangular distribution (Equation 79), and log-normal distribution (Equation 80). The triangles represent values that were calculated from the Cole-Cole distribution and used as trial data in determining the ranges of possible distribution functions.

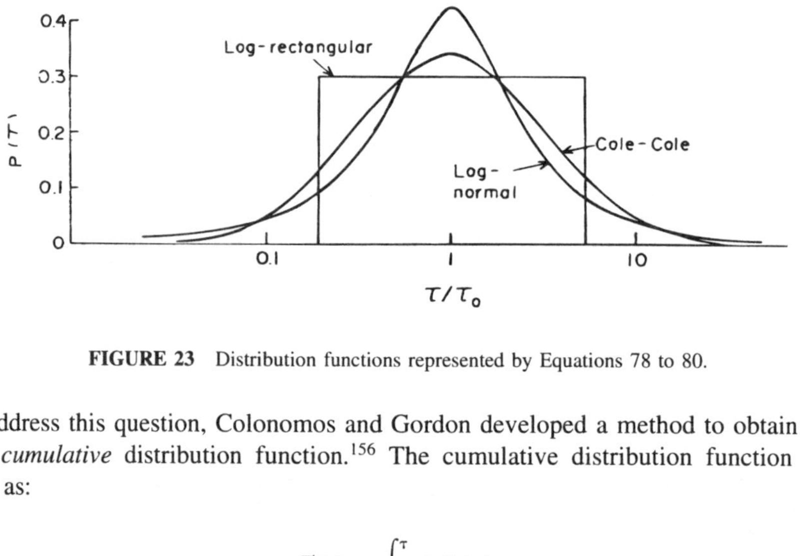

FIGURE 23 Distribution functions represented by Equations 78 to 80.

To address this question, Colonomos and Gordon developed a method to obtain bounds on the *cumulative* distribution function.[156] The cumulative distribution function $\Sigma(\tau)$ is defined as:

$$\Sigma(\tau) = \int_0^{\tau} p(\tau')d\tau' \tag{81}$$

The following constraints must be satisfied:

$$L'(\omega_k) = \int_0^{\infty} \frac{p(\tau)d\tau}{1 + (\omega_k\tau)^2} \tag{82}$$

$$L''(\omega_k) = \int_0^{\infty} \frac{p(\tau)\omega\tau d\tau}{1 + (\omega_k\tau)^2} \tag{83}$$

where $L'(\omega_k)$ and $L''(\omega_k)$ correspond the data at frequency ω_k, and $p(\tau)$ is the amplitude of the distribution at relaxation time τ. In the calculation discussed below, L' and L'' are just the real and imaginary parts of the function $(\epsilon^{\cdot} - \epsilon_\infty)/(\epsilon_s - \epsilon_\infty)$ at frequency ω_k. (Colonomos and Gordo analyzed the limits for the cumulative distribution of the microscopic relaxation times, which differ by a small factor from the macroscopic relaxation times of the bulk

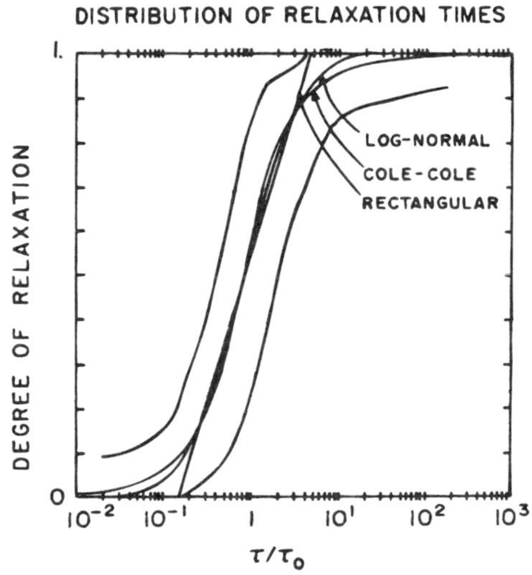

FIGURE 24 Cumulative distribution functions defined by Equation 81 for the three distribution functions (Equations 78 to 80), together with the upper and lower bounds calculated, using the linear programming technique. In these calculations, maximum relative errors of 5% were assumed for the complex permittivity values. The "degree of relaxation" is the cumulative distribution function R(τ).

material, and their equations include a factor in the definitions of L′ and L″ to correct for this difference.) Values for L′ and L″ can be chosen, and allowed to vary over a range (5% for the calculation discussed below), to simulate experimental uncertainties in the data. The maximum and minimum values of $\Sigma(\tau)$ can then be found using standard linear programming methods, treating $\Sigma(\tau)$ as the objective function subject to the constraints indicated.

Figure 24 shows the calculated bounds on $\Sigma(\tau)$, for the data shown in Figure 22, assuming experimental uncertainties of 5%. The vertical axis is the cumulative distribution function $\Sigma(\tau)$, whose values fall between the limits of 0 and 1. The figure also shows the cumulative functions corresponding to Equations 78–80.

The results of this calculation show:

1. A wide variety of continuous distribution functions can fit a given set of data if allowance is made for experimental error. The three continuous functions (Equations 78–80) would be difficult to distinguish experimentally. As the experimental uncertainties increase, the corresponding bounds quickly grow.
2. The data can also be fitted assuming that the distribution consists of discrete components whose time constants span an order of magnitude or more.

Clearly, if allowance is made for experimental error, a very large range of distribution functions can fit almost any set of experimental dielectric data. One might choose functions that provide the best fit, to compare different theoretical models. For example, one might determine whether a set of data is better represented by a Cole-Cole function, or by a sum of single time constant relaxations. However, this analysis is sensitive to the presence of any systematic error in the data. The analysis of Colonomos and Gordon provides a severe warning about the dangers of overinterpreting dielectric data. We will return to this point in the next section.

B. "BIOLOGICAL WATER"

Water is the major constituent in most tissues, and in several respects the dielectric properties of tissues reflect those of its water. At radiofrequencies, the conductivity of tissue is essentially that of its intracellular and extracellular fluids. At microwave frequencies the dielectric dispersion arises from the dipolar relaxation of the bulk tissue water. The properties of "biological" water have long been controversial, and it is interesting to study its behavior as reflected in these properties.

The approach is to measure the electrical properties in a suspension and compare it to the same properties of pure water. Inevitably the suspension shows a decrease in conductivity compared to that of the solvent water. This decrease can have two causes: an "obstruction effect" due to the requirement that ions must travel around nonconductive suspended matter, or a "hydration effect" from changes in the properties of the solvent due to its interaction with the suspended matter. The appropriate use of mixture theory can help to separate these two effects.

We note that the same mixture theory applies to a range of transport properties (thermal conductivity, magnetic susceptibility, and self-diffusion),[157] and comparisons of these properties can be fruitful. These properties reflect different motional properties of the solvent (more precisely, they reflect different combinations of correlation times); yet the geometric factors that give rise to the obstruction effect are the same for all of these properties. Thus, the motional properties of water in complex systems can be studied by comparing these various properties. We discuss below several of these transport properties measured in a range of biological and nonbiological materials.

1. Ionic Conductivity

The ionic conductivity of a variety of protein solutions was measured by Bull and Breese at 1 kHz.[158] From the decrease in the conductivity of the solutions (compared to that of the suspending electrolyte) they inferred the apparent volume of the suspended phase, assumed to be nonconductive, using the Maxwell-Fricke theory. Assuming that the volume fraction of the suspension consists of the protein plus its associated water of hydration, they calculated the "bound" water fraction to consist of 0.6 water per gram of protein. This value is close to, but somewhat above, the conventionally accepted values of 0.3 to 0.4 g water per gram protein obtained from thermodynamic studies of water binding in proteins.[159]

Using a similar approach, Pauly and Schwan[160] estimated the conductivity of the interior of the human erythrocyte to be 0.518 S/m at 25°C, compared with a value 1.45 S/m calculated from the known ionic composition of the cell. They concluded that this difference reflects, in part, the excluded volume in the solution. However, the decrease in conductivity was somewhat larger than would be expected assuming reasonable values of the protein hydration, which they attributed to a " 'frictional effect" or reduction in the mobility of ions near protein surfaces due to hydrodynamic effects.

This approach can be readily extended to tissues, provided that the conductivity is measured at a high frequency (>10 MHz) so that cell membranes do not significantly affect the conductivity. The conductivity of various soft, high water content canine tissues at 100 MHz is shown in Figure 19, as a function of the water content of the tissue. The "average" conductivity of the electrolyte in the tissue can be found by extrapolating the tissue data to zero solid content. Then the Maxwell-Fricke theory can be applied to find an apparent excluded volume p'; this value is compared with the experimental value p (obtained by drying the samples). The difference between p and p' presumably arises from hydration effects. This approach leads to hydration values that are similar to those obtained from protein solutions.

Finally, we recently reported the ionic conductivity of a simple polymer poly(ethylene) oxide in dilute electrolyte solution.[161] We interpreted the data using the Maxwell mixture theory, which led to conclusions about the amount and nature of water of hydration that were similar to those obtained from the biological materials.

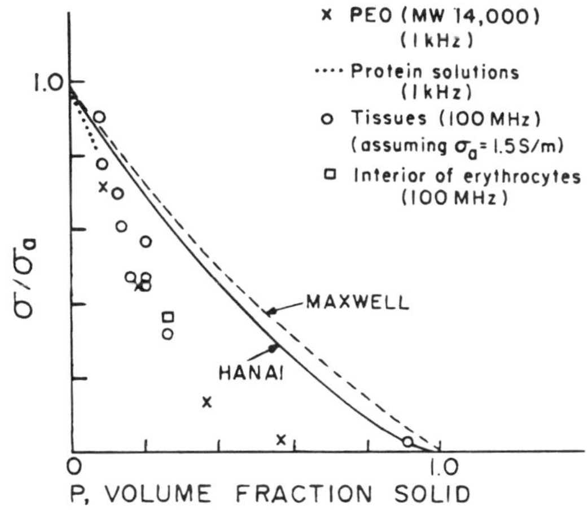

FIGURE 25 Conductivity of a variety of biological and nonbiological suspensions divided by the conductivity of the suspending electrolyte, showing similar hydration effects. The dashed line is the locus of the conductivity of several protein solutions (Ref. 158); other data are poly(ethylene) oxide (PEO) of 14,000 mol wt in 0.1 N KCI (Ref. 161) erythrocyte cytoplasm (Ref. 184) and tumor and normal tissues at 100 MHz (Ref. 120). The predictions of the Maxwell and Hanai relations (Equations 44 and 58) are also shown. A limiting conductivity of 1.5 S/m was assumed in representing the tissue, which was obtained from Figure 15 in the limit of zero solid content.

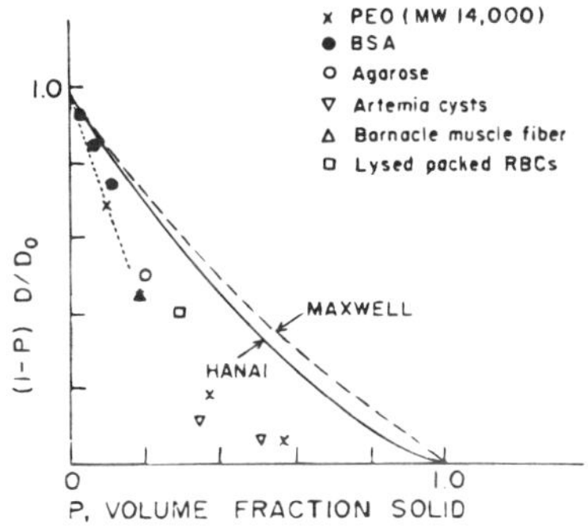

FIGURE 26 Water self-diffusion coefficients in several biological and nonbiological systems. The data from the tissues were obtained by a pulsed NMR technique, under conditions in which the outer cell membranes were considered unlikely to affect the water self-diffusion measurements. The theoretical curves are predictions from the Maxwell or Fricke mixture theories assuming that the suspended particles are impermeable to water. (From Blum, F. D., Pickup, S., and Foster, K. R., J. Colloid *Inter. Sci.* 113, 336, 1986. With permission).

Figure 25 shows the conductivity of these several materials as functions of the volume fractions of the suspension. Remarkably, all of these data fall on the same curve, if suitable assumptions are made about the conductivity of the suspending medium. The effects of suspended solids on the conductivity of biological and nonbiological materials are similar.

2. Water Self-Diffusion

A similar analysis applies to water self-diffusion, provided that care is taken to avoid the effects of cell membranes. Moreover, the measured self-diffusion coefficient must be multiplied by the factor $(1 - p)$ to correct for the differences in reference volumes in the diffusion vs. conductivity measurements.[162] Figure 26 compares self-diffusion coefficients of water in various biological and nonbiological systems. Again, the data all fall on the same curve. Astonishingly, similar results are even obtained for suspensions or solutions in nonaqueous solvents.[163] It seems that the manifestations "bound water" in biological and nonbiological materials are very much the same, even in nonaqueous systems!

3. Dipolar Effects

We also can investigate the properties of tissue water by comparing the dielectric dispersion of tissues at UHF-microwave frequencies with those of pure liquid water. Equation 11d is rewritten to represent the sum of a dipolar plus an ionic term

$$\sigma_s = \frac{2\pi f^2(\epsilon_s - \epsilon_\infty)\epsilon_0/f_c}{1 + (f/f_c)^2} + \sigma_s \tag{84}$$

where the parameters ϵ_∞, ϵ_s, and f_c pertain to pure water and σ_s is an ionic conductivity. Assuming that the solid content of tissues is essentially nonconductive, Equation 51 can be employed

$$\sigma \approx \frac{1 - p'}{1 + p'/2}\sigma_a + \frac{9p'}{(2 + p')^2}\sigma_i \tag{85}$$

where σ_i is the effective conductivity of the protein plus its water of hydration, and p' is the volume fraction of hydrated protein that is excluded from conduction. The excluded volume fraction p' is greater than p, the volume fraction of anhydrous protein; the difference $(p' - p)$ is by hypothesis the volume fraction of motionally altered water.

At high frequencies, the frequency dependence of the conductivity (σ) should reflect that of the bulk solvent water (σ_a) and variations in the conductivity of the suspended phase (σ_i) with frequency will become insignificant. Then, the measured conductivity will approach a linear function of $f^2/[1 + (f/f_c)^2]$ where f_c is the relaxation frequency of the bulk liquid. From the slope of this function, the effective excluded volume p' and consequently the volume fraction of hydration water can be obtained.

Figure 10 shows the conductivity of barnacle muscle at two temperatures, plotted vs. the quantity $f^2/[1 + (f/f_c)^2)]$, showing the delta dispersion at low gigahertz frequencies. The conductivity of a solution of poly(ethylene) oxide shows similar behavior[161] (Figure 27). Similar results are obtained for protein solutions,[164,165] and ionic and nonionic microemulsions.[166,167]

In all of these cases, the dielectric relaxation at microwave frequencies approaches single time constant behavior with the same relaxation time as that of pure liquid water. The amount of bulk-like water equals the total water content minus the fraction of motionally restricted water.

These various results lead to a consistent picture of the transport properties of the water in these materials. Some fraction of water exhibits a rotational and translational correlation time that is roughly ten times longer than that of bulk water. The remaining water is bulk-like in its dielectric and self-diffusion properties. The amount of motionally restricted water corresponds roughly to a monolayer of water surrounding the suspended molecules. It is comparable to the "bound" water estimated by thermodynamic measurements, and perhaps represents the same water fraction.

However, this identification is by no means certain. The dielectric and other transport measurements described here reflect *kinetic* properties quite different from the *thermodynamic* properties suggested by the term "bound water." The similarity of all the results suggests that the underlying mechanism is a physical effect (e.g. a "frictional effect" such as proposed by Pauly and Schwan[55] rather than a chemical effect (e.g. hydrogen bonding). The effects are not noticeably different in biological and nonbiological systems.

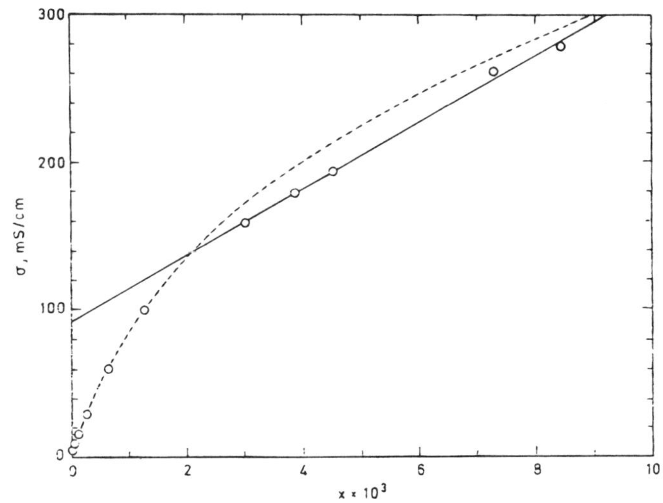

FIGURE 27 Plot of the conductivity vs. the quantity x for a 25% (volume) solution of poly(ethylene) oxide, showing the delta dispersion at low gigahertz frequencies that adds a total conductivity of ca. 9 S/m to the solution conductivity at microwave frequencies. This is to be compared with the similar plot (Figure 10) for muscle tissue, and shows the similar delta dispersion in these two materials. (From Foster, K. R., Cheever, E., Leonard, J. B., and Blum, F., *Biophys. J.*, 45, 975, 1984. With permission.) x is $f^2/[1 + (f/f_c)^2]$, where f_c is the relaxation frequency of pure water.

Some authors have argued that the mean dielectric relaxation time of the water in muscle tissue,[168] myoglobin solution,[169] and ocular and brain tissues,[170,171] is longer than that of the pure liquid water by a small factor of 1.5 or less. For example, Grant and colleagues[169,170,171] fitted the permittivity and conductivity data to the Cole-Cole function and other relaxation equations, and based their conclusions on the difference between the mean relaxation frequency and that of water. Masszi et al. fitted only the conductivity data over a limited frequency range (2 to 4 GHz) to the Debye equations to extract the "relaxation time." These authors considered data similar to those presented here; but they drew different conclusions about the properties of the major part of tissue water.

However, these conclusions all reflect implicit assumptions about the distribution of relaxation times; but different distributions might satisfactorily fit any practically obtainable data. Figures 10 and 27 suggest that the dielectric relaxation of tissue and (poly)ethylene oxide solutions approaches, at high frequencies, single time constant behavior similar to that of pure liquid water. This implies that the distribution is nonsymmetric, with one component of the water having a relaxation time similar to that of bulk water, and a smaller fraction being motionally restricted. The question of the "relaxation time" of tissue water is, more precisely, that of the distribution of relaxation times, and is a difficult one to settle by the experimental methods discussed here.

In spite of the obvious difficulties, a comparison of dielectric relaxation, conductivity, and diffusion is a useful way to study the properties of water in biological systems. Dielectric measurements at microwave frequencies reflect the motional properties of water on a short (picosecond) time scale. The other properties reflect the motion of the water molecules averaged over much longer times. But the macroscopic mixture theories for these different properties all involve similar approaches.

C. COUPLING CONSIDERATIONS

The dielectric properties of tissues and their underlying mechanisms provide insight into the interactions of electromagnetic fields with biological systems. Much is known about the interactions at the molecular and membrane levels. But to apply this knowledge we must translate field strength values such as exist outside of man or in the medium surrounding cells into the field values of interest. For example, field strength values might be desired inside the cell nucleus in order to judge the possibility of interaction with genetic apparatus, or across the membrane in order to evaluate possible membrane excitation phenomena.

This discussion is based on recent papers by Schwan.[96,172] We consider only objects of simple geometry to illustrate the magnitude of the effects; recent reviews of coupling considerations at the macroscopic level are by Stuchly and Stuchly,[173] Lin,[174] and Kaune and Anderson.[175]

Two coupling considerations are of prime interest: (1) How do fields external to the human body translate into *in situ* field strength values? (2) How are *in situ* field strengths related to membrane potentials and field values inside the cell or cell nucleus? Clearly, any quantitative biophysical analysis of the interaction of fields with membranes, cell components, and macromolecules must consider these coupling properties.

We first provide the necessary tools to address the above problems by solving Laplace's equation for a model cell consisting of a homogeneous spherical core, surrounded by a thin membrane, immersed in an infinite homogeneous medium. The solution to this will permit us to address the two coupling problems mentioned above. The following discussion is based on that by Schwan.[172]

We let the radius of the inner sphere be R_1, the thickness of the membrane $d = R - R_1$, and the subscripts 1, 2, and 3 refer to the external medium, membrane, and core respectively. After separation of variables, Laplace's equation then yields:

$$\varphi_1 = E(r - \alpha/r^2)\cos(\delta) \tag{86}$$

$$\varphi_2 = (Ar - B/r^2)\cos(\delta) \tag{87}$$

$$\varphi_3 = Cr\cos(\delta) \tag{88}$$

where E is the external unperturbed field, r is the radius vector, and δ is the angle between E and r. The coefficients A, B, C, and α are to be determined from the boundary conditions.

Each of Equations 86–88 gives a potential that is a sum of a homogeneous and a dipole field. The dipole term in Equation 88 vanishes to maintain a finite field at the center of the sphere. The homogeneous field term in Equation 86 is chosen to provide a field strength equal to E for large values of r. The four constants α, A, B, and C are then determined from the two sets of boundary conditions at the two interfaces between regions 1–2 and 2–3. The boundary conditions are: (1) the current density component normal to the surfaces is continuous, and (2) the potentials in the different media are equal at the interfaces.

Solving the resulting set of four linear equations yields:

$$A = 3\sigma_1^* E(\sigma_3^* + 2\sigma_2^*/D \tag{89}$$

$$B = 3\sigma_1^* E(\sigma_3^* - \sigma_2^*)/D \tag{90}$$

$$C = 9\sigma_1^*\sigma_2^* E/D \tag{91}$$

$$\alpha = [R^3(2\sigma_2^* + \sigma_3^*)(\sigma_2^* - \sigma_1^*) + R_1^3(2\sigma_2^* + \sigma_1^*)(\sigma_3^* - \sigma_2^*)]/D \tag{92}$$

with the common denominator D:

$$D = (\sigma_2^* + 2\sigma_1^*)(2\sigma_2^* + \sigma_3^*) + 2(R_1/R)^3(\sigma_2^* - \sigma_1^*)(\sigma_3^* - \sigma_2^*) \tag{93}$$

where

$$(R_1/R)^3 = (1 - d/R)^3 \approx 1 - 3d/R \tag{94}$$

since for cells $d << R$. Then,

$$D/3 \approx \sigma_2^*(\sigma_3^* + 2\sigma_1^*) - 2(\sigma_2^* - \sigma_3^*)(\sigma_1^* - \sigma_2^*)d/R \tag{95}$$

1. Coupling into Macroscopic Objects

We first consider the case of a homogeneous dielectric sphere to gain insight how fields surrounding a man couple into the body. In this case we set the shell thickness d to zero. Then, the internal fields strength is

$$E_i = \frac{3E_0\sigma_a^*}{\sigma_i^* + 2\sigma_a^*} \tag{96}$$

where E_o is the field strength outside the sphere and the complex conductivities (σ_i^*, σ_a^*) are those, respectively, inside and outside of the sphere.

This calculation applies in the quasistatic limit, where the wavelength of the radiation is much greater than the diameter of the sphere. If the outer medium is air while the sphere has dielectric properties typical of those of soft tissue, at low frequencies

and

$$\sigma_a = j\omega\epsilon_o \tag{97a}$$

$$E_i = \frac{j3E_o\epsilon_o\omega}{2\sigma_i}$$

$$j_i = \frac{j3E_o\epsilon_o\omega}{2} \tag{97b}$$

where j_i is the internal current density. The factor $j = \sqrt{-1}$ indicates, as usual, a 90° phase shift between applied field and induced current.

For typical tissues, $\sigma_i \approx 1$ S/m at 10 Hz, and the internal field strength is thus about seven orders of magnitude below that of the external field. We conclude that for objects coupled to an electric field in air:

1. At low frequencies the internal field strength is far smaller than the external field strength.

2. The internal field strength is proportional to the external field strength. Since the conductivity of most tissues increases by a factor of three or less between DC and the megahertz range, Equation 97a shows that internal field strength scales increases roughly proportional to the frequency.

These results can be extended to the case of homogeneous ellipsoids in an external field. The internal field is constant throughout the interior, but depends on the axial ratio of the ellipsoid and its orientation in the external field.

Equations 97 allows a simple estimate of the electric fields and current densities induced in an isolated animal or human situated in an external electric field. This estimate is valid only in the quasistatic limit, e.g. well below the electromagnetic resonance frequency of about 70 MHz for an erect human.

Clearly, below the radiofrequency range, the electric fields induced in a human located in an external electric field are far smaller than the external field. However, irregularities in the shape of the body determine the exact distribution of currents. For example, in a man standing in a field perpendicular to ground, the average current density in the legs is greater than in the trunk, by a factor that corresponds to the ratio of cross sectional areas of the trunk and leg.[176] Differences in the conductivity of the various tissues will also affect the current distribution. Finally, muscle is highly anisotropic at frequencies below 1 MHz, and currents will preferentially flow parallel to the fiber axis. Current will tend to flow around relatively nonconductive tissues such as lung, fat, and bone, and through relatively conductive tissues such as blood. However, the conclusions obtained from Equation 97b are still valid: the current densities induced by low-frequency electric fields external to the body are very small.

2. Coupling into Cells

We next determine various cellular potentials and field strength values.

a. Membrane Potential

The membrane potential is obtained by taking the difference in the two potentials at the inner and outer surfaces of the membrane. If the cytoplasm and the surrounding electrolyte are purely conductive, i.e. if $\sigma >> \omega\epsilon\epsilon_o$, and $d << R$, we obtain

$$V_m = \frac{1.5ER}{1 + (RG_m(\rho_i + 0.5\rho_a))} \frac{1}{1 + j\omega\tau} \cos(\delta) \tag{98}$$

where ρ_i and ρ_a are the internal and external resistivities (equal to $1/\sigma_i$ and $1/\sigma_a$, respectively), R is the cell radius, and G_m is the membrane conductance per unit surface area. The time constant τ has the value

$$\tau = \frac{RC_m(\rho_i + 0.5\rho_a)}{1 + RG_m(\rho_i + 0.5\rho_a)} \tag{99}$$

$$V_m = \sigma_i^* \sigma_a^* dE\cos(\delta)/D \tag{100}$$

and is equal to the time constant of the beta dispersion. Usually the membrane conductance is sufficiently low that the term containing RG_m is negligible in Equation 100. Then

$$|V_m| = \frac{1.5E\cos(\delta)}{\sqrt{1 + (\omega\tau)^2}} \tag{101}$$

which has a low frequency limit $V_o = 1.5\ ER\cos(\delta)$. The limit at high frequencies V_∞ is:

$$V_\infty = \frac{1.5E\cos(\delta)}{\omega C_m(\rho_i + 0.5\rho_a)} \tag{102}$$

These results apply only if the conductivity of the media outside and inside the cell are high enough that the Debye length of the medium is small compared to the cell diameter. With normal physiological media, this is certainly the case. However, when the cell is suspended in low-conductivity media (which is routinely done in studies of electrically-induced forces on cells and electroporation[172]), corrections are needed to take into account membrane conductance and space charge effects.[177]

In the limit of high frequencies, the impedance of the cell membrane becomes vanishingly small, and the density of current (J) through it is approximately the same as that through the external medium. In this case, the transmembrane potential that is induced by J is simply $J/\omega C_m$. At 1 GHz this would correspond to a membrane potential of a tenth of a microvolt, for a current density of 10 A/m^2 through the extracellular medium. (This discussion assumes that the resistivities of the cytoplasm and the external medium are the same; otherwise a small factor of the order of 1 must be included in this calculation; cf. Equation 102.)

A different situation exists when $\sigma << \omega\epsilon\epsilon_o$, for example when cells are suspended in media of low conductivity (as is the case in some electrorotation experiments). For $R >> d$ and $\epsilon_1 = \epsilon_3$, we obtain from Equations 95 and 98:

$$V_\infty = dE\cos(\delta)\epsilon_1/\epsilon_2 \tag{103}$$

if the displacement currents in media 1 and 3 exceed the conductive currents. Thus the field strengths in the membrane are different from those outside the cell by the factor ϵ_1/ϵ_2. Equations 101 and 103 suggest that the ratio V_o/V_m is about R/d if the permittivities ϵ_1 and ϵ_2 are comparable. Hence, the membrane potentials at microwave frequencies are a thousand times or so smaller than those induced by comparable field strengths at low frequencies.

Simple expressions are obtained for the average field strength within the membrane by dividing the total transmembrane potential by d. However, the field strength varies slightly in the radial direction throughout the membrane, as can be shown easily using the theory given above. We consider the maximum field strength that is obtained when $\cos(\delta) = 1$. From Equation 87

$$\begin{aligned} E_m &= \partial\varphi/\partial r \\ &= A + 2B/r^3 \end{aligned} \tag{104}$$

Introducing the values for the coefficients A and B yields the difference in field strengths between the inner and outer membrane surfaces

$$\begin{aligned} \Delta E &= E_1 - E_2 \\ &= 2\sigma_1^* E(\sigma_3^* - \sigma_2^*)d/RD \end{aligned} \tag{105}$$

For small σ_2^* (small membrane conductance and low frequencies) this reduces to

$$\Delta E = 3E \tag{106}$$

which is R/2d times smaller than the average membrane field strength $E_m = 1.5ER/d$ derived above in the limit of low frequencies. The strongest fields that are induced in the membrane are located at each pole of the cell (i.e. where $\delta = 0$ or π). By contrast, the induced fields along the equator (where $\delta = \pi/2$) are entirely trangential. It is easy to show that the tangential component of the field has a strength 1.5 E at the outer surface of the membrane, and is very small along the inner surface. These considerations apply at low frequencies and for poorly conducting membranes.

The electrical fields that act on the cell and its membrane result in mechanical forces on the cell. In particular, there is a compressional force on the membrane given by the product of the membrane permittivity and the square of the membrane field strength, $\epsilon_m E_m^2/2$. Since the field is slightly different at the inner and outer surfaces of the membrane, there will be a net outwards force. At the poles this force is $\epsilon_m(E_1^2 - E_2^2)/2 \approx 9\epsilon_m E^2 R/4d$. This force will elongate the cell in the direction of the field. A more exact analysis must include the tangential component of the field as well.[178]

b. Cytoplasmic Field Strengths

Similar considerations can be used to find the field strength within the cytoplasm. The total voltage drop across the cell is divided between the membrane and cytoplasm; however, there is a phase shift of nearly 90° because of the capacitive reactance of the membrane. Thus voltage developed across the cytoplasm and the induced membrane potential must be added vectorially. The internal field strength is obtained from Equations 88, 91, 93 and 95:

$$E_i = 9\sigma_1^*\sigma_2^* E/D \tag{107}$$

If displacement currents within the cytoplasm and the external electrolyte are negligible,

$$E_i = \frac{1.5ER\rho_i}{1 + RG_m(\rho_i + 0.5\rho_a)} \frac{G_m + j\omega C_m}{1 + j\omega\tau} \tag{108}$$

If the influence of the conductance of the membrane can also be neglected,

$$|E_i| = |E_i|_o \frac{\omega\tau}{\sqrt{1 + (\omega\tau)^2}} \tag{109}$$

with

$$|E_i|_o = 1.5E/(1 + 0.5\rho_a\rho_i) \tag{110}$$

the field strength in the cytoplasm in the limit of high frequencies. In the other limit (of low frequencies), the interior of the cell is shielded by its membrane: the field strength in the cytoplasm is zero if the conductance of the membrane is negligible. In this latter case, the entire potential that is applied to the cell is developed across its membrane, which results in an enhancement of the membrane field strength by the factor R/d. As the frequency is increased, the reactance of the membrane decreases until, in the limit of high frequencies, the entire applied potential is developed across the cytoplasm. These changes correspond to the beta dispersion in the dielectric properties of the cell in its surrounding medium.

These considerations can be readily extended to the case of a cell containing a nucleus or other membrane-bound organelle. In this case, Laplace's equation must be solved for the entire structure—which is straightforward but tedious.[179] A simpler approach, using the principle of superposition, anticipates important features of the exact solution, but ignores factors of order of magnitude unity. This approach assumes that the nucleus is small compared with the cell.

To estimate the field strength in the membrane, we assume that the field acting on the nucleus is given by Equation 108, and we obtain the nuclear membrane potential by replacing E by E_i in Equation 99:

$$V_n = \frac{2.25R_nE\rho_i\cos(\delta)}{[1 + R_nG_n(\rho_n + 0.5\rho_i)][1 + RG_m(\rho_i + 0.5\rho_a)]}\frac{G_m + j\omega C_m}{(1 + j\omega\tau)(1 + j\omega\tau_n)} \qquad (111)$$

where the subscript n refers to the nucleus and

$$\tau_n = \frac{R_nC_n(\rho_n + 0.5\rho_i)}{1 + R_nG_n(\rho_n + 0.5\rho_i)} \qquad (112)$$

is the time constant for the beta dispersion of a dilute suspension of the nuclei.

The field strength inside the nucleus is obtained from Equation 109 which must be applied twice, first to the cell in order to get the field strength external to the nucleus, and then to obtain the field strength within it:

$$E_n = \frac{2.25R_nRE\rho_i\rho_n}{[1 + R_nG_n(\rho_n + 0.5\rho_i)][1 + RG_m(\rho_i + 0.5\rho_a)]}\frac{(G_m + j\omega C_m)(G_n + j\omega C_n)}{(1 + jw\tau)(1 + j\omega\tau_n)} \qquad (113)$$

which reduces for small membrane conductances to

$$E_n = -\omega^2C_nC_m\frac{2.25R_nRE\rho_j\rho_n}{(1 + j\omega\tau)(1 + j\omega\tau_n)} \qquad (114)$$

It is of interest to compare the nuclear to the outer membrane potential. Thus

$$V_m/V_m = \frac{[1 + R_nG_n(\rho_n + 0.5\rho_i)](1 + j\omega\tau_n)}{1.5R_n\rho_i(G_m + j\omega C_m)} \qquad (115)$$

The high frequency limit of this is

$$(V_m/V_n)_\infty = (0.5 + \rho_n/\rho_i)C_n/1.5C_m \qquad (116)$$

This becomes simply C_n/C_m if $\rho_n = \rho_i$. Thus, the potential developed across the nuclear membrane in the limit of high frequencies might be higher or lower than that across the outer cell membrane depending on the two ratios ρ_n/ρ_i and C_n/C_m. For C_n equal to or greater than 1.5 C_m, the membrane potential V_m will be greater than the nuclear membrane potential V_n unless ρ_n is smaller than $0.5\rho_i$. At lower frequencies the nucleus is shielded by the cell membrane and its membrane potential will be lower than that of the outer cell membrane. We conclude that it will not be easy to induce significant potentials across the nuclear membrane without simultaneously inducing excessive potentials across the outer cell membrane.

The above results were obtained under the assumption that the displacement currents within the fluid compartments (the medium in which the cell is suspended, the cytoplasm, and the nuclear interior) can be neglected in comparison with the conduction currents. If this is not the case, as may happen when the cells are suspended in media of very low conductivity, simple closed form equations are no longer available and the full solution to Laplace's equation for the system must be obtained. Moreover, the above discussion is based on the (somewhat unrealistic) assumption that the nucleus is much smaller than the cell. Sauer and Schwan have calculated the field acting on a nucleus if this assumption is dropped.[180]

The physical meaning of these results can be easily summarized. At frequencies below the beta relaxation frequency of the cell, the outer membrane shields the interior of the cell and prevents the nuclear membrane from being polarized. Above the beta relaxation frequen-

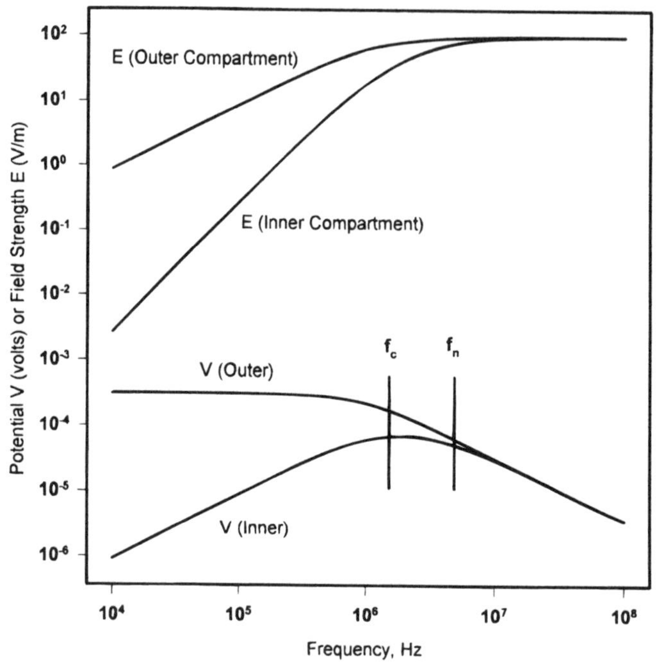

FIGURE 28 Internal field strengths and membrane potentials in a cell with a nucleus, calculated from Equations 86–94. The external field strength was assumed to be 100 V/m and the diameters of the cell and nucleus are, respectively, 2×10^{-6} and 6×10^{-6} m, with corresponding beta relaxation frequencies as indicated on the figure. The membrane capacitance is 10^{-2} F/m^2 and the conductivity of the inner and outer media are 1 S/m.

cies for the cell and nucleus, the reactances of both the outer and nuclear membranes become very small and the voltage drop across these membranes falls off as the inverse of the frequency. The potential induced across the nuclear membrane is largest at frequencies between the beta dispersion frequencies of the cell and nucleus, and, very roughly, is equal to the product of the external field strength times the nuclear radius. To illustrate typical frequency ranges, in liver (Table 3) the beta dispersion frequency of the outer cell membranes is roughly 0.7 MHz and for the nuclei two times higher. The resulting field strengths and membrane potentials calculated from Equations 91 to 94 are given in Figure 28.

More complicated effects are expected for organelles which connect with the outer membrane, such as the sarcoplasmic reticulum in the muscle fiber. The organelle could be charged by currents flowing into it from the electrolyte surrounding the cell, in which case the time constant would be that of the alpha dispersion, since the series component of the circuit in Figure 4 would be identified with an access impedance to the sarcotubular system. Alternately, the organelle could be charged through the cytoplasm, as would occur in an organelle not connecting with the external medium. It is expected that the maximum induced potential across the organelle would be larger for the low frequency process, and be comparable to the potential developed across the outer cell membrane. In the higher frequency process, the maximum potential developed would be roughly equal to the field strength times some distance that corresponds to the mean diameter of the organelle, a far smaller quantity.

Table 5 summarizes the results of this analysis, applied to a hypothetical cell containing one intracellular compartment. The two major conclusions are:

1. For a constant field strength in the electrolyte outside of the cell, the maximum induced membrane potentials are proportional to the diameter of the cell or subcellular structure. These induced membrane potentials fall off as the inverse of the frequency above the

TABLE 5
Summary of the Coupling Properties of External Fields to Cellular Membranes and Compartments

	$f < f_c$ (1 MHz)	$f_r < f < f_n$	$f > f_n$ (~10 MHz)
Cell			
Membranes	Polarized	Not polarized	Not polarized
Interior (proteins, etc.)	Doubly shielded	Shielded	Exposed
Organelles			
Membranes	Not polarized	Partly polarized	Not polarized
Interior (nucleic acids)	Doubly shielded	Shielded	Exposed
Connecting organelles			
Membranes	Polarized	Not polarized	Not polarized
Interior	Not exposed	Exposed	Exposed

Note: f_c is the beta dispersion frequency of the outer cell membrane; f_n that of the organelle.

Adapted from Schwan, H. P., *Biological Effects and Dosimetry of Static and ELF Electromagnetic Fields,* Grandolfo, M., Michaelson, S. M., and Rindi, A., Eds., Plenum Press, New York, 1985, 243.

respective beta dispersion frequency. For outer cell membranes this decline begins at radiofrequencies; for membranes of organelles this decline begins at audio frequencies if the organelle connects with the outer cell membrane or at radiofrequencies if not.

2. The interiors of cells and subcellular components are well shielded from the external field at low frequencies, and are increasingly exposed at radiofrequencies.

All this suggests that the fields induced within tissues by external fields depend greatly on the frequency of the field. The potential developed across internal membranes within a cell will be maximum at a frequency that is intermediate between the beta dispersion of the outer cell membrane and that of the organelle. These peaks will be quite broad, even in suspensions of cells, and will be broader still in tissues due to their structural heterogeneity.

This has important implications for understanding biological effects of electric fields. The frequency-dependence of the coupling will lead to a frequency dependence of any biological effect that arises at the level of cellular membranes or cytoplasm. (This is in addition to any frequency dependence that might arise from the kinetics of the response, such as the time-response of a gating mechanism.) One example is the well-known increase with frequency in the threshold for stimulating nerves by applied currents. Many other biological effects have been reported from applied fields, in some cases of high frequency, for which the mechanism is unknown and the effects themselves are of uncertain reproducibility. Such effects, if they can be replicated, should be studied over an extended frequency range with appropriate considerations for the prevailing potentials and field strengths at the cellular and subcellular levels.

VI. DIELECTRIC PROPERTIES OF TISSUES—TABULATED

Table 6 presents selected permittivity and conductivity data from various tissues. Where available, we present up to three values for the tissues, including measurements performed on excised tissues of various species and *in vivo*. The aim is to present primarily new data; in some cases we have included earlier data[181,182,183,184] to fill out the table. The large variability in reported properties of these tissues is illustrated in the table.

Other tabulations should be noted. Rajewsky[13] extensively reviewed data obtained before 1938; Schwan[1] extensively summarized many older data. Recent comprehensive reviews are

TABLE 6
Conductivity (S/m)

	Frequency	A Skeletal muscle parallel (nonoriented)	B Skeletal muscle perpendicular (nonoriented)	C Liver	D Lung	E Spleen	F Kidney	G Brain white matter	H Brain gray matter	I Bone	J Whole blood	K Fat
1	10 Hz	0.52	0.076	0.12	0.089							
4	100 Hz	0.52	0.076	0.13	0.092					0.0126	0.60	
7	1 kHz	0.52	0.08	0.13	0.096					0.0129		0.02–0.07
8										0.68		
10	10kHz	0.55	0.085	0.15	0.11					0.0133		
11						0.24 ± 0.03					0.68	
12				0.14		0.10	0.073					
13	100 kHz	0.65	0.40	0.15		0.62	0.24–0.25	0.12–0.15	0.17	0.0144	0.55	
14			0.56–0.59	0.16							0.68	
15			0.38–0.44	0.18		0.15	0.14					
16	1 MHz			0.27		0.63	0.37–0.39	0.14–0.19	0.21	0.0173		
17			0.83–0.85	0.30							0.71	
18			0.58–0.63	0.27		0.24	0.25					
19	10 MHz		0.86–0.87	0.47		0.84	0.64–0.68	0.21–0.28	0.35	0.0237		
20			0.92–0.96	0.46		0.55–0.53	0.50–0.57	0.30	0.38		1.11	
21			0.69–0.75	0.42–0.46		0.41	0.46	0.29–0.31	0.45–0.63			
22	100 MHz		0.95–0.99	0.72		1.05	0.94–1.05	0.36–0.48	0.69	0.0574	1.0	0.02–0.07
23			0.9 ± 0.08	0.70	0.53	0.73–0.76	0.66–0.72	0.45	0.7			
24			0.75–0.82	0.60–0.71		0.80 ± 0.02	0.75 ± 0.02	0.48–0.51	0.52–0.85		0.7–0.8	
25	1 GHz		1.38–1.45	0.98		1.2		0.89–0.94		0.05	1.4–1.6	
26			1.3	1.2	0.73	1.09–1.13	0.95–0.97	0.80	1.1		1.3	0.03–0.09
27			1.5	0.95–1.0		2.0	1.0	0.81–0.82	0.89–1.17			
28	3 GHz			2.0		2.5 ± 0.03		1.8–2.1		0.16	2.5–3.1	
29			2.7 ± 0.07	2.4				1.5	2.0			
30			2.8	2.8		2.7	2.3 ± 0.05					
31	10 GHz		8.3	5.8–6.7		6.5				0.5–1.7	9.1	0.3–0.4
32			7.7	10.0					10		10.5	
33			8.8			10.0	4.5–7.4	8				
34	35 GHz	39.9 ± 3.5					28.0 ± 2.5	38.9 ± 1.6		49.8 ± 0.4	2.7–0.8	

TABLE 6
Continued
Relative Permittivity

	Frequency	A Skeletal muscle parallel (nonoriented)	B Skeletal muscle perpendicular (nonoriented)	C Liver	D Lung	E Spleen	F Kidney	G Brain white matter	H Brain gray matter	I Bone	J Whole blood	K Fat
1	10 Hz	10^7	10^6	5×10^7	2.5×10^7							
4	100 Hz	1.1×10^6	3.2×10^5	8.5×10^5	4.5×10^5					3800		1.5×10^5
7	1 kHz	2.2×10^5	1.2×10^5	1.3×10^5	8.5×10^5					1000		5×104
8										2900		
10	10 kHz	8×10^4	7×10^4	5.5×10^4	2.5×10^4					640		2×10^4
11						1.4×10^4					2810	
12				1.8×10^4		3.4×10^4	4.8×10^4					
13	100 kHz	1.5×10^4	3×10^4	9760		3260	1.09×10^4–1.25×10^4	1960–3400	3800	280	4000	
14			2.5×10^4–2.7×10^4	1.4×10^4							2740	
15			1.4–1.6×10^4	6700		6300	8300					
16	1 MHz			1970		1450	2390–2690	543–827	1250	87		
17			2460–2530	1970							2040	
18			1900–2150	1000		1300	1500					
19	10 MHz		170–190	338		321	431–499	163–209	352	37		
20			187–204	300		352–410	190–204	200	380		200	
21			162–181	251–265		300	308	190–191	237–289			
22	100MHz		67–72	77		83	89–95	57–66	90	23		4.5–7.5
23			68 ± 2	79	35	71–76	56–62	65	90		67	
24			64–70	65–68		81 ± 3	85 ± 1	58–64	65–80		72–74	
25	1 GHz		57–59	46		54		40–44		8	58–62	4.3–7.5
26			58	55	35	50–51	43	5	45		63–67	3–6
27			48	47–49		50	46	38–39	47–51		63	
28	3 GHz			42		52 ± 0.6		35–41		7.5	55–56	4–7
29			52.5 ± 0.7	53				33	44			
30			46	42–43		46	47.5 ± 1					
31	10 GHz		40–42	38–38		42			8	50–52	3.5–4.0	
32			37	37				25	40		45	
33			35			38	38–37					
34	35 GHz		19.5 ± 1.5					17.8 ± 1.7	21.3 ± 0.9		23.6 ± 0.3	3.6 ± 0.3

TABLE 6
Continued

Coordinates	Tissues	Ref.	Coordinates	Tissues	Ref.
28 C	Bovine liver, 37°C	139	8 J	Rabbit blood, room temperature	180
23 D, J	Beef blood, excised human tissues (deflated lung)	132	11 J		
22 K	27°C (measurement frequencies 0.2–0.9 GHz)		14 J		
26 K			17 J		
26 D			20 J		
27 J			13 J	Normal human blood, hematocrit 40%, 21°C (50 kHz)	105
23 A, B	Various cat tissues, *in vivo,* (average ± SD) (55 measure-	185	25 C	Various tissues, dog, horse, 38°C (except 25 I, 28 I, 25°C);	
26 A, B	ments in 4 animals); value at 10 GHz extrapolated		30 C	measurements were made at 8.6 GHz and extrapolated to	
29 A, B	from 8.0 GHz; 1 GHz interpolated		25 K	10 GHz	
32 A, B			28 K		
24 E			25 J		
25 E			28 J		
28 E			25 I		
31 E			28 I		
24 F			31 A, B, I, J, K		
27 F			32 J	Human (9.4 GHz), 37°C	182
30 F			20 G, H	Dog brain, white ang gray matter, 37°C	187
33 F			23 G, H		
26 C	Dog liver, 37°C (value at 1, 3 GHz extrapolated)	120, 164	26 G, H		
29 C			29 G, H		
32 C			32 G, H		
41	Rat femur, 37°C, immersed in Hank's buffered saline,	129	4 J	Sheep blood, 18°C	184
71	radial direction (single sample)		47 J		
10 I			34 A, G, H, J, K	Various rabbit tissues (2 animals) 37°C	186
13 I			12 C, E, F	Various bovine tissues (data at 25:dgC adjusted to 37°C)	144
16 I			15 C, E, F	value at 10 KHz extrapolated from 20 KHz	
19 I			18 C, E, F		
22 I			21 E, F	Canine spleen, data adjusted from 25°C to 37°C	124
25 G	Mixed brain tissue, mouse, 37°C (value at 3 GHz	171	11 E		
28 G	interpolated)				

TABLE 6
Continued

Coordinates	Tissues	Ref.
1 A,B 4 A,B 7 A,B 10 A,B 13 A,B	Dog skeletal muscle, 37°C (average of 5 measurements) (SD~30%)	109
15 A,B 18 A,B 21 A,B 24 A,B	Nonoriented dog skeletal muscle, 37°C (range of 3 measurements)	116
14 A,B 17 A,B 20 A,B	Nonoriented rat skeletal muscle, 37°C (range of 2 measurements)	116
13 C 16 C 19 C 22 C	Dog liver, 37°C (single specimen)	116
14 C 17 C 20 C 23 C	Rabbit liver, 37°C (single specimen)	116
13 E 16 E 19 E 22 E	Dog spleen, 37°C (single specimen)	116
13 F 16F 19F 22 F	Dog kidney, 37°C (range of 2 measurements)	116
13 G,H 16 G,H 19 G,H 22 G,H	Dog brain, white and grey metter, 37°C (range of 2 measurements)	116

Coordinates	Tissues	Ref.
27 E 30 E 33 E	Dog spleen, 37°C (values at 1, 3 GHz interpolated) (single specimen)	120,164
27 A,B 30 A,B 33 A,B	Dog skeletal muscle, 37°C (values at 1, 3 GHz interpolated) (single specimen)	120,164
1 C 4 C 7 C 10 D	Dog lkiver, *in situ* (average of 20 measurements) (SD~25%)	135
1 D 4 D 7 D 10 D	Dog lung, inflated, *in situ* (average of 20 masurements) (SD~25%)	135
7 K 10 K	Dog fat, *in situ*	135
19 A,B 22 A,B 25 A,B	Cat skeletal muscle, 31°C, *in vivo* (range of 3 measurements)	146
20 E 23 E 26 E	Cat spleen, 35°C, *in vivo* (range of 3 measurements)	146
20 F 23 F 26 F	Cat kidney, *in vivo*, 35°C (range of 3 measurements)	146
21 G,H 24 G,H 27 G,H	Cat brain, *in vivo*, 33°C (range of 3 measurements)	146
21 C 24 C 27 C	Cat liver, *in vivo*, 35°C (range of 3 measurements)	146

by Geddes and Baker[108] and Stuchly and Stuchly.[9] Other primary research papers that provide extensive data are by Stoy,[116] Kraszewski and colleagues,[185] and Steel and Sheppard.[186]

All data pertain to tissues at body temperature (37 to 38°C). Temperature coefficients for some tissues are summarized in Figures 20 and 21.[187]

ACKNOWLEDGMENTS

We gratefully acknowledge the suggestions and helpful advice from many colleagues, in particular Drs. Edwin L. Carstensen, Constantino Grosse, and Charles Polk, and assistance with the numerical computations by Dr. Erik Cheever.

REFERENCES

1. **Schwan, H. P.,** Electrical properties of tissue and cell suspensions, in *Advances in Biological and Medical Physics,* Vol. 5, Academic Press, New York, 1957, 147.
2. **Foster, K. R. and Schwan, H. P.,** Dielectric properties of tissues—a review, in *Handbook of Biological Effects of Electromagnetic Radiation,* C. Polk and E. Postow, eds., CRC Press, 1986, pp. 27–96.
3. **Cole, K. S.,** *Membranes, Ions, and Impulses,* University of California Press, Berkeley, 1972.
4. **Pethig, R.,** *Dielectric and Electronic Properties of Biological Materials,* John Wiley & Sons, New York, 1979.
5. **Grant, E. H., Sheppard, R. J., and South, G. P.,** *Dielectric Behavior of Biological Molecules in Solution,* Oxford University Press, Oxford, 1978.
6. **Schanne, O. F. and P.-Ceretti, E. R.,** *Impedance Measurements in Biological Cells,* John Wiley & Sons, New York, 1978.
7. **Duck, F. A.,** *Physical Properties of Tissue,* Academic Press, London, 1990.
8. **Pethig, R. and Kell, D. B.,** The passive electrical properties of biological systems: their significance in physiology, biophysics and biotechnology, *Phys. Med. Biol.,* 32, 933, 1987.
9. **Stuchly, M. A. and Stuchly, S. S.,** Dielectric properties of biological substances—tabulated, *J. Microwave Power,* 15, 19, 1980.
10. **Schwan, H. P. and Foster, K. R.,** R-F field interactions with biological systems: electrical properties and biophysical mechanisms, Proc. IEEE, 68, 104, 1980.
11. **Stuchly, M.,** Interaction of radiofrequency and microwave radiation with living systems, *Radiat. Environ. Biophys.,* 16, 1, 1979.
12. **Gildemeister, M.,** Die passiven elektrischen Erscheinungen im Tier- und Pflanzenreich, in *Handbuch d. Norm. u. Pathol. Physiol.,* Vol. 8, Bethe, A. Ed., Berlin, Julius Springer, 1928, 657.
13. **Rajewsky, B.,** *Ergebnisse der Biophysikalischen Forschung,* Georg Thieme, Leipzig, 1938.
14. **Du Bois-Reymond, E.,** *Untersuchungen über Tierische Elektricität,* G. Reimer, Berlin, 1849.
15. **Peltier, J.,** C. A., cited in du Bois-Reymond, E., *Untersuchungen über Tierische Elektricität,* G. Reimer, Berlin, 1849.
16. **Hermann, L.,** Über eine Wirkung galvanischer Ströme auf Muskeln und Nerven, *Pflügers Arch. Gesamte Physiol.,* 5, 223, 1872.
17. **Du Bois, M.,** C. R. Acad. Sci., 126, 1790, 1898.
18. **Galler, H.,** Über den elektrischen Leitungswiderstand des tierischen Korpers, *Pflügers Arch. Gesamte Physiol.,* 149, 156, 1913.
19. **Gildemeister, M.,** Über elektrischen Widerstand, Kapazität und Polarisation der Haut, *Pflügers Arch. Gesamte Physiol.,* 176, 84, 1919.
20. **Einthoven, W. and Bijtel, J.,** Über Stromleitung durch den menschlichen Korper, *Pflügers Arch. Gesamte Physiol.,* 198, 439, 1923.
21. **Stewart, G. N.,** The relative volume or weight of corpuscles and plasma in blood, *J. Physiol.,* 24, 356, 1899.
22. **Maxwell, J. C.,** *Treatise on Electricity and Magnetism,* Oxford University Press, London, 1873.
23. **Höber, R.,** Eine Methode, die elektrische Leitfähigkeit im innern von Zellen zu messen, *Pflügers Arch. Gesamte Physiol.,* 133, 237, 1910.
24. **Höber, R.,** Ein zweites Verfahren die Leitfähigkeit im innern von Zellen zu messen, *Pflügers Arch. Gesamte Physiol.,* 148, 189, 1912.

25. **Geddes, L. A.,** A short history of electrical stimulation of excitable tissue, *The Physiologist,* 27, Suppl. 1, 1984.
26. **Licht, S.,** History of electrotherapy, in *Therapeutic Electricity and Ultraviolet Radiation,* 2nd ed., Licht, S., Ed., Waverly Press, Baltimore, 1967.
27. **Tesla, N.,** High frequency oscillators for electro-therapeutic and other purposes, lecture at 8th Annu. Meet. Am. Electro-Therapeutic Assoc., Buffalo, N. Y., September 1898, in *Nikola Tesla, Lectures, Patents, Articles,* Nikola Tesla Museum, Belgrade, Yugoslavia, 1956.
28. **Nernst, W.,** Zur Theorie des elektrischen Reizes, *Pflügers Arch. Gesamte Physiol.,* 122, 275, 1908.
29. **Coppee, G.,** La pararesonance dans l'excitation par les courants alternatifs sinusoidaux, *Arch. Int. Physiol.,* 90, 1, 1934.
30. **Philippson, M.,** Sur la resistance electrique des cellules et des tissus, *C. R. Soc. Belge Biol.,* 83, 1399, 1920.
31. **McClendon, J. F.,** Colloidal properties of the surface of the living cell. 11. Electric conductivity and capacity of blood to alternating currents of long duration and varying in frequency from 266 to 2,000,000 cycles per second, J. Biol. Chem., 69, 733, 1926.
32. **Osterhout, W. J. V.,** *Injury, Recovery and Death in Relation to Conductivity and Permeability,* Lippincott, Philadelphia, 1922.
33. **Dänzer, H.,** Über das Verhalten biologischer Körper im Hochfrequenzfeld, *Ann. Phys.,* 21(5), 463, 1934.
34. **Wachter, H.,** Fundamenta Radiologica (Berlin), 5, 1, 1939.
35. **Rajewsky, B. and Schwan, H. P.,** Die Dielektrizitätskonstante und Leitfahigkeit des Blutes bei ultrahohen Frequenzen, *Naturwissenschaften,* 35, 315, 1948.
36. **Cook, H. F.,** Dielectric behavior of human blood at microwave frequencies, *Nature,* 168, 247, 1951.
37. **Schwan, H. P.,** Determination of biological impedances, in *Physical Techniques in Biological Research,* Vol. 6, Oster, G. et al., Eds., Academic Press, New York, 1963, 323.
38. **Macdonald, J. R.,** Linear relaxation: distributions, thermal activation, structure, and ambiguity, *J. Appl. Phys.,* 62, R51, 1987.
39. **Morgan, F. D., and Lesmes, D. P.,** Inversion of dielectric relaxation spectra, *J. Chem. Phys.,* 100, 671, 1994.
40. **Cole, K. S. and Cole, R. H.,** Dispersion and absorption in dielectrics, I. Alternating current characteristics, *J. Chem. Phys.,* 9, 341, 1941.
41. **Cole, K. S. and Cole, R. H.,** Dispersion and absorption in dielectrics. II. Direct current characteristics, *J. Chem Phys.,* 10, 98, 1942.
42. **Markx, G. H., Davey, C. L., and Kell, D. B.,** To what extent is the magnitude of the Cole-Cole α of the β-dielectric dispersion of cell suspensions explicable in terms of the cell size distribution?, *Bioelectrochem. Bioenerg.,* 25, 195, 1991.
43. **Dissado, L.** A fractal interpretation of the dielectric response of animal tissues. *Phys. Med. Biol.,* 35, 1487, 1990.
44. **Hill, R. M. and Jonscher, A. K.,** The dielectric behavior of condensed matter and its many-body interpretations, *Contemp. Phys.,* 24, 75, 1983.
45. **Dyre, J. C.,** *J. Appl. Phya.,* 64, 2456, 1988.
46. **Kronig, R.,** de L., On the theory of dispersion of X-rays, J. Opt. Soc. Am., 12, 547, 1926.
47. **Grosse, C., and Greffe, J.-L.,** Permittivite statique des emulsions, J. Chimie *Physique,* 76, 305, 1979.
48. **Hanai, T.,** in *Emulsion Science,* Sherman, P., Ed., Academic Press, New York, 1968, chap. 5.
49. **Dukhin, S. S.,** Dielectric properties of disperse systems, *Surface Colloid Sci.,* 3, 83, 1971.
50. **Dukhin, S. S. and Shilov, V. N.,** *Dielectric Phenomena and the Double Layer in Disperse Systems,* John Wiley & Sons, New York, 1974.
51. **Wagner, K. W.,** *Arch. f. Elektrotechnik,* 2, 271, 1914.
52. **Fricke, H.,** A mathematical treatment of the electric conductivity and capacity of disperse systems, I. The electric conductivity of a suspension of homogeneous spheroids, *Phys. Rev.,* 24, 575, 1924.
53. **Fricke, H.,** A mathematical treatment of the electric conductivity and capacity of disperse systems. II. The capacity of a suspension of conducting spheroids surrounded by a non-conducting membrane for a current of low frequency, *Phys. Rev.,* 26, 678, 1925.
54. **Velick, S. and Gorin, M.,** The electrical conductance of suspensions of ellipsoids and its relation to the study of avian erythrocytes, *J. Gen. Physiol.,* 23, 753, 1940.
55. **Pauly, H. and Schwan, H. P.,** Über die Impedanz einer Suspension von kugelförmigen Teilchen mit einer Schale, *Z. Naturforsch.,* 14B, 125, 1959.
56. **Fricke, H.,** The complex conductivity of a suspension of stratified particles of spherical cylindrical form, *J. Phys. Chem.,* 59, 168, 1955.
57. **Irimajiri, A., Hanai, T., and Inouye, A.,** A dielectric theory of "multi-stratified shell" model with its application to a lymphoma cell, *J. Theor. Biol.,* 78, 251, 1979.
58. **Irimajiri, A., Doida, Y., Hanai, T., and Inouye, A.,** Passive electrical properties of cultured murine lymphoblast (L5178Y) with reference to its cytoplasmic membrane, nuclear envelope, and intracellular phases, *J. Membr. Biol.,* 38, 209, 1978.

59. **Bruggeman, D. A. G.,** Berechnung verschiedener physikalischer Konstanten von heterogen Substanzen. I. Dielektrizitätskonstanten und leitfähigkeiten der Mischorper aus isotropen Substanzen, *Ann. der Phys.,* 24, 636, 1935.
60. **Boned, C. and Peyrelasse, J.,** Etude de la permittivite complexe d'ellipsoides disperses dans un milieu continu. Analyses theorique et numerique, *Colloid Polym. Sci.,* 261, 600, 1983.
61. **Hanai, T. and Koizumi, N.,** Dielectric relaxation of W/O emulsions in particular reference to theories of interfacial polarization, *Bull. Inst. Chem. Res. Kyoto Univ.,* 53, 153, 1975.
62. **Hanai, T., Asami, K., and Koizumi, N.,** Dielectric theory of concentrated Suspensions of shell-sphere, in particular reference to the analysis of biological cell suspensions, *Bull. Inst. Chem. Res. Kyoto Univ.,* 57, 297, 1979.
63. **Chiew, Y. C. and Glandt, E. D.,** The effect of structure on the conductivity of a dispersion, *J. Colloid Interface Sci.,* 94, 90, 1983.
64. **Cole, K. S., Li, C., and Bak, A. F.,** Electrical analogues for tissues, *Exp. Neurol.,* 24, 459, 1969.
65. **Davey, C. L., Davey, H. M., and Kell, D. B.,** On the dielectric properties of cell suspensions at high volume fractions. *Bioelectrochem. and Bioenergetics,* 28, 319, 1992.
66. **Schwan, H. P. and Morowitz, H. J.,** Electrical properties of the membranes of the pleuropneumonia-like organism A 5969, *Biophys. J.,* 2, 395, 1962.
67. **Schwan, H. P., Takashima, S., Miyamoto, V. K., and Stoeckenius, W.,** Electrical properties of phospholipid vesicles, *Biophys. J.,* 10, 1102, 1970.
68. **Redwood, W. R., Takashima, S., Schwan, H. P., and Thompson, T. E.,** Dielectric studies on homogeneous phosphatidylcholine vesicles, *Biochim. Biophys. Acta,* 255, 557, 1972.
69. **Debye, P.,** *Polar Molecules,* Chemical Catalog Co., New York, 1929.
70. **Oncley, J. L.,** in *Proteins, Amino Acids, and Peptides,* Cohn, E. J. and Edsall, J. T., Eds., Reinhold, New York, 1943, 543.
71. **Takashima, S.,** Dielectric properties of proteins. 1. Dielectric relaxation, in *Physical Principles and Techniques of Protein Chemistry,* Academic Press, New York, 1969.
72. **Pennock. B. E. and Schwan, H. P.,** Further observations on the electrical properties of hemoglobin-bound water, *J. Phys. Chem.,* 73, 2600, 1969.
73. **Schwan, H. P., Sheppard, R. J., and Grant, E. H.,** Complex permittivity of water at 25°C, *J. Chem. Phys.,* 64, 2257, 1976.
74. **Grant, E. H., Szwarnowski, S., and Sheppard, R. J.,** Dielectric properties of water in the microwave and far-infrared regions, in *Biological Effects of Nonionizing Radiation,* (ACS Symposium Series 157), Illinger, K. H., Ed., American Chemical Society, Washington, DC, 1981.
75. **Buchanan, T. J., Haggis, G. H., Hasted, J. B., and Robinson, B. G.,** The dielectric estimation of protein hydration, Proc. R. Soc. London, Ser. A, 213, 379, 1952.
76. **Schwan, H. P.,** Electrical properties of bound water, Ann. N. Y. Acad. Sci., 125, 344, 1965.
77. **Grant, E. H.,** Electrical properties of bound water. Ann. N. Y. Acad. Sci., 125, 418, 1965.
78. **Schwan, H. P., Schwarz, G., Maczuk, J., and Pauly, H.,** On the low frequency dielectric dispersion of colloidal particles in electrolyte solution, *J. Phys. Chem.,* 66, 2626, 1962.
79. **Einolf, C. W. and Carstensen, E. L.,** Passive electrical properties of microorganisms, V. Low frequency dielectric dispersion in bacteria, *Biophys. J.,* 13, 8, 1973.
80. **Schwarz, G.,** Dielectric relaxation of biopolymers in solution, *Adv. Mol. Relax. Processes,* 3, 281, 1972.
81. **Mandel, M.,** Dielectric properties of charged linear macromolecules with particular reference to DNA, *Ann. N.Y. Acad. Sci.,* 303, 74, 1977.
82. **Schwarz, G.,** A theory of the low frequency dielectric dispersion of colloidal particles in electrolyte solution, *J. Phys. Chem.,* 66, 2636, 1962.
83. **Schwarz, G.,** Über die Dispersion des Orientierungsfeldeffektes von Polyelektrolyten in hochfrequenten elektrischen Feldern, *Z. Phys. Chem. (Frankfurt),* 19, 286, 1959.
84. **Mandel, M. and Odijk, T.,** Dielectric properties of polyelectrolyte solutions, *Annu. Rev. Phys. Chem.,* 35, 75, 1984.
85. **Fixman, M.,** Charged macromolecules in external fields. 1. The sphere, *J. Chem. Phys.,* 72, 5177, 1980.
86. **Fixman, M.,** Thin double layer approximation for electrophoresis and dielectric response, *J. Chem. Phys.,* 78, 1483, 1983.
87. **Chew, W. C. and Sen, P. N.,** Potential of a sphere in an ionic solution in thin double layer approximations, *J. Chem. Phys.,* 77, 2042, 1982.
88. **Chew, W. C., and Sen, P. N.,** Dielectric enhancement due to an electrochemical double layer: thin double layer approximation, *J. Chem. Phys.,* 77, 4683, 1982.
89. **Chew, W. C.,** Dielectric enhancement and electrophoresis due to an electrochemical double layer: a uniform approximation, *J. Chem. Phys.,* 81, 4541, 1984.
90. **Grosse, C. and Foster, K. R.,** Permittivity of a suspension of charged spherical particles in electrolyte solution, *J. Phys. Chem.,* 91, 3073, 1987.

91. **Grosse, C.,** Permittivity of a suspension of charged spherical particles in electrolyte solution. II. Influence of the surface conductivity and asymmetry of the electrolyte on the low and high frequency relaxations, *J. Phys. Chem.*, 92, 3905–3910, 1988.
92. **Schwan, H. P.** EM-field induced force effects. In *Interactions Between Electromagnetic Fields and Cells,* Chiabrera, A., Nicolini, C., and Schwan, H. P., eds. Plenum Press, New York, p. 371 (1985).
93. **Chizmadzhev, Yu. A., Kuzmin, P. I., Pastushenko, V. Ph.** Theory of the dielectrophoresis of vesicles and cells. (In Russian). *Biological Membranes (USSR)* 2, 1147, 1985.
94. **Bates, G. W., Saunders, J. A., and Sowers, A. E.,** Electrofusion, principles and applications, in *Cell Fusion,* A. E. Sowers, Ed., Plenum Press, N.Y., 1987, pp. 367–395.
95. **Schwan, H. P.,** Nonthermal cellular effects of electromagnetic fields: AC-field induced ponderomotoric forces, *Brit. J. Cancer,* 45 Suppl. V, 220, 1982.
96. **Schwan, H. P.,** Biophysical principles of the interaction of ELF fields with living matter: II. Coupling considerations and forces. in *Biological Effects and Dosimetry of Static and ELF Electromagnetic Fields,* M. Grandolfo, S. M. Michaelson, and A. Rindi, eds. Plenum Press, New York, 1985, p. 243.
97. **Schwan, H. P.,** Biophysical principles of the interaction of ELF fields with living matter: II. Coupling considerations and forces. in *Biological Effects and Dosimetry of Static and ELF Electromagnetic Fields,* M. Grandolfo, S. M. Michaelson, and A. Rindi, eds. Plenum Press, New York, 1985, p. 243.
98. **Rabinowitz, J. R.,** The effect of electric field induced perturbation of the distribution of ions near the cell surface on migration of charged membrane components, *J. Theor. Biol.* 99, 377, 1982.
99. **Woodward, A. M. and Kell, D. B.** On the nonlinear dielectric properties of biological systems *Saccharomyces cerevisiae. Bioelect. and Bioenerg.* 24, 83, 1990.
100. **Woodward, A. M., and Kell, D. B.** On the relationship between the nonlinear dielectric-properties and respiratory activity of the obligately aerobic bacterium *micrococcus-luteus. Bioelectrochem. and Bioenerg.,* 26, 423, 1991.
101. **Foster, K. R., Schepps, J. L., and Schwan, H. P.,** Microwave dielectric relaxation in tissue. A second look, *Biophys. J.,* 29, 271, 1980.
102. **Schwan, H. P.,** Electrical properties of cells: principles, some recent results and some unresolved problems, in *The Biophysical Approach to Excitable Systems,* Adelman, W. S. and Goldman, D., Eds., Plenum Press, New York, 1981, 3.
103. **Trautman, E. D. and Newbower, R. S.,** A practical analysis of the electrical conductivity of blood, *IEEE Trans. Biomed. Eng.,* 30, 141, 1983.
104. **Schwan, H. P.,** Electrical properties of blood and its constituents. Alternating current spectroscopy, *Blut,* 46, 185, 1983.
105. **Pfutzner, H.,** Dielectric analysis of blood by means of a raster-electrode technique, *Med. Biol. Eng. Computing,* 22, 142, 1984.
106. **Pethig, R.,** Dielectric properties of biological materials: biophysical and medical applications, *IEEE Trans. Elect. Insul.,* EI-19, 453, 1984.
107. **Alison, J. M. and Sheppard, R. J.,** Dielectric properties of human blood at microwave frequencies, *Phys. Med. Biol.,* 38, 971, 1993.
108. **Geddes, L. A. and Baker, L, E.,** The specific resistance of biological material—a compendium of data for the biomedical engineer and physiologist, *Med. Biol. Eng.,* 5, 271, 1967.
109. **Epstein, B. R. and Foster, K. R.,** Anisotropy in the dielectric properties of skeletal muscle, *Med. Biol. Eng. Comput.,* 21, 51, 1983.
110. **Schwan, H. P.,** Die elektrischen Eigenschaften von Muskelgewebe bei Niederfrequenz, *Z. Naturforsch.,* 9b, 245, 1954.
111. **Rush, S., Abildskov, J. A., and McFee, R.,** Resistivity of body tissues at low frequencies, *Circ. Res.,* 12, 40, 1963.
112. **Roth, B. J., Gielen, F. L. H., and Wikswo, J. P., Jr.,** Spatial and temporal frequency-dependent conductivities in volume-conduction calculations for skeletal muscle, *Math. Biosciences,* 88, 159, 1988.
113. **Fatt, P.,** An analysis of the transverse electrical impedance of striated muscle, *Proc R. Soc. London, Ser. B,* 159, 606, 1964.
114. **Falk, G. and Fatt, P.,** Linear electrical properties of striated muscle fibers observed with intracellular electrodes, *Proc R. Soc. London, Ser. B,* 160, 69, 1964.
115. **Takashima, S.,** Passive electrical properties and voltage dependent membrane capacitance of single skeletal muscle fibers, *Pflügers Arch. Gesamte Physiol.,* 403, 197, 1985.
116. **Stoy, R. D., Foster, K. R., and Schwan, H. P.,** Dielectric properties of mammalian tissues from 0.1 to 100 MHz: a summary of recent data, *Phys. Med. Biol.,* 27, 501, 1982.
117. **Foster, K. R. and Schepps, J. L.,** Dielectric properties of tumor and normal tissues at radio through microwave frequencies, J. Microwave Power, 16, 107, 1981.
118. **Schwan, H. P. and Kay, C. F.,** Capacitive properties of living tissues, *Circ. Res.,* 5, 439, 1957.
119. **Pauly, H., Packer, L., and Schwan, H. P.,** Electrical properties of mitochondrial membranes, *J. Biophys. Biochem. Cytol.,* 7, 589, 1960.

120. **Schepps, J. L. and Foster, K. R.,** The UHF and microwave dielectric properties of normal and tumor tissues: variation in dielectric properties with tissue water content, *Phys. Med. Biol.,* 25, 1149, 1980.
121. **Homburger, F. and Fishman, W. H.,** Eds., *The Physiopathology of Cancer,* Hoeber-Harper, New York, 1953, p. 554.
122. **Greenstein, J. P.,** *Biochemistry of Cancer,* Academic Press, New York, 1947, p. 246.
123. **Smith, S. R., Foster, K. R. and Wolf, J. L.,** Dielectric properties of VX-2 carcinoma vs. normal liver tissues, IEEE Trans. Biomed. Eng., BME-33, 522, 1986.
124. **Astbury, J. C., Goldschmidt, M. H., Evans, S., Neibauer, G. W., and Foster, K. R.,** The dielectric properties of canine normal and neoplastic splenic tissues. Proc. 14th Northeast Bioengineering Conference, Durham NH, March 1988.
125. **Fricke, H. and Morse, S.,** The electric capacity of tumors of the breast, *J. Cancer Res.,* 10, 340, 1926.
126. **Surowiec, A. J., Stuchly, S. S., Barr, J. R., and Swarup, A.,** Dielectric properties of breast carcinoma and the surrounding tissues, *IEEE Trans. Biomed. Eng.* BME-35, 257, 1988.
127. **Campbell, A. M. and Land, D. V.,** Dielectric properties of female human breast tissue measured *in vitro* at 3.2 GHz, *Phys. Med. Biol.,* 37, 193, 1992.
128. **Joines, W. T., Jirtle, R. L., Rafal, M. D., and Schaefer, D. J.,** Microwave power absorption differences between normal and malignant tissue, *Int. J. Radiat. Oncol. Biol. Phys.,* 6, 681, 1980.
129. **Kosterich, J. D., Foster, K. R., and Pollack, S. R.,** Dielectric permittivity and electrical conductivity of fluid saturated bone, *IEEE Trans. Biomed. Eng.,* BME-30, 81, 1983.
130. **Kosterich, J. D., Foster, K. R., and Pollack, S. R.,** Dielectric properties of fluid saturated bone: effect of variation in conductivity of immersion fluid, *IEEE Trans. Biomed. Eng.,* BME-31, 369, 1984.
131. **Kenyon, W. E.,** Texture effects on megahertz dielectric properties of calcite rock samples, *J. Appl. Phys.,* 55, 3153, 1984.
132. **Schwan, H. P. and Li, K.,** Capacity and conductivity of body tissues at ultrahigh frequencies, Proc. IRE, 41, 1735, 1953.
133. **Smith, S. R. and Foster, K. R.,** Dielectric properties of low-water-content tissues, *Phys. Med. Biol.,* 30, 965, 1985.
134. **Schwan, H. P. and Kay, C. F.,** Specific resistance of body tissues, *Circ. Res.,* 4, 664, 1956.
135. **Schwan, H. P. and Kay, C. F.,** The conductivity of living tissues, *Ann. N.Y. Acad. Sci.,* 65, 1007, 1957.
136. **Burdette, E. C., Cain, F. L., and Seals, J.,** In vivo probe measurement technique for determining dielectric properties at VHF through microwave frequencies, *IEEE Trans. Microwave Theory Tech.,* 18, 414, 1980.
137. **Magin, R. and Burdette, E. C.,** Electrical properties of tissue at microwave frequencies: a new approach to detection and treatment of abnormalities, in *Non-Invasive Physiological Measurements,* Vol. 2, Rolfe, P. M., Ed., Academic Press, New York, 1983.
138. **Athey, T. W., Stuchly, M. A., and Stuchly, S. S.,** Dielectric properties of biological substances at radiofrequencies. Part I. Measurement method, *IEEE Trans. Microwave Theory Tech.,* 30, 82, 1982.
139. **Brady, M. M., Symonds, S. A., and Stuchly, S. S.,** Dielectric behavior of selected animal tissues *in vitro* at frequencies from 2 to 4 GHz, *IEEE Trans. Biomed. Eng.,* BME-28, 305, 1981.
140. **Stuchly, M. A., Kraszewski, A., Stuchly, S. S., and Smith, A, M.,** Dielectric properties of animal tissues in vivo at radio and microwave frequencies: comparison between species, *Phys. Med Biol.,* 27, 927, 1982.
141. **Löfgren, B.,** The electrical impedance of a complex tissue and its relation to changes in volume and fluid distribution, *Acta Physiol. Scan.,* 23 Suppl. 81, 1, 1951.
142. **M. A. Fallert, M. S. Mirotznik, D. K. Bogen, E. B. Savage, K. R. Foster, and M. E. Josephson,** Myocardial electrical impedance mapping of ischemic sheep hearts and healing aneurisms. *Circulation,* 87, 188, 1993.
143. **Surowiec, A., Stuchly, S. S., and Swarup, A.,** Postmortem changes of the dielectric properties of bovine brain tissues at low radiofrequencies, *Bioelectromagnetics,* 7, 31, 1986.
144. **Surowiec, A., Stuchly, A. S., and Swarup, A.,** Radiofrequency dielectric properties of animal tissues as a function of time following death, *Phys. Med. Biol.,* 30, 1131, 1985.
145. **Surowiec, A., Stuchly, S. S., Keaney, M., and Swarup, A.,** *In vivo* and *in vitro* dielectric properties of feline tissues at low radiofrequencies, *Phys. Med. Biol.,* 31, 901, 1986.
146. **Stuchly, M. A., Athey, T. W., Stuchly, S. S., Samaras, G. M., and Taylor, G.,** Dielectric properties of animal tissues *in vivo* at frequencies 10 MHz–1 GHz. *Bioelectromagnetics,* 2, 93, 1981.
147. **Burdette, E. C., Cain, F. L., and Seals, J.,** *In situ* permittivity measurements: perspectives, techniques, results, in *Medical Applications of Microwave Imaging,* Krone, R., Ed., IEEE Press, New York, 1986, 13.
148. **Lippold, O. C. J. and Winton, F. R.,** *Human Physiology,* Little, Brown, Boston, 1968, p. 660.
149. **Burdette, E. C. and Karow, A. M.,** Kidney model for study of electromagnetic thawing, *Cryobiology,* 15, 142, 1978.
150. **Surowiec, A. J., Stuchly, S. S., Keaney, M., and Swarup, A.,** Dielectric polarization of animal lung at radio frequencies, *IEEE Trans. Biomed. Eng.,* BME-34, 62, 1987.

151. **Conway, J., Hawley, M. S., Seagar, A. D., Brown, B. H., and Barber, D. C.,** Applied potential tomography (APT) for noninvasive thermal imaging during hyperthermia treatment, *Elect. Let.,* 21, 886, 1985.
152. **Griffiths, H., and Ahmed, A.,** Applied potential tomography for noninvasive temperature mapping in hyperthermia, *Clin. Phys. Physiol. Meas. (GB),* 8 Suppl. A, 147, 1987.
153. **McRae, D. A. and M. A. Esrick,** The dielectric parameters of excised EMT-6 tumors and their change during hyperthermia, *Phys. Med. Biol.* 37, 2045, 1992.
154. **McRae, D. A. and M. A. Esrick,** Changes in electrical impedance of skeletal muscle measured during hyperthermia, *Int. J. Hyperthermia* 9, 247, 1993.
155. **Esrick, M. A. and D. A. McRae,** The effect of hyperthermia-induced tissue conductivity changes in electrical impedance temperature mapping, *Phys. Med. Biol.* 39, 133, 1994.
156. **Colonomos, P, and Gordon, R. G.,** Bounded error analysis of experimental distributions of relaxation times, *J. Chem. Phys.,* 71, 1159, 1979.
157. **Batchelor, G. K.,** Transport properties of two-phase materials with random structure, *Annu. Rev. Fluid Mech.,* 6, 227, 1974.
158. **Bull, H. B. and Breese, K.,** Electrical conductance of protein solutions, *J. Colloid. Interface Sci.,* 29, 492, 1969.
159. **Cook, H. F. and Kuntz, I. D.,** The properties of water in biological systems, *Annu. Rev. Biophys. Bioeng.,* 3, 95, 1974.
160. **Pauly, H. and Schwan, H, P.,** Dielectric properties and ion mobility in erythrocytes, *Biophys. J.,* 6, 621, 1966.
161. **Foster, K. R, Cheever, E., Leonard, J. B., and Blum, F.,** Transport properties of polymer solutions: a comparative approach, *Biophys. J.,* 45, 975, 1984.
162. **Clark, M. E., Burnell, E. E., Chapman, N. R., and Hinke, J. A. M.,** Water in barnacle muscle. IV. Factors contributing to the reduced self-diffusion, *Biophys. J.,* 39, 289, 1982.
163. **Blum, F. D., Pickup, S., and Foster, K. R.,** Solvent self-diffusion in polymer solutions, *J. Colloid Inter. Sci.,* 113, 336, 1986.
164. **Schepps, J. L.,** The Measurement and Analysis of the Dielectric Properties of Normal and Tumor Tissues at UHF and Microwave Frequencies, Ph.D. dissertation, University of Pennsylvania, Philadelphia, 1980.
165. **Foster, K. R., Schepps, J. L., and Epstein, B. R.,** Microwave dielectric studies on proteins, tissues, and heterogeneous suspensions, *Bioelectromagnetics,* 3, 29, 1982.
166. **Foster, K. R., Epstein, B. R., Jenin, P. C., and Mackay, R. A.,** Dielectric studies on nonionic microemulsions, *J. Colloid Interface Sci.,* 88, 233, 1982.
167. **Epstein, B. R., Foster, K. R., and Mackay, R. A.,** Microwave dielectric properties of ionic and nonionic microemulsions, *J. Colloid Interface Sci.,* 95, 218, 1983.
168. **Masszi, G., Szuarto, A., and Grof, P.,** Investigations on the ion- and water-binding of muscle by microwave measurements, *Acta Biochim. Biophys. Acad. Sci. Hung.,* 11, 129, 1976.
169. **Grant, E. H., Nightingale, N. R. V., Sheppard, R. J., and Gough, S. R.,** Dielectric properties of water in myoglobin solution, in *Biological Effects of Nonionizing Radiation,* (ACS Symposium Series 157), (Illinger, K. H., Ed., American Chemical Society, Washington, DC, 1981.
170. **Gabriel, C., Sheppard, R. J., and Grant, E. H.,** Dielectric properties of ocular tissues at 37 C, *Phys. Med. Biol.,* 28, 43, 1983.
171. **Nightingale, N. R. V., Goodridge, V. D., Sheppard, R. J., and Christie, J. L.,** The dielectric properties of cerebellum, cerebrum, and brain stem of mouse brain at radiowave and microwave frequencies, *Phys. Med. Biol.,* 28, 897, 1983.
172. **H. P. Schwan,** Dielectrophoresis and rotation of cells, in *Electroporation and Electrofusion in Cell Biology,* E. Neumann, A. Sowers, and C. Jordan, Eds., Plenum Press (1989).
173. **Stuchly, M. A. and Stuchly, S. S.,** Experimental radio and microwave dosimetry, in *CRC Handbook of Biological Effects of Electromagnetic Fields,* C. Polk and E. Postow, eds., CRC Press, Boca Raton FL, p. 229, 1986.
174. **Lin, J. C.,** Computer methods for field intensity predictions, in *CRC Handbook of Biological Effects of Electromagnetic Fields,* C. Polk and E. Postow, eds., CRC Press, Boca Raton FL, p. 273, 1986.
175. **Kaune, W. T.n and L. E. Anderson,** Physical aspects of ELF electric and magnetic field: measurement and dosimetry, in *Extremely Low Frequency Electromagnetic Fields: The Question of Cancer,* B. W. Wilson, R. G. Stevens, L. E. Anderson, eds., Battelle Press, Columbus, OH, p. 17, 1989.
176. **Guy, A. W., Davidow, S., Yang, G.-Y., and Chou, C.-K.,** Determination of electric current distributions in animals and humans exposed to a uniform 60 Hz high intensity electric field, *Bioelectromagnetics,* 3, 47, 1982.
177. **Grosse, C, and Schwan, H. P.,** Cellular membrane potentials induced by alternating fields, *Biophys. J.,* 63, 1632, 1992.
178. **K. R. Foster and A. E. Sowers,** Dielectrophoretic forces and potentials induced on pairs of cells in an electric field. *Biophysical J.,* 69(3), 1995.
179. **Drago, G. P., Marchesi, M., and Ridella, S.,** The frequency dependence of an analytical model of an electrically stimulated biological structure, *Bioelectromagnetics,* 5, 47, 1984.

180. **Sauer, F. and H. P. Schwan,** Rotation of spherical particles inside a cell by rotating electric fields, *J. Ferroelectricity* (in press).
181. **Fricke, H, and Curtis, H. J.,** The electric impedance of hemolyzed suspensions of mammalian erythrocytes, *J. Gen. Physiol.,* 18, 821, 1935.
182. **Herrick, J. F., Jelatis, D. G., and Lee, G. M.,** Dielectric properties of tissues important in microwave diathermy, *Fed. Proc. Fed. Am. Soc. Exp. Biol.,* 9, 60, 1950 (abstract only; data summarized in Ref. 1).
183. **England, T. S. and Sharples, N. A.,** Dielectric properties of the human body in the microwave region of the spectrum, *Nature,* 163, 487, 1949.
184. **Schwan, H. P.,** Über die Niederfrequenzleitfahigkeit von Bluten und Blutserum bei verschiedenen Temperaturen, *Z. Gesamte Exp. Med.,* 109, 531, 1941.
185. **Kraszewski, A., Stuchly, M. A., Stuchly, S. S., and Smith, A. M.,** *In vivo* and *in vitro* dielectric properties of animal tissues at radiofrequencies, *Bioelectromagnetics,* 3, 421, 1982.
186. **Steel, M. C. and Sheppard, R. J.,** The dielectric properties of rabbit tissue, pure water, and various liquids suitable for tissue phantoms at 35 GHz, *Phys. Med. Biol.,* 33, 467, 1988.
187. **Foster, K. R., Schepps, J. L., Stoy, R. D., and Schwan, H. P.,** Dielectric properties of brain tissue between 0.01 and 10 GHz, *Phys. Med. Biol.,* 24, 1177, 1979.

Chapter 2

INTERACTION OF DC AND ELF ELECTRIC FIELDS WITH BIOLOGICAL MATERIALS AND SYSTEMS

Frank S. Barnes

CONTENTS

I. INTRODUCTION

The fact that electrical currents can affect the behavior of biological systems has been known for more than 2000 years, and electric shocks have been used to treat a wide variety of ailments off and on ever since. However, it is surprising how incomplete our knowledge is today of the effects of direct current (DC) and low frequency voltages and currents on biological systems.

0-8493-0641-8/96/$0.00+$.50

Electrical signals are clearly important in the control of biological processes and in carrying information from one part of the body to another. Nerve cells propagate electrical signals from sensors of pressure, temperature, light, sound, etc., to the brain and return control signals to muscles and other tissue. Yet, if we choose to stimulate these processes with external electrical inputs, we have a relatively limited understanding of how a given electrical signal will affect various biological organs; what the safe limits of exposure are (particularly over extended periods of time); and how electrical signals are carried across cell membranes or propagated along nerves.

The purpose of this review is to bring together some of the physical concepts which underlie the interaction between electric fields and biological materials with the objective of providing background for determining safe levels of exposure and new applications for the use of electricity in therapy. It is organized to begin at the lowest level of complexity by examining some of the forces that are exerted on charged particles in fluids, and proceeds to some of the effects of electric fields on chemical reaction rates and membranes. At the next level of complexity, some effects of externally applied electric fields on currents through membranes and membrane nonlinearities are described. This is followed by a discussion of some long-term adaptive processes and secondary effects of current flow due to heating and by a description of a few effects on whole animals. This information is presented with the objective of specifying the general level or intensity of fields, currents, and temperatures where one can expect to observe a given class of biological responses. The next section contains data on the levels of typical naturally occurring and man-made fields and information on current safety standards. This section also includes information on both electrical signals and noise levels in the body. The final sections introduce the relationships of electric fields to tissue growth and applications to therapy.

II. PHYSICS OF THE INTERACTIONS OF ELECTRIC FIELDS WITH BIOLOGICAL MATERIALS

Biological systems consist of complex physical subsystems. In an attempt to understand them, we will start at the most elementary level. Perhaps the simplest level—which is already surprisingly complicated—is the effect of electric fields on biological fluids. These fluids contain a large number of components, including ions, polar molecules such as water, proteins, lipids, hormones, and colloidal particles. Current flow in these fluids is given by the sum of the drift and diffusion currents for each component. At low current densities the system is linear; however, at moderate to high current densities, nonlinearities are observed.[1] The next level of complexity involves the interaction of the fields with membranes that behave like porous solids for fields applied perpendicularly to their surface, and like viscous liquids for fields in the plane of the membrane.[2,3] Membranes are inhomogeneous so that different portions of them may be affected differently by the perturbing fields. Additionally, membranes are involved in active chemical reactions that change their porosity to various ions selectively so that both electrical potentials and chemical signals may change the membrane's conductivity by orders of magnitude. The next level of complexity occurs in the interactions between the biological fluids and the membranes in the presence of electric fields. Electric fields affect the selective transport of ions or molecules through the membrane. They change the build-up of charged ion layers at the surface and change the way new molecules are incorporated into the membrane or are bound to its surface. The result of changes in the transport of molecules or ions across cell membranes is changes in the performance of the cells and, in turn, of the organs of which they are a part. For example, a biasing electric current across a pacemaker cell in the heart will change its firing rate and thus the pumping rate of the heart.

The fundamental law describing forces on charged particles is

TABLE 1
Typical Values of Biological Coefficients

Particles	Mobilities	Ions	Mobilities
Proteins	$\mu = 10^{-10}$ to 10^{-8} m²/V sec	Ca^{++}	$\mu = 6.2 \times 10^{-8}$ m²/V sec
Na^+	$\mu = 5.2 \times 10^{-8}$ m²/V sec	Mg^{++}	$\mu = 5.4 \times 10^{-8}$ m²/V sec
K^+	$\mu = 7.6 \times 10^{-8}$ m²/V sec	Cl^-	$\mu = 7.9 \times 10^{-8}$ m²/V sec

$$\vec{F} = q(\vec{E} + \vec{v} \times \vec{B}) \tag{1}$$

where $\vec{F}$ is the force acting on a charged particle, q its effective charge, $\vec{E}$ the electric field, $\vec{v}$ the velocity of the particle, $\vec{B}$ the magnetic field, and $\rightarrow$ denotes a vector quantity. It is worth noting that $\vec{E}$ and $\vec{B}$ are coupled by Maxwell's equations so that a time-varying magnetic field generates an $\vec{E}$ field and vice versa. Additionally, if nonlinearities or time-varying impedances are present, alternating current (AC) fields can be rectified to produce DC. The direct magnetic field effects may occur through the term $(\vec{v} \times \vec{B})$. For example, for a Na^+ ion moving at a thermal velocity of 4×10^2 m/sec in the earth's magnetic field of about 5×10^{-5} T, this term has a magnitude of 2×10^{-2} V/m.

In addition to forces on charged particles, we can induce forces on polarizable neutral molecules or molecules with dipole moments in an electric field which varies in space. To first order these forces are described by

$$\vec{F_d} = (\vec{P} \cdot \vec{\nabla})\vec{E} \tag{2}$$

$$\vec{F_L} = \alpha V(\vec{E} \cdot \vec{\nabla})\vec{E} \tag{3}$$

where $\vec{F_d}$ is the force on a molecule with a permanent dipole moment $\vec{P}$, $\vec{\nabla}\vec{E}$ is the gradient of the electric field, and $\vec{F_L}$ is the force on a molecule with an induced dipole moment $\vec{p} = \alpha V\vec{E}$ where α is the tensor polarizability and V is the volume.[4]

Since most biological materials are highly inhomogeneous, there are nearly always important local variations in the electric field constituting a significant electric field gradient. This, in turn, means that when we are looking at the detail of the movement of molecules, membranes, and ions, we need to consider the field gradients as well as the fields.

The electric current density within a homogeneous fluid in an electric field is given by

$$\vec{J} = \sum_j q_j \mu_j C_j \vec{E} \tag{4}$$

where q_j is the charge on the ion of type j, μ_j is its electrical mobility, and C_j is its concentration. Typical ion mobilities are given in Table 1. The conductivity of biological fluid such as blood, which contains cells, is in the vicinity of $\sigma = 0.6$ Sm^{-1}, while that of physiological saline is approximately 1.4 Sm^{-1} (Note: see the paper by Schwan and Foster for a much more detailed treatment of conductivities of typical biological materials.) If the fluid channels between cells are relatively thick, the fluids are relatively good conductors and thus tend to cause a short circuit of voltages that might otherwise appear across membranes which typically have conductivities at least a thousand times smaller.

The forces applied by an electric field superimpose a drift velocity on top of the much larger random thermal velocity in opposite directions for positively and negatively charged particles. These forces can lead to a redistribution of ions or molecules as a result of the differential mobilities and to an increase in the concentration of ions at interfaces. The average drift velocities $\vec{v}$ are given by

$$\vec{v} = \mu_i \vec{E} \tag{5}$$

The separation of molecules as a result of the different velocities in a DC electric field is known as electrophoresis and is frequently used to identify large molecules or charged colloidal particles.[5] The separation of particles in an AC field gradient is known as dielectrophoresis.[4]

For a spherical particle in a homogeneous insulating fluid, the mobility μ_l is given by

$$\mu_l = \frac{q}{6\pi\eta a} \tag{6}$$

provided the particle is significantly larger than the background particles where η is the viscosity of the fluid and a is the radius of the particle. In a conducting medium, counterions, or ions with a charge of opposite sign to that of the particle, are attracted to it. They change the effective radius of the particle and partially shield its charge.[4,5] Additionally, small counterions may flow in the direction opposite to the particle motion, exerting a viscous drag. The theory for the motion of a rigid sphere through a conducting liquid is complicated if all these effects are taken into account. Often some of the parameters, including the charge on the sphere, are not measurable. However, a relatively simple expression for the electrophoretic mobility is often used:

$$\mu_l = \frac{\epsilon_l \zeta}{4\pi\eta} \tag{7}$$

ϵ_l and η are the dielectric permittivity and the viscosity of the fluid, and ζ is the electrical potential drop from the particle surface across the bound fluid, to the interface where the liquid begins to flow under the shear stress. Stated another way, the "zeta potential" ζ is the potential at the surface boundary between the stationary fluid and the liquid which is moving with the particle. It is to be noted that ζ is less than the total potential ψ across the charged double layer surrounding the charged particle.

In a uniform AC field a charged particle oscillates about its mean position, and the electrical energy added to the solution is largely converted to heat. If there is a gradient in the field, as is to be expected in biological materials which are highly inhomogeneous, then the gradient of the field leads to a net charge displacement. For example, if the driving field is given by

$$E(x,t) = f(x)\, g(t) = E_0(1 - \beta x) \cos \omega t, \tag{8}$$

one obtains by substitution of equation 8 into 5 and subsequent integration

$$x = \frac{1}{\beta} + \left(x_0 - \frac{1}{\beta}\right) \exp\left[-\frac{\mu\beta E_0}{\omega} \sin \omega t\right] \tag{9}$$

where x_0 is the initial position and β defines the slope for a linear variation of electric fields in space, and ω is the angular frequency.

If we set $x_0 = 0$, then for $x \le \frac{1}{\beta}$, and for large values of $\frac{\beta\mu E_0}{\omega}$ it oscillates between $x = \frac{1}{\beta}$ and

$$x = \frac{1}{\beta}\left[1 - e + \frac{\mu\beta E_0}{\omega}\right] \tag{10}$$

For large values of E, the velocity saturates and $\mu = \frac{\mu_0}{E}$. Thus equation 5 becomes

$$v = \frac{\partial x}{\partial t} = \mu E = \mu_0 \text{ and } x - x_0 = \mu_0(t - t_0)$$

so that the displacement grows with time.

Neutral particles with either permanent or induced dipole moments have different mobilities, depending upon their size and dipole moment. This can lead to changes in their distribution when they are placed in an electric field gradient.

In the case of an induced dipole moment for an ideal dielectric sphere with dielectric permittivity ϵ_2 and $\sigma_2 = 0$, placed in an ideal dielectric fluid with dielectric permittivity ϵ_1 and $\sigma_1 = 0$, and a nonuniform electric field prior to inserting the sphere, the force $\vec{F_d}$ acting on the sphere can be shown to be[4]

$$\vec{F_d} = \alpha V(\vec{E_1}\cdot\vec{\nabla})\vec{E_1} = 4\pi a^3\epsilon_1\frac{\epsilon_2 - \epsilon_1}{\epsilon_2 + 2\epsilon_1}(\vec{E_1}\cdot\vec{\nabla})\vec{E_1} \tag{11}$$

where $\vec{E_1}$ is the field in the fluid prior to insertion of the sphere. Written another way,

$$\vec{F_d} = \frac{3}{2}V\cdot\epsilon_1\frac{\epsilon_2 - \epsilon_1}{\epsilon_2 + 2\epsilon_1}\vec{\nabla}|\vec{E_1}|^2 \tag{12}$$

If we assume that the viscous drag on a spherical particle is given by Stokes' law, then

$$\vec{F_d} = 6\pi a\eta\vec{v} \tag{13}$$

The mobility μ_1 is given by

$$\mu_1 = \frac{2}{3}\frac{a^2}{\eta}\epsilon_1\frac{\epsilon_2 - \epsilon_1}{\epsilon_2 + 2\epsilon_1}\vec{\nabla}E_1 \tag{14}$$

In order for particles with dipole moments to change their distribution under the influence of an electric field gradient, the force $\vec{F_d}$ has to be large enough to overcome the other forces which may be binding them. One of the fundamental forces which must frequently be overcome is due to osmotic pressure or diffusion. The osmotic pressure Π due to a concentration of C_i particles per unit volume is given by $\Pi = C_i kT$, where k is the Boltzmann constant and T is the absolute temperature.[6] The osmotic pressure can be thought of as the force per unit area arising from diffusion or the random motion of the particles. The average differential force on a particle is proportional to the gradient of the osmotic pressure and is given by

$$\vec{F_{os}} = -\frac{1}{C_1}\vec{\nabla}\Pi = \frac{kT\vec{\nabla}C}{C_1} - k\vec{\nabla}T \tag{15}$$

If we consider the case of a spherical volume with a radial concentration gradient at constant temperature, the force is given by

$$\vec{F_{os}} = -kT\frac{\Delta C_i}{C_i}\cdot\frac{\vec{r'_o}}{\Delta r} \tag{16}$$

where $\vec{r}'_o$ is the unit vector, ΔC is the incremental change in concentration, and Δr is the incremental change in distance. The maximum change is

$$\frac{\Delta C_l}{C_l} = 1 \tag{17}$$

when the presence or absence of a particle occurs on the distance $\Delta r = 2a$, where a is the particle radius. In this case, we get the maximum force:

$$F_{os(max)} = -\frac{kT}{2a} \tag{18}$$

To get an idea of the size of these forces, consider a particle of fat with $a = 1$ μm in water. The maximum osmotic pressure at $T = 300$ K is $F_{os(max)} = 2 \times 10^{-15}$ N. The dielectric constant for water is approximately $\epsilon_l = 80\,\epsilon_o$ and for a fat particle $\epsilon_2 = 2\epsilon_o$ where $\epsilon_o = 8.854 \times 10^{-12}$ F/m. To get a dielectric force greater than the maximum osmotic force, we need a value of $\nabla|E^2| > 10^{12}$ V^2/m^3. This is given approximately by a voltage of 100 V across a 5-mm gap when the E field goes from zero to a peak value of 5×10^4 V/m over the same gap. For a particle with a single charge in a uniform field, we would need a field of $E = 1.3 \times 10^4$ V/m to get an equal force.

The electric current densities generated by a concentration gradient are given by

$$\vec{J_d} = -qD\vec{\nabla}C_l \tag{19}$$

where D is the diffusion constant and is given by

$$D = \upsilon KT \tag{20}$$

where υ is the hydrodynamic mobility with the dimensions of $\frac{\text{velocity}}{\text{force}}$ and D has the dimensions of meters squared per second.

There are four forces which may become important when considering the interaction between two particles in a fluid. These are the osmotic or diffusion force, electrostatic force, Van der Waals force, and hydration force.[6] These forces may all become important in considering the interaction between particles or bilipid membranes in an aqueous fluid. Electrostatic or coulomb forces between particles of like charge are repulsive. Because the charged particles attract free ions of the opposite sign, which produces a double layer, they are effectively shielded, or are screened by the charged ions of the opposite sign when immersed in a conducting fluid. This force decays exponentially or

$$F_c = F_o \exp\frac{-r}{\lambda_d} \tag{21}$$

where λ_d is known as the Debye screening length,[7]

$$\lambda_d = \left(\frac{2q^2n}{\epsilon kT}\right)^{1/2} \tag{22}$$

and n is the density of the ion species doing the shielding, q is the charge, ϵ is the dielectric constant of the solution, k is the Boltzmann constant, and T is the absolute temperature. For physiological saline solution of approximately 0.14 *M*, the Debye length is approximately 0.83 nm.[8] Thus the electrostatic forces are important only at very short ranges.

For like particles, the Van der Waals forces are repulsive at short distances (0.1 to 0.2 nm) and attractive at longer ranges. These forces may be thought of as being generated by transient electromagnetic fields due to fluctuations or displacements of the electric charges in the membranes or particles. These fluctuations occur because of thermal agitation or natural uncertainties in the position and momenta of the electrons and atomic nuclei. If one thinks of the local transient fluctuations in terms of the underlying contributions from oscillations at all possible frequencies, it can be shown that the strength of the contribution due to the local fluctuations at a given frequency is proportional to the absorption of light at that frequency by the material. For an individual atom these forces fall off very rapidly as $\frac{1}{r^7}$. However, when they are integrated over the surface of a membrane, which is thick compared to an atomic layer, they are correlated over many atoms, as the wavelengths are large compared to an atomic diameter.

A calculation of these fields has been done starting with quantum field theory.[9] The size of the forces and the rate at which they decay depends upon the distance between the membranes and on the difference of the bulk polarizability of the membrane and the aqueous gap in a complex way. All frequencies of charge fluctuations contribute to the attraction, and each gives rise to a different relationship between energy and the distance of separation.

One case of interest is for two membranes with a distance d_w across the aqueous gap between them. The thickness of the membranes is assumed to be large compared to the spacing. The force between these two membranes is approximately given by

$$F_w = \frac{H}{6d_w^3} \tag{23}$$

where H is the Hamaker coefficient.[8] In a typical situation, the distances over which the Van der Waals force is estimated to be important extend out to separations of 10 to 20 nm which is substantially longer than a Debye length or the rate of fall-off for the electrostatic or coulomb forces.

The hydration forces are repulsive forces that rise extremely rapidly as the membrane bilayers approach a separation distance of approximately an atomic spacing. Experimentally these forces can be expressed in the form[8]

$$P = P_o \exp \frac{-d_w}{\Lambda} \tag{24}$$

where Λ is a scaling constant. In the case of egg phosphomonoesterase bilayers, $P_o \simeq 7 \times 10^{-13}$ N/m^2 and $\Lambda = 0.256$ nm. This force may be important up to about 2 nm and is assumed to come about as a consequence of the work required to remove water from the hydrophilic surface of the membrane.

Recently, long-range attractive forces have been observed between hydrophobic surfaces.[10] These forces are proportional to the contact area and may be orders of magnitudes larger then the Van der Waals forces. They may also have decay lengths up to 25 nm. The magnitude and range of these forces depends on the temperature and the length of the surfactant's chain, and they appear only when the chains are in a fixed ordered state. These forces appear to be generated when the fields emanating from one surface induce a larger polarization in the other surface layer than in the intervening medium. Some relatively complex expressions for these attractive forces have been worked out.[10] These forces are important in the self-organizing processes such as protein folding, ligand binding to hydrophobic receptor sites, and the transformation of membrane structures.

All these forces act over relatively short ranges in typical biological fluids. The ion densities are so large that charge neutrality is maintained everywhere except very close to the charged surfaces.

The effect of an electric field, or an electric field gradient in a fluid, is to superimpose a small drift velocity on a relatively large random thermal velocity. For example, if we apply an electric field of 10^3 V/m to a Na^+ ion, we would expect a drift velocity of about 5×10^{-5} m/sec as compared to a thermal velocity of about 4×10^2 m/sec. For a protein, we would expect a drift velocity approximately one tenth of the speed of the Na^+ ion, although at higher fields. This means that if we are to transport proteins or other small charged particles over appreciable distances of a few millimeters, we can expect it to take periods of time of minutes or longer. In the case of bacteria, we have measured drift velocities of 10^{-6} m/sec at 100 Hz in fields of about 10^4 V/m and gradients of 5×10^6 V/m^2 or about 0.2% of the velocity of the Na^+ ions in the same field.[4,11]

This drift velocity may also change chemical reaction rates if the rate is limited by the availability of one of the charged components. Consider the case of a chemical reaction which takes place in a homogeneous fluid if the chemical reaction has the form of

$$A + B \underset{k_2}{\overset{k_1}{\leftrightarrow}} C \tag{25}$$

where (A) and (B) are the concentrations of the two input chemical reactants, and where k_1 is the reaction rate for A + B to C and k_2 is the rate for the back reaction of C to A. If we simplify the system and let k_2 be small, then the initial reaction rate may take the form[12,13]

$$k_1 = k\,(A)^n(B)^m \tag{26}$$

where n and m are the order of the reaction. In order to find the values of n and m, one can make the concentration of one of the reactants small so that k_1 takes the form

$$k_1 = k\,(A_o)^n\,(B)^m \tag{27}$$

where we have made the concentration of (A_o) large enough so that it is approximately constant, and the changes in the reaction rate can be measured by varying (B). The value of k is given by

$$k = zpe^{-E/RT} \tag{28}$$

where z is the collision frequency and p is the steric factor which is less than one and reflects the fact that not all collisions occur with the right orientation of the molecules to react. E is the activation energy, R is the gas constant, and T is the absolute temperature. For many cases the collision frequency is proportional to the current density, and thus there are terms which are proportional to both the drift and diffusion currents. For example, consider the case of an enzyme reaction on a charged substrate such as a biological membrane. The total current density for a given ion in the fluid incident on the membrane is given by the sum of equations 4 and 19 so that

$$\vec{J} = q\mu C\vec{E} - qD\vec{\nabla}C \tag{29}$$

If the chemical reaction rate is limited by the number of ions arriving at the membrane surface with an energy large enough to overcome some barrier, then a DC drift current may either add to or subtract from the diffusion current. If the field direction is such as to prevent the ion from reaching the surface, then the chemical reaction is blocked. If field direction is reversed, the rate can grow exponentially. These changes in chemical reaction rates with the direction of the electric field are likely to be responsible for changes in growth and reabsorption of neurites.[14] It is also likely to be involved in the motility of cells such as leukocytes and fibroblasts.[15–16]

An AC drift current will add to the diffusion current. For the AC fields, the drift current can be thought of as increasing the volume covered by a particle executing a random walk as a result of Brownian motion. Thus, an electric field oscillating in the x direction increases the number of particles which will strike the a y-z plane in a fixed period of time and can increase the chemical reaction rate for a catalytic reaction at the y-z plane. Seto and Hsieh show that AC fields as low as 5 V/m can increase enzyme reaction rates by a factor of 5.[17] For a 60-Hz field, the peak to peak displacement for Ca^{++} resulting from this field is estimated to be about 1.6 nm or about twice the thickness of the Debye layer. AC magnetic fields have also been shown to change the growth rate of corn roots at fields levels of 5×10^{-3}T. It is likely that these fields are inducing significant currents.[18]

III. THE EFFECTS OF ELECTRIC FIELDS ON CELL MEMBRANES

Electric fields play a very important role in the normal biological function of membranes. It would be surprising if externally applied fields did not affect the membrane behavior. First, an electric field exerts a mechanical force on a membrane by means of the force exerted on charges in the Debye layer on either side of it. The effects from fields in the plane of the membrane, where large molecules such as proteins are free to move as in a viscous fluid, are significantly different from the effects of fields in the transverse direction where the membrane components are bound in a layer typically 5 to 15 nm thick.

The interiors of cells are normally negatively biased in relation to the surrounding fluid by 50 to 150 mV, which leads to transverse electric fields up to tens of millions of volts per meter.[19] The effective membrane resistance (R_m) per unit area takes on values of 0.14 to 15 Ω/m^2 in the transverse direction. This corresponds to resistivities in the range of $\rho_m = 10^7\ \Omega m$ to $\rho_m = 10^9\ \Omega m$. Both the surrounding fluid and the interior of a cell have resistivities (ρ_f) of about $\rho_f = 2\ \Omega m$. This means that the cell membrane tends to shield the interior of a cell very effectively from externally applied fields.

Consider the case of a hypothetical rectangular cell as shown in Figure 1. An external field, $\vec{E}$, causes a current density, $\vec{J}_f = \vec{E}/\rho_f$, to flow in the external medium, where ρ_f is the resistivity of the fluid. The corresponding voltage drop is $V = E\ell = J_f\rho_f\ell$ which we can consider to be applied to the cell. This voltage is distributed across the cell length as

$$V = [\rho_m(2t) + \rho_f(\ell - 2t)]J_m \tag{30}$$

where J_m is the current density through the cell and ρ_m is the resistivity of the membranes.

Typical cell membrane thicknesses are 6 to 10 nm and typical dimensions are 10 to 150 μm. Setting $\ell = 100\ \mu$m and

$$\rho_m 2t = 10\ \Omega m^2 \tag{31}$$

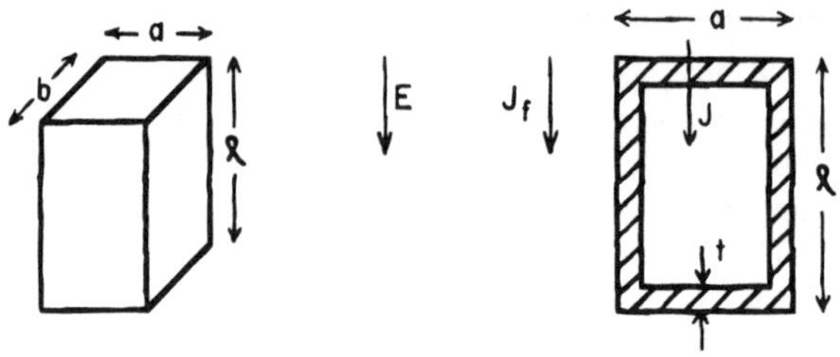

FIGURE 1 Current distribution in a hypothetical rectangular cell.

we get

$$V = (10\ \Omega m^2 + 2 \times 10^{-d}\ \Omega m^2)J_m \tag{32}$$

This shows that essentially all of the transverse voltage drop occurs across the membrane, and the interior of the cell is almost completely shielded from external fields. A more complete theory for long cells which accounts for the internal resistance of the cell is given by Cooper et al.[20] It is to be noted that muscles and a number of other cells may be much longer than 100μm. Additionally, blood vessels form long, low resistance paths which may concentrate currents due to externally applied fields. The anisotropic characteristics of cells are reflected in an anisotropy of the dielectric and conductive properties of tissue which may give variations of as much as 10 to 1 in conductivity, depending on direction of measurement relative to cell orientation.[21]

A number of fish, including sharks, have been shown to use very long cells and to sum signals from cells in both series and parallel to increase the voltage drop across a sensitive membrane in order to sense fields as low as 10^{-6} V/m.[22–23] The long cell may be thought of as an antenna that concentrates the field across a very thin, voltage-sensitive detector membrane. This membrane appears to have a built-in amplifier that allows detection of signals which are only a little above the natural electrical noise.

Membranes are not just simple linear resistors but usually are nonlinear and, in the case of nerve cells, are time-varying resistors. For a passive membrane in which the membrane potential is primarily determined by the concentration gradient of a single ion such as K^+, the Nernst equation predicts a diode rectifying characteristic of the form[19]

$$I = I_0\left[\exp\left(\frac{V_m}{\eta V_T}\right) - 1\right] \tag{33}$$

where V_T is given by

$$V_T = \frac{kT}{q} = 0.026\ V \tag{34}$$

at

$$T = 300\ K \tag{35}$$

q is the charge on the electron, V_m is the voltage across the membrane, and I_0 is the current for ideal back-biased current in amperes and η is a dimensionless constant. Thus, for currents flowing through a membrane in one direction, the current is nearly constant while in the other it increases exponentially with voltage. In addition to passive currents, cells also use the energy from metabolic processes for the active transport of ions against the fields established by the concentration gradients. These processes are usually modeled as current sources and described as pumps.[19] A thermodynamic approach to pumping shows that ions can be pumped if they form a compound with a material which can flow through the membrane and which is created on one side of the membrane and destroyed on the other.[24] A large variety of models have been generated to characterize the effects of externally applied fields on the transport of ions through membranes.[25–27] However, the details of the pumping process are not well understood. In addition to ion transport, electrical fields can change the binding of ions or molecules to the membrane surface.

In the case of pacemaker cells, there are also feedback processes that lead to an oscillating membrane potential and a membrane resistance which is a function of time. The current flow for these cells is described empirically by the Hodgkin Huxley equation.[19] An alternate approach which treats the nerve pulse like a plasma instability has been proposed by Triffet and Green and Vaccaro and Green.[28–29] Na^+ and K^+ currents are the dominant carriers for the propagation of nerve impulses along a cell. It is generally believed that the Na^+ and K^+ currents, which flow through the membrane in opposite directions, are carried through separate channels. Ca^{++} ion currents are involved in the activation of at least a portion of the K^+ currents and are voltage-gated. By activating the K^+ currents, the Ca^{++} ions shorten the length of time the cell is depolarized and thus speed up the firing cycle.[30] A statistical approach to the formation of protein channels in the membrane by Baumann and Easton predicts many of the observed characteristics.[31–33]

During the firing of a nerve cell, the Na^+ current pulse precedes the K^+ current pulse which returns the cell to its resting potential.[19] The overall concentration balances are maintained by active ion pumps. Cl^-, Mg^{++} and possibly OH^- and H^+ ions may also be involved in the current flow across a cell membrane.

The firing of a nerve cell typically involves voltage spikes of 10^{-1} V and peak current densities of 1.5 A/m^2. Changes in the firing rate can be induced by the injection of charge through a micro electrode of $< 10^{-9}$ A for a few milliseconds. However, in cases where electrodes are used to stimulate muscles or to control epilepsy, the current is injected through a series of cell membrane fluid boundaries at a distance from the controlling nerve fiber. Thus, typical injected currents to produce behavioral changes in cells are milliamps, and current densities are 10 A/m^2 or higher.

For fields parallel to the plane of the membrane, it is possible to obtain electrophoresis or a rearrangement of charged particles. This has been shown by Poo in striking fashion in cultured embryonic *Xenopus* myotomal muscle cells.[34–35] Receptors on the surface of the cell were labeled with a fluorescent dye and allowed to uniformly distribute themselves. Exposures to electric fields of 10^2 to 10^3 V/m were sufficient to concentrate the fluorescent labeled receptors on the side of the anode in about 10 min. After shutting off the field, diffusion returned the dye to its uniform distribution in about 2 hr. This corresponds to an in-plane diffusion constant of about 3×10^{-12} m^2/sec. The force on the receptor molecules or particles in the membrane includes not only $q\vec{E}$ but also any viscous drag which may be generated by the flow of ions of the opposite sign moving along the surface in the opposite direction. The direction of motion for a given charged particle seems to depend on whether it has a larger or smaller zeta (ζ) potential than the potential across the charged double layer at the interface between the cell surface and the fluid (see equation 7).

Additional work has shown that the distribution of acetylcholine (ACh) receptors is changed by external fields.[34–35] These receptors are concentrated on the cathode-facing surface of the cell in fields of 10^3 V/m over a period of 30 min by literally rearranging channels already existing in the cell membrane. The concentration or clustering persists for at least 5 hr after the field has been turned off, indicating that the clustering is relatively stable.

Single channel patch measurements show both a higher density of ACh channels in the clusters near the cathode and a longer mean duration of the pulses through the transmembrane channels. The length of the current pulse near the anode does not differ from the controls, indicating that the field itself does not have a direct effect on the channel kinetics. The lateral diffusion coefficient, D, of ACh receptors in the plasma membrane of cultured *Xenopus* embryonic muscle cells is estimated to be 2.6×10^{-6} m^2/sec at 22°C.[34] Lateral concentration gradients in lipid monolayers have been shown to be induced by externally applied electric field gradients. For binary mixtures of dihydrocholesterol and dimyristoylphosphatidylcholine, the application of an electric field gradient at pressures below the critical pressure produces a liquid-liquid phase separation in a monolayer that is otherwise homogenous.[36] This separation occurs at field levels on the order of 10^7 volts/m and gradients of 10^9 to 10^{11} volts/m^2.

IV. NONLINEAR EFFECTS OF AC FIELDS ON CELLS

A. INTRODUCTION

The application of an AC electric field to nonlinear systems which can be described by either a nonlinear resistance or capacitance leads to at least partial rectification of the input signal and the generation of harmonics. If two or more signal frequencies are applied, it also leads to frequency mixing of the form

$$f_0 = \pm m f_1 \pm n f_2 \tag{36}$$

where f_0 is the output frequency, f_1 and f_2 are input frequencies, and m and n are integers. The rectified component of the AC current can, in turn, lead to ion accumulation at interfaces, which results in changes of ion concentration.[37–38] These changes in ion concentration, in turn, can affect biological function. Another important additional effect is the dependence of the dielectric constant on frequency. Thus, the electrophoretic or dielectrophoretic forces become both size- and frequency-dependent. A third—possibly important—additional effect is excitation of frequency-sensitive biological systems in a resonant manner. By driving such systems near their resonant frequency, we may change the effective amplitude of the stimulating signal and change the frequency of the nerve cells firing.

In this section we will review the rectification process at the cell membrane in some detail. Additionally, we will show that cell nonlinearities lead to frequency-dependent effects such as injection phase locking of pacemaker cells. We will also briefly examine some problems associated with safety, exposure of cells to very low ELF fields, and application of large ELF fields to biological systems.

B. RECTIFICATION BY CELL MEMBRANES

The rectification of currents flowing across membranes has been studied by many authors beginning with Katz in 1949. Much of this work is referenced by Hayashi and Fishman in their paper on the inward rectifier K^+ channel kinetics.[37]

For many passive cell membranes, an approximate relation for the transmembrane current can be derived from the Nernst equation as given in equation 33.[19] If we apply an AC signal across the membrane of the form $V_M = V_0 + V_1 \cos\omega t$, the resulting current can be approximated for small values of V_M (i.e., $qV_M < \eta kT$) by a Taylor series yielding

$$I = \frac{I_0}{\eta V_T}\left(V_0 + \frac{V_1^2}{2\eta V_T} + V_1\cos\omega t + \frac{V_0 V_1 \cos\omega t}{2\eta V_T} + \frac{V_1^2}{4\eta V_T}\cos 2\omega t + \ldots\right) \tag{37}$$

It is to be noted that the second term in the expression is the first approximation to the fraction of the applied AC voltage V_1, which yields a DC current component ΔI,

$$\Delta I = \frac{I_0}{4}\left(\frac{V_1}{\eta V_T}\right)^2 \tag{38}$$

or an offset voltage V_{DC} given by

$$V_{DC} \approx \frac{I_0}{4}\left(\frac{V_1}{\eta V_T}\right)^2 R_m \tag{39}$$

where R_m is the membrane impedance. This predicted voltage offset for an applied AC current has been measured by Montaigne and Pickard.[39] In their experiments, an AC signal was applied to a large plant cell by a strip line, and the measured voltage shift was obtained through micro electrodes located outside the applied AC fields. For an applied AC field of about 0.2 V, they measured a DC offset of 1 to 2×10^{-4} V. For frequencies above 2.5 kHz, the effects of the membrane capacitance must be taken into account, and the effective driving voltage is reduced to

$$(V_1)\text{eff} = \frac{\sqrt{2}\, a\sigma_e E_{Irma}}{[(\sigma_e + aG)^2 + (a\omega C)^2]^{1/2}} \tag{40}$$

where a is the cell radius, σ_e is the conductivity of the medium, E_{Irms} is the electric field strength in the medium surrounding the cell, G is the membrane conductance per unit area, ω is the frequency, and C is the membrane capacitance per unit area.[40] This leads to the usual roll-off in the measured DC offset with increasing frequency. It is to be noted that the DC effect gets still smaller at higher frequencies (above 1 MHz) because of transit time limitations for ion flow across the membrane.[41] The relaxation times for a typical K+ channel in an *Aplysia* membrane has been measured to be from 2 to 8 ms.[37] Rectification has also been demonstrated in thin lipid membranes.[42] In these systems, both the conductivity of the membrane and the ion concentration differences across it can be controlled. The Nernst equation was shown to apply to the I vs. V curve over a range of voltages from −60 to +40 mV. Depending on the ion concentration and membrane doping, the values of η ranged from 1 to 0.25.

A different treatment of the nonlinear response of passive cell membranes to an applied AC field has been carried out by Franceschetti and Pinto and by Casaleggio et al.[43–44] Both these groups have expanded the Nernst equation in a Volterra series which takes into account memory of the preceding state of the cell. They have also treated the cell in spherical rather than plane geometry. The inclusion of a spherical cell requires that the total current into and out of the cell be equal to zero, and thus loops are formed circulating through the cell membrane (see Figure 2). All of the theoretical treatments predict a DC component which varies as the square of the input signal V_I and tends to hyperpolarize the cell or make the interior of the cell more negative.

Cain has considered the effects of an AC field on nonlinearities of the nerve cell by numerical analysis of the Hodgkin Huxley equation.[*45] He applied a voltage

$$V_m = V_0 + V_1 \cos\omega t\,[u(t) - u(t - \tau)]$$

across the membrane where u(t) and u(t − τ) are unit step functions which define an AC pulse of length τ. For the case when the AC frequency is large compared to the reciprocal

* Bisceglia and Pinto have applied a Volterra series expansion to the Hodgkin Huxley equations. This approach gives an alternate method to Cain's of computing the current shifts resulting from applied AC signals.[46]

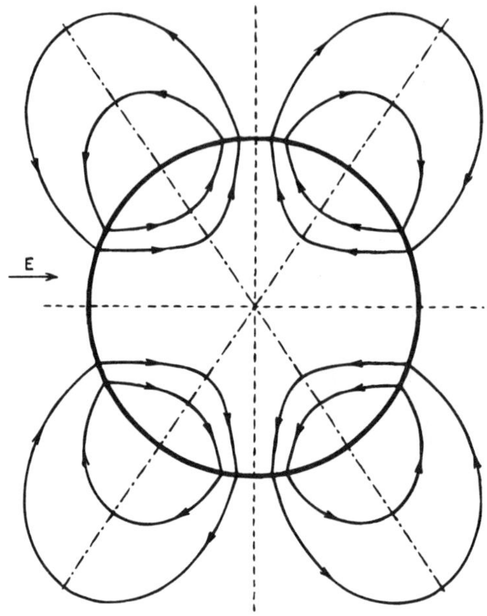

FIGURE 2 Induced DC current distribution in a spherical cell. (Bisceglia and Pinto, I., "Volterra Series Solution of Hodgkin-Huxley Equation," personal communication, 1984.)

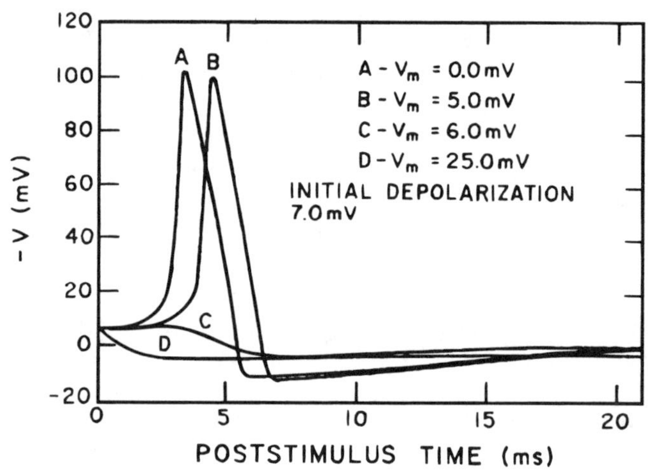

FIGURE 3 Computed membrane action potentials in response to an initial membrane depolarization of 7 mV for different values of V_m. Curves are solutions to the Hodgkin-Huxley equations. (Cain, C. A., "Biological Effects of Oscillating Electric Fields: Role of Voltage Sensitive Ion Channels, *Bioelectromagnetics,* 2, 23, 1981.)

of the pulse length, if a 7-mV depolarizing pulse is also applied to the membrane, the action potential is obtained as shown in Figure 3. He has assumed coefficients appropriate to the giant squid axon. Increasing V_1 first delays and then suppresses the action potential. If no depolarizing pulse is applied, the predicted changes in g_{NA} and g_K and the deviation V from the resting potential are shown in Figure 4 for a 10 msec AC pulse with $V_1 = 25$ mV. Note that the applied AC frequency is assumed high enough not to be resolved in these figures. The predictions for the changes in the Na^+ and K^+ conductances g_{NA} and g_K are shown in Figure 4 for a 10 msec pulse with $V_1 = 25$ mV. From these results, it is clear that AC signals can induce substantial changes in the operating characteristics of nerve cells at moderate to

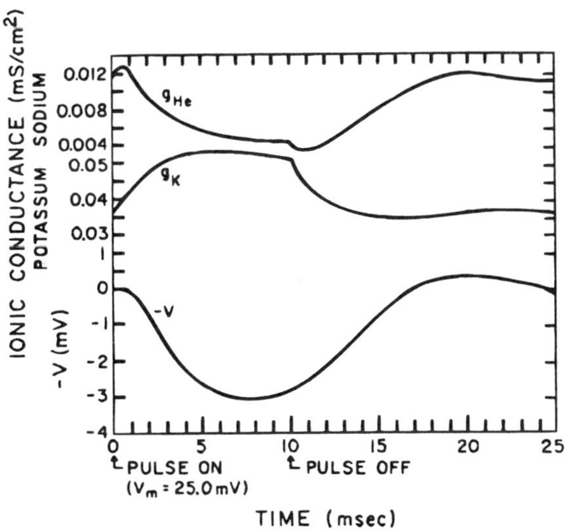

FIGURE 4 Response of model axon to a pulsed oscillating component of membrane electric field (10-msec pulse, V_m = 25 mV). The membrane potential and the sodium and potassium conductances are shown. These curves are solutions of the Hodgkin Huxley equations. (Note: 1 mS/cm² = 10 S/m².) (Cain, C. A., "Biological Effects of Oscillating Electric Fields: Role of Voltage Sensitive Ion Channels, *Bioelectromagnetics,* 2, 23, 1981.)

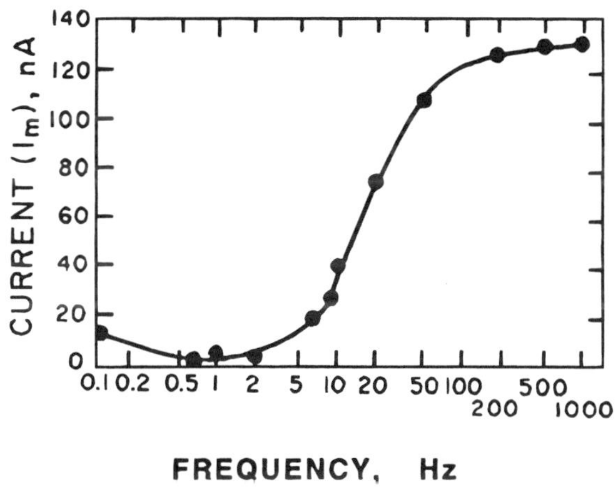

FIGURE 5 Intracellular (transmembrane) currents I_m (in nA) needed at different frequencies to produce firing-pattern changes (in pacemaker neuron). Note the detectable changes take on different forms at different frequencies. (Wachtel, H., "Firing-pattern Changes and Transmembrane Currents Produced by Low Frequency Fields in Pacemaker Neurons," *Proceedings of the 18th Annual Hansford Life Science Symposium,* Technical Information Center, U.S. Dept. of Energy, Richard, Washington, USA, 132, 1978.)

high levels of applied voltage. Although the appropriate coefficients were not measured in order to make a direct comparison between theory and experiments, Wachtel's results on *Aplysia* at frequencies above the lock-in range would appear to support Cain's theoretical predictions.[47]

Wachtel has made a series of measurements which demonstrate the nonlinear characteristics of pacemaker cells from *Aplysia.*[47] First, he measured the current input through a micro electrode which changed the firing rate of the cell. The current threshold for a minimum detectable change was approximately 6×10^{-10} A at frequencies between 0.8 and 1 Hz (see Figure 5). The natural firing rate for this cell is about 0.8 Hz, and an increasing current is required to synchronize the cell to the injected signal as the frequency deviates from the natural firing rate. A theory for injection locking of electronic oscillators predicts that the signal required for locking an oscillator to an external signal increases linearly as the difference between the two frequencies $\Delta\omega$ increases.[48] The signal required for lock-in according to this theory is given by

$$I_l \approx |A\ \Delta\omega_0| I \tag{41}$$

where I_l, is the injected signal current and I is the peak unperturbed oscillator current. $A = \frac{d\phi}{d\omega}$ is the rate of change of phase with respect to frequency in the unperturbed oscillator. $\Delta\omega_0$ is equal to the difference between the frequency of the free-running oscillator and the injected signal. This expression is applicable as long as

$$\Delta\omega_0 \leq \frac{2\pi}{\tau} \tag{42}$$

where τ is the time constant for adjusting the gain of the circuit. The time constant τ for the *Aplysia* cells varied between $\tau = 0.1$ to 0.5 sec, and this corresponds to a maximum measured lock-in frequency of about 10 Hz. The results shown in Figure 5 show the threshold for 1-to-1 locking up to about 2 Hz. In the range from 2 to 10 Hz, Wachtel observed a lower threshold for subharmonic locking than 1-to-1 locking. At frequencies above 80 Hz, he observed a constant shift in the firing rate of the neuron in response to the injected transmembrane AC. The natural firing rate would be restored by also injecting a transmembrane DC equal in amplitude to about 1% of the peak-to-peak value of the AC current. This DC current was in the depolarizing direction, making the exterior of the cell more negative with respect to the cell cytoplasm to increase the firing rate (i.e., to restore it to its natural value). Apparently the applied transmembrane AC current was partially rectified so as to hyperpolarize the membrane (making the interior of the cell more negative with respect to the external fluid). The details of how the applied field modifies the ion flow are only partially understood, but one characteristic is an increase in the conductivity for K^+ which increases its flow out of the cell. Wachtel also injected low frequency currents into the sea water surrounding the cell preparation through external electrodes.[47] In this case, the minimum current densities flowing in the vicinity of the cell preparation for injection-locking were estimated to be about 10^{-2} A/m^2, and there was about a 10 to 1 variation between the maximum and minimum sensitivity for changes in angle between the applied field and the cells. At frequencies above 100 Hz, a minimum of about 0.35 A/m^2 was necessary to get a detectable change in firing rate.

These studies have been extended by Barnes, et al., and injection locking at harmonic and subharmonics has been shown to occur. It is suggested that phase locking may provide a mechanism for narrow banding or time averaging so that a weak coherent signal may be distinguished from noise by a cell.[49] Extensive modeling of phase locking for a squid axon using two versions of the Hodgkin Huxley equations has been carried out by Fohlmeister, et al.[50] They show that phase locking can occur for a wide variety of frequencies with AM modulated signals at injected current densities greater than 0.1 A/m^2.[50] For natural oscillation frequencies less than the externally applied signal, the system may be treated as a parametric process. For parametric amplification, a phase stability such that

$$d\Phi/dt < \Delta\omega - KV_s \tag{43}$$

is required for injection locking of the frequency of oscillation to an external signal, where

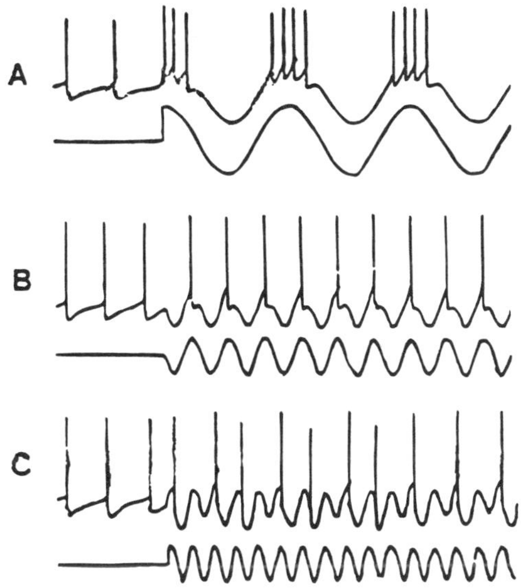

FIGURE 6 Examples of several modes of synchrony between an imposed ELF field and neuronal patterns. In each case the ELF current is shown below the transmembrane potential recording. A. for ELF frequencies well below FR_0, several nerve impulses (spikes) are locked to each ELF half cycle. B: ELF frequencies slightly above FR_0 are effective in phase locking the neuronal spikes on a one for one basis. C: For ELF frequencies several times greater than FR_0, phase locking can take the form of spikes occurring on alternate cycles (2 for 1) synchrony. (*Bioelectromagnetics Supplement,* Barnes, F.S., "Some Engineering Models for Interactions of Electric and Magnetic Fields with Biological Systems," 67–85, 1, 1992.)

$d\Phi/dt$ is the rate of change of the phase, $\Delta\omega$ is the frequency offset, K is the linear control characteristic in units of $\left(\frac{2\pi Hz}{volt}\right)$ and is closely related to the loop gain, and V_s is the injected signal.[51] Stated in words, this equation requires that the amplified signal, KV_s, be large enough to correct for the random frequency fluctuations $d\Phi/dt$ generated by the noise for the system to become phase locked to a signal which is displaced by $\Delta\omega$. (See Figure 6 for some examples of injection locking of pacemaker cells to an external signal.)[52] A related phenomenon is stochastic resonance. In this case, a bistable system such as a pacemaker cell is driven by both noise and a periodic signal. It is shown that the signal-to-noise ratio can be enhanced by the addition of noise to a weak periodic signal and that power can extracted from the noise. A strong periodic signal can be generated at signal-to-noise ratios less than one.[53–54] This phenomenon occurs when two energy states are separated by a barrier, and the probability of a transition increases exponentially with increasing noise power. For periodic signals which are insufficient to cause a transition over the barrier but periodically increase the energy of the particle, the transition rate at the signal frequency first increases with increased noise power up to some maximum. When the noise power is increased above this level, the output signal becomes more random.

An increase in the sensitivity to electromagnetic fields has also been shown in isolated frog hearts for signals which approach the natural resonant frequency or firing rate.[55] In these experiments, the heart firing rate was shown to increase as much as 30% when a signal in the vicinity of 10 to 20 V/m was applied through the Ringer's solution to the isolated frog hearts at a frequency between 0.5 and 1 Hz. The natural firing rate of these excised hearts started out at approximately 1 Hz and dropped to about 0.5 Hz over a period of 2 hours where they remained stable for at least 5 hours. To get a 30% increase in firing rate at 60 Hz, it was necessary to apply field strengths of 60 to 80 V/m. Thus, we have additional

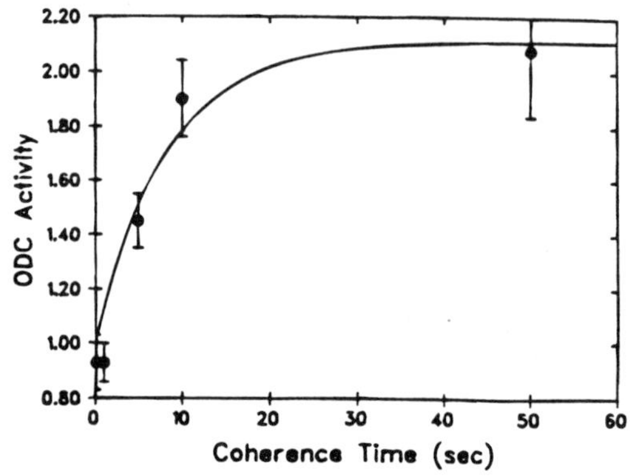

FIGURE 7 Plot of the enhancement of ODC activity (exposed/control) as a function of the coherence time, τ_{coh}, of the applied field. The solid line is the best fit to the mathematical function given by Eq. 44 where τ_{cell} is found to be 8.2 sec. The experimental points shown represent a minimum of six different exposures. (© Academic Press, Litovitz, T.A., Krause, D., Mullins, J.M., "Effects of Coherence Time of the Applied Magnetic Field on Ornithine Decarboxylase Activity," *Biochemical and Biophysical Research Communications,* 862–865, 178, 3, 1991.)

evidence that electric fields with repetition rates near the natural biological signaling frequencies are more likely to induce changes than those of higher frequencies, and that signal strengths required for a given shift increase approximately linearly up to some cutoff, as shown in Figure 5.

For weak fields, it has been shown that cells can respond differently to signals which are both space and time coherent than they do for signals which look like the background noise. T. Litovitz, et al. have shown that the application of 10 μT magnetic fields at either 55 or 65 Hz doubles the specific activity of ornithine decarboxylase (ODC) in L929 cells if the signals are coherent for periods of 10 sec or longer during the course of a 4-hr exposure.[56] The applied signal and the corresponding ODC response as a function of the coherence time are shown in Figure 7. The ODC response of the cell can be fitted to an exponential curve of the form

$$(\text{ODC}) = 1 + 1.26\left[1 - \exp\left(\frac{-\tau_{coh}}{\tau_{cell}}\right)\right] \tag{44}$$

where τ_{coh} is the length of the time between shifts in frequency and the introduction of a random phase shift, and τ_{cell} is the effective time constant of the cell.[56] τ_{cell} has a value of about 8 secs for these cells. If a spatially coherent noise signal with a power spectral density ranging from 30 to 90 Hz is superimposed on the coherent signal, the increased ODC response decreases with a decreasing signal to noise ratio and is less than 10% at a signal-to-noise ratio of one.[57]

For the exposure geometry used in these experiments, the magnetic field induced a corresponding electric field of 4μ V/m. This signal is well below the calculated thermal noise field of 0.02 V/m for a 20 μm cell diameter. The combination of the two experiments cited above indicate that both space and time coherence may be used by cells to separate useful signals from larger natural background noise signals. For example, to get a significant biological response, some threshold number of channels or receptor molecules may need to be activated within a given period of time; this, in turn, requires nearly simultaneous activation over a significant fraction of the cell surface. Similar results have been obtained for developing chick embryos, where weak coherent signals lead an increased incidence of abnormalities.[58] In this work, Litovitz and his colleagues show an increase in the incidence of abnormalities of approximately a factor of 3 for white leg horns incubated in periodic magnetic fields with peak field strength of 1 μT (100 Hz repetition rate, 500 μs pulse duration, 2 μs rise and decay times) when compared with the controls. This increased rate of the incidence of abnormalities was nearly eliminated with the addition of band-filtered noise with a spectrum running from 30 Hz to 100 Hz and an RMS value of 1 μT. Thus, Litovitz makes a strong case for a requirement of both space and time coherence for biological systems to detect signals below the natural noise environment.

A number of experiments indicate that at least two mechanisms are involved in the effects of low-level time-varying magnetic fields on membrane transport. The first of these is through Faraday's law or the induced electric field which in turn induces electric currents. In these experiments, one would expect to get the same effects by introducing electric fields with electrodes at levels that induce the same current densities. The second group of experiments indicates that the background DC magnetic field is also important and that the combined effects of AC and DC magnetic fields are observed.

The initial experiments by Walleczek and Liburdy showed an enhanced uptake of Ca^{++} in Con A activated rat thymocytes with exposures of one hour to 60-Hz magnetic fields of 22 mT and induced current densities of 0.16 A/m^2.[59] In this paper the exposure system consisted of concentric rings on cell culture plates which in turn were placed in a water cooled solenoid which produced a uniform magnetic field. This was followed by a group of experiments by Liburdy on Ca^{++} transport across mitogen-activated lymphocyte membranes.[60] In these experiments, both the DC and AC magnetic fields where controlled so that the DC geomagnetic field and the ambient 60-Hz fields were perpendicular to the exposed and control plates. The results show an increase in the Ca^{++} influx during the plateau phase of the calcium signaling for Con A activated lymphocytes, which was a function of the induced electric field and which could be reproduced by applying the electric fields across the cells with a salt bridge at levels between 0.1 V/m and 0.17 V/m. This corresponds to induced current levels of 0.168 to 0.28 A/m^2 in the fluid surrounding the cells which had a conductivity, $\sigma = 1.68$ S/m, which is approximately a hundred times larger than the current densities observed around growing cells. Thus, the approximately 20 to 25% increase in the initial Ca^{++} uptake is the result of a relatively large external current. In other experiments it was also shown that the response is dependent on the age of the animals from which cells are taken.[61]

Most other reported experiments have not been done in a way to sort out the differences between possible direct effects of the magnetic fields and the induced electric fields. Yost and Liburdy have also done experiments in this same system which show a direct dependence of the calcium uptake on the DC magnetic field.[62]

The experiments by McLeod, Lee and Ehrlich show both a frequency dependence and a dependence on the electric field strength across the cell membrane.[63] They exposed neonatal bovine fibroblast cells to electric fields in culture through a media bridge. The fibroblasts populated a collagen matrix that enabled the cells to be grown with a dominant orientation and exposed to a well defined current. An estimate of newly synthesized protein was made by measuring the incorporation of (^{3}H) proline into macromolecules after a 12-hr exposure to current densities ranging from 10^{-3} to 10 A/m^2 and frequencies from 0.1 Hz to 1 KHz. The results in Figure 8 show an approximately 30% reduction in the ^{3}H counts with current densities as low as 10^{-2} A/m^2. This reduction is interpreted as a reduction in the incorporation of newly synthesized protein into the extracellular matrix rather than as a change in the cell number. The frequency specificity for this threshold is shown in Figure 8; the peak sensitivity was recorded at 5×10^{-3} A/m^2 and 10 Hz. The corresponding peak electric field intensity was 4.5 mV/m. The fractional change in the (^{3}H) proline was nearly independent of the current density for increases in current density up to two orders of magnitude above 10^{-2}A/m^2. The cell membranes have a resistance many times higher than the resistance of the matrix

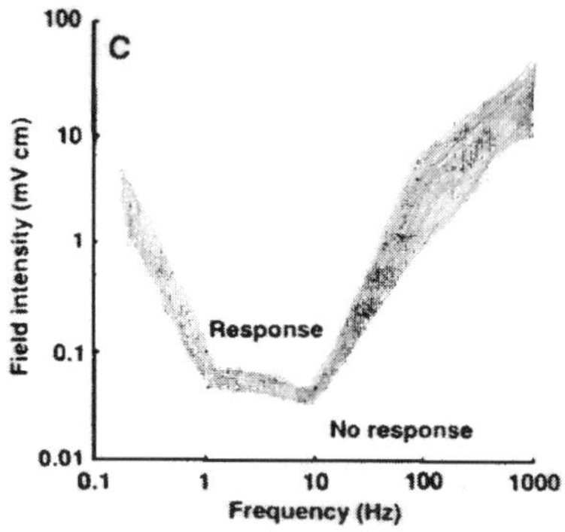

FIGURE 8 Minimum field intensity for a detectable response. Summary of results for all tested frequencies and current densities. Current densities were converted to peak field intensities by using the measured media resistivity of 65 ohm-cm. The lower boundary of the gray region represents the highest field intensity at which no significant change in extracellular protein accumulation was detected; the upper boundary represents the lowest intensity evoking a statistically significant change (n = 6). (McLeod, L.J., Lee, R.C., Ehrlich, H.P., "Frequency Dependence of Electric Field Modulation of Fibroblast Protein Synthesis," *Science,* 1465–1469, 136, 1987.)

as a whole. The cells are also asymmetric with a ratio of major to minor axis of from about 7 to 10. Thus, the current through the cell membranes would be expected to be at a maximum when the long axes of the cells were parallel to the applied field. For randomly oriented cells, current densities of 3 mA/m^2 produced no significant effect on the rate of proline incorporation. However, when the cells were oriented parallel to the electric field that was estimated at 2 mV/m and at 10 Hz, a little more than a 30% reduction was observed. The estimated transmembrane potential was 0.5 μV. With the cell oriented perpendicular to the field, no significant change in proline incorporation was measured at 6 mA/m^2.

In addition to the nonlinear conductances associated with the Na^+ and K^+ currents, membranes also exhibit nonlinear (i.e., potential-dependent) and frequency-dependent capacitances and inductances. It is sometimes useful to think of these effects in terms of a phasor diagram as shown in Figure 9, where the electric field vector $\vec{E}$ is rotating at velocity ω and ϕ is the phase angle between $\vec{E}$ and the current density $\vec{J}$. If there is, for example a fixed time delay between the field activation of a current gate and the current flow, then, depending on the

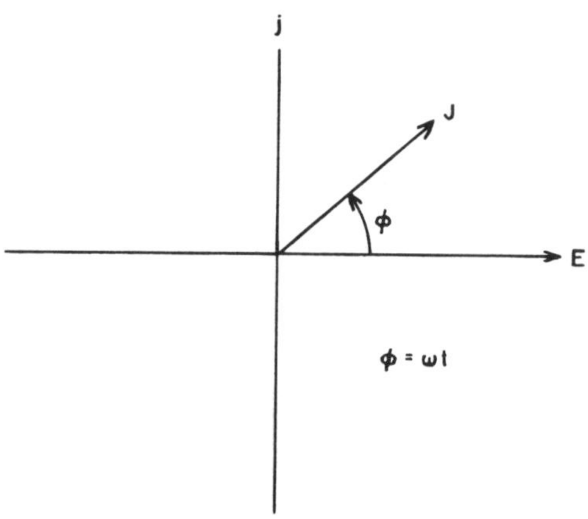

FIGURE 9 Steady state vector characterization of electric fields $\vec{E}$ vs. current density $\vec{J}$ in polar form. ϕ is the phase angle between the sinusoidal electric field $\vec{E}$ and the resulting current density $\vec{J}$.

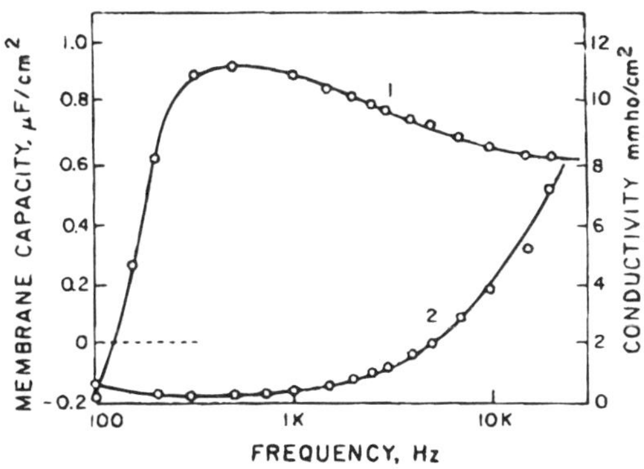

FIGURE 10 Membrane capacitance (Curve 1) and conductivity (Curve 2) of squid giant axon at various frequencies. Note anomalous behavior at low frequencies. (note: 1 μF/cm² = 0.01 F/m².) (Takashima, S., "Frequency Response of Membrane Components in Nerve Axon," *Biological Effects of Nonionizing Radiation,* Illinger, K.H., Ed., ACS Symposium Series, No. 157, 133–145, 1981.)

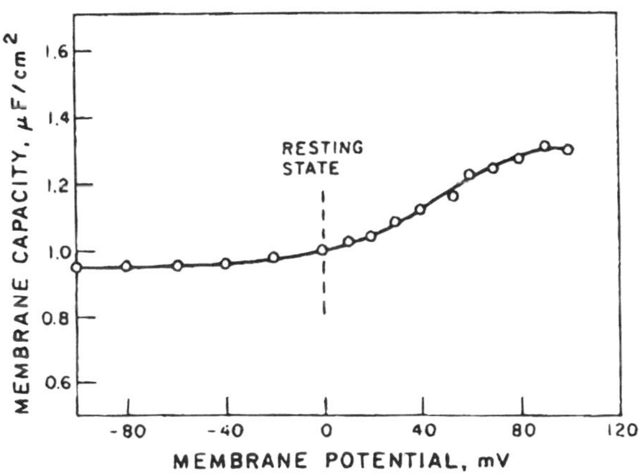

FIGURE 11 Membrane capacitance of squid giant axon at various membrane potentials. Membrane potential was shifted by injecting currents. The abscissa shows actual potential across the membrane in mV. (Takashima, S., "Frequency Response of Membrane Components in Nerve Axon," *Biological Effects of Nonionizing Radiation,* Illinger, K.H., Ed., ACS Symposium Series, No. 157, 133–145, 1981.)

frequency, $\vec{J}$ may be in any of the four quadrants and appear capacitive or inductive or even present a negative resistance to an external driving source.

Nonlinear inductive effects seem to be associated with the time delay for the onset of the K^+ currents under excitation in a typical excitable membrane, and they have been studied in the giant squid axon.[64] The nonlinear capacitive effects are difficult to measure at frequencies below a few kilohertz. Extra care needs to be exercised to minimize the series resistance and the end effects of the wire being used to measure the capacitance or inductance. Additionally, corrections must be made in the calculations of the membrane capacitance to take into account the appropriate variations in the frequency response which these terms introduce. However, when this is done, it can be shown that the membrane capacitance has both frequency- and voltage-dependent terms. The capacitance of giant squid axons is shown in Figures 10 and 11 as a function of frequency and membrane voltage.

Variation of the capacitance of these membranes with frequency and amplitude differs from that of a simple bilipid membrane which has nearly constant capacitance. The variation appears to be associated with changes in the conformation of the proteins associated with the Na^+ conductance channels. Nonlinearity in conductance and capacitance can be induced into a bilipid membrane by the addition of Alamethicin. The nonlinear inductance or capacitance may also generate both sum and difference frequencies if two signals are applied. For the case of the single signal, a DC term is added to the current density which is proportional to membrane potential and the square of the applied AC signal.[65] The effects due to nonlinear membrane capacitance thus far observed are small. They appear likely to be more important in providing an understanding of the possible gating mechanism in membranes than as a mechanism for introducing rectification.

Another form of nonlinearity in the electrical response of cells comes about in what is often described as adaptive processes. For example, we found that repetitive exposures of pacemaker cells, taken from the ganglion of an *Aplysia*, to microwave pulses resulted in a decreasing reduction in the firing rate by successive pulses. This kind of change has also been shown to occur in neurons which have been conditioned with repetitive stimulation. Studies of conditioning have shown decreases in potassium ion conductance through membranes, thus raising the internal potential and enhancing the excitability.[66–67] The decrease in resistance between adjacent cells can occur in two ways. First, the resistance of gap junctions may be reduced by repetitive electrical stimulation which increases the electrical coupling between the cells by amounts ranging up to 62%. Second, repetitive electrical stimulation can modify the chemical excitatory postsynaptic potential by amounts ranging from 31–140%.[68] This change is associated with the movement of protein kinase C from the interior of the cell into the membrane. An accompanying change in Ca^{++} concentrations and the movement of a second messenger, diacylglycerol, into the membrane reduces the potassium ion flow. This enhanced excitability reduces the voltage or the charge required to initiate an action potential. If charge is transferred efficiently between cells, either actively or passively, cell length is effectively multiplied in the linear model by the number of cells in the chain; this in turn reduces the external electric field required to generate a given voltage across a terminating membrane.

An interesting speculation which is raised by these adaptive processes is whether or not a neural network can be trained to identify a repetitive signal such as 60 Hz in the presence of larger electric fields generated by the surrounding biological material. To test this hypothesis we programmed a computer to simulate a neural network as shown in Figure 12.[52] Using a back-propagation algorithm to adjust the connecting weights between neurons, a sigmoidal summing junction to model the neurons, and a pseudorandom noise generator, we measured the number of runs required to train the network to recognize a 60-Hz signal with 97%

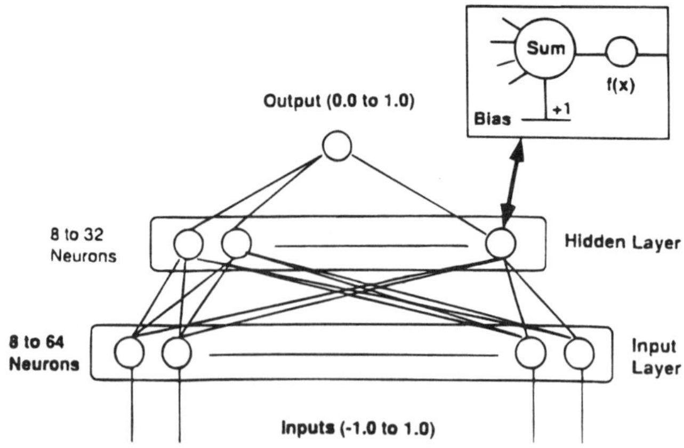

FIGURE 12 Back-propagation neural network. (*Bioelectromagnetics Supplement,* Barnes, F.S., "Some Engineering Models for Interactions of Electric and Magnetic Fields with Biological Systems," 67–85, 1, 1992.)

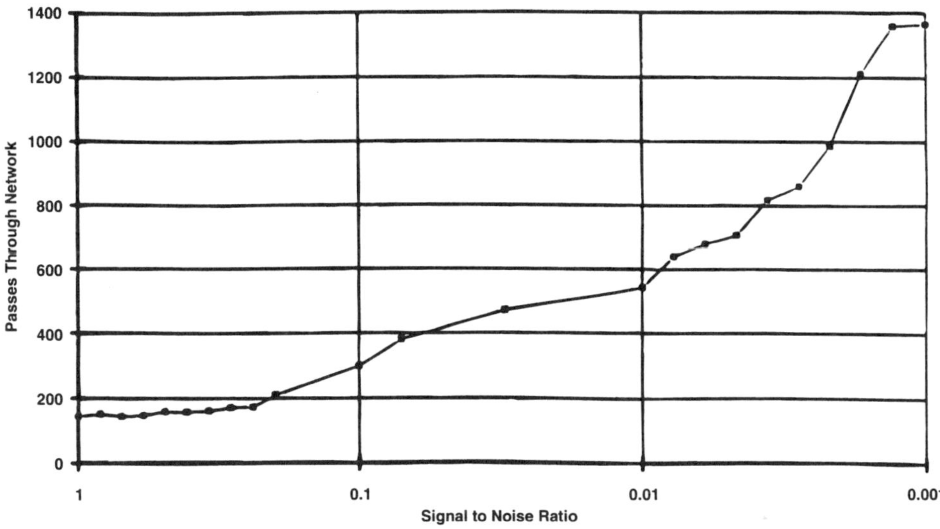

FIGURE 13 The learning response of a neural network with 64 input neurons, 8 neurons in the hidden layer, and one output neuron to a 60-Hz input signal and a pseudo random noise signal with a decreasing signal-to-noise ratio. (*Bioelectromagnetics Supplement*, Barnes, F.S., "Some Engineering Models for Interactions of Electric and Magnetic Fields with Biological Systems," 1, 67–85, 1992.)

accuracy as a function of the input signal-to-noise ratio. The results shown in Figure 13 show that the training time increased from about 200 runs to about 1400 runs as the signal-to-noise ratio decreased from 1 to 0.001. The way the noise is presented to this network during the training makes a difference. For example, if you want the network to separate 59 Hz from 60 Hz, it helps to tell the network that 59 Hz is noise. This computer network model is clearly too simple to describe a biological nervous system, but it may provide a clue to one way in which a collection of cells may be able to respond to weak externally applied electric fields in a way in which a single cell would not.

V. THERMAL EFFECTS

One important effect of current flow due to electric fields is heating. The power input to a given volume of material can be expressed by $P' = I^2R$ where I is the total current and R is the resistance of the sample. For many calculations, a more useful expression is given by the power per unit volume or $P = \sigma E^2 = \rho J^2$ where σ is the conductivity, E is the electric field intensity, ρ is the resistivity, and J is the current density. The temperature rise resulting from this heat input is determined by the thermal capacity of the volume and the mechanisms for carrying the heat energy away. Typically these thermal loss mechanisms include a combination of conduction and convection processes. For short current pulses, the heat dissipation is usually dominated by thermal conduction, and the basic equation for the rate of change of temperature is given by

$$\frac{\partial T}{\partial t} = \frac{P}{\rho' C_p} - \frac{T - T_0}{\tau_c} \tag{45}$$

where T is the temperature, T_0 is the initial temperature, t is time and p is the power supplied per unit volume, ρ' is the density of the material in Kg/m^3, C_p, is the specific heat under constant pressure, and τ_c is the thermal relaxation time.

If we consider a homogeneous sphere of radius a immersed in an infinite fluid, the thermal conductive relaxation time is approximately given by

$$\tau_C = \frac{a^2}{4\overline{K}} \tag{46}$$

where $\overline{K}$ is the thermal diffusivity and is measured in meters squared per second.[69] The thermal diffusivity is given by

$$\overline{K} = \frac{K'}{\rho' C_p} \tag{47}$$

where K' is the thermal conductivity in Cal/m s °C, ρ' is the material density in kg/m^3, and C_p is the thermal capacity in Cal/°C kg. If an applied current pulse is short compared to τ_c, the maximum temperature change is given by

$$\Delta T_{max} = \left(\frac{3}{2\pi e}\right)^{3/2} \frac{\overline{H}}{\rho' C_p a^2} \tag{48}$$

where $\overline{H}$ is the total input energy in calories and e is the base of natural logarithms.[69] For current inputs that are long compared to the thermal relaxation time τ_c, the peak temperature is determined by a balance between the input power and the dissipation processes controlled by conduction and convection. For many situations where the volume involved is a cubic millimeter or larger, the thermal time constant is controlled by the amount of blood flowing through the volume. In these cases, temperatures are often more easily measured than calculated since a complicated thermal and electrical boundary value problem would have to be solved analytically to calculate the temperature rise. This is particularly true since the viscosity η and other thermal and electrical parameters such as ρ, C_p, $\overline{K}$, etc. are functions of temperature. For example, C_p for an artificial bilipid membrane is shown in Figure 14.[70] Another example of the importance of change in temperature is the conductivity of saline,

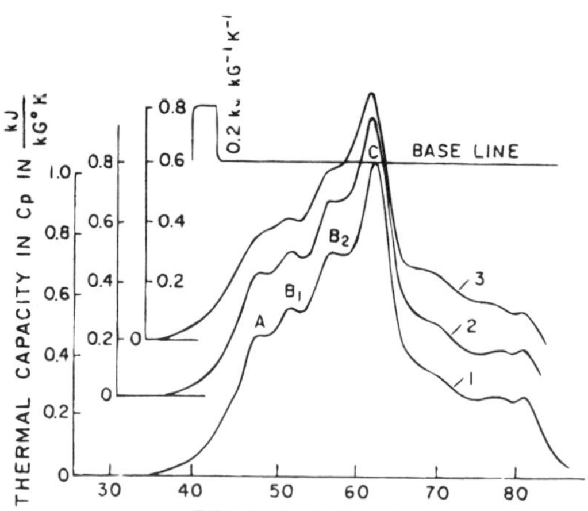

FIGURE 14 Differential changes in the heat capacity Cp erythrocyte membranes as a function of temperature in 5 mmol/*l* sodium phosphate with pH = 7.4 and a concentration of 5 mg protein per milliliter. The changes at A, B, B$_2$, and C correspond to changes in the structure of the membrane with temperature and are irreversible. Curve (1) intact membranes; (2) irradiated at 330 mHz for 5 min (SAR 9 m 2/g); (3) irradiated at 330 mHz for 30 min (SAR 9 mw/g). (Shuyrou, V.L., Zhodan, G.G. and Akornv, I.G., "Effects of 330 MHz Radiofrequency," private communication, 1984.)

$$\sigma \approx C_l \left[10^{\left(\frac{1}{T} + \alpha\right)\left(\frac{1}{b}\right)} \right] \times 10^{-4} \text{ S/m} \tag{49}$$

where C_l is the concentration of NaCl in milligram equivalents per liter, T is the absolute temperature, $\alpha \approx -6.23 \times 10^{-3}$ degrees^{-1}, and $b \approx -1.4 \times 10^{-3}$ degrees^{-1}.[71] In the range around 37.5°C, this means that a 5°C change in temperature corresponds to a little less than 9% change in conductivity.[71]

Changes in temperature are important, not only because they change transport properties such as viscosity η, mobility μ, and diffusion coefficient D, but also because they change chemical reaction rates. Typical biochemical reactions can be described by an equation of the form

$$\frac{dS}{dt} = -K'S \tag{50}$$

where S is the fraction of the material which has undergone the chemical reaction, t is the time, and K′ is the reaction rate.[72] K′ is often given by

$$K' = \frac{kT}{h} \exp\left(\frac{+\Delta H' - T\Delta S'}{R'T} \right) \tag{51}$$

where k is the Boltzmann constant, T is the absolute temperature, H′ is the free energy, S′ is the entropy, h is Planck's constant, and R′ is the gas constant. The significant feature is that the reaction rate K′ varies exponentially with temperature, and $\Delta H'$ and $T\Delta S'$ are large numbers. Thus, very small changes in temperature can lead to big changes in chemical reaction rates.

In addition to chemical reaction rate changes, there may be changes in the binding of proteins to cell membranes which lead to a shedding of proteins with a small increase in temperature. An exponential temperature dependence of the binding to membrane receptors is to be expected just as it is for chemical reactions.[73]

A rule of thumb which the author uses to estimate whether or not significant biological changes are likely is to see if ΔT is > 10°C for 10^{-6} sec, or 5°C for 1 sec, or 2°C for hours. If the ΔTs are larger, then they can be expected to lead to important changes in the biological system. Typical mammalian temperature regulatory systems will hold the internal body temperature constant to better than 0.5°C.

In addition to the magnitude of the temperature change, it can be shown that the rate of temperature rise, $\frac{dT}{dt}$, is important and can induce current to flow across membranes. Changes in the firing rate of pacemaker cells from the ganglion of *Aplysia* have been induced by total temperature changes of as little as 1/10°C when the rates of change are about 1°C/sec.[74] This change of the firing rate corresponds to the injection of approximately 1 nA to the cell. By taking the time derivative of the Nernst equation, which describes the passive equilibrium potential across a membrane for a single ion, it can be shown that a current proportional to the temperature derivative is to be expected, or

$$I = -qV_l'C_l\left(\frac{\phi}{\phi_T}\right)\left(\frac{\dot{\phi}}{\phi} - \frac{\dot{T}}{T}\right) \tag{52}$$

where q is the charge on the ion, V_l' is the volume of the cell, C_l is the concentration of ions inside the cell, ϕ is the resting potential, ϕ_T is given by $\phi_T = \frac{kT}{q}$, $\dot{\phi}$ is the derivative of the membrane potential with respect to time, T is the temperature, and $\dot{T}$ is the temperature derivative with respect to time.[75]

Bol'shakov and Alekseyev have observed similar changes in the firing rate of pacemaker cells taken from the large parietal ganglion of the central nervous system of *Limnea stagnalis.*[76] In their experiments they observed a slow increase in temperature, a degree per minute or slower, to increase the firing rate of pacemaker cell and a rapid increase in temperature, 0.1°C/sec or faster, to decrease or stop the firing. They ascribe these changes to changes in the Na^+ pump as the rapid temperature effect was completely blocked by adding ouabaine to the solution. In addition to the changes in the Na^+ currents, Ca^+ currents have been shown to be sensitive to rapid changes in temperature.[77] The rate of rise has also been shown to be significant in exciting a brain slice from a mouse with pulses of 10^{-3} sec and peak temperature rises of less than 0.5°C.[78]

Temperature rises also lead to thermal expansion, and rapid temperature rises lead to the generation of acoustic waves.[79] These acoustic waves in turn can affect stretch receptors in nerve cells and other tissue and thus generate a biological response which may be at a considerable distance from the electrical heating.[80]

To get an idea of the magnitudes of both heating (as described in equation 45) and the effect of the rate of rise (as given by equation 52), consider the case of liver tissue with $\sigma = 0.14$ S/m and a field strength in the tissue of 2×10^3 V/m. The rate of temperature rise is approximately 13°C/sec assuming no conduction or diffusion heat losses and the thermal capacity of water. For this high field, a significant temperature rise requires about 1/2 sec. However, the rate of rise has been shown to be significant in exciting a brain slice from a mouse with pulses of 10^{-3} sec.[81]

VI. NATURAL FIELDS AND MAN-MADE FIELDS

It is of interest to compare man-made with naturally occurring fields. First, we would like to know the approximate magnitudes of the fields that occur in nature outside man or the biological system of interest. Second, we would like to have values for the internal or physiological fields.

The natural electric fields at the surface of the earth have both DC and AC components.[82] One may think of the earth as a spherical capacitor where the surface is negatively charged with respect to an electrical conducting ionosphere which is about 50 km above the surface. This capacitor is being continuously charged by about 100 lightning strokes per second from thunderstorms worldwide. Since the atmosphere is a finite conductor, it also discharges with an RC time constant of about 18 sec. the result is an average electric field of about 130 V/m. This field is not uniform with height and typically falls off to 30 V/m at 1 km above the surface. The local values vary widely with temperature and humidity. In the Sahara during dust storms caused by winds in the dry season, a field of 1500 V/m has been measured with the polarity reversed from the normal. In thunderstorms, fields of up to 3000 V/m have been measured without lightning, and the polarity has been known to reverse in minutes. Storms as far as 50 km away have been shown to affect local fields.

The atmosphere is a relatively poor conductor and as such will suspend a significant number of charged ions, dust particles, etc. This helps to contribute to local field variations of 20 to 50% over the course of the day and is a normal characteristic of our environment. The level of natural AC fields in the atmosphere falls very rapidly from a DC value of about 130 V/m.[82] The average value of the vertical component of the electric field above 1 Hz has a typical value of 10^{-4} V/M $Hz^{1/2}$. However, this value fluctuates widely with the time of day, the season of the year, and location. Additionally, the Schumann resonances impose multiple cavity resonances on this spectrum with a periodicity of about 10 Hz. These resonances may be explained in terms of standing waves in a cavity formed by the earth and atmosphere. These very low levels of the natural fields are one of the reasons why electronic communications in the ELF band are useful for ships at sea and submarines. However, because of the very low level of the natural atmospheric fields at frequencies above a few Hertz, there is very little reason for biological organisms to develop natural protection against perturbations at these frequencies. It also means that biological systems could communicate internally at these frequencies using very low signal power levels and still maintain a good signal-to-noise ratio.

The signals generated within the body are the result of nerve firing and other cell activity. A typical nerve cell fires with an action potential of 50 to 100 mV and transmits a current pulse about 0.4 msec long.[83] The rise time for this current spike is approximately 0.1 msec, and the fall time is about 0.5 msec. Each pulse is followed by a refractory period which is typically on the order of 1 to 3 msec. The longitudinal fields along the exterior of a nerve cell membrane are estimated to have a maximum value of about 5×10^{-2} V/m during an action potential when the cell is surrounded by a relatively high conductivity fluid of 5 S/m.[83] If we look at these signals closely, it will be noted that the interspike interval along any given nerve cell fluctuates in time. Additionally, variations in the beat to beat intervals for the ECG are random or chaotic, and the period can vary up to 30%. This is frequently seen, particularly at slow heart rates.

In looking at the natural fields in the body, we have two concerns. The first is how large an external signal it takes to perturb the on-going natural signal which is being used to communicate or control some biological process.[84] The second is how much of the signal field typically leaks away from active nerve fibers or bundles to form a background noise environment for surrounding tissue and processes. Regarding the first of these questions, it is interesting to look on the microscopic level at the electrical noise, i.e., the fluctuations that occur fundamentally as a result of the electrical process itself.

The first of several sources of noise which is always present is blackbody radiation, or Johnson noise, which is given by

$$P_n = kTB \tag{53}$$

P_n is the noise power, k is the Boltzmann constant, T is the absolute temperature, and B is the bandwidth.[85–87] The voltage equivalent of this noise power which can be delivered to a matched load, or the mean squared voltage fluctuation $\overline{V_n^2}$ across a resistance R, is given by[86–87]

$$\overline{V_n^2} = 4kTBR \tag{54}$$

or

$$\overline{i_n^2} = \frac{4kTB}{R} \tag{55}$$

Johnson noise applies to systems at thermodynamic equilibrium. Living systems are not at thermodynamic equilibrium. Thus, the foregoing expressions must be applied with caution to only those portions of biological systems where thermodynamic equilibrium is a good approximation. In the case of lasers, the spontaneous emission noise associated with the nonequilibrium population inversion of the energy levels can be obtained from Planck's radiation law by defining a negative temperature which assumes a Boltzmann distribution of atoms with N_2 atoms in the excited energy level E_2 which is greater than the N_1 atoms in the energy level E_l such that $\frac{N_2}{N_1} = \exp - \left[\frac{E_2 - E_1}{kT}\right]$. In this case, the spontaneous emission noise $P_n = h\nu B$, where h is Planck's constant and ν is the frequency of the radiation corresponding to a transition from E_2 to E_1.[85] The noise implications of the nonequilibrium conditions of living systems have, to the author's knowledge, only begun to be studied. However, in those situations where the nonequilibrium characteristic may be described by an amplifier which can be modeled by a negative resistor or by energy storage in an inverted population distribution, the concept of a negative temperature may be a useful approach.

The second source of noise which is also present is the shot noise, which is given by

$$\bar{i}_n^2 = 2qI_{DC}B \tag{56}$$

where $\bar{i}_n^2$ is the mean-squared current fluctuation. This noise comes about because of the discreteness of the electronic charge q. It gives an AC fluctuation $\bar{i}_n^2$ which is proportional to the average value of the current I_{DC}. A third source of noise is (1/f) noise. This noise may be generated by many processes. However, a critical characteristic of models which lead to noise of this kind is that they require memory of about one state variable per decade of time over which the model is used to generate noise with a power density spectrum s(f) = (C/f), where C is a constant.[88] This means that we can expect to find this kind of noise for processes which evolve with time. 1/f noise is the dominant source of fluctuations at low frequencies in such diverse phenomena as transistors, quartz crystal oscillators, and the weather. It is also generated by the flow of ion currents through an orifice and thus is a fundamental part of the transport of current through channels in membranes.[87] Measurements of the noise voltage across a 10 μm hole in a 6 μm mylar film showed that for a wide range of ionic concentrations the voltage noise spectral density S(f) is given by

$$\frac{S_\phi(f)}{\phi^2} = \frac{a}{bnr^3f} \tag{57}$$

where b is a numerical geometric factor, n is the density of ions in the solution, r is the radius of the hole, a is a constant, and ϕ is the applied voltage. The data showed $2.5 < a < 40$ with a mean value of 10 for a wide range of solutions including HCl, KCl, $AgNO_3$, etc. with concentrations from 0.05 to 5 mol.

For natural membranes this noise has been shown to take on the form of

$$S_E(f) = \frac{C_E}{f^\alpha} \tag{58}$$

where $0.7 < \alpha < 1.2$ with a mean close to $\alpha = 1$. For the frog node of *Ranvier*, the noise is a function of the membrane voltage as shown in Figure 15.[91] The dominant source of this noise appears to be the K^+ current, and it has a minimum when the membrane is biased so that this K^+ current, is biased to zero.

To get an estimate of the size of these noise sources, let us consider a pacemaker cell from the abdominal ganglion of *Aplysia*. This cell fires 20-msec pulses at about 1 Hz/sec. It has a resting voltage of about 50 mV and a resistance R measured with a micro electrode between the inside of the cell and the surrounding solution of approximately 10^6 Ω. If we assume a system bandwidth of 100 Hz and T = 300 K, the Johnson noise voltage would be $\bar{V}_n \approx 3 \times 10^{-6}$ V. This gives a resting potential to noise $\bar{V}_n$ ratio of about 4×10^4. The peak current flow in these cells is estimated to be about 10^{-7} A, and thus the estimated shot

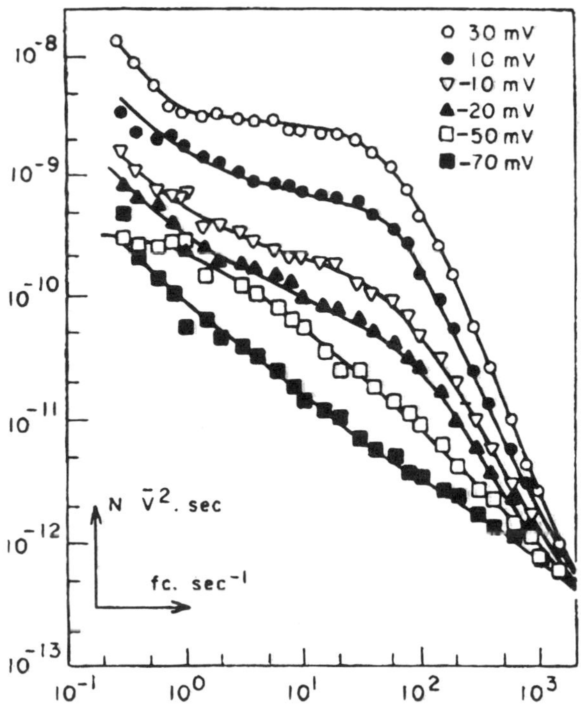

FIGURE 15 Voltage noise spectra of a frog node of Ranvier at different levels of membrane potential. (Sichenga, E., Verveen, A.A., "The Dependence of the 1/f Noise Intensity of the Diode of Ranvier on Membrane Potential," *Proceedings of the First European Biophysics Congress,* 5, Verlag Wiener Medizinischen Akademic, Vienna, Austria, 219, 1971.)

noise current is $\bar{i}_n \approx 2 \times 10^{-12}$A, and the ratio of the peak current to the noise current is about 2×10^4. We do not have available the value S(f) for the *Aplysia*. However, $\bar{v}_\phi = \sqrt{S(f)B}$, where B is the bandwidth. If it is assumed that the maximum value of the noise is the same as that of the frog node of *Ranvier,* then for a bandwidth of 1 Hz we get $\bar{v}_\phi = 1.4 \times 10^{-5}$ V at a center frequency of 1 Hz from the curve for -50 mV of Figure 15. This is about a factor of 10^3 greater than the Johnson noise. It is likely that (1/f) noise is the largest source of noise at the cell membranes for frequencies below 160 Hz.[89–90] It is to be noted that these fundamental sources of noise which are generated by random fluctuations in the position of ions and their transport through channels are spatially incoherent.[58]

In addition to the electrical noise generated by currents and voltages which are part of the single cell operation, electrical signals propagate through the body as the result of the incomplete confinement of electrical signals propagating in nerve cells. In some sense, these signals may be thought of as noise if they are not pertinent to the activity in that portion of the body through which they are propagating. If, on the other hand, they are used by tissue within the organism at some distance from the source, they must be thought of as signals. In the brain, the fraction of these signals which reach the scalp is called the EEG (electroencephalogram). The EEG is obtained by placing two or more contacting electrodes on the scalp and measuring the voltage between them. For electrodes placed 5 cm apart, the peak-to-peak voltages range up to 30 μV.[93] The author views this voltage as the integral of the vector sum of the leakage fields from the firing of the nerve cells in the brain between the two electrodes. Since there is a very large number of cells firing, most of the 50-mV signals from an isolated nerve are canceled by summing over many like cells firing at different times and by the attenuation caused by propagation through the lossy tissue. Estimates of surface potential gradients along a nerve fiber range from 3×10^{-4} to 5×10^{-2} V/m, and the corresponding current densities external to the nerve cells range from 5×10^{-2} to 4 A/m^2.[93–94] The EEG voltage has a strong periodic component (particularly during sleep) near 10 Hz, which is known as the alpha (α) wave. A peak amplitude of this component may be as large as 50 μV when measured at the surface of the scalp.

At the surface of the chest, a signal may be recorded between two electrodes known as the ECG or EKG (electrocardiogram). This signal results from the highly coordinated firing of the cells in the heart and has a definite wave shape which is closely related to the operation of the heart. The peaks of the so-called R wave in this signal may range up to 2.5 mV and are typically 0.5 to 1.5 mV, depending on the placement of the electrodes, the amount of body fat, etc. The pulse repetition rate is usually in the range of from 1 to 2 Hz, and the "QRS spike" of the typical cardiogram is 40 msec long. Again, the signal measured at the surface of the skin is the result of leakage from electrically active cells located at a distance. The estimated current density near the firing heart cell ranges up to 1 A/m^2.[95] In this case the shape of the signal reaching the skin is so closely related to the activity of the heart that it provides detailed information on heart function.

One result of electric discharges in the atmosphere, as well as natural ionizing radiation, is the creation of small positive and negative ions in the atmosphere. In clean country or mountain air, the typical ion density is about 10^{10}/m^3 with an average ion lifetime of a few minutes.[96] During a hot dry wind, positive ions created by the shearing forces can increase in concentration significantly. It has been shown that increases in the negative ion concentration reduce the amount of serotonin (5-HT) in mice and rabbits, possibly by accelerating the enzymatic oxidation process.[96] A similar result has been demonstrated in the oxidation of cytochrome *c*. Positive ions appear to block monoamine oxidase action, thus raising the concentration of free 5-HT.[96] Changes in 5-HT levels produce significant changes in the central nervous system, with high levels of positive ions raising the anxiety levels under stress. Other effects of increased positive ion concentration include a decrease in the survival rate of mice exposed to a measured dose of influenza virus, while an increase of negative ions reduced the mortality rate.[96]

The significance of these results is that it is relatively easy to change the ion concentration in air using high-voltage DC systems where a leakage current of 1 μA from a burr or other sharp point would correspond to the generation of about 10^{12} ions per second.

Relatively few high-voltage DC transmission lines are in use today for distribution of power. Because the shocks resulting from a short contact across a high DC voltage are so painful and obviously dangerous, these systems are nearly always shielded. Thus, one is rarely exposed to DC electric fields $> 10^3$ V/m. An additional feature of this exposure is that air is such a good insulator that the DC currents through the body in a noncontacting situation are very small, as explained in the Introduction. For example, 1000 V across a 1cm gap would yield a current density of approximately 10^{-7} A/m^2 flowing across the air gap. Thus the principal hazards from DC fields occur when parts of the body make contact with a conductor.

VII. MAN-MADE FIELDS IN TYPICAL LIVING ENVIRONMENT

Most of us are exposed to low-level man-made ELF fields on a continuous basis. We live a large part of our lives surrounded by a grid of wires that delivers the energy we need to power lights, electric motors, and many of the conveniences that make modern living possible. The majority of our exposure comes from fringing fields generated by the power distribution systems. In a typical suburban house, these background 60-Hz fields range from 1 to 20 V/m and may be considerably larger near appliances, where they may range up to 250 V/m.

Table 2 shows some typical values for appliances.[97] The corresponding current densities induced in the body for some of these fields are given in Table 3.[97] It is also interesting to note that fields on the order of 150 to 500 V/m have been measured in a 1/4-inch spacing between an electric blanket and the human body (measurements made by the author). An important characteristic of the power line fields which distinguish them from biological signals is the stability of the frequency at either 50 or 60 Hz.

The scaling from external fields to fields inside the body depends on the geometry and the electrical properties of the body. Quantitative calculations of the fields and induced current densities have been carried out for a variety of approximate models for man in different configurations.[99–101] These models all assume that the currents flow along the surface of the body and show the induced current densities to be inhomogeneous and geometry dependent for the application of uniform external fields. They are sufficiently good to estimate the total of induced currents in a standing human and to estimate the current distributions along the surface of the skin. (See Figure 16.) In a field of 10 kV/m at 60 Hz, the total current in a 1.7 m human is estimated to be about 150 μA.[103] Additionally, typical induced body currents in the legs are 15 A/kV/m of external field. The fields internally are generally reduced by about a factor of 10^{-7} from the externally applied field, and rarely exceed a factor of 10^{-4}.

Electric fields are also induced in the body by time-varying magnetic fields such that

$$\int_{\ell} \vec{E} \cdot d\vec{\ell} = \frac{\partial}{\partial t} \int_{S} \vec{B} \cdot d\vec{S} \tag{59}$$

where ℓ is the length of the closed path encircling the area S and t is the time. These fields lead to quite different current distributions and are much more likely to generate significant current densities in the interior of the body, as shown in Figure 17. Upon exposure to either a vertical or horizontal magnetic field of 10^{-7} T at 60 Hz, the maximum induced current density is approximately 4 $\mu A/m^2$.[103] The models are not yet sufficiently detailed to take

TABLE 2
Summary of Representative Electric Fields & Currents from Household Appliances

Appliance	A Field	B Field	Leakage Currents	Induced Currents
Electric blanket	250 V/m			
Hair dryer	40 V/m	10^{-3} to 2.5×10^{-3}T		
Electric train	60 V/m	10^{-6} to 10^{-5}T		
Food mixer	50 V/m	10^{-4} to 5×10^{-4}T		
ANSI standard fixed appliance			750μA	
ANSI standard cord connected			500μA	
Coffee mill			380μA	
Refrigerator			40μA	
Sewing machine			34μA	
Coffee maker			6μA	
Heating pad[a]				18μA
Electric blanket[a]				7 to 27μA

[a] Electric fields were measured 30cm from the appliance, magnetic fields at a few centimeters; induced current was measured through grounded arm, and leakage current through body to ground.

TABLE 3
Comparison of Body Current and Related Current Densities in Terms of Equivalent Electric Fields

Source	Current (μA)	Current Densities (10^{-2}A/m^2)	Electric Fields (k/Vm)
Sinusoidal waveforms			
Cord-connected household appliance	20 to 500	0.5 to 12[a]	1.5 to 38
Man in 8kV/m field	120	3[a]	8
Electric blanket	7 to 25	2 to 40[b]	0.5 to 1.7
Man in 0.16k V/m field	2.2	0.05[a]	0.16
Non-sinusoidal waveforms—medical devices			
Electric anesthesia (100 Hz square-wave)	10,000	71,000[c]	670
Pacemaker electrode in myocardium[d,e]	6,000	20,000	400
Pacemaker electrode implanted in abdomen[d,f]	6,000	300	400

[a] Passing through a 40cm^2 ankle; [b] 0.25inch from electric wire in blanket [c] Next to electrode;
[d] Peak pulse current $\sim 10^{-3}$sec duration repeated every 0.8sec; [e] 0.3cm^2 electrode area;
[f] 20cm^2 electrode area.

into account the complexities and inhomogeneous nature of the body including low resistance blood vessels, tissue, and high resistance bones and fat, etc. Thus the fields at particular cell membranes or organs cannot be accurately calculated in many cases of interest. Rough estimates can be made by using detailed models of the anatomy using numerical techniques (see Chapt. 9).

VIII. THE EFFECTS OF ELECTRIC FIELDS ON BIOLOGICAL SYSTEMS

Many of the effects of exposure to DC and low frequency fields on whole plants and animals are difficult to assess—particularly at low levels. This is fundamentally true for at least two reasons: biological systems are very complex, and they contain a great deal of negative feedback that tends to null out perturbations which would damage them or change their performance. For example, increased energy dissipation, which would raise man's internal temperature, is compensated for by increased blood flow and sweating, which reduces the measurable temperature changes to a small fraction of what one would calculate assuming a constant energy dissipation system. Thus, the fact that many experiments on the effects of electric fields have not shown positive changes in growth, reproduction, blood chemistry, etc.[104] does not mean that we are not disturbing the system. It is possible that we have perturbed the system only a little less than the ability of the body to compensate. However, we can get some guidance concerning the effects of electric fields on complex living systems by working up from the cell level to more complex structures. For example, the slime mold *Physarium polycephalum* exposed to 75-Hz fields of 0.2 μT and 0.7 V/m shows a lengthening of the mitotic cycle by as much as 20%.[105] This effect was very slow to develop (80 to 100 days at 63 Hz and 100 to 120 days at 75 Hz). Recovery to normal mitotic cycles occurs after 30 to 60 days. These are very slowly developing changes in growth characteristics.

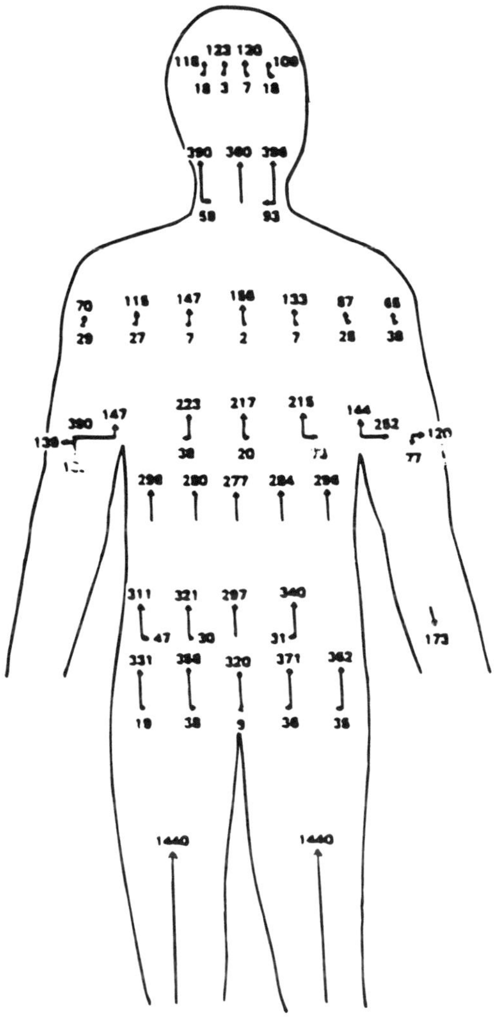

FIGURE 16 Current densities measured in the midfrontal plane of a saline model of a standing man exposed to a 60 Hz, 10 kV/m electric field. Induced r.m.s. current densities are given in units of nA/cm^2. The model was grounded with equal currents passing through both feet. The vertical and horizontal arrows represent the axial and radial components of the induced current densities, respectively. (© Williams & Wilkins, T.S. Tenforde and W.T. Kaune, "Interaction of Extremely Low Frequency Electric and Magnetic Fields with Humans," *Health Physics,* 53, 6, 1987.)

At the genetic level, it appears that electric and magnetic fields act like other physiological stresses.[105] Exposure to electric fields at the level of about 3×10^{-3} V/m applied to HL-60 cells at 60 Hz for 20 minutes has been reported to increase the levels for oncogenes such as C-myc and hsp 70 transcripts. The corresponding current density was about 10^{-2} A/m^2. The maximum sensitivity occurred at 45 Hz and was about 4 times greater than at 60 Hz. The same kind of results were obtained with a magnetically induced electric field in a toroidal tube at a level of 6 mV/m with a corresponding stray magnetic field level between 1 and 2×10^{-7} T at the center of the toroid when the dominant part of the magnetic field was confined to an iron core inside the toroid. Direct exposure to time-varying magnetic fields at 8×10^{-6} T and induced electric fields of 1.1×10^{-5} V/m in a pair of Helmohotz coils yield similar results. All these results are very similar to those obtained with a heat

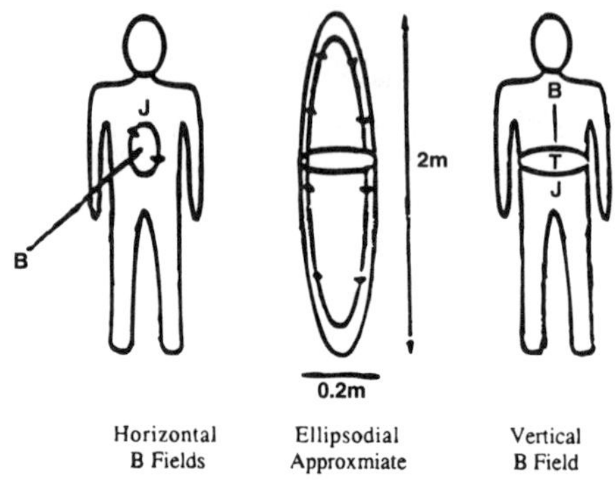

FIGURE 17 An elliptical approximation for the induced currents in man. (© IEEE, 1990, F.S. Barnes, "The Effects of Time Varying Magnetic Fields on Biological Effects," *IEEE Transactions on Magnetics,* 26, 5.)

shock with $\Delta T = 13°C$ raising the cell culture from 20°C to approximately 33°C as measured by an increase in the fractional distribution of proteins synthesized with a molecular weight at 30 KD or by the fraction of material with an isoelectric point near 6.[107] This similarity of the electric and field responses to heat shock response indicates that stimulation of cells with these fields has many similarities to other physiological stresses and that we should expect to find related responses in the immune system and other parts of the biological systems that are known to be affected by stress. Some preliminary work by Korneva indicates changes in the leukocyte count and their activity for the exposure of mice at 50 Hz and 2×10^{-5} T.[108] In this case the calculated E fields predicted peak currents of 10^{-5} A/m^2.

At higher levels of exposure a variety of effects have been shown. An examination of nerves taken from rats which had been exposed to fields of 100 kV/m for 30 days at 60 Hz has shown increases in the C-fiber conduction velocity, a decreased rate of fatigue, and shifts in the strength duration curve which suggest that the exposure had increased neural excitability.[109]

In another study, experiments on dogs showed a significant change in blood chemistry for exposures at 25 kV/m and 50 Hz for 8 hr/day over a period of 6 weeks.[110] These changes included an increase in reticulocytes, a decrease in hemoglobin concentration, an increase in serum protein, a decrease in neutrophils, and an increase in lymphocytes. In separate experiments on rats, the growth rate was slowed upon exposure to fields of both 25 and 100 kV/m. These rats also showed an increase in neutrophils and a decrease in lymphocytes.

In experiments carried out on honeybees whose hives were located under a 765-kV transmission line and exposed to fields of 0.6 to 4.5 kV/m, important deleterious effects were shown, including a 64% increase in the over-winter mortality rate by comparison to shielded hives in the same area.[111] In the same study, those hives with exposures ranging from 1.1 to 8.3 kV/m showed almost no honey production over the summer and showed a loss of bee weight. Furthermore, all hives exposed at this level failed to survive the following winter. More recent experiments show that electrical shocks on entering the hive were the source of these effects. This whole group of experiments showed the ability of large kilovolt-per-meter ELF electric fields to disrupt biological systems when applied for extended periods of time.

At levels as low as 2.5 V/m, 10-Hz signals have been reported to reduce the Circadian rhythm in man by more than 0.6 hr in a shielded environment. These signals were also shown to prevent the internal desynchronization and to resynchronize the length of sleep and activity periods.[112] Experiments on monkeys trained to estimate the passage of a time interval of 5 sec showed that exposure to 7-Hz fields at a level above 10 V/m caused the monkeys to underestimate the delay.[113] These results coupled with the Ca^{++} efflux experiments suggest the ability of very weak fields at frequencies close to natural biological system frequencies to influence behavior.[114]

In all of the above experiments, exposure to ELF electric fields (and in one case simultaneous exposure to magnetic fields) produced currents inside various organisms which did not make electrical contact with any external source of power. The fields inside the organisms were therefore, in general, much smaller than the external fields, as explained in the Introduction to this book. When direct electrical contact with a conductor is established, the situation becomes very different, because the large series resistance of air no longer limits current flow through the biological system. In view of the obvious safety problems associated with shocks and burns from contact with 60-Hz lines, a number of studies have been done to set standards for insulation. For direct contact with a metal wire connected to a 60-Hz generator, the minimum detectable current by humans is about 0.1 mA.[115] The average value at 60 Hz for just perceptible current is 1.1 mA for men and 0.7 mA for women. Dry skin has a resistance per unit area of about 10 to 30 Ω/m^2. About 100 V is needed for a detectable sensation with a contact resistance of 105 Ω. However, if the skin is moist, the resistance drops to about 0.1 Ω/m^2, and for a break in the skin, a typical body impedance for current from one major extremity to another is between 100 and 500 Ω. At voltages higher than 200 V, the skin resistance breaks down and the value of 500 Ω is often assumed for calculating the current flow through the body. Currents strong enough to freeze muscle, or "let go currents," range upward from 5 mA and average 16 mA and 10.5 mA for women. The maximum safe current for men is taken to be 9 mA and for women 6 mA. At still higher currents of 3 to 6 A, violent muscle contractions occur which may include the arrest of both heart function and breathing.

For very short, high voltage pulses, the energy threshold for burns is estimated at 50 J. Electrical burns are estimated to constitute about 4% of the burns in the U.S. (About 2400 cases were seen in hospital emergency rooms in 1977.)[116] The burns occurred in the areas where the current density is highest. Current densities tend to be the highest in regions of small cross-section which are often adjacent to one of the contacts. The relatively high resistance of bone tends to exclude the current and direct it into lower resistance arteries, nerves, and tissues. Tissue resistivities are functions of both temperature and current density. In general, the resistivity increases with temperature for muscle, nerves, and arteries and decreases for fat. Burn damage tends to occur first along blood vessels and nerves.

As a result of this experimental background, the American National Standards Institute (ANSI) in November 1970 issued a standard of 0.5 mA allowable leakage current for two-wire portable appliances and 0.75 mA for heavy cord connected major appliances such as stoves, refrigerators, etc. This factor of 10 to 20 below the "let go current" seems adequate for normal use. However, it is clearly above what may be a safe operating level in a hospital situation where the leakage current might couple into a catheter supplying fluids to a patient. For patients with pacemakers or catheters in the vicinity of the heart, currents which are picked up by pacemaker leads or induced into a catheter fluid in the range of 20 to 800 μA may cause ventricular fibrillation. Thus, the standards for hospital equipment should be at least 100 times lower than for the general population.[115]

IX. MEDICAL APPLICATIONS

The use of electricity to treat a great variety of ills dates back at least 2000 years. In early times, the electric shock was obtained from electric eels or torpedo fish and was used to treat such things as gout and hemorrhoids.[117] Until the 1920s, various kinds of electric shocks were reputed to have all the curative powers of a patent medicine and were used to treat a great variety of ills. Even Benjamin Franklin is reputed to have cured a girl of convulsions by giving her four shocks, morning and evening, for a period of about 3 months.

The first use in modern medicine seems to have been in pacemakers for defective hearts in 1960. The pacemaker electrodes are placed in contact with the heart muscle and pulses of 0.1 to 10 mA, a few milliseconds long, are used to synchronize the firing of the cells. The original pacemaker operated at a fixed rate; however, more recent devices can supply pulses on demand.[112] A second use is for defibrillation or to synchronize the firing of the heart cells by overwhelming the natural signals which have gotten out of sequence with a large current pulse. Defibrillation pulses may be as large as 20 A at 6000 V and last for a few milliseconds. Typical energies for adults are 200 to 400 J.[119–120] Still another use of external electrical signals includes the stimulation of muscles in which the control nerves communicating with the brain have been severed. Typical current pulses for stimulating muscles in the leg range from 1 to 10 mA and, in some cases, as much as 20 mA depending on location and the efficiency of coupling from the electrode to the tissue to be activated.[117] These pulses can be from 10 to 60 μsec long, and the rate of firing determines the strength of the muscle response. Typical values are from 10 to 35 pulses per second. Current experiments include an attempt to use computer sequencing of muscle stimulation to permit walking by quadriplegics.

Still another example of the application of electrical stimulation has been the treatment of epilepsy.[121] In one study, rectangular millisecond pulses were coupled at a 10-Hz rate through electrodes placed over the anterior lobe of the cerebellum and automatically switched between two areas at 8-min intervals. After an exposure of 24 hr/day for 30 days at a level just below that which would induce headaches, a significant reduction in the frequency of epileptic seizures was shown in five out of six chronic patients. Seizure occurrence rates were reduced by factors of from 2 to 10 and, in one case, from 100/month to none.

The current level was not reported for these treatments, but it seems likely that the peak electrode currents were a few milliamps. A supporting study on cat brains showed that the effect of the current pulses was first to stimulate and then to suppress the firing rate of cerebral neurons. Over the period of days involved in this study, it seems likely that there was a change in the level of neurotransmitter agents as a result of electrical stimulation. A related use of electricity is electroconvulsive therapy.[122] In this treatment, electrodes may be applied to either the temple or forward part of the scalp and voltages from 70 to 130 V at 60 Hz are applied for from 0.1 to 0.5 sec. The corresponding currents are between 0.2 and 0.6 A. The objective of the current pulses is to induce convulsions which in turn release chemicals that modify brain behavior. The treatment has been most successful in relieving severe depression where the potential for suicide is high. However, because of the possible side effects which may induce partial loss of memory and brain damage, this form of treatment is controversial.[123]

It is clear that we still have much to learn about how and where to apply electric signals to the brain. We have still not reached the stage of technology where we can separately contact individual nerve fibers in a bundle and supply pulses to them in a controlled fashion over extended periods of time. When we succeed, there will be many exciting applications. For example, accident victims may be enabled to regain the use of their limbs.

Another important application of electric fields is the modification of cells. A technique developed by Zimmermann, et al.[124–128] uses dielectrophoresis, or the forces exerted on an induced dipole moment of the cell in a large electric field gradient, to line up the cells into "pearl chains" to be fused. A theory worked out by Schwan, and also by Hu, shows that particles with a dielectric permittivity ϵ_1 immersed in a fluid of dielectric permittivity ϵ_2 take on the lowest energy configuration in an alternating electric field.[129–130] At low frequencies, this means that the cells form long chains upon the application of field strengths of 1 to 2×10^4 V/m. Depending on the cell size and the suspending fluid, the optimum frequency is usually in the range of 10 kHz to 100 MHz; however, we have also successfully used frequencies from 60 Hz to 1 kHz to move bacteria.

X. THE EFFECTS OF ELECTRIC FIELDS ON GROWTH

It has been shown that electric fields can be used either to inhibit or to stimulate growth under appropriate conditions. This is currently a field of active research.[131–132] Studies of wound healing, for example, have shown that application of a negatively biased electrode to an ulcer suppressed both healing and infection, while a positive bias stimulated growth. Thus, a method of treatment has evolved in which the wound is first biased negatively until it can be considered clinically noninfected, and then the potential is reversed to stimulate growth. The net result for a set of 14 ulcers was about a fourfold increase in the healing rate. The current levels for these experiments were in the range of 200 μA to 1 mA.[15] The processes for growth occur over extended periods of time and typically involve currents which are several orders of magnitude smaller than those used to stimulate nerve or muscle cells.

Naturally growing plant cells typically show an inward flowing current to the growth region of tens of picoamps and current densities in the range 0.01 to 1 A/m^2. In lily pollen grain, a relatively steady transcellular current begins to flow hours before germination or the initiation of visible growth.[133] The interior of the growth area is negatively biased, which implies an inward flow of K^+ and other positive ions. In the case of barley roots, the dominant current has been shown to be H^+.[134] This was shown by raising the pH from 5.5 to 7.1 and by insensitivity to changes in Na^+, K^+, Ca^{++} and Cl^- concentrations. On top of the gross inward current are superimposed micro current loops through each growing root hair. These micro currents enter the growing root hair tip with current densities of about 2×10^{-2} A/m^2 and correlate well with the tip growth. Nongrowing hairs did not show the inward currents. The current return path for both the macro- and micro currents includes a pathway through the xylem and possibly through the epidermal layer between the zones of elongation and the root hair zone. There are also apparently alternating bands of alkalinity and acidity near growing barley roots with strong steady currents entering the alkaline zones. Peak current densities are about 8×10^{-2} A/m^2.[134] In *Nitella* cells, the elongation or growth occurs in the acidic region or regions of current exit.[134] Growth in these regions was stimulated by external acidification and inhibited by alkalinization. This pattern is contrary to the general rule of cell growth at current leaks or points of current entry.

Ca^{++} currents appear to precede visible tip growth in various plant cells. There is direct evidence with radioactive $^{45}Ca^{++}$ for leaks in both fucoid eggs and in lily pollen.[134] In developing plants of the water mold *Blastocladiella*, Ca^{++} tip currents of 10^{-2} A/m^2 have been observed.[134] These currents are blocked by La^3+, and Ca^{++} has been shown to be both necessary and sufficient to maintain the current. The La^{3+} also blocks spore growth.

In animal egg cells between fertilization and the first cleavage, steady currents are found to enter at the animal pole and leave at the vegetal pole. In frog and sea urchin eggs, the current first enters the cell near the area where the cell eventually divides. Shortly afterwards, it reverses to flow out of the cleavage region. These currents may be accompanied by voltage drops across the interior of the cell of up to 10 mV. In the case of root cells, potential differences of up to 60 mV have been measured along the surface of the cell corresponding to currents of 0.1 to 0.4 μA and current densities of 0.1 A/m^2.

On a larger scale, strong electrical currents have been shown to flow out of the primitive streak of growing chick embryos.[135] Current densities as high as 1 A/m^2 leave the streak and return over a broader area. A current of about 10^{-4} A flows out of the streak. The component of the current flowing along the streak reverses direction about 2×10^{-4} m behind Hensen's node with current flowing away from this epicenter in all directions. Figures

18 and 19 illustrate these currents. These currents are believed to be strong enough to help control the organization of the embryo growth.

The application of external fields to growing cells can modify both the growth rate and the orientation of the growth. Patel and Poo have shown that fields in the range of 10^{-3} to 10^{-1} V/m have a large effect on the growth of neurite from single, dissociated *Xenopus* neurons in culture.[136] Upon exposure to fields of 2.5×10^{-2} V/m for 6 hours, the growth rate toward the cathode becomes approximately double that toward the anode. Additional neurites which were initially growing at right angles to the field bend to grow toward the cathode. These effects are reversible for exposures of up to at least 2 hours, with the asymmetry in the growth disappearing several hours after the field is turned off. No asymmetry in the growth is seen with application of alternating fields. The asymmetry in growth seems to be associated with the concentration of receptor for concanavalin A (Con A) under the influence of the field toward the cathode and can be blocked by binding tetramethylrodamine (R-Con A) to the neurons prior to the application of the electric field. This binding of R-Con A presumably cross-links and immobilizes the Con A receptors on the cell surface. The overall implication is that the underlying mechanism for the field-induced orientation of neurite growth toward the cathode is the influence of the electric field upon the distribution of growth-controlling glycoproteins.

A similar result of directed nerve growth toward the cathode and suppression on the anode side is reported for explants of embryonic chick dorsal root ganglia exposed to electric fields of 50 V/m for 3 to 29 hours.[133]

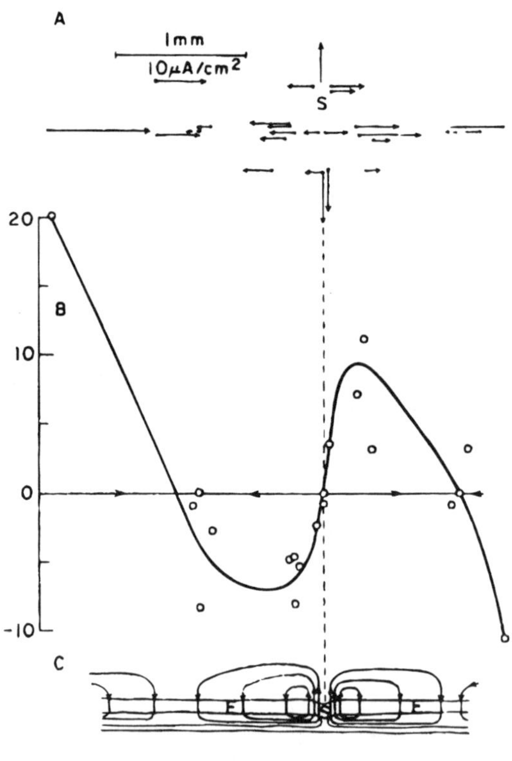

FIGURE 18 The pattern of current that flows across the streak of the embryo, measured in a horizontal plane 0.2 mm above the epiblast. (A) Map of a scan across the middle of the streak together with all measurement made nearby. (B) Graph of the main scan. As the arrowheads on the abscissa indicate, positive values indicate current going right and negative ones show current going left. (C) Diagram of inferred current pattern in a cross-section during the streak. The thickness of the embryo is exaggerated. (Jaffee, L.F., "Strong Electrical Currents Leave the Primitive Streak of Chick Embryos," *Science*, 206, 569, 1979.)

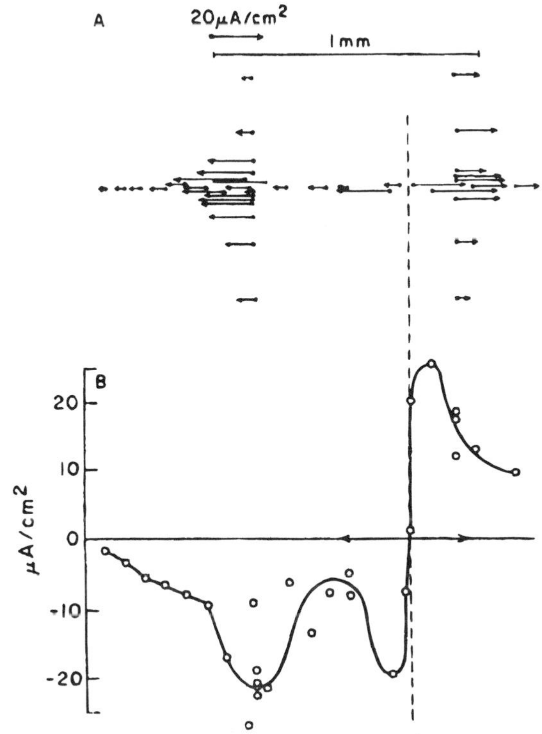

FIGURE 19 The pattern of current that flows along the streak of the embryo, measured in a plane 0.2 mm above the epiblast. (A) Map of a scan along the streak together with all measurements made nearby; (B) graph of the main scan. Positive values indicate current going right (toward the front end of the embryo) and negative ones show current going left. The dashed line goes through the central reversal point, or epicenter. (Jaffee, L.F., "Strong Electrical Currents Leave the Primitive Streak of Chick Embryos," *Science*, 206, 569, 1979.)

The work on nerve regrowth also includes a study of enhanced spinal cord regeneration in lamprey when electric fields of approximately 10 V/m are imposed across the neural break.[137] With a total delivered current of about 10 μA for a period of 5 or 6 days and current densities of 1 A/m^2, about 73% of the spinal cords would propagate action potentials in both directions across the lesions. This compared with 23% for the untreated controls. Additionally, microscopic examination showed twice as many swollen tips and tips with neurites for the spinal cord breaks exposed to the electric fields as those which were left to heal on their own.

The mechanism by which the regeneration takes place is not understood, but it is to be noted that the imposed current is in the opposite direction to the natural injury current. The injury current density starts out at about 5 A/m^2 and drops to 4 or 5 $\times$ 10^{-2} A/m^{-2} over the first 36 hours and continues at about this level for at least 4 days.

In another example, Yamamoto has shown that unfertilized eggs of the medaka fish can be activated by a field of 200 V/m applied for 20 to 60 sec.[138] Activation occurs at either electrode and twice as rapidly at the anode. This field should produce a voltage drop of 360 mV across the 1.2-mm egg. It is likely that the nonlinear characteristics of the cell membrane lead to a hyperpolarization of the cell and a large influx of Ca^{++}, which is known to be a central event in the activation of these eggs.[133]

In a second example, when the sign is opposite to that for the *Xenopus* neurites, eggs from the algae *Pelvetia* normally tend to grow on the side nearest a positive electrode under the influence of fields of 30 to 60 V/m, corresponding to a voltage drop across the cell of

3 to 6 mV.[133] These cells also grow toward the highest concentration in a Ca^{++} gradient, and it is likely that externally applied fields lead to an increased flux of Ca^{++} ions.

In summary, there are strong natural ionic currents associated with cell growth, and there is evidence that these currents may be driven by a separation of the leaks and pumps within the cell membranes.[131] The voltage gradients (2.5 V/m) developed by these currents in a semi-infinite surrounding medium would seem to be too small to be important in organizing the cell growth. However, when these currents are confined to narrow intercellular spaces, voltage drops from 0.1 to several millivolt may be generated over the length of a cell, and these voltages have been shown to be large enough to rearrange channels in membranes and thus affect cell growth. It is also likely that there is positive feedback so that the currents grow in time up to some limiting value, and current generated by one cell, contributes to inducing the current in adjacent cells. The evidence for the effects of naturally generated currents on cell growth and patterning has been reviewed in detail by Jaffe.[139–140]

XI. DISCUSSION AND SUMMARY

In this review, we have examined some of the mechanisms by which time-varying or slowly pulsed electric fields affect biological systems. An attempt has been made to list a few typical values of electric field strength and current densities that are known to affect transport properties so that the reader can make his or her own estimates of the probability that a given exposure will be significant. In very rough terms, both the medical and safety literature indicate threshold levels of concern for total currents leaving and entering the body above a few milliamperes or current densities of 10^{-1} A/m^2, and electric field strengths inside the body in excess of several volts per meter for short periods of time, on the order of seconds. For longer exposures to coherent excitations, the level of concern drops to 10^{-2} A/m^2 and about 10 V/m or less. Our understanding of the details of how changes in electric fields affect the transport of various ions and molecules across membranes and along their surfaces, and of the various feedback systems that control their biological function is still incomplete. Thus, it becomes very difficult to set safety standards. Standards which are too restrictive will result in very large costs for unnecessary insulation and limit the use of electrical instruments and appliances. Standards set too high may permit unnecessary biological damage. At the present time, our understanding is most incomplete with respect to the effects of long-term, low-level fields. This is not surprising, since the necessary experiments are the most difficult to do and the most subject to uncontrolled changes in the biological subject and the environment.

ACKNOWLEDGMENTS

The author wishes to express appreciation to Mikhail Zhadin, Maria Stuchly, Ross Adey, Mike Marron, Elliot Postow, and Charles Polk for their many helpful comments and suggestions. He also wishes to express his appreciation to the Office of Naval Research under Contract N00014-81-K-0387 and the University of Colorado for financial support of his work in this area.

REFERENCES

1. **Barnes, F. S., Hu, C. L.,** Nonlinear interactions of electromagnetic waves with the biological materials, in *Nonlinear Electromagnetics*, Uslenghi, P.L.L., Ed., Academic Press, New York, 1980, 391–426.

2. **Singer, S. J., Nicolson, G. L.,** The fluid mosaic model of the structure of cell membranes, *Science,* 175, 720–731, 1972.
3. **Poo, M.,** In situ electrophoresis or membrane components, in *Annual Review of Biophysics and Bioengineering,* Vol. 10, Mullins. L. J., Ed., Academic Press, New York, 1981, 245–276.
4. **Pohl, H. A.,** Dielectrophoresis, the behavior of neutral matter, in *Nonuniform Electric Fields,* Cambridge University Press, Cambridge, 1978.
5. **Bier, M., Ed.,** *Electrophoresis: Theory, Methods, and Applications,* Vol. 2, Academic Press, New York, 1967.
6. **Rand, E. P.,** Interacting phospholipids bilayers: measured forces and induced structural changes, in *Annual Review of Biophysics and Bioengineering,* Vol. 10, Mullins, L. J., Ed., Academic Press, New York, 1981, 277–314.
7. **Pethig, R.,** *Dielectric and Electronic Properties of Biological Materials,* John Wiley & Sons, New York, 1979, 184.
8. **Parsegian, J. L.,** Long range physical forces in the biological milieu, in *Annual Review of Biophysics and Bioengineering,* Mullins, L. J., Ed., Academic Press, New York, 1973, 221.
9. **Lifshitz, E. M.,** *Zh. Eksp. Teor. Fiz.,* 29.95, 1955: *Sov. Phys. JETP,* 2, 73, 1956.
10. **Tasao, Y. T., Evans, D. F., Wennerstrom, H.,** Long-range attractive force between hydrophobic surfaces observed by atomic force microscopy, *Science,* 262, 1993.
11. **Barnes, F. S., Ginley, R., Shulls, W.,** AC electric field effects on bacteria, presented at Third Annual Conference, Bioelectromagnetics Society (BEMS), Washington, D.C., 1981.
12. **Zumdahl, S. S.,** *Chemistry,* 2nd ed., D.C. Health Co.:Lexington, Massachusetts, USA, 546, 1989.
13. **Atkins, P. W.,** *Physical Chemistry,* 4th ed., W. H. Freeman Co.: New York, New York, USA, Chapter 26, 1990.
14. **McCaig, C. D.,** Spinal neurite reabsorption and regrowth *in vitro* depend on the polarity of an applied electric field, *Development,* 100, 31–41, 1987.
15. **Nuccitelli, R.,** Physiological electric fields can influence cell motility, growth, and polarity, *Adv. in Cell Biol.,* 2, 213–233, 1988.
16. **Erickson, C. A., Nuccitelli, R.,** Embryonic fibroblast motility and orientation can be influenced by physiological electric fields, *J. of Cell Biol.,* 98, 296–307, 1984.
17. **Seto, Y. J., Hsieh, S. T.,** Electromagnetic induced kinetic effects on charged substrates in localized enzyme systems, *Biotechnology and Bioengineering,* XVIII, 813–837, 1976.
18. **Muraji, M., Nishimura, M., Tatebe, W., Fujii, T.,** Effect of Alternating Magnetic Field on the growth of the primary root of corn, *IEEE Trans. on Magnetics,* 28, 4, 1996–2000, 1992.
19. **MacGregor, R. J., Lewis, E. R.,** *Neural Modelling,* Plenum Press, New York, 1977.
20. **Cooper, M. S., Miller, J. P., Fraser, S. E.,** Electropheretic repatterning of charged cytoplasmic molecules within tissues coupled by gap junctions by externally applied electric fields, *Dev. Biol.,* 132, 170–188, 1989.
21. **Epstein, B. R. and Foster, K. R.,** Anisotropy in the dielectric properties of skeletal muscle, *Med. Biol. Eng. Comp.,* 21, 1983.
22. **McCleave, J. D., Romell, A. S.,** Cathcart, Weak electric and magnetic fields in fish orientation, in *Orientation: Sensory Basis,* 188, Adler, H. E., Ed., New York Academy of Sciences, 1971, 270.
23. **Kalmijn, A. D.,** Detection of weak electric fields, in *Sensory Biology of Aquatic Animals,* Atema, E.J., et.al., Eds., Springer, Verlag, Berlin, Germany, 1987, Chapter 6.
24. **Ransom, B. P., Eyring, H.,** Membrane permeability and electrical potential, in *Ion Transport Across Membranes,* Clarke, H. J., Nachmansohn, D., Eds., Academic Press, New York, 1954, 103.
25. **Tsong, T. Y., Liu, D. S., Chauvin, R.,** Electroconformational coupling (ECC): An electric field induced enzyme oscillation for cellular energy and signal transductions, *Bioelectrochemistry and Bioenergetics,* 21, 319–331, 1989. Elsevier, Sequoia, S.A., Lausanne, Switzerland, A section of *J. Electrical. Chem. and Constituting,* Vol. 275, 1985.
26. **Liu, D. S., Astumian, R. D., Tsong, T. Y.,** Activation of Na^+ and K^+ Pumping Modes of (Na, K)-ATPase by an Oscillating Electric Field, *J. of Biological Chemistry,* 265, 13, 7260–7267, 1990.
27. **Horn, L. W.,** A novel method of the observation of membrane transporter dynamics, *Biophys. J.,* 64, 281–289, 1993
28. **Triffet, T., Green, H. S.,** Information and energy flow in a simple nervous system, *J. Theor. Biol.,* 86, 3, 1980.
29. **Vaccaro, S. R., Green, H. S.,** Ionic processes in excitable membranes, *J. Theor. Biol.,* 81, 771, 1979.
30. **Eckert, R., Ewald, D.,** Residual calcium ions depress activation of calcium-dependent current, *Science,* 216, 730, 1982.
31. **Baumann, G., Easton, G.,** Modeling state-dependent sodium conductance data by memoryless random process, *Math Biosci.,* 60, 265, 1982.
32. **Schauf, C. L., Baumann, G.,** Experimental evidence consistent with aggregation kinetics in sodium current of myxicola giant axons, *Biophys. J., 35,* 707, 1981.
33. **Baumann, G., Eastman, G. S.,** Charge immobilization linked to inactivation in the aggregation model of channel gating, *J. Theor. Biol.,* 99, 249, 1982.

34. **Poo, M.**, Rapid lateral diffusion of functional ACh receptors in embryonic muscle cell membrane, *Nature,* 295, 332, 1982.
35. **Poo, M. M., Poo, W. J., Lam, J. W.**, Lateral electrophoresis and diffusion of concanavalin A receptors in the membrane of embryonic muscle cell, *J. Cell Biol.,* 76, 483, 1978.
36. **Lee, K. Y. C., Klingler, J. B., McConnell, H. M.**, Electric field-induced concentration gradients in lipid monolayers, *Science,* 263, 655, 1994.
37. **Hayashi, H., Fishman, H.**, Inward rectifier K^+ channel from analysis of the complex conductance of aplysia neuronal membrane, *Biophys. J.,* 53, 747–757, 1988.
38. **Barnes, F. S., Hu, C. L.**, Model for some non thermal effects of radio and microwave fields on biological membranes, *IEEE Trans. Microwave Theory and Technology,* 25, 742, 1977.
39. **Montaigne, K., Pickard, W. F.**, Offset of the vascular potential of characean cells in response to electromagnetic radiation over the range of 250Hz to 250 kHz, *Bioelectromagnetics,* 5, 31, 1984.
40. **Pickard, W. F., Rosenbaum, F. J.**, Biological effects of microwaves at membrane level: Two possible athermal electrophysiological mechanisms and a proposed experimental test, *Math. Biosci.,* 39, 235, 1978.
41. **Pickard, W. F., Barsoum, Y. H.**, Radio-frequency bioeffects at the membrane level: Separation of thermal and athermal contributions in the Characeae, *J. Membrane Biol.,* 61, 39, 1981.
42. **Kalkwarf, D. R., Frasco, D. L., Brattain, W. H.**, Current rectification and action potentials across thin lipid membranes, *Physical Principles of Biological Membranes,* Snell, F., Wolken, J., Iverson, G., Lam, J., Eds., Gordon & Breach, New York, 1970.
43. **Franceschetti, G., Pinto, I.**, Cell membrane nonlinear response to applied electromagnetic field, *IEEE Transactions on Microwave Theory and Technology,* 32, 7, 1984.
44. **Casaleggio, A., Marconi, L., Morgavi, G., Ridella, S., Rolando, C.**, Current flow in a cell, with a nonlinear membrane, stimulated by an electric field, personal communication, 1985.
45. **Cain, C. A.**, Biological effects of oscillating electric fields: Role of voltage sensitive ion channels, *Bioelectromagnetics,* 2, 23, 1981.
46. **Bisceglia, Pinto, I.**, Volterra series solution of Hodgkin-Huxley equation, *Bioelectromagnetics,* personal communication, 1984.
47. **Wachtel, H.**, Firing-pattern changes and transmembrane currents produced by low frequency fields in pacemaker neurons, in *Proc. 18th Annu. Hansford Life Sci. Symp.,* Technical Information Center, U.S. Department of Energy, Richland, Washington, 132, 1978.
48. **Adler, R.**, A study of locking phenomena in oscillators, *Proc.,* Institute of Radio Engineering (IRE), 34, 351, 1946.
49. **Barnes F. S., Smoller, A, Sheppard, A.**, Injection locking of pacemaker cells, *Proc. of the 9th Annu. Conf.,* IEEE Engineering in Medicine and Biology Society, 1986.
50. **Fohlmeister, J. F., Adelman, W. J., Jr., Poppele, R. E.**, Excitation properties of the squid axon membrane and model systems with current stimulation, *Biophys J.,* 30, 79–98.
51. **Kroupa, V. F.**, *Frequency Synthesis,* John Wiley & Sons, New York, 1973.
52. **Barnes, F. S.**, Some engineering models for interactions of electric and magnetic fields with biological systems, *Bioelectromagnetics Supplement,* 1, 67–85, 1992.
53. **Moss, F.**, Stochastic resonance, *Ber. Bunsenges. Phys. Chem.,* 95, 3, 302–311, 1991.
54. **McNamara, B.**, Wiesenfeld, Theory of stochastic resonance, *Physical Review A,* 39, 9, 4854–4869, 1989.
55. **Koss, D. A., Carstensen, E. L.**, Effects of ELF electric fields on the isolated frog heart, *IEEE Trans. on Biomed. Eng.,* 30, 347, 1983.
56. **Litovitz, T. A., Krause, D., Mullins, J. M.**, Effect of coherence time of the applied magnetic field on ornithine decarboxylase activity, *Biochemical and Biophysical Research Communications,* 178, 3, 862–865, 1991.
57. **Mullins, J. M., Krause, D., Litovitz, T. A.**, Simultaneous application of a spatially coherent noise field blocks the response of cell cultures to a 60 Hz electromagnetic field, *Proc. of the 1992 World Congress on Bioelectromagnetism.,* Bioelectromagnetics society, Orlando, Florida, 1992.
58. **Litovitz, T. A., Montrose, C. J., Doinov, P., Brown, K. M., Barber, M.**, Superimposing spatially coherent electromagnetic noise inhibits field induced abnormalities in developing chick embryos, *Bioelectromagnetics,* 15, 2, 105–113, 1994.
59. **Walleczek, J., Liburdy, R. P.**, Nonthermal 60 Hz sinusoidal magnetic-field exposure enhances $^{45}Ca^{2+}$ uptake in rat hymocytes: Dependence on mitogen activation, *Federation of European Biochemical Societies Letters,* 271, 1/2, 157–160, 1990.
60. **Liburdy, R. P.**, Calcium signaling in lymphocytes and ELF fields, *Federation of European Biochemical Societies,* 301, 1, 53–59, 1992.
61. **Liburdy, R. P.**, Biological interactions of cellular systems with time-varying magnetic fields, *Annals New York Academy of Sciences,* 649, 74–94, 1992.
62. **Yost, M. G., Liburdy, R. P.**, Time-varying and static magnetic fields act combination to alter calcium signal transduction in the lymphocyte, *Federation of European Biochemical Societies,* 296, 2, 17–122, 1992.

63. **McLeod, K. J., Lee, R. C., and Ehrlich, H. P.,** Frequency dependence of electric field modulation of fibroblast protein synthesis, *Science,* 136, 1465–1469, 1987.
64. **Takashima, S.,** Non-linear properties of nerve membranes, *Biophys. Chem.,* 11, 447, 1980.
65. **Berkowitz, G. C., Barnes, F. S.,** The effects of nonlinear membrane capacity on the interaction of microwave and radio frequencies with biological materials, *IEEE Trans. Microwave Theory and Technology,* 27, 204, 1979.
66. **Alkon, D. L.,** Memory storage and neural systems, *Sci. Am.,* 261, 42–50, 1989.
67. **Alkon, D. L., Rasmussen, H.,** A spatial temporal model of cell activation, *Science,* 26, 998–1004, 1988.
68. **Yang, X., Korn, H., Faver, A. S.,** Long term potentiation of electronic coupling at mixed synapses, *Nature,* 348, 542–545, 1990.
69. **Hu, C. L., Barnes, F. S.,** The thermal-chemical damage in biological materials under laser irradiation, *IEEE Trans. Biomed. Eng.,* 17, 220, 1970.
70. **Shuyrou, V. L., Zhodan, G. G., Akorv, I. G.,** Effects of 330 MHz radio frequency, Academy of Science, Institute of Biophysics, Pushchino, Moscow Region, Russia, personal communication, 1984.
71. **Trautman, E. D., Newbower, R. S.,** A practical analysis of the electrical conductivity of blood, *IEEE Trans. Biomed. Eng.,* 30, 141, 1983.
72. **Johnson, F. H., Eyring, H., Stover, B. J.,** *The Theory of Rate Processes in Biology and Medicine,* John Wiley & Sons, New York, 1974.
73. **Liburdy, R. P., Penn, A.,** Microwave bioeffect in erythrocyte are temperature and pO dependent action permeability and protein shedding occur at the membrane phase transition, *Bioelectromagnetics,* 5, 283–291, 1984.
74. **Chalker, R.,** The effect of microwave absorption and associated temperature dynamics on nerve cell activity in Aplysia, M.S. thesis, University of Colorado, Boulder, 1982.
75. **Barnes, F. S.,** Cell membrane temperature rate sensitivity predicted from the Nernst equation, *J. Bioelectromagnetics,* 5, 113, 1964.
76. **Bol'shakov, M. A., Alekseyev, S. I.,** Change in the electrical activity of the pacemaker neurons of *L. Stagnalis* with the rate of their heating, *Biophysics,* Vol. 31, 569–571, 1986.
77. **Alekseyev, S. I., IL'in, V. I., Tyazhelov, V. V.,** Effect of electromagnetic radiation in decimeter wavelength range on calcium current of mollusk neurons, *Biophysics, 31, 2,* 290–295,1986.
78. **Philippova, T. M., Novoselov, V. I., Bystrova, M. F., Alekseev, S. I.,** Microwave effect on camphor binding to rat olfactory epithelium, *Bioelectromagnetics,* 9, 347–354, 1988.
79. **Bushanam, S., Barnes, F. S.,** Laser generated thermoelastic shock waves in liquids, *J. of App. Physics,* 46, 5, 2074–2082, May 1975.
80. **Mihran, R. T., Barnes, F. S., Wachtel, H.,** Temporally-specific modification of myelinated axon excitability in vitro following a single ultrasound pulse, *Ultrasound in Medicine and Biology,* 16, 3, 297–309, 1990.
81. **Adey, G., McNaughton, B. L., Wachtel, H.,** A system for recording microwave effects on isolated mammalian brain slices, presented at the 5th Annual Conference, Bioelectromagnetics Society, Boulder, Colorado, 1983.
82. **Polk, C.,** Sources, propagation amplitude and temporal variations of extremely low frequency (0–100 Hz) electromagnetic fields, in *Biological and Clinical Effects of Low-Frequency Magnetic and Electric Fields,* Llaurado, J. G., Sances, A., Jr., Battocletti, J. H., Eds., Charles C Thomas, Springfield, Illinois, 1974, 21.
83. **Plonsey, R.,** *Bioelectric Phenomena,* McGraw-Hill, New York, 1969.
84. **Barnes, F. S., Seyed-Madani, M.,** Some possible limits on the minimum electrical signals of biological significance, in *Mechanistic Approaches to Interactions of Electric and Electromagnetic Fields with Living Systems,* Blank, M., Findl, E., Eds., Plenum Press, New York, 1987.
85. **Yariv, A.,** *Introduction to Optical Electronics,* Holt, Reinhart & Winston, New York, 1976.
86. **MacDonald, D. K. C.,** *Noise and Fluctuations: An Introduction,* John Wiley & Sons, New York, 1962.
87. **Beck, A. H. W.,** *Statistical Mechanics, Fluctuations, and Noise,* Edward Arnold, London, England, 1967.
88. **Keshner, M. S.,** 1/f noise, *Proc. IEEE,* 70, 3, 212–218, 1982.
89. **Dorset, D. L., Fishman, H. M.,** Excess electrical noise during current flow through porous membranes separating ionic solutions, *Journal of Membrane Biology,* 21, 291–301, 1975.
90. **Verveen, A. A., DeFelice, L. J.,** Membrane noise, *Prog. Biophys. Mol. Biol.,* 28, 189, 1974.
91. **Sichenga, E., Verveen, A. A.,** The dependence of the 1/f noise intensity of the diode of Ranvier on membrane potential, *Proc. 1st. Eur. Biophys. Congr.,* 5, Verlag Wiener Medizinischen Akademic, Vienna, Austria, 1971, 219.
92. **Verveen, A. A., Derksen, H. E.,** Fluctuation phenomena in nerve membrane, *Proc. IEEE,* 56, 906, 1968.
93. **Nuney, P. L., Katznelson, R. D.,** *Electric Fields of the Brain—the Neurophysics of EEG,* Oxford University Press, New York, 1981.
94. **Bernhardt, J.,** The direct influence of electromagnetic fields on nerve and muscle cells of man within the frequency range of 1 Hz to 30 MHz, *Radiat. Environ. Biophys.,* 16, 309, 1979.
95. **Beckwith, J. R., McGuire, L. B.,** *Basic Electrocardiography and Vector Cardiography,* Raven Press, New York, 1982.

96. **Krueger, A., Reed, E.,** Biological impact of small air ions, *Science,* 193, 1209, 1976.
97. **Bridges, J. E., Preache, M.,** Biological influences of power frequency electric fields: A tutorial review from a physical and experimental viewpoint, *Proc. IEEE,* 69, 1092, 1981.
98. **Buffler, P. A.,** Future Epidemiologic Studies of Health Effects of Electric and Magnetic Fields, *Proc.,* Electric Power Research Institute, Palo Alto, California, September 1992.
99. **Chen, K., Chuang, H., Lin, C.,** Quantification of interaction between ELF-LF electric fields and human bodies, *IEEE Trans on Biomedical Engineering,* 33, 746–756, 1986.
100. **Kaune, W. T., Phillips, R. D.,** Comparison of the coupling of grounded humans, swine and rats to vertical, 60Hz electric fields, *Bioelectromagnetics,* 1, 117–129, 1980.
101. **Kaune, W. T., Forsythe, W. C.,** Current densities measured in human models exposed to 60 Hz electric fields, *Bioelectromagnetics,* 6., 13–32, 1985.
102. **Takuma, T., Kawamoto, T., Isaka, K., Yoko, Y.,** A three-dimensional method for calculating currents induced in bodies by extremely low-frequency electric fields, *Bioelectromagnetics,* 11, 71–89, 1990.
103. **Tenforde, T. S., Kaune W. T.,** Interaction of extremely low frequency electric and magnetic fields with humans, *Health Physics, Special Section: Non-Ionizing Radiation,* 53, 6, 583–606, 1987.
104. **Sheppard, A. R., Eisenbud, M.,** *Biological Effects of Electric and Magnetic Fields of Extremely Low Frequency,* New York University Press, New York, 1977.
105. **Greenbaum, E., Goodman, E. M., Marron, M. T.,** Long-term effects of weak 45–75 Hz electromagnetic fields on the slime mold physarum polycephalum, *Proc. Biol. Eff.,* Union of Radio Scientists International/URSI Annual Symposium, Boulder, Colorado, 1975.
106. **Blank, M., Khorkova, O., Goodman, R.,** Changes in polypeptide distributions stimulated by different levels of electromagnetic and thermal stress, to be published, *Bioelectromagnetics,* 1994.
107. **Goodman, R., Blank M., Lin, H., Dai, R., Khorkova, O., Soo, L., Weisbrot, D., Henderson, A.,** Increased levels of HSP70 transcripts are induced when cells are exposed to low frequency electromagnetic fields, *Bioelectrochem. Bioenerg.,* in press, 1994.
108. **Korneva, E.,** Head, Dept. of General Pathology and Pathophysiology, Institute for Experimental Medicine, RAMS, St. Petersburg, Russia, private communication, 1993.
109. **Jaffe, R. A., Phillips, R. D., Kaune, W. T.,** Effects of chronic exposure to a 60 Hz electrical field on synaptic transmission and peripheral nerve function, biological effects of extremely low frequency electromagnetic fields, *Proc. 18th Annu. Hansford Life Sci. Symp.,* Technical Information Center, U.S. Department of Energy, Richland, Washington, 277, 1978.
110. **Ceretelli, P., Veicsteinas, A., Margonato, V., Cantone, A., Viola, D., Malagati, C., Previ, A.,** 1000 kV project research on the biological effects of 50 Hz electric fields in Italy, biological effects of extremely low frequency electromagnetic fields, *Proc. 18th Annu. Hansford Life Sci. Symp.,* Technical Information Center, U.S. Department of Energy, Richland, Washington, 241, 1978.
111. **Greenberg, B., Kunich, J. C., Bindokas, V. P.,** Effects of high voltage transmission lines on honeybees, biological effects of extremely low frequency electromagnetic fields, *Proc. 18th Annu. Hansford Life Sci. Symp.,* Technical Information Center, U.S. Department of Energy, Richland, Washington, 74, 1978.
112. **Weber, R. A.,** *The Circadian System of Man—Results of Experiments and Temporal Isolation,* Springer-Verlag, Basel, Switzerland, 1979.
113. **Gavalas, R. J., Walter, D. O., Hamer, J., Adey, W. R.,** Effect of low-level low-frequency electric fields in EEG behavior, *Macaca nemestrina, Brain Res.,* 18, 491, 1970.
114. **Adey, W. R.,** Evidence for cooperative mechanisms in the susceptibility of cerebral tissue to environmental and intrinsic electric fields, in *Functional Linkage in Biomolecular Systems,* Schmitt, F. O., et al., Eds., Raven Press, New York, 1975.
115. **Dalziel, C. F.,** Electric shock, in *Advances in Biomedical Engineering,* 3, Kenedi, R. M., Ed., Academic Press, New York, 223, 1973.
116. **Mehn, W. H.,** The human considerations in bioeffects of electric fields, biological effects of extremely low frequency electromagnetic fields, *Proc. 18th Annu. Hansford Life Sci. Symp.,* Technical Information Center, U.S. Department of Energy, Richland, Washington, 21, 1978.
117. **Peckham, P., Martimer, J. T.,** Restoration of hand function in the quadriplegic through electrical stimulation, in *Biomedical Engineering and Instrumentation,* 3, Hambrech, F. T. and Rewick, J. R., Eds., Marcel Dekker, New York, 1977.
118. **Schaldach, M., Furman, S., Hein, F., Thull, R., Eds.,** *Advances in Pacemaker Technology,* Springer-Verlag, Basel, Switzerland, 1975.
119. **Tacker, W. A., Jr.,** Geddes, L. A., Fibrillation and defibrillation, *Electrical Defibrillation,* CRC Press, Boca Raton, Florida, 1980, Chapter 1.
120. **Tacker, W. A., Jr., Galioto, F. M., Jr., Giuliani, E., Geddes, L. A., McNamara, D. G.,** Energy dosage for human transchest electrical ventricular defibrillation, *N. Engl. J. Med.,* 214, 1974.

121. **Gilman, S., Daully, G., Tennyson, V., Kramzer, L. T., Defenlini, R., Carroll, J.,** Clinical morphological, biochemical and physiological effects cerebellar stimulation, *Biomedical Engineering and Instrumentation,* 3, Hambrech, F. T., Rewick. J. R., Eds., Marcel Dekker, New York, 1977.
122. **Maletzky, B. M.,** *Multiple-Monitored Electroconvlsive Therapy,* CRC Press, Boca Raton, Florida, 1981.
123. **Breggin, P. R.,** *Electroshock—Its Brain-Disabling Effects,* Springer-Verlag, Basel, Switzerland, 1979.
124. **Benz, R., Zimmermann, U.,** Pulse length dependence of the electrical breakdown in lipid bilayer membranes, *Biochem. Biophys., Acta,* 597, 637, 1980.
125. **Vienken, J., Ganzer, R., Hampp, R., Zimmermann, U.,** Electric field induced fusion of isolated vacuoles and protoplasts and metabolic provenience, *Physio. Plant,* 52, 153, 1981.
126. **Vienken, J., Zimmermann, U.,** Electric field-induced fusion: Electro-hydraulic procedure for production of hetrokaryon cells in high yield, *Federation of European Biochemical Societies Lettters,* 137, 11, 1982.
127. **Zimmermann, U.,** Electric field-mediated fusion and related electrical phenomena, *Biochimim. Biophys., Acta,* 694, 227, 1982.
128. **Zimmermann, U., Scheurich, P.,** High frequency fusion of plant protoplasts by electric fields, *Planta,* 151, 26, 1981.
129. **Schwan, H. P., Sher, L. D.,** Alternating current field-induced forces and their biological implications, *J. Electrochem. Soc.,* 116, 170, 1969.
130. **Hu, C. L., Barnes, F. S.,** A simplified theory of pearl chain effects, *Radiat. Environ. Biophys.,* 12, 1975.
131. **Jaffe, L. F.,** Control of development of ion currents, in *Biological Structures and Coupled Flows,* Optlaka, A., Balaban, M., Eds., Academic Press, New York, 445, 1983.
132. **Rowley, B., McKenna, J. M., Chase, G. R., Wolcott, L. E.,** The influence of electrical current on an infecting microorganism in wounds, in *Electrically Mediated Growth Mechanisms in Living Systems,* 238, Liboff, A. R., Rinaldi, R. A., Eds., New York Academy of Sciences, New York, 1974.
133. **Jaffe, L. F., Nuccitelli, R.,** Electrical controls of development, *Annual Review of Biophysics and Bioengineering,* 6, Academic Press, New York, 445–476, 1977.
134. **Jaffe, L. F.,** Control of plant development by steady ionic currents, in *Plant Membrane Transport: Current Conceptual Issues,* Spanswick, R. M., Lucas, W. J., Dainty, J., Eds., Elsevier/North-Holland, Amsterdam, The Netherlands, 381, 1980.
135. **Jaffe, L. F.,** Strong electrical currents leave the primitive streak of chick embryos, *Science,* 206, 569, 1979.
136. **Patel, N., Poo, M.,** Orientation of neurite growth by extracellular electric fields, *J. Neurosci.,* 2, 483–496, 1982.
137. **Borgens, R. B., Roederer, E., Cohen, M. J.,** Enhanced spinal cord regeneration in lamprey by applied electric fields, *Science,* 213, 611, 1981.
138. **Yamamoto, T.,** Physiological studies on fertilization and activation of fish eggs. III. The activation of the unfertilized egg with electric current, *Cytologia,* 14, 219, 1947.
139. **Jaffe, L. F.,** The role of ionic currents in establishing developmental pattern, *Philos. Trans. R. Soc London, S er. B.,* 295, 553, 1981.
140. **Jaffe, L. F.,** Developmental currents, voltages, and gradients, in *Development Order: Its Origin and Regulation,* Alan R. Liss, New York, 183, 1982.

Chapter 3

BIOLOGICAL EFFECTS OF STATIC MAGNETIC FIELDS

Richard B. Frankel and Robert P. Liburdy

CONTENTS

I. INTRODUCTION

Recent advances in medical instrumentation and concerns about exposure have stimulated research into the effects of magnetic fields on organisms and biological materials.[1–3] There has also been a great increase in knowledge of how organisms use and are affected by the geomagnetic field.[4–7] Many organisms are also known to possess inclusions of magnetic material.[8–11, 299]

There are many aspects to the effects of magnetic fields in biology. In order to make the treatment tractable, time-varying magnetic field effects have been arbitrarily separated from static magnetic field effects. Static field effects will be covered here; time-varying field phenomena including magneto-phosphenes and effects of induced currents will be covered in a companion chapter. An especially important area of growing interest concerns the biological effects and safety aspects of nuclear magnetic imaging and spectroscopy.[12–15] Static magnetic fields of up to 2.0 Tesla are now used in current FDA-approved imaging systems, and several million people a year experience these fields.

Even the subject of static magnetic field effects has many aspects. These include the use of magnetic fields in spectroscopies of biological material, including nuclear magnetic resonance (NMR), mentioned above, electron paramagnetic resonance (EPR), recoilless nuclear gamma resonance (Mossbauer effect), and magnetic susceptibility and magnetization

0-8493-0641-8/96/$0.00+$.50

measurements. Magnetic fields have also been used to orient and separate cells or cell fragments in suspension. Applications of magnetism in physiology and clinical medicine include magnetic resonance imaging (MRI), magnetic targeting and modulation of drug delivery, magnetic separation of biological materials, use of magnetism in surgical procedures, and noninvasive measurement of blood flow. An important topic area spanning AC and DC magnetic field regimes is the measurement of magnetic fields generated by the human body, and the use of those measurements in medicine and physiology. A large topic area involves mutagenic, mitogenic, metabolic, morphological, and developmental effects of exposure of organisms or biological materials to intense DC magnetic fields or to null field conditions. Another important topic area includes behavioral effects of magnetic fields, including effects on orientation, migration, and homing and the involvement of the geomagnetic field in the activities of organisms. Biomineralization of magnetic materials is especially significant for this latter topic, but could also play a role in other interactions of organisms with electromagnetic fields.

II. PHYSIOLOGICAL EFFECTS AND MEDICAL APPLICATIONS OF MAGNETISM

Magnetic properties and spectroscopy of biological materials have been extensively studied. Especially significant advances in NMR have been made.[16] These include increased resolution and sensitivity, pulsed programming, Fourier decomposition of complex spectra allowing study of whole organisms or perfused organs, and observation of the time development of chemical intermediates, e.g., metabolites containing phosphorus, in metabolic pathways.[17–19] Multidimensional NMR has been applied to the determination of protein structure in solution.[20–22]

Superconducting magnets with bores large enough to accommodate human bodies have allowed development of NMR imaging systems with resolution comparable and even superior to X-ray and positron computerized tomography and ultrasound techniques.[23,24] The basis of the method is that when a magnetic field gradient is superposed on a static, homogeneous magnetic field, nuclei such as protons will resonate at different frequencies at each point along the gradient. The amplitude of the signals at each frequency is proportional to the number of protons at that point. By switching the gradient to different directions and recording the spectra, a two- or three-dimensional proton density map can be reconstructed by computer. Measurement of relaxation times also allows discrimination of chemical differences, e.g., different types of tissue. Fields of the order of 5×10^{-2} to 2.0 T with gradients of the order of 10^{-2} T/m are used for periods of the order of 1/2 hr. Switching of the gradients can involve rates of magnetic field change as high as 2 T/sec.

Major safety issues associated with NMR imaging have been identified by Kanal and colleagues.[15] Biological effects associated with static magnetic fields is one of eight safety concerns relevant to NMR imaging; others are ferromagnetic attractive "projectile" effects, biological effects of the time-varying magnetic fields (mainly induced currents),[25,26] biological effects of the RF field (mainly RF heating),[27,28] auditory noise caused by the pulsed magnetic fields, concerns about the cryogens used in the superconducting magnets, claustrophobia and anxiety, and complications due to the NMR contrast agents. The specific problems associated with static magnetic fields and ferromagnetic objects have been reviewed in detail.[29] An associated safety issue relates to potential health risks to NMR imaging workers who are exposed to static and time-varying magnetic fields used in NMR. Epidemiological data was recently published in which women workers in more than 90% of clinical NMR facilities in the U.S. were evaluated for menstrual-reproductive experiences and work activities.[30] No statistically significant differences were observed between NMR workers and the control cohort group in spontaneous abortions, conception taking more than 12 months, delivery before 39 weeks, birth weight below 5.5 lb, or male gender of the offspring. These results suggest that a typical NMR worker environment is not a risk factor for these common adverse reproductive outcomes. A smaller but related study of the health of 11 human volunteers who experienced varying degrees of exposure to a 4.0 Tesla whole-body NMR imaging device over a 12 month period also reported no major adverse health problems associated with exposures.[31] The only positive findings related to mild sensations of vertigo, nausea, metallic taste, and magnetophosphene production all associated with motion within the 4.0 Tesla static magnetic field.

The above findings of mild sensory effects in humans exposed to 4.0 Tesla fields may have relevance to behavioral alterations reported in rats exposed to 4.0 Tesla static fields.[32] A simple T-maze was used to evaluate behavioral effects of rats in 1.5 and 4.0 Tesla fields. Rats were observed to enter the bore of the magnet at 0 and 1.5 Tesla freely. At 4.0 Tesla 97% of the rats would not enter the magnet and most of the decisions to turn around were made at the edge of the magnet in a region of a strong gradient field (13 T/m).

Several studies have investigated central nervous system activity in human subjects before and after exposure to NMR imaging fields up to 2.0 Tesla. Several groups have studied auditory evoked potentials in the human brainstem, in which subjects were assessed before and after routine NMR imaging at 1.5 or 2.0 Tesla, or during stepwise increments in field strength to 1.5 or 2.0 Tesla.[33–35] No long-lasting, significant differences in interpeak latencies were noticed before or after NMR imaging, or during the stepwise field intensity tests. A third report describes measurements of somato-sensory evoked potentials excited from median nerve stimulation in normal human subjects before, during, and after short-term exposure to a 1.5 Tesla static magnetic field.[36] No changes in interpeak latencies were observed. The above studies suggest that nerve conduction and synaptic transmission are within normal limits in human subjects exposed to such fields. These findings are consistent with studies discussed below (Section IV) in which static magnetic fields of 4.7 Tesla do not alter visually evoked potentials in the cat brain.

Human cognition has been investigated in 157 volunteer, neurologically-normal, subjects who were randomly assigned to a typical NMR treatment (0.15 Tesla), sham treatment, or control treatment groups.[37] Subjects were tested in a double-blind, prospective format at pre-treatment, post-treatment, and follow-up time periods. Six different tests were administered that assessed visual retention, mental rotation (3D visualization), memory scanning, and anxiety. No significant differences were reported between treatment groups. These studies are consistent with negative findings from studies employing rats in which spatial memory, open field behavior, and passive avoidance was investigated in animals exposed to a 0.15 Tesla NMR imaging procedure.[38,39] Human studies have also been conducted to assess whether hormone secretion is altered due to NMR fields.[40] Four healthy adult male volunteers (22–35 yrs) participated in the study which involved obtaining nine hourly blood samples between 2000 and 0400hrs on two different nights, one week apart. Between 2400 and 0200 hrs the subjects were exposed to a 0.15 Tesla NMR procedure or sham treatment. Hormone levels of melatonin peak at night and lighting was maintained well below threshold levels that trigger melatonin secretion. In addition, prolactin and growth hormone were also monitored and no statistically significant alteration in these three hormone levels was reported.

A number of animal studies have been conducted using NMR imaging fields. Rats were exposed to a 0.15 Tesla NMR procedure for 23 minutes for five successive days or for twenty-one successive days and 13 to 23 months. Thereafter animals were examined for abnormalities in body weight, spleen, heart, thymus, adrenal weights, white blood cell counts, hemoglobin, adrenocorticotrophin, and cortisol levels.[41] No evidence for changes in long-term survivability or stress was reported. A series of studies in mice and snails has shown that 0.15 Tesla NMR fields attenuate morphine-induced, and fentanyl-induced analgesia in mice.[42–45] Of interest in these studies is that the time-varying magnetic field component was nearly as effective in eliciting the response by itself as the complete NMR field environment.[42,45] The static magnetic field, therefore, is not strongly implicated as a causative factor in these studies. The authors note that these responses are small and appear to be reversible, and may operate through alterations in second messengers such as calcium and in protein kinase C activity.

Several animal studies exist which suggest a potential effect of NMR imaging fields on embryo development and cell metabolism. Concerns about NMR imaging fields influencing pregnancy outcomes was a major factor in initiating the Kanal epidemiology study, discussed above, which dealt with technical staff that experienced fringe fields during NMR imaging procedures. In contrast, patients experience the full complement of all three field components in the magnet bore. Perhaps the greatest difference in patient exposures is the increased likelihood for localized heating, a known teratogen, from the focused RF fields and time-varying magnetic fields in the magnet bore.[46,47] In this regard, the combination of localized heating plus X-ray diagnostic procedures which may precede NMR imaging scans is a potential factor for consideration. When pregnant mice were exposed for 16 hours beginning on gestation day 8.75 to 0.35 Tesla NMR imaging fields, a significant reduction in crown-to-rump length was reported, with no change in resorptions, stillbirths, fetal weight, or homeotic shifts.[48] In studies reported by Tyndall, the teratogenic potential of NMR imaging fields on development of the eye in mice was investigated.[49] Mice were exposed to NMR imaging fields at the isocenter of the magnet (1.5 Tesla) or the entrance (0.4 Tesla) on day 7 of gestation, the most susceptible stage of eye development. On day 14, concepti were removed and eye formation was evaluated by veterinary pathological examination and by computerized morphometric analysis. Both exposure locations to the NMR fields resulted in a similar statistically significant increase in abnormal eye formations compared to controls. It is of interest that both exposure locations lead to the same increase in affected fetuses. Exposure to static magnetic, RF, and time-varying magnetic fields are significantly different at the isocenter compared to the entrance of a NMR system. In particular, the static magnetic field at the isocenter is spatially uniform whereas at the entrance it has the characteristics of a gradient field. In other studies, using the chicken egg as a model system, NMR imaging fields of 1.0 or 4.0 Tesla did not result in alterations of embryo mortality, hatching rate or vitality of the chicken.[50] NMR imaging fields at 0.15 Tesla are reported to increase the synthesis of collagen in the dentin and bone of mice, but not synthesis in alveolar and tibial bone tissue.[51]

The blood-brain barrier (BBB) has been investigated in animals exposed to NMR imaging fields. At least seven laboratories have reported on this question, and, as such, it is one of the most widely investigated topics across laboratories. In addition, there has been an attempt to define which of the three field components is responsible for an effect on the BBB. In general, rats have been employed and the animals are chemorestrained so that an agent, such as a radiotracer-labeled compound, is injected into the circulation and the brain assessed for uptake of this compound following NMR imaging procedures. Technical differences such as the use or non-use of anesthesia as a chemo-restraint, the molecular size of the BBB marker compound, whether an acoustic control was employed or whether the animals were conditioned for handling,[52] and whether the rat circulatory system was perfused after the experiment to flush out excess tracer, do not appear to correlate with an effect on BBB permeability. The only ostensible correlate appears to be that NMR imaging systems that operate at < 0.15 Tesla show a small, reproducible alteration in the blood-brain barrier.[53–56] In contrast, negative findings have been reported for studies employing higher strength static magnetic fields up to 4.7 Tesla.[56–59] The biological basis for such alterations in the BBB may relate to functional changes in pinocytotic activity or capillary endothelium function, or to alterations in brain circulation. Since these alterations appear to be small and reversible, the clinical significance of these changes are as yet unclear.

Understanding the physical basis for these results is a challenging task since three different electromagnetic fields are present in the NMR imaging environment. In addition, the time-varying magnetic field is pulsed and may possess extremely-low-frequency (ELF) components raising the possibility that ELF magnetic fields associated with the time-varying field may be a factor in observed bioeffects.[12,25,42,60] Two of the above BBB investigations have attempted to experimentally separate the field components to directly test whether the static magnetic field, time-varying magnetic field, or RF field is the operative component. Oldendorf and colleagues reported that NMR imaging procedures led to a slight, but statistically significant, increase in 3H-mannitol uptake in the brain at 0.3 and 0.5 Tesla, but not at 1.5 Tesla.[56] Moreover, when field components were studied separately they report that no changes in BBB were observed for the 0.3 Tesla static field alone or the RF field alone. However, the time-varying magnetic field alone led to changes that could account for changes due to the complete field. Persson and colleagues report that the BBB is increased in rats exposed to 0.08 Tesla NMR imaging fields, but not in those exposed to 2.35 Tesla.[53] To address the question of which field component was operative at 0.08 Tesla they performed separate exposures to static and time-varying magnetic fields and the RF field. Although the number of animals was small (5–6 per group) they reported a statistically significant increase in BBB permeability, as assessed by Evan's Blue dye extravasation, for rats treated with all three components separately, however, the RF component was most prominent and similar to the complete field treatment. Studies at 915 MHz RF (CW and modulated at 8–215 Hz) using a larger cohort of rats (20–35 per group) led to statistically significant increases in the number of animals showing endogenous albumin leakage into the brain assessed by fluorescent antibody techniques (15%, controls vs. 67% exposed, $p = 0.0001$). The reversibility or clinical significance of these changes is at present unknown.

A small number of studies of the effects of NMR imaging fields on cellular model systems have been reported. The major advantage to using well-defined cellular systems in bioeffects studies is that biologically-based interaction mechanisms can be addressed directly using simple systems. Exposure of cells to NMR imaging fields requires special consideration of temperature control and isolation from vibrations during exposures.[2]

Two groups reported effects involving calcium metabolism in lymphocytic cells. These studies are of interest since calcium is a second messenger and plays a major role in cell growth and differentiation.[25,61,62] Alteration in calcium signaling across the cell membrane is an attractive biological framework for understanding magnetic field effects on fundamental events in cells involved in the signal transduction cascade such as RNA, DNA, and protein synthesis and cell proliferation. The T-lymphocyte is one of the best characterized model systems for investigating calcium signaling.[25]

When human peripheral blood lymphocytes or rat thymic lymphocytes were placed at the end positions of a 2.35 Tesla NMR imaging magnet, a significant increase in calcium-45 uptake was reported for cells undergoing signal transduction during mitogen activation.[25]

For both human and rat lymphocytes the mitogen employed was Con-A used at suboptimal concentrations so that the mitogen itself did not significantly elevate calcium uptake. However, for both cells the NMR fields acted synergistically to enhance calcium uptake in the presence of suboptimal levels of mitogen by approximately 30%. Thus, the NMR fields acted as a co-mitogen to amplify calcium signaling. This concept of field enhancement of mitogen activation represents a synergistic interaction that has biological significance. The effective threshold for mitogen activation was shifted to lower mitogen concentrations in the presence of the NMR fields and this suggests that the NMR fields enhanced the cells responsiveness to mitogen. Further studies with this model cell system were reported in which two field components, the 2.35 Tesla static magnetic field and the RF field, were evaluated separately; neither influenced calcium influx during mitogen activation.[63] This suggests that neither the RF nor the static magnetic field was involved in the field interaction. Additional exposures to static magnetic fields were reported at 7.5 Tesla with no effect on mitogen-activated calcium influx, which is consistent with findings at 2.35 Tesla, but exposures at 9.0 Tesla resulted in a statistically significant decrease in mitogen-activated calcium uptake.[63] The latter effect is in the opposite direction to that reported for 2.35 Tesla NMR imaging fields discussed above. These decreases in calcium influx in a 9.0 Tesla magnetic field may represent a static field interaction on cell membrane components that have a threshold between 7.5 and 9.0 Tesla. Calcium influx may lead to changes in free intracellular calcium,[25] and intracellular calcium was recently reported to be elevated by approximately 30% in HL-60 lymphocytic tumor cells exposed to 0.15 Tesla NMR imaging fields.[64] It is of interest that the field component associated with this field effect was reported to be the time-varying magnetic field. Thus the static magnetic field was not implicated. The increase in intracellular calcium observed in the complete 0.15 Tesla NMR field is consistent with an increase in calcium influx observed for lymphocytes exposed to the 2.35 Tesla NMR imaging field discussed above.

The effects of a static gradient magnetic field on cell cycle and growth rate were reported for mammalian FM3A cells.[65] Exposure to 0.6 Tesla/m resulted in a 5% reduction in growth rate, a decrease in cells in G1 phase up to eight hours post-exposure, and a drop in cell viability.

A number of studies have reported on the effects of static and ELF magnetic fields on the pineal gland and melatonin levels in rats and other organisms.[66–71] Reuss and Olcese[68] reported that light stimulation was necessary for the sensitivity of the pineal gland and melatonin content to magnetic fields. Lerchl et al.[69] has reported that the magnetic sensitivity of the pineal gland results from induced currents. However, no connection between these effects and behavioral responses to magnetic fields have been established.

The development of superconducting quantum interference devices (SQUIDS) with very high sensitivity has spurred the study of the very weak magnetic fields generated by electrical processes in the human body.[72–77] Applications include magnetocardiography, magnetoencephalography, measurement of pulmonary activity, and detection of body iron stores due to asbestos inhalation or diseases such as Thalassemia, which result in hemochromatosis.[75]

Magnetic devices have been used in several surgical procedures including repair of giant retinal tears,[78] bougienage of esophogeal atresia in infants,[79] and in the treatment of aneurysms.[80] Magnetic agitation has been used to modulate the release of macromolecules such as insulin from polymers with magnetic inclusions which are implanted in the body.[81]

The voltage induced when an electrolyte flows perpendicular to a magnetic field, known as the magnetohydrodynamic effect, has been used noninvasively to measure blood flow.[82,83] The voltage induced across a blood vessel is proportional to the vessel diameter, the magnetic flux density and the flow velocity. A velocity of 0.6 m/sec in the human aorta and 1 Tesla result in a 15 mV potential difference. These potentials also show up in electrocardiograms of animals in magnetic fields[84] (see Section IV).

Magnetic microcarriers have been used to target drugs to specific locations in the body and in cell separation and immunological assays.[85,86] These applications are based on the translation of paramagnetic and ferromagnetic particles in the direction of increasing magnetic field, relative to translation of the diamagnetic fluid background in the opposite direction. The particle velocity due to magnetic forces depends on the nature of the material, the magnitude of the field gradient, and the size of the particle. The size of magnetic microcarriers is limited by the need for a large surface-to-volume ratio, and by the tendency of large particles to coagulate due to interparticle forces.

To employ magnetic drug microcarriers, large magnetic field gradients must be generated outside the body by suitably shaped, magnetized pole pieces. In the *in vitro* procedures, the microcarriers, to which specific antibodies or antigens are chemically attached, can be separated by the use of high gradient magnetic separation (HGMS) filters consisting of fine stainless steel wires in a magnetic field strong enough to magnetize the wires. Gradients as high as 10^5 to 10^6 T/m can be generated within a few diameters of the wires. As the fluid flows through the filter the magnetic particles and anything attached to them are trapped on the wires. This method has been applied to the separation of cells, proteins and nucleic acids[87–91] and to removal of microorganisms from water.[92,93]

Magnetic fields have been used to release drugs from liposome vesicles and this suggests a novel means for noninvasively triggering drug delivery.[94] Liposomes are used to carry pharmaceutical agents or chemicals in the lipid bilayer or in the interior aqueous phase of the vesicle;[95,96] encapsulation of highly toxic drugs is advantageous since sequesteration prevents damage to normal tissue. The challenge is to direct the liposome vesicle to the appropriate target tissue and to trigger release of the agent into the target cell. Achieving the desired biodistribution has been approached in different ways; targeting by antibodies to cell surface markers,[97] specialized lipid compositions to promote prolonged circulation in the blood stream,[98] and modification of liposome size to preferentially enhance target tissue specificity.[99] Recently, static magnetic fields and time-varying fields have been shown to trigger the release of drugs from liposome vesicles.[94,99–103] This technique permits the controlled delivery of drugs using a noninvasive physical agent such as a magnetic field; liposome depots placed subcutaneously in different areas of the body can be treated with magnetic fields to release drugs in a controlled manner into the bloodstream.[103] Controlled, on-demand, temporal delivery of drugs is an evolving area of medicine that seeks to optimize the timing of drug delivery for maximal drug effectiveness.[104] Mechanisms of magnetic field interaction with the synthetic phospholipid bilayer of liposome vesicles is discussed in Section III.

A novel medical application of static magnetic fields is the use permanent magnet devices for retaining prosthetic appliances. In 1972 the rare earth cobalt magnet was developed; this led to its application in dentistry.[105] Sm-Co magnets have excellent magnetic properties but are not easily castable, and three new alloys were developed composed of palladium and cobalt (Pd-Co).[106] Biocompatibility tests showed least cytotoxicity with a nickel composition (Pd-Co-Ni).[107] Optimization schemes and design considerations have been developed,[108] as well as construction techniques[109] and specific dental prostheses.[110,111]

III. CELLS, BIOMOLECULES AND CHEMICAL REACTIONS IN MAGNETIC FIELDS

Alignment of molecular aggregates with sufficient diamagnetic anisotropy will occur in intense magnetic fields (see Appendix). Experimental results have been reviewed by Maret and Dransfeld,[112,113] and by Gretz et al.[114] Muscle fibers,[115] chloroplasts,[116] retinal elements,[117,118] sickled erythrocytes,[119,120] bacteriophage fibers,[121] membranes,[122,123] and macromolecules including nucleic acids[112] have diamagnetic anisotropy and have been aligned in intense, homogeneous magnetic fields. Highly oriented structures can result from polymerization in an intense field. Torbet et al.[124] and Freyssinet et al.[125] produced oriented fibrin gels from fibrin monomers in a field of 11 T. The fibrin monomers were produced by enzymatic cleavage of fibrinogen. Murthy and Yannas[126] prepared oriented collagen films by heat precipitation in a magnetic field. Collagen in solution dissociates into monomers at low temperatures and precipitates at higher temperature ($\sim 37°C$). The monomers are apparently not aligned. Alignment of dimers, trimers, etc. occurs as they form from monomers with increasing temperature in the field. Conversely, magnetic fields of 0.5–1.8 T were reported to disrupt cellulose polymerization in oat leaf fibrils and in the bacterium *Acetobacter xylinum*.[114]

In addition to polymer alignment, intense magnetic fields may interact with biomaterials in other ways. Sperber et al.[127] observed oriented growth of pollen tubes in intense fields and suggested a redistribution of membrane proteins that regulate intracellular concentrations of calcium. Audus[128,129] previously observed oriented growth, or magnetotropism, in oat shoots and cress roots in inhomogeneous magnetic fields, with growth occurring in the direction of decreasing field intensity. Labes[130] and Aceto et al.[131] proposed interaction of magnetic fields with cell membranes as a plausible mechanism for physiological effects. They noted that membranes have liquid crystal-like properties and are close to phase transitions at physiological temperatures. Magnetic fields could affect membrane fluidity or other properties. Magnetic orientation of diamagnetically anisotropic domains in artificial phospholipid bilayers has been reported.[132] Magnetic fields also affect the fluid-gel transition in agarose gels,[133] possibly by alignment of the monomers in the fluid phase.

There are reports of alteration of enzyme activity *in vitro* by magnetic fields.[134,135] For example, Haberditzl[134] reported that fields up to 7.8 T diminished the activity of glutamic dehydrogenase, while a 6 T field enhanced the activity of catalase. Nonuniform fields produced larger effects than uniform fields. Weissbluth and co-workers[136] reported no effects of intense magnetic fields up to 22 T on the activities of several enzymes.

Because of the important role that the cell membrane plays in mediating interactions with static magnetic fields, a number of studies have been undertaken to investigate how simple synthetic lipid bilayers interact with static magnetic fields. In these studies, the liposome vesicle is used as a model system for the natural cell membrane since the phospholipid bilayer represents a simple bilayer membrane structure that can be made more complex by incorporating, for example, protein receptor ensembles and ion channels structures. Unilammelar liposome systems are the simplest bilayer systems for use in laboratory investigations; An important feature of such single bilayer systems is that they display well-defined structural changes in organization during phase transitions.[101] This appears to be important to static magnetic field interactions since profound structural changes in phospholipid organization, e.g. changes in packing density of hydrophobic hydrocarbon tails in the bilayer and of polar headgroups at the membrane surface, as well as lateral compression of the bilayer, will influence the magnetic susceptibility tensor which determines the magnitude of the static field interaction.[2,101] Such changes in organization of the bilayer are relevant to functional aspects of natural cell membranes.[137]

Previous studies of dipalmitoyl lecithin vesicles in 3.8 Tesla static magnetic fields have reported changes in magnetically induced birefringence at temperatures above the phase transition temperature, T_c.[138] This was interpreted as reflecting lipid orientation in the bilayers. Light scattering studies on liposomes of various compositions have also reported that changes in turbidity occur during exposure to static fields greater that 0.2 Tesla at temperatures near or at the phase transition. These studies were also interpreted as reflecting membrane phospholipid orientation.[139] In addition, other studies at 9.3 Tesla indicate that phospholipid orientation at temperatures below the phase transition in the gel phase does not occur.[140] The above studies raise an interesting question: are field effects on structural changes such as phospholipid orientation, which are dependent on T_c, able to change membrane transport properties of the bilayer? This relates directly to a functional property of natural membranes.

This question was addressed in experiments in which unilammellar liposome vesicles were loaded with a radiolabeled aqueous phase marker, tritiated arabinofuranoside (3H-ARA-C), in the interior of the liposome, and solute release was assessed during exposure to a range of static magnetic fields and temperatures.[94] These liposomes exhibited enhanced solute release during brief (15 minute) exposures to 7.5 Tesla magnetic fields at temperatures near but lower than the characteristic phase transition temperature of ~41C (Figure 1). Kinetic studies at 7.5 Tesla revealed that the magnetic field approximately doubled the rate of solute leakage across the bilayer essentially instantaneously and that a slow rise in this permeability increase occurred over a 40–minute period (Figure 2).

These data indicated that magnetic field-induced, membrane permeability changes are possible in simple phospholipid bilayers, and that these field interactions display a dependence on the structural organization of the membrane.

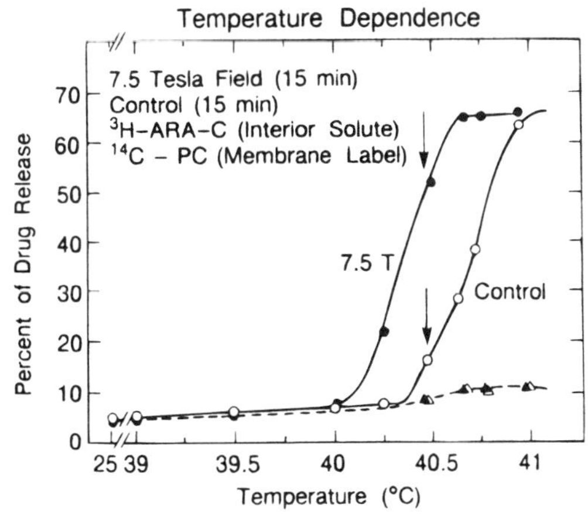

FIGURE 1 Temperature dependence of drug release from liposomes loaded with 3H-arabinofuranoside (3H-ARA-C; aqueous phase) and 14C-dipalmitylphosphatidlycholine (14C-PC; membrane label) in a 7.5 Tesla magnetic field.

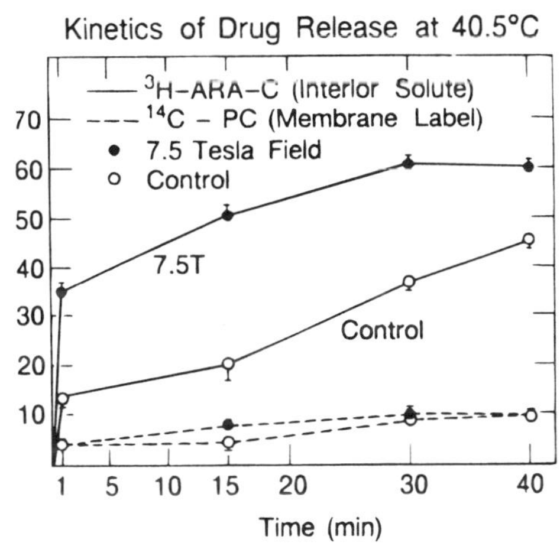

FIGURE 2 Time course of drug release from liposomes loaded with 3H-arabinofuranoside (3H-ARA-C; aqueous phase) and 14C-dipalmitylphosphatidlycholine (14C-PC; membrane label) in a 7.5 Tesla magnetic field.

An approach for analyzing magnetic field interactions with cell membranes, based on molecular dynamics, takes into account changes in the local curvature of the bilayer. The groundwork for this approach was originally developed by Helfrich[141] and was later applied to liposomes.[101] This approach rests on the observation that membrane structures undergo oscillations or vibrations (dynamic fluctuations) at the molecular level that are dependent on density packing of phospholipid elements in the aggregate structure.[142] Molecular dynamic fluctuations spontaneously produce local regions of curvature in the bilayer.[143] Such structures have been identified in liposomes and these local areas of molecular curvature in the bilayer are known as ripple structures which are formed spontaneously at pretransition temperatures.[144]

The idea of a singularity at which the cell membrane becomes unstable and is extremely sensitive to physical deformation by the application of an external magnetic field provides a theoretical basis for low field intensity interactions which are not addressed by classical thermal interaction analysis. Natural cells change shape during growth and differentiation. These changes in morphology may potentially predispose the cell to magnetic field deformation; such an interaction would be expected to exhibit cell-cycle dependence. In addition, natural cell membranes exhibit phase separations and lipid clustering that give rise to the presence of lipid domains.[94,137,145–148] These changes in membrane organization or phase transitions are associated with profound changes in enzyme activity and ion transport, usually at or near physiological temperatures. Local changes in lipid-protein packing, bilayer compressibility, and membrane fluidity would all play a role in modulating the local, spontaneous curvature of the cell membrane and, thus, the environment of the enzyme or ion channel.

At the surface of the cell a number of important processes occur that are critical to cellular function such as receptor binding events and the transport of second messenger ions, such as calcium, through ion channel structures in the bilayer. The effects of static magnetic fields have been reported on these processes in a number of model systems. Static magnetic fields of 2.0 Tesla, comparable to that used in NMR imaging systems, were used to investigate the binding of ligand to the acetlycholine receptor (AchR) *in vitro.*[149] Binding of the competitive antagonist, alpha-Bungarotoxin, to AchR in the magnetic field was significantly reduced at 45 minutes compared to control treatment. This effect was reversible and did not decrease the total number of available binding sites since maximal binding (as in controls) was achieved after 2.75 hours in the field. Pretreatment of AchR in the magnetic field for 2.5 hours had no effect on ligand binding. Thus, this interaction required the presence of both ligand and receptor, and apparently influenced the initial rate of receptor-ligand complex formation. Aoki and colleagues studied the effects of a 0.4 Tesla static magnetic field on adriamycin (ADR) movement across the cell membranes of cultured TALL-1 cells.[150] A fifteen minute field exposure was reported to result in an increase of less than 5% of ADR accumulation observed in the sham-treated cells. Experiments were carried out within a temperature range of 41–46C corresponding to the phase transition temperature, suggesting that lipid ordering and bilayer structural organization, as discussed above, play a role in this interaction. Other studies implicate calcium movement across the cell membrane as a factor in interactions with static magnetic fields.[151] Human polymorphonuclear leukocytes were reported to exhibit decreased migration, decreased release of lysozyme and lactate dehydrogenase, and decreased cell viability when exposed to a 0.1 Tesla field for times as short as 30 minutes. This loss in cell viability may explain the loss of functional parameters observed. Of interest is the fact that three calcium channel antagonists, diltiazem, nifedipine, and verapamil, were able to protect the cells from these effects. Since nifedipine was the best protective agent, the site of interaction of the magnetic field may be at the extracellular surface component of the calcium channel where this drug acts.

Cellular function has been investigated in intense static magnetic fields. Freshly isolated human peripheral blood T-lymphocytes were exposed to static fields up to 6.3 Tesla and evaluated for cell growth and viability.[152] No changes were noted for lymphocytes under standard cell-culture conditions. However, when the T-cells were stimulated with the mitogen PHA, which triggers cell proliferation, cell growth was significantly inhibited compared to controls. Dose response data indicate that this effect was detectable for fields between 4.0–6.3 Tesla, but not for fields below 2.0 Tesla. It is relevant that calcium influx across the cell membrane is required as a second messenger during mitogen activation in the lymphocyte, and a plausible interaction mechanism for this effect on the PHA-treated cells is the inhibition of calcium influx across the cell membrane. These findings are also consistent with the reports dealing with calcium transport of mitogen-activated lymphocytes and calcium-mediated cell function discussed above.[25,63] A normal human fibroblast cell, DMD-D, and a human malignant melanoma cell line, PS1273, were also assessed for cell growth and viability in intense static magnetic fields.[153] A 4.7 Tesla magnetic field did not influence cell growth or cell viability of these two cell lines over a 72 hour growth period. However, the authors reported that a significant detachment of the melanoma cells was observed by 12 hours and by 72 hours only a few percent of the cells remained attached; cells could reattach when returned to a nonmagnetic environment. A dose-dependence was observed with fields as low as 0.5

Tesla resulting in 75 % detachment at 72 hours. Since cell viability and cell growth were not affected, this field interaction specifically involves cell-surface coupling which mediates cellular adhesion. Rat fibroblasts and osteoblasts were assessed for thymidine and for proline incorporation in the presence of a 0.61 Tesla static magnetic field.[154] Fibroblasts, but not osteoblasts, showed a statistically significant increase in both thymidine and proline incorporation of approximately 0.5–1.0 fold throughout the 10-day exposure protocol; increases in these indices reflect increases in total DNA synthesis and protein synthesis, respectively. In contrast, when explanted neonatal rat calvaria tissue, which was the source of the fibroblasts and the osteoblasts, was used intact in similar tests an inhibition of these indices was detected. The above studies indicate that different cellular model systems can be influenced and can exhibit different sensitivities to static magnetic fields.

Magnetic fields of the order of 10^{-3} to 10^{-2} T can affect chemical reactions by influencing the electronic spin states of reaction intermediates.[155–161] Such effects have the potential to lead to biological consequences,[156–158, 162–165] however, it should be cautioned that a magnetic field effect on chemical reaction intermediates in biological systems has not yet been demonstrated under actual physiological condition.[3]

A relatively simple chemical illustration of the effect involves homolytic cleavage of a chemical bond to produce two radicals.* Since the electrons in the chemical bond are spin-paired in an $S = 0$ or singlet state,[166] these electrons on the nascent radicals will also have overall singlet character as the radicals separate. Separation is a diffusion-controlled process and there is a high probability that the two radicals will re-encounter each other. If the electrons retain their overall singlet character, a re-encounter is likely to produce recombination. If the electrons have overall triplet ($S = I$) character, the bond will not reform and the radicals will eventually separate and perhaps participate in other chemical reactions. The transition from singlet to triplet can result from the interaction of the odd electrons of the radicals with the nuclear magnetic moment(s) of the atom(s) on which they have high probability density. This interaction—the magnetic hyperfine interaction—is equivalent to a local magnetic field produced by the nuclei at the location of the electrons. Different local magnetic fields cause the electrons on the radicals to precess at different rates, which destroys singlet phasing and results in triplet formation. However, an applied magnetic field will decouple the electrons and the nuclei, suppressing formation of the triplet state. This enhances the recombination rate and suppresses the other chemical reactions. The decoupling of the electrons and the nuclei will occur when the intensity of the applied field exceeds the effective magnetic fields produced by the hyperfine interactions. Then the electrons will precess in phase about the applied field rather than at different rates about the local fields. This condition is typically satisfied for fields of the order of 10^{-3} to 10^{-2} T.

A variation of this scheme is proposed to account for the effects of magnetic fields on electron transport in photosynthetic, purple bacterial membranes.[162,164,165] The effects are observed when elements of the transport chain are electrochemically reduced, which is nonphysiological, forcing back transfer of the photoexcited electron. The electron transport sequence can be summarized as follows:

1. $^{S}(A)$ + photon $\sim$ $^{S}(A)^*$
2. $^{S}(A)^* + (B)$—$^{S}(A^+ + B^-)$
3. $^{S}(A^+ + B^-)$—$^{T}(A^+ + B^-)$
4. $^{S}(A^+ + B^-)$—$^{S}(A) + (B)$
5. $^{T}(A + B)$—$^{T}(A) + (B)$

* Homolytic cleavage is the breaking of a covalent, single bond so that one electron from the bond is left on each fragment, resulting in two free radicals with single, unpaired electrons.

(A) corresponds to bacteriochlorophyll dimer and (B) corresponds to bacteriopheophytin. S and T stand for singlet and triplet, respectively, and * stands for the photoexcited state. (1) Bacteriochlorophyll absorbs a photon resulting in electron excitation. (2) Electron transfer to bacteriopheophytin occurs, resulting in positive and negative charges on donor and acceptor, respectively. (3) The positive ion-negative ion pair are initially in a singlet state which can evolve into a triplet state via the hyperfine interaction mechanism. In the blocked transport chain, the ion pair decays by back transfer of the electron to a less energetic state of bacteriochlorophyll. (4) If the ion pair is in the singlet state, back transfer populates the singlet ground state of bacteriochlorophyll. (5) If the ion pair is in the triplet state, back transfer populates an intermediate energy triplet state of bacteriochlorophyll which is detected by a delayed fluorescence method. It was found that the amount of bacteriochlorophyll triplet produced following flash excitation is magnetic field-dependent.[165] Because the photosynthetic apparatus is highly structured and membrane-bound, exchange interactions between the ions also play a role in the formation of the triplet state in (3). Although the experimental conditions cited above are certainly nonphysiological, mechanisms of this kind could conceivably play a role in electron transport in viable biological systems. However, the fact that free radical production in cells, for example, is extremely tightly regulated by multiple redundant systems argues against magnetic field effects on reaction intermediates leading to any adverse, long-lasting consequences.

The action of an intense static magnetic field on the movement of blood or the diffusion of ions, plasma proteins, and paramagnetic molecules has been studied. These investigations address the question of a magnetohydrodynamic effect on flow processes (Figure 3).

As noted in section II, the flow of a conductive element in the presence of a static magnetic field will lead to induced electrical potentials, and the magnetohydrodynamic effect has been predicted on theoretical grounds to produce a reduction in the flow velocity and a compensatory elevation in blood pressure to retain a constant volume flow rate.[167] The occurrence of changes in dynamic pressure could, in principle, impose cardiovascular stress. This might pose a potential problem particularly in biologically-compromised patients during NMR imaging procedures. Monkeys (*Macaca cynomolgous*) were exposed to a 1.5 Tesla static magnetic field and no measurable alterations in blood pressure were reported.[168] However, an instantaneous, field strength-dependent increase in the EEG signal amplitude at the T wave locus was observed in fields > 0.1 Tesla; this was reversible. This is consistent with

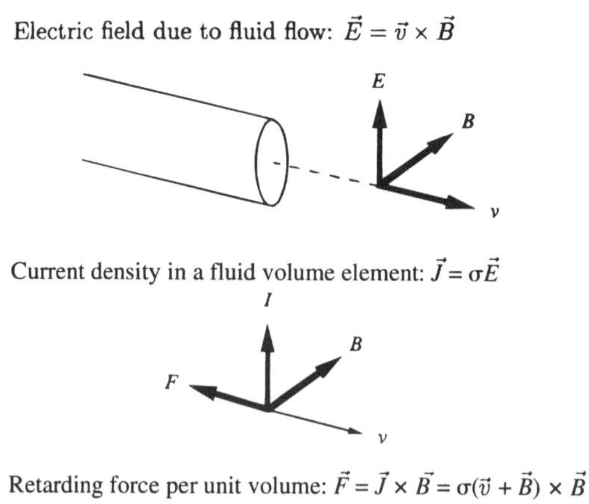

FIGURE 3 Illustration of the magnetohydrodynamic effect, i.e., the drag force generated in a conducting fluid moving in a static magnetic field.

the presence of a magnetically-induced aortic blood flow potential superimposed on a normal T-wave signal. In studies employing humans exposed to static magnetic fields up to 1.0 Tesla for 10-minute sessions, there was no evidence of alterations in local blood flow at the skin of the thumb or at the forearm that were attributable to the applied magnetic fields.[169] Theoretical studies have estimated the magnetohydrodynamic effects that would be observed at very high intensity fields. A 10 Tesla magnetic field is predicted to change the vascular pressure in a model of the human vasculature by less than 0.2%, and experimental results for the retardation of 15% NaCl flowing transverse to a 2.3 or a 4.7 Tesla field are in general agreement with these predictions.[170] In addition, the issue of magnetic fields affecting the diffusion of ions and proteins in solution has been addressed theoretically.[171] A Lorentz force interaction is predicted for charged ions such as Na^+, K^+, Ca^{2+}, Cl^-, and for plasma proteins with a threshold at more than 10^6 Tesla. In gradient magnetic fields, Maxwell stress interactions are predicted for gradients more than 100,000 T^2/m for paramagnetic molecules such as $FeCl_3$ and O_2, also for plasma proteins. Typical gradient fields in MRI devices are usually less than 100 T^2/m. However, the movement of deoxygenated erythrocytes and $FeCl_3$ colloids (greater than 1000 molecules) are easily influenced by normal gradient fields due to a volume effect.

IV. MUTAGENIC, MITOGENIC, MORPHOLOGICAL, AND DEVELOPMENTAL EFFECTS OF MAGNETIC FIELDS

A large number of papers have been published on this topic and a number of bibliographies, reviews, and symposia have appeared.[1–3, 172–189] Moreover, a number of interaction mechanisms have been proposed.[190–196] However, this area is still an empirical science with little elucidation of effects in terms of mechanism. Only some of the more recent reports will be cited here.

Mutagenic effects of chronic exposure to DC magnetic fields have been investigated. A recent review on the genotoxic potential of static magnetic fields assessed a wide range of recent studies. In these studies, the authors conclude that the preponderance of evidence suggests that static magnetic fields do not have a clearly demonstrated potential to cause genotoxic effects.[197] Mahlum et al.[198] exposed male mice to a magnetic field of up to lT for 28 days. The mice were subsequently mated to two females per week for up to 8 weeks and the resulting embryos were assayed for viability 10 days later. No significant differences were reported between exposed and sham-exposed (control) groups. Kale and Baum[199] exposed fruitfly (*Drosophila melanogaster*) male eggs, larvae, pupae, and adults to fields up to 3.7 T for up to 7 days. After mating with females, broods were tested for induction of genetic damage by the sex-linked, recessive lethal test. No evidence for induction of mutations under the conditions of exposure were reported. Skopek et al.[200] exposed Salmonella and cultured human lymphoblasts to 10 T magnetic fields for 4 hr. Cells were surveyed for toxic and mutagenic effects with a forward mutation assay. No effects of magnetic field exposure were found for either cell type when compared with sham-exposed controls.

Morphological and developmental effects have been investigated. S. Ueno and colleagues studied the embryonic development of the frog (*Xenopus Laevis*) by exposing fertilized eggs in Ringers solution to static magnetic fields up to 6.5 Tesla for varying lengths of time.[201] The time course of early, very rapid cleavage was assessed to determine if static fields affected cell multiplication and differentiation during a 7 hour exposure. No changes were detected in cleavage rates or in the percentage of embryos reaching stage 21 (late neural stage, 22 hour) between control and exposed embryos. In addition, no appreciable defects were observed between control and exposed embryos developing into tadpoles (stage 42) at day 4. In this organism the processes of DNA replication and cell division during early gastulation, followed by tissue differentiation and organogenesis appears to be insensitive to intense static magnetic fields. Earlier studies have also reported negative findings on early stages of development of the frog (*Rana pipiens*) exposed to a 0.7 Tesla static magnetic field in a NMR system.[202] Drosophila larvae, in contrast, have been reported to show abnormalities in embryogenesis during exposures to static magnetic fields as low as 1 mTesla.[203] In these studies M.-W. Ho and colleagues confirmed earlier reports that exposure of Drosophila to mTesla fields decreases hatching rates,[204] but they extended these observations to specific morphological abnormalities. They reported that exposures for 30 minutes were as effective as 24 hours, and that a flux-density dependence in abnormalities was reported over the range 0–3 mTesla with no further change in abnormalities between 3–9 mTesla. The authors state that such weak magnetic fields probably do not influence transembryonic ionic currents in the Drosophila, or do not exert orientation effects on cell membrane components, but that they may be acting via a high degree of cooperativity among molecules involved in the processes of pattern recognition. These recent studies, as well as previous studies, emphasize the important fact that different animal species can display sensitivity to static magnetic fields.

Sikov et al.[205] reported no effects on the development of mice after intrauterine exposure to 1 T during gestation. An earlier study by Nahas et al.[206] had indicated that exposure of rodents to 0.02–0.12 T fields for 1 month resulted in no toxic or histopathological effects. Brewer[207] studied guppies (*Lebistes reticulatus*) chronically exposed to a 0.05 T magnetic field and reported reduction in spawn rate and gestation period in successive generations exposed to the field. However, field effects were not permanent; reproduction eventually returned to normal when fish were removed from the magnetic field. Mild et al.[208] studied development of frog (*Xenopus laevis*) embryos exposed at 0.25 T for periods up to 1 week, at temperatures just above the threshold for development in the embryos. If the effect of the magnetic field is equivalent to a reduction in temperature, exposed embryos should not have developed. However, no differences between development of exposed embryos and unexposed controls were reported. Previous studies had indicated effects of magnetic field exposure on development of frog embryos.[209] Strand et al.[210] reported enhancement of fertilization following exposure of trout sperm, ova, or both to 1 T magnetic fields.

Frazier et al.[211] investigated mammalian cell cultures continuously exposed to magnetic fields of 0.1 or 0.3 T through 80 cell doublings. No mitogenic effects of the field were reported when doubling times of exposed cells were compared to controls. Differences in plating efficiencies between exposed cells and controls were cited and ascribed to an as yet unexplained increased clumping of exposed cells. Exposure of frozen cells at 1 T did not result in changes in cell morphology[211] as reported earlier.[212] Cultured cells from human bronchogenic carcinoma and from Burkitt Lymphoma were exposed to DC magnetic fields up to 1.15 T by Chandra and Stefani.[213] They reported that growth characteristics were unaffected by exposure. *In vivo* exposure of mouse tumors did not cause retardation of growth of the tumor. Leitmannova et al.[214] reported changes in morphology of aged red blood cells in magnetic fields.

Moore[215] studied growth of five species of bacteria and a yeast in DC and slowly varying magnetic fields up to 0.09 T. He reported stimulation or retardation of growth depending on the field strength, frequency, and organism. A number of previous studies had indicated that growth of bacteria and yeasts could be altered by static magnetic fields.[216–218].

Electrophysiological effects of static magnetic fields have been investigated. Blatt and Kuo[219] reported no change in the action potential in the interpodal cells of the fresh water *alga Nitella* exposed to fields up to 2 T. These cells have bioelectrical activity and previous studies[220] had indicated a reduction of the action potential in magnetic fields. Extended exposure at 1.6 T was not toxic for cells. Edelman et al.[221] reported an increase with time in the amplitude of the compound action potential of stimulated frog sciatic nerve when fields up to 0.71 T were applied perpendicular to the nerve. Fields applied parallel to the nerve produced no changes. However, Gaffey and Tenforde[222] reported no changes in electrical conduction in frog sciatic nerve in fields up to 2 T and suggested that results of Edelman et al. were due to changes in temperature. Semm et al.[223] reported electrical changes in cells in the pineal glands of guinea pigs when exposed to magnetic fields of the order of 10^{-4} T. Raybourn[224] reported that 10^{-3} to 10^{-2} T fields acutely reduce electroretinographic responses in turtle retina, but do not reduce retinal sensitivity. This effect might involve magnetic field effects on chemical reactions (see Section III).

Studies by Rosen indicate that a significantly weaker magnetic field of 0.12 T can decrease both the amplitude and the variability of a visually evoked response in the adult cat striate cortex.[226] This effect was observed to have a latency period since the effect developed slowly 50 seconds after the field was turned on and persisted for minutes after the field was turned off. This group also reported that 0.12 T magnetic fields altered the spontaneous discharge frequency and discharge patterns of principal cells in the cat's lateral geniculate body.[227] Decreases in frequency and in short interspike intervals were detected, with a gradual onset of the effects. The authors hypothesized in both reports that calcium ions may be involved in this response since latency is consistent with an effect on neurostransmitter release which is calcium dependent. In contrast to these results, when the somatosensory evoked potentials of the cat were studied in fields up to 4.7 T no statistically significant changes were observed in EEG intensity, or EEG frequency index that correlated with field magnitude although substantial variations in these parameters were observed during experiments[228] The authors interpreted this as reflecting spontaneous fluctuations of vigilance and not as effects of the static magnetic field.

Studies on miniature end plate potentials (MEPPs) in 0.12 T static magnetic fields have been reported.[229] Murine neuromuscular junction preparations maintained in a 0.12 T magnetic field exhibited a modest increase in frequency at temperatures below 34C and a prominent decrease in frequency at temperature above 35C. When calcium was removed from the media this temperature-dependent field response was not observed. The author interpreted the data to indicate that a phase transition was operating and that the applied magnetic field was sufficient to orient diamagnetic elements of the presynaptic membrane and alter calcium transfer. Further studies with this model system indicate that a minimum of 50 seconds and a maximum of 150 seconds of field exposure is associated with MEPP inhibition at 35C.[230] This interpretation is consistent with a slow orientation of diamagnetic domains that might be coupled to neurotransmitter release in this system.

Static magnetic fields affect electrocardiograms in mammals.[167,168] Alterations in the T wave of rats, rabbits, and baboons are reported at exposures above 0.3 T, and are due to the potentials associated with the flow of blood in the magnetic field, as discussed above. There are apparently no chronic physiological effects associated with this phenomenon.

Ripamonti et al.[231] studied the effect of magnetic field exposures up to 12.5 T on the responses of the contractile protozoan *Spirostomum ambiguum* to the toxic substance 2,2′ dipyridyldisulfide. Magnetic field exposure reportedly diminished the ability of the organism to survive the drug and lengthened the extension phase of the contraction cycle. It was hypothesized that interactions of the magnetic field with cellular membranes alters the regulation of calcium transients. There were no toxic effects of exposure to magnetic fields without the drug. Bucking et al.[232] had previously reported that magnetic field exposure affected the force of contraction of isolated muscle, which also involves regulation of intracellular calcium.

deLorge[233] reported that low-intensity magnetic fields have no measurable effects on operant behavior in monkeys. However, experiments at high magnetic fields showed a suppression of a learned response above a threshold between 4.6 and 7.0 T. Davis et al.[234] reported no behavioral alterations in mice exposed to 1.5 T magnetic fields. Further data on the effects of very intense magnetic fields come from NMR studies on perfused, whole organisms. Fossell et al.[9] noted that exposure of perfused rat hearts at 6.4 T did not alter either pressure development or heart rate.

A number of studies report effects in animals exposed to static magnetic fields. When mice were chemoimmobilized and exposed to a 1.4 Tesla static magnetic field the activity of thymidine kinase in bone marrow cells was influenced.[235,236] The effect depended on animal body temperature with an increase in activity at 27–29C, and a decrease in activity at 37–39.5C. A dose-response was observed from 0.2–1.4 Tesla for the low temperature effect. A thirty-minute exposure was required for full expression of the effect at 1.4 Tesla, which was reversible within 5–10 minutes. Of interest is the observation that these effects were not observed in moving rats, in cell suspensions, or in enzymes in solution. The authors interpreted these findings to indicate that a complex, intact structure such as the cell membrane is likely involved in this interaction in contrast to a magnetic field effect on individual free enzymes or other molecules in solution such as free radicals. Another study employed immobilized rats, but a ventilated restraining chamber was used instead of a chemorestraint during exposure to a 1.0 Tesla static magnetic field.[237] Rats were exposed to the magnetic field for 30 minutes on each of ten consecutive days. No significant differences were reported in blood alkaline phosphatase, acid phosphatase, calcium ion concentration, and phosphate, ion concentration between control, sham-exposed, and magnetic field-exposed animals. The authors suggested that these results support the idea that short-term exposures to 1.0 Tesla fields do not alter physiological mechanism of bone mineralization.

The effects of static magnetic fields on *in vivo* immunity has been studied in rats implanted with micromagnets (600 Gauss, 60 mTesla) in the skull in the frontal, parietal and occipital regions.[237a] After 20 days, rats were either challenged with antigen or not challenged; both implanted-micromagnet and control animal groups were subsequently assessed for immune status up to 34 days. The implanted micromagnets were reported to immunopotentiate both humoral and cell-mediated immune responses, with highest performance obtained from occipital exposure for a total period of 24 days. The use of implanted magnets in bioeffects studies is a recent development; there is a limited database for comparison of results. Several such studies have been reported in the area of permanent magnet devices for retaining dental prosthetic appliances, discussed above. The use of implanted magnets raise challenging questions regarding biocompatibility and *in vivo* dosimetry to target tissues.

A survey of workers exposed to intense magnetic fields was conducted by Beischer[238] who found no adverse effects of short exposures to fields up to 0.5 T. Epidemiological studies of magnetic field effects in humans are being conducted presently by Budinger et al.[239] and by the National Radiological Protection Board in the U.K. Reviews concerned especially with potential hazards of magnetic field exposure associated with NMR imaging have been published,[12–14] as discussed above.

Finally, nature has conducted an experiment over geologic times on life in a substantial magnetic field. Magnetotactic bacteria[240] are sediment-dwelling bacteria that contain particles of Fe_3O_4 that cause them to be oriented in the geomagnetic field (see Section V). These particles produce strong intracytoplasmic magnetic fields and field gradients in the bacterium. The fields can be as large as several tenths of a Tesla near the surface of the particles. Thus, these bacteria carry out all of their cellular and metabolic functions in intracellularly generated magnetic fields, and have presumably done so for billions of years.

V. THE GEOMAGNETIC FIELD IN THE ORIENTATION AND HOMING OF ORGANISMS

At the end of the last century, Kreidl[241] published a report describing magnetic field effects on orientation in crabs. The experimental design was contrived to produce an effect and so does not test behavior in the natural environment of crabs, but could provide a paradigm for effects in other organisms. Crabs periodically molt and form a new exoskeleton. In the process of molting, they also lose their statoliths, the dense particles that form part of the vestibular system. They subsequently pick up particles of sand to serve as new statoliths. Kreidl took newly molted crabs and placed them in an aquarium in which only magnetic particles were available. The crabs indeed placed magnetic particles in their ear labyrinths. When the crabs were subsequently approached with a bar magnet, they adopted an orientation that could be correlated with the resultant of the magnetic and gravitational forces on the particles. Electrophysiological responses to magnetic field stimulation have been recorded in crayfish with ferrite statoliths.[242]

Since Kreidl's experiment, magnetic field effects on orientaion and homing have been reported for a very diverse group of normal organisms, including bacteria,[240] algae,[243] snails,[244] planaria,[245] honeybees,[246] fish,[247] mollusks,[248] amphibians,[249] homing pigeons,[250] migratory birds,[5,251] mammals,[252] and humans.[253–255] In addition, conditioning experiments on pigeons,[256] skates,[257] tuna[258] and honeybees[259] have demonstrated the ability of these organisms to sense magnetic fields. Cetacean strandings have been correlated with geomagnetic field anomalies.[260,261] In addition to behavioral effects, electrophysiological responses to changes in magnetic field intensity or direction have been reported in several organisms.[262,263]

These observations imply existence of an organ or organelle that transduces magnetic field intensity, direction or gradient infromation. Three interaction mechanisms have been proposed:

1. detection by the organism of the electric field induced by Faraday effect as the organism moves through the magnetic field;
2. interaction of the magnetic field with magnetic material in the organism;
3. effect of the magnetic field on chemical reactions or absorption of photons.

The first mechanism is apparently operative in marine sharks, skates, and rays,[257] which are sensitive to electric fields as low as 5×10^{-7} V/m in sea water.[264,265] The animals detect the electric fields through the *ampullae of Lorenzini,* which are long, conductive channels that connect electrically sensitive cells in the snout with pores on the skin. Flowing ocean currents or motion of the animal through the geomagnetic field induce voltage gradients with sign and magnitude depending on orientation, which are, in general, above the sensitivity threshold of the animal. Kalmijn[257] demonstrated that skates could be trained to use magnetic fields of the order of the geomagnetic field as an orientational cue. Brown et al.[266] used electrophysiological measurements to show that the *ampullae of Lorenzini* can detect variations in the geomagnetic field. Jungerman and Rosenblum[267] have considered the possibility of the magnetic induction mechanism for an animal in air. They concluded that a circular, electrically conducting loop millimeters in diameter, would be required to overcome thermal noise, with voltages induced by changes in magnetic flux in the loop as the animal changes its heading.

Evidence for orientation by the second mechanism was obtained for homing pigeons in the classical experiment of Keeton.[268] Keeton glued small bar magnets to the backs of the heads of a group of homing pigeons and compared their homing ability with that of a group of control birds carrying brass weights. Under sunny skies, both groups oriented and homed equally well when released from unfamiliar sites many miles from the home loft, but, under overcast skies, when the birds could not see the sun, the orientation of the birds carrying magnets was disrupted, whereas control birds oriented normally. Subsequently, Walcott and Green[269] used Helmholtz coils attached to the heads of pigeons to change the orientation of the birds under overcast conditions. The orientation depended on the direction of the magnetic field, as determined by the direction of current in the coils. Further studies suggest that the magnetic compass sense in pigeons and migratory birds is sensitive to the inclination of the geomagnetic field.

Pigeon orientation is also affected by magnetic anomalies and magnetic storms.[270,271] The experimental results suggest that in addition to a magnetic compass, a homing pigeon may have a "map"[272,273] or position finding system that includes magnetic cues. The results have been reviewed by Walcott,[250] Gould,[246,272] Able,[274] Griffin,[275] and Wiltschko.[5,276] Although attempts to observe magnetic sensitivity in pigeons by cardiac response have not been successful, Bookman[256] reported training pigeons to detect the presence of magnetic fields in a flight cage.

Walcott et al.[277] dissected pigeons with nonmagnetic tools and found magnetic material in head and neck sections. Most of the magnetic material was localized in a piece of tissue between the dura and the skull. Each pigeon had an inducible, remanent moment* of 10^{-8} to 10^{-9} $A{\cdot}^{m2}$, which disappeared at 575°C, indicating Fe_3O_4. Presti and Pettigrew[278] found magnetic material in the neck musculature of pigeons and migratory, white-crowned sparrows but did not find localized magnetic materials in the heads. Thus, the connection between the magnetic material and magnetic sensitivity was not definitely established. Yorke,[279] Kirschvink and Gould,[280] and Presti and Pettigrew,[278] have speculated on the possible role of magnetite, Fe_3O_4 in a magnetic sensor. Yorke pointed out that if a pigeon could somehow measure the total magnetization of its ensemble of magnetic particles, there is enough magnetic material present to indicate the field direction with high accuracy.

Migratory birds are also able to use the geomagnetic field as a compass to find and maintain direction (see reviews by Able[274] and by Wiltschko[5,276]). As in pigeons, the compass is an inclination compass referring to dip angle, rather than polarity. Beason studied the bobolink (*Dolichonyx oryzivorus*),[281] a nocturnal migrant that integrates information from several sources to determine the preferred direction. Magnetic and histological studies revealed the presence of magnetic material, probably magnetite, in the ethmoidal region. Electrophysiological studies have revealed that the trigeminal nerve system of the boblink responds to changes in magnetic field intensity and direction.[263,282] Responses to magnetic field changes were also recorded in the optic tectum and pineal gland. It has been suggested that there are two separate magentic systems, with at least one involving magnetite for transduction of magnetic field information.

A possible connection between magnetic material and magnetic field effects on behavior was also found in honeybees. The behavioral effects have been reviewed by von Frisch,[283] Martin and Lindauer[284,285] and Gould.[246] Honeybee workers communicate the location of a food source to other workers in a hive by means of a "waggle dance" on a vertical honeycomb. The angle between the direction of the dance and the vertical direction indicates the angle between the food source and the sun. Consistent errors in the dance angle occur which vanish when the magnetic field in the hive is nulled by means of external coils. In anomalous situations where bees are made to dance on horizontal surfaces, after an initial period of disorientation they dance along the eight magnetic, compass directions (N, NE, E, SE, etc.).[246,284] If the geomagnetic field is nulled, the dances become disoriented. Evidence has also been reported that bees can use the diurnal variations in the geomagnetic field to set their circadian rhythms.[284]

Walker and Bitterman have reported conditioning experiments that demonstrate the sensitivity of free-flying bees but not stationary bees to local magnetic anomalies.[259] The threshold sensitivity was determined to be ca. 0.6% of the background geomagnetic field.[286] Magnetized wires attached to the bees abdomen disrupted the ability of the bees to detect the anomalies.[286]

Gould et al.[287] have found that honeybees contain magnetic material. They measured an average, induced remanent moment of about 2×10^{-9} $A{\cdot}m^2$ per bee, distributed between single-magnetic-domain and superparamagnetic particles. The magnetic material was mostly localized in the abdomen. Kuterbach et al.[288] found bands of cells around the abdominal segments in honeybees that contained numerous iron-rich granules. The granules were primarily a hydrous iron oxide, which can be a precursor in the precipitation of Fe_3O_4.[289]

Evidence for magnetic sensitivity has been obtained for bony fish including several species of salmon and trout. Walker observed unconditioned and conditioned responses to local magnetic field anomalies in actively swimming yellowfin tuna (*Thunnus albacares*).[258] Magnetic studies of tuna indicated the presence of magnetite particles associated with the dermethmoid tissue in the head.[291] Magnetite particles extracted from the ethmoid region in sockeye salmon were found to be single magnetic domains arranged in strings.[292] Single magnetic domain particles have also been found in chinook salmon.[293]

In addition to the cases cited above, magnetic inclusions, principally Fe_3O_4, occur widely in the biological world.[8,10] Magnetic material has been reported in organisms as diverse as dolphins,[294] butterflies,[295] tuna,[291] green turtles,[296] marine crustacea,[297] bacteria,[298] and humans.[299] The first identification of Fe_3O_4 in an organism was by Lowenstam,[300] who found it in the tooth denticles on the *radulae* of chitons, a group of mollusks. Fe_3O_4 is very hard as well as magnetic which is advantageous to chitons since they scrape algae off rocks. Kirschvink et al.[299] have recently reported the detection and identification of magnetite particles in human brain tissue. The concentration is ca. 5 nanograms per gram of tissue, out of a total iron concentration of ca. 200 micrograms per gram tissue. The neurophysiological function or significance is not known, but a considerable fraction is in the form of single-magnetic domain cubo-octahedra.

The best documented connection between magnetically sensitive behavior and the presence of magnetic material is for motile, aquatic bacteria that orient and migrate along geomagnetic field lines.[4,240,301,302] This behavior, magnetotaxis, is exhibited by a number of freshwater and marine bacteria; the diversity of morphological types suggests that the phenomenon is a feature of a number of bacterial species.[301,303] Magnetotactic bacteria are common in the sediments of almost any aquatic environment. They are also localized in horizontal plates at specific depths in water columns with vertical chemical and redox gradients, principally at the oxic-anoxic transition zone.[304] Several magnetotactic bacteria have been cultured axenically.[305,306]

Individual magnetotactic bacterial cells contain intracytoplasmic mineral particles of magnetite (Fe_3O_4),[298,306,307] or greigite (Fe_3S_4),[308–310] usually arranged in chains. Greigite is isostructural with magnetite and is also magnetically ordered at ambient temperatures. The particles in a given species or strain are characterized by a narrow size distribution and a specific crystalline habit.[310] A number of particle habits for both magnetite and greigite have been elucidated by high resolution transmission electron microscopy and electron diffraction studies.[310,311] Some of the habits are equilibrium forms, i.e., they preserve the symmetry of the fcc unit cell, while others are of lower symmetry, implying the relative acceleration or retardation of the growth of certain cystal faces. While most magnetotactic bacteria contain only one particle type, one organism, a multicellular prokaryote,[312] contains particles of nonmagnetic pyrite, FeS_2, as well as greigite.[308] Another organism, a large marine rod, contains particles of magnetite and greigite with different morphologies, arranged within the same chain of particles.[313]

The mineral particles in magnetotactic bacteria are enclosed in a membrane within the cell;[314] a particle and its enveloping membrane is known as a magnetosome.[315] In *Aquaspirillum magnetotacticum,* the membrane has a protein profile that is different than that of the cytoplasmic membrane. The membranes form or are attached to an ultrastructural entity that organizes the magnetosomes into chains, and holds the particles in a fixed position within the cell.

While the details of magnetosome chain formation in magnetotactic bacteria are unknown, it is thought to involve two separately controlled processes, an ultrastructural process associated with chain assembly, and a mineralization process associated with particle formation.[313]

Phylogenetic analyses of magnetite- and greigite-forming bacteria show that the two groups are distantly related, suggesting that that different mineralization processes are involved in the two mineral particle types.[316]

The magnetosome chain is a hierarchical structure that constitutes a permanent magnetic dipole within each cell.[317] The magnetosome particles are typically with the single-magnetic-domain size ranges for magnetite and greigite. When arranged in chains, the individual particle moments are oriented parallel to each other along the chain axis. The orientation of the net cellular dipole moment by the geomagnetic or ambient magnetic field causes the cell to migrate along the field lines. Thus magnetotaxis is a passive process and each cell is, in effect, a motile biomagnetic compass needle. Cells can use other sensory responses, such as aerotaxis in conjunction with magnetotaxis efficiently to find and maintain position in chemical gradients.[318]

The best documented behavioral evidence for the involvment of light in magnetic orientation is for Eastern red-spotted newts (*Notophthalmus viridescens*). The ability of these organisms to use the geomagnetic field for shoreward orientation was studied by Phillips,[319,320] who investigated whether the magnetoreception mechanism of shoreward-orienting newts is axial or polar, i.e., whether it determines the magnetic axis, or determines a preferred direction along the magnetic axis. This was done by studying the orientation responses of the animals following inversion of the vertical component of the ambient magnetic field. It was found that newts that had been trained to associate shoreward orientation with the ambient magnetic field changed their orientation by about 180 degrees in response to inversion of the vertical component of the field, compared to controls for which the vertical component was not inverted. Since inversion of the vertical component left the horizontal component of the ambient field unchanged, if the magnetoreception mechanism were polar, the orientation of the animals should not have changed. The behavioral response of the newts therefore suggests that the magnetoreceptor mechanism in shoreward orientation is axial, and that the animal relies on the magnetic field inclination to discriminate between the two directions along the horizontal projection of the magnetic axis.

Newts also use geomagnetic field cues in homing behavior when displaced long distances. In this case, the mechanism appears to be polar, because inversion of the vertical component of the ambient field did not affect homing orientation. Thus newts apparently have two magnetoreception modalities.

Phillips subsequently studied red-spotted newts which were trained in natural (full spectrum) light to maintain a consistent orientation with respect to the ambient magnetic field, by association with orientation toward an artificial shore. They were subsequently tested in an arena where the horizontal component of the ambient magnetic field, as well as the wavelength composition of the ambient light, could be varied. Under full spectrum illumination, newts maintained the same orientation to the horizontal component of the ambient magnetic field as in the training regime. In the dark, the newts were randomly oriented. Under monochromatic short wavelength light (400 or 450 nm), the newts were oriented as under full spectrum light. However, under long wavelength light (500, 550 or 600 nm), the newts were oriented at 90 degrees (anti-clockwise) to the full spectrum direction. Under 475 nm light, newts were oriented at random. When newts were trained under long wavelength (>500 nm) light, they maintained the same orientation with respect to the ambient magnetic field when tested under long wavelength light, but were oriented at 90 degrees (clockwise) to the long-wavelength direction when tested under full spectrum conditions. The experimental results can be understood in terms of magnetic field modulation of photoreceptors by light. Phillips has hypothesized two types of light dependent receptors, or two spectral mechanisms in the same cell, with long- and short-wavelength response respectively, that are sensitive to the ambient magnetic field. The two receptor types, or mechanisms, give antagonistic neural inputs, but the short-wavelength receptor, or mechanism, has greater light-intensity sensitivity and predominates under full spectrum conditions.

Evidence for wavelength-dependent effects of light on magnetic compass orientation in fruit-flies (*Drosophila melanogaster*)[324] and in migratory birds[251,325] has also been reported.

On the other hand, Lohmann has reported[326] that loggerhead and leatherback sea turtle hatchlings can orient to the geomagnetic field in complete darkness. Magnetic orientation in darkness also occurs in other organisms. In turtles, magnetic orientation is axial as in newts. Lohmann has proposed that magnetoreception in darkness could still involve photoreceptors based on magnetic field dependent biochemical reactions. However, there is no current evidence for magnetic field modulation of light independent biochemical reactions in the retina or elsewhere.

Lohmann and Willows[248] studied the spontaneous orientation of a marine mollusk, the nudibranch *Tritonia diomedea,* under dark conditions. The animals spontaneously adopted a significant, preferential orientation with respect to the geomagnetic field. The orientation direction varied with the lunar phase. When the horizontal component of the geomagnetic field was cancelled with a coil system, the animals' preferential orientation vanished. Further experiments with migration in a maze showed that *T. diomedea* can use the geomagnetic field for directional cues. The organism has large, individually identifiable neurons, and intracellular electrical recordings[262] from at least one neuron has shown electrophysiological responses to changes in the ambient (earth-strength) magnetic field, such as rotation of the field. It has not yet been determined whether this neuron is the primary magnetoreceptor or part of the signal pathway from another source.

In conclusion, a magnetoreception organ or even mechanism has not been clearly identified in any organism except for magnetotactic bacteria[4] and skates and rays.[257] However, the recent results on the involvement of light in magnetoreception, and on electrophysiological responses to changes in the ambient magnetic field are very promising for finding a magnetoreceptor and elucidating a magnetoreception mechanism. It may turn out that there is more than one magnetoreception modality, based on magnetic particles,[280] magnetic effects on chemical reactions,[156,157] light absorption[327] or some combination of the three.

ACKNOWLEDGMENTS

We thank R. C. Beason, M. E. Bitterman, J. B. Phillips, K. J. Lohmann, C. Walcott, and A. O. D. Willows for discussions and communications. RBF was partially supported by the Office of Naval Research under contract N00014-91-J-1290. RPL was partially supported by the Office of Environmental Health Research and the the Office of Energy Management Systems of the Department of Energy, under contract DE-AC03-76SF000098, and from the N.I.H. under contract CA53711 from the N.C.I.

APPENDIX: ELECTROMAGNETIC UNITS AND DEFINITIONS

Purcell[328] gives an excellent discussion of magnetic field concepts; for a review of magnetic measurements see Morrish[329] and Foner.[330] In discussing the interactions of static magnetic fields with materials most workers use the centimeter-gram-second (cgs)-electromagnetic units (emu). In the SI system, the magnetic flux density is defined by

$$\mathbf{B} = \mu_0\mathbf{H}$$

and

$$\mathbf{B} = \mu_0(\mathbf{H} + \mathbf{M}) \tag{A1}$$

in vacuum and in a material medium, respectively. **H** is the magnetic field intensity and **M** is the magnetization per unit volume. The permeability of free space is given by

$$\mu_0 = 4\pi \times 10^{-7} \text{ H/m} \tag{A2}$$

where H/m (henry per meter) is equivalent to weber per meter per ampere. The volume magnetization of diamagnetic and paramagnetic materials is related to the magnetic field intensity by the magnetic susceptibility χ

$$\mathbf{M} = \chi\mathbf{H} \tag{A3}$$

In the SI system, χ is a dimensionless quantity. In magnetically ordered materials, **M** is a complex function of **H** and can have finite values even at **H** = 0.

In the cgs-emu system, $\mu_0 = 1$ and

$$\mathbf{B} = \mathbf{H} + 4\pi\mathbf{M} \tag{A4}$$

Some of the relations between emu and SI units are as follows:

Magnetic flux density B:	1 gauss = 10^{-4} Tesla
Magnetic field intensity H:	1 oersted = $(10^3/4\pi)$ amp/m
Magnetic moment μ:	1 emu = 1 erg/gauss = 10^{-3} amp·m^2
Magnetization M:	1 emu/cm^3 = 10^3 amp/m
Magnetic susceptibility χ:	χ(emu) = $(1/4\pi)$ χ(MKS)

In the emu system, Faraday's law of magnetic induction is

$$\epsilon = -10^{-8}\, d\Phi/dt \tag{A5}$$

where the emf (electromotive force) ϵ (volts) is induced in a conducting loop of area A normal to B, and the magnetic flux

$$\Phi = BA \tag{A6}$$

The emf can be produced by a time-varying field in a stationary loop or by a loop whose area perpendicular to the field direction is changing with respect to a static magnetic field.

The magnetic moment, magnetization, and magnetic susceptibility of materials are expressed on a unit weight, unit volume, unit mole, unit atom, or unit molecule basis. Magnetic moments and magnetization have the units emu, gauss, cgs, or ergs per gauss in the emu system, with 1 emu = 1, G = 1, cgs = 1, erg/G. For example, the saturation magnetic moment per unit volume, or magnetization, of Fe_3O_4 is 480 emu/cm^3, or 92 emu/gram. The conversion factor is the density. Magnetic moments of atoms and molecules are often expressed as bohr magnetons (μ_B) with 1 $\mu_B = 0.927 \times 10^{-20}$ ergs/G. The electron has a magnetic moment of 1 μ_B. Fe_3O_4 has a magnetic moment of 4 μ_B per formula unit.

The free energy of magnetic dipoles or of materials with permanent, macroscopic magnetization M oriented at angle Θ in a magnetic field of flux density B is

$$E_m = -\mathbf{M} \cdot B = -\mathbf{M}B \cos\Theta \tag{A7}$$

In a homogeneous magnetic field the free energy is a minimum when $\cos\Theta = 1$, i.e., **M** is parallel to **B**. In an inhomogeneous magnetic field, additional lowering of the free energy comes from translational motion of the material toward increasing field strength. The translational force along the gradient is

$$F_x = M\, dB/dx \tag{A8}$$

where the magnetic field gradient is taken along the x axis.

For materials without a permanent dipole moment, a magnetic field will induce a magnetization **M**, where

$$\mathbf{M} = \chi \mathbf{H} \tag{A9}$$

χ is the susceptibility tensor with units emu/G or ergs/G^2. For diamagnetic materials, χ is small and negative and is generally independent of temperature. For example, H_2O has an isotropic volume magnetic susceptibility $\chi_v = -0.4 \times 10^{-7}$ emu/(G·cm^3). In paramagnetic materials, i.e., materials with unpaired electrons, c is larger in magnitude, positive in sign, and is generally temperature-dependent. At ambient temperatures, typical paramagnets have susceptibilities that follow the Curie Law:

$$\chi = N\,(\mu_{eff})^2/3k_BT \tag{A10}$$

where N is the number density of paramagnetic atoms, μ_{eff} is the effective magnetic moment per atom and k_BT is Boltzmann's constant times temperature. According to quantum mechanics, the saturation magnetic moment of an atom or molecule is proportional to the total angular momentum

$$\mu = g\,\mu_B\,J \tag{A11}$$

where μB is the Bohr magneton, g is the proportionality constant known as the g-factor, and J is the angular momentum quantum number. If we consider spin angular momentum only, g = 2. The effective magnetic moment is then

$$\mu_{eff} = [g^2\,S\,(S + 1)\,\mu_B{}^2]^{1/2} \tag{A12}$$

In a hypothetical example, if every water molecule had an unpaired electron spin (S = 1/2), the magnetic moment per molecule would be 1 μ_B and the paramagnetic susceptibility per cubic centimeter at 300 K would be 2.2×10^{-6} emu/G. The total susceptibility would be the sum of paramagnetic and diamagnetic contributions, giving 2.16×10^{-6} emu/G. In some cases the diamagnetic susceptibility contribution can be larger than the paramagnetic contributions. This is often the case in large biological molecules that have a single or a few paramagnetic atoms.

In a magnetic field the magnetic energy is given by

$$E_m = -(1/2)\,\mathbf{H} \cdot \chi \cdot \mathbf{H} \tag{A13}$$

where the magnitude and direction of the induced moment depend on the orientation of the molecule in the field. If the susceptibility is isotropic, the induced moment is always parallel to **H** and

$$E_m = -(1/2)\,\chi\,H^2 \qquad \text{(A14)}$$

If χ is isotropic, there are no translational forces in a homogeneous magnetic field. In an inhomogeneous magnetic field, the material will experience a translational force in the direction of increasing or decreasing field strength depending on the sign of χ

$$F_x = \chi\,V\,H\,dH/dx \qquad \text{(A15)}$$

where χ is the susceptibility per unit volume and V is the volume. Diamagnetic materials move in the direction of decreasing field strength while paramagnetic (and ferro or ferrimagnetic) materials move toward increasing field strength. As discussed in Section II, this is the basis of a method for separating diamagnetic from paramagnetic materials. In suspensions or solutions, the force depends on the difference in susceptibility between the material and the medium.

Anisotropic materials require two or, at most, three independent parameters to specify the magnetic susceptibility. In the most general case,

$$E_m = -(1/2)\,(\chi_x H_x^2 + \chi_y H_y^2 + \chi_z H_z^2 \qquad \text{(A16)}$$

where x, y, and z denote the eigenvectors of the diagonalized susceptibility tensor. These three vectors often correspond to molecular symmetry axes. In addition to translational forces in inhomogeneous fields, there are rotational forces in homogeneous fields. For example, benzene has an inplane susceptibility $\chi\| = -4.5 \times 10^{-7}$ emu/G·cm^3 ($= -5.7 \times 10^{-5}$ in SI units) and a susceptibility normal to the plane $\chi\perp = -12 \times 10^{-7}$ emu/G·cm^3 ($= -1.5 \times 10^{-5}$ in SI units). The molecule will experience a torque, tending to align its plane parallel to the field direction. In general, any molecule or molecular assembly with anisotropic diamagnetism will tend to align so that the least negative susceptibility direction is parallel to the applied field.

In anisotropic paramagnets the highest (positive) susceptibility direction will tend to align parallel to the field. If $\chi\|$ is the susceptibility along the minimum energy direction and Θ is the angle between that direction and the applied field

$$E_m = -\,(1/2)\,[\chi\|H^2 - \Delta\chi\,H^2\cos^2\Theta]$$

$$\Delta\chi = \chi\| - \chi\perp \qquad \text{(A17)}$$

where by definition $\Delta\chi > 0$ and E_m is minimized when $\Theta = 0$. The degree of orientation of an ensemble of molecules at a given field strength and temperature can be calculated from statistical mechanics. The angular distribution function $F(\Theta)$, which specifies the probability that a molecule has an equilibrium orientation at angle Θ with respect to H, can in general be expanded in terms of $\cos^n\Theta$. For molecules with cylindrical symmetry, odd terms in n vanish. Then a convenient measure of the degree of orientation is the average value of the second Legendre polynomial

$$\langle P_2(\cos\Theta)\rangle = \langle (3/2)\cos^2\Theta - (1/2)\rangle \qquad \text{(A18)}$$

From statistical mechanics,

$$\langle P_2(\cos\Theta)\rangle = [\textstyle\int (3/2\cos^2\Theta - 1/2)\exp(-1/2\,\Delta\chi H^2\cos^2\Theta/k_BT)\,d\cos\Theta]/Z \qquad \text{(A19)}$$

where Z is the partition function. Complete alignment means $\langle\cos^2\Theta\rangle = 1$ and $\langle P_2(\cos\Theta\rangle =$

1. For random orientation in three dimensions $\langle \cos^2\Theta \rangle = 1/3$, hence $\langle P_2(\cos\Theta) \rangle = 0$. Even for molecules such as benzene with large diamagnetic anisotropies, thermal agitation will overcome magnetic alignment and the equilibrium alignment will be small. However, if N molecules are contained in an ordered array or aggregate, $\Delta\chi$ for the single molecule would be replaced by $N\Delta\chi$ in the exponential in Equation A19, resulting in substantial enhancement of alignment.

REFERENCES

1. **Tenforde, T. S.** Ed., *Magnetic Field Effects on Biological Systems,* Plenum Press, New York, 1979.
2. **Maret, G., Kiepenheuer, J., Boccara, N., Eds.,** *Biophysical effects of steady magnetic fields,* Springer Verlag, New York, 1986.
3. **Simon, N. J.,** Biological Effects of Static Magnetic Fields, International Cryogenic Materials Committee, Boulder, CO, 1992.
4. **Frankel, R. B., and Blakemore, R. P.,** Magnetite and magnetotaxis in microorganisms, *Bioelectromagnetics,* 10, 223, 1989.
5. **Wiltschko, W. and Wiltschko, R.,** Magnetic orientation in birds, *Current Ornithology, Vol. 5,* Ed. Johnston, R. F., Plenum, New York, 1988, 67.
6. **Tenforde, T. S.,** Electroreception and magnetic reception in simple and complex organisms, *Bioelectromagnetics,* 10, 215, 1989.
7. **Blakemore, R. P.,** Biomagnetism and geomagnetic field detection by organisms, *Encyclopedia of Applied Physics, Vol. 2,* VCH Publishers, Berlin, 1991, 421.
8. **Frankel, R. B., and Blakemore, R. B.,** *Iron Biominerals,* Plenum Press, New York, 1991.
9. **Mann, S., Webb, J., Williams, R. J. P., Eds.,** *Biomineralization: Chemical, and Biochemical Perspectives,* VCH, New York, 1989.
10. **Kirschvink, J. L., Jones, D. S., and MacFadden, B. J., Eds.,** *Magnetic Biomineralization and Magnetoreception in Organisms,* Plenum Press, New York, 1985.
11. **Kirschvink, J. L.,** Magnetite biomineralization and geomagnetic sensitivity in higher animals; an update and recommendation for future study, *Bioelectromagnetics,* 10, 239, 1989.
12. **Magin, R. L., Liburdy, R. P. and B. Persson (Eds.)** Biological Effects and Safety Aspects of Nuclear Magnetic Resonance Imaging and Spectroscopy, *Annals of the N.Y. Acad. Sci,* 649, 1992.
13. **Persson, B. R. R. and Stahlberg, F.** *Health and Safety of Clinical NMR Examinations,* CRC Press, Inc., Boca Raton, FL, 1989.
14. **Beers, G. J.,** Biological Effects of Weak Electromagnetic Fields From 0 Hz to 200 MHz: A Survey of the literature with Special Emphasis on Possible Magnetic Resonance Effects. *Mag. Res. Imag.* 7, 309, 1989.
15. **Kanal, E. Shellock, F. G. and Talagala, L.** Safety Consideration in MR Imaging, *Radiology,* 176, 593, 1990.
16. **Shulman, R. G.,** *Biological Applications of Magnetic Resonance,* Academic Press, New York, 1979.
17. **Shulman, R. G.,** NMR spectroscopy of living cells, *Sci. Am.,* 248(1), 86, 1983.
18. **Ackerman, J. H., Grove, T. H., Wong, G. G., Gadion, D. G., and Radda, G. K.,** Mapping of metabolites in whole animals by 31P NMR using surface coils, *Nature,* 283, 167, 1980.
19. **Fossell, E. T., Morgan, H. E., and Ingwall, J. S.,** Measurement of changes in high-energy phosphates in the cardiac cycle using gated ^{31}P nuclear magnetic resonance, *Proc. Natl. Acad. Sci. U.S.A.,* 77, 3654, 1980.
20. **Wutrich, K.** Protein determination in solution by nuclear magnetic resonance spectroscopy, *Science* 243, 45, 1989.
21. **Clore, G. M., and Gronenborn, A. M.,** Structures of large proteins in solution: three and four dimensional heteronuclear NMR spectroscopy, *Science* 252, 1390, 1991.
22. **Shaanan, B., Gronenborn, A. M., Cohen, G. H., Gilliland, G. L., Veerapandian, B., Davies, D. R. and Clore, G. M.,** Combining experimental information from crystal and solution studies: joint x-ray and NMR refinement, *Science* 257, 961, 1992.
23. **Mansfield, P. and Morris, P. G.,** *NMR Imaging in Biomedicine,* Academic Press, New York, 1982.
24. **Kaufman, L., Crooks, L. E., and Margulis, A. R., Eds.,** *Nuclear Magnetic Resonance Imaging in Medicine,* Igaku-Shoin, New York, 1982.
25. **Liburdy, R. P.** Biological Interactions of Cellular Systems with Time-Varying Magnetic Fields, *Annals New York Acad. Sci,* 649, 74, 1992.
26. **Reilly, J. P.** Principles of Nerve and Heart Excitation by Time-Varying Magnetic Fields, *Annals New York Acad. Sci,* 649, 96, 1982.

27. **Shellock, F. S.** Thermal Responses of Human Subjects Exposed to Magnetic Resonance Imaging, *Annals New York Acad. Sci,* 649, 260, 1992.
28. **Gordon, C. J.,** Local and Global Thermoregulatory Responses to MRI Fields, *Annals New York Acad. Sci,* 649, 273, 1992.
29. **Shellock, F. G. and Curtis, J. S.** MR Imaging and Biomedical Implants, Materials, and Devices: An Updated Review., *Radiology,* 180, 541, 1991.
30. **Kanal, E., Gillen, J, Evans, J. A., Savitz, D. A. and Shellock, F. G.** Survey of Reporductive Health Among Female MR Workers. *Radiology,* 187, 395, 1993.
31. **Schenk, J. F., Dumoulin, C. L., Redington, R. W., Kressel, H. Y., Elliot, R. T. and McDougall, I. L.** Human Exposure to 4.0 Tesla Magnetic Fields in a Whole-Body Scanner. *Medical Physics,* 19, 1089, 1992.
32. **Weiss, J. Herrick, R. C., Taber, K. H., Contant, C. and Plishker, G. A.** Bio-Effects of High Magnetic Fields: A Study Using a Simple Animal Model, *Magentic Resonance Imaging,* 10, 689, 1992.
33. **Muller, S. and Hotz, M.** Human Brainstem Auditory Evoked Potentials (BAEP) Before and After MR Examinations. *Magnetic Resonance in Medicine,* 16, 476, 1990.
34. **Hotz, M. A., Muller, S., Allum, J. H. and Pfaltz, C. R.,** Human Auditory-Evoked Potentials Before and After Magnetic Resonance Imaging, *European Archives of Oto-Rhino-Laryngology,* 249, 85, 1992.
35. **Buetnner, U. W.,** Human Interactions with Ultra High Fields, *Annals New York Acad. Sci,* 649, 59, 1992.
36. **Hong, C. Z. and Shellock, F. G.** Short-Term Exposure to a 1.5 Tesla Static Magnetic Field Does Not Affect Somato-Sensory-Evoked Potentials in Man. *Magnetic Resonance Imaging,* 8, 65, 1990.
37. **Sweetland, J., Kertesz, A., Prato, F. S. and Nantau, K.,** The Effect of Magnetic Resonance Imaging on Human Cognition. *Magnetic Resonance Imaging,* 5, 129, 1987.
38. **Ossenkopp, K.-P., Innis, N., Prato, F. and Sestini, E.,** Behavorial Effects of Exposure to Nuclear Magnetic Resonance Imaging: I. Open Field Behavior and Passive Avoidance Learning in Rats. *Magnetic Resonance Imaging,* 4, 275, 1986.
39. **Innis, N., Ossenkopp. K.-P., Prato, F. and Sestini, E.,** Behavioral Effects of Exposure to Nuclear Magnetic Resonance Imaging: II. Spatial Memory Tests. *Magnetic Resonance Imaging,* 4 281, 1986.
40. **Prato, F. S., Ossenkopp, K.-P., Kavaliers, M. Uksik, P., Nicholson, R. L., Drost, D. and Sestini, E. A.,** Effects of Exposure to Magnetic Resonance Imaging on Noctural Serum Melatonin and Other Hormone Levels in Adult Males: Preliminary Findings., *J. Bioelectricity,* 7, 169, 1989.
41. **Teskey, G. C., Ossenkopp, K.-P., Prato, F. and Sestini, E.,** Survivability and Long-Term Stress Reactivity Levels Following Repeated Exposure to Nuclear Magnetic Resonance Imaging Procedures in Rats, *Physiological Chemistry & Physics & Medical NMR,* 19 43, 1987.
42. **Prato, F. S. Kavaliers, M., Ossenkopp, K.-P., Carson, J. J. L., Drost, D. J. and Frappier, R. H.,** Extremely Low Frequency Magnetic Field Exposure from MRI/MRS Procedures. *Annals New York Acad. Sci,* 649, 44, 1992.
43. **Prato, F. S., Wills, J. M., Kavaliers, M., Galea, L. A. M., Saksida, L. M. and Ossenkopp, K.-P.** Magnetic Resonance Imaging at 1.5 Tesla Attenuates Opoid Analgesia: Evidence From Morphine- and an Enkephalinase Inhibitor-Induced Analgesia in Snails. In *Electricity and Magnetism in Biology and Medicine* (M. Blank, Ed), 1993, 752.
44. **Teskey, G. C., Prato, F. S., Ossenkopp, K.-P. and Kavaliers, M.,** Exposure to Time Varying Magnetic Fields Associated with Magnetic Resonance Imaging Reduces Fentanyl-Induced Analgesi in Mice. *Bioelectromagnetics,* 9, 167, 1988.
45. **Prato, F. S., Ossenkopp, K.-P., Kavaliers, M., Sestini, E. and Teskey, G. C.,** Attemuation of Morphine-Induced Analgesi in Mice by Exposure to Magnetic Resonance Imaging: Separate Effects of the Static, Radiofrequency and Time-Varying Magnetic Fields. *Magnetic Resonance Imaging,* 5, 9, 1987.
46. **Johnson, C. C. and Guy, A. W.** Non-ionizing Electromagnetic Wave Effects in Biological Materials and Systems. *Proc. I.E.E.E.* 60, 692, 1972.
47. **Michaelson, S. M.** Microwave and Radiofrequency Radiation. In *Ionizing Radiation Protections.* Copenhagen, WHO, 1982, 97.
48. **Heinrichs, W. L., Fong, P., Flannery, M., Heinrichs,** Gestational Exposure of Pregnant Balb/c Mice to Magnetic Resonance Imaging Conditions. *Magnetic Resonance Imaging,* 6,305, 1988.
49. **Tyndall, D. A.** Magnetic Resonance Imaging and X-Ray Effects on Eye Development in C57BL/6J Mice, *Annals. N.Y. Acad. Sci,* 649, 363, 1992.
50. **Behr, K. P., Tiffe, H. W., Luders, H., Friederichs, M. and Hundeshagen, H.,** The Effect of Magnetic Resonance treatment on Chicken Embryos, *Dtw. Deutsche Tierarztliche Wochenschroft,* 98, 149, 1991.
51. **Kwong-Hing, A., Sandhu, H. S., Prato, F. S., Frappier, J. R. H. and Kavaliers, M.,** Effects of Magnetic Resonance Imaging (MRI) on the Formation of Mouse Dentin and Bone, *J. of Experimental Zoology,* 252, 53, 1989.
52. **Hayes, R. L, Pechura, C. M., Povishock, J. T. and Becher, D. P.** Changes in Blood-Brain Barrier Function Associated with Conditioned Fear, In *Brain Oedema* (Eds. Y. Inaka & M. Spatz), Springer-Verlag, Berlin, 1985, 224.

53. **Persson, B. R. R., Salford, L. G, Brun, A., Eberhardt, J. L. and Malmgren,** Increased Permeabiliity of the Blood-Brain Barrier Induced by Magnetic and Electromagnetic Fields, *Annals N.Y. Acad. Sci.*, 649, 356, 1992.
54. **Shivers, R. R., Kavaliers, M., Teskey, G. C., Prato, F. S. and Pelletier, R.-M.,** Magnetic Resonance Imaging Temporarily Alters Blood-Brain Barrier Permeability in the Rat. Neuroscience Letters, 76, 25 (1987).
55. **Prato, F. S., Frappier, J. R. H., Shivers, R. R., Kavaliers, M., Zabel, P., Drost, D. and Lee, T. Y.,** Magnetic Resonance Imaging Increases the Blood-Brain Barrier Permeability to 153-Gadolinium Diethylene-triaminepentaacetic Acid in Rats, *Brain Research,* 523, 301, 1990.
56. **Garber, H. J., Oldendorf, W. H., Braun, L. D. and Lufkin, R. B.** MRI Gradient Fields Increase Brain Mannitol Space, *Magnetic Resonance Imaging,* 7, 605, 1989.
57. **Adzamli, I. K., Jolesz, F. A. and Blau, M.** An Assessment of Blood-Brain Barrier Integrity Under MRI Conditions: Brain Uptake of Radiolabeled Gd-DTPA and In-DTPA-IgG. *Journal of Nuclear Maedicine,* 30, 839, 1989 (Abstract).
58. **Liburdy, R. P., de Manincor, D. J., Roos, M. S. and Brennan, K. M.,** Permeability of the Blood-Brain Barrier of the Rat is not Significantly Altered by NMR Exposure. *Annals N.Y. Acad. Sci.*, 649, 345, 1992.
59. **Preston, E., Butler, K. and Hass, N.,** Does Magnetic Resonance Imaging Compromise Integrity of the Blood-Brain Barrier, *Neuroscience Letters,* 101, 46, 1989.
60. **Carson, J. J. L., Prato, F. S., Drost, D. J., Diesbourg, L. D. and Dixon, S. J.,** Time-Varying Magnetic Fields Increase Cytosolic Free Calcium in HL-60 Cells. *Amer. J. Physiol. (Cell Physiol.)* 28, C687, 1990.
61. **Liburdy, R. P.** Calcium Signaling in Lymphocytes and ELF Fields: Evidence for an Electric Field Metric and a Site of Interaction Involving the Calcium Ion Channel, *FEBS Letters,* 301, 53, 1992.
62. **Yost, M. G. and Liburdy, R. P.** Time-Varying and Static Magnetic Fields Act in Combination to Alter Calcium Signal Tansduction in the Lymphocyte, *FEBS Letters,* 296, 117, 1992.
63. **Liburdy, R. P., Falk, M. S. and de Manincor, D. J.** Ultrahigh Static Magnetic Fields and Signal Transduction in the Lymphocyte. In *Electricity and Magnetism in Biology & Medicine* Blank, M., Ed., San Francisco Press, San Francisco, 1993, 749.
64. **Carson, J. J. L., Prato, F. S., Drost, D. J., Diesbourg, L. D. and Dixon, S. J.,** Time-Varying Magnetic Fields Increase Cytosolic Free Calcium in HL-60 Cells, *Am. J. Physiol.* 259 (Cell Physiol. 28), C687, 1990.
65. **Kobayashi, H. and Sakuma, S.,** Biological Effects of static Magnetic Field on Cultured Mammalian Cells and Combined Effects with 60Co Gamma-Rays., *Nippon Igaku Hoshasen Gakkai Zasshi. Nippon Acta Radioligica,* 52, 1679, 1992.
66. **Welker, H. A., Semm, P., Willing, R. P., Commentz, J. C., Wilthschko, W., Vollrath, L,** Effects of an artificial magnetic field on serotonin N-acetyltransferase activity and melatonin content of the rat pineal gland, Exp. Brain Res. 50, 426, 1983.
67. **Semm, P.,** Neurobiological investigations on the magnetic sensitivity of the pineal gland in rodents and pigeons, *Comp. Biochem. Physiol.*, 76A, 683, 1983.
68. **Reuss S, and Olcese, J.,** Magnetic field effects on the rat pineal gland: Role of retinal activation by light, Neurosci. Lett. 64, 97, 1986.
69. **Lerchl, A, Nonaka, K. O., and Reiter, R. J.,** Pineal gland magnetosensitivity to static magnetic fields is a consequense of induced electic currents (eddy currents), J. Pineal Res. 10, 109, 1991.
70. **Stehle, J., Reuss, S., Schroder, H., Henschel, M,. Vollrath, L.,** Magnetic field effects on pineal N-acetyltransferase activity and melatonin content in the gerbil: Role of pigmentaion and sex. Physiol. Behav. 44, 91, 1988.
71. **Cremer-Bartels, G., Krause, K., and Kuchle, H. J.,** Influence of low magnetic field strength variations on retinal and pineal glands of quail and humans, *Graefe's Arch. Clin. Exp. Ophthalmol.*, 220, 248, 1983.
72. **Cohen, D.,** Magnetic fields of the human body, *Phys. Today,* 28, 34, 1975.
73. **Williamson, S. J. and Kaufman, L.,** Magnetic fields of the human brain, *J. Magn. Magn. Mater.*, 15–18, 1548, 1980.
74. **Erne, S. N., Hahlebohm, H. D., and Lubbig, H., Eds.,** *Biomagnetism,* de Gruter, Berlin, 1982.
75. **Farrell, D. E., Tripp, J. H., Zanzucchi, P. E., Harns, J. W., Brittenham, G. M., and Muir, W. A.,** Magnetic measurement of human iron stores, *IEEE Trans. Magn.*, 16, 1818, 1980.
76. **Williamson, S. J., Romani, G. L., Kaufman, L., and Modena, I., Eds.,** *Biomagnetism,* Plenum Press, New York, 1983.
77. **Weinberg, H., Stroink, G., and Katila, T., Eds.,** *Biomagnetism: Applications and Theory,* Pergamon Press, New York, 1985.
78. **Lobel, D., Hale, J. R., and Montgomery, D. B.,** A new magnetic technique for the treatment of giant retinal tears, *Am. J. Ophthalmol.*, 85, 699, 1978.
79. **Hendren, W. H. and Hale, J. R.,** Electromagnetic bougienage to lengthen esophageal segments in congenital esophageal atresia, *N. Engl. J. Med.*, 293, 428, 1975.

80. **Hale, J. R.,** Medical applications of magnetic devices, *IEEE Trans. Magn.,* 11, 1405, 1975.
81. **Hsieh, D. S. T., Langer, R., and Folkman, J.,** Magnetic modulation of release of macromolecules from polymers, *Proc. Natl. Acad. Sci. U.S.A.,* 78, 1863, 1981.
82. **Kolin, A.,** Magnetic fields in biology, *Phys. Today,* 21, 39, 1968.
83. **Salles-Chuna, S. X., Battocletti, J. H., and Sances, A.,** Steady magnetic fields in non-invasive electromagnetic flowmetry, *Proc. IEEE,* 68, 149, 1980.
84. **Gaffey, C. T. and Tenforde, T. S.,** Alterations in the rat electrocardiogram induced by stationary magnetic fields, *Bioelectromagnetics,* 2, 357, 1981.
85. **Widder, K. J., Senyei, A. E., and Sears, B.,** Experimental methods in cancer therapeutics, *J. Pharm. Sci.,* 71, 379, 1982.
86. **Hirschbein, B. L., Brown, D. W., and Whitesides, G. M.,** Magnetic separations in chemistry and biochemistry, *Chem. Tech.,* March, 172, 1982.
87. **Owen, C. S.,** High gradient magnetic capture of cells and ferritin-bound particles, *IEEE Trans. Magn.,* 18, 1514, 1982.
88. **Owen,, C. S., Babu, V. M., Cohen, S. W., and Maurer, P. H.,** Magnetic enrichment of antibody secreting cells, *J. Immunol. Methods,* 51, 171, 1982.
89. **Tagle, D. A., Swaroop, M., Lovett, M., and Collins, F. S.,** Magnetic bead capture of expressed sequences encoded within large genomic segments, *Nature,* 361, 751, 1993.
90. **Melville, D., Paul, F., and Roath, S.,** Direct separation of red cells from whole blood, *Nature,* 255, 706, 1975; Fractionation of blood components using high gradient magnetic separation, *IEEE Trans. Magn.* 18, 1680, 1982.
91. **Kronick, P. L., Campbell, G. L., and Joseph, K.,** Magnetic microspheres prepared by redox polymerization used in a cell separation based on gangliosides, *Science,* 200, 1074, 1978.
92. **Kurinobu, S. and Uchiyama, S.,** Recovery of plankton from red tide by HGMS, *IEEE Trans. Magn.,* 18, 1526, 1982.
93. **deLatour, C.,** Magnetic separation in water pollution control, *IEEE Trans. Magn.,* 9, 314, 1973.
94. **Liburdy, R. P., Tenforde, T. S. and Magin, R. L.** Magnetic Field-Induced Drug Permeability in Liposome Vesicles, *Radiation Research,* 108, 102, 1986.
95. **Gregoriadis, G. and Patel, H. M. (Eds.)** *Liposomes in Drug Delivery,* Harwood Publ., 1992.
96. **Gregoriadis, G. (Ed.),** *Liposome Technology, Vol. I, II, III* CRC Press, Inc,. Bacon Raton, FL., 1984.
97. **Gregoriadis, G., Florence, A. T. and Patel, H. M. (Eds.),** Drug Targeting and Delivery. Vol.2: Liposomes in Drug Delivery, Harwood Publ., U.K., 1992.
98. **Harwood, J. L.** Understanding Liposomal Properties to Aid Their Clinical Usage, *Trends. in Biochem. Sci.* 17, 203, 1992.
99. **Liburdy, R. P.** *Electromagnetic Field Triggered Drug and Chemical Delivery Via Liposomes,* U.S. Patent #5, 190,761, 1993.
100. **Saalman, E., Norden, B., Arvidsson, L., Hammnerius, Y., Hojevik, P., Connell, K. E. and Kurucsev, T.** Effect of 2.45 GHz Microwave Radiation on Permeability of Unilamellar Liposomes to 5(6)-Carboxyfluorescein. Evidence of Non-thermal Leakage. *Biochim. et Biophys. Acta,* 1064, 124, 1991.
101. **Tenforde, T. S. and Liburdy, R. P.,** Magnetic Deformation of Phospholipid Bilayers: Effects on Liposome Shape and Solute Permeability at Prephase Transition Temperatures, *J. Theoretical Biology,* 133, 385, 1988.
102. **Liburdy, R. P. and Magin R. L.,** Microwave-Stimulated Drug Release from Liposomes, *Radiation Res.,* 103, 266, 1985.
103. **Liburdy, R. P.,** The Influence of Oscillating Electromagnetic Fields on Membrane Structure and Function: Synthetic Liposomes and the Lymphocyte cell Membrane with Direct Application to the Controlled Delivery of Chemical Agents, In Interaction Mechanisms of Low-Level Electyromagnetic Fields in Living Systems, (B. Norden and C. Ramel, Eds.) Oxford University Press, New York, 261–279, 1992.
104. **Hrushesky, W. J. M., Langer, R. and Theeuwes, F.(Eds.)** *Temporal Control of Drug Delivery,* Annals New York Acad. Sci, 618, 1991.
105. **Tsutsui H., Kinouchi, Y., Sasaki, H., Shiota, M. and Ushita, T.,** Studies on the SM-Co Magnet as a Dental Material, *J. Dental Research,* 58, 1597, 1979.
106. **Kinouchi, Y., Ushita, T., Tsutsui, H., Yoshida, Y., Sasaki, H. and Miyazaki, T.,** Pd-Co Dental Casting Ferromagnetic Alloys, *J. Dental Res.* 60, 50, 1981.
107. **Kawata, Y., Shiota, M., Tsutsui, H., Yoshida, Y., Sasaki, H. and Kinouchi, Y.** Cytotoxicity of Pd-Co Dental Casting Ferromagnetic Alloys. *J. Dental Res.,* 60, 1403, 1981.
108. **Kinouchi, Y., Ushita., T., Tsutsui, H. and Yoshida, Y.,** Optimization and Design of Permanent Magnet Devices for Retaining Dental Prosthetic Appliances, *IEEE Trans. Biomed. Engr.* BME-30, 201, 1983.
109. **Yoshida, Y., Kinouci, Y., and Nabeshima, T.** A Technique for Constructing Magnetic Attachments, *Quintessence of Dental Technology,* 9, 589, 1985.
110. **Sasaki, H., Kinouchi, Y., Tsutsui, H., Yoshida, Y. and Ushita, T.,** A Magnetic Attachment for Overdentures, *J. of Prosthetic Dentristry,* 51, 450, 1984.

111. **Sasaki, H. Kinouchi, Y., Tsutsui, H., Yoshida, Y., Karv, M. and Ushita, T.** Sectional Prostheses Connector By Samarium-Cobalt Magnets J. of Prostethic Dentristry, 52, 556, 1984.
112. **Maret, G. and Dransfeld, K.**, Macromolecules and membranes in high magnetic fields, *Physica,* 86–88B, 1077, 1977.
113. **Maret, G. and Dransfeld, K.**, Biomolecules and polymers in high steady magnetic fields, in *Applications of Strong and Ultrastrong Magnetic Fields,* Herlach, F., Ed., Springer-Verlag, Basel, 1981.
114. **Gretz, M. R., Folsom, D. B., and Brown, R. M., Jr.**, Cellulose biogenesis in bacteria and higher plants is disrupted by magnetic fields, Naturwissenschaften, 76, 380, 1989.
115. **Arnold, W., Steele, R., and Mueller, H.**, On the magnetic asymmetry of muscle fibers, *Proc. Natl. Acad. Sci. U.S.A.,* 44, 1, 1958.
116. **Geacintov, N. E., van Nostrand, F., Pope, M., and Tinkel, J. B.**, Magnetic field effects on the chlorophyll fluorescence in *Chlorella, Biochim. Biophys. Acta,* 226, 486, 1971.
117. **Chagneux, R., Chagneux, H., and Chalazonitis, N.**, Decrease in magnetic anisotropy of external segments of the retinal rods after a total photolysis. Biophvs. J., 18, 125, 1977.
118. **Chambre, M.**, Diamagnetic anisotropy and orientation of a-helix in frog rhodopsin and meta 11 intermediate, *Proc. Natl. Acad. Sci. U.S.A.,* 75, 5471, 1978.
119. **Murayama, M.**, Orientation of sickled erythrocytes in a magnetic field, *Nature,* 206, 420, 1965.
120. **Brody, A. S., Embury, S. H., Mentzer, W. C., Winkler, M. L., and Gooding, C. A.**, Preservation of sickle cell blood flow patterns during MR imaging: an in vivo study, Amer. J. Roentgenol. 151, 139, 1988.
121. **Torbert, J. and Maret, G.**, Fibers of highly oriented Pfl bacteriophage produced in a strong magnetic field, *J. Mol. Biol.,* 134, 843, 1979.
122. **Hong, F. T.**, Photoelectric and magneto-orientation effects in pigmented biologic membranes, *J. Coll. Interface Sci.,* 58, 471, 1977.
123. **Neugebauer, D. C., Blaurock, A. E., and Worcester, D. L.**, Magnetic orientation of purple membranes demonstrated by optical measurements and neutron scattering, *FEBS Lett.,* 78, 31, 1977.
124. **Torbet, J., Freyssinet, J. M., and Hudry-Clergon, G.**, Oriented fibrin gels formed by polymerization in strong magnetic fields, *Nature,* 289, 91, 1981.
125. **Freyssinet, J. M., Torbet, J., Hudry-Clergeon, G., and Mater, G.**, Fibrinogen and fibrin structure and fibrin formation measured by using magnetic orientation. Proc. Natl. Acad. Sci. (USA), 80, 1616, 1983.
126. **Murthy, N. S. and Yannas, I. V.**, Liquid crystallinity in collagen solutions and formation of collagen fibrils, private communication, 1982.
127. **Sperber, D., Dransfeld, K., Maret, G., and Weisenseel, M. H.**, Oriented growth of pollen tubes in strong magnetic fields, *Naturwissenschaften,* 68, 40, 1981.
128. **Audus, L. J.**, Magnetotropism: a new plant-growth response, *Nature,* 185, 132, 1960.
129. **Audus, L. J. and Whish, J. C.**, Magnetrotropism, in *Biological Effects of Magnetic Fields,* Barnothy, M. F., Ed., Plenum Press, New York, 1964, 170.
130. **Labes, M. M.**, A possible explanation for the effect of magnetic fields on biological systems, *Nature,* 211, 968, 1966.
131. **Aceto, H., Jr., Tobias, C. A., and Silver, I. L.**, Some studies on the biological effects of magnetic fields, *IEEE Trans. Magn.,* 6, 368, 1970.
132. **Gaffney, B. J. and McConnell, H. M.**, Effect of magnetic field on phospholipid membranes, *Chem. Phys. Lett.,* 24, 310, 1974.
133. **Kalkwarf, D. R. and Langford, J. C.**, Response of agarose solutions to magnetic fields, in *Biological Effects of Extreme Low Frequency Electromagnetic Fields,* Phillips, R. D. and Gillis, M. F., Eds., Conf. 781016, U.S. Department of Energy, 1979, 408.
134. **Haberditzl, W.**, Enzyme activity in high magnetic fields, *Nature,* 213, 72, 1967.
135. **Akoyunoglou, G.**, Effect of a magnetic field on carboxydismutase, *Nature,* 202, 452, 1964.
136. **Weissbluth, M.**, Enzyme-substrate reactions in high magnetic fields, in *Magnetic Field Effects in Biological Systems,* Tenforde, T., Ed., Plenum Press, New York, 1979, 44
137. **Stuchly, M. A., Stuchly, S. S., Liburdy, R. P. and Rousseau, D. A.** Dielectric properties of liposome vesicles at the phase transition, *Phys. Med. Biol.* 33, 1309, 1988.
138. **Maret, G. and Dransfeld, K.**, Biomolecules and polymers in high magnetic fields, *Top. Appl. Phys.* 57, 143, 1985.
139. **Braganza, L. F., Blott, B. H., Coe, T. J. and Melville, D.**, The superdiamagnetic effect on magnetic fields on one and two component multilamellar systems, *Biochim. Biophys. Acta,* 801, 66, 1985.
140. **Speyer, J. B., Sripada, P. K., Gupta, S. K. D., Shipley, G. G. and Griffin, R. G.**, Magnetic orientation of sphingomyelin-lecithin bilayers. *Biophys. J.,* 51, 687, 1987.
141. **Helfrich, W.**, Elastic properties of lipid bilayer: theory and possible experiments, *Z. Naturforschung,* 28C, 693, 1973.
142. **Smit, B. Hilbers, P. A. J., Esselink, K., Rupert, L. A. M., van Os, N. M. and Schlijper, A. G.**, Computer simulations of a water/oil interface in the presence of micelles, *Nature,* 348, 624, 1990.

143. **Israelachvilli, J. N and Wennerstrom, H.**, Entropic forces between ampiphilic surfaces in liquids, *J. Phys. Chem.*, 96, 520, 1992.
144. **Zasadzinski, J. A., Schneir, J. Gurley, J., Elings, V. and Hansma, P. K.** Scanning tunneling microscopy of freeze-fracture replicas of biomembranes, *Science,* 239, 1013, 1988.
145. **Liburdy, R. P. and Vanek, P. F.** Microwaves and the cell membrane. II. temperature, plasma, and oxygen mediate microwave-induced membrane permeability in the erythrocyte. *Radition Res.* 102, 190, 1985.
146. **Liburdy, R. P. and Vanek, P. F.** Microwaves and the cell membrane. III. protein shedding is oxygen and temperature dependent: evidence for cation bridge involvement. *Radiation Res.* 103, 382, 1987.
147. **Liburdy, R. P., Rowe, A. W. and Vanek, P. F.,** Microwaves and the cell membrane. IV. protein shedding in the human erythrocyte: quantitative analysis by high-performance liquid chromatography, 114, 500, 1988.
148. **Zakim, D., Kavecansky, J. and Scarlata, S.** Are membrane enzymes regulated by the viscosity of the membrane environment? *Biochemistry,* 31, 11589, 1992.
149. **Chiles, C., Hawrot, E., Gore, J. and Byck, R.,** Magnetic field modulation of receptor binding. Magn. Res. Me 10, 241–245, 1989.
150. **Aoki, H., Yamazaki, H. and Akagi, T.** Effects of static magnetic fields on membrane permeability of a cultured cell line, *Res. Communications in Chemical Pathology and Pharmacology* 69, 103, 1990.
151. **Papatheofanis, F.** Use of calcium channel antagonists as magnetoprotective agents, *Radiation Research,* 122, 24, 1990.
152. **Norimura, T., Imada, H.,, Kunugita, N., Yoshida, N. and Nikaido, M.** Effects of strong magnetic fields on cell growth and radiation response of human T-lymphocytes in culture, *Sangyo Ika Daigaku Zasshi* 15, 103, 1993.
153. **Short, W. O., Goodwill, L., Taylor, C. W., Job, C., Arthur, M. E. and Cress, A. E.** Alteration of human tumor cell adhesion by high-strength static magnetic fields, *Invest. Radiology* 27, 836, 1992.
154. **McDonald, F.** Effect of static magnetic fields on osteoblasts and fibroblasts *in vitro. Bioelectromagnetics,* 14, 187, 1993.
155. **Atkins, P. W.,** Magnetic field effects, *Chem. Br.,* 12, 214, 1976.
156. **Schulten, K.,** Magnetic field effects in chemistry and biology, *Adv. Solid State Phys.,* 22, 61, 1982.
157. **Schulten, K. and Windemuth, A.,** Model for a physiological magnetic compass, in Ref. 2.
158. **Steiner, U. E. and Ulrich, T.** Magnetic field effects in chemical kinetics and related phenomena, *Chem. Rev.* 89, 51, 1989.
159. **McLauchlan, K. A. and Steiner, U.,** The spin correlated radical pair as a reaction intermediate, *Molecular Physics* 73, 241, 1991.
160. **McLauchlan K. A.,** The effects of magnetic fields on chemical reactions, *Sci. Prog. Oxf.* 67, 509, 1981.
161. **Polk, C.** Dosimetry of extremely-low-frequency magnetic fields, *Bioelectromagnetics,* Supplement #1, 209, 1992.
162. **Bube, W., Haberkorn, R., and Michel-Beyerle, M. E.,** Magnetic field and isotope effects induced by hyperfine interactions in a steady-state photochemical experiment, *J. Am. Chem. Soc.,* 100, 5953, 1978.
163. **Swenberg, C. E.,** Theoretical remarks on low magnetic field interaction with biological systems, in *Magnetic Field Effects in Bioloigcal Systems,* Tenforde, T., Ed., Plenum Press, New York, 1979, 88.
164. **Hoff, A. J.,** Magnetic field effects on photosynthetic reactions, *Q. Rev. Biophys.,* 14, 599, 1981.
165. **Blankenship, R. E., Schaafsma, T. J., and Parson, W. W.,** Magnetic field effects on radical pair intermediates in bacterial photosynthesis, *Biochim. Biophys. Acta,* 461, 297, 1977.
166. **Pohl, H. A.,** *Quantum Mechanics for Science and Engineering,* Prentice-Hall, Englewood Cliffs, N.J., 1967, 81.
167. **Gaffey, C. T., Tenforde, T. S., and Dean, E. E.,** Alterations in the electrocardiograms of baboons exposed to dc magnetic fields, *Bioelectromagnetics, I,* 209, 1980
168. **Tenforde, T. S., Gaffey, C. T., Moyer, B. R., and Budinger, T. F.,** Cardiovascular alterations in *Macaca* monkeys exposed to stationary magnetic fields: experimental observations and theoretical analysis, *Bioelectromagnetics,* 4, 1, 1983.
169. **Stick, C., Hinkelman, K., Eggert, P., and Wendhausen, H.,** Do strong magnetic fields in NMR tomography modify tissue perfusion? Nuklearmedizin, 154, 326, 1991.
170. **Keltner, J. R., Roos, M. S., Brakeman, P. R. and Budinger, T. F.** Magnetohydrodynamics of blood flow, *Mag. Resonance in Med.* 16, 139, 1990.
171. **Kinouchi, Y., Tanimoto, T., Sato, K., Yamaguchi, H. and Miyamoto, H.** Effects of static magnetic fields in diffusion in solutions, *bioelectromagnetics* 9, 159, 1988.
172. **Barnothy, M. F., Ed.,** *Biological Effects of Magnetic Fields, Vol. I,* Plenum Press, New York, 1964; *Biological Effects of Magnetic Fields, Vol. 2,* Plenum Press, New York, 1969.
173. **Kholadov, Y. A.,** *Magnetism in Biology,* Nauka, Moscow, 1970.
174. **Presman, A. S.,** *Electromagnetic Fields and Life,* Plenum Press, New York, 1970.
175. **Kaufman, G. E. and Michaelson, S. M.,** Critical review of the biological effects of electric and magnetic fields, in *Biological and Clinical Effects of Low-Frequency Magnetic and Electric Fields,* Llaurado, J. G., Sances, A., Jr., and Battocletti, J. H., Eds., Charles C. Thomas, Springfield, IL. 1974, 49.

176. **Persinger, M. A.,** Ed., *ELF and VLF Electromagnetic Field Effects,* Plenum Press, New York, 1974.
177. **Silver, I. L. and Tobias, C. A.,** Magnetic fields and their biological effects, in *Space Radiation Biology and Related Topics,* Tobias, C. A. and Todd, P. W., Eds., Academic Press, New York, 1979, 258.
178. **Sheppard, A. R. and Eisenbud, M.,** *Biological Effect of Electric and Magnetic Fields of Extremely Low Frequency,* New York University Press, 1977.
179. **Committee on Biosphere Effects of Extremely-Low-Frequency-Radiation,** Biological effects of electric and magnetic fields associated with proposed project Seafarer, U.S. National Academy of Sciences, Washington, D.C., 1977.
180. **Dubrov, A. P.,** *The Geomagnetic Field and Life (Geomagnetobiologyl,* Plenum Press, New York, 1978.
181. **Ketchen, E. E., Porter, W. E., and Bolton, N. E.,** The biological effects of magnetic fields on man, *Am. Ind. Hyg. Assoc. J.,* 39, 1, 1978.
182. **St. Lorant, S. J.,** Biomagnetism: a review, in *Proc. 6th Int. Conf. Magn. Technol., Vol 6,* Alpha Press, Bratislava, 1978, 337.
183. **Taylor, L. S. and Cheung, A. Y.,** Eds., *The Physical Basis of Electromagnetic Interactions with Biological Systems,* Food and Drug Administration, U.S. Department of Health, Education and Welfare, Washington, D. C., 78–8055, 1978.
184. **Lie, R.,** *The Biological Effecfs of Magnetic Fields, a Bibliography,* Los Alamos Scientific Laboratory, Los Alamos, N.M. 7723-Ms, 1979.
185. **Phillips, R. D., Gillis, M. F., Kaune, W. T., and Mahlum, D. D., Eds.,** Biological effects of extremely low frequency electromagnetic fields, Conf. 781016, U.S. Department of Energy Symposium Series 50, 1979.
186. **Kinn, J. B. and Postow, E.,** Index of publications on biological effects of electromagnetic radiation (0–100 GHz), U.S. Environmental Protection Agency, 600/9-81-011, 1981.
187. **Kleinstein, B. H.,** Biological effects of non-ionizing electromagnetic radiation: a digest of current literature, Vols. I—5, U.S. Department of Commerce, 1974 to 1982.
188. **Budinger, T. F.,** Potential medical effects and hazards of human NMR studies, in *Nuclear Magnetic Resonance Imaging in Medicine,* Kaufman, L., Crooks, L. E., and Margulis, A. R., Eds., Igaku-Shoin, New York, 1982, 207.
189. **Easterly, C. E.,** Biological effects of static magnetic fields: a selective review with emphasis on risk assessment, Oak Ridge National Laboratory Report, Oak Ridge, Tenn., TM-7860, 1982.
190. **Liboff, R. L.,** A biomagnetic hypothesis, *Biophys. J., 5,* 845, 1965; Neuromagentic thresholds, J. *Theor. Biol.,* 83, 427, 1980.
191. **Wikswo, J. P., Jr. and Barach, J. P.,** An estimate of the steady magnetic field strength required to influence nerve conduction, *IEEE Trans. Biomed. Eng.,* 27, 722, 1980.
192. **Kim, Y. S.,** Some possible effects of static magnetic fields on cancer (review), *T. I. T. J. Life Sci.,* 6, 11, 1976.
193. **Cope, F. W.,** Biological sensitivity to weak magnetic fields due to biological superconductive Josephson junctions?, *Physiol. Chem. Phys., 5,* 173, 1973.
194. **Neurath, P. W.,** Simple theroetical models for magnetic interactions with biological units, in *Biological Effects of Magnetic Fields, Vol. 1,* Barnothy, M. F., Ed., Plenum Press, New York, 1964, 25.
195. **Roberts, A. M.,** Motion of *Paramecium* in static electric and magnetic fields, *J. Theor. Biol.,* 27, 97, 1970.
196. **Leask, M. J. M.,** A physiochemical mechanism for magnetic field detection by migratory birds and homing pigeons, *Nature,* 267, 144, 1977.
197. **McCann, J., Dietrich, F., Raferty, C. and Martin, A. O.,** A critical review of the genotoxic potential of electric and magnetic fields, *Mutation Research,* 297, 61 (1993).
198. **Mahlum, D. D., Sikov, M. R., and Decker, J. R.,** Dominant lethal studies in mice exposed to direct current magnetic fields, in *Biological Effects of Extremely Low Frequency Electromagnetic Fields,* Phillips, R. D. and Cillis, M. F., Eds., Conf. 781016, U.S. Department of Energy, 1979, 474.
199. **Kale, P. G. and Baum, J. W.,** Genetic effects of strong magnetic fields in *Drosophila melanogaster:* homogenous fields ranging from 13,000 to 37,000 gauss, *Environ. Mutagen., 1,* 371, 1979.
200. **Skopek, T., Fairfax, S., and Thilly, W. G.,** unpublished results.
201. **Ueno, S., Shiokawa, K. and Iwamoto,** Embryonic development of *xenopus laevis* under static magnetic fields up to 6.34 Tesla, J. Appl. Phys. 67, 5841, 1990.
202. **Prasad, N., Wright, D. A. and Forster, J. D.,** Effect of NMR on early stages of amphibian rana pipiens development, *Magn. Reson. Imaging,* 1, 35, 1982.
203. **Ho, M.-W., Stone, T. A., Jerman, I., Bolton, J., Bolton, H., Goodwin, B. C., Saunders, P. T. and Robertson, F.,** Brief exposures to weak static magnetic fields during early embryogenesis cause circular pattern abnormalities in drosophila larvae. *Phys. Med. Biol.* 37, 1171, 1992.
204. **Ramirez, E., Monteagudo, J. L, Garcia-Garcia, M. and Delgado, J. M. R.,** Oviposition and development of drosophila modified by magnetic fields. *Bioelectromagnetics,* 4, 315, 1983.
205. **Sikov, M. R., Malhum, D. D., Montgomery, L. D., and Decker, J. R.,** Development of mice after intrauterine

exposure to direct current magnetic fields, in *Biological Effects of Extremely Low frequency Electromagnetic Fields,* Phillips, R. D. and Gillis, M. F., Eds., U.S. Department of Energy, 1979, 402.

206. **Nahas, G. G., Boccalon, H., Berryer, P., and Wagner, B.,** Effects on rodents of a one month exposure to magnetic fields (200–1200 gauss), *Adv. Space Environ. Med.,* 46, 1161, 1975.
207. **Brewer, B. B.,** Some preliminary studies of the effects of a static magnetic field on the life cycle of *Lebistes reticulatus* (guppy), *Biophys. J.,* 28, 305, 1979.
208. **Mild, K. H., Sandstrom, M., and Lovtrup, S.** Development of *Xenopius laevis* embryos in a static magnetic field, *Bioelectromagnetics,* 2, 199, 1981.
209. **Neurath, P. W.,** High gradient magentic field inhibits embryonic development of frogs, *Nature,* 219, 1358, 1968; *Biological Effects of Magnetic Fields,* Barnothy, M. F., Ed., Plenum Press, New York, 1969, 177.
210. **Strand, J. A., Abernethy, C. S., Skalski, J. R., and Genoway, R. G.,** Effects of magnetic field exposure on fenilization success in rainbow trout, *Salmo gairdneri, Bioelecrromagnetics,* 4, 295, 1983.
211. **Frazier, M. E., Andrews, T. K., and Thompson, B. B.,** *In vitro* evaluation of biomagnetic effects, in *Biological Effects of Extremely Low Frequency Electromagnetic Fields,* Phillips R. D. and Gillis, M. F., Eds., Conf. 781016, U.S. Depanment of Energy, 1979, 417.
212. **Malinin, G. I., Gregory, W. D., and Morelli, L.,** Evidence of morphological and physiological transformation of mammalian cells by strong magnetic fields, *Science,* 194, 844, 1976.
213. **Chandra, S. and Stefani, S.,** Effect of constant and alternating magnetic fields on tumor cells in vitro and in vivo, in *Biological Effecrs of Extremely Low Frequency Electromagnetic Fields,* Phillips, R. D. and Gillis, M. F., Eds., Conf. 781016, U.S. Department of Energy, 1979, 436.
214. **Leitmannova, A., Stosser, R., and Glaser, R.,** Changes in the shape of human erythrocytes under the influence of a static homogeneous magnetic field, *Acta Biol.* Med. *Ger.,* 36, 931, 1977.
215. **Moore, R. L.,** Biological effects of magnetic fields: studies with microorganisms, *Can. J. Microbiol.,* 25, 1145, 1979.
216. **Gerencer, V. F., Barnothy, M. F., and Barnothy, J. M.,** Inhibition of bacterial growth by magnetic fields; *Nature,* 196, 539, 1962.
217. **Varga, A.,** Proteinbiosynthese bei Mikroorganismen unter Einwirkung von ausseren electromagnetischen Felden, *Fortschr. Exp. Theor. Biophys.,* 20, 1, 1976.
218. **Schaarschmidt, B., Lamprecht, I., and Muller, K.,** Influence of a magnetic field on the UV sensitivity in yeast, Z. *Naturforsch.,* 290, 447, 1974.
219. **Blatt, F. J. and Kuo, Y.,** Absence of biomagnetic effects in *Nitella. Biophys.* J., 16, 441, 1976.
220. **Arajs, S., Yehlin, L. C. L., and Farrington, T. E.,** Behavior of the action potential of *Nitella clavata* cells in the presence of uniform magnetic fields, *Proc. Int. Conf. Magn. Magn. Mater.,* American Institute of Physics, New York, 1975, 759.
221. **Edelman, A., Teulon, J., and Puchalska, I. B.,** Influence of the magnetic field on frog sciatic nerve, *Biochem. Biophys. Res. Commun., 91,* 118, 1979.
222. **Gaffey, C. T and Tenforde, T. S.,** Electrical properties of conducting frog sciatic nerve exposed to high dc magnetic fields, *Bioelectromagnetics, 1,* 208, 1980; Bioelectric properties of frog sciatic nerves during exposure to stationary magnetic fields, *Radiat. Environ. Biophys.,* in press, 1983.
223. **Semm, P., Schneider, T., and Vollrath, L.,** Effects of an earth-strength magnetic field on electrical activity of pineal cells, *Nature,* 288, 607, 1980.
224. **Raybourn, M. S.,** The effects of dc magnetic fields on vertebrate photoreception, Science, 220, 715, 1983.
225. **Beischer, D. E. and Knepton, J. C.,** Influence of strong magnetic fields on the electro-cardiogram of squirrel monkeys, *Aerosp. Med.,* 35, 939, 1964.
226. **Rosen, A. D. and Lubowsky, J.** Magnetic Field Influence on Central Nervous System Function, Experimental *Neurology,* 95, 679, 1987.
227. **Rosen, A. D. and Lubowsky, J.** Modification of Spontaneous Unit Discharge in the Lateral Geniculate Body by a Magnetic Field, *Experimental Neurology,* 108, 261 1990.
228. **Kloiber, O., Okada, Y. and Hassmann, K. A.** A 4.7 Tesla static magnetic field has no effect on the electrical activity of the brain in cats, *Eeg-Emg Zeitschrift fur Elektroenzephalographie, Elektromyographic und Verwandte Gebiete,* 21, 229, 1990.
229. **Rosen, A. D.** Magnetic field influence on acetylcholine release at the neuromuscular junction, Am. *J. Physiol. (Cell Physiol.)* 31, C1418, 1992.
230. **Rosen, A. D.** Membrane response to static magnetic fields: effect of exposure duration. *Biochim. et Biophys. Acta* 1148, 317, 1993.
231. **Ripamonti, A., Ettienne, E. M., and Frankel, R. B.,** Effect of homogeneous magnetic fields on responses to toxic stimulation in *Spirostomum ambiguum, Bioelectromagnetics,* 2, 187, 1981; *Bioelectromagnetics,* 3, 391, 1982.
232. **Bucking, J., Herbst, M., and Pointer, P.,** The influence of a strong magnetic field on muscular contraction, *Radiat. Environ. Biophys., 11,* 79, 1974.

233. **deLorge, J.**, Effects of magnetic fields on behavior in nonhuman primates, in *Magnetic Field Effects in Biological Systems*, Tenforde, T., Ed., Plenum, Press, New York, 1979, 32.
234. **Davis, H. P., Mizumori, S. J. Y., Allen, H., Rosenzweig, M. R., Bennett, E. L., and Tenforde, T. S.**, Behavioral studies with mice exposed to dc and 60-Hz magnetic fields, *Bioelectromagnetics*, 5, 147, 1984.
235. **Feinendegen, L. E. and Muhlensiepen, H.** Magnetic field affects thymidine kinase in *vivo, Int. J. Radiat. Biol.* 47, 723, 1985.
236. **Feinendegen, L. E. and Muhlensiepen, H.** *In vivo* enzyme control through a strong stationary magnetic field—the case of thymidine kinase in mouse bone marrow cells, *Int. J. Radiat. Biol.* 52, 469, 1987.
237. **Papatheofanis, F. J. and Papatheofanis, B. J.** Short-term effect of exposure to intense magnetic fields on hematologic indices of bone metabolism. *Invest. Radiol.* 24, 221, 1989.
237a. **Jankovic, B. D., Maric, D., Ranin, J. and Veljic, J.**, Magnetic fields, brain and immunity: effect on humoral and cell-mediated immune reponses, *Intern. J. Neurosci.*, 59, 25, 1991.
238. **Beischer, D. E.**, Human tolerance to magnetic fields, *Astronautics*, 7, 24, 1962.
239. **Budinger, T. F., Wong, P. D. C., and Yen, C. K.**, Magnetic field effects on humans: epidemiological design study, in *Biological Effects of Eltremely Low Frequency Electromagnetic Fields*, Phillips, R. D. and Gillis, M. F., Eds., Conf. 781016, U.S. Department of Energy, 1979, 379.
240. **Blakemore, R. P.**, Magnetotactic bacteria, *Science*, 190, 377, 1975.
241. **Kreidl, A.**, Weitere Beitrage zur Physiologie des Ohrlabyrinthes: Versuche an Krebsen, *Sitzungsber. Akad. Wiss. Wien, Math. Naturwiss. K.*, 102, 149, 1893.
242. **Ozeki, M., Takahata, M., and Hisada, M.**, Afferent response pattern of the crayfish statocyst with ferrite grain statolith to magnetic field stimulation, *J. Comp. Physiol.*, 123, 1, 1978.
243. **Torres de Araujo, F. F., Pires, M. A., Frankel, R. B., and Bicudo, C. E. M.**, Magnetite and magnetotaxis in algae, Biophys. J. 50, 375, 1986.
244. **Brown, F. A., Jr., Barnwell, F. H., and Webb, H. M.**, Adaptation of the magneto-receptive mechanism of mud snails to geomagnetic field strength, *Biol. Bull.*, 127, 221, 1964.
245. **Brown, F. A., Jr. and Park, Y. H.**, Duration of an after effect in planarians following a reversed horizontal magnetic vector, *Biol. Bull.*, 128, 347, 1965.
246. **Gould, J. L.**, The case for magnetic sensitivity in birds and bees (such as it is), *Am. Sci.*, 68, 256, 1980.
247. **Quinn, T. P.**, Evidence for celestial and magnetic compass orientation in lake migrating sockeye salmon fry, J. *Comp. Physiol.*, 137, 243, 1980.
248. **Lohmann, K. J., and Willows, A. O. D.**, Lunar-modulated geomagnetic orientation by a marine mollusk, *Science*, 235, 331, 1987.
249. **Phillips, J. B.**, Use of the earth's magnetic field by orienting cave salamander *(Eurycea lucifuga), J. Comp Physiol.*, 121A, 273, 1977.
250. **Walcott, C.**, Magnetic orientation in homing pigeons, *IEEE Trans. Magn.*, 16, 1008, 1980.
251. **Wiltschko, W. and Wilschko, R.**, Magnetic compass of European robins, *Science*, 176, 62, 1972.
252. **Mather, J. and Baker, R.**, Magnetic sense of direction in woodmice for route-based navigation, *Nature*, 291, 152, 1981
253. **Baker, R. R.**, Goal orientation by blindfolded humans after long-distance displacement: possible involvement of a magnetic sense, *Science*, 210, 555, 1980
254. **Baker, R. R.**, *Human Navigation and the Sixth Sense*, Hodder and Stoughton, London, 1981
255. **Gould, J. L. and Able, K. P.**, Human homing: an elusive phenonenon, *Science*, 212, 1061, 1981.
256. **Bookman, M. A.**, Sensitivity of the homing pigeon to an earth-strength magnetic field, in *Animal Migration, Navigation and Homing*, Schmidt-Koenig, K. and Keeton, W. T, Eds., Springer-Verlag, Basel, 1978, 127.
257. **Kalmijn, A. J.**, Electric and magnetic field detection in elasmobranch fishes, *Science*, 218, 916, 1982; Biophysics of geomagnetic field detection, *IEEE Trans. Magn.*, 17, 1113, 1981.
258. **Walker, M. M.**, Learned magnetic field discrimination in yellowfin tuna, Thunnus albacares, *J. Comp. Physiol. A*, 155, 673, 1984.
259. **Walker, M. M. and Bitterman, M. E.**, Conditioning analysis of magnetoreception in honeybees, *Bioelectromagnetics*, 10, 261, 1989.
260. **Klinowska, M.**, Cetacean live stranding sites relate to geomagnetic topography, *Aquatic Mammals*, 11, 27, 1985.
261. **Kirschvink, J. L., Dizon, A. E., Westphal, J. A.**, Evidence from strandings for geomagnetic sensitivity in cetaceans, *J. Exp. Biol.*, 120, 1, 1986.
262. **Lohmann, K. J., Willows, A. O. D., and Pinter, R. B.**, An identifiable molluscan neuron responds to changes in earth-strength magnetic fields, *J. Exptl. Biol.*, 161, 1, 1990.
263. **Semm, P. and Beason, R. C.**, Responses to small magnetic variations by the trigeminal system of the bobolink, *Brain Res. Bull.*, 25, 735, 1990.
264. **Bullock, T. H. and Heiligenberg, W., Eds.**, *Electroreception*, Wiley, New York, 1986.
265. **Kalmijn, A. J.**, The detection of electric fields from inanimate and animate sources other than electric organs, in *Handbook of Sensory Physiology, Vol.* 3, Fessard, A., Ed, Springer-Verlag, Basel, 1974, 147.

266. **Brown, H. R. and Ilyinsky, O. B.**, The ampullae of Lorenzini in the magnetic field, *J. Comp. Physiol.*, 126, 333, 1978; **Brown, H. R., Ilyinsky, O. B., Muravejko, V. M., Corshkov, E. S., and Fonarev, G. A.**, Evidence that geomagnetic variations can be detected by Lorenzini ampullae, *Nature*, 277, 648, 1979
267. **Jungerman, R. L. and Rosenblum, B.**, Magnetic induction for the sensing of magnetic fields by animals—an analysis, *J. Theor. Biol.*, 87, 25, 1980.
268. **Keeton, W. T.**, Magnets interfere with pigeon homing, *Proc. Natl. Acad. Sci. U.S.A.*, 68, 102, 1971
269. **Walcott, C. and Green, R.**, Orientation of homing pigeons altered by a change in the direction of an applied magnetic field, *Science*, 184, 180, 1974.
270. **Keeton, W. T., Larkin, T. S., and Windsor, D. M.**, Normal fluctuations in the earth's magnetic field influence pigeon orientation, *J. Comp. Physiol.*, 95, 95, 1974.
271. **Walcott, C.**, Anomalies in the earth's magnetic field increase the scatter of pigeons vanishing bearings, in *Animal Migration, Navigarion and Homing*, Schmidt-Koenig, K. and Keeton, W. T., Eds, SpringerVerlag, Basel, 1978.
272. **Gould, J. L.**, The map sense of pigeons, *Nature*, 296, 205, 1981.
273. **Walcott, C.**, Magnetic maps in pigeons, *Orientation in Birds*, ed. P. Berthold, Birkhauser, Basel, 1991, 38–51.
274. **Able, K. P.**, Mechanisms of orientation, navigation and homing, in *Animal Migration, Orientation and Navigation*, Gauthreaux, S. A, Jr., Ed., Academic Press, New York, 1980, 283.
275. **Griffin, D. R.**, Ecology of migration: is magnetic orientation a reality?, Q. *Rev. Biol.*, 57, 293, 1982.
276. **Wiltschko, W.**, Further analysis of the magnetic compass of migratory birds, in *Animal Migration, Navigation and Homing*, Schmidt-Koenig, K. and Keeton, W T., Eds, Springer-Verlag, Basel, 1978, 302
277. **Walcott, C., Gould, J. L., and Kirschvink, J. L.**, Pigeons have magnets, *Science*, 205, 1027, 1979.
278. **Presti, D, and Pettigrew, J. D.**, Ferromagnetic coupling to muscle receptors as a basis for geomagnetic field sensitivity in animals, *Nature*, 285, 99, 1980
279. **Yorke, E. D.**, Two consequences of magnetic material found in pigeons, *J. Theor. Biol.*, 89, 533, 1981.
280. **Kirschvink, J. L. and Gould, J. L.**, Biogenic magnetite as a basis for magnetic field detection in animals, *Biosystems*, 13, 181, 1981.
281. **Beason, R. C., Nichols, J. E.**, Magnetic orientation and magnetically sensitive material in a transequatorial migratory bird, *Nature*, 309, 151, 1984.
282. **Semm, P., and Beason, R. C.**, Sensory basis of bird orientation, *Experientia*, 46, 372, 1990.
283. **von Frisch, K.**, *The Dance Language and Orientation of Bees*, Harvard University Press, Cambridge, MA, 1967.
284. **Martin, H. and Lindauer, M.**, Der Einfluss der Erdmagnetfeld und die Schwereorientierung der Honigbiene, *J. Comp. Physiol.*, 122, 145, 1977.
285. **Martin, H., Korall, H., Forster, B.**, Magnetic field effects on activity and ageing in honeybees, *J. Comp. Physiol.* 164, 423, 1989.
286. **Walker, M. M. and Bitterman, M. E.**, Honeybees can be trained to respond to very small changes in geomagnetic field intensity, *J. Exp. Biol.* 1445, 489, 1989; Attached magnets impair field discrimination by honeybees, *J. Exp. Biol.*, 141, 447, 1989.
287. **Gould, J. L., Kirschvink, J. L., and Deffeyes, K. S.**, Bees have magnetic remanence, *Science*, 201, 1026, 1978
288. **Kuterbach, D. A., Walcott, B., Reeder, R. J., and Frankel, R. B.**, Iron-containing cells in the honeybee *(Apis mellifere)*, *Science*, 218, 695, 1982.
289. **Lowenstam, J. A.**, Minerals formed by organisms, *Science*, 211, 1126, 1981.
290. **Chew, G. L., and Brown, G. E.**, Orientation of rainbow trout *(Salmo gairdneri)* in normal and null magnetic fields, *Can. J. Zool.*, 67, 641, 1989.
291. **Walker, M. M., Kirschvink, J. L., Chang, S. B. R., and Dizon, A. E.**, A candidate magnetic sense organ in the yellowfin tuna, *Thunnus albacares Science*, 224, 751, 1984.
292. **Mann, S., Sparks, N. H. C., Walker, M. M., and Kirschvink, J. L.**, Ultrastructure, mophology, and organization of biogenic magnetite from sockeye salmon, *Oncorhynchus nerka* implications for magnetoreception, J. Exp. Biol., 140, 35, 1988.
293. **Kirschvink, J. L., Walker, M. M., Chang, S-B. R., Dizon, A. E., and Peterson, K. A.**, Chains of single-domain magnetite particles in chinook salmon, *Oncorhynchus tshawytscha*, J. Comp. Physiol. A, 157, 375, 1985.
294. **Zoeger, J., Dunn, J. R., and Fuller, M.**, Magnetic material in the head of the common Pacific dolphin, *Science*, 213, 892, 1981
295. **Jones, D, S. and MacFadden, B. J.**, Induced magnetization in the monarch butterfly *(Danus plexippus)*, *J. Exp. Biol.*, 96, 1, 1982.
296. **Perry, A., Bauer, G. B., and Dizon, A. S.**, Magnetite in the green turtle, *Trans. Am. Geophys. Un.*, 62, 850, 1981
297. **Buskirk, R. E.**, Magnetic material in marine crustacea, *Trans. Am. Geophys. Un.*, 62, 850, 1981.
298. **Frankel, R, B., Blakemore, R. P., and Wolfe, R. S.**, Magnetite in freshwater magnetotactic bacteria, *Science*, 203, 1355, 1979.

299. **Kirschvink, J. L., Kobayashi-Kirschvink, A. and Woodford, B. J.,** Magnetite biomineralization in the human brain, *Proc. Natl. Acad. Sci. USA,* 89, 7683, 1992.
300. **Lowenstam, H. A.,** Magnetite in the denticle capping in recent, chitons *(Polyplacophora), Geol. Soc. Am. Bull.,* 73, 435, 1962.
301. **Blakemore, R. P.,** Magnetotactic bacteria, *Annu. Rev. Microbiol.,* 36, 217, 1982.
302. **Moench, T. T. and Konetzka,** A novel method for the isolation and study of magnetotactic bacterium, *Arch. Microbiol.,* 119, 203, 1978.
303. **Blakemore, R. B., Blakemore, N. A., Bazylinski, D. A., and Moench, T. T.,** The magnetotactic bacteria, in *Bergy's Manual of Systematic Bacteriology, Vol. 3,* Staley, J. T., Bryant, M. P., Pfennig, N., and Holt, J. G., Eds., Williams and Wilkins, Baltimore, 1989, 1882.
304. **Stoltz, J. F.,** Magnetotactic bacteria: biomineralization, ecology, sediment magnetism, environmental indicator, in: *Biological and Mineralogical Processes of Fe and Mn in Contemporary Environments,* Fitzpatrick, R. W., and Skinner, H. C. W., Eds., Catena, Braunschweig, 1992, 133.
305. **Blakemore, R. P., Maratea, D., and Wolfe, R. S.,** Isolation and pure ulture of a freshwater magnetic spirillum in chemically defined medium, *J. Bacteriol.,* 140, 720, 1979.
306. **Bazylinski, D. A., Frankel, R. B., Jannasch, H. W.,** Anaerobic magnetite production by a marine, magnetotactic bacterium, *Nature,* 334, 518, 1988.
307. **Towe, K. M. and Moench, T. T.,** Electron-optical characterization of bacterial magnetite, *Earth Planet Sci. Lett.,* 52, 213, 1981.
308. **Mann, S., Sparks, N. H. C., Frankel, R. B., Bazylinski, D. A., and Jannasch, H. W.,** Biomineralization of ferrimagnetic greigite (Fe_3S_4) and iron pyrite (FeS_2) in a magnetotactic bacterium, *Nature,* 343, 258, 1990.
309. **Heywood, B. R., Bazylinski, D. A., Garratt-Reed, A., Mann, S. and Frankel, R. B.,** Controlled biosynthesis of greigite (Fe3S4) in magnetotactic bacteria, *Naturwissenschaften,* 77, 536, 1990.
310. **Heywood, B. R., Mann, S., and Frankel, R. B.,** Structure, morphology and growth of biogenic greigite, *Mat. Res. Soc. Symp. Proc.* 218, 93, 1991.
311. **Mann, S. and Frankel, R. B.,** Magnetite biomineralization in unicellular organisms, in Ref. 9, 389.
312. **Rodgers, F. G., Blakemore, R. P., Blakemore, N. A., Frankel, R. B., Bazylinski, D. A., Maratea, D., Rodgers, C.,** Intercellular structure in a many-celled magnetotactic prokaryote, *Arch. Microbiol.,* 154, 18, 1990.
313. **Bazylinski, D. A., Heywood, B. R., Mann, S. and Frankel, R. B.,** Fe_3O_4 and Fe_3S_4 in a bacterium, *Nature,* 366, 218, 1993.
314. **Gorby, Y. A., Beveridge, T. J., and Blakemore, R. B.,** Characterization of the bacterial magnetosome membrane, *J. Bacteriol.,* 170, 834, 1987.
315. **Balkwill, D. L., Maratea, D., snd Blakemore, R. P.,** Ultrastructure of a magnetotactic spirillum, *J. Bacteriol.,* 141, 1399, 1980.
316. **DeLong, E. F., Frankel, R. B., and Bazylinski, D. A.,** Multiple evolutionary origins of magnetotaxis in bacteria, *Science* 259, 803, 1993.
317. **Frankel, R. B.,** Magnetic guidance of organisms, *Annu. Rev. Biophys. Bioeng.,* 13, 85, 1984.
318. **Frankel, R. B. and Bazylinski, D. A.,** Magnetotaxis and magnetic particles in bacteria, *Hyperfine Interactions,* in press, 1994.
319. **Phillips, J. B.,** Magnetic compass orientaion in the Eastern red-spotted newt *(Notophthalmus viridescens), J. Comp. Physiol.* 158, 103, 1986.
320. **Phillips, J. B.,** Homing orientation in the Eastern Red-spotted newt *(Notophthalmus viridescens), J. Exp. Biol.* 131, 215, 1987.
321. **Beason, R. C.,** Use of an inclination compass during migratory orientation by the bobolink *(dolichonyx oryzivorus), Ethology* 81, 291, 1989
322. **Phillips, J. B.,** Two magnetoreception pathways in a migratory slamander, *Science,* 233, 765, 1986
323. **Phillips, J. B. and Borland, S. C.,** Behavioral evidence for use of a light-dependent magnetoreception mechanism by a vertebrate, *Nature* 359, 142, 1992
324. **Phillips, J. B., and Sayeed, O.,** Wavelength-dependent effects of light on magnetic compass orientation in Drosophila melanogaster, *J. Comp. Physiol. A.* 172, 1993.
325. **Wiltschko, W., Munro, U., Ford, H., and Wiltschko, R.,** Red light disrupts magnetic orientation of migrating birds, *Nature,* 364, 525, 1993.
326. **Lohmann, K. J.,** Magnetic orientation by hatchling sea turtles *(Caretta caretta), J. Exp. Biol.* 155, 37, 1991.
327. **Leask, M. J. M.,** A physicochemical mechanism for magnetic field detection by migratory birds and homing pigeons, *Nature,* 267, 144, 1978.
328. **Purcell, E. M.,** *Electricity and Magnetism* (Berkeley Physics Course) Vol. 2, McGraw-Hill,. New York, 1965.
329. **Morrish, A. H.,** *The Physical Principles of Magnetism,* John Wiley and Sons, New York, 1968.
330. **Foner, S.,** Review of magnetometry, *IEEE Trans. Magn.,* 17, 3358, 1981.

Chapter 4

INTERACTION OF ELF MAGNETIC FIELDS WITH LIVING SYSTEMS

T. S. Tenforde

CONTENTS

I. INTRODUCTION

The primary topics to be considered in this chapter are the mechanisms of interaction, biological effects, and potential human health effects of time-varying magnetic fields in the

0-8493-0641-8/96/$0.00+$.50

extremely-low-frequency (ELF) range below 300 Hz. This frequency band therefore contains the 50 Hz and 60 Hz frequencies used throughout the world for electric power transmission, distribution, and consumer applications.

The possible relationship between exposure to ELF magnetic fields from power lines, electrical appliances, machinery, and other common sources and an elevated risk of adverse health effects has become a subject of considerable public interest and concern during the past decade. A principal objective of this chapter will be to describe the experimental evidence for interactions of ELF magnetic fields with living systems, and the implications of these observations in the context of potential human health effects.

II. HUMAN EXPOSURES TO ELF MAGNETIC FIELDS

There are numerous sources of environmental ELF magnetic fields, both naturally occurring and man-made, to which humans are exposed.[1] Large intermittent, time-varying magnetic fields are produced in the atmosphere as a result of intense solar activity and thunderstorms, and can reach levels up to 0.5 μT (T = tesla = 10^4 gauss) during magnetic storms.[2] Diurnally varying magnetic fields with flux densities of about 30 nT are produced as a result of solar and lunar influences on ion currents in the upper atmosphere. Superimposed on the slowly time-varying fields associated with irregular atmospheric events are weak fields that are generated by lightning discharges and propagate in the resonant cavity formed by the earth's surface and the lower boundary of the ionosphere.[3] This effect is known as the Schumann resonance phenomenon, and the fields have a periodic frequency dependence with the lowest five resonances occurring at 8, 14.1, 20.3, 26.4 and 32.5 Hz. At 8 Hz the measured value of the horizontal magnetic field associated with the Schumann resonance effect is, per unit frequency bandwidth, 0.6 $nT/Hz^{1/2}$. The intensity of the field decreases at higher resonant frequencies.[3]

Numerous man-made sources of ELF magnetic fields are present in the home, workplace, and public places such as urban transit systems. These sources include electric power transmission and distribution lines, long-range ELF military communication systems, and a wide variety of appliances, tools, and machines used in offices and various industries.[1,4–6] Of considerable interest in the context of possible health effects are the fields associated with high-voltage transmission lines and residential power distribution lines.[7–10] For three-phase transmission lines operated at 765 kV and carrying a load current of 2 kA, the maximum magnetic field at 1 m above the ground surface is 33 μT. This field level decreases to 28 μT for a 500-kV line with a load current of 1.5 kA, and to 19 μT for a 345-kV line with a load current of 1.0 kA. Transmission lines generally terminate at substations, where the voltage is stepped down by transformers to levels of less than 35 kV for local power distribution on primary lines that are typically less than a few km in length. At locations that are close to the end users, the voltage is again stepped down by transformers to levels of 110 to 480 V and transported over short distances on secondary distribution lines. Because of their lower voltages and load currents, the electric and magnetic fields at ground level in the vicinity of distribution lines are substantially less than those near high-voltage transmission lines.[10] The strength of the 50/60 Hz magnetic fields below power distribution lines are typically less than 2 μT.

The largest source of human exposure to ELF magnetic fields in the home is from the fields generated by many common household appliances and tools. Several of these devices produce local magnetic fields that exceed 1 mT at a distance of 30 cm from their surfaces.[11] Although the fields drop off rapidly with distance due to the localized nature of the source, the residual fields at typical distances from the body during appliance use (~15–30 cm) are still large compared to those from other sources of magnetic fields in the home. Exposures to local fields from appliances and tools, however, are usually of short duration (minutes per day) and the field levels are large only in regions of the body closest to the device.[12,13] One example of a prolonged residential exposure to a relatively high 50/60-Hz magnetic field is provided by the older style of electric blankets,[12] in which the fields at the surface typically reach levels of 2–5 μT. An interesting aspect of many personal appliances, especially among older models, is the production of harmonic fields that extend into the MHz frequency range.[14] At locations in homes away from appliances, the ambient ELF magnetic field levels are generally less than 0.3 μT and typical values of 0.1–0.2 μT have been found in a number of studies.[15–19]

In an office environment, the typical levels of ELF magnetic fields exceed those commonly found in homes, and several devices such as visual display units (VDUs), photocopiers, and microfilm readers produce local fields that frequently exceed 1 μT.[20] The VDUs present in most offices and many homes produce local ELF magnetic fields with flux densities up to 5 μT, although the typical exposure level at the operator's location is less than 1 μT.[21,22]

Several studies published in the last few years have shown convincingly that many classes of electrical occupations involve exposures to time-weighted-average 50/60-Hz magnetic field levels that are significantly greater than those encountered in residences and other non-occupational settings.[19,22–29] Average exposures during the workday ranging from 0.5–2.5 μT have been measured with personal dosimeters for several classes of electrical workers. Some of the highest levels of exposure have been recorded for workers at electric power stations, live-line workers, electricians, cable splicers, and welders. For these classes of workers, mean exposure levels generally range from 1–2.5 μT.

A number of industrial heating processes produce ELF magnetic fields of high intensity within the occupational environment. For example, based on a survey of electrosteel and welding industries in Sweden, it was reported that the local fields near 50-Hz ladle furnaces reached a level of 8 mT, and flux densities as high as 0.07 T were measured near induction heating devices operating in the 50 Hz to 10 kHz range.[30] ELF magnetic fields with high intensities have also been recorded near specialized equipment such as welding machines and demagnetizers;[31,32] at worker locations the field levels are typically in the range 0.1–1.0 mT.

A source of exposure to both workers and the general public that has been a subject of several studies during the past three years are electrified urban transit systems.[33] The fundamental ELF operating frequencies used for these systems in the United States, Europe, and other locations worldwide are typically 16.67, 25, 50, or 60 Hz. However, significant harmonic fields extending to frequencies of several kHz have been measured for several different electrified rail systems. The strength of the ELF fields, which are the dominant source of exposure, reach levels of 10 μT in some sections of the train operator's compartment, with typical levels of 1–5 μT at locations where the operators are seated in the various types of trains. Average ELF fields of a similar magnitude have been measured in the passenger compartments of several electrified rail systems, with the maximum levels reaching 60 μT in the frequency range of 5–40 Hz on the Washington, DC Metro and the New Jersey AMTRAC trains.

As discussed in Chapter 5, time-varying magnetic fields are also used in several types of medical procedures, including the stimulation of bone fracture reunion.[34–36] The magnetic resonance imaging procedure produces time-varying magnetic fields as a result of the switching at ELF frequencies of the magnetic field gradients used for the localization in tissue of nuclei with magnetic moments, such as protons, ^{13}C, ^{15}N, and ^{31}P.[37–39]

III. MECHANISMS OF INTERACTION OF ELF MAGNETIC FIELDS

ELF magnetic fields interact with living systems by two well established mechanisms: induced electric fields produced in accord with Faraday's law of magnetic induction, and direct magnetic field effects on magnetic particles such as the crystals of magnetite (Fe_3O_4) that have been found in a number of organisms. A large number of other possible interaction mechanisms, including resonance effects, have been proposed. However, the evidence for the existence of these interactions in living systems is not well established. The physical basis and biological implications of these various interactions are discussed in the following sections of this chapter.

A. MAGNETICALLY-INDUCED ELECTRIC FIELDS

ELF magnetic fields induce electrical currents in tissue that circulate in loops within planes that are orthogonal to the direction of incidence of the field. This relationship between a time-varying magnetic field and the circulating electric field that it induces is expressed formally by Faraday's law:

$$\partial \mathbf{B}/\partial t = -\nabla \times \mathbf{E} \tag{1}$$

where $\nabla \times \mathbf{E}$ is the curl of the electric field vector. A magnetically induced electric field gives rise to currents that are predicted from Ohm's law, $\mathbf{J} = \sigma\mathbf{E}$, where $\mathbf{J}$ is the induced current density vector, expressed in the MKS/SI system in A/m^2, and σ is the tissue conductivity in S/m.

For the specific case of a circular loop of tissue with radius R intersected by a spatially uniform, time-varying magnetic field orthogonal to the loop, Faraday's law gives the magnitude of the peak electric field tangent to the loop as:

$$E_{peak} = \frac{dB}{dt}(R/2) \tag{2}$$

If the magnetic field is sinusoidal with amplitude B_o and frequency f, then $B = B_o\sin(2\pi ft)$ and from Equation 2 and Ohm's law, the peak current density induced in tissue with an average conductivity σ is given by

$$J_{peak} = \pi fRB_o\sigma\sin(2\pi ft) \tag{3}$$

The simple relation between the induced current density and the applied magnetic field given in Equation 3 can be used to calculate the magnitude of an ELF magnetic field that would be expected to perturb the function of critical biological tissues such as the heart and the central nervous system (CNS). Bernhardt[40] has estimated that the endogenous current densities associated with electrical activity of the brain and heart have lower limits of 1 and 10 mA/m^2, respectively. Perturbations of normal biological functions might be expected to occur in the presence of ELF magnetic fields that induce tissue current densities above these levels. Using values of $R = 0.1$ m and $\sigma = 0.1$ S/m appropriate to the human brain, a calculation based on Equation 3 leads to the conclusion that a peak current density of 1 mA/m^2 would be induced by a 60-Hz magnetic field with an amplitude of 0.5 mT. A similar calculation for the human heart, with $R = 0.06$ m and $\sigma = 0.2$ S/m, leads to the result that a peak current density of 10 mA/m^2 would be induced by a 60-Hz field with an amplitude of 4.4 mT. Because ELF magnetic fields with flux densities of several mT are present near the surfaces of several types of appliances and industrial machinery, the induction of tissue currents at levels that could potentially perturb biological functions in electrically excitable tissues is therefore possible.

The magnitude of the induced current density can also be calculated easily from Equation 1 and Ohm's law for several more complex geometries. Consider, for example, a model of the human body as a uniformly conductive ellipsoid of revolution with the major axis, z, parallel to the long axis of the body. If a sinusoidal magnetic field with an amplitude, B_o, is incident along the z axis, then the peak amplitude of the induced current density in a plane defined by the orthogonal x and y coordinate axes is given by:

$$J_{peak} = \frac{2\pi f B_o \sigma}{a^2 + b^2}(b^4x^2 + a^4y^2)^{1/2} \tag{4}$$

In Equation 4, a and b are the semi-axes along the x and y coordinates, respectively. The induced currents circulate in closed loops within a plane defined by the x and y coordinates (orthogonal to the z axis).

Using Equation 4 the magnetically-induced current densities in an erect human can be modeled to a good approximation. A typical adult male has a height of 1.7 m, a mass of 70 kg and a ratio of body width to thickness of about 2. An ellipsoid with semi-axes of 0.85 m, 0.20 m and 0.10 m has the same height, the same width-to-thickness ratio, and a body volume of 7.1×10^{-2} m^3. The maximum electric field and current density induced in this ellipsoidal model occur when the lines of magnetic flux are horizontal and perpendicular to the front of the body. Using Equation 4 with a = 0.20 m and b = 0.85 m, the calculated value of the peak current density is 1.2 $fB_o\sigma$. As an example, for a human with an average tissue conductivity of 0.2 S/m in a 60-Hz, 33-μT field (the value at ground level under a 765-kV transmission line), J_{peak} = 0.5 mA/m^2. For a typical field level of 0.2 μT in a home, the peak induced current density would be approximately 3 $\mu A/m^2$. From Ohm's law, the corresponding peak electric field induced in the body would be approximately 15 μV/m.

Although an initial physical mechanism of interaction of time-varying magnetic fields with living tissues is the induction of electric currents, the triggering of biological effects by these electrical signals must involve subsequent electrochemical processes that influence cellular functional properties. There is now a relatively strong body of evidence suggesting that the transduction processes through which induced electric signals influence cellular properties involve interactions at the level of the cell membrane.[41,42] Because the membrane's electrical conductivity is smaller by approximately 5–6 orders of magnitude than that of the extracellular fluid, it forms an excellent electrical barrier that mediates interactions of a cell with pericellular electric fields in the external medium. A growing body of evidence indicates that induced ELF fields and currents circulating in the extracellular medium can alter ion binding to membrane macromolecules, influence ion transport across the membrane, and modify ligand-receptor interactions at the cell membrane surface (e.g., the binding of hormones, mitogens, etc.).[41–43] These changes in membrane properties are envisioned as setting up transmembrane signaling events, possibly mediated by Ca^{++} or cyclic nucleotides, that trigger abnormal biochemical and cell growth states. It has been demonstrated, for example, that a magnetically-induced electric field of 0.1 V/m in the extracellular fluids can significantly increase Ca^{++} uptake in mitogen-activated thymocytes.[44] A number of other membrane effects and cellular manifestations of ELF field interactions with cell membranes—many of which occur at threshold levels of induced fields below 0.1 V/m are discussed in Section IV.B of this chapter.

The observation of several cellular and membrane responses to weak ELF fields has raised the question of how the magnitudes of these signals compare with the intrinsic electrical noise present in cell membranes. The four major sources of electrical noise in biological membranes are:[45]

1. Johnson-Nyquist thermally-generated electrical noise, which produces a 3 μV transmembrane voltage shift at physiological temperatures;

2. 1/f noise associated with ion current flows through membrane channels, which typically produces a 10 μV transmembrane voltage shift;
3. "shot" noise, which results from the discrete nature of ionic charge carriers and is a minor source of membrane electrical noise; and
4. endogenous biological background fields produced by electrically active organs such as the heart, muscles, and the nervous system, which can exceed the contribution of physical noise sources by an order of magnitude or more.

Weaver and Astumian[46] have addressed the question of how the weak signals generated by ELF fields compare with the thermally-generated electrical noise present in cell membranes. For fields in the ELF range, the minimum signal strength required to exceed the Johnson-Nyquist membrane noise was estimated to be approximately 0.1 V/m for the case of an elongated cell with a cylindrical geometry (such as a fibroblast, neuron, or muscle cell). Larger threshold ELF field levels (greater than 1 V/m) were predicted for small spherical cells such as lymphocytes. Further increases in the threshold field level are expected if other sources of electrical noise within the cell membrane are taken into account. However, a lower threshold of 4 mV/m was estimated by Polk,[47] and 10 mV/m by Tenforde,[43] if the ELF field effects depend upon a redistribution of charge in the counterion layer at the cell surface, rather than an effect on the transmembrane potential that exceeds the Johnson-Nyquist noise level.

In other theoretical analyses, both Adair[48] and Bennett[49] have argued that biological effects at the cellular level could not occur in response to the weak environmental ELF fields to which humans are routinely exposed. They argue that the fields induced in tissue at the level of individual cells would be too weak to overcome the effects of Boltzmann thermal noise or electrical noise in cell membranes. However, these theoretical treatments do not adequately address the considerable signal amplification that can occur in large arrays of electrically coupled cells in tissue. Weaver and Astumian[50] have considered the signal-to-noise ratio in cell aggregates with junctional coupling that provides a pathway for electrical communication between cells. They presented arguments that, for an aggregate of N electrically coupled cells, the threshold ELF field level required to achieve a signal-to-noise ratio of 1 is lower than the threshold for single cells by approximately a factor of $N^{-5/6}$. The physical basis for this predicted reduction in the field level that would exceed membrane electrical noise was: (i) field amplification due to the larger size of the aggregate of cells, and (ii) reduction of membrane voltage noise as a result of the larger capacitance of the aggregate. This model is highly idealized, and does not take into account the fact that junctions between membranes represent only a small fraction of the total membrane area and may, in addition, have a relatively high impedance. Nevertheless, it is physically plausible that the threshold field level that exceeds intrinsic membrane electrical noise will be reduced, and perhaps substantially reduced, in large aggregates of electrically coupled cells relative to the threshold field level for single cells.

Another aspect of the signal-to-noise issue that has been addressed by Litovitz et al.[51,52] is the temporal incoherence of electrical noise fields relative to the field induced by an ELF signal at a fixed frequency. By switching the frequency of an applied magnetic field between 55 and 65 Hz at random time intervals, these investigators demonstrated that the frequency must be maintained at a fixed value for a period in excess of 1 sec in order to produce an elevation in the activity of the cytoplasmic enzyme ornithine decarboxylase in a rodent fibroblast cell line. A maximum effect on enzyme activity was found when the switching interval exceeded 10 sec. Similar results were obtained for both sinusoidal ELF fields and microwave fields that were amplitude-modulated at ELF frequencies. These results were interpreted by the authors as evidence that temporal coherence of ELF fields must be maintained for a period of seconds in order to achieve signal recognition above the incoherent background electrical noise at the cellular level.

B. DIRECT MAGNETIC FIELD INTERACTIONS

The presence of biogenic magnetite particles in the tissues of a large number of organisms, including several mammalian species,[53,54] has led to the suggestion that oscillatory magneto-mechanical forces and torques on these particles could provide a mechanism for direct interactions of ELF magnetic fields with cellular constituents. Of particular interest is the recent demonstration of magnetite crystals in various anatomic locations within the human brain.[55] Kirschvink et al.[56] have proposed a model in which oscillatory magnetic forces on magnetite particles at ELF frequencies are visualized as producing the opening and closing of pressure-sensitive ion channels in membranes. The 60-Hz field level required to overcome the effects of Brownian motion is predicted in this theoretical model to be on the order of 0.1 mT, which is within the range of field levels to which humans are exposed at locations close to various types of appliances and machinery. An independent theoretical treatment by Adair[57] of the torque on a magnetite particle in the cytoplasm of a cell led to the prediction that a 5-μT, 60-Hz magnetic field, which he takes to be a high flux density relative to typical human exposures, has an interaction energy considerably below the Boltzmann thermal noise level. In a more comprehensive analysis of a similar physical model, Polk[58] has argued that significant interactions could occur between a 5-μT, 60-Hz field and clusters of magnetite particles within the cytoplasm of a cell.

One difficulty in modeling possible biological effects in the human brain mediated by direct ELF magnetic field interactions with magnetite particles is the scarcity of magnetite crystals relative to the number of cells in brain tissue. Kirschvink et al.[55] have reported that human brain tissue contains a few million magnetite crystals per gram, distributed in 0.5–1.0 million discrete clusters. The number of cells in brain tissue thus exceeds the number of magnetite crystals by a factor of about 100. It is therefore difficult to envision how oscillating magnetomechanical interactions of an ELF magnetic field with magnetite crystals could affect a significant number of pressure-sensitive ion channels in the brain. However, the effects of such interactions on neural signaling in localized brain regions could possibly result in a biological response, although there is no evidence at present to support this hypothesis. The possibility also exists that some subpopulations of cells may have extremely high concentrations of magnetite that experience large forces and torques in a magnetic field of moderate strength. Evidence for such cells was recently obtained in studies on magnetite in human leukemic leukocytes.[59]

C. RESONANCE INTERACTIONS

Several resonance models that involve the combined action of an ELF magnetic field and the static geomagnetic field have received considerable attention during the past several years. The most widely discussed of this class of models is the ion cyclotron resonance (ICR) model proposed by Liboff[60] as a mechanism that could facilitate the movement of ions such as Ca^{++} through membrane channels in the presence of the geomagnetic field and a weak ELF magnetic field tuned to the ICR frequency. The resonance condition is formally expressed by the relationship $f_c = QB/2\pi m$, where f_c is the ICR frequency, Q is the ion charge, and m is the ion mass. For the typical range of the geomagnetic field over the surface of the earth (30–70 μT), the resonant frequencies of many biologically important ions such as Na^+, K^+, and Ca^{++} fall within the ELF range. Although some data on Ca^{++} uptake by lymphocytes[61] and diatoms[62] have been cited as lending support to this model, there have also been a number of negative experimental findings.[63–69] In addition, a number of fundamental physical arguments can be raised against the ICR model,[2] the most serious of which is the collisional damping of resonant ion motion that is expected to occur in a condensed phase.[70]

Two other resonance models that have been proposed recently are the "quantum beats" model of Lednev,[71] in which combined static and ELF fields affect vibrational energy levels and transition probabilities of bound ions (e.g., Ca^{++} ions bound to calmodulin), and the model of Zhadin and Fesenko,[72] in which the rotational energy levels of bound ions are pumped by combined static and ELF magnetic fields. Shuvalova et al.[73] have obtained data on the effect of combined fields on the rate of calmodulin-dependent phosphorylation of myosin that appear to support the predictions of the Lednev model. However, another study in which an optical technique was used to study Ca^{++} binding to calmodulin and to metallochromic dyes failed to find any effects of combined static and time-varying fields under the resonance conditions predicted by Lednev's model.[74] In a recent theoretical analysis, Blanchard and Blackman[75] claimed that the original Lednev model contained mathematical errors that led to an incorrect prediction for the ion transition probability between energy levels. Using their revision of the Lednev model, these investigators obtained a reasonably good fit to data on combined static and ELF field inhibitory effects on neurite outgrowth in cultured PC-12 cells stimulated by nerve growth factor.[76] In view of the significant differences in the original and revised Lednev models, there is a clear need for further investigations on both the theoretical basis and biological implications of this resonance model for the action of combined static and ELF magnetic fields. However, regardless of the details of the model, Adair[77] has pointed out that the Lednev model of resonant field effects is improbable because of the long lifetime of the excited vibrational states (approximately 8 sec), during which de-excitation would occur as a result of collisional damping. Similar arguments can be raised against the type of resonant field effects envisioned in the Zhadin and Fesenko[72] model. In addition, the rate of transition of a bound ion to an excited state predicted by this model is so low that a transition would occur only once in several months at typical environmental magnetic field levels.

Another resonance phenomenon that has recently been proposed as a possible mechanism through which weak ELF fields could be detected in biological systems is the stochastic resonance effect. This form of resonance is a nonlinear cooperative effect in which a weak periodic stimulus entrains noise fluctuations with a resulting enhancement of the periodic signal. Preliminary experiments with mechanoreceptor hair cells from crayfish have given evidence for the enhanced detection of weak stimuli in the presence of random noise.[78] It has been suggested that stochastic resonance is the basis for this effect, and perhaps underlies many other examples of weak signal detection by sensory systems.[79] Arguments have also been made that stochastic resonance is a plausible mechanism for the detection of periodic electromagnetic fields under conditions in which the signal-to-noise ratio is less than one.[80] This hypothesis clearly merits further evaluation and experimental testing in biological systems that exhibit reproducible responses to weak ELF fields.

D. OTHER PROPOSED INTERACTION MECHANISMS

A wide variety of physical and chemical interactions have been proposed as possible mechanisms through which weak ELF fields could exert biological effects under conditions in which the interaction of the imposed signal is weaker than the Boltzmann thermal energy or the various sources of intrinsic electrical noise discussed in Section III.A.[9,41,81–84] One example of these theoretical models is the suggestion that dissipative instabilities involving cooperative transitions of allosteric membrane proteins occur in the presence of an applied field. It has also been proposed that critical-state instabilities in membrane physical properties near the phase transition temperature can be amplified by an applied electromagnetic field. Other models involve coherent field interactions that lead to the excitation of oscillating dipolar modes in membrane proteins or the production of nonlinear oscillations (solitons) that facilitate vibrational energy transfer in macromolecular structures. In many cases it is difficult to evaluate the relevance of these interaction mechanisms to biological systems because of the lack of adequate experimental data with which to test predictions made on the basis of the theoretical models. Some progress in this regard has been made in the development of a predictive model of the effects of applied fields on nonequilibrium membrane behavior near phase transition temperatures.[85] This model accurately predicts the changes in solute permeability observed in the membranes of liposomes and eukaryotic cells exposed to magnetic and electromagnetic fields at temperatures near critical phase transitions.

Another model of ELF magnetic field interactions that has received recent attention is the possible effect of these fields on the triplet state yield in radical-mediated charge transfer processes. There are numerous examples of biological processes that involve the transfer of electrons from a donor to an acceptor molecule, which exist in a radical-pair intermediate state during the charge transfer process.[86,87] If the lifetime of the intermediate state that exists prior to radical recombination is sufficiently long, on the order of nanosec, the electrons undergoing transfer evolve from a singlet to a triplet state as a result of hyperfine interactions with the fields from neighboring magnetic nuclei. The application of a magnetic field with a flux density exceeding approximately 1 mT can influence the triplet-singlet interconversion and the triplet state yield via Zeeman interactions that remove degeneracies in the energy sublevels of the triplet-state radical pairs.

As predicted by McLaughlan[88] and demonstrated experimentally by Scaiano et al.[89], a slowly time-varying magnetic field in the ELF frequency range can have similar effects to those of static fields of comparable magnitude since the period of oscillation of the imposed ELF field is extremely long compared to the lifetime of the radical pair intermediate state. The relevance of this type of magnetic field interaction to the functions of living cells is not clear at the present time, although it has been proposed that certain enzyme systems and DNA could serve as hosts for magnetic field-dependent radical processes.[89] It is clear, however, that the amplitude of an applied ELF magnetic field would have to be on the order of 1 mT in order to have significant effects on the radical pair intermediate state.

IV. EXPERIMENTAL STUDIES ON ELF MAGNETIC FIELD EFFECTS

A large body of published literature exists on the biological interactions of ELF magnetic fields at the molecular, cellular, tissue, and animal levels. Several comprehensive reviews have been published on this subject,[2,41,42,90–93] and the focus of the following summary will be on describing research results published during the last few years. The implications of recent findings in relation to possible human health effects of ELF magnetic fields will also be discussed.

A. INTERACTIONS WITH ELECTRICALLY EXCITABLE TISSUES AND BEHAVIORAL EFFECTS

1. Visual, Nervous System, and Cardiovascular System Effects

An effect of time-varying magnetic fields that was first described by d'Arsonval[94] in 1896 is the induction of a flickering illumination within the visual field known as magnetophosphenes. As observed in many independent studies with human subjects, this phenomenon occurs as an immediate response to stimulation by either pulsed or sinusoidal magnetic fields with frequencies less than 100 Hz, and the effect is completely reversible with no apparent influence on visual acuity. The properties and biological basis of phosphenes have been studied extensively in several laboratories in the United States and Europe during the past century.[90] In general, there has been good agreement among different investigators on the characteristics of phosphenes, which have been shown to be induced both by inductively coupled ELF magnetic fields and capacitatively coupled ELF electric fields applied to the head in the region of the eyes.

The maximum visual sensitivity to sinusoidal magnetic fields has been reported to occur at a frequency of 20 Hz in human subjects with normal vision. At this frequency, the threshold magnetic field flux density found by Lovsund et al.[95] to induce phosphenes is approximately 10 mT. The threshold field level increases rapidly as a function of frequency below and above 20 Hz. The corresponding time rate of change of the sinusoidal field is 1.26 T/sec. In other studies Silny[96] has observed a threshold for magnetophosphene perception in human subjects as low as 5 mT with 18-Hz sinusoidal fields. In studies with pulsed fields having a rise time of 2 ms and a repetition rate of 15 Hz, the threshold values of dB/dt for eliciting phosphenes ranged from 1.3 to 1.9 T/sec in five adult subjects.[97] A trend in the data suggested that the threshold was lower among younger subjects. In related studies it was also observed that the stimulus duration is an important parameter, because pulses of 0.9-ms duration with dB/dt = 12 T/sec did not evoke phosphenes.

Several types of experimental evidence indicate that the magnetic field interaction leading to magnetophosphenes occurs in the retina: (i) magnetophosphenes are produced by time-varying magnetic fields applied near the eyes, and not by fields directed toward the visual cortex in the occipital region of the brain;[98] (ii) pressure on the eyeball abolishes sensitivity to magnetophosphenes;[98] (iii) the threshold magnetic field flux density required to elicit magnetophosphenes in human subjects with defects in color vision was found to have a different dependence on the field frequency than that observed for subjects with normal color vision;[95] and (iv) in a patient in whom both eyes had been removed as the result of severe glaucoma, phosphenes could not be induced by time-varying magnetic fields, thereby precluding the possibility that magnetophosphenes can be initiated directly in the visual pathways of the brain.[95]

Although experimental evidence has clearly implicated the retina as the site of magnetic field action leading to phosphenes, it is not as yet resolved whether the photoreceptors or the neuronal elements of the retina are the sensitive substrates that respond to the field. In a series of experiments on *in vitro* frog retinal preparations, extracellular electrical recordings were made from the ganglion cell layer of the retina immediately following termination of exposure to a 20-Hz, 60-mT field.[99] It was found that the average latency time for response of the ganglion cells to a photic stimulus increased by 5 msec in the presence of the magnetic field, and that this was a statistically significant change. In addition, the ganglion cells that exhibited electrical activity during photic stimulation ("on" cells) ceased their activity during magnetic field stimulation (i.e., they became "off" cells). The converse behavior of ganglion cells was also observed. These observations indicate that stimulation of the retina by light and by a time-varying magnetic field elicits responses in similar post-synaptic neural pathways.

The direct involvement of electric currents induced in the retina by ELF magnetic fields in the stimulation of postsynaptic neurons is also suggested by the close similarity of electrophosphenes and magnetophosphenes. As demonstrated by Lovsund et al.,[100] the frequency response of the human eye to electrically-generated phosphenes is similar to that observed for magnetically-induced phosphenes. Carstensen et al.[101] found that electrophosphenes were generated by a 25-Hz electric current of 4 mA passing along a path from the left eye to the right hand. At 60 Hz the current required to generate electrophosphenes increased by a factor of 5 relative to that required at 25 Hz. These investigators estimated that the threshold 60-Hz electric field that must be induced in the retina to produce phosphenes is approximately 1 V/m. Based on a finite-element analysis of the field distribution in the head, the threshold 60-Hz current density required to induce electrophosphenes was estimated to be on the order of 10 mA/m^2.

Several phenomena related to the sensitivity of the visuosensory system to time-varying magnetic fields have been studied by Silny.[96] In experiments with human subjects it was found that the distinct flickering characteristic of phosphenes could be elicited in the visual field by sinusoidal magnetic fields with frequencies ranging from 5 to 60 Hz. The threshold field intensity varied with the field frequency and background light level, but was as low as 5 mT under optimal conditions. Alterations in visually evoked potentials (VEP) were also reported to occur in sinusoidal magnetic fields at intensity levels that are 5 to 10 times greater than thsoe which produce magnetophosphenes.[96] The change in VEP was characterized by a reversal of polarity and a decreased amplitude of the three major evoked potentials. These effects were observed within 3 min after onset of the magnetic field exposure, and the VEP returned to normal only after a recovery period of approximately 30–70 min following termination of the exposure. The relationship of these changes in the VEP to the mechanism of magnetophosphene induction is not clear from the evidence that is presently available.

Time-varying magnetic fields that induce current densities above 1 A/m^2 in tissue lead to neural excitation and are capable of producing irreversible biological effects such as cardiac fibrillation[9,102] Several investigators have achieved direct neural stimulation using pulsed or sinusoidal magnetic fields that induced tissue current densities in the range of 1–10 A/m^2. In a study involving electromyographic recordings from the human arm,[103] it was found that a pulsed field with dB/dt greater than 10^4 T/sec was required to stimulate the median nerve trunk. The duration of the magnetic stimulus has also been found to be an important parameter in the excitation of nerve and nerve-muscle specimens. Using a 20-kHz sinusoidal field applied in bursts of 0.5 to 50 msec duration, Oberg[104] found that a progressive increase in the magnetic flux density was required to stimulate the frog gastrocnemius neuromuscular preparation when the burst duration was reduced to less than 2–5 msec. A similar rise in threshold stimulus strength has been observed for frog neuromuscular stimulation using pulsed magnetic fields with pulse durations less than approximately 1 msec.[105,106] Bernhardt[107] has described evidence obtained from studies on the electrical activity of isolated nerve preparations which indicates that the threshold current density for neural stimulation may be as low as 100 mA/m^2.

The effects of ELF magnetic fields on the CNS have been studied in both human subjects and laboratory animals. In a study on 14 human subjects, Bell et al.[108] observed changes in the electroencephalogram (EEG) in one-half of the subjects exposed to either a 60-Hz, 25-μT field (3 subjects) or a 60-Hz, 50-μT field (4 subjects). In all cases, the EEG power expressed in $(\mu V)^2$ was decreased at various brain wave frequencies ranging from 1 to 15 Hz. Similar observations were made by the same group of investigators in laboratory studies on rabbits exposed to a 5-Hz, 100-μT field.[109] Silny[96] has reported changes in the EEG power density spectrum during 5-hr exposures of cats to 50-Hz, 20-mT fields, but the response was evident only for a brief period at the beginning of the exposure.

Evoked brain potentials have been studied in both humans and monkeys exposed to combined 60-Hz electric and magnetic fields at levels that simulate the environment of an electric power transmission line. The early and mid-latency periods of auditory, visual, and somatosensory evoked potentials were unaffected during exposure of monkeys to 60-Hz fields at levels up to 30 kV/m and 90 μT.[110] However, the late components of somatosensory evoked potentials were decreased in amplitude in monkeys exposed to combined field levels of 10 kV/m and 30 μT or 30 kV/m and 90 μT. An increase in the latency of auditory evoked potentials and an increase in the P300 component of the evoked response was observed in humans exposed to combined 60-Hz electric and magnetic fields at levels of 6 kV/m and 10 μT, 9 kV/m and 20 μT, and 12 kV/m and 30 μT.[111–113] Significant effects were observed only at the intermediate level of combined fields, i.e., 9 kV/m and 20 μT. However, there were no significant changes in measures of CNS performance such as the estimation of time intervals by subjects exposed to fields of this magnitude. The interpretation of these studies is complicated by the fact that both electric and magnetic fields were present simultaneously. However, the overall set of findings suggest that there are few, if any, reproducible effects on the peripheral and central afferent sensory pathways of ELF fields with intensities comparable to those in the vicinity of high-voltage transmission lines.

Cardiac performance has also been studied in humans exposed to 50-Hz and 60-Hz fields at high intensities. In electrocardiogram (ECG) studies on a large number of subjects directly exposed to the fields from 50-Hz transmission lines, no significant exposure-related effects were observed on cardiac performance indices such as the heart rate.[114,115] The fields used in these studies varied over a wide range, with the highest levels of the electric and magnetic field components being 10.2 kV/m and 15.4 μT, respectively. In another study conducted with humans in a laboratory environment, a statistically significant slowing of the heart rate by a few percent was observed to occur in subjects exposed to combined 60-Hz electric and magnetic fields at levels of 9 kV/m and 20 μT, respectively.[111–113,116] A similar effect was not observed at either higher (12 kV/m and 30 μT) or lower (6 kV/m and 10 μT) field levels. The investigators speculated that there may be "windows" of field intensities to which the cardiovascular system may be the most responsive, although there is no known mechanism to explain a differential sensitivity to fields at various levels of intensity.

Responses to pulsed magnetic fields of high intensity have been reported to occur in both cardiac and nerve tissues. An increase in the beating rate of isolated rat hearts was observed in response to 4.5–6 Hz magnetic field pulses with flux densities that exceeded 10 mT.[117] Similarly, pulsed fields of comparable strength and repetition rates ranging from 20–55 Hz were found to increase the contraction rate of excised canine cardiac tissues.[117] An analgesic effect and long-term pain relief produced by stimulation with high-intensity magnetic pulses was reported on the basis of clinical trials with subjects suffering from chronic pain associated with peripheral nerve injury or trauma incurred as a result of surgical procedures.[118] The treatment procedure involved applying a 1.17-T field to the body surface as a pulse with a repetition frequency of 40/min. Accelerated healing of a crush lesion in rat sciatic nerves was reported to occur following treatment with a pulsed magnetic field.[119] The field had a flux density of 300 μT, and was applied at a rate of 2 Hz for 4 hr/day over a period of 1–4 days, or for 15 min/day over a 2-day interval.[119] In a study with rabbits in which cerebral ischemia was initiated by arterial occlusion, it was observed that subsequent application of a pulsed magnetic field during a 4-hr reperfusion period attenuated the extent of cortical ischemic edema.[120] The pulsed field was applied to the head using a pair of coils, each of which generated a magnetic field with a maximum flux density of 2.8 mT. The impulse duration and repetition frequency were, respectively, 1.3 msec and 75 Hz.

2. Animal Behavior

Extensive reviews of the literature on behavioral alterations in animals exposed to ELF magnetic fields have revealed that nearly all of the positive findings were obtained in studies involving relatively low field strengths, whereas no behavioral effects were observed in studies with rodents, monkeys, and humans exposed to higher field levels.[90,92] Although the detection and behavioral responses to weak ELF fields exhibited by several animal species, including bees and avians, may be attributable to interaction of the field with deposits of biogenic magnetite (see discussion in Section III.B of this chapter and in Chapter 3), there is no clear biological basis for a difference in behavioral sensitivity to low versus high levels of ELF magnetic fields in any of the animal species that have been studied.

Several recent studies with rodents and humans have confirmed the general observation that few, if any, behavioral effects result from exposure to ELF magnetic fields with flux densities exceeding 1 mT. In a shuttlebox test on rats exposed to 60-Hz, 3-mT fields, the rodents were found to exhibit no avoidance of exposure to the field.[121] No significant effects on open-field behavior or locomotor activity were observed in rats exposed to a 50-Hz, 18-mT field for periods of 2 hr/day over 20 consecutive days.[122] In these experiments, however, a reproducible decrease in the avoidance of aversive stimuli was exhibited by the exposed animals relative to controls. In a study on the EEG and performance measures in humans exposed either continuously or intermittently (1 sec on, 1 sec off) for 1 hr to a sinusoidal 45-Hz, 1.26-mT field, changes were observed in brain wave activity.[123] These changes were observed primarily with the intermittent field exposure regimen. However, there was no significant effect of exposure to the field on reaction time performance.[123]

In studies related to neurological and behavioral effects of ELF fields on humans, two groups of investigators have analyzed the occurrence of symptoms of depression among electrical workers and women living in homes adjacent to transmission lines. In a comparison of several indices of depression among electrical and nonelectrical workers, Savitz et al.[124] concluded that there was no reliable indication of an increased risk of depression among the electrical workers. In a study on the effects of residential exposure to 60-Hz magnetic fields from transmission lines, no elevation in the risk of depression was found among women living in homes next to transmission lines as compared to a control group of women that lived one block away from the nearest transmission line.[125] Based on measurements with a portable exposure meter, it was found that the average 60-Hz magnetic field level at the front door of homes of women living next to transmission lines was 0.486 μT. In contrast, the average 60-Hz magnetic field level at the front door of homes of women in the comparison group was only 0.068 μT. These studies were carefully controlled for factors other than 60-Hz fields that could have influenced the results.

B. CELLULAR AND TISSUE INTERACTIONS

1. Molecular, Cellular, and Tissue Effects

During the past 25 years numerous studies have been published on biological effects of ELF fields at the tissue, cellular, and subcellular levels.[84,90,93,126] Among the effects of ELF magnetic fields that have been reported are alterations in cell membrane properties (e.g., ion transport and interactions of mitogens with cell-surface receptors), changes in cellular functions and growth properties (e.g., increased proliferation and alterations in metabolism, gene expression, protein biosynthesis, and enzyme activities), and modifications in various functional properties of organs and tissues (e.g., altered hormone production and changes in immune responsiveness to antigenic stimulation). In many cases, these effects were associated with the application of fields that induced current densities on the order of 1–10 mA/m^2 or higher in the tissue or cellular system under study, which is the typical range of endogenous current densities to which most living tissues are exposed *in vivo*.[40,87] However, there have also been a number of effects reported to occur in response to weak ELF magnetic fields that induce current densities below the endogenous background level. Examples of such effects observed in eukaryotic cells are alterations in the binding of Ca^{++} to cell surfaces[127–129] and changes in gene expression.[130–132] In the following discussion, an emphasis will be placed on describing recent laboratory findings that have provided insights into possible mechanisms through which ELF magnetic field signals are transduced, and the possible implications of these interactions on the regulation of cellular and tissue functions and growth properties.

(a) DNA and Chromatin Structure

Because of their extremely weak interactions with tissue, it is not expected that ELF fields at typical environmental levels would disrupt chemical bonds or significantly perturb the secondary and tertiary structures of biological macromolecules. It is therefore not surprising that exposure to 60-Hz magnetic fields with flux densities up to 2 mT, either alone or in combination with a 60-Hz, 38-V/m electric field, did not introduce strand breaks into the DNA of Chinese hamster ovary cells.[133] In other studies with human peripheral lymphocytes, it was found that pulsed magnetic fields with a peak intensity of 2.5 mT, a pulse duration of 2 msec, and a 50-Hz repetition rate had no effect on DNA structure as judged by measurements of unscheduled DNA synthesis.[134] In two series of experiments, no effects of ELF magnetic fields were found on the cellular repair of DNA damage induced by ionizing radiation.[134,135]

As expected from the lack of effects on DNA structure, experimental tests of chromosome damage, genetic mutations, and cell transformation have had negative outcomes. No evidence was obtained for either sister chromatid exchanges or chromosome instability in studies on normal human lymphocytes and a variety of tissue cells exposed *in vitro* to 60-Hz, 0.2-mT magnetic fields, either alone or in combination with 60-Hz, 2.25-V/m electric fields produced by electrodes in contact with the medium.[136,137] Continuous exposure for 72 hr to a 50-Hz, 300-μT field was found to have no effect on the frequency of chromosome aberrations in human amnionic cells.[138] However, in the same study a small, but significant, 2% increase in chromosome aberration frequency was observed in response to an intermittent exposure (15 sec on, 15 sec off) for 72 hr to a 50-Hz, 30-μT field. At present there is no explanation for the different results obtained with intermittent versus continuous exposures to an ELF magnetic field. In another study, human lymphocytes exposed to 60-Hz, 1.4-mT magnetic fields following ionizing radiation were observed to have an increased frequency of near-tetraploid chromosomes compared to cells exposed to ionizing radiation alone.[139] This finding suggests that the magnetic field exposure may have influenced chromosome structure during the premitotic block in the cell division cycle that is introduced by ionizing radiation. In an extensive series of *in vitro* experiments using mouse C3H/10T1/2 cells, no evidence was obtained for cell transformation following exposure to 60-Hz magnetic fields with flux densities up to 0.6 mT.[140]

(b) Gene Expression, DNA and Protein Synthesis, and Cell Proliferation

A large number of publications have appeared during the past decade in which an increase in DNA transcription was reported to occur in response to either pulsed or sinusoidal ELF magnetic field exposures. Goodman, Henderson, and their associates have observed significant effects on messenger RNA (mRNA) levels, gene transcription, and protein synthesis in dipteran salivary gland cells in response to a wide variety of pulsed and sinusoidal magnetic fields with frequencies in the ELF range.[141–144] Similar effects have been observed with cultured human HL-60 leukemic cells.[130–132] The changes observed in RNA transcripts exhibited a strong dependence on the amplitude and frequency of the applied ELF magnetic field, and were transient in nature with a rise and subsequent decline of the specific mRNA over a period of less than 1 hr. A possible explanation of these observations is that the field alters the rate constant for one (or more) of the intermediate sequential reactions that are involved in RNA synthesis and degradation.[145,146] The threshold level of the magnetically-induced electric field capable of producing effects on gene transcription and translation appears to be on the order of 1 mV/m or less.[130,131] One other interesting aspect of this work is that the expression of a broad spectrum of specific genes appears to be activated by exposure to ELF magnetic fields, including several oncogenes (e.g., c-*myc* and c-*fos*) and a number of stress-related genes that transcribe proteins characteristic of the cellular response to thermal stress.[147,148] It should be noted, however, that the mechanism through which heat-stress proteins such as HSP70 are activated by weak ELF magnetic fields is very likely different from that involved in the activation of these genes by thermal stress. This suggestion is based on the fact that the weak electric currents induced by the ELF magnetic fields used in these experiments are several orders of magnitude below the level that would produce measurable heating of the cell culture medium.

Several other laboratories have analyzed changes in the RNA content and expression of specific genes in various lines of cultured eukaryotic cells. Both pulsed and sinusoidal ELF magnetic fields have been found to increase the total RNA content of HL-60 cells and Daudi cells, a human lymphoma cell line, by Ning et al.[149] Similarly, Greene et al.[150] observed a transient increase in total RNA levels of HL-60 cells exposed to a 60-Hz, 1.0-mT field.

Phillips and McChesney[151] observed increases in both total RNA and mRNA in CCRF-CEM cells exposed to a pulsed magnetic field with a peak flux density of 3.5 mT and a repetition rate of 72 Hz. Phillips et al.[152] also observed transient increases in the transcript levels of several genes, including c-*fos,* c-*jun,* and c-*myc* and protein kinase C, in CEM-CM3 T-lymphoblastoid cells exposed for 15–120 min to a sinusoidal 60-Hz, 0.1-mT field.

In contrast to these findings, neither Lacy-Hulbert et al.[153] or Saffer and Thurston[154] were able to detect changes in c-*myc* or β-actin transcript levels in HL-60 or Daudi cells.[154] Parker and Winters[155] also reported that the expression of four oncogenes and the heat-stress protein HSP70 in several mouse and human cell lines were unaffected by exposure to a 60-Hz, 0.1-mT field. This study, however, involved exposure of the cells for 24 hr, and the transient effects of ELF fields on mRNA transcript levels reported by other investigators may not have been observable following an exposure of this duration.

Several studies have been conducted to detect changes in DNA synthesis and cell proliferation in cells exposed to ELF magnetic fields. In experiments with cultured human fibroblast cells, Liboff et al.[156] observed an elevation in DNA synthesis in response to exposure to sinusoidal magnetic fields with frequencies ranging from 15 Hz to 4 kHz and flux densities of 2.3–500 μT. The most interesting aspect of this finding was that the elevated DNA synthesis rate was independent of the current density induced in the cell culture medium. In contrast, Cridland and Saunders[157] observed no change in DNA synthesis rate in human fibroblasts exposed to a 50-Hz, 2-mT field. Phillips and McChesney[151] found no change in DNA synthesis in lymphoblastoid cells exposed to a pulsed magnetic field under conditions in which significant changes were observed in the cellular content of mRNA and total RNA.

As reviewed by Cridland,[126] a large number of studies have been conducted on the stimulation of DNA synthesis and cell proliferation in lymphocyte cultures exposed to ELF magnetic fields, either alone or in combination with chemical mitogens. Although some of these studies have shown no direct effect of either pulsed or sinusoidal ELF magnetic fields on DNA synthesis rate, several reports have presented evidence that DNA synthesis and proliferative responses of lymphocytes to mitogens are influenced by magnetic field exposure. The reported effects, however, have been highly variable, with claims of both enhanced and diminished mitogen effects as a result of magnetic field exposure. A more consistent set of findings on the stimulation of bone cell growth and bone fracture repair by pulsed and sinusoidal magnetic fields has been reported in the literature, as discusssed in Chapter 5.

(c) Membrane Properties and ELF Signal Transduction

A growing body of experimental evidence has implicated the cell membrane as an important site of interaction of ELF fields, as discussed in Section III.A. In the specific case of ELF magnetic fields, several of the field-related changes in membrane functional and structural properties that have been reported include alterations in Ca^{++} binding to fixed-charge sites on the cell surface,[127–129] increased uptake of Ca^{++} in mitogen-activated cells,[44,158,159] changes in the activity of the membrane-associated adenylate cyclase enzyme and its response to the binding of parathyroid hormone (PTH) molecules at receptor sites on the cell surface,[160–164] alterations in the expression of specific antigens and transferrin receptors at the surfaces of tumor cells,[165,166] changes in membrane surface charge,[167] and modifications in membrane microfluidity and other properties.[168] In the case of magnetic field effects on the process that leads to adenylate cyclase activation by PTH binding to the cell membrane, there is evidence that the action of the field reduces the ability of the PTH-receptor complex to activate the alpha subunit of membrane-associated G protein.[169] This effect of the field interferes with the sequence of events that subsequently activate adenylate cyclase at the cytoplasmic membrane surface. Another interesting aspect of ELF magnetic field interactions with membranes is the general observation that the membranes must be activated by mitogen binding before significant field-related effects occur on the transport of cations such as Ca^{++}. For example, an increased uptake of Ca^{++} by lymphoid cells has been observed only when they were exposed to the plant lectin Concanavalin A (Con-A).[44,158] In several different cell types that were not stimulated by mitogens, no effect of pulsed or sinusoidal ELF fields was observed on the transport of Ca^{++} or other cations.[170,171] One exception to this set of observations is the report of Lyle et al.[172] that a sinusoidal 13.6-Hz, 20-μT field increased the uptake of Ca^{++} by normal lymphocytes and lymphoma cells.

The membrane interactions that lead to effects of ELF magnetic fields on the kinase-C signaling pathway are of considerable interest because this pathway is known to be activated by the binding of tumor promoters such as phorbol esters. Activation of this pathway by the binding of a mitogen leads to a cascade of events that produce an increase in kinase-C activity and an increase in the concentration of free cytosolic Ca^{++} ions. A significant 70% increase in kinase-C activity has been observed in human HL-60 cells exposed to a pulsed magnetic field.[173] The field effect was considerably reduced by the addition of EGTA (a Ca^{++} chelator) to the culture medium. This observation is consistent with the known dependence of kinase-C activation on the mobilization of Ca^{++} ions in the cytosol. Another relevant study is the observation described above that the stimulation of Ca^{++} uptake into rat thymocytes by the mitogen Con-A is significantly increased by exposure to a sinusoidal 60-Hz magnetic field. In lymphoid cells the binding of Con-A to surface receptors triggers cytoplasmic signaling events involved in the activation of kinase-C. It was also shown in this study that the enhancement of the Con-A effect depends upon the strength of the 60-Hz electric field induced in the extracellular medium.[158] The threshold induced electric field level that produced an increase in Ca^{++} uptake by the Con-A stimulated cells was 0.1 V/m.

(d) Hematology and Immunology

Several extensive studies with rodents, monkeys and human subjects have found no significant effects of ELF magnetic fields on the hematological system.[111,174–177] In these studies none of the major classes of blood cells were found to be affected, even as a result of long-term exposure to 60-Hz fields with flux densities as high as 110 mT. Similarly, no significant exposure-related changes were observed in serum chemistry.

Two comprehensive studies on immune responsiveness in rodents exposed chronically to 60-Hz magnetic fields have not revealed any significant field-related effects.[179,180] In mice exposed to a 60-Hz, 2-mT field for 6 hr/day over a period of 21 weeks, no changes were observed in the level of natural killer cell activity in lymphocyte populations isolated from the blood and the spleen.[181] deSeze et al.[182] found no alteration in the antibody response to foreign antigen by spleen cells from mice that were exposed 10 hr/day for 5 days to a square-wave pulsed magnetic field with a repetition frequency of 0.8 Hz and a peak flux density of 120 mT. However, in these experiments an increase in natural killer cell activity was observed in spleen lymphocytes obtained from animals exposed to pulsed fields with peak flux densities greater than 30 mT. This effect may be related to the large instantaneous current densities induced in the tissues of the exposed rodents during application of the strong magnetic pulses, and merits further study.

(e) Endocrine Functions

Several reports have described alterations in hormonal regulation in both experimental animals and humans exposed to ELF magnetic fields. Studies on rats exposed to 50-Hz, 20-mT fields for periods of 1–7 days provided evidence for an increase in adrenal 11-hydroxy corticosteroids and a mobilization of adrenal catecholamines, both of which may reflect stress-related responses to the exposure.[183,184]

In a study on rats exposed to rapid variations in the strength of the static geomagnetic field, it was observed that both the nocturnal pineal melatonin content and the activity of serotonin N-acetyltransferase, an enzyme involved in melatonin biosynthesis, were reduced.[185]

An independent study with avians provided evidence that the activity of hydroxyindole-O-methyl transferase, which catalyzes the transfer of a methyl group to N-acetylserotonin to form melatonin, is reduced in both the retina and pineal gland as a consequence of varying the local geomagnetic field.[186] In addition to these neurochemical effects, changes in the electrical activity of pineal gland cells were observed to occur in response to changes in the local geomagnetic field.[187,188] Later studies by Olcese et al.[189] and Lerchl et al.[190] confirmed that the nocturnal biosynthesis of melatonin in the rat pineal gland is depressed in response to artificial variations in the local geomagnetic field. Lerchl et al.[191] have also presented evidence that this effect depends upon the induced electric currents that result from rapid time variations in the static magnetic field strength.

Yellon et al.[192] have shown that the nocturnal production of pineal melatonin in Djungarian hamsters is extremely sensitive to a brief 15 min exposure to a sinusoidal 60-Hz, 0.1-mT magnetic field. When the field was applied 2 hr before the onset of darkness, the nighttime rise in pineal melatonin was heavily damped and the peak production of this hormone was phase shifted by approximately 2 hr to a later part of the dark period. This finding is similar to earlier observations of a damping and phase shifting in the circadian rhythm of melatonin production in rats that were chronically exposed to 60-Hz electric fields.[193,194]

In a recent series of publications, Kato et al.[195,196] have reported that a 6-week exposure of rats to a 50-Hz, 1-μT circularly polarized magnetic field led to a significant reduction in the nighttime concentrations of melatonin in the blood and the pineal gland. The levels of melatonin returned to normal within one week following cessation of the magnetic field exposure.[197] It is of interest that the same group of investigators found no effect of 50-Hz, 1-μT linearly-polarized magnetic fields, both horizontally and vertically oriented, on the pineal or blood levels of melatonin.[198] The effects on pineal melatonin synthesis of the circularly polarized relative to the linearly polarized fields could be related to the larger induced currents with continuously varying directions produced in the body of a rat by the circularly polarized fields. Kato et al.[199] also observed no effect of 50-Hz circularly polarized fields at flux densities of 1, 5, and 50 μT on the blood level of testosterone in rats. They concluded from this observation that the effect of circularly polarized ELF magnetic fields on the pineal production of melatonin does not reflect a generalized suppression of endocrine functions in the hypothalmic-pituitary-gonadal axis. Several investigators have presented evidence that the effect of ELF magnetic fields on nocturnal melatonin synthesis in the pineal gland is mediated by effects of these fields on visual pathways, although the exact mechanism by which this occurs has not been identified.[189,200,201] It has also been suggested that magnetic field effects on the pineal gland could be mediated through effects on β-adrenergic receptors for norepinephrine on the membranes of pinealocytes.[202] These interactions could underlie the observed reduction in cAMP levels of rat pinealocytes following exposure to perturbations in the geomagnetic field level. The lower cAMP levels could, in turn, explain the reduced serotonin N-acetyltransferase activity and depressed pineal melatonin production observed in a number of studies.

In a study with humans exposed at night to 60-Hz magnetic fields produced by electric blankets, Wilson et al.[203] found that 7 of the 28 subjects who used a continuous polymer wire (CPW) type of electric blanket showed a significant decrease in the overnight excretion of 6-hydroxymelatonin sulfate (6-OHMS), a urinary metabolite of melatonin. The average flux density of the 60-Hz fields produced by the CPW blankets was estimated to be 0.4 μT at the location of the head of a user. In this study it was also found that the largest changes in urinary 6-OHMS levels occurred immediately following the onset or cessation of exposure to fields from the electric blankets. The investigators interpreted this set of findings as preliminary evidence for a possible influence of weak ELF magnetic fields on nocturnal levels of melatonin in humans.

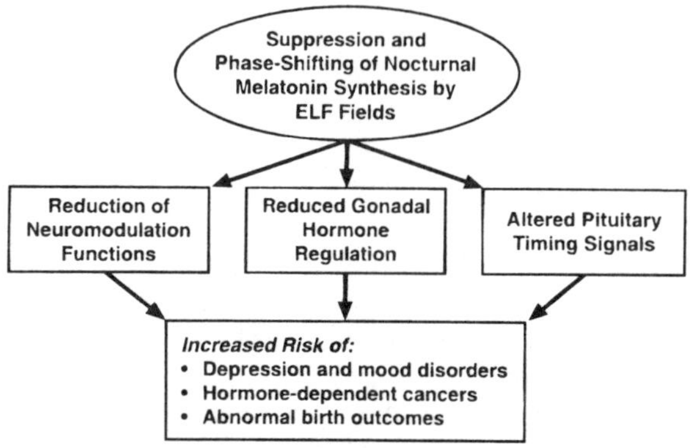

FIGURE 1 Illustration of possible physiological and psychological effects of suppressing and phase-shifting pineal melatonin synthesis by ELF magnetic field exposure.

Because of the broad range of physiological regulatory functions exerted by pineal melatonin, several authors have speculated on the possible health implications of the reduced nocturnal levels of this hormone that result from ELF magnetic field exposure.[204–207] The major functions of melatonin include: (i) circadian regulatory effects on numerous physiological variables; (ii) reproductive effects mediated through melatonin's regulatory effect on gonadal hormones; (iii) effects on mood and behavior associated with melatonin's neuroendocrine modulation functions; (iv) melatonin effects on the immune system influence antibody production in response to challenge with a foreign antigen; and (v) melatonin exerts direct effects on the growth of breast, prostate, and colon carcinomas. Strong evidence exists for all of these effects of melatonin, and the possible physiological and behavioral consequences of suppressing the magnitude and the circadian phase of nocturnal pineal melatonin synthesis by exposure to ELF magnetic fields is illustrated in Figure 1. Although there is no convincing evidence for effects of typical levels of human exposure to ELF magnetic fields on mood or reproductive outcomes (see Sections IV.A.2, IV.B.2, V.A, and V.B), the possible role of melatonin suppression on the risk of breast cancer and other tumors is currently a subject of active investigation (see Section IV.C).

The possible link between exposure to ELF magnetic fields and an elevated risk of breast cancer was originally suggested by Stevens,[208] who noted that reduced concentrations of melatonin in the circulation leads to increased concentrations of steroid hormones (estrogen and testosterone) which, in turn, can stimulate the proliferation of breast tissue. Through this mechanism small foci of tumor cells with high levels of steroid hormone receptors could enter a rapid growth state. This hypothesis is supported by the results of rodent studies in which it was demonstrated that melatonin injection suppresses mammany carcinogenesis and pinealectomy increases breast tumor formation.[209]

An alternative mechanism through which the suppression of circulating melatonin levels could increase the risk of breast cancer has been proposed by Liburdy et al.[210] The hypothesis he proposed is that ELF magnetic field interactions could inhibit the oncostatic action of melatonin in estrogen-receptor-positive (ER+) breast cancer cells. Using the human ER+ breast cancer cell line MCF-7, the growth inhibition of physiological concentrations of melatonin was tested in the presence and absence of weak 60-Hz magnetic fields with flux densities ranging from 0.2–1.2 μT. At the 1.2-μT magnetic field level it was observed that the oncostatic effect of 10^{-9} M melatonin was inhibited. This promising finding merits further study to identify the site of magnetic field action that interferes with melatonin's inhibitory effect on the proliferation of breast cancer cells.

2. Effects on Reproduction and Development

A large number of studies have been conducted on reproductive outcomes and postnatal development in both mammalian and nonmammalian (primarily avian) species exposed to ELF magnetic fields.[211,212] From experiments on mammals exposed to either sinusoidal or pulsed magnetic fields with ELF frequencies, there is little evidence to indicate significant teratological effects. Huuskonen et al.[213] exposed female rats from day 0 to day 20 of pregnancy to a sinusoidal 50-Hz, 35.6-μT field or a 20-kHz sawtooth field (typical of VDU fields) with a peak flux density of 15 μT. The pregnancy rates, incidence of resorptions, late fetal deaths, fetal body masses, and incidence of major malformations were found not to differ significantly between the exposed and control animals. There was, however, a small increase in the incidence of minor skeletal abnormalities in the rats exposed to either the sinusoidal 50-Hz field or the sawtooth 20-kHz field relative to controls. In a study on the fetal development of mice exposed to a sinusoidal 50-Hz, 20-mT field during the first 17 days of gestation, Kowalczuk et al.[214] found no significant field-associated effects on embryo survival or the incidence of visceral or skeletal abnormalities. In a subsequent study by the same group of investigators,[215] no major changes were observed in the postnatal development or behavior of the exposed mice relative to controls. In a large study on rats exposed to a sinusoidal 60-Hz, 1-mT field throughout gestation, no significant effects of the exposure were observed on litter size, fetal body weight, or the incidence of malformations.[216]

Several studies on the fetal development of rodents following exposure to pulsed magnetic fields with either ELF or very-low-frequency (VLF: 300–30,000 Hz) repetition frequencies have given variable results. Tribukait et al.[217,218] exposed mice from day 0 to day 14 of pregnancy to either 100-Hz pulsed fields with a rectangular waveform and peak flux densities of 1 μT or 15 μT, or to 20-kHz sawtooth pulses of the same magnitude. A statistically significant increase in the percentage of malformed fetuses was observed in the mice exposed to the 15-μT sawtooth field, but no teratological effects were found with the other pulsed fields that were studied. In a similar study using the same sawtooth pulsed field but a different strain of mice, Frolen et al.[219,220] did not observe a statistically significant increase in the incidence of fetal malformations. However, these investigators observed a significant increase in the percentage of placental resorptions in the exposed group relative to control mice. In an extensive study on rats exposed throughout gestation to 18-kHz sawtooth fields with peak flux densities of 5.7, 23, or 66 μT, Stuchly et al.[221] observed no significant field-related effects on the number of implantations, fetal resorptions, fetal body weight, or the incidence of major malformations. However, an increase in the number of minor skeletal anomalies was found in the group of mice exposed to the highest field level. In another extensive study involving several thousand exposed mice, Wiley et al.[222] observed no effect of exposure throughout the gestational period to 26-kHz sawtooth pulsed fields with peak flux densities up to 0.2 mT.

The possibility that pulsed magnetic fields have significant effects on embryonic development has been tested in a large number of studies on chicken embryos, as discussed in several reviews.[41,211,212] The first significant study designed to detect teratological effects in this species was conducted by Delgado et al.[223,224] with pulsed magnetic fields. These studies involved the exposure of embryos during the first 48 hr of development to square-wave pulses with a duration of 500 μsec, a 2-μsec pulse rise time, and peak flux densities of 0.12, 1.2, and 12 μT. Pulses with repetition frequencies of 10, 100, and 1000 Hz were tested for effects on embryo development. The most significant effects on embryo development were observed with the 1.2-μT pulsed field repeated at a 100-Hz frequency. The authors interpreted the fact that fields at higher and lower intensities and frequencies had a smaller effect as a demonstration of a possible "window" effect, but offered no explanation in terms of possible biological mechanism(s) that could explain such an effect.

Subsequent to the early reports from Delgado's laboratory, a large number of studies have been conducted throughout the world to detect teratogenic effects of both sinusoidal and pulsed magnetic fields with a wide variety of waveforms.[225–241] The results of these studies have been highly variable, with several groups of investigators finding no effects of fields that were similar to those used by Delgado and his associates.[226,228,229,236,238] In a major effort to determine whether the original set of findings by Delgado et al.[223,224] could be replicated consistently, the Office of Naval Research and the U.S. Environmental Portection Agency sponsored an international cooperative effort involving six independent laboratories.[238] Laboratories located in Spain, Sweden, Canada, and the United States were equipped with identical pulsed magnetic field exposure systems that had been constructed and tested by the same engineering team. The pulse parameters chosen for this study were a 100-Hz repetition frequency, a 500-μsec pulse duration, 2-μsec pulse rise and fall times, and a peak magnetic flux density of 1.0 μT. The waveform was therefore very similar to the square-wave pulsed waveform used in Delgado's laboratory. Each of the six laboratories conducted 10 separate experiments with 20 chick eggs, 10 of which were exposed to the pulsed field for the first 48 hr of incubation while the remaining 10 eggs were sham-exposed for the same time interval. Two of the six laboratories observed a statistically significant increase in the proportion of abnormal embryos ($p < 0.001$ and $p = 0.03$), while the other four laboratories did not observe a significant difference between the exposed and sham-exposed embryos. The overall data from the six different laboratories, however, did show a statistically significant increase in the proportion of abnormal embryos in the exposed groups of eggs. The interlaboratory variations observed in this series of experiments are indicative of the difficulty often encountered in replicating the results of biological studies of ELF field effects, even when exceptional efforts are made to control the relevant experimental variables. In a recent publication, Handcock and Kolassa[242] were critical of the statistical methods used to analyze the results of the interlaboratory study conducted by Berman et al.[238] Using a linear logistic modeling technique, these investigators concluded that the interlaboratory variations in the study by Berman et al.[238] were at least as large as any effects of the pulsed magnetic field observed in individual laboratories. Several sources of interlaboratory variability existed in the study by Berman et al.,[238] including the use of a different strain of chicken in one laboratory and differences in the ages of the eggs at the time of exposure to the pulsed magnetic field.

C. CANCER-RELATED STUDIES

As discussed in several recent reviews, environmental levels of ELF fields are sufficiently weak that direct effects on DNA and chromatin structure leading to mutation and neoplastic transformation are not expected to occur.[2,40,42,93,107,126,243,244] This expectation has generally been supported by the results of laboratory studies designed to detect DNA and chromosomal damage, mutational events, and increased cell transformation frequency in response to ELF magnetic field exposures (see Section IV.B.1.a). The possibility exists, however, that epigenetic effects of these fields (for example, on cell signaling pathways and gene expression) could influence the growth of cancer cells. Although effects of this nature have been reported in the literature, attempts at replication have led to both positive and negative findings (see Section IV.B.1.b). The following is a summary of studies with cellular and animal models that were designed to detect effects of ELF magnetic fields on cancer cell proliferation. A particular focus of these studies has been the detection of possible effects of ELF magnetic fields on the promotion and progression of tumors following initiation with carcinogenic chemicals.

Studies on *in vitro* tumor cell growth, transplanted tumors in rodents, and rodents that are genetically susceptible to cancer have not provided significant insight into the possible carcinogenic effects of exposure to ELF magnetic fields. For example, studies with pulsed or sinusoidal magnetic fields have led to findings of an increased growth rate,[245] a depressed growth rate,[246] and an unchanged growth rate[151,247] of transformed cell lines exposed *in vitro*. Bellossi[248] reported no effect on cancer development of exposure to pulsed magnetic fields throughout the lifespan of five consecutive generations of leukemia-prone AKR mice. The mice were exposed twice weekly for 30 min to pulsed magnetic fields with a peak flux density of 6 mT and repetition frequencies of 12 or 460 Hz. Thompson et al.[249] found no effect of exposing mice to 60-Hz magnetic fields with flux densities of 1.4, 200, or 500 μT on the growth of transplanted P388 leukemia cells. Mice were exposed 6 hr/day, 5 days/week throughout the survival period after injection of the tumor cells.

Several studies of more direct relevance to the possible effects of ELF magnetic fields on tumor development in humans have involved *in vivo* experiments with skin, mammary, liver, and brain tumor models in rodents. In experiments on skin tumor promotion, three studies in mice have shown no promoting effect of chronic exposure to ELF magnetic fields with flux densities ranging from 0.05 to 2 mT on the development of skin papillomas induced by topical application of 10 nanomole of the carcinogen, 7,12-dimethyl-benz(a)anthracene (DMBA).[181,250,251] In positive control groups of animals, skin tumor development was observed when the DMBA-initiated mice were treated twice weekly with topical application of the phorbol ester, 12-O-tetradecanoyl-phorbol-13-acetate (TPA). In one experiment, the effect of an intermittently applied 50-Hz magnetic field (15 sec on/15 sec off) was tested for tumor-promoting activity in comparison with a continuously applied field.[251] A small increase in the number of skin tumors was observed in the mice exposed to intermittent versus continuous fields, but neither group of animals showed a statistically significant increase relative to control mice treated with DMBA alone. Hence there was no indication of a tumor-promoting effect with either type of magnetic field exposure.

Another *in vivo* study of particular interest in the context of possible tumor-promoting effects of ELF fields is the finding by Stuchly et al.[252] that a 60-Hz, 2-mT magnetic field exerts a copromoting effect on mouse skin carcinogenesis. In these experiments skin tumors were initiated by the topical application of 10 nanomole of DMBA, and were then promoted by the application of 4.9 nanomole of the phorbol ester TPA, once per week for a total of 23 weeks. In the group of mice that received DMBA and TPA plus exposure to the 60-Hz field for 6 hr/day, 5 day/wk throughout the tumor promotion phase, both the percentage of mice with tumors and the number of tumors per mouse increased more rapidly with time than in the group of mice that received DMBA plus TPA alone. For example, at week 18 the percentage of mice with tumors in these two groups were, respectively, 25% and 8% and the mean number of tumors per mouse were 1.90 ± 0.69 (SE) and 0.65 ± 0.46. By week 23 the differences between the mice exposed to the 60-Hz magnetic field and the nonexposed group were no longer statistically significant. Other studies by the same group of investigators had previously shown that a 60-Hz, 2-mT magnetic field acting alone on DMBA-initiated skin cells does not have a tumor-promoting effect.[181] It would be of considerable interest to conduct further experiments in which comparative measurements are made of kinase-C activity in the affected skin cells throughout the tumor promotion phase in the field-exposed and nonexposed groups of mice. Such experiments would provide a test of the hypothesis that ELF fields may exert a copromoting effect on tumor development through an influence on the kinase-C signaling pathway activated by phorbol esters.

Three studies of mammary cancer development in rodents exposed to ELF magnetic fields following tumor initiation with a chemical initiator have provided some evidence for a tumor-promoting effect of the fields.[253–255] In the first of these studies, rats were injected intravenously with a 50 mg/kg dose of nitrosomethylurea (NMU), which produced mammary tumors in 54% of the rats.[253] This percentage increased to 86% in a group of rats initiated with NMU and exposed to a 50-Hz, 0.2-mT magnetic field for 3 hr/day for 5 weeks. The latency time to tumor development was also found to decrease significantly in the rats exposed to both NMU and the magnetic field relative to rats that received only NMU. In a second study, rats that were initiated by oral doses of DMBA (20 mg total dose) were exposed for 23 weeks to a 50-Hz, 30-mT field.[254] Although the overall tumor incidence did not differ significantly between the field-exposed and control groups, there was a statistically significant increase in the number of tumors per rat in the group exposed to the 30-mT field. A third study was conducted with the same protocol as the second study described above, except that a considerably smaller field level of 0.1 mT was used.[255] In this experiment the overall mammary tumor incidence was observed to be 50% higher in the field-exposed rats relative to the control group of animals.

These observations of increased tumor incidence in rats exposed to ELF magnetic fields may be related to a field-induced suppression of pineal melatonin and a resultant elevation in steroid hormone levels and breast cancer risk (see Section IV.B.1.d). Further research on the effects of ELF fields on the development of chemically-induced mammary tumors in rodents, including parallel measurements of field effects on the level of pineal melatonin in the exposed animals, would provide useful insights into the pathway by which these fields may exert carcinogenic effects.

Two series of experiments have been designed to study the effects of chronic exposure to 50-Hz magnetic fields on the development of chemically-induced liver tumors in partially hepatectomized rats.[256,257] In both studies liver tumor development was initiated by a 30-mg dose of diethylnitrosamine (DENA) administered intraperitoneally 24 hr following partial removal of liver tissue. Tumor development was promoted with phenobarbital as a positive control. The development of transformed foci of liver cells was assayed by histochemical staining for the enzymes gamma-glutamyl transpeptidase and glutathione S-transferase. The development of tumor foci in the livers of the chemically-initiated rats was not significantly influenced by a 3-month exposure to 50-Hz magnetic fields with flux densities ranging from 0.5 μT to 500 μT, thus indicating a lack of tumor-promoting effect by the fields. Similarly, no evidence was obtained for a copromoting effect in rats exposed to both phenobarbital and 50-Hz fields following tumor initiation with DENA.

A recently completed study provided evidence that chemically-induced brain tumors in rats are not promoted by exposure to 50-Hz magnetic fields.[258] Brain tumors of various histological types, including gliomas and astrocytomas, were induced by intravenous injection of 50 mg/kg ethylnitrosourea (ENU) into female rats on the 19th day of pregnancy. After weaning, the offspring were chronically exposed to fields of 1, 10, and 100 μT and studied for survival time. Neither male or female offspring showed a statistically significant difference in mean survival time relative to control rats that were exposed to ENU but not to 50-Hz magnetic fields.

Overall, with the possible exception of mammary tumors, the available evidence is not strong for a promoting or copromoting effect of ELF fields on tumor development in rodents. However, further studies are needed to clarify the possible influence of ELF fields on second messenger signaling and endocrine regulation in exposed animals, both of which are factors that could influence the development of tumors by promoting the proliferation of initiated cells.

V. HUMAN HEALTH STUDIES

A subject of considerable contemporary interest and debate is the possible relationship between human exposure to ELF magnetic fields and adverse health effects. A comprehensive review of the epidemiological literature on the possible health effects of exposure to ELF fields in residential and occupational settings is presented in Chapter 7. The focus of these studies, however, has been on the biological interactions and potential health effects of the magnetic, rather than the electric, component of ELF fields from residential power distribution lines, appliances, machinery, and other common sources. There are two primary reasons for focusing on ELF magnetic fields as the major source of human exposure: (i) nearly complete shielding from the ELF electric fields of power distribution lines is provided in homes, offices, and other buildings by intervening trees, walls, and other structures, and (ii) the intensities of the ELF electric fields that emanate from many common electrical devices used in the home and workplace are small, and couple to the body more weakly than the ELF magnetic fields produced by these devices. Because of the emphasis on ELF magnetic fields in nearly all of the human health assessments conducted to date, a brief summary and critique of the literature on this subject is presented in this chapter.

A. REPRODUCTIVE OUTCOMES AND CANCER RISK IN RELATION TO ELECTRIC APPLIANCE USE

In 1986, Wertheimer and Leeper[259] published evidence that seasonal changes in fetal growth and abortion rates may be linked to the use of electrically-heated waterbeds and electric blankets during the winter months. The authors contended that these adverse effects on fetal development could result from exposure of the fetus to the 60-Hz fields present at the surfaces of the electrically-heated beds and electric blankets. They pointed out, however, that the potentially harmful effect of excessive heat on fetal development cannot be excluded on the basis of their data. The methods used in this epidemiological study and the conclusions drawn by the authors have been criticized by Chernoff et al.,[211] primarily on the basis of the small sample size and the unconventional procedures used for selecting subjects and analyzing data on pregnancy outcomes. In a subsequent study conducted in Springfield and Eugene, Oregon, where electric ceiling cables are frequently used to provide heat in homes during the winter months, Wertheimer and Leeper[260] also found evidence suggesting a higher rate of fetal loss during the winter months. They again suggested that the effect was associated with an elevated level of 60-Hz fields from the ceiling cables, which they determined from measurements to be approximately 1 μT and 10–50 V/m when the ceiling cables are turned on. These values of the magnetic flux density and electric field intensity are about 5–10 times the typical levels observed in homes without ceiling cables. In evaluating this study, it should be noted that the rate of abortions among subjects that lived in homes with ceiling cables was the same as that among women who lived in homes with other types of heating. The only trend in the data that supported the conclusions of Wertheimer and Leeper[260] was the seasonal variation in abortion rate among women in the cohort of subjects living in homes with ceiling cables. In another study on fetal development in relation to exposure to ELF fields from electrically-heated waterbeds and electric blankets, Dlugosz et al.[261] found no evidence for an elevated rate of congenital malformations in the offspring of women who used electrically-heated bedding.

Two recent studies evaluated the occurrence of adverse pregnancy outcomes in relation to the average level of ELF magnetic fields in the residential environment. Juutilainen et al.[262] reported that the risk of early fetal loss was increased among women who lived in homes with the highest levels of ambient ELF magnetic fields measured at the front door. In homes with typical levels of ELF fields, there was no significant elevation in risk. Savitz and Ananth[263] subsequently found no evidence for an association between residential levels of ELF magnetic fields and the risk of miscarriage, low birth weight, or preterm delivery. Both of these studies involved a small number of subjects, thus limiting the statistical power of their conclusions. In addition, the studies were designed to detect a possible correlation between adverse pregnancy outcomes and residential levels of ELF magnetic fields, including contributions from neighboring power lines, and did not specifically consider the effects of fields from appliances or other sources within the home. Nevertheless, the general lack of a significant relationship between adverse reproductive effects and residential field levels in these two studies is of interest in the broader context of potential health effects from exposure to ELF magnetic fields produced within the home by a variety of sources such as electric appliances.

Several studies have addressed the issue of cancer risk in relation to exposure to ELF magnetic fields from electric blankets and other personal appliances used within the home. Three of these studies found no evidence for an elevated risk of myelogenous leukemia,[264] testicular cancer,[265] or breast cancer[266] among users of electric blankets. In a preliminary study with a small number of cases and controls, Lovely et al.[267] reported that the use of massage units was correlated with an elevated risk of nonlymphocytic leukemia. A similar association was not observed between leukemia risk and the use of electric razors or hair dryers. Overall, the use of one or more of these appliances was not associated with a statistically significant increase in the risk of nonlymphocytic leukemia. The findings in this group of studies support the general conclusion that there is no substantial evidence for a link between cancer risk and exposure to the ELF magnetic fields from appliances used in the home.

B. REPRODUCTIVE EFFECTS AMONG OPERATORS OF VISUAL DISPLAY UNITS

A large number of epidemiological surveys have been conducted to detect adverse reproductive outcomes among women who spend a majority of their time at work operating VDUs. As discussed in several recent reviews,[211,212,268–270] most of these studies have found no significant evidence for an increased risk of abnormal pregnancy outcomes associated with the use of a VDU. However, in two studies an increased risk of spontaneous abortion was observed under certain conditions of VDU usage. In the first of these studies, Goldhaber et al.[271] reported that the rate of miscarriages increased by approximately 80% among women who worked on VDUs for more than 20 hr/week, as compared to women who did similar work without the use of these devices. No statistically significant elevation in the risk of miscarriage was found among women who worked at VDUs for less than 20 hr/week. Overall, the increase in the rate of birth defects was about 40% for women who worked at VDUs for more than 5 hr/week, but this increase was not statistically significant. Although the results of this study suggest that the fields from VDUs may enhance the risk of miscarriage, the possible role of job stress or other unidentified factors cannot be excluded on the basis of the information collected by Goldhaber et al.[271] In a second study with positive findings, Lindbohm et al.[272] reported that the risk of spontaneous abortion was significantly elevated among a small subset of women who used VDUs with the highest level of exposure to ELF magnetic fields (greater than 0.9 μT). However, in this study there was no statistically significant elevation in the overall risk of adverse pregnancy outcomes. Apart from the results of these two epidemiological surveys, 15 other studies have found no evidence for a significant elevation in the risk of spontaneous abortion or birth defects as a result of prolonged exposure during pregnancy to the fields from VDUs.[273–287] In other studies on the general health of VDU operators, no major health problems have been identified.[288] Certain adverse reactions to prolonged use of VDUs, such as skin rashes or eye discomfort, have been attributed to stressful working conditions or ergonomic factors.[288]

C. RESIDENTIAL FIELDS AND CANCER RISK

Since the initial report by Wertheimer and Leeper[289] of a possible association between residential exposure to ELF magnetic fields from power distribution lines and the risk of childhood cancer, 16 epidemiological studies of a similar nature on both childhood and adult cancer have been conducted in the United States, Europe, and Asia. The following is a brief summary of these studies, several of which suggest that some relationship may exist between the proximity of residences to power lines and the risk of childhood cancer. Little evidence has been acquired to suggest a similar association between power line fields and adult cancer, nor has it been established that the ELF magnetic fields from power lines are causally linked to cancer risk among children.

In their initial study, Wertheimer and Leeper[289] found that cancer deaths among children less than 19 years of age in the Denver, Colorado, area were correlated with the presence of high-current primary and secondary wiring configurations in the vicinity of their residences. This retrospective epidemiological study was based on 344 fatal childhood cancer cases during the period 1950 to 1973 and an equal number of age-matched controls chosen from birth records. The electric power lines near the birth and death residences of the cancer cases and the residences of the controls were inspected and classified as being either high-current configurations (HCC) or low-current configurations (LCC). This "wire code" was assumed to reflect the local strength of the 60-Hz magnetic field within the homes of the subjects. The percentage of the cancer cases whose birth and death residences were near HCC was found to be significantly greater than for the residences of the control subjects, from which Wertheimer and Leeper[289] concluded that an association may exist between the strength of magnetic fields from the residential power distribution lines and the frequency of childhood cancer.

In a subsequent study, Fulton et al.[290] used methodology that was matched as closely as possible to that of Wertheimer and Leeper,[289] including the designation of HCC and LCC power lines. This study involved 119 leukemia patients with ages of onset from 0 to 20 years, whose address histories were obtained from medical records at Rhode Island Hospital, and 240 control subjects chosen from Rhode Island birth certificates. In their study, Fulton et al.[290] concluded that no statistically significant correlation existed between the incidence of leukemia and the residential power line configuration. Wertheimer and Leeper[291] were critical of the study by Fulton et al.[290] on the basis that the control and case groups had not been matched for interstate migration, for years of occupancy of residences, or for the ages of the children at the time their residential addresses were determined from birth records and hospital medical records. In a subsequent analysis of the data obtained by Fulton et al.,[290] Wertheimer and Leeper[291] excluded cases and controls aged 8 and above in order to define a complete residential history for the remaining subjects (53 cases and 71 controls). In this subset of the total population studied by Fulton and associates, Wertheimer and Leeper[291] found a weakly significant correlation ($p \sim 0.05$) between the incidence of leukemia and residential HCC wiring configurations.

Two case/control studies were conducted in England to evaluate childhood cancer risk in relation to residential proximity to overhead transmission and distribution lines. The first of these studies by Myers et al.[292] involved 376 cancer cases and 590 controls in the Yorkshire Health Region in England. Exposure was assessed by estimating the distances of residences from overhead power lines, and also from estimates of the 50-Hz magnetic fields in homes based on historical records of the current loads carried by neighboring power lines. No statistically significant elevation in the incidence of fatal leukemias or solid tumors was observed among the subjects living closest to overhead power lines. A second study in England of childhood leukemia risk in relation to living near overhead power lines was conducted by Coleman et al.[293] In this study 771 childhood leukemia cases were identified from the Thames Cancer Registry during the period 1965–1980, and matched with 1432 controls that were registered with a solid tumor (excluding lymphoma) or 231 general population controls. Exposures to 50-Hz magnetic fields were assessed from the distance, type and power loading of electrical supply equipment and power lines within 100 m of each subject's home. None of the comparisons between cases and controls indicated a statistically significant elevation in leukemia risk for children living in close proximity to either overhead power lines or substations.

Another study of childhood cancer incidence was conducted in the County of Stockholm by Tomenius et al.,[294] who analyzed the residential 50-Hz magnetic fields for 716 cases that had a stable address from the time of birth to the time of cancer diagnosis, and for 716 controls that were matched for age, sex, and birth location. These investigators evaluated the electrical wiring configurations near the residences of the study population, and also made measurements at the entrance door to each residence of the magnetic field flux density in the frequency range above 30 Hz. Among the residences where the magnetic field exceeded 0.3 μT, the incidence of cancer was greater by a statistically significant amount than the expected level. This elevation in cancer risk was primarily associated with an excess of nervous tissue tumors. It is interesting to note from the data presented by Tomenius et al.[294] that the arithmetic mean of the magnetic flux density for the residences of all the cancer cases did not exceed the mean value for the residences of the control group.

A study of considerable importance was conducted by Savitz et al.,[295] in which an attempt was made to replicate the original findings of Wertheimer and Leeper[289] on childhood cancer in the Denver, Colorado, area. In this study, both the wire code developed by Wertheimer and Leeper and spot measurements of 60-Hz magnetic fields in homes were used as a basis for comparison of cancer risk among cases and controls. The cases were children less than 14 years of age diagnosed as having cancer during the period from 1976 to 1983, which did not overlap the 1950 to 1973 period studied by Wertheimer and Leeper.[289] In a comparison of cancer risk associated with HCC versus LCC wiring configurations, Savitz et al.[295] observed a statistically significant odds ratio of 1.5. In contrast, no significant correlation was found between point-in-time measurements of magnetic fields in homes and the risk of cancer among the most highly exposed subjects. This set of observations thus provided some support for the original findings of Wertheimer and Leeper,[289] but also indicated that the 60-Hz magnetic fields in homes of cancer cases may not be the causal factor underlying an elevation in cancer risk. Several possible confounding factors such as socioeconomic status, *in utero* exposure to potential carcinogens, and residential traffic density were examined in relation to the relative cancer risk of cases living near HCC wiring configurations. Although factors such as maternal smoking during pregnancy and high residential traffic density were found to be associated with a greater occurrence of fatal childhood cancers, adjustment of the odds ratio to account for these factors did not significantly modify the estimate of cancer risk associated with residing near HCC power lines.

A later study that provided results similar to those of the Savitz et al.[295] report was conducted by London et al.[296] in Los Angeles County, California. This study was carried out on 232 childhood leukemia cases obtained from a population-based tumor registry during the period 1980–1987, and an equal number of controls matched by age, sex, and ethnicity. A statistically significant two-fold elevation in the risk of childhood leukemia was found to be associated with residence near HCC power lines. However, no significant association between leukemia risk and residential 60-Hz fields was observed, either on the basis of point-in-time measurements or continuous 24-hr measurements of the magnetic field in the child's bedroom. Potential confounding associated with exposure to fields from household and personal appliances, exposure to toxic chemicals, and drug use was evaluated by London et al.[296] Although the use of some personal appliances and chemical agents were found to be associated with an increased risk of childhood leukemia, the adjustment of cancer odds ratios for these factors did not significantly influence the observed association between residence near HCC power lines and the risk of childhood leukemia. Similar to the earlier study by Savitz et al.,[295] the results of the epidemiological survey by London et al.[296] thus supports an association of childhood cancer risk with high-current power line configurations, but not with directly measured 60-Hz fields in the home environment.

Four other studies have been conducted to evaluate childhood cancer risk in relation to residential power line fields. None of these epidemiological surveys used the Wertheimer and Leeper HCC and LCC power line configurations for estimating residential exposures, but instead used proximity to high-voltage power lines or models of residential magnetic fields from power lines for estimating the exposures of childhood cancer cases and controls. In a case-control study conducted in Taiwan in the late 1980s, Lin and Lu[297] reported a small, statistically insignificant elevation of all cancers, leukemias, lymphomas, and brain tumors among children who were assumed to have a higher exposure as a consequence of residing within 50 m of a high-voltage power line (greater than 20 kV), transformer, or substation.

Three Scandanavian studies on childhood cancer risk were published in 1993, in which both proximity to high-voltage power lines and historical models of residential fields were used as bases for estimating exposure to 50-Hz magnetic fields. The first of these three studies was conducted by Feychting and Ahlbom,[298,299] who used the comprehensive Swedish Cancer Registry as the basis for a case-control study of childhood cancer in relation to exposure to ELF magnetic fields. The cases were children diagnosed with all types of cancer during the period 1960–1985. Exposures were estimated from proximity of the residences of case and control subjects to 200- and 400-kV transmission lines, spot measurements of magnetic fields in homes, and a model of historical exposure based on the distance, configuration, and power load (current flow) on the power lines nearest to each home over a time period relevant to the epidemiological study. Predictions based on this model gave results that correlated reasonably well with contemporary measurements of 50-Hz magnetic fields in residences. A statistically significant association was found between estimated historical exposure to 50-Hz magnetic fields from power lines and the risk of childhood leukemia, but the association was not significant for other types of cancer. However, consistent with the results of earlier studies by Savitz et al.[295] and London et al.,[296] there was not a significant correlation between childhood leukemia risk and 50-Hz magnetic fields that were determined by short-term measurements in residences.

In response to a question by Stevens,[300] Feychting and Ahlbom[301] calculated that 62% of the variance in their estimates of the historical fields in residences is explained by distance from the nearest power lines. The exposure model used by Feychting and Ahlbom[298,299] that incorporates both distance and the load on nearby power lines is therefore superior to an estimate based only on distance. In their study Feychting and Ahlbom[298,299] also found that the association between childhood leukemia risk and estimated exposure to 50-Hz magnetic fields was stronger for children living in single-family homes than for those living in apartment buildings. They speculated that this result may be attributable to the lower accuracy of exposure estimates for children living in apartment buildings, which could lead to errors in the stratification of subjects by exposure groups.

The second and third Scandanavian studies were performed by Olsen et al.[302] in Denmark and by Verkasalo et al.[303] in Finland. The Danish case-control study had a design similar to the Swedish study discussed above, and was based upon all cases of childhood leukemia, lymphomas, other reticuloendothelial cancers, and central nervous system tumors registered during the period 1968–1986. Again, a model of power line contributions to residential 50-Hz magnetic fields and historical records of power loads were used to estimate exposure of the subjects in this study. No association was observed between any of the types of cancer recorded in this study and exposure to power line fields, although a statistically insignificant elevation was observed in the relative risk of leukemia and lymphomas.

In the study conducted in Finland,[303] no statistically significant elevation in the risk of all cancers, leukemia, lymphomas, or nervous tissue tumors was observed among a cohort of children who were estimated to be the most highly exposed to 50-Hz residential magnetic fields from power lines. Similar to the other recent Scandanavian studies, exposure of children to 50-Hz magnetic fields in the home was estimated from a model based on the distance and load currents carried by the nearest 110-, 220-, and 400-kV power lines. Subjects were identified from cancer cases recorded in the Finnish cancer registry during the period 1970–1989.

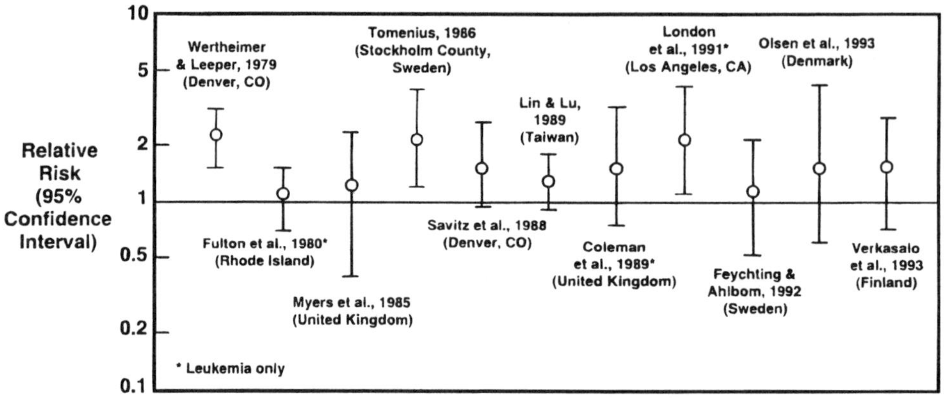

FIGURE 2 Graphical summary of studies on childhood cancer in relation to living in the proximity of electric power lines. Data are represented as the relative cancer risk based on the following exposure metrics: Wertheimer and Leeper wire codes [Wertheimer and Leeper,[289] Fulton et al.,[290] Savitz et al.,[295] and London et al.[296]] or estimates of 50/60-Hz magnetic field exposure based on either residential proximity to power lines or models incorporating the distance and current load of neighboring power lines [Myers et al.,[292] Tomenius,[294] Lin and Lu,[297] Coleman et al.,[293] Feychting and Ahlbom,[298,299] Olsen et al.,[302] and Verkasalo et al.[303]]. In several studies,[295,296,298,299] directly measured 50/60-Hz magnetic fields in residences were found not to be associated with an elevated cancer risk (data not shown). Three of the studies measured the relative risk only for childhood leukemia,[290,293,296] while the other studies determined the relative risk for all types of cancer. In the Feychting and Ahlbom study,[298,299] the relative risk was determined both for all types of cancer (data shown in the figure) and for subsets of children who had died of leukemia, lymphoma, and CNS tumors. Only the relative risk of leukemia exhibited a statistically significant level for the most highly exposed children. For the cancer cases exposed to residential fields exceeding 0.25 μT, the relative risk was 3.3 with a 95% confidence interval of 1.3 to 7.9. The relative risks of lymphomas, CNS tumors, or total cancers were not statistically significant in the Feychting and Ahlbom study.[298,299]

Although each of the three recent Scandanavian studies involved extensive cancer registries for the entire population of children residing in Sweden, Denmark, and Finland during periods of time ranging from 18 to 25 years, the total number of cancer cases among the most highly exposed subjects was only 12, 6, and 11 for the three studies, respectively. For this reason, an effort was made to increase the statistical power of the results from these three studies by pooling the data.[304] Based on this analysis, it was concluded that the overall risk of leukemia was significantly increased by approximately two-fold in the most highly exposed group of children. The overall risk of nervous tissue tumors, lymphomas, and all cancers combined was not found to be elevated by a statistically significant amount.

Figure 2 displays graphically the results of the 11 major studies on the risk of childhood cancer in relation to power line configuration (HCC versus LCC), proximity of residences to power lines, or estimates of residential exposure to magnetic fields based on models that use historical data on current loads carried by power lines. There is a clear trend in the data toward a small elevation in the risk of cancer among the most highly exposed children, although this effect was not statistically significant in a majority of the individual studies. As discussed above, none of the epidemiological surveys that involved measurements of ELF fields in homes over short time intervals (a few minutes to one day) found a correlation between the field levels and cancer risk among children.

A total of six epidemiological studies have been carried out to assess whether an association exists between exposure to residential power line fields and the risk of cancer among adults. The first of these studies was conducted by Wertheimer and Leeper,[305] and was based on 1179 cancer cases (78% fatal cancers) in the Denver, Boulder, and Longmont, Colorado, area during the period 1967–1977. As in the original study on childhood cancer in the same geographic area, a statistically significant elevation in cancer risk was observed among adults living in residences near HCC configurations of power lines.

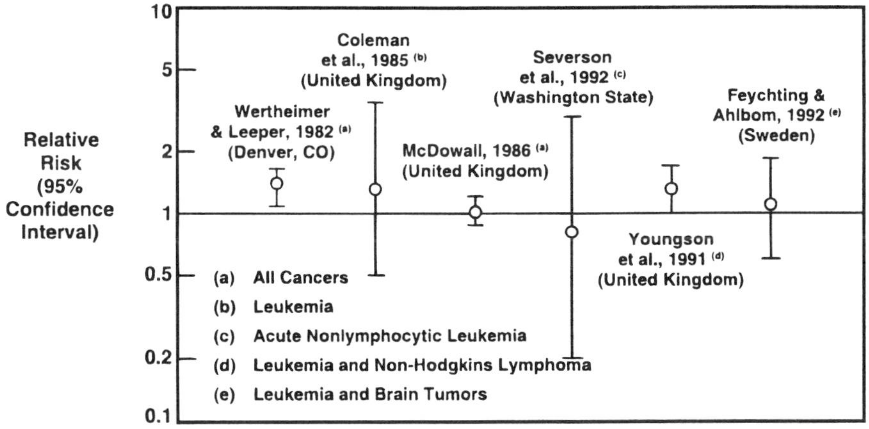

FIGURE 3 Graphical summary of studies on adult cancer in relation to living in the proximity of electric power lines. Data are presented on relative cancer risk based on either the Wertheimer and Leeper wire code [Wertheimer and Leeper[305] and Severson et al.[309]], or estimates of 50/60-Hz magnetic field exposure based on either residential proximity to power lines or models incorporating the distance and current load of neighboring power lines [Coleman et al,[306] McDowall,[307] Youngson et al.,[308] and Feychting and Ahlbom[298]]. In the studies by Severson et al.[309] and Feychting and Ahlbom,[298] directly measured 50/60-Hz magnetic fields in residences were found not to be significantly associated with an elevated cancer risk (data not shown).

Five subsequent epidemiological studies observed no correlation between adult cancer risk and exposure to power line fields estimated on the basis of residential proximity to power lines,[306–308] the Wertheimer/Leeper wire code and direct measurements of fields in the homes of cases and controls,[309] or exposure estimates based on a model of fields in homes that takes into account proximity to power lines and their historical current loads.[298] The primary focus of these studies was on leukemia, although other hematological malignancies, breast cancer, and brain tumors were considered in some of the studies.

The relative risk of adult cancer estimated in the six published studies is represented graphically in Figure 3. Based on an evaluation of the composite body of data acquired in the original Wertheimer and Leeper[305] study and the five subsequent studies on adult cancer, an advisory committee to the National Radiological Protection Board in England concluded that there was no convincing evidence of an association between the ELF fields from residential power lines and the risk of cancer among adults.[310]

D. OCCUPATIONAL EXPOSURE AND CANCER RISK

Controversy surrounding the issue of exposure to ELF fields and cancer risk has been increased by a large number of reports on an apparent association between employment in various electrical occupations and an elevated cancer risk. Following the original publication by Milham[311] in 1982, in which he observed an elevated risk of leukemia among electrical workers relative to the general population of adult males in the State of Washington, forty reports on this subject have appeared in the published literature. More than two-thirds of these publications have reported an elevation in the risk of certain types of cancer among electrical workers, primarily leukemia and nervous tissue tumors. In an effort to strengthen the conclusions drawn from individual studies, in 1987 Savitz and Calle[312] collated the data from 11 published reports to estimate the average relative risk of all leukemias, acute leukemias, and acute myelogenous leukemias among workers in 12 different classes of electrical occupations. The overall relative risk and 95% confidence limits for leukemia mortality were found to be the following: (i) total leukemias: 1.2 (1.1–1.3), (ii) acute leukemias: 1.4 (1.2–1.6); and (iii) acute myelogenous leukemias: 1.5 (1.2–1.8). Similar results were obtained by Coleman and Beral,[313] who also conducted a meta-analysis of data on leukemia risk among electrical workers that had been published prior to 1988. A similar conclusion of a small elevation in the risk of leukemia among electrical workers was drawn in several later evaluations of the epidemiological data on electrical workers.[310,314,315] In addition, a growing body of information on cancer risk among electrical workers has indicated that this occupational group has an elevated risk of brain tumors.[307,311,312] Limited evidence also suggests an elevated risk of breast cancer among electrical workers.[316–319] Three of these epidemiological studies were conducted with male electrical workers as subjects, among whom breast cancer is extremely rare, and the fourth study was on female workers. The finding of an elevated risk for breast cancer is especially interesting in the context of the hypothesis that exposure to ELF fields may depress the nocturnal synthesis of pineal melatonin, thereby leading to an increase in the circulating level of steroid hormones and a resultant increase in the proliferation of hormone-dependent breast cancer cells (see Section IV.B.1.e).

Assessment of the possible link between occupational exposure to ELF fields and cancer risk is complicated by several factors, including the wide range of variability of results obtained in different studies and the lack of quantitative exposure assessment in nearly all of the studies published prior to 1990. In the last few years there have been several notable attempts to overcome these deficiencies and to increase the statistical power of each individual epidemiological study by analyzing cancer data on large cohorts of exposed workers. Unfortunately, the results of these studies have yielded such a wide range of risk estimates for leukemia and nervous tissue tumors that they have not provided a clear answer to the question of whether electrical workers are at increased risk for specific types of cancer. This conclusion is illustrated by the following brief summary of the major epidemiological studies on electrical workers published in the last few years.

In a major study on 243,000 workers at electric power companies in Canada and France, Thieriault et al.[29] observed a two-fold to three-fold elevation in the risk of leukemia, including both nonlymphoid leukemia and acute myelogenous leukemia, among workers who had more than the median cumulative exposure to ELF magnetic fields based on direct measurements using personal dosimeters. No exposure-related increases were observed in 29 other types of cancer, including brain tumors and breast tumors. Some support for this finding was obtained in a registry-based study in Los Angeles County, California, by London et al.[28] These investigators found a statistically significant trend in the risk of leukemia and acute myelogenous leukemia of 1.2 and 1.6, respectively, per microtesla increase in 60-Hz magnetic field exposure determined with personal dosimeters. In contrast, another study on California electrical workers that also used personal dosimeters for exposure assessment did not find a statistically significant elevation in the risk of leukemia, lymphomas, brain tumors, or all cancers combined.[27]

Two other large studies that involved the use of personal dosimeters for exposure assessment have failed to observe a statistically significant elevation in the risk of leukemia among the most highly exposed groups of electrical workers. The first of these was a study on telephone lineworkers whose exposure to ELF fields was a result of installing telephone lines on poles shared by electric power lines.[320] Although the overall risk of leukemia was not significantly elevated among these workers, the data suggested an increasing trend of risk with increasing levels of exposure. Another major study that led to negative findings on the total cancer risk and the specific risk of leukemia was conducted by Savitz and Loomis,[321] who analyzed morality data from the period 1950–1988 among 139,000 workers at five electric power companies in the United States. In this study, however, a statistically significant increase was observed in the risk of brain tumors among the most highly exposed workers. The relative risk of brain cancer and the 95% confidence interval obtained for workers employed from 5 to 20 yr were 1.87 and 1.20–2.92, respectively. For workers employed more than 20 yr, the relative risk and 95% confidence interval were 1.45 and 0.83–2.53. The value for the relative brain tumor risk among workers employed more than

20 yr approaches, but does not reach, statistical significance. The results of this study therefore differ from the findings of Thieriault et al.[29] and Sahl et al.,[27] who found no elevation in the risk of brain tumors among electrical workers. It is interesting to note that the results of this recent study by Savitz and Loomis[321] are similar to those obtained by the same investigators in an earlier case-control study that was based on an analysis of occupational mortality data from the United States vital statistics registry during the period 1985–1986.[322] In this epidemiological survey men in electrical occupations were found to have no increase in leukemia risk relative to men in all other occupations, whereas the risk for brain cancer among the electrical workers was significantly elevated with an odds ratio of 1.4. In a relatively small study on male workers in electrical occupations in Los Angeles County, California, Mack et al.[323] found a significant elevation in the risk of astrocytomas among subjects employed for more than 10 years. However, the risk of all brain tumors, meningiomas, and gliomas was not elevated significantly.

A set of epidemiological studies on cancer risk among electrical workers in three Scandanavian countries have also been published since 1990. In the first of these studies, Juutilainen et al.[324] reported small, but statistically significant, elevations in the relative risk of leukemia (1.42) and brain tumors (1.29) among Finnish electrical workers in job categories that were judged to involve excess exposure to ELF magnetic fields. The second study was conducted on Norwegian electrical workers, and led to the finding of a statistically significant 40% elevation in the incidence of leukemia among exposed workers who had been employed in electrical trades for more than 10 years.[325] The incidence of brain tumors was not elevated significantly among exposed workers with long durations of employment. As in an earlier report by the same group of investigators,[319] a significant twofold increase in breast cancer incidence was observed in this study. The third study in this group from Scandanavian countries was conducted by Floderus et al.[326] These investigators found a statistically significant two-fold to three-fold elevation in the relative risk of leukemia among the oldest workers (greater than 60 years of age at diagnosis) who routinely worked in 50-Hz magnetic fields with flux densities greater than 0.2 μT. The elevated risk of leukemia was primarily associated with chronic lymphocytic leukemia, and not acute myelogenous leukemia. This result therefore differs from earlier findings[28,29,312,313] in terms of the subclass of leukemia to which electrical workers are at greatest risk. In the Swedish study, the overall risk of brain tumors was not significantly elevated, although there was a trend of increased risk among the most highly exposed categories of electrical workers. Overall, a large variability in the results of independent epidemiological studies on different populations of electrical workers is evident in the Scandanavian studies, as well as in the studies conducted in the United States. Although many of the recent studies have involved estimates of the effects of confounding variables such as cigarette smoking and workplace exposures to carcinogenic chemicals, correction of relative risk values for the influence of these variables has, in general, had little effect on the results obtained in individual studies or reduced the variability among results obtained in different studies.

E. CRITIQUE OF EPIDEMIOLOGICAL STUDIES

The overall set of information obtained to date from epidemiological studies has not demonstrated a strong association between cancer risk and residential exposure to ELF fields. In the case of adult cancer, there is no convincing evidence for a correlation between residential exposures to ELF magnetic fields and the risk of any type of cancer. A similar conclusion cannot be drawn for childhood cancer, as indicated by the fact that a recent meta-analysis of published data using residential proximity to power lines as an exposure metric led to the conclusion that there is a small, but statistically significant, increase in the risk of leukemia and nervous system tumors among the most highly exposed children.[327]

One of the most puzzling findings in recent studies on childhood cancer risk is the fact that ELF magnetic fields measured in homes have consistently shown a much weaker association with cancer risk than have various surrogate measures of exposure such as the Wertheimer/Leeper wire code. One possible interpretation of this finding is that wire codes may be better indicators of historical exposure to ELF magnetic fields in the home environment than point-in-time measurements of the field levels. This hypothesis has not been supported by a recent study which showed that measurements of 60-Hz magnetic fields in the homes of subjects studied by Savitz et al.[295] were remarkably stable over a period of 5 yr.[328] Another possible interpretation is that wire codes may be reliable indicators of carcinogenic factors other than ELF fields in the residential environment. Although the role of confounding factors such as exposure to carcinogenic chemical substances has been analyzed in several recent studies on childhood cancer, the number of exposed subjects was generally small and corrections for the influence of confounding variables had little effect on the overall estimate of cancer risk. Nevertheless, the association of childhood cancer risk with confounding variables has, in some cases, been found to be stronger than the association with power lines or measured values of ELF magnetic fields in the home. For example, Savitz and Feingold[329] found that residential traffic density was strongly associated with childhood cancer, especially leukemia, among the same study population that had previously been reported to show an association between cancer risk and high-current power line configurations.[295] The authors speculated that benzene, a known leukemogen, in automobile exhaust fumes may have been a contributing factor to the elevated incidence of childhood cancer. Such findings suggest that it would be worthwhile to evaluate the possible impact of confounding variables on the relative risk of childhood cancer using pooled data from several independent epidemiology studies as a means of increasing the sample size and the statistical power of the risk estimates. The importance of identifying and introducing corrections for confounding variables in epidemiological studies has been emphasized by Savitz and Baron.[330]

Another factor in the childhood cancer studies that merits further attention is the possible role of exposure to ELF magnetic fields from appliances, ground currents, and other sources of fields within the home environment. Two extensive studies have shown that only 17 to 21% of the variance in residential ELF magnetic field levels can be accounted for by differences in power line configurations (for example, the proximity of high-current versus low-current power lines in the Wertheimer/Leeper classification).[17,331] Other sources of exposure have not been completely analyzed, and may not have been fully accounted for, in studies that utilized personal dosimeters or fixed-site measurements to characterize ELF field levels in homes. It should also be noted that both the time resolution and the bandwidth of the personal dosimeters used in these studies were limited. As a result, brief exposures to high-amplitude magnetic field transients and to high-frequency harmonic fields were not captured in the personal exposure histories recorded with these dosimeters. The characterization of these exposures clearly merits further attention.

The large body of published literature on occupational exposure to ELF magnetic fields and cancer risk is similarly difficult to interpret. Although the majority of studies conducted to date have shown an elevation in the risk of various types of cancer, especially leukemia, nervous tissue tumors, and breast cancer, the lack of uniformity in the results obtained in different studies is troublesome. Attempts to improve the statistical reliability of the conclusions drawn from studies on cancer risk among electrical workers, either by increasing the sample size in individual studies or by conducting a meta-analysis of pooled data, have contributed little to our understanding of the true level of risk. It is becoming increasingly evident, however, that any excess in the risk of various types of cancer among electrical workers compared to other occupational groups is small, and very difficult to detect in a reliable manner by conventional epidemiological methods. Efforts to analyze cancer risk for electrical workers in relation to their exposure to ELF fields have also been complicated by the lack of an adequate assessment of workplace exposure characteristics and the possible role of confounding variables. Many of the same problems discussed above for epidemiological studies on residential cancer risk pertain to the study of excess cancer risk among electrical workers.

A final important issue that has complicated all of the epidemiological studies conducted to date is the lack of knowledge of a biologically plausible mechanism (or mechanisms) through which exposure to weak ELF fields could influence the risk of cancer. As a consequence, there is no clear information to guide the design of future epidemiological studies in terms of the most appropriate exposure parameters and biological endpoints to measure.

VI. SUMMARY AND CONCLUSIONS

The literature summary presented in this chapter illustrates the wide variety of effects of ELF magnetic fields that have been reported to occur at the animal, tissue, cellular, and subcellular levels. Several effects, such as the induction of magnetophosphenes in the visual system, are associated with high field levels and have been successfully replicated in different laboratories. Many other effects, especially at low field levels, have not been independently verified, or in some cases, replication efforts have led to conflicting results. An example of the latter situation is provided by the results of a large number of independent studies on the transport of Ca^{++} into cells during exposure to combined static and ELF magnetic fields under resonance conditions.

Evidence is growing for a central role of cell membranes in the reception, transduction, and amplification of signals imposed by ELF magnetic fields. A major challenge in future laboratory-based research will be the elucidation of specific physical mechanisms and molecular pathways through which these fields can influence transmembrane signaling events and affect the functional and proliferative states of cells and organized tissues. Of particular importance will be studies on possible mechanisms through which ELF fields may play a role in the development of tumors. At the present time there is little evidence for a promoting or copromoting effect of ELF magnetic fields on tumor development, with the possible exception of mammary tumors in which endocrine alterations resulting from field exposure may play an important role. Further research is needed to gain an understanding of the ELF signal characteristics that are the most biologically effective, and to define the threshold field parameters above which predictable biological responses occur. Recent laboratory studies have provided a number of insights on the pathways through which ELF fields may interact at the tissue, cellular, and subcellular levels. However, a great deal of research remains to be done in order to fully characterize the molecular substrates of ELF field interactions and the resultant cascade of electrical and biochemical signals that lead to cellular and tissue responses, including possible carcinogenic effects.

Numerous epidemiological studies on the possible health effects of ELF magnetic fields have contributed information on adverse pregnancy outcomes and cancer risk among the most highly exposed groups of individuals. In the case of field-related effects on fetal development, there is little evidence for an elevated risk of spontaneous abortion or fetal malformations among women exposed to fields from VDUs or appliances such as electric blankets. A less clear picture has emerged from the large number of studies conducted over the last 15 years to detect an elevation in cancer risk associated with either residential or occupational exposure to ELF magnetic fields. Broadly speaking, these studies have provided some indication of a small elevation in the risk of leukemia, nervous tissue tumors, and breast cancer among the most highly exposed individuals. However, a lack of consistency in the results obtained in different epidemiological studies, together with an inadequate characterization of the exposure environment and the possible role of confounding variables, have made it impossible to determine with accuracy the excess cancer risk associated with exposure to ELF fields. The design and interpretation of meaningful epidemiological studies has been further complicated by the lack of information on biologically plausible mechanisms through which exposure to weak ELF fields could elevate the risk of cancer. There is a critical need to define such mechanisms on the basis of laboratory research, and to determine the most appropriate exposure metric(s) for assessing the potential health effects of ELF fields. Without this information it is likely that future epidemiological studies conducted by traditional methods will continue to yield variable results of limited usefulness for evaluating whether ELF magnetic fields exert measurable carcinogenic effects at typical levels of human exposure.

ACKNOWLEDGMENTS

Research support is received from the U.S. Department of Energy under Contract DE-AC06-76RLO 1830 with the Pacific Northwest Laboratory. The Pacific Northwest Laboratory is operated for the U.S. Department of Energy by the Battelle Memorial Institute.

REFERENCES

1. **Tenforde, T. S.,** Spectrum and intensity of environmental electromagnetic fields from natural and man-made sources, in *Biological Effects of Environmental Electromagnetic Fields.* Blank, M., Ed., American Chemical Society Press, Washington, DC, in press.
2. **Tenforde, T. S.,** Biological interactions and potential health effects of extremely-low-frequency magnetic fields from power lines and other common sources, *Annu. Rev. Public Health,* 13, 173, 1992.
3. **Polk, C.,** Schumann resonances, in *CRC Handbook of Atmospherics. Vol. 1.* Volland, H., Ed., CRC Press, Boca Raton, FL, 1982, 111.
4. **Stuchly, M. A.,** Human exposure to static and time-varying magnetic fields, *Health Phys.,* 51, 215, 1986.
5. **Tenforde, T. S.,** Magnetic field applications in modern technology and medicine, in *Biological Effects of Static and Extremely-Low-Frequency Magnetic Fields.* Bernhardt, J. H., Ed., MMV Medizin Verlag, Munich, Germany, 1986, 23.
6. **Bernhardt, J. H. and Matthes, R.,** ELF and RF electromagnetic sources, in *Non-Ionizing Radiation.* Greene, M. W., Ed., UBC Press, Vancouver, British Columbia, Canada, 1992, 59.
7. **Bracken, T. D.,** Field measurements and calculations of electrostatic effects of overhead transmission lines, *IEEE Trans. Power Appar. Syst.,* 95, 494, 1976.
8. *Transmission Line Reference Book—345 kV and Above, 2nd ed.,* Electric Power Research Institute, Palo Alto, CA, 1982.
9. **Tenforde, T. S. and Kaune, W. T.,** Interaction of extremely-low-frequency electric and magnetic fields with humans, *Health Phys.,* 53, 585, 1987.
10. **Kaune, W. T.,** Assessing human exposure to power-frequency electric and magnetic fields, *Environ. Health Perspect.,* 101 (Suppl. 4), 121, 1994.
11. **Gauger, J. R.,** Household Appliance Magnetic Field Survey, IIT Res. Inst. Rep. No. E06549-3, U.S. Naval Electron. Syst. Command, Washington, DC, 1984.
12. **Delpizzo, V.,** A model to assess personal exposure to ELF magnetic fields from common household sources, *Bioelectromagnetics,* 11, 139, 1990.
13. **Mader, D. L. and Peralta, S. B.,** Residential exposure to 60-Hz magnetic fields from appliances, *Bioelectromagnetics,* 13, 287, 1992.
14. **Wilson, B. W., Hansen, N. H., and Davis, K. C.,** Magnetic-field flux density and spectral characteristics of motor-driven personal appliances, *Bioelectromagnetics,* 15, 439, 1994.
15. **Caola, R. J., Deno, D. W., and Dymek, V. S. W.,** Measurements of electric and magnetic fields in and around homes near a 500 kV transmission line, *IEEE Trans. Power Appl. Syst.,* 102, 3338, 1983.
16. **Kaune, W. T., Stevens, R. G., Callahan, N. J., Severson, R. K., and Thomas, D. B.,** Residential magnetic and electric fields, *Bioelectromagnetics,* 8, 315, 1987.
17. **Barnes, F., Wachtel, H., Savitz, D., and Fuller, J.,** Use of wiring configuration and wiring codes for estimating externally generated electric and magnetic fields, *Bioelectromagnetics,* 10, 13, 1989.
18. **Silva, M., Hummon, N., Rutter, D., and Hooper, Co.,** Power frequency magnetic fields in the home, *IEEE Trans. Power Deliv.,* 4, 465, 1989.

19. **Bracken, T. D., Kheifets, L. I., and Sussman, S. S.,** Exposure assessment for power frequency electric and magnetic fields (EMF) and its application to epidemiologic studies, *J. Exp. Analy. Environ. Epidemiol., 3,* 1, 1993.
20. **Breysse, P., Lees, P. S. J., McDiarmid, M. A., and Curbow, B.,** ELF magnetic field exposures in an office environment, *Am. J. Indust. Med.,* 25, 177, 1994.
21. **Stuchly, M. A., Lecuyer, D. W., and Mann, R. D.,** Extremely low frequency electromagnetic emissions from video display terminals and other devices, *Health Phys.,* 45, 713, 1983.
22. **Tofani, S. and D'Amore, G.,** Extremely-low-frequency and very-low-frequency magnetic fields emitted by video display units, *Bioelectromagnetics,* 12, 35, 1991.
23. **Bowman, J. D., Garabrant, D. H., Sobel, E., and Peters, J. M.,** Exposures to extremely-low-frequency (ELF) electromagnetic fields in occupations with elevated leukemia rates, *Appl. Ind. Hyg.,* 3, 189, 1988.
24. **Deadman, J. E., Camus, M., Armstrong, B. G., Heroux, P., Cyr, D., Plante, M., and Thérlault, G.,** Occupational and residential 60-Hz electromagnetic fields and high-frequency electric transients: Exposure assessment using a new dosimeter, *Am. Ind. Hyg. Assoc. J.,* 49, 409, 1988.
25. **Hayashi, N., Isaka, K., and Yokoi, Y.,** ELF electromagnetic environment in power substations, *Bioelectromagnetics,* 10, 51, 1989.
26. **Bracken, T. D.,** Exposure assessment for power frequency electric and magnetic fields, *Am. Ind. Hyg. Assoc. J.,* 54, 165, 1993.
27. **Sahl, J. D., Kelsh, M. A., and Greenland, S.,** Cohort and nested case-control studies of hematopoietic cancers and brain cancer among electric utility workers, *Epidemiol.,* 4, 104, 1993.
28. **London, S. J., Bowman, J. D., Sobel, E., Thomas, D. C., Garabrant, D. H., Pearce, N., Bernstein, L., and Peters, J. M.,** Exposure to magnetic fields among electrical workers in relation to leukemia risk in Los Angeles County, *Am. J. Ind. Med.,* 26, 47, 1994.
29. **Thériault, G., Goldberg, M., Miller, A. B., Armstrong, B., Guenel, P., Deadman, J., Imbernon, E., To, T., Chevalier, A., Cyr, D., and Wall, C.,** Cancer risks associated with occupational exposure to magnetic fields among electric utility workers in Ontario and Quebec, Canada, and France: 1970–1989, *Am. J. Epidemiol.,* 139, 550, 1994.
30. **Lövsund, P., Öberg, P. A., and Nilsson, S. E. G.,** ELF magnetic fields in electrosteel and welding industries, *Radio Sci.,* 17(5S), 35S, 1982.
31. **Lövsund, P.,** Biological Effects of Alternating Magnetic Fields with Special Reference to the Visual System, Medical Dissertation No. 92, Linkoping University, Linkoping, Sweden, 1980.
32. **Stuchly, M. A. and Lecuyer, D. W.,** Exposure to electromagnetic fields in arc welding, *Health Phys.,* 56, 297, 1989.
33. **Goellner, D., Inge, T., Gillette, L., Hankin, N., and Hostage, B.,** Safety of High Speed. Guided Ground Transportation Systems: EMF Exposure Environments Summary Report, Rep. No. DOT/FRA/ORD-93/28, U.S. Department of Transportation, Washington, DC, 1993.
34. **Basset, C. A. L., Mitchell, S. N., and Gaston, S. R.,** Pulsing electromagnetic field treatment in ununited fractures and failed arthrodeses, *J. Am. Med. Assoc.,* 247, 623, 1982.
35. **Hinsenkamp, M., Tuerlinckx, B., and Rooze, M.,** Effect of ELF fields on bone growth and fracture repair, in *Biological Effects and Dosimetry of Static and ELF Electromagnetic Fields,* Grandolfo, M., Michaelson, S. M., and Rindi, A., Eds., Plenum, New York, 1985, 441.
36. **Pilla, A. A.,** State of the art in electromagnetic therapeutics, in *Electricity and Magnetism in Biology and Medicine,* Blank, M., Ed., San Francisco Press, San Francisco, CA, 1993, 17.
37. **Budinger, T. F. and Lauterbur, P. C.,** Nuclear magnetic resonance technology for medical studies, *Science,* 226, 288, 1984.
38. **Tenforde, T. S. and Budinger, T. F.,** Biological effects and physical safety aspects of NMR imaging and *in vivo* spectroscopy, in *NMR in Medicine: The Instrumentation and Clinical Applications,* Thomas, S. R. and Dixon, R. L., Eds., Amer. Assoc. Phys. Med., Med. Monograph No. 14, Am. Inst. Physics, New York, 1986, 493.
39. **Cohen, S. M., Ed.,** *Physiological NMR Spectroscopy: From Isolated Cells to Man,* NY Acad. Sci., New York, 1987, Vol. 508.
40. **Bernhardt, J.,** The direct influence of electromagnetic fields on nerve and muscle cells of man within the frequency range of 1 Hz to 30 MHz, *Radiat. Environ. Biophys.,* 16, 309, 1979.
41. **Adey, W. R.,** Tissue interactions with nonionizing electromagnetic fields, *Physiol. Rev.,* 61, 435, 1981.
42. **Tenforde, T. S.,** Biological responses to static and time-varying magnetic fields, in: *Extremely-Low-Frequency Electromagnetic Fields: The Question of Cancer,* Wilson, B. W., Stevens, R. G., and Anderson, L. E., Eds., Battelle Press, Columbus, OH, 1990, 291.
43. **Tenforde, T. S.,** Cellular and molecular pathways of extremely-low-frequency electromagnetic field interactions with living systems, in: *Electricity and Magnetism in Biology and Medicine,* Blank, M., Ed., San Francisco Press, San Francisco, CA, 1993, 1.
44. **Wallaczek, J. and Liburdy, R. P.,** Nonthermal 60-Hz sinusoidal magnetic field exposure enhances $^{45}Ca^{2+}$ uptake in rat thymocytes: Dependence on mitogen activation, *FEBS Lett.,* 271, 157, 1990.

45. **Fishman, H. M. and Leuchtag, H. R.,** Electrical noise in physics and biology, *Curr. Top. Membr. Transport,* 37, 3, 1990.
46. **Weaver, J. C. and Astumian, R. D.,** The response of cells to very weak electric fields: The thermal noise limit, *Science,* 247, 459, 1990.
47. **Polk, C.,** Physical mechanisms for biological effects of ELF low-intensity electric and magnetic fields: Thermal noise limit and counterion polarization, in: *Electricity and Magnetism in Biology and Medicine,* Blank, M., Ed., San Francisco Press, San Francisco, CA, 1993, 543.
48. **Adair, R. K.,** Constraints on biological effects of weak extremely-low-frequency electromagnetic fields, *Phys. Rev.,* A43, 1039, 1991.
49. **Bennett, W. R., Jr.,** Cancer and power lines, *Phys. Today,* April 1994, 23.
50. **Weaver, J. C. and Astumian, R. D.,** ELF biological threshold estimates: An approach for electric fields based on noise and mechanism models, *Bioelectromagnetics,* Suppl. 1, 119, 1992.
51. **Litovitz, R., Krause, D., and Mullins, J. M.,** Effect of coherence time of the applied magnetic field on ornithine decarboxylase activity, *Biochem. Biophys. Res. Comm.,* 178, 862, 1991.
52. **Litovitz, T. A., Krause, D., Penafiel, M., Elson, E. C., and Mullins, J. M.,** The role of coherence time in the effect of microwaves on ornithine decarboxylase activity, *Bioelectromagnetics,* 14, 395, 1993.
53. **Kirschvink, J. L., Jones, D. S., and MacFadden, B. J., Eds.,** *Magnetite Biomineralization and Magnetoreception in Animals: A New Biomagnetism,* Plenum, New York, 1985.
54. **Kirschvink, J. L.,** Biogenic magnetite and magnetoreception, *Bioelectromagnetics,* 10, 239, 1989.
55. **Kirschvink, J. L., Kobayashi-Kirschvink, A., and Woodford, F. B.,** Magnetite biomineralization of the human brain, *Proc. Natl. Acad. Sci. (USA),* 89, 7683, 1992.
56. **Kirschvink, J. L., Kobayashi-Kirschvink, A., Diaz-Ricci, J., and Kirschvink, S. J.,** Magnetite in human tissues: A mechanism for the biological effects of weak ELF magnetic fields, *Bioelectromagnetics,* Suppl. 1, 101, 1992.
57. **Adair, R. K.,** Effects of ELF magnetic fields on biological magnetite, *Bioelectromagnetics,* 14, 1, 1993.
58. **Polk, C.,** Effects of extremely-low-frequency magnetic fields on biological magnetite, *Bioelectromagnetics,* 15, 261, 1994.
59. **Kirschvink, J. L. and Kobayashi-Kirschvink, A.,** Magnetite biomineralization in human tissues, Paper K1–8, Abstracts of XXIVth General Assembly of the International Union of Radio Science, Kyoto, Japan, Aug. 25–Sept. 2, 1993, 512.
60. **Liboff, A. R.,** Geomagnetic cyclotron resonance in living cells, *J. Biol. Phys.,* 13, 99, 1985.
61. **Liboff, A. R., Rozek, R. J., Sherman, M. L., McLeod, B. R., and Smith, S. D.,** $^{45}Ca^{++}$ cyclotron resonance in human lymphocytes, *J. Bioelectr.,* 6, 13, 1987.
62. **Smith, S. D., McLeod, B. R., Liboff, A. R., and Cooksey, K.,** Calcium cyclotron resonance and diatom mobility, *Bioelectromagnetics,* 8, 215, 1987.
63. **Parkinson, W. C. and Hanks, C. T.,** Search for cyclotron resonance in cells *in vitro, Bioelectromagnetics,* 10, 129, 1989.
64. **Liboff, A. R. and Parkinson, W. C.,** Search for ion-cyclotron resonance in a Na+ transport system, *Bioelectromagnetics,* 12, 77, 1991.
65. **Prasad, A. V., Miller, M. W., Carstensen, E. L., Cox, Ch., Azadniv, M., and Brayman, A. A.,** Failure to reproduce increased calcium uptake in human lymphocytes at purported cyclotron resonance exposure conditions, *Radiat Environ. Biophys.,* 30, 305, 1991.
66. **Parkinson, W. C. and Sulik, G. L.,** Diatom response to extremely-low-frequency magnetic fields, *Radiat. Res.,* 130, 319, 1992.
67. **Durney, C. H., Kaminski, M., Anderson, A. A., Bruckner-Lea, C., Janata, J., and Rappaport, C.,** Investigation of AC-DC magnetic field effects in planar phospholipid bilayers, *Bioelectromagnetics,* 13, 19, 1992.
68. **Galt, S., Sandblom, J., Mamnerius, Y., Hojevik, P., Saalman, E., and Norden, B.,** Experimental search for combined AC and DC magnetic field effects on ion channels, *Bioelectromagnetics,* 14, 315, 1993.
69. **Coulton, L. A., and Barker, A. T.,** Magnetic fields and intracellular calcium: Effects on lymphocytes exposed to conditions for 'cyclotron resonance', *Phys. Med. Biol.,* 38, 347, 1993.
70. **Halle, B.,** On the cyclotron resonance mechanism for magnetic field effects on transmembrane ion conductivity, *Bioelectromagnetics,* 9, 381, 1988.
71. **Lednev, V. V.,** Possible mechanism for the influence of weak magnetic fields on biological systems, *Bioelectromagnetics,* 12, 71, 1991.
72. **Zhadin, M. N. and Fesenko, E. E.,** Ionic cyclotron resonance in biomolecules, *Biomed. Sci.,* 1, 245, 1990.
73. **Shuvalova, L. A., Ostrovskaya, M. V., Sosunov, V. A., and Lednev, V. V.,** Influence of a weak magnetic field under conditions of paramagnetic resonance on the rate of calmodulin-dependent phosphorylation of myosin in solution, *Proc. Natl. Acad. Sci. (USSR),* 317, 227, 1991.

74. **Bruckner-Lea, C., Durney, C. H., Janata, J., Rappaport, C., and Kaminski, M.,** Calcium binding to metallochromic dyes and calmodulin in the presence of combined AC-DC magnetic fields, *Bioelectromagnetics,* 13, 147, 1992.
75. **Blanchard, J. P. and Blackman, C. F.,** Clarification and application of an ion parametric resonance model for magnetic field interactions with biological systems, *Bioelectromagnetics,* 15, 217, 1994.
76. **Blackman, C. F., Blanchard, J. P., Benane, S. G., and House, D. E.,** Empirical test of an ion parametric resonance model for magnetic field interactions with PC-12 cells, *Bioelectromagnetics,* 15, 239, 1994.
77. **Adair, R. K.,** Criticism of Lednev's mechanism for the influence of weak magnetic fields on biological systems, *Bioelectromagnetics,* 13, 231, 1992.
78. **Wiesenfeld, K. and Moss, F.,** Stochastic resonance and the benefits of noise: from ice ages to crayfish and SQUIDs, *Nature,* 373, 33, 1995.
79. **Douglas, J. K., Wilkins, L, Pantazelou, E., and Moss, F.,** Noise enhancement of information transfer in crayfish mechanoreceptors by stochastic resonance, *Nature,* 365, 337, 1993.
80. **Kruglikov, I. L. and Dertinger, H.,** Stochastic resonance as a possible mechanism of amplification of weak electric signals in living cells, *Bioelectromagnetics,* 15, 539, 1994.
81. **Taylor, L. S.,** The mechanisms of athermal microwave biological effects, *Bioelectromagnetics,* 2, 259, 1981.
82. **Postow, E. and Swicord, M. L.,** Modulated fields and 'window' effects, in: *Handbook of Biological Effects of Electromagnetic Fields,* Polk, C. and Postow, E., Eds., CRC Press, Boca Raton, FL, 1986, 425.
83. **Tenforde, T. S.,** Microscopic dosimetry of extremely-low-frequency electric and magnetic fields, *Bioelectromagnetics,* Suppl. 1, 61, 1992.
84. **Adey, W. R.,** Nonlinear electrodynamics in cell membrane transductive coupling, in: *Membrane Transport and Information Storage, Vol. 4,* Aloia, R. C., Curtain, C. C., and Gordon, L. M., Eds., Wiley-Liss, New York, 1990, 1.
85. **Tenforde, T. S. and Liburdy, R. P.,** Magnetic deformation of phospholipid bilayers: Effects on liposome shape and solute permeability at prephase transition temperatures, *J. Theor. Biol.,* 133, 385, 1988.
86. **Hoff, A. F.,** Magnetic field effects on photosynthetic reactions, *Quart. Rev. Biophys.,* 14, 599, 1981.
87. **Tenforde, T. S.,** Biological responses to static and time-varying magnetic fields, in: *Electromagnetic Interaction with Biological Systems,* Lin, J. C., Ed., Plenum, New York, 1989, 83.
88. **McLauchlan, K.,** Are environmental magnetic fields dangerous? *Phys. World,* Jan. 1992, 41.
89. **Scalano, J. C., Mohtat, N., Cozens, F. L., McLean, J., and Thansandote, A.,** Application of the radical pair mechanism to free radicals in organized systems: Can the effects of 60 Hz be predicted from studies under static fields? *Bioelectromagnetics,* 15, 549, 1994.
90. **Tenforde, T. S.,** Interaction of ELF magnetic fields with living matter, in: *Handbook of Biological Effects of Electromagnetic Radiation,* Polk, C. and Postow, E., Eds., CRC Press, Boca Raton, FL, 1986, 197.
91. **Nair, I., Morgan, M. G., and Florig, H. K.,** Biological Effects of Power Frequency Electric and Magnetic Fields, Rep. No. OTA-BP-E-53, Office of Technology Assessment, U.S. Government Printing Office, Washington, DC, 1989.
92. **Anderson, L. E.,** Biological effects of extremely-low-frequency electromagnetic fields: In vivo studies, *Am. Ind. Hyg. Assoc. J.,* 54, 186, 1993.
93. **Sienkiewicz, Z. J., Cridland, N. A., Kowalczuk, C. I., and Saunders, R. D.,** Biological effects of electromagnetic fields and radiation, in: *The Review of Radio Science: 1990–1992,* Stone, W. R. and Hyde, G., Eds., Oxford Univ. Press, Oxford, UK, 1993, 737.
94. **d'Arsonval, M. A.,** Dispositifs pour la mesure des courants alternatifs à toutes frequences, *C. R. Soc. Biol. (Paris),* 3(100 Ser.), 450, 1896.
95. **Lövsund, P., Öberg, P. A., Nilsson, S. E. G., and Reuter, T.,** Magnetophosphenes: A quantitative analysis of thresholds, *Med. Biol. Eng. Comput.,* 18, 326, 1980.
96. **Silny, J.,** The influence threshold of a time-varying magnetic field in the human organism, in: *Biological Effects of Static and Extremely-Low-Frequency Magnetic Fields,* Bernhardt, J. H., Ed., MMV Medizin Verlag, Munich, Germany, 1986, 105.
97. **Budinger, T. F., Cullander, C., and Bordow, R.,** Switched magnetic field thresholds for the induction of magnetophosphenes, Abstracts of Society of Magnetic Resonance in Medicine 3rd Annual Meeting, New York, Aug. 13–17, 1984, 118.
98. **Barlow, H. B., Kohn, H. I., and Walsh, E. G.,** Visual sensations aroused by magnetic fields, *Am. J. Physiol.,* 148, 372, 1947.
99. **Lövsund, P., Nilsson, S. E. G., and Öberg, P. A.,** Influence on frog retina of alternating magnetic fields with special reference to ganglion cell activity, *Med. Biol. Eng. Comput.,* 19, 679, 1981.
100. **Lövsund, P., Öberg, P. A., and Nilsson, S. E. G.,** Magneto- and electrophosphenes: A comparative study, *Med. Biol. Eng. Comput.,* 18, 758, 1980.
101. **Carstensen, E. L., Buiettner, A., Genberg, V. L., and Miller, M. W.,** Sensitivity of the human eye to power frequency electric fields, *IEEE Trans. Biomed. Eng.,* 32, 561, 1985.

102. **Reilly, J. P.,** Peripheral nerve stimulation by induced electric currents: Exposure to time-varying magnetic fields, *Med. Biol. Eng. Comp.,* 27, 101, 1989.
103. **Polson, M. J. R., Barker, A. T., and Freeston, I. L.,** Stimulation of nerve trunks with time-varying magnetic fields, *Med. Biol. Eng. Comput.,* 20, 243, 1982.
104. **Öberg, P. A.,** Magnetic stimulation of nerve tissue, *Med. Biol. Eng.,* 11, 5, 1973.
105. **Ueno, S., Matsumoto, S., Harada, K., and Oomura, Y.,** Capacitive stimulatory effect in magnetic stimulation of nerve tissue, *IEEE Trans. Mag.,* 14, 958, 1978.
106. **Ueno, S., Harada, K., Ji, C., and Oomura, Y.,** Magnetic nerve stimulation without interlinkage between nerve and magnetic flux, *IEEE Trans. Mag.,* 20, 1660, 1984.
107. **Bernhardt, J. H.,** Non-ionizing radiation safety: Radiofrequency radiation, electric and magnetic fields, *Phys. Med. Biol.,* 37, 807, 1992.
108. **Bell, G. B., Marino, A. A., Chesson, A. L., and Struve, F. A.,** Human sensitivity to weak magnetic fields, *Lancet,* 338, 1521, 1991.
109. **Bell, G., Marino, A., Chesson, A., and Struve, F.,** Electrical states in the rabbit can be altered by light and electromagnetic fields, *Brain Res.,* 570, 307, 1992.
110. **Dowman, R., Wolpaw, J. R., Segal, R. F., and Satya-Murti, S.,** Chronic exposure of primates to 60-Hz electric and magnetic fields: Ill. Neurophysiologic effects. *Bioelectromagnetics,* 10, 303, 1989.
111. **Graham, C., Cook, M. R., and Cohen, H. D.,** Immunological and Biochemical Effects of 60-Hz Electric and Magnetic Fields in Humans, Midwest Research Institute Final Report for Contract No. DE-FC01-84-CE-76246 (Order No. DE90006671), U.S. Department of Energy Office of Scientific and Technical Information, Oak Ridge, TN, 1990.
112. **Cook, M. R., Graham, C., Cohen, H. D., and Gerkovich, M. M.,** A replication study of human exposure to 60-Hz fields: Effects on neurobehavioral measures, *Bioelectromagnetics,* 13, 261, 1992.
113. **Graham, C., Cook, M. R., Cohen, H. D., and Gerkovich, M. M.,** Dose response study of human exposure to 60-Hz electric and magnetic fields, *Bioelectromagnetics,* 15, 447, 1994.
114. **Korpinen, L., Partanen, J., and Uusitalo, A.,** Influence of 50-Hz electric and magnetic fields on the human heart, *Bioelectromagnetics,* 14, 329, 1993.
115. **Korpinen, L. and Partanen, J.,** Influence of 50-Hz electric and magnetic fields on the pulse rate of human heart, *Bioelectromagnetics,* 15, 503, 1994.
116. **Maresh, C. M., Cook, M. R., Cohen, H. D., Graham, C., and Gunn, W. S.,** Exercise testing in the evaluation of human responses to powerline frequency fields, *Aviat., Space, Environ. Med.,* 59, 1139, 1988.
117. **Ramon, C. and Powell, M. R.,** Preliminary report: Modification of cardiac contraction rate by pulsed magnetic fields, *Bioelectromagnetics,* 13, 303, 1992.
118. **Ellis, W. V.,** Pain control using high-intensity pulsed magnetic stimulation, *Bioelectromagnetics,* 14, 553, 1993.
119. **Grant, G., Cadossi, R., and Steinberg, G.,** Protection against focal cerebral ischemia following exposure to a pulsed electromagnetic field, *Bioelectromagnetics,* 15, 205, 1994.
120. **Kanje, M., Rusovan, A., Sisken, B., and Lundborg, G.,** Pretreatment of rats with pulsed electromagnetic fields enhances regeneration of the sciatic nerve, *Bioelectromagnetics,* 14, 353, 1993.
121. **Lovely, R. H., Creim, J. A., Kaune, W. T., Miller, M. C., Phillips, R. D., and Anderson, L. E.,** Rats are not aversive when exposed to 60-Hz magnetic fields at 3.03 mT, *Bioelectromagnetics,* 13, 351, 1992.
122. **Trzeciak, H. I., Grzesik, J., Bortel, M., Kuska, R., Duda, D., Michnik, J., and Malecki, A.,** Behavioral effects of long-term exposure to magnetic fields in rats, *Bioelectromagnetics,* 14, 287, 1993.
123. **Lyskov, E. B., Juutilainen, J., Jousmäki, V., Partanen, J., Medvedev, S., and Hänninen, O.,** Effects of 45-Hz magnetic fields on the functional state of the human brain, *Bioelectromagnetics,* 14, 87, 1993.
124. **Savitz, D. A., Boyle, C. A., and Holmgreen, P.,** Prevalence of depression among electrical workers, *Am. J. Ind. Med.,* 25, 165, 1994.
125. **McMahan, S., Ericson, J., and Meyer, J.,** Residential proximity to high-voltage transmission lines and depressive symptomatology, in: *Electricity and Magnetism in Biology and Medicine,* Blank, M., Ed., San Francisco Press, San Francisco, CA, 1993, 453.
126. **Cridland, N. A.,** Electromagnetic Fields and Cancer: A Review of Relevant Cellular Studies, Rep. No. NRPB-R256, National Radiological Protection Board, Chilton, Didcot, Oxon, UK, 1993.
127. **Blackman, C. F., Benane, S. G., Rabinowitz, J. R., House, D. E., and Joines, W. T.,** A role for the magnetic field in the radiation-induced efflux of calcium ions from brain tissue *in vitro, Bioelectromagnetics,* 6, 327, 1985.
128. **Blackman, C. F., Benane, S. G., Elliott, D. J., House, D. E., and Pollock, M. M.,** Influence of electromagnetic fields on the efflux of calcium ions from brain tissue *in vitro*: A three-model analysis consistent with the frequency response up to 510 Hz, *Bioelectromagnetics,* 9, 215, 1988.
129. **Blackman, C. F., Benane, S. G., House, D. E., and Elliott, D. J.,** Importance of alignment between local DC magnetic field and an oscillating magnetic field in responses of brain tissue *in vitro* and *in vivo*, *Bioelectromagnetics,* 11, 159, 1990.

130. **Goodman, R., Wei, L.-X., Xu, J.-C., and Henderson, A.,** Exposure of human cells to low frequency electromagnetic fields results in quantitative changes in transcripts, *Biochim. Biophys. Acta,* 1009, 216, 1990.
131. **Wei, L.-X., Goodman, R., and Henderson, A.,** Changes in c-*myc* and histone H2B following exposure of cells to low-frequency sinusoidal electromagnetic fields: Evidence for a window effect, *Bioelectromagnetics,* 11, 269, 1990.
132. **Goodman, R. and Henderson, A. S.,** Transcription and translation in cells exposed to extremely-low-frequency electromagnetic fields, *Bioelectrochem. Bioenerg.,* 25, 335, 1991.
133. **Reese, J. A., Jostes, R. F., and Frazier, M. E.,** Exposure of mammalian cells to 60-Hz magnetic or electric fields: Analysis for DNA single-strand breaks, *Bioelectromagnetics,* 9, 237, 1988.
134. **Cossarizza, A., Monti, D., Sola, P., Moschini, G., Cadossi, R., Bersani, F., and Franceschi, C.,** DNA repair after gamma irradiation in lymphocytes exposed to low-frequency pulsed electromagnetic fields, *Radiat. Res.* 118, 161, 1989.
135. **Frazier, M. E., Reese, J. A., Morris, J. E., Jostes, R. F., and Miller, D. L.,** Exposure of mammalian cells to 60-Hz magnetic or electric fields: Analysis of DNA repair of induced, single-strand breaks, *Bioelectromagnetics,* 11, 229, 1990.
136. **Cohen, M. M., Kunska, A., Astemborski, A., McCulloch, D., and Paskewitz, D. A.,** Effect of low-level, 60-Hz electromagnetic fields on human lymphoid cells: I. Mitotic rate and chromosome breakage in human peripheral lymphocytes, *Bioelectromagnetics,* 7, 415, 1986.
137. **Cohen, M. M., Kunska, A., Astemborski, J. A., and McCulloch, D.,** The effect of low-level electromagnetic fields on human lymphoid cells: II. Sister chromatid exchanges in peripheral lymphocytes and lymphoblastoid cell lines, *Mutat. Res.,* 172, 177, 1986.
138. **Nordenson, I., Mild, K. H., Andersson, G., and Sandström, M.,** Chromosomal aberrations in human amniotic cells after intermittent exposure to fifty hertz magnetic fields, *Bioelectromagnetics,* 15, 293, 1994.
139. **Hintenlang, D. E.,** Synergistic effects of ionizing radiation and 60 Hz magnetic fields, *Bioelectromagnetics,* 14, 545, 1993.
140. **Frazier, M. E., Reese, J. A., and Morris, J. E.,** Effect of 60-Hz electromagnetic fields on growth rates and transformation frequencies of C3H10T1/2 cells, Abstracts of Bioelectromagnetics Society 12th Annual Meeting, San Antonio, TX, June 10–14, 1990, 9.
141. **Goodman, R., Bassett, C. A. L., and Henderson, A. S.,** Pulsing electromagnetic fields induce cellular transcription, *Science,* 220, 1283, 1983.
142. **Goodman, R. and Henderson, A. S.,** Sine waves enhance cellular transcription, *Bioelectromagnetics,* 7, 23, 1986.
143. **Goodman, R. and Henderson, A. S.,** Exposure of salivary gland cells to low frequency fields alters polypeptide synthesis, *Proc. Natl. Acad. Sci. (USA),* 85, 3928, 1988.
144. **Goodman, R., Weisbrot, D., Uluc, A., and Henderson, A.,** Transcription in *Drosophila melanogaster* salivary gland cells is altered following exposure to low-frequency electromagnetic fields: Analysis of chromosome 3R, *Bioelectromagnetics,* 13, 111, 1992.
145. **Litovitz, T. A., Montrose, C. J., Goodman, R., and Elson, E. C.,** Amplitude windows and transiently augmented transcription from exposure to electromagnetic fields, *Bioelectromagnetics,* 11, 197, 1990.
146. **Litovitz, T. A., Montrose, C. J., and Wang, W.,** Dose-response implications of the transient nature of electromagnetic-field-induced bioeffects: Theoretical hypotheses and predictions, *Bioelectromagnetics,* Suppl. 1, 237, 1992.
147. **Blank, M., Khorkova, O., and Goodman, R.,** Similarities in the proteins synthesized by *Sciara* salivary glands in response to electromagnetic fields and to heat shock, *Bioelectrochem. Bioenerget.,* 31, 27, 1993.
148. **Blank, M., Khorkova, O., and Goodman, R.,** Changes in polypeptide distribution stimulated by different levels of electromagnetic and thermal stress, *Bioelectrochem. Bioenerget.,* 33, 109, 1994.
149. **Ning, J., Casamento, J., Czerska, E., Swicord, M., Al-Barazi, H., Davis, C., and Elson, E.,** Comparison of the effect of ELF on total RNA content in normal and transformed human cells, *Ann. N. Y. Acad. Sci.,* 649, 353, 1992.
150. **Greene, J. J., Skowronski, W. J., Mullins, J. M., and Nardone, R. M.,** Delineation of electric and magnetic field effects of extremely low frequency electromagnetic radiation on transcription, *Biochem. Biophys. Res. Comm.,* 124, 742, 1991.
151. **Phillips, J. L. and McChesney, L.,** Effect of 72 Hz pulsed magnetic field exposure on macromolecular synthesis in CCRF-CEM cells, *Cancer Biochem. Biophys.,* 12, 1, 1991.
152. **Phillips, J. L., Haggren, W., Thomas, W. J., Ishida-Jones, T., and Adey, W. R.,** Magnetic field-induced changes in specific gene transcription, *Biochim. Biophys. Acta,* 1132, 140, 1992.
153. **Lacy-Hulbert, A., Wilkins, R. C., Hesketh, T. R., and Metcalfe, J. C.,** Cancer risk and electromagnetic fields [letter to the Editor], *Nature,* 375, 23, 1995.

154. **Saffer, J. D. and Thurston, S. J.,** Cancer risk and electromagnetic fields [letter to the Editor], *Nature*, 375, 22, 1995.
155. **Parker, J. E. and Winters, W.,** Expression of gene-specific RNA in cultured cells exposed to rotating 60-Hz magnetic fields, *Biochem. Cell Biol.*, 70, 237, 1992.
156. **Liboff, A. R., Williams, T., Jr., Strong, D. M., and Wistar, R. Jr.,** Time-varying magnetic fields: Effect on DNA synthesis, *Science*, 23, 818, 1984.
157. **Cridland, N. A. and Saunders, R. D.,** Proliferative responses of human fibroblasts exposed to ELF magnetic fields, in: *Electricity and Magnetism in Biology and Medicine*, Blank, M., San Francisco Press, San Francisco, CA, 1993, 628.
158. **Liburdy, R. P.,** Calcium signaling in lymphocytes and ELF fields, *FEBS Lett.*, 301, 53, 1992.
159. **Walleczek, J.,** Electromagnetic field effects on cells of the immune system: The role of calcium signaling, *FASEB J.*, 6, 3177, 1992.
160. **Luben, R. A., Cain, C. D., Chen, M. C.-Y., Rosen, D. M., and Adey, W. R.,** Effects of electromagnetic stimuli on bone cells *in vitro*: Inhibition of responses to parathyroid hormone by low-energy low-frequency fields, *Proc. Natl. Acad. Sci. (USA)*, 79, 4180, 1982.
161. **Jones, D. B.,** The effect of pulsed magnetic fields on cyclic AMP metabolism in organ cultures of chick embryo tibiae, *J. Bioelectr.*, 3, 427, 1984.
162. **Murray, J. C. and Farndale, R. W.,** Modulation of collagen production in cultured fibroblasts by a low-frequency, pulsed magnetic field, *Biochim. Biophys. Acta*, 838, 98, 1985.
163. **Cain, C. D., Adey, R., and Luben, R. A.,** Evidence that pulsed electromagnetic fields inhibit coupling of adenylate cyclase by parathyroid hormone in bone cells, *J. Bone Miner. Res.*, 2, 437, 1987.
164. **Hiraki, Y., Endo, N., Takigawa, M., Asada, A., Takahashi, H., and Suzuki, F.,** Enhanced responsiveness to parathyroid hormone and induction of functional differentiation of cultured rabbit costal chondrocytes by a pulsed electromagnetic field, *Biochim. Biophys. Acta*, 931, 94, 1987.
165. **Phillips, J. L., Winters, W. D., and Rutledge, L.,** *In vitro* exposure to electromagnetic field: Changes in tumor cell properties, *Int. J. Radiat. Biol.*, 49, 463, 1986.
166. **Phillips, J. L., Rutledge, L., and Winters, W. D.,** Transferrin binding to two human colon carcinoma cell lines: Characterization and effect of 60-Hz electromagnetic fields, *Cancer Res.*, 46, 239, 1986.
167. **Smith, O. M., Goodman, E. M., Greenebaum, B., and Tipnis, P.,** An increase in the negative surface charge of U937 cells exposed to a pulsed magnetic field, *Bioelectromagnetics*, 12, 197, 1991.
168. **Parola, A. H., Porat, N., and Kiesow, L. A.,** Chicken embryo fibroblasts exposed to weak, time-varying magnetic fields share cell proliferation, adenosine deaminase activity, and membrane characteristics of transformed cells, *Bioelectromagnetics*, 14, 215, 1993.
169. **Luben, R. A.,** Effects of low-energy electromagnetic fields (pulsed and DC) on membrane signal transduction processes in biological systems, *Health Phys.*, 61, 15, 1990.
170. **Farndale, R. W., Maroudas, A., and Marsland, T. P.,** Effects of low-amplitude pulsed magnetic fields on cellular ion transport, *Bioelectromagnetics*, 8, 199, 1987.
171. **Garcia-Sancho, J., Montero, M., Alvarez, J., Fonteriz, R. I., and Sanchez, A.,** Effects of extremely-low-frequency electromagnetic fields on ion transport in several mammalian cells, *Bioelectromagnetics*, 15, 579, 1994.
172. **Lyle, D. B., Wang, X., Ayotte, R. D., Sheppard, A. R., and Adey, W. R.,** Calcium uptake by leukemic and normal T-lymphocytes exposed to low frequency magnetic fields, *Bioelectromagnetics*, 12, 145, 1991.
173. **Monti, M. G., Pernecco, L., Moruzzi, M. S., Battini, R., Zaniol, P., and Barbiroli, B.,** Effects of ELF pulsed electromagnetic fields on protein kinase C activation process in HL-60 leukemia cells, *J. Bioelectr.*, 10, 119, 1991.
174. **Beischer, D. E., Grissett, J. D., and Mitchell, R. R.,** Exposure of Man to Magnetic Fields Alternating at Extremely Low Frequency, Rep. No. NAMRL-1180, Nav. Aerosp. Med. Res. Lab., Pensacola, FL, 1973.
175. **deLorge, J.,** Effects of magnetic fields on behavior in nonhuman primates, in *Magnetic Field Effects on Biological Systems*, Tenforde, T. S., Ed., Plenum Press, New York, 1979, 37.
176. **Fam, W. Z.,** Biological effects of 60-Hz magnetic field on mice, *IEEE Trans. Mag.*, 17, 1510, 1981.
177. **Sander, R., Brinkmann, J., and Kuhne, B.,** Laboratory Studies on Animals and Human Beings Exposed to 50-Hz Electric and Magnetic Fields, Paper 36-01, Proc. Int. Congr. Large High Voltage Elec. Syst., Paris, France, September 1–9, 1982.
178. **Lorimore, S. A., Kowalczuk, C. I., Saunders, R. D., and Wright, E. G.,** Lack of acute effects of 20 mT, 50 Hz magnetic fields on murine hemopoiesis, *Int. J. Radiat. Biol.*, 58, 713, 1990.
179. **Putinas, J. and Michaelson, S. M.,** Humoral responsiveness of mice exposed to 500 μT, 60 Hz magnetic field, *Bioelectrochem. Bioenerget.*, 24, 371, 1990.
180. **Morris, J. E., Sasser, L. B., Buschbom, R. L., and Anderson, L. E.,** Natural immunity in rats exposed to 60-Hz magnetic fields, Abstracts of the DOE Annual Review of Research on Biological Effects of 50 and 60 Hz Electric and Magnetic Fields, Denver, CO, November 4–8, 1990, A-23.
181. **McLean, J. R. N., Stuchly, M. A., Mitchel, R. E. J., Wilkinson, D., Yang, H., Goddard, M., Lecuyer,**

D. W., Schunk, M., Callary, E., and Morrison, D., Cancer promotion in a mouse-skin model by a 60-Hz magnetic field: II. Tumor development and immune response, *Bioelectromagnetics,* 12, 273, 1991.

182. **deSeze, R., Bouthet, C., Tuffet, S., Deschaux, P., Caristan, A., Moreau, J. M., and Veyret, B.,** Effects of time-varying uniform magnetic fields on natural killer cell activity and antibody response in mice, *Bioelectromagnetics,* 14, 405, 1993.
183. **Udintsev, N. A. and Moroz, V. V.,** Response of the pituitary-adrenal system to the action of a variable magnetic field, *Byull. Eksp. Biol. Med.,* 77, 51, 1974.
184. **Sakharova, S. A., Ryzhov, A. I., and Udintsev, N. A.,** Mechanism of the sympathoadrenal system's response to the one time action of a variable magnetic field, *Kosm. Biol. Aviakosmicheskaya Med.,* 15, 52, 1981.
185. **Welker, H. A., Semm, P., Willig, R. P., Commentz, J. C., Wiltschko, W., and Vollrath, L.,** Effects of an artificial magnetic field on serotonin N-acetyltransferase activity and melatonin content of the rat pineal gland, *Exp. Brain Res.,* 50, 426, 1983.
186. **Cremer-Bartels, G., Krause, K., and Kuchle, H. J.,** Influence of low magnetic-field-strength variations on the retina and pineal gland of quail and humans, *Graetes Arch. Clin. Exp. Ophthalmol.,* 220, 248, 1983.
187. **Semm, P., Schneider, T., and Vollrath, L.,** Effects of an earth-strength magnetic field on electrical activity of pineal cells, *Nature,* 288, 607, 1980.
188. **Semm, P.,** Neurobiological investigations on the magnetic sensitivity of the pineal gland in rodents and pigeons, *Comp. Biochem. Physiol.,* 76A, 683, 1983.
189. **Olcese, J., Reuss, S., and Vollrath, L.,** Evidence of the involvement of the visual system in mediating magnetic field effects on pineal melatonin synthesis in the rat, *Brain Res.,* 333, 382, 1985.
190. **Lerchl, A., Nonaka, K. O., Stokkan, K.-A., and Reiter, R. J.,** Marked rapid alterations in nocturnal pineal serotonin metabolism in mice and rats exposed to weak intermittent magnetic fields, *Biochem. Biophys. Res. Comm.,* 169, 102, 1990.
191. **Lerchl, A., Honaka, K. O., and Reiter, R. J.,** Pineal gland "magnetosensitivity" to static magnetic fields is a consequence of induced electric currents (eddy currents), *J. Pineal Res.,* 10, 109, 1991.
192. **Yellon, S. M.,** An acute 60-Hz magnetic field exposure suppresses the nighttime melatonin rise in the pineal and circulation of the adult Djungarian hamster, Abstracts of the Annual Review of Research on Biological Effects of 50 and 60 Hz Electric and Magnetic Fields, Milwaukee, WI, November 3–7, 1991, A-25.
193. **Wilson, B. W., Anderson, L. E., Hilton, D. I., and Phillips, R. D.,** Chronic exposure to 60-Hz electric fields: Effects on pineal function in the rat, *Bioelectromagnetics,* 2, 371, 1981 [erratum: 4, 293, 1983].
194. **Reiter, R. J., Anderson, L. E., Buschbom, R. L., and Wilson, B. W.,** Reduction of the nocturnal rise in pineal melatonin levels in rats exposed to 60-Hz electric fields in utero and for 23 days after birth, *Life Sci.,* 42, 2203, 1988.
195. **Kato, M., Honma, K.-I., Shigemitsu, T., and Shiga, Y.,** Effects of exposure to a circularly polarized 50-Hz magnetic field on plasma and pineal melatonin levels in rats, *Bioelectromagnetics,* 14, 97, 1993.
196. **Kato, M., Honma, K., Shigemitsu, T., and Shiga, Y.,** Circularly polarized 50-Hz magnetic field exposure reduces pineal gland and blood melatonin concentrations of Long-Evans rats, *Neurosci. Lett.,* 166, 59, 1994.
197. **Kato, M., Honma, K.-I., Shigemitsu, T., and Shiga, Y.,** Recovery of nocturnal melatonin concentration takes place within one week following cessation of 50-Hz circularly polarized magnetic field exposure for six weeks, *Bioelectromagnetics,* 15, 489, 1994.
198. **Kato, M., Honma, K., Shigemitsu, T., and Shiga, Y.,** Horizontal or vertical 50-Hz, 1-μT magnetic fields have no effect on pineal gland or plasma melatonin concentration of albino rats, *Neurosci. Lett.,* 168, 205, 1994.
199. **Kato, M., Honma, K., Shigemitsu, T., and Shiga, Y.,** Circularly polarized, sinusoidal, 50-Hz magnetic field exposure does not influence plasma testosterone levels of rats, *Bioelectromagnetics,* 15, 513, 1994.
200. **Reuss, S. and Olcese, J.,** Magnetic field effects on the rat pineal gland: Role of retinal activation by light, *Neurosci. Lett.,* 64, 97, 1986.
201. **Reiter, R. J.,** Static and extremely low frequency electromagnetic field exposure: Reported effects on the circadian production of melatonin, *J. Cell. Biochem.,* 51, 394, 1993.
202. **Rudolph, K., Wirz-Justice, A., Krauchli, K., and Feere, H.,** Static magnetic fields decrease nocturnal cAMP in the rat, *Brain Res.,* 446, 159, 1988.
203. **Wilson, B. W., Wright, C. W., Morris, J. E., Buschbom, R. L., Brown, D. P., Miller, D. L., Sommers-Flannigan, R., and Anderson, L. E.,** Evidence for an effect of ELF electromagnetic fields on human pineal gland function, *J. Pineal Res.,* 9, 259, 1990.
204. **Wilson, B. W., Stevens, R. G., and Anderson, L. E.,** Neuroendocrine mediated effects of electromagnetic-field exposure: Possible role of the pineal gland, *Life Sci.,* 45, 1319, 1989.
205. **Stevens, R. G., Davis, S., Thomas, D. B., Anderson, L. E., and Wilson, B. W.,** Electric power, pineal function, and the risk of breast cancer, *FASEB J.,* 6, 853, 1992.
206. **Reiter, R. J. and Richardson, B. A.,** Magnetic field effects on pineal indoleamine metabolism and possible biological consequences, *FASEB J.,* 6, 2283, 1992.
207. **Reiter, R. J.,** A review of neuroendocrine and neurochemical changes associated with static and extremely low frequency electromagnetic field exposure, *Integr. Physiol. Behav. Sci.,* 28, 57, 1993.

208. **Stevens, R. G.,** Electric power use and breast cancer: A hypothesis, *Am. J. Epidemiol.,* 125, 556, 1987.
209. **Tamarkin, L., Cohen, M., Roselle, D., Reichert, C., Lippman, M., and Chabner, B.,** Melatonin inhibition and pinealectomy enhancement of 7,12-dimethylbenz(a)anthracene-induced mammary tumors in the rat, *Cancer Res.,* 41, 4432, 1981.
210. **Liburdy, R. P., Sloma, T. R., Sokolic, R., and Yaswen, P.,** ELF magnetic fields, breast cancer, and melatonin: 60-Hz fields block melatonin's oncostatic action on ER+ breast cancer cell proliferation, *J. Pineal Res.,* 14, 89, 1993.
211. **Chernoff, N., Rogers, J. M., and Kavet, R.,** A review of the literature on potential reproductive and developmental toxicity of electric and magnetic fields, *Toxicol.,* 74, 91, 1992.
212. **Brent, R. L., Gordon, W. E., Bennett, W. R., and Beckman, D. A.,** Reproductive and teratologic effects of electromagnetic fields, *Reproduct. Toxicol.,* 7, 535, 1993.
213. **Huuskonen, H., Juutilainen, J., and Komulainen, H.,** Effects of low-frequency magnetic fields on fetal development in rats, *Bioelectromagnetics,* 14, 205, 1993.
214. **Kowalczuk, C. I., Robbins, L., Thomas, J. M., and Saunders, R. D.,** Effects of prenatal exposure to 50-Hz magnetic fields on development in mice: I. Implantation rate and fetal development, *Bioelectromagnetics,* 15, 349, 1994.
215. **Sienkiewicz, Z. J., Robbins, L., Haylock, R. G. E., and Saunders, R. D.,** Effects of prenatal exposure to 50-Hz magnetic fields on development in mice: II. Postnatal development and behavior, *Bioelectromagnetics,* 15, 363, 1994.
216. **Rommereim, D. L., Rommereim, R. L., Miller, D. L., Buschbom, R. L., and Anderson, L. E.,** Developmental toxicology evaluation of 60-Hz, horizontal magnetic fields in rats, Abstracts of the Thirtieth Hanford Symposium on Health and the Environment, Richland, WA, October 29–November 1, 1991, 65.
217. **Tribukait, B., Cekan, E., and Paulsson, L.-E.,** Effects of pulsed magnetic fields on embryonic development in mice, Abstracts of the International Scientific Conference: Work with Display Units, Stockholm, Sweden, May 12–15, 1986, 68.
218. **Tribukait, B., Cekan, E., and Paulsson, L.-E.,** Effects of pulsed magnetic fields on embryonic development in mice, in *Work with Display Units 86,* Knave, B. and Wideback, P.-G., Eds., Elsevier, New York, 1986, 129.
219. **Frölen, H., Svedenstål, B. M., Bierke, P., and Fellner-Feldegg, H.,** Repetition of a Study on the Effects of Pulsed Magnetic Fields on the Development of Fetuses in Mice, Final Rep. for SSI Project No. 346.86, National Institute of Radiation Protection, Stockholm, Sweden, 1987.
220. **Frölen, H., Svedenstål, B.-M., and Paulsson, L.-E.,** Effects of pulsed magnetic fields on the developing mouse embryo, *Bioelectromagnetics,* 14, 197, 1993. [Erratum: *Bioelectromagnetics,* 14, 393, 1993.]
221. **Stuchly, M. A., Ruddick, J., Villeneuve, D., Robinson, K., Reed, B., Lecuyer, D. W., Tan, K., and Wong, J.,** Teratological assessment of exposure to time-varying magnetic field, *Teratol.,* 38, 461, 1988.
222. **Wiley, M. J., Corey, P., Kavet, R., Charry, J., Agnew, D., Harvey, S., and Walsh, M.,** The effects of continuous exposure to pulsed 20 kHz sawtoothed magnetic fields on the litters of CD-1 mice, *Teratol.,* 46, 391, 1992.
223. **Delgado, J. M. R., Monteagudo, J. L., Garcia, M. G., and Leal, J.,** Teratogenic effects of weak magnetic fields, *ICRS Med. Sci.,* 9, 392, 1981.
224. **Delgado, J. M. R., Leal, J., Monteagudo, J. L., and Garcia, M. G.,** Embryological changes induced by weak, extremely low frequency electromagnetic fields, *J. Anat.,* 1344, 533, 1982.
225. **Ubeda, A., Leal, J., Trillo, M. A., Jimenez, M. A., and Delgado, J. M. R.,** Pulse shape of magnetic fields influence chick embryogenesis, *J. Anat.,* 137, 513, 1983. [Erratum: *J. Anat.,* 140, 721, 1985.]
226. **Maffeo, S., Miller, M., and Carstensen, E. L.,** Lack of effect of weak low frequency electromagnetic fields on chick embryogenesis, *J. Anat.,* 139, 613, 1984.
227. **Martucci, G., Gailey, P. C., and Tell, R. A.,** Investigations of possible effects of weak pulsed magnetic fields on the chick embryo, Abstracts of the 6th Annual Meeting of the Bioelectromagnetics Society, Atlanta, GA, July 15–19, 1984, 1.
228. **Rooze, M. and Hinsenkamp, M.,** *In vivo* modifications induced by electromagnetic stimulation of chicken embryos, *Reconstr. Surg. Traumatol.,* 19, 87, 1985.
229. **Siskin, B. F., Fowler, L., Mayand, C., Ryaby, J. P., and Pilla, A. A.,** Pulsed electromagnetic fields and normal chick development, *J. Bioelectr.,* 5, 25, 1986.
230. **Bardasano, J. L., Meyer, A. J., and Picazo, L.,** Pineal cells with multipolar spindles in chicken embryos exposed to magnetic fields—first trials, Z. *Mikrosk. Anat. Forsch. Leipzig,* 100, 85, 1986.
231. **Juutilainen, J., Harri, M., Saali, K., and Lahtinen, T.,** Effects of 100-Hz magnetic fields with various waveforms on the development of chick embryos, *Radiat. Environ. Biophys.,* 25, 65, 1986.
232. **Juutilainen, J. and Saali, K.,** Development of chick embryos in 1 Hz to 100 kHz magnetic fields, *Radiat. Environ. Biophys.,* 25, 135, 1986.
233. **Juutilainen, J., Läärä E., and Saali, K.,** Relationship between field strength and abnormal development in chick embryos exposed to 50 Hz magnetic fields, *Int. J. Radiat. Biol.,* 52, 787, 1987.

234. **Leal, J., Trillo, M. A., Ubeda, A., Abraira, V., Shamsaifar, K., and Chacon, L.,** Magnetic environment and embryonic development: A role of the earth's field, *IRCS Med. Sci.,* 14, 1145, 1986.
235. **Sandström, M., Mild, K. H., and Løvtrup, S.,** Effects of weak pulsed magnetic fields on chick embryogenesis, in *Work with Display Units 86,* Knave, B. and Wideback, P.-G., Eds., Elsevier, New York, 1987, 135.
236. **Maffeo, S., Brayman, A. A., Miller, M. W., Carstensen, E. L., Ciaravino, V., and Cox, C.,** Weak low frequency electromagnetic fields and chick embryogenesis: Failure to reproduce positive findings, *J. Anat.,* 157, 101, 1988.
237. **Martin, A. H.,** Magnetic fields and time dependent effects on development, *Bioelectromagnetics,* 9, 393, 1988.
238. **Berman, E., Chacon, L., House, D., Koch, B. A., Koch, W. E., Leal, J., Løvtrup, S., Mantiply, E., Martin, A. H., Martucci, G. I., Mild, K. H., Monahan, J. C., Sandström, M., Shamsaifar, K., Tell, R., Trillo, M. A., Ubeda, A., and Wagner, P.,** Development of chicken embryos in a pulsed magnetic field, *Bioelectromagnetics,* 11, 169, 1990.
239. **Martin, A. H.,** Development of chicken embryos following exposure to 60-Hz magnetic fields with differing waveforms, *Bioelectromagnetics,* 13, 223, 1992.
240. **Litovitz, R. A., Montrose, C. J., Doinov, P., Brown, K. M., and Barber, M.,** Superimposing spatially coherent electromagnetic noise inhibits field-induced abnormalities in developing chick embryos, *Bioelectromagnetics,* 15, 105, 1994.
241. **Ubeda, A., Trillo, M. A., Chacon, L., Blanco, M. J., and Leal, J.,** Chick embryo development can be irreversibly altered by early exposure to weak extremely-low-frequency magnetic fields, *Bioelectromagnetics,* 15, 385, 1994.
242. **Handcock, M. S. and Kolassa, J. E.,** Statistical review of the henhouse experiments: The effects of a pulsed magnetic field on chick embryos, *Bioelectromagnetics,* 13, 429, 1992.
243. **Tenforde, T. S.,** Biological interactions of extremely-low-frequency electric and magnetic fields, *Bioelectrochem. Bioenerget.,* 25, 1, 1991.
244. **Foster, K. R.,** Health effects of low-level electromagnetic fields: Phantom or not-so-phantom risk? *Health Phys.,* 62, 429, 1991.
245. **Akamine, T., Muramatsu, H., Hamada, H., and Sakou, T.,** Effects of pulsed electromagnetic field on growth and differentiation of embryonal carcinoma cells, *J. Cell. Physiol.,* 124, 247, 1985.
246. **Rodemann, H. P., Bayreuther, K., and Pfleiderer, G.,** The differentiation of normal and transformed human fibroblasts *in vitro* is influenced by electromagnetic fields, *Exp. Cell Res.,* 182, 610, 1989.
247. **Florani, M., Cantoni, O., Sestili, P., Conti, R., Nicolini, P., Vetrano, F., and Dacha, M.,** Electric and/or magnetic field effects on DNA structure and function in cultured human cells, *Mutat. Res.,* 282, 25, 1992.
248. **Bellossi, A.,** Effect of pulsed magnetic fields on leukemia-prone AKR mice: No effect on mortality through five generations, *Leuk. Res.,* 15, 899, 1991.
249. **Thomson, R. A. E., Michaelson, S. M., and Nguyen, Q. A.,** Influence of 60-Hertz magnetic fields on leukemia, *Bioelectromagnetics,* 9, 149, 1988.
250. **Rannug, A., Ekström, T., Mild, K. H., Holmberg, B., Gimenez-Conti, I., and Slaga, T. J.,** A study on skin tumor formation in mice with 50-Hz magnetic field exposure, *Carcinogenesis,* 14, 573, 1993.
251. **Rannug, A., Holmberg, B., Ekström, T., Mild, K. H., Gimenez-Conti, I., and Slaga, T. J.,** Intermittent 50-Hz magnetic field and skin tumor promotion in SENCAR mice, *Carcinogenesis,* 15, 153, 1994.
252. **Stuchly, M. A., McLean, J. R. N., Burnett, R., Goddard, M., Lecuyer, D. W., and Mitchel, R. E. J.,** Modification of tumor promotion in the mouse skin by exposure to an alternating magnetic field, *Cancer Lett.,* 65, 1, 1992.
253. **Beniashvili, D. Sh., Bilanishvili, V. G., and Menabde, M. Z.,** Low-frequency electromagnetic radiation enhances the induction of rat mammary tumors by nitrosomethylurea, *Cancer Lett.,* 61, 75, 1991.
254. **Mevissen, M., Stamm, A., Buntenkotter, S., Zwingelbert, R., Wahnschaffe, U., and Löscher, W.,** Effects of magnetic fields on mammary tumor development induced by 7,12-dimethylbenz(a)anthracene in rats, *Bioelectromagnetics,* 14, 131, 1993.
255. **Löscher, W., Mevissen, M., Lehmacher, W., and Stamm, A.,** Tumor promotion in a breast cancer model by exposure to a weak alternating magnetic field, *Cancer Lett.,* 71, 75, 1993.
256. **Rannug, A., Holmberg, B., Ekström, T., and Mild, K. H.,** Rat liver loci study on coexposure with 50-Hz magnetic fields and known carcinogens, *Bioelectromagnetics,* 14, 17, 1993.
257. **Rannug, A., Holmberg, B., and Mild, K. H.,** A rat liver foci promotion study with 50-Hz magnetic fields, *Environ. Res.,* 62, 223, 1993.
258. **Brugere, H., Pupin, F., and Lambrozo, J.,** Effects of 50-Hz magnetic fields on ENU-induced brain tumors in rats, Abstracts of the Annual Review of Research on Biological Effects of Electric and Magnetic Fields from the Generation, Delivery, and Use of Electricity, Savannah, GA, October 31–November 4, 1993, 74.
259. **Wertheimer, N. and Leeper, E.,** Possible effects of electric blankets and heated waterbeds on fetal development, *Bioelectromagnetics,* 7, 13, 1986.
260. **Wertheimer, N. and Leeper, E.,** Fetal loss associated with two seasonal sources of electromagnetic field exposure, *Am. J. Epidemiol.,* 120, 220, 1989.

261. **Dlugosz, L., Vena, J., Zielezny, M., Byers, T., Roth, G., and Marshall, E.,** Congenital malformations and periconceptional use of electric blankets and heated waterbeds: A register-based case-control study, *Am. J. Epidemiol.,* 132, 763, 1990.
262. **Juutilainen, J., Matilainen, P., Saarikoski, S., Läärä, E., and Suonio, S.,** Early pregnancy loss and exposure to 50-Hz magnetic fields, *Bioelectromagnetics,* 14, 229, 1993.
263. **Savitz, D. A. and Ananth, C. V.,** Residential magnetic fields, wire codes, and pregnancy outcome, *Bioelectromagnetics,* 15, 271, 1994.
264. **Preston-Martin, S., Peters, J. M., Yu, M. C., Garabrant, D. H., and Bowman, J. D.,** Myelogenous leukemia and electric blanket use, *Bioelectromagnetics,* 9, 207, 1988.
265. **Verreault, R., Weiss, N. S., Hollenbach, K. A., Strader, C. H., and Daling, J. R.,** Use of electric blankets and risk of testicular cancer, *Am. J. Epidemiol.,* 131, 759, 1990.
266. **Vena, J. E., Graham, S., Hellmann, R., Swanson, M., and Brasure, J.,** Use of electric blankets and risk of postmenopausal breast cancer, *Am. J. Epidemiol,* 134, 180, 1991.
267. **Lovely, R. H., Buschbom, R. L., Slavich, A. L., Anderson, L. E., Hansen, N. H., and Wilson, B. W.,** Adult leukemia risk and personal appliance use: A preliminary study, *Am. J. Epidemiol.,* 140, 510, 1994.
268. **Marcus, M.,** Epidemiologic studies of VDT use and pregnancy outcome, *Reprod. Toxicol.,* 4, 51, 1990.
269. *Health Effects Related to the Use of Visual Display Units,* National Radiological Protection Board, Chilton, Didcot, Oxon, England, 1994.
270. **Delpizzo, V.,** Epidemiological studies of work with video display terminals and adverse pregnancy outcomes (1984–1992), *Am. J. Indust. Med.,* 26, 465, 1994.
271. **Goldhaber, M. K., Polen, M. R., and Hiatt, R. A.,** The risk of miscarriage and birth defects among women who use video display terminals during pregnancy, *Am. J. Indust. Med.,* 13, 695, 1988.
272. **Lindbohm, M.-L., Hietanen, M., Kyyrönen, P., Sallmen, M., von Nandelstadh, P., Ylikoski, M., and Hemminki, K.,** Magnetic fields of video display terminals and spontaneous abortion, *Am. J. Epidemiol.,* 136, 1041, 1992.
273. **Kurppa, K., Holmberg, P. C., Rantala, K., Nurminen, T., and Saxen, L.,** Birth defects and exposure to video display terminals during pregnancy, *Scand. J. Work Environ. Health,* 11, 353, 1985.
274. **Butler, W. J. and Brix, K. A.,** Video display terminal work and pregnancy outcome in Michigan clerical workers, in *Allegations of Reproductive Hazards from VDUs,* Pearce, B. G., Ed., Humane Technology, Nottingham, England, 1986, 31.
275. **Ericson, A. and Kallen, B.,** An epidemiological study of work with video display screens and pregnancy outcome: I. A registry study, *Am. J. Indust. Med.,* 9, 447, 1986.
276. **Ericson, A. and Kallen, B.,** An epidemiological study of work with video display screens and pregnancy outcome: II. A case-control study, *Am. J. Indust. Med.,* 9, 459, 1986.
277. **McDonald, A., Cherry, N., Delorme, C., and McDonald, J. D.,** Visual display units and pregnancy: Evidence from the Montreal survey, *J. Occup. Med.,* 28, 1126, 1986.
278. **Bjerkedal, T. and Egenaes, J.,** Video display terminals and birth defects, in *Work with Display Units 86,* Knave, B. and Wideback, P.-G., Eds., Elsevier, New York, 1986, 111.
279. **Westerholm, P. and Ericson, A.,** Pregnancy outcome and VDU work in a cohort of insurance clerks, in *Work with Display Units 86,* Knave, B. and Wideback, P.-G., Eds., Elsevier, New York, 1986, 104.
280. **McDonald, A. D., McDonald, J. C., Armstrong, B., Cherry, N., Nolin, A. D., and Robert, D.,** Work with visual display units in pregnancy, *Br. J. Indust. Med.,* 45, 509, 1988.
281. **Nurminen, T. and Kurppa, K.,** Office equipment, work with video display terminals, and course of pregnancy: Reference mothers' experience from a Finnish case-referent study of birth defects, *Scand. J. Work Environ. Health,* 14, 293, 1988.
282. **Bryant, H. and Love, E.,** Video display terminal use and spontaneous abortion risk, *Int. J. Epidemiol.,* 15, 132, 1989.
283. **Neilsen, C. V. and Brandt, L.,** Spontaneous abortion among women using video display terminals, *Scand. J. Work Environ. Health,* 16, 323, 1990.
284. **Brandt, L. and Neilsen, C. V.,** Congenital malformations among children of women working with video display terminals, *Scand. J. Work Environ. Health,* 16, 329, 1990.
285. **Windham, G. C., Fenster, L., Swan, S. H., and Neutra, R. R.,** Use of video display terminals during pregnancy and the risk of spontaneous abortion, low birth weight, or intrauterine growth retardation, *Am. J. Indust. Med.,* 18, 675, 1990.
286. **Schnorr, T. M., Grajewski, B. A., Hornung, R. W., Thun, M. J., Egeland, G. M., Murray, W. E., Conover, D. L., and Halperin, W. E.,** Video display terminals and the risk of spontaneous abortion, *New Engl. J. Med.,* 324, 727, 1991.
287. **Roman, E., Beral, V., Pelerin, M., and Hermon, C.,** Spontaneous abortion and work with visual display units, *Br. J. Indust. Med.,* 49, 507, 1992.
288. **Bergqvist, U.,** *Health Problems During Work with Visual Display Terminals,* Doctoral Dissertation, Department of Neuromedicine, National Institute of Occupational Health, Solna, Sweden, 1993.

289. **Wertheimer, N. and Leeper, E.,** Electrical wiring configurations and childhood cancer, *Am. J. Epidemiol.,* 109, 2743, 1979.
290. **Fulton, J. P., Cobb, S., Preble, L., Leone, L., and Forman, E.,** Electrical wiring configurations and childhood leukemia in Rhode Island, *Am. J. Epidemiol.,* 111, 292, 1980.
291. **Wertheimer, N. and Leeper, E.,** RE: Electrical wiring configurations and childhood leukemia in Rhode Island, *Am. J. Epidemiol.,* 111, 461, 1980.
292. **Myers, A., Cartwright, R. A., Bonnell, J. A., Male, J. C., and Cartwright, S. C.,** Overhead power lines and childhood cancer, IEE Conf. Publ. No. 257, International Conference of Electric and Magnetic Fields in Medicine and Biology, London, England, December 4–5, 1985, 126.
293. **Coleman, M. P., Bell, C. M. J., Taylor, H.-L., and Primic-Zakelj, M.,** Leukemia and residence near electricity transmission equipment: A case-control study, *Br. J. Cancer,* 60, 793, 1989.
294. **Tomenius, L.,** 50-Hz electromagnetic environment and the incidence of childhood tumors in Stockholm County, *Bioelectromagnetics,* 7, 191, 1986.
295. **Savitz, D. A., Wachtel, H., Barnes, F. A., John, E. M., and Tvrdik, J. G.,** Case-control study of childhood cancer and exposure to 60-Hz magnetic fields, *Am. J. Epidemiol.,* 128, 21, 1988.
296. **London, S. J., Thomas, D. C., Bowman, J. D., Sobel, E., Cheng, T.-C., and Peters, J. M.,** Exposure to residential electric and magnetic fields and risk of childhood leukemia, *Am. J. Epidemiol.,* 134, 923, 1991.
297. **Lin, S. R. and Lu, P. Y.,** An epidemiologic study of childhood cancer in relation to residential exposure to electromagnetic fields, presented at Annual DOE/EPRI Contractors Review on Biological Effects from Electric and Magnetic Fields, Portland, Oregon, Nov. 13–16, 1989.
298. **Feychting, M. and Ahlbom, A.,** Magnetic Fields and Cancer in People Residing near Swedish High Voltage Power Lines, Rep. No. 6/92, Institutet for Miljomedicin, Karolinska Institutet, Stockholm, Sweden, 1992.
299. **Feychting, M. and Ahlbom, A.,** Magnetic fields and cancer in children residing near Swedish high-voltage power lines, *Am. J. Epidemiol.,* 138, 467, 1993.
300. **Stevens, R. G.,** RE: Magnetic fields and cancer in children residing near Swedish high-voltage power lines, *Am. J. Epidemiol.,* 140, 75, 1994.
301. **Feychting, M. and Ahlbom, A.,** The authors reply (to comments by R. G. Stevens), *Am. J. Epidemiol.,* 140, 75, 1994.
302. **Olsen, J. H., Nielsen, A., and Schulgen, G.,** Residence near high-voltage facilities and the risk of cancer in children, *Br. Med. J.,* 307, 891, 1993.
303. **Verkasalo, P. K., Pukkala, E., Hongisto, M. Y., Valjus, J. E., Järvinen, P. J., Heikkilä, K. V., and Koskenvuo, M.,** Risk of cancer in Finnish children living close to power lines, *Br. Med. J.,* 307, 895, 1993.
304. **Ahlbom, A., Feychting, M., Koskenvuo, M., Olsen, J. H., Pukkala, E., Schulgen, G., and Verkasalo, P.,** Electromagnetic fields and childhood cancer, *Lancet,* 342, 1296, 1993.
305. **Wertheimer, N. and Leeper, E.,** Adult cancer related to electrical wires near the home, *Int. J. Epidemiol.,* 11, 345, 1982.
306. **Coleman, M., Bell, C. M. J., Taylor, H.-L., and Thornton-Jones, H.,** Leukemia and electromagnetic fields: A case-control study, IEE Conf. Publ. No. 257, International Conference of Electric and Magnetic Fields in Medicine and Biology, London, England, December 4–5, 1985, 122.
307. **McDowall, M. E.,** Mortality of persons resident in the vicinity of electricity transmission facilities, *Br. J. Cancer,* 53, 271, 1986.
308. **Youngson, J. H. A. M., Clayden, A. D., Myers, A., and Cartwright, R. A.,** A case/control study of adult hematological malignancies in relation to overhead powerlines, *Br. J. Cancer,* 63, 977, 1991.
309. **Severson, R. K., Stevens, R. G., Kaune, W. T., Thomas, D. B., Henser, L., Davis, S., and Sever, L. E.,** Acute nonlymphocytic leukemia and residential exposure to power frequency magnetic fields, *Am. J. Epidemiol.,* 128, 10, 1988.
310. *Electromagnetic Fields and the Risk of Cancer,* National Radiological Protection Board, Chilton, Didcot, Oxon, England, 1992.
311. **Milham, S., Jr.,** Mortality in workers exposed to electromagnetic fields, *Environ. Health Perspect.,* 62, 297, 1985.
312. **Savitz, D. A. and Calle, E. E.,** Leukemia and occupational exposure to electromagnetic fields: Review of epidemiologic surveys, *J. Occup. Med.,* 29, 47, 1987.
313. **Coleman, M. and Beral, V.,** A review of epidemiological studies of the health effects of living near or working with electricity generation and transmission equipment, *Int. J. Epidemiol.,* 17, 1, 1988.
314. **Thériault, G.,** Health effects of electromagnetic radiation on workers: Epidemiologic studies, in: *Proc. Scientific Workshop on the Health Effects of Electric and Magnetic Fields on Workers,* Bierbaum, P. J. and Peters, J. M., Eds., Publ. No. 91-111, U. S. Dept. Health and Human Services (NIOSH), Cincinnati, OH, 1991, 91.
315. *Evaluation of the Potential Carcinogenicity of Electromagnetic Fields,* Rep. No. EPA/600/6-90/005A, U. S. Environmental Protection Agency, Washington, DC, 1990.

316. **Tynes, T. and Andersen, A.**, Electromagnetic fields and male breast cancer, *Lancet,* 36, 1596, 1990.
317. **Matanoski, G. M., Breysse, P. N., and Elliott, E. A.**, Electromagnetic field exposure and male breast cancer, *Lancet,* 337, 737, 1991.
318. **Demers, P. A., Thomas, D. B., Rosenblatt, K. A., Jimenez, L. M., McTiernan, A., Stalsberg, H., Stemhagen, A., Thompson, W. D., McCrea-Curnen, M. F., Satariano, W., Austin, D. F., Isacson, P., Greenberg, R. S., Key, C., Kolonel, L. N., and West, D. W.**, Occupational exposure to electromagnetic fields and breast cancer in men, *Am. J. Epidemiol.,* 13, 340, 1991.
319. **Loomis, D. P., Savitz, D. A., and Ananth, C. V.**, Breast cancer mortality among female electrical workers in the United States, *J. Natl. Cancer Inst.,* 86, 921, 1994.
320. **Matanoski, G. M., Elliott, E. A., Breysse, P. N., and Lynberg, M. C.**, Leukemia in telephone linemen, *Am. J. Epidemiol.,* 137, 609, 1993.
321. **Savitz, D. A. and Loomis, D. P.**, Magnetic field exposure in relation to leukemia and brain cancer mortality among electric utility workers, *Am. J. Epidemiol.,* 141, 123, 1995.
322. **Loomis, D. P. and Savitz, D. A.**, Mortality from brain cancer and leukemia among electrical workers, *Br. J. Indust. Med.,* 47, 633, 1990.
323. **Mack, W., Preston-Martin, S., and Peters, J. M.**, Astrocytoma risk related to job exposure to electric and magnetic fields, *Bioelectromagnetics,* 12, 57, 1991.
324. **Juutilainen, J., Läärä, E., and Pukkala, E.**, Incidence of leukemia and brain tumors in Finnish workers exposed to ELF magnetic fields, *Int. Arch. Occup. Environ. Health,* 62, 289, 1990.
325. **Tynes, T., Andersen, A., and Langmark, F.**, Incidence of cancer in Norwegian workers potentially exposed to electromagnetic fields, *Am. J. Epidemiol.,* 136, 81, 1992.
326. **Floderus, B., Persson, T., Stenlund, C., Linder, G., Johansson, C., Kiviranta, J., Parsman, H., Lindblom, M., Wennberg, A., Öst, A., and Knave, B.**, Occupational exposure to electromagnetic fields in relation to leukemia and brain tumors: A case-control study, *Cancer Causes and Cont.,* 4, 465, 1993.
327. **Washburn, E. P., Orza, M. J., Berlin, J. A., Nicholson, W. J., Todd, A. C., Frumkin, H., and Chalmers, T. C.**, Residential proximity to electricity transmission and distribution equipment and risk of childhood leukemia, childhood lymphoma, and childhood nervous system tumors: Systematic review, evaluation, and meta-analysis, *Cancer Causes and Cont.,* 5, 299, 1994.
328. **Dovan, T., Kaune, W. T., and Savitz, D. A.**, Repeatability of measurements of residential magnetic fields and wire codes, *Bioelectromagnetics,* 14, 145, 1993.
329. **Savitz, D. A. and Feingold, L.**, Association of childhood cancer with residential traffic density, *Scand. J. Work Envir. Health,* 15, 360, 1989.
330. **Savitz, D. A. and Baron, A. E.**, Estimating and correcting for confounder misclassification, *Am. J. Epidemiol.,* 129, 1062, 1989.
331. **Bracken, T. D.**, EMDEX Project Residential Study: Interim Report, Rep. No. TR-102011, Electric Power Research Institute, Palo Alto, CA, 1993.

Chapter 5

ELECTRIC AND MAGNETIC FIELDS FOR BONE AND SOFT TISSUE REPAIR

Charles Polk

CONTENTS

I. HISTORY

As early as 1962 in the US[2]—even earlier in Japan, 1957[1]—it was shown that electric potential differences appear across both living and dead bone subjected to mechanical stress. C.A.L. Bassett and R.O. Becker observed that these stress-generated electrical signals decayed very slowly in comparison with similarly initiated signals in piezo-electric crystals and concluded that piezo-electric phenomena "while probably present, were not the sole cause of these potentials."[2] Later analysis and experiments established that the observed signals were primarily due to ion displacement within the porous regions and multiple fluid filled channels present in all bone. The early observations already suggested that direct application of an externally generated voltage might have an effect on bone development. This was shown to be the case by Bassett, Pawluk and Becker[3] who found that a DC current of the order of 1 μA (corresponding to a current density of approximately 0.01 A/m^2) produced massive osteogenesis near the cathode when electrodes were implanted into the femur of living dogs.

Having shown that application of a DC electric field to non-excitable, connective-tissue cells can produce effects similar to those elicited by mechanical stress, C.A.L. Bassett and his co-workers realized that clinical exploitation of these phenomena would require surgical implantation of electrodes with the attendant danger of infection. Therefore, they proceeded to explore whether non-invasive, inductive coupling, that gave wave forms similar to those endogenously produced by mechanical stress, could lead to beneficial bone development. Favorable results obtained with pulsed electromagnetic fields on dogs were reported by them in 1974.[4] Signals of this type have generally been identified as "PEMF" in the orthopedics/electrical stimulation community for the last twenty years and have been applied successfully in a large number of cases for the repair of non-unions.[5] In Germany, relatively large

0-8493-0641-8/96/$0.00+$.50

amplitude, low frequency (< 20 Hz) sinusoidal magnetic fields have been used for both bone repair and wound healing.[6] More recently sinusoidal fields at selected frequencies below 100 Hz, but of much lower intensity, were found to be effective for bone repair by research groups in the U.S. when applied simultaneously with a static (DC) magnetic field of less than 50 μT.[7,8]

Although the non-invasive, time-varying magnetic field treatment for non-unions (fractures which fail to heal) became—at least in the US—the most widely used clinical application of sub-radio frequency fields, several investigators pursued the application of DC electric fields through implanted electrodes and the application of higher frequency currents through electrode contacts placed on the skin surface to enhance bone repair.[9] At the same time laboratory investigations, *in vitro* and on animals, explored the application of all three modalities—PEMF, implanted DC electrodes and higher frequency coupling through skin electrodes—to produce blood-vessel regeneration (*angiogenesis*), soft-tissue healing, nerve repair or regeneration and regression of tumors. The use of time varying magnetic fields for the treatment of arthritis was also explored.[10] Motivated by the suggested clinical applications, a large number of basic science investigations have been initiated, and are continuing today; the objective is understanding the mechanisms through which low intensity electric and magnetic fields affect cells and living tissue. Some of these are reviewed briefly in Section IV below.

II. BONE AND CARTILAGE REPAIR

In the US, medical devices are approved for clinical use only after it has been shown to the satisfaction of the US Food and Drug Administration (F.D.A.) that they are not only safe, but also effective. This is a much more stringent requirement than the mere demonstration of safety demanded presently in most European countries (although some EEC countries are considering moving towards approval criteria that are similar to those used in the US). The US laws governing the sale of medical devices for clinical use are also much more restrictive than the controls over implementation of new surgical procedures which involve only informal medical peer review. Even organized multi-patient clinical trials require no more than approval by a local hospital–based institutional review board. The F.D.A. Office of Device Evaluation does approve some new devices, after careful review, for clearly limited clinical trials. However, information on such limited, temporary approval is not made available. Table 1 therefore shows only those devices which are presently (as of March 1994) approved and does not include experimental systems which currently may be undergoing limited clinical trials. Some of the latter are discussed further on, based only on information furnished by manufacturers or available from the published medical literature.

The devices listed in Table 1 are approved for one of three applications: The treatment of non-unions (fractures which have failed to heal after standard treatment involving setting and stabilization with casts), congenital pseudoarthroses[11] and promotion of spinal fusion. Although many animal experiments (and possibly a few human trials, especially in Europe) have evaluated the application of electric or magnetic fields for acceleration of fresh fracture healing, for reversal of osteoporosis and for the treatment of arthritis, no devices are currently approved in the US for these purposes.

Classified by electrical and mechanical characteristics, the devices in Table 1 are either:

noninvasive:

a) generating time varying magnetic fields applied by coils to the affected body part (I and "PEMF" devices A,B,C,D)

b) generating time varying electric fields applied through skin-surface electrodes ("capacitively" coupled) (E)

invasive or semi-invasive:

DC applied from an implanted battery (F, G) or (semi-invasive) DC applied with percutaneous pins (H).

A signal typical for some PEMF (A,C,D) devices is illustrated in Figure 1. Figure 1A shows the magnetic field versus time and 1B the corresponding electric field induced into a linear, isotropic medium. The wave form shown on Figure 1B can be measured by a probe coil having a sufficiently large number of turns. The frequency spectrum of the electric field is shown in Figures 2A and 2B. Signals used by the different manufacturers are protected by patents and FDA approval is for particular signal parameters within specified tolerances on time and amplitude. The pseudoarthrosis signal used by EBI (B in Table 1) consists of single pulses repeated at a rate of 72 pps rather than the pulse bursts illustrated in Figures 1 and 2. Each magnetic field pulse increases from zero to 3.5 mT in 380 μs and then decreases slowly to zero in approximately 4.5 ms. The signals that are now in use have evolved

TABLE 1
Electrical Bone Growth Stimulators Approved by the US F.D.A. as of March 1994

Manufacturer	Device	Indications	Technology	Approval Date	Text Ref.
Electro-Biology, Inc.	EBI Bone Healing System*	non-union; congenital pseudoarthroses, failed fusions	noninvasive pulsed e.m. field (PEMF)	Nov. 1979	A
					B
American Medical Electronics, Inc.	Physio-Stim	nonunions (excluding vertebrae and flat bones)	noninvasive PEMF	Feb. 1986	C
American Medical Electronics, Inc	Spinal-Stim	to promote spinal fusion as an adjunct to surgery or as non-operative treatment when 9 mos. have elapsed since the last surgery	noninvasive PEMF	Feb. 1990	D
Orthologic Corp.	Orthologic 1000	Nonunions (excluding vertebrae and flat bones)	Noninvasive DC + sinusoidal magnetic fields	March 1994	I
Bioelectron, Inc.	Orthopak BGS System	nonunions (excluding vertebrae and flat bones)	noninvasive/capac. coupled	Feb. 1986	E
Electro-Biology, Inc.	Orthogen/ Osteogen	nonunion of long bones	Implantable DC	Jan. 1980	F
Electro-Biology, Inc	Sp F-4 (2) Implantable BGS**	Spinal fusion adjunct	Implantable DC	April 1987	G
Bioelectron, Inc.	Zimmer Direct Current Bone Growth Stimulator (DCGBS)	Nonunion fractures	Semi-invasive DC with percutaneous pins	Nov. 1979	H

* Also known as Bi-Osteogen Systems 204.
** Also known as BGS-Osteostim HS 11.

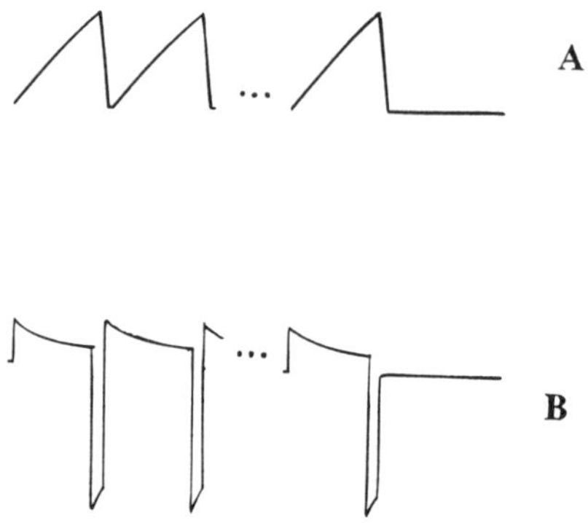

FIGURE 1 A typical PEMF signal (signal A of Table 1). Signal consists of 15 pulse bursts per second. Each burst is 4.5 ms long and contains 20 magnetic field pulses. In each pulse the magnetic field increases from 0 to approx. 2 mT during 200 μs, decreases to 0 again during 23 μs and is equal to 0 for 2 μs before the next 225 μs sequence begins. 1A: Magnetic field vers. time. 1B: Electric field ($\propto \frac{\partial B}{\partial t}$) versus time.

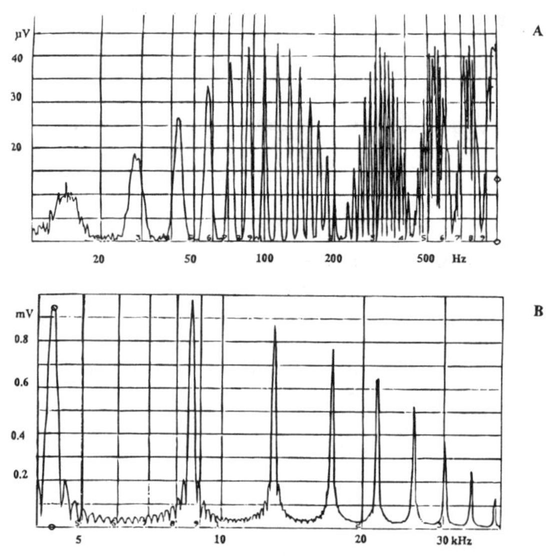

FIGURE 2 Electric field spectrum $|E(\omega)|$ of signal in Fig. 1 as measured by the output from an air-core coil (self-inductance 0.147 mH, resistance 2.13 Ω, mean radius 1.5 cm). 2A: 10 Hz to 1 kHz (Frequency resolution: 3 Hz). 100 μV correspond to 0.6 mT/sHz in a 4.5 ms pulse burst. 2B: 1 kHz to 10 kHz (Frequency resolution: 10 Hz). 1 mV corresponds to 6 mT/sHz in a 4.5 ms pulse burst.

considerably from those employed in the initial studies and some have little resemblance to the endogenous electrical signals elicited by mechanical stress.

The PEMF signals employed by the various manufacturers in the US and in Europe can have several different pulse shapes, rise and decay times, pulse widths, pulse repetition rates and amplitudes. Since it has been shown (see IV below) that all of these variables can have a profound effect on the biological action of a particular signal, it is essential that reports on effectiveness or lack of effectiveness of "PEMF" give an exact description of the signal that was used. Unfortunately the medical literature is replete with examples where this information is either incomplete or completely absent. It is particularly important that a PEMF signal not be described simply by its "frequency," when what is meant is the pulse repetition frequency. Referring to the legends of Figure 1 and Figure 2, it is clear that this particular signal should not be identified as a "15-Hz" signal since it has significant frequency

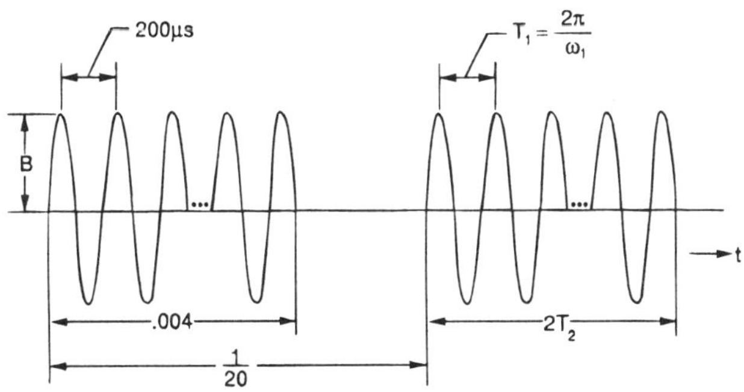

FIGURE 3 "Biphasic" pulse burst repeated at frequency $(\omega_0/2\pi) = 20$ Hz. Time t in seconds.

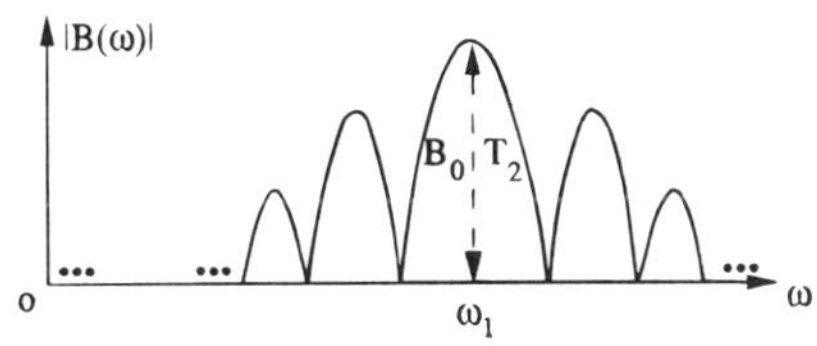

FIGURE 4 Spectral envelope of the signal of Fig. 3 (If a single burst is replaced by bursts repeated at frequency f_0, the continuous sepstrum becomes the envelope of spectral lines spaced at intervals f_0).

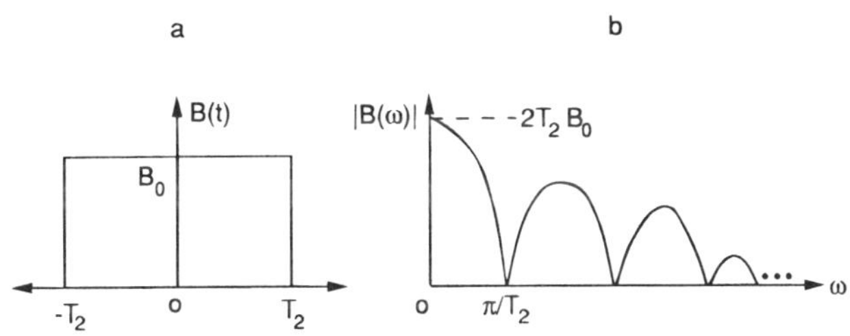

FIGURE 5 DC pulse (a) and Fourier spectrum (b).

content well into the kilo Hertz region. Only the spacing between the many different spectral lines is 15 Hz as is apparent from panel A of Figure 2.

As a further illustration of the importance of differentiating between simply "frequency" and pulse repetition frequency, we consider the hypothetical signal of Figure 3 which is, however, similar to that of some therapeutic signals. That signal would be correctly described as having a pulse-repetition frequency of 20 Hz. However it should not simply be called a "20-Hz" signal or an "ELF" signal, because very little of its energy is at 20 Hz or even below 1000 Hz. Fourier integral analysis of such a pulse burst[12,13] gives

$$F_1(\omega) = 2\pi \sum_{n=-\infty}^{n=\infty} BT_2\left[\frac{\sin(\omega - \omega_1)T_2}{(\omega - \omega_1)T_2} + \frac{\sin(\omega + \omega_1)T_2}{(\omega + \omega_1)T_2}\right]\delta(\omega - n\omega_0) \tag{1}$$

where the delta function $\delta\ (\omega - n\omega_0)$ indicates discrete spectral lines at multiples of the pulse repetition frequency $(\omega_0/2\pi)$, which is 20 Hz in the selected example. The amplitudes of the spectral lines are given by the expression in the bracket. The spectral peak for positive frequencies occurs, as illustrated in Figure 4, where $[\sin(\omega - \omega_1)T_2]/(\omega - \omega_1)T_2]$ is equal to one, which occurs when $\omega = \omega_1$ or at the frequency of the sine wave within each individual burst. In the present example this is $[10^6/200] = 5$ kHz! The duration $2T_2$ of the pulse burst

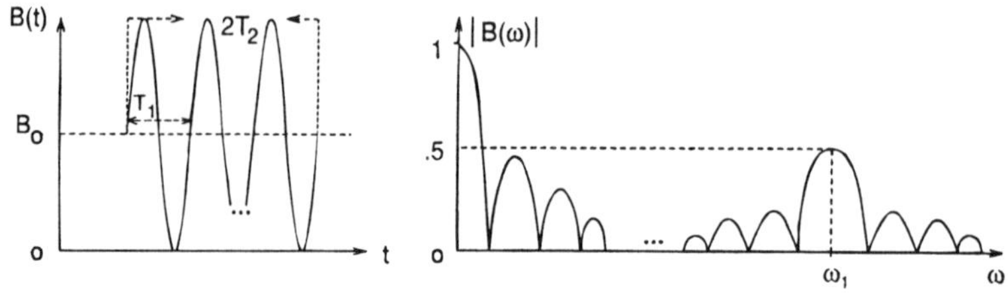

FIGURE 6 Unidirectional pulse and spectral envelope. If burst is repeated at frequency f_0, the spectrum consists of discrete spectral lines at f_0 intervals. The illustrated continuous spectrum then becomes the envelope indicating relative amplitudes of the spectral lines.

determines the widths of the spectral lobes in Fig. 4 as indicated by Equation 1. The spectrum of a unidirectional ("DC") pulse (Figure 5a) is shown on Figure 5b. Only if such a pulse is combined with the "biphasic" signal of Figure 3, so as to give the signal of Figure 6, will the Fourier spectrum have significant energy content at low frequencies. The example serves to illustrate the difference in frequency content—rather than only DC component—between unidirectional and "biphasic" pulsed signals.

Details of shape, orientation and location of the application coil or coils are also important, since these parameters, together with pulse amplitude and shape, determine the nature of the magnetic and electric field at the location of the injured tissue. If the amplitude of the axially directed magnetic flux density B is constant over some region of radius R within the cross-section of a circular cylinder, the induced electric field is

$$\bar{E} = -\left(\frac{\partial B}{\partial t}\right)\frac{r}{2}\hat{\phi} \tag{2}$$

provided the material of the cylinder is electrically *homogenous* and *isotropic.* In Equation (2) $r < R$ is the distance from the center of the cylinder and $\hat{\phi}$ a unit vector in the circumferential direction. For magnetic fields varying sinusoidally as $B_0 \cos \omega t$, $\bar{E} = \omega B_0(r/2) \sin \omega t\hat{\phi}$. Since most biological objects are neither homogeneous nor isotropic, the actual induced electric fields at various points in the tissue or cells may deviate substantially from the values given by Equation 2.[14,15] Equation 2 is useful only for *estimating* the spatial *average* value of the induced electric field, which depends in the bone environment on the point-to-point variation of the electrical properties of muscle, fat, cartilage, periosteum (outer bone membrane) and bone marrow.

Current pulses of the PEMF devices are usually produced by the discharge of capacitor banks controlled by a timing network. The applicator coil cannot be interchanged among different devices because its inductance and resistance are a part of the discharge network. While earlier bone growth stimulators employed Helmholtz coil-pair arrangements, most present devices have single coils which can be custom-shaped for particular limbs. Figure 7 shows a typical system sold by EBI (Electrobiology, Inc.). This unit is driven by rechargeable batteries and the control unit includes an elapsed-time clock to measure the total time of stimulation of the fracture being treated. A typical treatment time can be between two and ten hours per day over a period of six months and is 30 minutes per day with some newer low field intensity systems.

The so-called capacitively coupled device (E on Table 1) generates a continuous sine wave at a frequency of 60 kHz. The total current through the skin contains a conduction component since conductive contact is made between the applicator electrodes and the skin which represents a very "leaky" capacitor. Electric fields produced at the tissue level by this device are between 1 V/m and 50 V/m.[16] These levels are very much higher than the average

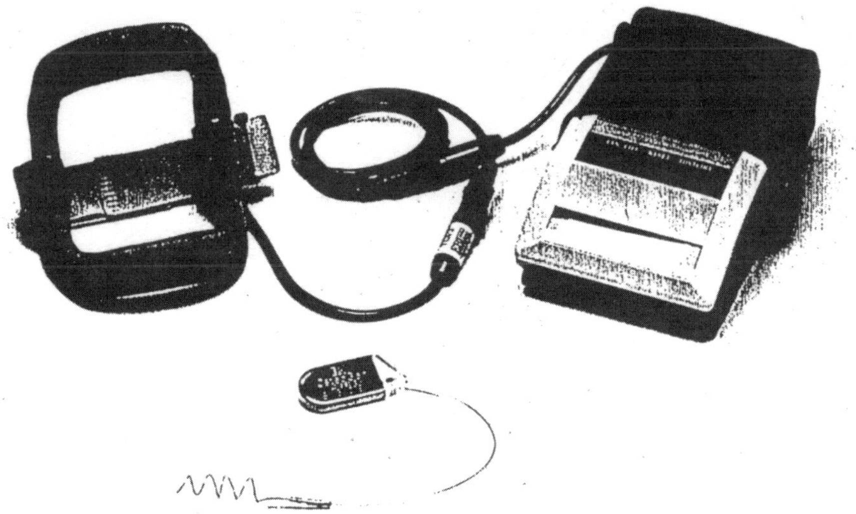

FIGURE 7 PEMF applicator and control unit and implantable DC stimulating device (battery with wire electrode) manufactured by Electrobiology Inc. Systems A and G of Table 1. (Photograph courtesy of Electrobiology Inc.)

amplitude of the electric fields produced in tissue by PEMF devices and are also higher than the instantaneous peak values produced by some of the PEMF systems. It is interesting to note here that *in vitro* experiments with truly capacitively coupled fields showed enhancement of bone cell proliferation at very much lower amplitude (10^{-5} V/m) when the frequency of the continuous sine wave was 10 Hz rather than 60 kHz.[17]

An invasive (implantable) DC device (F,G in Table 1) is also shown in Fig. 7. The small (approximately $4 \times 2 \times 0.5$ cm) titanium case contains a long-life battery that is connected to it. The case acts as the anode and two (shown) or four titanium wires act as the cathode. The amplitude of the continuous current is between 5 μA (for some spinal fusion applications) and 20μA (for nonunion of long bones). The cathodes are placed at the location where bone growth is to be enhanced (for example at the vertebrae which are surgically fused) while the case is placed in a convenient location at some distance from the bone. Treatment details and success rates, in comparison with surgical procedures without use of DC stimulation, are discussed in the medical literature.[18]

Although not used or approved for use in the US, the German "Magnetodyn" system[6] is interesting not only because it employs sinusoidal magnetic fields between 2 Hz and 20 Hz, but also because it relies on metallic implants which are used for fixation of the bone, to act as the "secondary" of a "transformer" whose primary is the external applicator. Sometimes an implanted pick-up coil ("secondary") is connected to fixation bars or to screws (electrically insulated from the bars) on each side of a pseudoarthrotic gap. With this system peak electric fields of 40 V/m and current densities of 5 A/m^2 have been produced in the gap.

A PEMF signal (similar to B in Table 1) has also been used experimentally to arrest osteonecrosis (bone death, possibly due to vascular impairment or toxic agents) of the femoral head. Application for 8 hours/day over 12 months gave substantially better results than the standard surgical ("core decompression") treatment.[19]

The most recent addition to non-invasive systems for the repair of non-union fractures (I in Table 1) employs an ELF time varying magnetic field of less than 100 μT in combination with a DC magnetic field of comparable amplitude. The field vectors have the same orientation and are generated by the same coils. A clinically used applicator is shown on Figure 8. The rationale for using an AC/DC "resonance" combination is discussed further on under "Mechanisms." The company that markets this equipment, Orthologic, states that the system

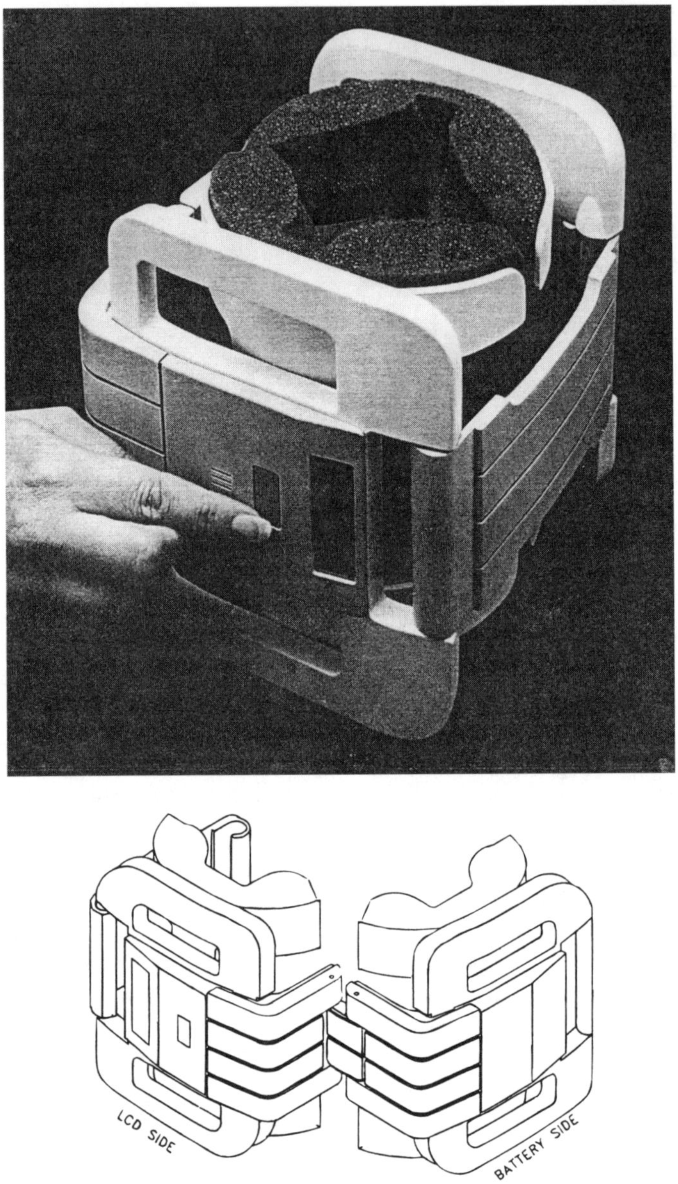

FIGURE 8 Applicator of the Orthologic 1000 DC/AC equipment, system I of Table 1. (Photograph courtesy of Orthologic Inc.)

focuses on "the use and control of cellular signaling, or the movement of ions across the cell membrane, as a signal or switch to initiate bone regeneration."[20]

III. SOFT-TISSUE REPAIR AND NERVE REGENERATION

No electric or magnetic system to aid nerve regeneration or soft tissue repair is approved by the F.D.A. at the present time for non-experimental therapy. However, considerable animal and *in vitro* experimentation in this country and abroad suggests the clinical usefulness of electric currents for soft tissue repair[21] and possibly also to enhance repair of nerve fibers which have sustained crush or transsection injury.[22,23] Since there is a great variety of soft-tissue pathologies which could respond to electric or magnetic fields, the volume of application in this area could in the future become larger than in orthopedics, provided field-tissue and field-cell interactions become better understood and clinical benefits for specific injuries and diseases are established.

Beneficial effects of time varying electric fields in wound healing are most likely related to promotion of angiogenesis. Wound healing consists of several stages, the first being inflammation when changes in vascular permeability occur, infiltration of leucocytes and macrophages takes place and cells migrate, synthesize granulation tissue, collagen and proteoglycans, and initiate formation of capillaries. This is followed by transitional repair and remodelling phases. Electrical currents are probably only important in the first two stages[21] and the experimental clinical trials performed thus far involve therapy for inflammation. One was a double blind study of persistent rotator cuff tendinitis[24] which showed beneficial effects of a PEMF very similar to the pseudarthrosis signal used by EBI (B in Table 1, described above in II). Cuff tendinitis is probably due to partial interruption of blood supply to the rotator cuff tendons of the shoulder by compression of vessels between adjacent bones. Another double blind study[25] employing a PEMF signal indicated beneficial effects in the treatment of skin ulcers.

Clinical trials[26] and animal experiments also involved irradiation of wounds with pulsed radio frequencies between 3 MHz and 44 MHz, at power levels from 7 μW to 15W, employing pulse widths between 65 μs and 100 μs and pulse repetition frequencies between 200 Hz and 1 kHz. A commonly used animal model for wound healing is the McFarlane Skin flap involving partial excision of a rectangular skin section on the back of the rat. Survival of the flap depends mainly on blood supply, with vascularization of the skin flap being an indirect measure of treatment success. The EBI signal shown on Figure 1 [A in Table 1] is reported to have decreased skin flap necrosis when exposure was for 6 hours per day for 3 days, while exposure for 18 hours on the first day after injury had no observable effect.[27] Exposure with a triangular, symmetric and almost continuous magnetic field (18 ms triangular pulse, followed by 2 ms pause) at a frequency of 50 Hz (8 mT peak to peak value) produced a significant increase in wound contraction on rats (in comparison with controls) who were exposed to the field for 30 minutes immediately after surgery and for the same period thereafter every 12 hours. In an effort to determine what type of signal is most beneficial for the acceleration of wound healing, skin flap necrosis was observed under exposure to sinusoidal magnetic fields at constant ($\partial B/\partial t$) of 0.5 T/s using frequencies of 20, 72 and 500 Hz.[28] While signals at 20 Hz and 72 Hz significantly decreased necrosis after 7 days, the 500 Hz signal was ineffective.

Very soon after some types of bone injury and pathology were first treated with PEMF, the effects of PEMF on peripheral nerve regeneration became subjects of investigation. Improved neural function that appeared as an unintended "side-effect" in the clinical treatment of non-unions lead Kort, Ito and Bassett to a systematic investigation of PEMF effects on neural regeneration in rats.[29] Other investigators employed other animal models and also compared PEMF with DC current as agents for neural regeneration. More recent work[23] employed a PEMF signal consisting of 20 ms pulses at a repetition rate of 2 pulses per second with exponential rise and decay times of, respectively, 0.5 ms and 1 ms. Amplitudes were 0.3 mT for experiments with crush lesions of the sciatic nerve in rats and 0.05 mT for stimulation *in vitro* of neurite outgrowth in dorsal root ganglia. Estimated values for the induced electric field pulses were 5 mV/m in the animal experiments and 0.7 mV/m in the 60 mm diameter culture dishes of the *in vitro* work. Stimulation for two hours per day of the *in vitro* cultures produced approximately 50 per cent enhancement of neurite outgrowth compared with controls after two days.

The *in vivo* experiments using the 0.3 mT pulse produced "a 22 per cent increase in the rate of regeneration relative to controls" as measured by a standardized test of reflex response.

A very interesting observation was that animals exposed to the 0.3 mT field for four hours per day for seven days *prior* to the crush injury, who received *no* treatment after injury, responded similarly to those treated post injury. Analysis of extracts from sciatic nerve segments after sacrifice of the animals showed that treatment with PEMF changed the molecular weight distribution of synthesized polypeptides.[23]

A report on a very limited (13 subjects) clinical trial[30] of spinal nerve stimulation in para- and quadriplegics employing pulsed electric current introduced by needle electrodes produced "encouraging results" in terms of increased sensory perception and motor function. The signal obtained from a Chinese multipurpose therapy apparatus was described as follows[30]: "The pulsed-wave generator produced a biphasic wave form of 2 ms duration with an initial slow positive deflection followed by an 80-μs rise time to its maximal negative deflection and subsequent asymptotic decay. This wave shape was pulsed in ramped bursts of 200 pulses per second for 1.5 s with 0.75 s rest time before the next ramped burst. Peak-to-peak voltages of approximately 30 V were common, and current flow was on the order of 1 mA."

IV. MECHANISMS

Although clinical effects of electric currents, PEMF and sinusoidal electric and magnetic fields are well documented, and although some specific biochemical results have been obtained *in vitro* or in animal experiments that suggest explanation of bone and soft tissue effects, the mechanism of field-to-cell or field-to-protein transduction is presently not understood. As a consequence optimum "dose" (what field magnitude for how long) and optimum wave shapes or frequencies for particular clinical applications are unknown and dosimetry relies largely on trial and error methods.

It is known that electrokinetic or streaming potentials rather than piezoelectricity make the principal contribution to electric potentials generated by mechanically stressed bone.[31,32] Thus potential differences appear when mechanical loading displaces fluid that contains "counter-ions" which normally reside opposite ions fixed to cell or intercellular matrix surfaces. These potentials are likely to play a role in intercellular signaling and in bone as well as cartilage and soft tissue development. While the original intent of electrical bone therapy was simply to mimic endogenously generated fields, a much wider range of signals were found to be clinically useful. Furthermore, it was found later that some weak ionic currents ($\approx 5 \cdot 10^{-2}$ A/m^2)[32] appear endogenously without mechanical stress and that extremely weak sinusoidal electric fields can produce profound effects on cells *in vitro*. For example, "transcription" (information transfer from DNA to RNA) and "translation" (formation of proteins initiated by RNA) have possibly been modified by electric fields estimated at 10^{-5} V/m which were generated inductively by a sinusoidal 60-Hz magnetic field of 5.7 μT.[33] Similarly, calcium metabolism was affected significantly in mitogen-activated lymphocytes by a 42 μT, 16-Hz magnetic field (in the presence of a 23.4 μT DC magnetic field) which induced an average electric field of about (2) 10^{-5} V/m.[34] Sinusoidal 15-Hz magnetic fields at the 0.5 mT level, giving an estimated mean electric field in the affected tissue of less than 10^{-3} V/m, significantly affected cartilage development in immature rats.[35]

It is likely that the mechanism involved when DC currents are directly applied to injured bone (or other tissue) differs from the transduction sequence that must be acting when low intensity alternating fields are employed. Even the 5 μA continuous current of the implantable devices (F, G in Table 1) when distributed over an (estimated) 5 cm^2 area corresponds to a steady electric field of 1 V/m in bone tissue of 0.01 S/m conductivity. This value is large compared with the average (but not the peak) fields induced by some PEMF devices and very large compared with the average fields induced in tissue by other PEMF devices[36] or the mV/m ELF sinusoidal fields that affect cartilage and bone development.[35,37] For example, if one assumes a mean radius of 4 cm for a particular human bone fracture, one obtains from Equation 2 the electric field peak values between 0.18 and 1.74 V/m shown on Table 2. When electrodes are implanted, as for the DC signals, chemical reactions at the electrodes may play a role in bone and cartilage formation. For example, the reaction at a stainless steel cathode involves consumption of dissolved oxygen and increase in local pH.[32]

If the mechanism of interaction were to involve simple charge transfer by the applied electric field, it would be useful to compare the magnitude of the charge transferred by a single pulse, within a specified volume, with the random charge fluctuation due to thermal excitation during the pulse. An equation due to Einstein[38] gives the mean square value of the charge fluctuation δq in terms of Boltzman's constant k, the absolute temperature T, the conductance G of the current carrying region and the observation time t:

$$\langle \delta q^2 \rangle = 2G\, kTt \tag{3}$$

It is then easy to find the electric field required to transfer during time t charge at least equal to δq over the length of a conductance of volume v [v = length × cross-section area] and uniform conductivity σ. One obtains

$$E \geq \left[\frac{2kT}{v\sigma t}\right]^{1/2} \tag{4}$$

Assuming bone tissue with conductivity 10^{-2} S/m, a physiological temperature of 37°C, an interaction volume of 10^{-14} m^3 (about equal to the volume of a cell with 10 μm radius) and an observation time of 200 μs, equal to the duration of the positive pulse of device A, one obtains from Equation 4 a minimum value of 6.6 V/m. Unless v and σ can be assumed to be much larger, this would indicate that the values given in Table 2 should be below thermal noise. If one considers instead of charge transfer over some as yet unknown path, the voltage induced by the applied field across the membrane of the idealized spherical cell with 10 μm radius, and compares the energy of the repetitive pulse below 100 Hz—where biological action apparently occurs[37]—with the thermal noise voltage ("Johnson noise") given by

$$\langle V_n^2 \rangle = 4\, kTR\, (\Delta f) \tag{5}$$

where (Δf) is the bandwidth and R the transmembrane resistance, one finds again that the electric field due to the PEMF devices would be at best only marginally above thermal noise. The induced electric fields of the *in vitro* experiments and of the "Orthologic" device (I in Table 1) are also clearly below thermal noise.

TABLE 2
Electric Field (V/m) Induced by Bone Therapy Signals at a Radius of 4 cm Into an Electrically Uniform Medium. B Perpendicular to Plane in Which Radius is Defined

	Positive Peak	Negative Peak	Average of Rectified Signal
PEMF device A (Table 1, Figs. 1,2,7)	0.2	1.74	0.024
PEMF device B (Table 1)	0.18	0.015	$(1.4)\ 10^{-4}$
AC-DC device I (Table 1, Fig. 8)	76.6 Hz sinusoid, peak	$\pm\ (1.9)\ 10^{-4}$	$(1.2)\ 10^{-4}$

It is possible to obtain somewhat better signal to noise ratios if one considers either larger interaction volumes (assuming electrical phenomena involving the inter-cell volume), enlongated cells, or cells connected by gap junctions.[39] It is also important to note that in the extremely inhomogeneous biological system, the actual electric field at a particular point can be considerably larger or smaller than its spatial average. Nevertheless very substantial improvement of signal to noise ratios would require signal averaging and limitation of bandwidth by resonance phenomena.[40,41] Weak steady and time varying magnetic fields could also be detected above thermal equilibrium noise be ferrimagnetic single domain particles which have recently been detected in the human brain.[42] However in view of the experimental evidence for biological effects of weak time varying magnetic fields (for examples see references 7, 8, 17, 33, 34), as well as on theoretical grounds, the simple application of Equations 3 to 5 to living systems cannot be justified. These equations were derived for passive systems, i.e. for systems without continuous energy input in thermal equilibrium. Living cells—as contrasted with dead tissue—are not in thermal equilibrium. They depend upon continuous input of chemical signals, as well as energy input by ongoing chemical processes. Living systems also contain non-linear oscillators[43,44] as evidenced by periodic variations (usually "spike" trains rather than sinusoidal) in the concentration of various ions. Some molecules inside cells may at times also be involved in systematic and guided rather than random thermal motion.[45]

Mathematical models of such processes[46,47] predict transitions between different stable states, e.g. from oscillation to no oscillation, or from oscillation at one frequency to oscillation at a different frequency. In the presence of noise the phenomenon of "stochastic resonance" may then become important[48,49,50] when an exogenous, periodic, driving force, such as a weak ELF electric or magnetic field, is applied. If a bistable or multistable nonlinear system is driven by random fluctuations ("noise") as well as by a coherent signal, the output, i.e. the transition between states, involves interaction between noise and input signal. As a consequence, the ratio of the output signal (which, in general, can contain different frequencies than the input signal) to the output noise can become larger than the input signal-to-noise-ratio. What is missing in this picture at the present time, however, is the knowledge of the exact transduction mechanism that converts the time varying magnetic field initially into a chemical signal which is then magnified many times in enzyme assisted chemical reactions.

One interesting suggestion, supported by some experimental evidence, is due to Litovitz et al.[51] who postulate that endogenous noise due to ion motion in the vicinity of chemical receptors at the cell surface is spatially incoherent. Thus, identical receptors at different locations on the cell surface would "see" different noise. An externally applied field would, however, be necessarily coherent over the entire cell (and in fact over many cells and organs). Since the typical cell has not only many different types of receptors, but thousands of each type,[52] a spatially coherent signal recognized by a particular type would be amplified thousand fold upon transmission to the cell interior. The experimental support for this idea is that at least some effects of a weak external single frequency sinusoidal signals can be cancelled by application of a weak external noise signal that appears spatially (but not temporally) coherent at the cell membrane.[51]

Several attempts have been made to construct theoretical models which would explain narrow band resonances in biological systems. Experiments to confirm or reject these hypotheses have thus far given ambiguous results. One theory assumes that ion transfer through cell membranes is affected by cyclotron resonance.[53] It is based on the fact that the cyclotron resonance frequency $\omega_c = 2\pi f_c$ of several physiologically important ions of charge Q and mass m, in the steady magnetic field B_o of the earth, falls into the ELF range:

$$\omega_c = \frac{QB_0}{m} \tag{6}$$

For example, the Ca^{2+} ion has a resonance frequency of 16 Hz in a DC field of 21.0 μT. However, it has been pointed out that the collision frequency in the physiological environment would be very much larger than the cyclotron frequency and would therefore wipe out any resonance motion, that the usually hydrated ions would have a total mass larger than m, and that the energy gain caused by an alternating field of frequency ω_c (as in a cyclotron) would require an orbital radius larger by many orders of magnitude than a typical cell radius.[54] Another theory, due to Lednev, postulates that the binding of Ca^{2+} to the protein calmodulin (ubiquitous in all vertebrates) should be affected by magnetic fields at frequencies ω_c and ω_c/n (where n is an integer).[55–57] This mechanism involves Zeeman splitting at ELF, due to B_0, of infrared vibrational modes that are chemically or thermally excited. Some experiments showing a resonant cell response at the frequency given by equ. (6) could not be replicated while others were performed only over a very narrow frequency range, thereby only confirming other work showing that low intensity magnetic fields at frequencies below 50 Hz affect bone development.[35,37] Nevertheless, apparently successful attempts have been made to stimulate bone cell proliferation with "resonant" field combinations[58] leading to the development of the "Orthologic" device (I in Table 1).

Recently Lednev's "parametric resonance" model was examined in experiments that considered as biological endpoint the number of "PC-12" cells with neurite outgrowths in a magnetic field exposed cell population.[59] The Lednev model,[55,57] as well as a modification proposed by Blanchard and Blackman,[60] predict that the affected biological result should depend—at the resonance frequency in a specified manner (i.e. in proportion to the amplitude of a Bessel function) upon the amplitude ratio of the AC to DC field (which appears as the argument of the Bessel function). The experiments by Blackman et al.[59] seemed to indicate agreement with their version of the theory over a rather wide range of amplitude ratios and at two resonant frequencies (25 and 45 Hz), provided the affected ions were Mg^{2+} and/or V^{4+} rather than Ca^{2+}. The parametric resonance model was also employed to quantitatively fit amplitde and frequency dependence of DC/AC magnetic field elicited changes in opioid mediated analgesia of snails.[65] The results of these experiments provided support for the original Lednev model at DC field magnitudes above 30 μT.

To clarify the sequence of biological events which occur when PEMF signals are applied to developing bone, the following *in vivo* experiment was performed.[61] Twenty five mg of demineralized rat bone matrix in powdered form was implanted along the thoracic musculature of immature rats. This powder recruits cells from the surrounding tissue leading to the formation of cartilage within 6 to 10 days. Thereafter, progressive calcification occurs leading to the formation of fibrous particles by days 12–14 and the formation of a small bone ("ossicle"). These developments were compared in a large number of paired rats, with equal numbers unexposed and exposed (8 hours/day) to the PEMF signal illustrated on Figures 1 and 2 (A in Table 1). Estimates of the mean electric fields in the exposed tissues give values equal to about one fourth of those listed on line one in Table 2. Chemical and histological analysis of ossicles harvested from animals, sacrificed on every second day, showed that exposure to this PEMF signal at the applied level significantly increased both rate and quantity of cartilage formation and enhanced maturation of the subsequent bone. The experimenters concluded that field exposure either enhanced recruitment or proliferation of cartilage precursor cells, increased differentiation of precursor to cartilage cells, or accelerated maturation of cartilage cells.

Getting even closer to fundamental events at the cellular level, both signals A and B (Table 1) were used to expose cultured mouse bone cells and mouse skin fibroblasts, as well as explanted mouse pineal cells in organ culture.[62] In all three cases various chemical procedures were employed to examine "beta-adrenergic receptors." These are cell surface protein strands that span the cell membrane and emerge from it; they mediate cell response to agents such as epinephrine (= adrenaline) or norepinephrine through so-called G proteins which act essentially as molecular amplifiers at the interior surface of the cell. Other G proteins are involved in the response to growth factors. The arrangement of the exposure coils and culture plates was such as to give mean electric fields equal to about one tenth of the values shown in Table 2. Exposures were of four hour duration. Specific types of G-proteins were stimulated by the A-signal, others by the B-signal. The total number of binding sites on the cells was not affected, but the affinity of the receptors for specific hormones was changed, suggesting a change in receptor conformation. It is interesting to compare this work, which employed peak values of the order of 10^{-2} V/m and time average values not greater than 2 (10^{-3}) V/m, with other *in vitro* experiments showing effects on enzyme activity at the cell surface by ELF sinusoidal fields between 5 (10^{-4}) V/m and 30 V/m.[63] Related experiments at higher field intensities and theory of field effects on catalysis[64] also show that electric field action at the exterior of the cell surface can be translated via enzyme catalyzed chemical reactions to the cell interior.

REFERENCES

1. **Fukada, E. and Yasuda, I.,** On the piezoelectric effect in bone, *J. Phys. Soc. Japan,* 12, 1158, 1957.
2. **Bassett, C. A. L. and Becker, R. O.,** Generation of electric potentials by bone in response to mechanical stress, *Science,* 137, 1063, 1963.
3. **Bassett, C. A. L., Pawluk, R. J. and Becker, R. O.,** Effects of electric currents on bone *in vivo,* Nature, 204, 652, 1964.
4. **Bassett, C. A. L., Pawluck, R. J. and Pilla, A. A.,** Augmentation of bone repair by inductively coupled electromagnetic fields, *Science* 184, 575–577, 1974.
5. **Gossling, H. R., Bernstein, R. A. and Abbott, J.,** Treatment of ununited tibial fractures, a comparison of surgery and pulsed electromagnetic fields (PEMF) *Orthopedics* 15-6, 711–719, 1992.
6. **Kraus, W.,** Magnetfeld therapie and magnetisch induzierte Elecktrostimulation in der Orghopadie, *Orthopade,* 13, 78–92, 1984.
7. **Ryaby, J. T., Magee, F. P. and Weinstein, A. M.,** The effect of combined AC/DC magnetic fields on resting articular cartilage metabolism, in *Electricity and Magnetism in Biology and Medicine* Blank, M. Ed., San Francisco Press, San Francisco, 1993. 371–374.
8. **Horton, P., Ryaby, J. T., Magee, F. P. and Weinstein, A. M.,** Stimulation of neural differentiation proteins in PC-12 cells by combined AC/DC magnetic fields, in *Electricity and Magnetism in Biology and Medicine,* Blank M., Ed., San Francisco Press, San Francisco, 1993, 619–622.
9. **Brighton, C. T., Friederberg, Z. B. and Black, J.,** Evaluation of the use of constant direct current in the treatment of nonunion, pp. 519–545 in *Electrical Properties of Bone and Cartilage,* Brighton C. T., Blank J., Pollack S. R., eds., Grune and Stratton, New York, London, Toronto, Sydney, 1979.
10. **Trock, D. H., Bollet, A. J., Dyer, R. H., Fielding, L. P., Miner, W. K. and Markoll, R.,** A double-blind trial of the clinical effect of pulsed electromagnetic fields in osteoarthritis, *J. of Rheumatology,* 20:3, 456–460, 1993.
11. **Bassett, C. A. L.,** Biology of fracture repair, nonunion and pseudoarthrosis, in *Compilations of Fracture Management,* Goosling H.R. and Pillsbury, S.L., eds., J.B. Lippincott, New York, pp 1–8, 1984.
12. **Papoulis, A.,** *The Fourier Integral and its Applications,* McGraw Hill Book Co. Inc., New York, London, Toronto, 1962.
13. **Hsu, H. P.,** *Outline of Fourier Analysis,* Unitech Outlines, Unitech Division, Simon and Schuster Inc.: New York, NY 1967. pp 108, 117, 146–151.
14. **Polk, C. and Song, J. H.,** Electric fields induced by low frequency magnetic fields in inhomogeneous biological structures that are surrounded by an electric insulator, *Bioelectromagnetics,* 11, 235–249, 1990.
15. **Van Amelsfort, A. M. J.,** An analytical algorithm for solving inhomogeneous electromagnetic boundary-value problems for a set of coaxial circular cylinders, James Clerk Maxwell Foundation, Eindhoven, The Netherlands, 1990.
16. **Pollack, S. R. and Brighton, C. T.,** Dosimetry in electrical stimulation, *Bioelectric Repair and Growth Society Transactions IX-40,* 1989.
17. **Fitzsimmons, R. J., Farley, J., Adey, W. R. and Baylink, D. I.,** Embrionic bone matrix formation is increased after exposure to a low amplitude capacitively coupled electric field *in vitro,* Biochimica et Biophysica Acta, 882, 51–56, 1989.

18. **Nerubay, J., Marganit, B., Bubis, J. J., Tadmar, A. and Katzelson, A.,** Stimulation of bone formation by electrical current on spinal fusion, *Spine,* 11, 167, 1986.
19. **Aaron, R. K. and Steinberg, E.,** Electrical stimulation of the femoral head, *Seminars in Arthroplasty,* 2–3, 214–224, 1991.
20. **Zoltan, J. D. and Ryaby, Y. T.,** Exogenous signal generators—a review of the electrical stimulation of bone, *Int. J. Orthopaedic Trauma,* 2:25–30, 1992.
21. **Canady, D. J. and Lee, R. C.,** Scientific basis for clinical applications of electric fields in soft tissue repair, in *Electromagnetics in Medicine and Biology,* Brighton C.T. Pollack S.R., eds, San Francisco Press, 275–280, 1991.
22. **Ito, H. and Bassett, C. A. L.,** Effect of weak, pulsing electromagnetic fields on neural regeneration in the rat, *Clinical Orthopaedics,* 181, 283–290, 1983.
23. **Sisken, B. F., Kanje, M., Lundborg, G. and Kurtz, W.,** Pulsed electromagnetic fields stimulate nerve regeneration *in vitro* and *in vivo, Restorative Neurology and Neuroscience,* 1, 303–309, 1990.
24. **Binder, A., Parr, G., Hazelman, B. and Fitton-Jackson S.,** Pulsed electromagnetic field therapy of persistent rotator cuff tendinitis. A double-blind controlled clinical assessment, *Lancet,* March 31, 1984, 695–698.
25. **Ieran, M., Zaffuto, S., Bagnacani, M., Annovi, M., Moratti, A. and Cadossi, R.,** Effect of low frequency pulsing electromagnetic fields on skin ulcers of venous origin in humans; A double-blind study, *J. Orthop Research,* 8-2, 276–282, 1990.
26. **Bentall, R. H. C.,** Low-level pulsed radiofrequency fields and the treatment of soft-tissue injuries, *Bioelectrochemistry and Bioenergetics,* 16, 531–548, 1986.
27. **Luce, E. A. and Bryant, G. C.,** Dose-response of electromagnetic field current in rat skin flap survival, *Trans. Bioelectrical Repair and Growth Society,* 6, 72, 1986.
28. **Sisken, B. F. and Herbst, E.,** Wound healing: electrical and electromagnetic fields, *Proc. 12th Ann. Intern. Conf. IEEE Engineering in Medicine and Biology Society,* 4/5, 1533, 1990.
29. **Kort, J., Ito H., Bassett, C. A. L.,** Effects of pulsing electromagnetic fields on peripheral nerve regeneration, *J. Bone Jt. Surg. Orthop. Trans.,* 4, 238, 1980.
30. **Ellis, W.,** Pulsed subcutaneous electrical stimulation in spinal cord injury: preliminary results, *Bioelectromagnetics,* 8, 159–164, 1987.
31. **Gross, D. and Williams, W. S.,** Streaming potential and the electromechanical response of physiologically moist bone, *J. Biomechanics,* 15, 227–295, 1982.
32. **Lavine, L. S. and Grodzinsky, A. J.,** Electrical Stimulation of Repair of Bone, *J. Bone Joint Surgery,* 69-4, 626–630, 1987.
33. **Goodman, R. and Henderson, S. A.,** Transcription and translation in cells exposed to extremely low frequency electromagnetic fields, *Bioelectrochemistry and Bioenergetics,* 25, 335–355. 1991.
34. **Yost, M. G. and Liburdy, R. P.,** Time-varying and static magnetic fields act in combination to alter calcium signal transduction in the lymphocyte, *FEBS,* 296-2, 117–122, 1992.
35. **McK Ciombor, D., Aaron, R., Fisher, H., Polk, C., Gautreau, D, and Cherlin, D.,** Effect of 15-Hz sinusoidal magnetic field on cartilage development *in vivo* depends non-linearly on duration of daily stimulation, Project resumes, The Annual Review of Research on Biological Effects of 50 and 60 Hz Electric and Magnetic Fields. (U.S. Dept. of Energy, Office of Energy Management), P-8, 1991.
36. **Rubin, C. T., McLeod, K. J. and Lanyon, L. E.,** Prevention of osteoporosis by pulsed electromagnetic fields, *J. Bone and Joint Surgery,* 71-A (3), 411–418, 1989.
37. **McLeod, K. J. and Rubin, C. T.,** Frequency specific modulation of bone adaptation by induced electric fields, *J. Theor. Biol.,* 145, 385–396, 1990.
38. **Einstein, A.,** *Investigation on the Theory of Brownian Movement,* M.R. Furth and A.D. Cowper, eds. Dover Publications, New York, p 33, 1956 (originally published 1905 in German).
39. **Polk, C.,** Dosimetric extrapolations across biological systems: dosimetry of ELF magnetic fields, *Bioelectromagnetics,* 13-S1, 1992.
40. **Weaver, J. C. and Astumian, R. D.,** The response of living cells to very weak electric fields; the thermal noise limit, *Science,* 247, 459–562 (26 January), 1990.
41. **Adair, K. R.,** Constraints on biological effects of weak extremely low frequency electromagnetic fields, *Physical Review,* A, 1039–1048, 1991.
42. **Kirschvink, J. L., Kobayashi-Kirschvink, A. and Woodford, B. J.,** Magnetic biomineralization in the human brain, *Proc. Natl. Acad. Sci. USA 89 (16),* 7683–7687, 1992.
43. **Hartoonian, A. T., Kao, J. P. Y. Paranjape, S. and Tsien, R. Y.,** Generation of calcium oscillations in fibroblasts by positive feedback between calcium and IP_3. *Science,* 251, 75–77, 4 January, 1991.
44. **Jacobs, R.,** Calcium oscillation in electrically non-excitable cells, *Biochimica et Biophysica Acta,* 1052, 427–438, 1990.
45. **Hoffman, M.,** Motor molecules on the move, *Science,* 256 (26 June), 1758–1760, 1992.
46. **Meyer, T, and Stryer, L.,** Molecular model for receptor-stimulated calcium spiking, *Proc. Natl. Acad. Sci. USA,* 85:5051–5055 (Biochemistry), July 1988.

47. **Eichwald, C., Kaiser, F.,** Model for receptor-controlled cytosolic calcium oscillations and for external influences on the signal pathway, *Biophys.* J., 65, 2047–2058, Nov. 1993.
48. **Benzi, R., Parisi, G., Sutera, A., Vulpiani, A.,** Stochastic resonance in climatic change, *Tellus,* 34, 10–16, 1982.
49. **McNamara, B., Wiesenfeld, K.,** Theory of stochastic resonance, *Phys. Rev.,* A 39 (9) 4854–4869, 1989.
50. **Wiesenfeld, K., Moss, F.,** Stochastic resonance and the benefits of noise: from ice ages to crayfish and squids. Nature 373, 33–36, Jan. 5, 1995.
51. **Litovitz, T. A., Montrose, C. J., Doinov, P., Brown, K. M. and Barber, M.,** Superimposing spatially coherent electromagnetic noise inhibits field-induced abnormalities in developing chick embryos, *Bioelectomagenetics,* 15, 1994, 105–113.
52. **Lauffenburger, D. A. and Linderman, J. J.,** *Receptors, Models for Binding, Trafficking and Signaling,* Oxford University Press, New York, Oxford, 1993, p. 170.
53. **Liboff, A. R. and McLeod, B. R.,** Kinetics of channelized membrane ions in magnetic fields, *Bioelectromagnetics,* 9, 39–51 (1988).
54. **Polk, C.,** Physical mechanisms by which low-Frequency magnetic fields can affect the distribution of counterions on cylindrical biological cell surfaces, *J. Biol. Phys.,* 14, 3–8, 1986.
55. **Lednev, V. V.,** Possible mechanism for the influence of weak magnetic fields on biological systems, *Bioelectromagnetics,* 12, 71–75, 1991.
56. **Adair, R. K.,** Criticism of Lednev's mechanism for the influence of weak magnetic fields on biological systems, *Bioelectromagnetics,* 13, 231–235, 1992.
57. **Lednev, V. V.,** Possible mechanism for the effect of weak magnetic fields on biological systems: correction of the basic expression and its consequences, in *Electricity and Magnetism in Biology and Medicine,* Blank M., Ed., San Francisco Press, San Francisco, 1993. 550–552.
58. **Fitzsimmons, R., Baylink, D., Magee, F. P. and Weinstein, A. M.,** Electromagnetic field stimulated bone cell proliferation (Abstract), *J. Bone and Mineral Research,* 6 (Supplement 1), August 1991.
59. **Blackman, C. F., Blanchard, J. P., Benane, S. G., House D. E.,** Empirical test of an ion parametric resonance model for magnetic field interactions with PC-12 cells, *Bioelectromagnetics,* 15, (4), 1994.
60. **Blanchard, J. P. and Blackman, C. F.,** Clarification and application of an ion parametric resonance model for magnetic field interaction with biological systems, *Bioelectomagnetics,* 15, (4), 1994.
61. **Aaron, R. K., Ciombor, D. M. and Grant, J.,** Stimulation of experimental endochondral ossification by low-energy pulsing electromagnetic fields, *J. Bone and Mineral Research,* 4-2, 227–233, 1989.
62. **Luben, R. A.,** Effects of low energy electromagnetic fields on signal transduction by G protein linked receptors, in *Electricity and Magnetism in Biology and Medicine,* Blank M., Ed. San Francisco Press, San Francisco, 1993. 57–62.
63. **Blank, M.,** Na, K-AT Pase function in alternating electric fields, *FASEB Journal,* 6, 2434–2438 (April), 1992.
64. **Robertson, B. and Astumian, R. D.,** Frequency dependence of catalyzed reactions in a weak oscillating field, *J. Chem. Phys.,* 94, 7414–7419, 1991.
65. **Prato, F. S., Carson, J. J. L., Ossenkopp, K.-P., Kavaliers, M.,** Possible mechanisms by which extremely low frequency magnetic fields affect opioid function, FASEB Journal, 9, 807–813, June, 1995.

ADDITIONAL SOURCES OF INFORMATION

Adey, W. R., Electromagnetic fields, cell membrane amplification, and cancer promotion, pp. 211–249 in *Extremely Low Frequency Electromagnetic Fields: The Question of Cancer,* Wilson B. W., Stevens R. G., Anderson L. E., eds. Battelle Press, Columbus, OH 1990.

Bioelectromagnetics, the bi-monthly (formerly quarterly) journal of the Bioelectromagnetics Society, Published by Wiley-Liss, New York, since 1980.

Black, J., *Electrical Stimulation,* Praeger Inc. New York, 1987.

Branden, C. and Tooze J., *Introduction to Protein Structure,* Garland Publishing Inc., New York 1991.

Brighton C. T. Black, J. and Pollack, S. (eds.), *Electrical Properties of Bone and Cartilage,* Grune and Stratton, New York, London, Toronto, Sidney, 1979.

Brighton C. T. and Pollack, S. R. (eds.), *Electromagnetics in Medicine and Biology,* San Francisco Press, 1991.

Transactions of the Bioelectric Repair and Growth Society, Vols. I through XIII, published annually since 1980 by the Bioelectric Repair and Growth Society, P.O. Box 64, Dresher, Pennsylvania 19025.

Chapter 6

ELECTROPORATION

James C. Weaver and Yuri Chizmadzhev

CONTENTS

I. BACKGROUND

Nature has evolved significant barriers to the movement of ions and charged molecules in biological systems. One of the most fundamental is the phospholipid bilayer-based cell membrane, which separates the intra- and extracellular regions. Other significant barriers consist of one or more layers of cells connected by tight junctions around bladders and ducts (retention of specialized fluids), and the tough, flexible skin of animals (prevention of water loss, entry of toxic molecules and infectious agents). The phenomenon of electroporation

0-8493-0641-8/96/$0.00+$.50

results in an essentially universal physical reduction of such barriers, by creating aqueous pathways (large dielectric constant $K_w \approx 80$) across lipid- containing (small dielectric constant $K_m \approx 2$) barriers. Almost all of the electroporation studies to date have focused on bilayer membranes, both artificial planar bilayer membranes and cell membranes. In bilayer membranes, electroporation occurs under biochemically mild conditions, usually with a small temperature rise. For most easily observable phenomena such as cell transfection and molecular uptake, electroporation-related phenomena are believed to depend non-linearly on the transmembrane voltage, U(t). A duration-dependent onset within the range $0.2 < U < 1$ V is usually found for single short (10^{-6} to 10^{-3} sec) pulses. Both dramatic electrical behavior ("reversible electrical breakdown" = REB for cells) and significant molecular transport occur. Most interest to date has focused on the electropermeabilization that occurs under REB conditions, which involves a large increase in membrane permeability to essentially all ions and molecules. Other consequences of electroporation are electrofusion of cells to other cells or tissue, and electro-insertion of membrane proteins into cell membranes.

The ability of a lipid-containing sheet to exclude ions and charged molecules is fundamental, and is a result of the change in "Born energy" associated with moving a charge from a high dielectric medium (e.g. water) to a low dielectric region (e.g. lipid). The Born energy is the electrostatic energy required to assemble a particular configuration of charge, and is written here in terms of the electric field, E, and the permittivity, $\epsilon = K\epsilon_0$ (K is the dielectric constant and $\epsilon_0 = 8.85 \times 10^{-12}$ F/m).[1]

$$W_{Born} = \int_{\text{all space excluding ion}} 0.5\,\epsilon E^2\, dV \tag{1}$$

The essential barrier function of cell membranes can be represented by a thin sheet of lipid. This allows the magnitude of the Born energy barrier, ΔW_{Born} to be computed by the following process: a charged sphere (representing the ion or molecule) is initially located in water far from the lipid sheet ($W_{Born,i}$) and then moved to the center of the sheet ($W_{Born,f}$), which requires the expenditure of energy. The barrier height, $\Delta W \equiv W_{Born,f} - W_{Born,i}$ is large for small monovalent ions (e.g. Na^+ and Cl^-), is larger for multivalent charged molecules, and depends on the membrane thickness, sphere radius and the amount and distribution of charge within the sphere.

For a single, isolated charge such as a small ion (e.g. Na^+ ion), the largest contribution to W_{Born} arises from the region close to the ion. The small ion diameter ($2r_s \approx 0.4$ nm) is significantly smaller than a typical membrane thickness ($d \approx 6$ nm). In this case W_{Born} can be estimated by calculating the change in energy to move the ion from bulk water ($K_w \approx 80$) to bulk lipid ($K_m \approx 2$)

$$W_{Born} \approx \frac{e^2}{8\pi\epsilon_0 r_s}\left[\frac{1}{K_m} - \frac{1}{K_w}\right] \approx 65kT \tag{2}$$

where T = 37°C = 310 K was used, and $e = 1.6 \times 10^{-19}$ C. A numerical solution as a function of position for a thin low dielectric constant sheet agrees with this simple estimate (Figure 1). Spontaneous barrier crossing is therefore negligible. An early calculation considered not only the "intact sheet" case, but also the cases of a mobile aqueous cavity and a fixed cylindrical pore.[2] Both aqueous configurations lowered ΔW, but the greater reduction was achieved by the pore. This basic structure, penetration of the lipid membrane by an aqueous pathway, is of course the essence of gated channels. It is also the basis of electroporation.

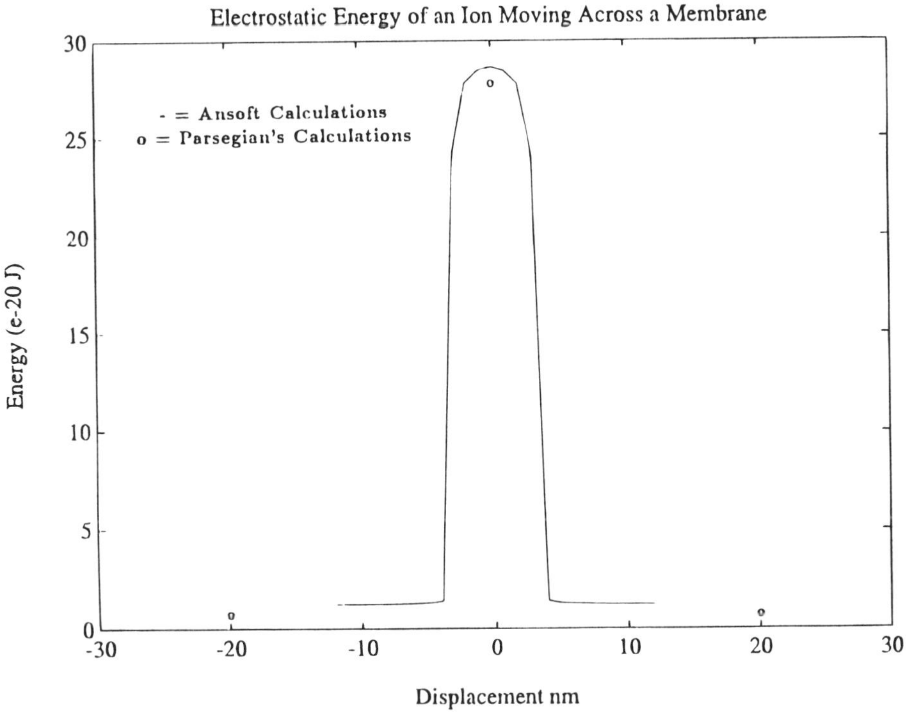

FIGURE 1 The electrostatic energy, W_{Born}, as a function of ion displacement for a small ion which is represented by a 0.2 nm radius sphere with charge $q = e = 1.6 \times 10^{-19}$ C.[90] The pore-free membrane is represented by a circular patch of low dielectric constant ($K_m = 2$) material with thickness $d = 7$ nm. This result for the Born energy barrier was obtained by using commercially available software (Ansoft, Inc., Pittsburgh, PA) to solve Poisson's equation for a continuum model consisting of a circular patch of this low dielectric constant material immersed in water ($K_w = 80$). The values plotted here as a solid line ("—Ansoft Calculations") are in good agreement with Parsegian's original calculation **o** ("Parsegian's Calculations."[2]) As predicted by the simple estimate of equation 2, the barrier is large, viz. $\Delta W \approx 2.8 \times 10^{-19}$ J $\approx$ 65 kT. The appearance of aqueous pathways ("pores"; Fig. 2) provides a large reduction in this barrier. (Reproduced with permission from reference 90).

To our knowledge, the first experimental observation of electroporation-related phenomena was the "breakdown" of the excitable membrane of the node of Ranvier.[3] Almost a decade later the nonthermal killing of microorganisms by electric field pulses was reported,[4] followed by the observation of a large molecular permeability increase in natural vesicles.[5] Subsequently there were increasing numbers of reports which described dramatic electrical behavior,[6–12] and indications of significantly increased molecular transport.[13–15] Since this early period, there has been a steady increase in imaginative manipulations of cells by electroporation, far too many to cite individually. Recent books and reviews should be consulted to appreciate the scope of recent work.[16–21]

An important feature of artificial and cell membranes is that they concentrate potential drops, and by this means amplify applied electric fields. In the case of artificial planar bilayer membranes, the membrane completely blocks the current pathway, so that the entire voltage across the electrodes appears (after a characteristic charging time) across the membrane. In the case of cells, the situation is more complicated, but an approximate guide is obtained for the case of a spherical cell for which $U_{max} = 1.5Er_{cell}$, which shows that the change in transmembrane voltage is given approximately by the product of the electric field and the cell size.

The usual pulse conditions involve large electric field magnitudes and relatively low frequencies. The field pulses typically have magnitudes of 10^5 to 10^6 V/m in aqueous

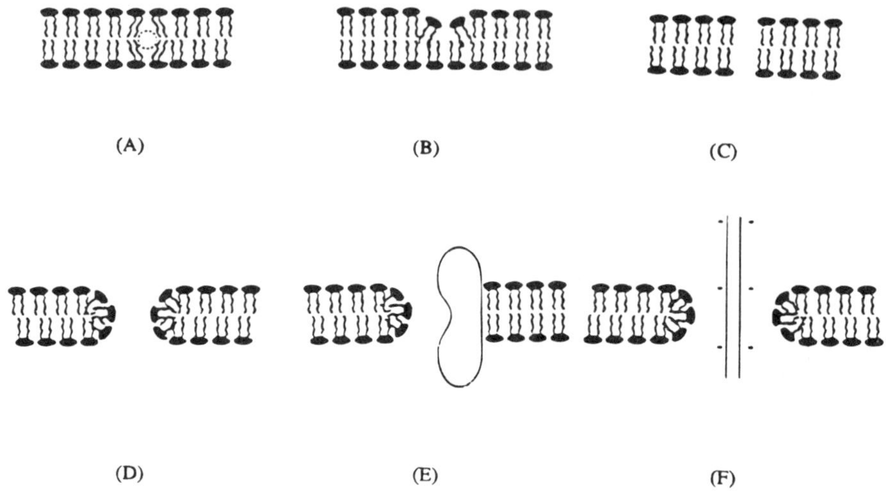

FIGURE 2 Drawings of hypothetical structures of both transient and metastable membrane conformations that may be involved in electroporation.[20] A: membrane free volume fluctuation[143], B: aqueous protrusion into the membrane ("dimple")[144,145] C: hydrophobic pore transient structure,[6] D: hydrophilic pore transient structure,[6,37,38] these are usually regarded as the "primary pores" through which ion and molecules pass, E: composite pore with one or more proteins at the pore's inner edge,[40] F: composite pore with "foot-in-the-door" charged macromolecule inserted into a hydrophilic pore.[141] Although the actual transitions are not known, the transient aqueous pore model assumes that transitions from A → B → C or D occur with increasing frequency as U is increased. Type F may form by entry of a tethered macromolecule during the time that U is significantly elevated, and then persist after U has decayed to a small value because of pore conduction. These hypothetical structures have not been directly observed, and indeed it is unlikely that they can be visualized by known forms of microscopy.[141] Instead, evidence for them comes from interpretation of a variety of experiments involving electrical, optical, mechanical and molecular transport behavior. (Reproduced with permission from reference 20).

extracellular media. As discussed below, most cells have characteristic linear dimensions of $L_{cell} \approx 10^{-6}$ to 10^{-5} m, so that such pulses result in the transmembrane voltage U(t), reaching 0.5 to 1.5 V. Exponential pulses ($E = E_0\, e^{-t/\tau}$) and square pulses ($E = E_0$ or 0) with a duration τ_{pulse} of 10^{-4} sec and longer are predominantly used. Neither other waveforms (e.g. gated radio frequency pulses) nor multiple pulses have received nearly as much attention.

II. PLANAR MEMBRANE ELECTROMECHANICAL PHENOMENA

Artificial planar bilayer membranes are widely used to investigate basic aspects of the bilayer portion of cell membranes,[22] and provide an electric field amplification by the factor $L_{electrode}/d$, where $L_{electrode}$ is the electrode separation and d is the membrane thickness. Within a charging transient characterized by a time constant $\tau_{CHG} = R_E C_m$, the electric field within the membrane reaches $E_m = U/d$. Here R_E is the combined resistance of the electrolyte and electrodes, and C_m is the BLM capacitance. For example, if one-half volt is applied to the electrodes, U reaches 0.5 V within τ_{CHG}. An advantage of such membranes for mechanism studies is that several measurements can be made on the same preparation.

Creation of these planar membranes involves use of relatively large amounts of organic solvent, or use of previously formed monolayers on aqueous solutions.[23] Unlike their cellular counterparts, these membranes are essentially macroscopic, often spanning a circular aperture ($D_{aperture} \approx 10^{-3}$ m; membrane area is $A_m = \pi D^2_{aperature}/4 \approx 7 \times 10^{-7}$ m^2) in an electrically insulating septum. The convenient access to the chambers on both sides of the septum allows electrical measurements to be readily made, and the macroscopic chamber size results in a small electrical resistance ($R_E \approx 100$ ohm) between the membrane and macroscopic electrodes located about 10^{-2}m away. Bilayer membranes have a capacitance per area of about $C_{area} \approx 10^{-2}$ F/m^2, and therefore a typical BLM has a capacitance of order $C_M \approx 3 \times 10^{-8}$ F. As a result, the charging time constant is small, $\tau_{CHG} = R_E C_m \approx 3 \times 10^{-6}$ sec. For this reason, BLM experiments have excellent time resolution for electrical measurements.

Measurements of the transmembrane current, I(t) following a step change in the applied voltage, V_0 allows the induced membrane conductance to be followed. In such experiments it is found that during its lifetime a membrane can pass through up to four distinguishable stages. The first stage is a simple charging of the membrane, governed by the τ_{CHG} given above. The second stage is characterized by a constant voltage and associated (small) current. The third stage has a fluctuating current, but sometimes in this stage the membrane reverts to a quiet, steady current state. If a large voltage is applied to the membrane for a short time, then the intact membrane often makes a transition into a peculiar long-lived (tens of minutes) excited state that is characterized by a large conductance and pronounced current fluctuations, even at low U. This peculiar state was termed a "stress state."[24] During a fourth stage is often reached, in which I(t) exhibits a drastic irreversible increase to saturation. Visual inspection shows that at saturation the membrane has ruptured, i.e. the membrane material no longer spans the aperture, but instead has collected on the aperture rim.

If a new membrane is made and the experiment repeated, then the next I(t) is reproduced qualitatively, but the membrane lifetime, τ_m (defined as the time from the step voltage to the onset of a drastic current rise), is usually different. The character of the fluctuations, and the duration of the fluctuating stage, are also different. In short, the lifetime exhibits stochastic behavior. As discussed below, this stochastic behavior can be explained by the hypothesis of transient aqueous pores.

To elucidate the mechanism of membrane rupture, the dependence of the mean membrane lifetime, $\bar{\tau}_m$, on various factors was studied.[6–12] Of these, the effect of varying U was most revealing. In the range 300 to 600 mV the dependence of $\log_{10} \bar{\tau}$ on U is usually almost linear, such that an increase in U by 100 mV causes a tenfold decrease in $\bar{\tau}$. At larger transmembrane voltages the decrease in $\bar{\tau}$ with increasing U becomes less pronounced. For example, if U is increased from 200 mV to 1.4 V, the mean lifetime changes by more than six orders of magnitude, decreasing to about 10μsec. This highly nonlinear dependence on U is not expected for electrostrictive mechanisms, and provides key support for the transient aqueous pore hypothesis.

Rupture (irreversible breakdown) was also studied by the charge injection technique.[7] In these experiments, the membrane was charged quickly (≈400 nsec) to about 0.4 V. After 300 to 400 μsec, U declines to zero over an interval of about 50μsec because of rupture. Many features of rupture following charge injection can be described quantitatively by a transient aqueous pore model.[25]

One of the most dramatic aspects of electroporation is so-called "reversible electrical breakdown" (REB), which is actually a high conductance state that acts to protect the membrane through a rapid discharge. Planar membranes made of oxidized cholesterol have been studied using the "charge injection" method.[7] If such membranes are rapidly (400 ns) charged to about 1 V, the membrane resistance reversibly decreases by nearly nine orders of magnitude. In a typical experiment the membrane voltage reached 0.9 and 1.2 V. Significantly, the value U ≈ 1.2 V cannot be exceeded by increasing the charging pulse amplitude further. After a rapid discharge, the membrane survives, remains mechanically stable, and can be recharged. This was the first observation of REB in planar bilayer membranes. Later it was also studied using a voltage-clamp method.[26] It is very important to recognize that electroporation of cells also exhibits reversible behavior, and that is similar to the behavior of a planar membrane with oxidized cholesterol[7] or UO_2^{++}.[12]

III. CELL MEMBRANE ELECTROMECHANICAL PHENOMENA

It is an understatement that a cell membrane is much more complex than a planar bilayer membrane. But even if the detailed biochemical composition is neglected, there are still important physical differences. These include the membrane's non-planar shape, the closed membrane topology, the non-zero transmembrane voltage due to pump activity, and the presence of membrane proteins, and membrane cytoskeleton interactions. Although real cells range from spherical to elongated cylinders to irregular three dimensional shapes, the portion of the membrane which experiences most of electroporation is often believed to be a subsystem which can be reasonably approximated by a simpler geometry. Consider the most commonly treated cell shape (spherical) for the small electric field case. As is well known, for small fields a spherical cell develops a change in transmembrane voltage, U(t), which varies in position over the membrane and also with time:[27]

$$U(t) = \frac{1.5 E r_{cell}}{1 + r_{cell} G(\rho_{in} + 0.5\rho_{out})} [1 - \exp(-t/\tau_{cell})] \cos\theta \tag{3}$$

Here r_{cell} is the cell's radius, ρ_{in} is the intracellular medium resistivity, ρ_{out} is the intracellular medium resistivity, G_m = constant is the average (no significant transient aqueous pores) membrane conductance, θ is the angle between the applied electric field, $E_e = \vec{E}_0 e^{jwt}$, and the site on the cell membrane at which U is measured, and the time constant associated with charging the cell membrane is

$$\tau_{cell} = \frac{r_{cell} G_m(\rho_{in} + 0.5\rho_{out})}{1 + r_{cell} G_m(\rho_{in} + 0.5\rho_{out})} \tag{4}$$

Typical mammalian cells with $r_{cell} \approx 10^{-5}$ m have $\tau_{cell} \approx 10^{-6}$ sec. However, it should be strongly emphasized that equation 3 does not apply if the membrane conductivity varies with position or time. Yet this is what occurs during electroporation; G_m varies dramatically with time at the sites of large U, but hardly at all at other sites. This has been convincingly established experimentally by sub-microsecond fluorescence measurements that respond to the transmembrane voltage.[28] Moreover, deformation of spherical membranes is expected,[29] and have been observed.[30] For these fundamental reasons, theoretical models which persist in using equation 3 under electroporation conditions should be viewed with suspicion.

Two different membrane destruction processes have been identified. In the first, a mechanically bounded portion of the cell membrane, e.g. a region bounded by cytoskeletal elements, may behave like a small planar membrane, and may therefore exhibit prompt rupture. In this case the behavior of a BLM is relevant. A topologically closed membrane (e.g. a vesicle) is expected to not rupture globally, as there is no boundary at which membrane phospholipids can accumulate upon expansion of a large pore.[31]

The effective tension, Γ_{eff}, within the cell membrane is also important, and is expected to be small for some cells or vesicles, but large for others, particularly if an osmotic pressure difference, $\Delta\pi$, exists across the membrane. In the case of BLM the tension is due to the meniscus at the aperture, and is a constant during pore expansion. For cells, however, the membrane tension is associated with osmotic pressure through Laplace's Law. Following electroporation, $\Delta\pi$ (if it exists) goes to zero, and the tension, Γ, also goes to zero. In this case the evolution of pores is more complicated, and involves the laws of colloid osmotic lysis. In the second, reversible electroporation may lead to significant molecular transport between the extra- and intracellular volumes, with the ensuing chemical imbalance leading to cell stress and eventual lysis. Instead, large pores are favored for large transmembrane voltages, but as the membrane discharges through the pores (REB), the vesicular membrane returns to its initial state.[31] This inability to promptly rupture via critical pores probably applies generally to cell membranes.

Cell membranes are believed to exhibit reversible electrical breakdown, i.e. a tremendous increase in electrical conduction which can be measured, and which is believed due to ionic conduction through transient aqueous pores. Electric field pulses which cause U to rapidly rise into the range 0.5 to 1.5 V cause onset of reversible electrical breakdown, with essentially no dependence on membrane composition. In short, the onset of electroporation involving REB is universal, as it has been observed for a wide variety of cell types. The behavior of the transmembrane voltage, U(t), during membrane charging, and the subsequent appearance and evolution of a pore population, is intimately connected with the number and size of the pores. The success of a transient aqueous pore model in providing a quantitative description of U(t) under these conditions provides confidence that electroporation is a valid concept. Significantly, for essentially the same conditions that cause REB, a large increase in molecular transport across cell membranes is found.

IV. NON-PORE THEORIES OF ELECTROMECHANICAL PHENOMENA

The first theoretical approach to the problem of planar membrane stability at elevated U was based on an electromechanical collapse due to compression of membrane.[32] This model has several important attributes. First, it is deterministic, predicting a critical transmembrane voltage, U_c, above which rupture occurs. For realistic values for the membrane compressibility of solvent-free membranes, this model predicts $U_c \approx 5000$ mV, which is about an order of magnitude too large.[33] Further, the absence of a marked change in C before rupture argues strongly against large scale electro-compression, as an increase in C is an inevitable consequence of a decrease in membrane thickness, h. Moreover, the observed stochastic nature of rupture is in direct conflict with the concept of a deterministic critical voltage, U_c. Neither the stochastic nature, nor the strong lifetime dependence on U, are expected for an electromechanical mechanism. Finally, the fate of the membrane in this model depends only on U, but experiments also show a dependence on pulse duration.

Related models based on viscoelastic behavior have also been advanced[34,35] but are essentially the same as the electrocompression approach.[32] The only difference is that the electrocompression model attributes the increase in system energy to elastic compression energy of the membrane, whereas the electrohydrodynamic model it corresponds to the work required to form new membrane surface. In both models the development of instability ("electric breakdown" or "rupture") represents a nonlocal process that occurs simultaneously over a large area of the membrane. In contrast, pore models involved localized events that involve only a small fraction of the total membrane area.[36]

V. PORE THEORIES OF ELECTROMECHANICAL PHENOMENA

The entry of water and aqueous electrolytes (large dielectric constant) into a fluid bilayer membrane (small dielectric constant) is increasingly favored as the electric field within the lipid region of the bilayer becomes larger. This is the basis of the transient aqueous pore theoretical model, originally introduced in a series of papers, of which only the first is cited here.[6] This involved an estimate of the free energy change, $\Delta W(r, U)$, associated with formation of a large dielectric constant aqueous pore within the small dielectric constant lipid portion of the membrane. The initial, simplest form neglected the "spreading resistance" and "Born energy repulsion" which were included in some later versions, and gave

$$\Delta W(r, U) = 2\pi\gamma r - \pi\Gamma r^2 - 0.5C_{LW}U^2\pi r^2 \quad (5)$$

where γ is the linear energy density associated with a pore edge, Γ is the area energy density of a pore-free planar membrane, and $C_{LW} = \epsilon_0[K_w - K_m]/d$ is the difference in specific capacitance associated with replacement of lipid by water. At zero transmembrane voltage, $\Delta W(r)$ is a parabola with a large maximum, more than 100 kT, consistent with infrequent spontaneous rupture. The initial suggestion that pore formation could occur and lead to membrane rupture did not involve electrical behavior. Instead, the possibility of rupture due to spontaneous pore formation was independently suggested by two groups.[37,38] The basic idea was that a "cookie cutter" model for a pore formation energy, $\Delta W(r)$ could be used, based on a gain in "edge energy," $2\pi r\gamma$, as a pore is created, and a simultaneous reduction in energy, $\pi r^2\Gamma$, due to the loss of a circular region of membrane (Fig. 3). The interpretation is simple: a pore-free membrane is envisioned, then a circular region is cut out of the membrane, and the difference in energy between these two states calculated, and identified

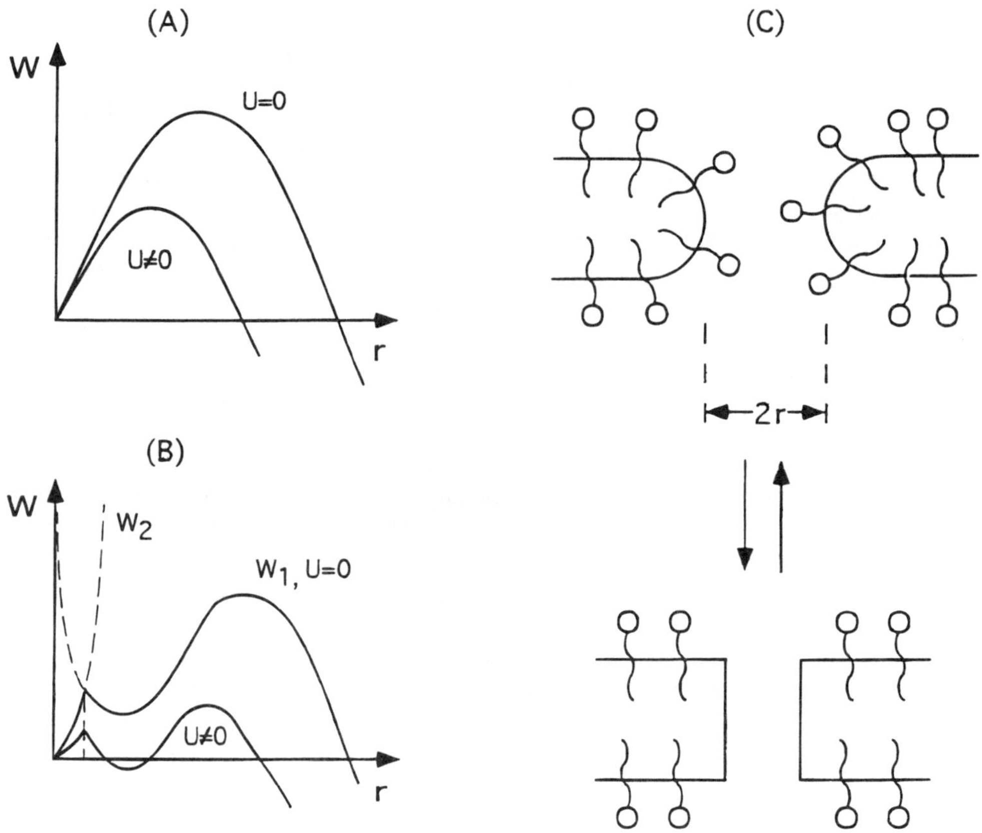

FIGURE 3 Formation energy $\Delta W(U,r)$, of a membrane containing a single pore of radius r, with both plotted in arbitrary units. (A) Simplified version of $\Delta W(U,r)$ for a hydrophilic pore for zero transmembrane voltage ($U = 0$) and an elevated value ($U \neq 0$) for which the energy is reduced. (B) Simplified illustration of two branches for pore energies. The curve ΔW_1 corresponds to a hydrophilic pore, and curve ΔW_2 to a hydrophobic pore. Again the cases of $U = 0$ and $U \neq 0$ are illustrated. (C) Schematic illustration of the two basic pore types, a hydrophilic pore (upper) and hydrophobic pore (lower). Detailed discussions of these basic concepts can be found in the text and in the cited papers.

as ΔW. Here γ is the energy per length along the circumference needed to form the pore, and Γ is the surface energy per area of a flat, pore-free membrane.

The change in pore energy due to an elevated transmembrane voltage is contained in the third term of the right-hand side of equation 5. The membrane is regarded as a capacitor, with this term representing the change in electrical energy resulting from replacing membrane material with dielectric constant K_m by water with dielectric constant K_w in a cylindrical pore of radius "r." In other words, creation of a hydrophilic pore (hereafter simply called "pore") is described as a change in specific capacitance, C_{LW}, in the region occupied by the pore.[6] It was recognized that the existence of ions within small pores in a low dielectric membrane is improbable, and for this reason the pore can be represented by a water-filled, rather than electrolyte-filled, capacitor. However, for small hydrophilic pores, even if bulk electrolyte exists within the pores, the permittivity would be $\epsilon \approx 70\epsilon_0$, only about ten percent smaller than that of pure water. Moreover, the pore resistance would still be large, $R_p = \rho_e h/\pi r^2$, compared to the spreading resistance (see below), so the voltage across the pore would be very close to U.

Although each of the "ingredients" of the pore model are plausible, if they are combined to represent a complete experimental situation, the overall model is too complex to solve in closed form. However, such models can be run as computer simulations that allow predictions to be generated, and these results can be compared with experimental behavior.[25,39–41] To date simulations for artificial planar bilayer membranes have been developed. They produce reasonable, but not completely correct, quantitative descriptions of measurable quantities such as the transmembrane voltage, U(t), the membrane conductance, G(t), which increases tremendously during REB, and the molecular transport of small charged molecules. A single simulation run with only different pulse magnitudes predicts the four distinguishable outcomes that are found experimentally for an oxidized cholesterol planar membrane stimulated with a 0.4μsecond square charging pulse: (1) simple charging of the membrane capacitance for small pulses, (2) charging to about 0.5 V, followed by a sigmoidal decay of U(t) as the membrane ruptures for a moderate pulse, (3) incomplete reversible electrical breakdown in which the membrane achieves a high conductance state, but not enough to discharge the membrane fully to zero, for a larger pulse, and (4) reversible electrical breakdown due to a still larger but transient high conductance that discharges the membrane completely, for still larger pulses. This complex behavior reflects the interplay between the rapidly changing electrical conductance and the mechanical expansion and contraction of the individual pores in a heterogeneous, dynamic pore population.

Moreover, an extension of the single artificial planar bilayer membrane to two identical membranes in series can be used as a very simple model of a "cubic cell." By using a very low value of Γ the inability of a spherical membrane to rupture can be represented, and this allows the behavior of cell membranes to longer pulses to be explored. Such an approach is ongoing, but generates predictions of U(t) and the total transport of small charged molecules, N_s, that are reasonable.[40,41] This approach has also generated a surprise, the prediction of an approximate plateau in N_s as a function of exponential pulse amplitude for small molecules such as calcein.[41] While too simple to fully represent cellular electroporation, the prediction of this experimentally observed behavior (see "Molecular Transport" section) is encouraging.

The various theoretical models support the idea that the reversibility of electroporation in planar membranes is due to several factors: (1) the chemical composition of the membrane, (2) the pulse protocol, and (3) feedback at the pore level that is associated with the spreading resistance, such that as a pore expands, the local transmembrane voltage across the pore decreases. Overall, considerable progress has been made in devising theoretical models that can quantitatively describe some aspects of electroporation, while others are very poorly understood. Mechanical and electrical aspects of electroporation of a relatively simple artificial planar bilayer membrane do account for: (1) the approximate magnitude of U at which membrane rupture (irreversible breakdown) occurs, (2) the steep dependence of membrane lifetime on U, and (3) the stochastic nature of rupture. Moreover, a reasonable but not exact quantitative description of U(t) is given by a transient aqueous pore theory, and the maximum fractional aqueous area of the membrane during reversible electroporation is correctly explained. There has been some progress in predicting molecular transport, but much remains to be done, and the mechanisms of membrane recovery, cellular stress and the determinants of cell survival or death are essentially not understood. Moreover, single pulse electroporation has been emphasized to date, and relatively little is known about the importance of pulse shape, repetition and spacing of multiple pulses.

VI. MOLECULAR TRANSPORT

Greatly enhanced molecular transport is the basis of most applications of electroporation, as delivery of biochemicals and indicators into the otherwise inaccessible regions of cells and tissues is a fundamental manipulation, with many potential applications in biology, biotechnology and medicine. Qualitatively, the basic idea is simple: an elevated U leads to pore formation in the bilayer portion of membranes, and water soluble molecules move though pores under the influence of local driving forces (Figure 4). Quantitatively, however, far less is known.

Early studies reporting molecular release from biological vesicles[5] or molecular uptake into cells used large numbers of cells, and determined the average molecular transport by measuring a total cell population effect. These studies showed that greatly enhanced molecular transport occurred for approximately the same conditions that cause REB. This was an

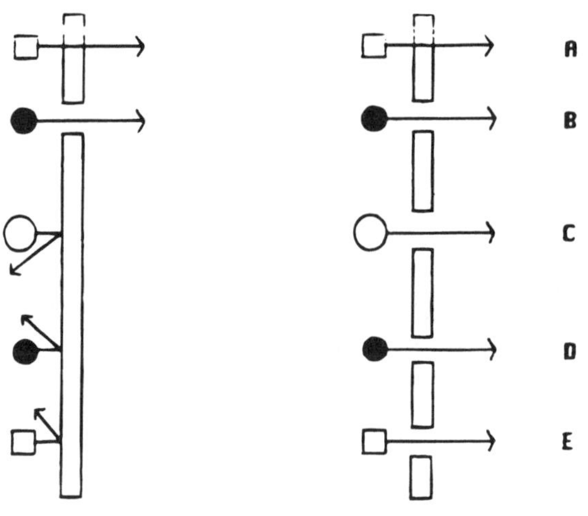

FIGURE 4 Schematicized drawing of molecular transport pathways through pre-existing and new aqueous pathways in lipid-containing barriers such as artificial planar bilayer membranes, cell membranes, cell monolayers and the stratum corneum (SC) of mammalian skin. Left: Unpulsed lipid-based barrier shown with [A] a lipid soluble molecule with some water solubility (□), which passes through the barrier by dissolution and diffusion, and redissolves in the aqueous region on the right side of the barrier. [B] is a pre-existing aqueous pathway, which can accommodate a small charged molecule (●) but not a larger charged molecule (○); it might also accommodate a small lipid soluble molecule that has some water solubility [not shown]. Right: Pulsed lipid-based barrier in which two newly created aqueous pathways are shown: [C] is larger than [B], a pre-existing aqueous pathway, and [D] is the same size as [B]. [E] provides increased transport of a lipid soluble molecule that is also water soluble. Measurement of two fluxes simultaneously for the same barrier of molecules with different sizes, charge and water and lipid solubilities should reveal and partly characterize both pre-existing and new aqueous pathways. Reproduced with permission from Weaver, J. C. (1994) Abstract Proced. Intern. Symp. Control. Ref. Bioact. Mater. 21 Controlled Release Society pp 15–16.

important finding, but did not then address the question of how much molecular transport occurred for various pulsing conditions, nor was the question of cell-cell variation considered. Some subsequent experiments used appropriate calibrations to obtain quantitative net transport results from total population measurements, i.e. these experiments determined the average transport ($\bar{N}_s$) per cell.[44–46] Still other subsequent experiments showed that significant cell-cell variation in molecular transport occur, based on large numbers of individual cell measurements that respond to the number of molecules taken up per cell.[47–50] One particularly interesting study showed that uptake of a first fluorescence-labeled molecule (fluorescence-labeled dextran; 19 kD) was a reliable indicator for the uptake of a second, larger molecule (phycoerythrin, 240 kD).[51] This observation is consistent with the view that electroporative molecular transport is a physical process (movement through aqueous pathways) which is relatively non-specific if the pathway is not too close to the molecule's size.

The question of cell-cell variability is important, as considerable heterogeneity is in fact expected. As discussed earlier, studies with artificial planar bilayer membranes show that electroporation depends on the history of the transmembrane voltage, U(t). From this it follows that physical factors which govern ΔU over the extent of the membrane should be important, i.e. the shape, size and orientation of cells with respect to the local electric field. In addition, biological systems generally display cell-cell variation in many properties ("biological variability"). Finally, the microscopic basis of electroporation itself is believed to involve both a deterministic contribution (electric field energy within the membrane) and a stochastic contribution (thermal fluctuations), so that the overall process of forming pores is fundamentally stochastic, and will therefore display fluctuations on some scale. For all of these reasons, electroporation-related phenomena are expected to exhibit significant cell-cell variability, and this suggests that quantitative molecular transport determinations at the individual cell level are appropriate.

Of the many possible molecular measurement possibilities, only optical methods have adequate sensitivity and speed. Specifically, both image analysis (quantitative microscopy) and flow cytometry make quantitative measurements at the individual cell level. This allows measurement of the number of molecules present by light absorbence, fluorescence or luminescence. With appropriate corrections for background and non-specific binding to external cell surfaces, this allows determination of the number of molecules taken up (or released), N_s, from each measured cell. However, their capabilities are presently complimentary: image analysis excels at providing measurements with spatial resolution (of order 1 μm) and temporal resolution (of order 10 msec)[52–57] to less than a μsecond[28,58] of one or a few cells within a field of view, whereas flow cytometry can make spatially-unresolved individual cell measurements at rates of 10^2 to 10^4 cells sec^{-1}, and routinely analyze large cell populations.[59–61]

Insight into the kinetics of molecular transport can be obtained using image analysis of a few cells within a field of view. This is most readily accomplished for fluorescent molecular release into an extracellular region devoid of fluorescent molecules[53,55] as there is a general limitation for molecular uptake determinations. Unless the molecule entering the cell exhibits a measurable change in optical property (e.g. fluorescence that increases upon binding within a cell), then the extracellular medium containing the molecule must be replaced before a measurement can be made. For example, it was possible to observe slow (seconds) increases in spatially localized fluorescence associated with the uptake of ethidium bromide (394 D, $z = +1$) by NIH3T3 cells due to different pulse protocols.[56] There is, however, a possible exception to the need to remove the molecule-containing extracellular medium: if confocal microscopy is used with sufficiently large cells that a portion of the intracellular region can be volumetrically resolved, then molecular transport into an intracellular volume might be quantitatively monitored during the uptake process.[62,63] Without such schemes, however, the extracellular medium must be replaced before a measurement, and this is limited by the time needed to replace solutions. As a consequence only the late stages ($\approx$10 to 100 sec or longer) of uptake processes can be measured.

The spatial resolution of image analysis also provides information regarding the localization of molecular transport. This has allowed observation of the degree of uptake symmetry in cells which are close to spherical. The instantaneous transmembrane voltage, U(t), at the two poles of a cell ($\theta = 0, \pi$ in equation 3) suggests that U will be increased at one pole and diminished at the other. Qualitatively, it would be expected that the "resting transmembrane voltage," U_0, adds at one pole and subtracts at the other. Quantitatively, however, this effect should probably be considered by treating U_0 as due to a metabolically driven current, $I_{metabolic}$ that ordinarily passes through leaks to establish U_0, but which during electroporation also is shunted through the pores. The experimental evidence is presently equivocal,[52,53,55,56] emphasizing that molecular transport is still not well understood.

As an example of quantitative individual cell level molecular transport determinations, yeast cells in 80 μM propidium iodide (PI; 668 D, z = +2) exhibit negligible uptake if not pulsed, which is expected because of the large Born energy barrier (equation 2) for small charged molecules. In fact, PI is a well established "membrane exclusion dye" that is routinely used to test for cell membrane integrity.[59] However a single square electric field pulse ($E = 5 \times 10^5$ V/m for $t_{pulse} = 5 \times 10^{-5}$ sec) results in significant average uptake, $\overline{N}_{PI} \approx 10^8$ molecules cell^{-1}, with PI binding to doubly stranded nucleic acids which significantly increases the fluorescence of PI.[47] As a second example, red blood cell ghosts suspended in a medium with 1 mM calcein (623 D, z = −4; fluorescent) acquire only very slight fluorescence if unpulsed (attributed to weak surface binding and autofluorescence), but if a single exponential pulse $E(t) = E_0 \exp(-t/\tau)$ with $E_0 = 3 \times 10^5$ V/m or larger, and $\tau \approx 10^{-3}$ sec, then a plateau in uptake $\overline{N}_{calcein} \approx 7 \times 10^4$ molecules/ghost is found.[48] For the same conditions, but with 10μM lactalbumin (14.5 kD, z ≈ −15 e) 10μM bovine serum albumin (BSA; 68 kD, z ≈ −25 e), a plateau is also found ($\overline{N}_s \approx 7 \times 10^4$ molecules/ghost. Surprisingly, for the single exponential pulse experiments, the plateau uptake per ghost is well below the estimated equilibrium value obtained by multiplying the supplied extracellular concentration and the ghost volume, $\overline{N}_{equi} = c_{extra} V_{ghost}$. Extension of these experiments to multiple pulses for BSA uptake reveals that the plateau value for N_s increases as the pulse number is increased, but is still well below $\overline{N}_{equi}$.[49]

The average molecular transport per cell is a reasonable quantity to consider for cell populations without subpopulation responses. For yeast uptake of PI by a 50μsec square pulse and ghost uptake of calcein, lactalbumin and BSA by an exponential pulse with a 1 msec time constant, a single population with a very broad distribution in the amount of uptake (N_s) is found, i.e. there were no distinct subpopulations. In contrast to these two experiments, molecular uptake of calcein by a suspension of yeast (*Saacharomyces cerevisiae*) subjected to a single exponential pulse with a 1 msec time constant exhibits several subpopulations that change markedly as the pulse amplitude is increased from an approximate threshold $E_0 \approx 1 \times 10^5$ V/m up to 7×10^5 V/m.[50] In spite of this significant heterogeneity the average uptake per cell, $\overline{N}_{calcein}$, also plateaued, with separate plateaus found for both a "major uptake subpopulation" and a "minor uptake subpopulation." Surprisingly, for these single exponential pulse experiments with yeast (elongated, budding) the plateau uptake per cell is well below (more than an order of magnitude) than $\overline{N}_{equi}$, even after a determination of the intra-/extracellular partition coefficient is included. Other studies with yeast found that the uptake of a fluorescence-labeled macro-molecules bovine surum albumin also exhibit a plateau.[64]

At this early phase of quantitative molecular transport investigation, several single pulse experiments have found non-equilibrium uptake. A square pulse experiment using only one pulse magnitude found non-equilibrium uptake for both fluorescence-labeled dextran and for PI. The exponential pulse experiments used a series of pulse magnitudes, and found a plateau in $\overline{N}_s$ that is far below $V_{cells}[s]_{ext}$. This latter result suggests that the plateau may be a feature of molecular transport driven by local electric field forces (in this case electrophoresis) for this type of pulse.[48–50,64] Interestingly, an approximate plateau is predicted by a theoretical model that estimates only the molecular transport due to local electrical driving force (electrophoresis) within the pores of a rapidly changing pore population.[41]

At this stage of electroporation investigation, quantitative molecular transport measurements at the individual cell level have been made on only two systems: (1) red blood cell ghosts, for which the uptake per ghost of calcein, lactalbumin and BSA because of a single exponential pulse with a one msec time constant, and (2) intact yeast (*S. cerevisiae*) cells, for which the uptake per cell was determined for propidium iodide because of a single 50μsec pulse, and calcein and BSA because of a exponential pulse with a one msec time constant. Thus, it is not established whether non-equilibrium uptake is a general feature. Likewise, it is not yet known whether multiple pulses, pulses of other shapes or certain features of the transported molecules are particularly important.

The mechanism of molecular transport is not yet understood in detail. For small molecules the creation of a pore population probably occurs independently of molecular transport,[40,41] but for larger molecules, particularly highly charged macromolecules such as DNA, electrically created pores may be significantly enlarged by through interaction with approaching, field-driven molecules.[65,66] In addition to electrophoresis through pores in response to local, inhomogeneous electric fields, diffusion and electro-osmosis[53,55] may play a role, as may stimulated endocytosis.[67] Externally supplied supplemental driving forces such as a pressure difference due to simultaneous or subsequent centrifugation may act to both enlarge primary (electric field-created) pores, and to drive flow through the enlarged pores.[68] However the strongest evidence to date supports involvement of electrical drift (electrophoresis) for moving charged molecules through pores. Interpretation of experiments which introduce highly charged DNA are consistent with local electrophoresis.[65,66] Moreover, a transient aqueous pore theoretical model involving only electrophoretic transport through a dynamic pore population created by a single exponential pulse predicts some of the plateau behavior found for red blood cell ghosts and yeast cells. A tentative interpretation is that electroporation involves a rapid interaction between the membrane conductance due to pores, and the transmembrane voltage, U(t). A slight increase in U results in rapid creation of many more pores, which by a voltage divider effect tends to reduce U. This tendency to a nearly constant U seems to prevail during much, but not all, of the pulse duration; this is qualitatively consistent with a local transmembrane electrophoretic transport that is independent of the pulse magnitude once a significant number of pores exist. Given the central importance of molecular transport to applications, considerably more development of basic insight into molecular transport mechanism is needed.

VII. ELECTROTRANSFECTION

Molecular transport of DNA into cells by electric field pulsing was first reported in 1976, but was a demonstration of principle, as it did not use active genetic material.[69] The first electrically induced gene transfer with subsequent expression was demonstrated in 1982,[70] and was followed by a large number of papers demonstrating the applicability of electrotransfection to mammalian, plant, bacterial and yeast protoplasts and cells.[16,18] Electrotransfection has significant advantages: it is physical and therefore universal, effective, and often is the only method which gives positive results. For this reason electrotransfection is considered as promising for human gene therapy, based on two general approaches: (1) cells treated *in vitro*, with subsequent reimplantation into the body,[16,18] and (2) cells are transfected within tissue, either in isolated tissues (e.g. plant tissue)[71] or mammalian tissue *in vivo*.[72] For maximum utility, the efficiency of introduction of active DNA and cell survival must be simultaneously optimized, but our understanding of mechanism is presently inadequate.

With this in mind, we summarize what is now known about electroporative DNA introduction. Many investigators have reported a correlation between electroporation and electrotransfection. This could be interpreted as support for the role of electropores as the actual pathways for DNA transport. But most primary pores caused solely by an increased U[25] are too small to accommodate DNA. The frequency-of-occurrence of large pores, and the possibility that entering DNA forces enlargement of a pore must therefore both be considered. In any case, it is clear that DNA uptake is not due to passive diffusion. Instead, the following driving forces were considered: (1) flow of water occurring due to the colloid-osmotic cell swelling following electropermeabilization,[73] and (2) electroosmotic flow through a pore with a charged interior surface,[55] (3) electrophoretic transfer of DNA through pores,[74,75] and (4) electrophoretically driven adsorption of DNA onto the membrane, followed by membrane invagination which engulfs the DNA (endocytosis-like mechanism).[76]

However, more recent experiments support the mechanism of electrophoresis by local electric fields focused (by the spreading resistance) through pores, such that DNA is carried into and through a pore which is actively enlarged by the electrically driven DNA.[65,66] The evidence for this process comes from *in vitro* transformation studies using plasmid DNA with a β-galactosidase gene, the efficiency of transfection determined by transient expression of this gene. If the DNA was introduced immediately after pulsing, the transfection efficiency was small compared to that for prepulse addition. Addition of a DNA-degrading enzyme (DNAase) to the suspension three seconds after a pulse did not decrease transfection efficiency. For this reason, the time of DNA translocation was concluded to be less than 3 seconds. If osmotic conditions were used that prevented cell swelling, no change in transfection efficiency was found. But this observation eliminates two possibilities: (1) that DNA passively diffuses through long-lived pores, and (2) that osmotically driven flow carries DNA through pores. Another experimental test exploited a spatial asymmetry: cells were cultured in a monolayer on a porous film. This allowed DNA to be provide on one side of the cell monolayer, and the direction of the electric field chosen to drive the DNA toward or away from the cells. For the pulse polarity that drove DNA electrophoretically towards the cell, the transfection efficiency was ten-fold larger than for the reverse polarity. Moreover, addition of Ficoll-400 (up to 10% w/v), which increases the viscosity reduced the transfection by two to three-fold, as did addition of Mg^{++} ions (10 nM), which decrease the effective charge of DNA. Together these observations constitute support for significant involvement of DNA electrophoresis.

Introduction of DNA by electroporation also increases the uptake of dextrans (MW $<$ 20 kD) supplied after pulsing. This occurs even if the dextran is added a few minutes after pulsing, implying a pore life-time of 10–100 seconds. Moreover, the larger the DNA fragment, the greater the increase in permeability to the dextrans. The interpretation is that the use of a two-pulse technique (a large first pulse to create pores, a second long but much smaller to cause electrophoresis) provides separation of the primary electroporation process and subsequent field-driven transport through long lifetime pores.

An associated theory predicts that DNA in the electroporating electric field exerts high pressure on the membrane, which can then induce structural rearrangements of the membrane.[77] In this view, electrophoresis is the transport mechanism for DNA uptake, as it brings DNA to the cell surface and then drags DNA through the membrane, initially entering pre-existing pores, which are then enlarged and deformed in course of translocation. This is an example of processes leading to secondary pores, which are significantly different than the primary pores which arise solely by the interaction of the electric field with the membrane.

VIII. MEMBRANE RECOVERY

Both reversible and irreversible electroporation lead rapidly (typically 10^{-6} to 10^{-4} sec) to a high conduction state, which discharges the membrane if pulsing has ceased, or which provides a small membrane resistance that participates in a voltage divider effect to reduce the transmembrane voltage even during the pulse.[25,40,41,78,79] By definition, irreversible behavior has no physical membrane recovery, and recovery by biological repair of significant membrane openings due to large magnitude or long lasting pulses has not yet been investigated. In considering what is known about membrane recovery, it is noted that both electrical and molecular assessments of membrane barrier function can be made.

Early experiments used electrical measurements and planar membranes to show that after a decrease in the voltage there is a very rapid (<5 μsec) decrease in the membrane conductance.[10,12] Initially this result was interpreted as reflecting rapid resealing of pores.[10] However, subsequently it was shown that a rapid decrease in conductance is associated with the nonlinear behavior of a pore's current-voltage curve.[26,80–82] The number and the mean radius of the pores were unchanged on a time-scale of μsec, but in the time-scale of 1 to 10msec the decrease in conductance is governed by a change in the size of the pores. For artificial planar bilayer membranes, pores disappear from the membrane on a time-scale of 10 to 100 sec. For cell membranes the time-scale for disappearance of pores is much longer, ranging from second to hours.[82]

Although electrical assessment of recovery has the advantage of speed and convenience, testing for restoration of membrane barrier function by probing with molecules of different size and charge may fundamentally be more revealing as to the extent and kinetics of recovery. Indeed, a few studies have reported the use of delayed addition of probe molecules to pulsed cells Not surprisingly, the qualitative observation from experiments using total population measurements is that significantly less uptake occurs if molecules are added to the extracellular medium after the pulse. Typically the added molecules are small,[13,83] but sometimes macromolecules, including DNA, were used.[66] An important finding is that if analytical techniques with individual cell level capability (microscopy, flow cytometry, image analysis) are used, then subpopulations with uptake of delayed addition molecules are found.[50,84–86] The factors that affect membrane recovery kinetics have not yet been thoroughly investigated. However, a number of experiments show that reduced temperatures are associated with much longer membrane recovery times.[87] This is not surprising, as the diffusive pore shrinking that is expected as the membrane discharges is expected to have this qualitative temperature dependence. Moreover, most physical and chemical processes of multi-molecular systems are characterized by one or more rates that have Boltzmann factors, with their strong, non-linear reduction in rate at lower temperature.

After electrical treatment leading to pore formation and evolution, a dynamic pore population is expected. If so, the change with time of the membrane's ability to exclude molecules of different sizes and charge should provide additional insight into the membrane recovery process. Indeed, experiments have shown that if molecules are first presented to cells at various time delays after pulsing, then a decreasing fraction of cells have a persistent ability to take up molecules.[84] This suggests that a subpopulation of cells has metastable pores with a distribution of pore lifetimes.

Enhanced recovery by adding certain surfactants has been demonstrated.[88] Not only does this challenge our understanding of what membrane recovery is, it also offers promise as a therapy following electrical injury that causes membrane damage by otherwise irreversible electroporation.

IX. CELL STRESS AND SURVIVAL

A general trend is observed when many cell experiments are considered: for a given pulse shape, as the magnitude is increased, phenomena such as molecular uptake become significant at a "fuzzy threshold" (dependent on pulse duration), and then increase. However, at about the same "fuzzy threshold" cell viability begins to drop, with almost all cells being killed as the electric field is increased. Although the mechanism of cell killing by electroporation is not definitely established, a plausible hypothesis is that cell survival or death is mainly due to the degree of molecular exchange between cells and their environment, and the cell stress due to chemical imbalances that results.[84,89,90] This is consistent with the observation that cell killing can occur without significant heating,[4,91,92] and that a tremendous increase in membrane permeability and associated molecular transport occurs.

Both the transient pore population, and possibly a small number of metastable pores, may contribute. In the case of cells, relatively non-specific molecular exchange between the intra- and extracellular volumes probably occurs, and can lead to chemical imbalances. Depending on the ratio of intra-and extracellular volume, the composition of the extracellular medium and the cell type, the cell may not recover from the associated stress, and will therefore die.

Both reversible and irreversible electroporation result in transient openings (perforations) of the membrane, which are often large enought the molecular transport is expected to be non-specific. In the case of irreversible electroporation, it is plausible that a portion of the cell membrane behave much like a small planar membrane, and undergoes rupture. In the case of reversible electroporation, significant molecular transport between the intra- and extracellular volumes may lead to a significant chemical imbalance. If this imbalance is too large, recovery may not occur, with cell death the result. Here it is hypothesized that the volumetric ratio

$$F_{vol} \equiv \frac{V_{extracellular}}{V_{intracellular}} \tag{6}$$

may correlate with cell survival or death. Specifically, for a given cell type and extracellular medium composition, then $F_{vol} \gg 1$ (typical of *in vitro* conditions such as cell suspensions and anchorage dependent cell culture) should tend to favor cell death, while the other extreme $F_{vol} \ll 1$ (typical of *in vivo* tissue conditions) should tend to favor cell survival. This hypothesis therefore suggests that for the same degree of electroporation, significantly less damage may occur in tissue than in body fluids or under most *in vitro* conditions.

Caution should be exercised in assessing cell viability after electroporation protocols. Specifically, there is little *a priori* justification for using membrane-exclusion dyes (e.g. trypan blue [960 D, z = −4], propidium iodide [668 D, z − +2]), as electroporation causes membrane permeability to significantly increase by unknown amounts for unknown times. Membrane-exclusion dyes are generally larger than critical ions, such as Ca^{++}, so their exchange may occur even if membrane recovery has occurred to the degree that membrane-exclusion dyes are excluded by the membrane. Thus, unless an independent validation is made for a given cell type in particular medium for each pulse protocol, it cannot be assured that the cell membrane has recovered sufficiently that fatal chemical imbalances are avoided.

Instead, relevant functional tests should be considered. For cells that can be cultured, clonal growth is probably the most stringent functional test available for electrically pulsed cells.[93] Clonal growth involves many biochemical pathways, and has long been used as the basis of cytotoxicity testing for chemical and viral challenges. For differentiated cells and tissue, other specialized functional tests may suffice, e.g. the ability to develop a normal open circuit voltage by isolated viable epithelia such as frog skin.[94] If electroporation is to be optimally used, a significantly better understanding of the mechanism of cell death will be needed.

X. ELECTROFUSION

Electrofusion is a promising method of somatic hybridization of cells[95]which has been used for increasing numbers of applications.[16,18] However, the mechanism of cell electrofusion has not yet been fully elucidated. Experiments using model systems such as two planar lipid bilayers have provided significant insight. After the application of a pressure the bilayers pass into the state of close contact. This stage usually is necessary also for cell electrofusion. The distance between bilayers is determined by the equilibrium of all acting forces: molecular, electrostatic, hydrostatic pressure and hydrational. After a lag time, one or more bridges (stalks) form spontaneously between two adjacent monolayers. Subsequently, these bridges expand, leading to a so called trilaminar structure, which contains a single bilayer in contact with two bilayers along a perimeter. A trilaminar structure is rather stable, but after a voltage pulse is applied, electroporation occurs, followed by irreversible rupture of the contact bilayer with its transition to so called membrane tube. This stage of the process results in fusion because now there are joined lipid phase and water compartments. It is important to note that two different structural rearrangements take place: (1) stalk (or bridge) formation between adjacent monolayers, and (2) pore formation in contact bilayer—with measurable time lag. Note that the first stage is essentially spontaneous, and that the electric field is involved only during second stage of the fusion process.

Additional insight follows from noting that the probability of stalk formation depends on its energy, which in turn is determined by the favorable molecular geometry of phospholipid molecules. From the viewpoint of the mechanics of continuous media this implies a dependence on the spontaneous curvature of monolayers. The above theory was compared with the results of experiments in which lag times were measured for membranes with different spontaneous curvatures. Good agreement was obtained, supporting the view that the stalk mechanism is involved in lipid bilayers fusion.

Cell fusion is much more complicated. Attempts to find evidence for trilaminar structures in electrically pulsed cell suspensions have not yet been successful. For this reason another fusion mechanism was considered, the interaction of coaxial electropores at adjacent membranes.[96] In this case it was shown that attractive forces due to the Maxwell tension between two parallel bilayers with coaxial pores are large enough to almost bend the membranes, and to bring the pores together until a fusion of pore edges is achieved. These two mechanisms differ in their sequence of events: (1) mono-layer fusion and followed by electroporation (the first case; see above), and (2) electroporation followed by fusion of pore edges (the second case). Significantly, it has been experimentally demonstrated[97,98] that fusion can also proceed by using a protocol in which pulsing precedes cell contact. This supports the mechanism of fusion through electropores. Further elucidation of electrofusion may thus also depend on further developments in the understanding of electroporation.

Extension of cell/cell electrofusion *in vitro* to cell/tissue *in vivo* has been briefly described,[99,100] and offers the prospect of many applications in research and medicine. In this case critical issues involve contacting the introduced cells or vesicles to the target tissue, and then causing a high probability of fusion while retaining viability of the both the introduced cells and the target tissue.

XI. PROTEIN INSERTION INTO MEMBRANES

The intriguing observation that membrane proteins can be electrically inserted into the membrane by pulses similar to those used to cause electroporation not only offers still another cell manipulation possibility,[101–104] it presents a significant challenge in that it is necessary to understand why such molecules do not pass all the way through the membrane. A basic feature of membrane proteins is that they have hydrophilic regions separated by hydrophobic regions, a feature which is qualitatively consistent with the stable residence of such molecules in a bilayer membrane. In the electro-insertion experiments, the molecule to be inserted is apparently first adsorbed onto the cell membrane, and then the cells are electrically pulsed. A qualitative hypothesis is that the adsorption process effectively tethers one end of the molecule to the membrane surface, such that if a pore is created nearby, it may be drawn into the pore but with one end tethered to the membrane. Then upon pore shrinkage around the tethered molecule the process is complete, and insertion has been accomplished. Clearly relatively little of this aspect of electroporation is understood.

XII. TISSUE ELECTROPORATION

A fundamental attribute of cells and tissues is that applied electric fields result in local electric fields that are largest in regions made of highly resistive material, usually lipids. This is the basis of the "amplification" associated with equation 3. Although well established at the cellular level, the corresponding concentration of electric fields in tissues has received less attention. Here we emphasize that this concentrating effect means that electroporation of other lipid-based barriers should occur preferentially, but with nearby tissue low resistance tissue experiencing far less electroporation. For preferable electroporation two features should be sought: (1) tissue barriers comprised mainly of lipids, and (2) mechanical deformability (compliance) of the lipids, so that the electrostatically favored entry of water into the lipid can result in creation of aqueous pathways.

Generalized tissue can be regarded as an aggregate of cells, with different types of cell-cell inter-connections. In many cases there are significant spaces between the cells, such that for small electric fields the electrical currents at low frequencies flow mainly between the cells. In this case, tissue electroporation consists of electroporation of individual cells whose behavior is expected in many respects to be similar to the electroporation of isolated cells, but with two major differences: (1) the local extracellular electric field depends in a complicated way on the many other neighboring cells, and (2) the ratio of the extra- to intracellular volume (equation 7) is usually small, just the opposite of most *in vitro* electroporation conditions. This means that if chemical exchange between the intra- and extracellular volumes is the main cause of cell stress, and therefore cell death, tissue electroporation may be intrinsically less damaging than most *in vitro* electroporation conditions.

Tumor tissue is an important example of tissue for which many cells have intercellular aqueous pathways, but a significant physiological resistance to entry of anticancer drugs nevertheless exists because of limited blood perfusion, elevated interstitial pressure and relatively large distances to blood vessels.[105] Tissue electroporation, which can be caused by localized electric field pulses, should reduce all of these resistances. Under ordinary cancer chemotherapy conditions blood perfusion delivers the anticancer drug to sites within a tumor, and the drug then diffuses across the avascular regions. Local tissue electroporation should create aqueous pathways that assists drug movement, and that also releaves pressure. This means that with reduced pressure the perfusion should increase, and the diffusional barriers are also reduced because of the new aqueous pathways. Striking results have, in fact, been reported in experimental tumor systems, mainly using the drug bleomycin, which is a potent drug that ordinarily does not cross the membrane readily.[106–110] Initial clinical applications have also been reported.[111–113] Most of these initial results were obtained by using surface tumors. For more general applicability, it will be necessary to localize the electric field pulses to the less accessible tumor volumes. As a partial guide, the use of electrical impedance measurements, or even electrical impedance imaging[114] can be considered. Interestingly, at the present time there may be a somewhat artificial barrier to rapidly developing this use of tissue electroporation. Essentially by definition, almost all approved drugs readily cross the cell membrane, actively or passively. Thus, even though it should be attractive to use tissue electroporation with cell membrane-impermeant drugs (e.g. an established drug with several charge groups added that are quickly cleaved by intracellular enzymes) to greatly reduce side effects, the modified drug will be regarded as new, and the expensive and lengthy approval process for new drugs is a major disincentive.

Another general tissue type is that comprised of one or more monolayers of cells in which the cells are connected by tight junctions. Such tissues form the linings of organs and other specialized structures, such as the lining of sweat gland ducts.[115] Qualitatively, the main barrier of each cell mono-layer consists of two cell membranes in series, and can therefore be regarded approximately as two planar bilayer membranes in series. In fact, some tissues such as frog skin and toad bladder are plausible approximations to an ideal single cell monolayer tissue, in that a single cell monolayer is the main barrier. For this reason, such tissues provide a convenient experimental system that allows investigation of some aspects of tissue electroporation.[94]

The mammalian skin is a more complicated tissue, as it has both a specialized barrier structure, the stratum corneum (SC) and appendages (sweat ducts and hair folicles) which perforate the SC.[116,117] The SC is the main barrier, which is a dead tissue whose essential features a often described by the "brick wall model."[118] The "bricks" are corneocytes, which contain cross-linked keratin and increase in volume by up to a factor of five if the SC is hydrated, the usual condition for transdermal drug delivery. The "mortar" is the intervening lipid which surrounds the corneocytes, and exists mainly as mostly parallel bilayers made of special (non-phospholipid) lipids, with about five bilayers between each of the approximately twenty rows of corneocytes. The SC is mechanically flexible but tough, essentially impenetrable to most infectious microorganisms and very impermeable to water soluble molecules, particularly charged molecules, as the lipids are present as about one-hundred bilayers in series. The corneocytes are residual, left over from cells which migrated outwards to form the SC, with the multilamellar lipid bilayer providing most of the barrier function of the skin. Clearly this very effective barrier has evolved to provide protection, but it also presents a major barrier to potential medical interventions, viz. transdermal drug delivery[119,120] and extraction of chemical analytes such as subcutaneous interstitial fluid glucose.[121]

The appendages are usually lined with an extension of the SC, or, deeper within the skin, by a double layer of living cells connected by tight junctions.[115] Both the appendage barriers and the SC barrier are candidates for electroporation, but at very different barrier voltages, $U_{barrier}$. The double cell lining of sweat ducts should experience electroporation at about $U_{barrier} \approx 2$ to 4 volts, but the approximately one-hundred multilamellar bilayers of the SC need $U_{barrier} \approx 50$ to 100 V for 100 μsecond to 1 msecond pulses, i.e. about 0.5 to 1 V per bilayer.[122]

Electroporation of skin and other tissue can be studied *in vitro* using an experimental system based on a well established apparatus, a permeation chamber,[123] which is a version of the Ussing chamber originally used for molecular flux and electrical studies of epithelial tissues.[124] Two- and four-electrode systems can be used for electrical stimulation and measurement. In addition, molecular transport measurements can be made by supplying molecules in the donor chamber, and removing aliquots at intervals from the acceptor chamber for chemical assay. Experiments of this type with human skin show that if exponential pulses with $U_{SC,0} \approx 50$ to 300 V and time constant $\tau \approx 10^{-3}$ second are applied every five seconds for one hour, then there is an enhancement by up to a factor of 10^4 in the flux of charged molecules of up to about 1 kD.[125–128] Companion electrical impedance measurements show a rapid ($\lesssim$ 25 msecond) decrease in skin resistance;[129,130] both molecular flux and electrical measurements show that either reversible or irreversible behavior occurs, depending on the pulse amplitude, $U_{SC,0}$. Exploratory *in vivo* experiments have also been reported, using the hairless rat, which show that significant transdermal fluxes into the blood can be achieved, with insignificant damage revealed by visual inspection.[127]

An extension of the above *in vitro* method to continuous molecular flux measurement uses a sampling stream and quantitative, computer-controlled spectrofluorimetry to measurement two fluorescent molecule concentrations at alternating intervals, so that with deconvolution two molecular fluxes and passive electrical properties can be determined for each skin preparation.[131] A faster responding, single flux version of the same method allowed demonstration of both the large magnitude and rapid onset and cessation of the flux of calcein (623 D, $z = -4$), and supports the idea that rapid electrical control may be another important attribute of skin electroporation.[132] This experimental method provides a flexible approach to characterization of the new aqueous pathways due to skin electroporation.

Presently several potential applications of tissue electroporation have been identified. In addition to enhanced tumor chemotherapy (ECT = electrochemotherapy) and skin electroporation for transdermal drug delivery and chemical analyte extraction, it should be possible to use tissue electroporation for localized gene therapy. In this case, if a therapy does not require that all cells in a tissue be modified, local tissue electroporation and introduction of a virus deficient in its cell entry mechanism might be used. In principle, this would allow viral control of DNA integration, but with far less risk of reversion to a pathogenic virus. A partial demonstration based on subcutaneous injection of DNA followed by surface electrode electroporation leading to transformation of some cells *in vivo* has already been reported.[72]

As with all potential technologies, undesired side effects must be considered along with possible uses. Thus, the possibility of tissue damage must also be considered. Fortunately, major advances in understanding the role of electroporation in tissue damage has already begun, through a series of important studies motivated by electrical injury.[133–138] Although electrical injury to tissue has in the past been interpreted as a thermal denaturation phenomenon, more recent studies make a convincing case that electroporation can be a significant factor, particularly for the larger cells such as skeletal muscle cells and nerve cellS.[137,138] Moreover, the tantalizing discovery that certain surfactants can significantly accelerate membrane recovery has been suggested as the basis of a possible therapy for electrical injury,[88] and might also be considered as a damage-minimizing substance for use with tissue electroporation.

Finally, the possible role of tissue electroporation in cardiac defibrillation has been considered. In this case, the widespread use of electrical pulses to stimulate fibrillating hearts has led to investigation of a possible relationship between this medical intervention and electroporation.[139,140] Based on these studies, it is likely that tissue electroporation does not occur during ordinary defibrillation, but may occur for cardiac tissue immediately adjacent the defibrillating electrodes if a catheter-based system is used.

Tissue electroporation thus has compelling potential advantages. A general feature is that it achieves chemical results (transport of molecules into desired sites) without leaving other chemical residue (e.g. transdermal drug delivery enhancers such as dimethyl sulfoxide). Perhaps even more important, however, is the fact that electroporation is caused electrically, and the tissue experiencing electroporation can be rapidly assessed electrically (e.g. by impedance measurements). This means that much better control of molecular transport for drug delivery and analyte extraction may be possible. It is also clear, however, that a better fundamental understanding of aqueous pathway creation is needed. Further, the characteristics of such pathways need to be quantitatively established, in terms of both molecular transport and electrical properties. Finally, the potential side effects of possible tissue damage must be confronted, and the trade off between desired and undesired consequences understood.

XIII. ELECTROPORATION, BREAKDOWN AND ELECTROPERMEABILIZATION

The terms electroporation, breakdown and electropermeabilization appear frequently in the literature. In the early literature, "breakdown" was used as "dielectric breakdown," and often was qualified as being irreversible or reversible. The idea of breakdown in the sense of a sudden, drastic change is appealing, but not at the molecular level. The traditional meaning of dielectric breakdown involves creation of charge pairs that then accelerate and lead to further molecular ionization. But this traditional dielectric breakdown cannot occur, as there is not enough energy (zeU) at $U \approx 1$ V to ionize most molecules, even though the electric field within the membrane is very large, of order $E \approx 2 \times 10^8$ V m^{-1}. For this reason, it now appears better to think of a high conductance state due to the appearance of conducting aqueous pores, rather than traditional breakdown, as causing the large decrease in membrane resistance.

In the case of the term "electropermeabilization," there is indeed much evidence that a tremendous increase in permeability occurs if U is significantly elevated. However, "permeability" is usually intended to be a phenomonological parameter, which provides an operational description of molecular transport, such as diffusive permeation (dissolution and diffusion). Specifically, a permeating steady-state flux of molecular species "s" is $J_s = A_m P_m \Delta[s]$, where A_m is the membrane area, P_m the diffusive permeability and $\Delta[s]$ is the concentration difference. Thus, the usual view is that $P_m = g_s D_s/d$ where g_s, is the partition coefficient

TABLE 1
Notation used here and in many previous papers.

A_m	membrane area
C_m	capacitance of membrane
C_{LW}	difference in specific capacitance upon replacement of lipid by water
D_p	diffusion constant in pore radius space
D_s	diffusion constant of molecule of type "s"
E(t)	electric field as function of time
e	charge on electron
F_{vol}	ratio of extracellular to intracellular volumes
G_m	membrane conductance
g_s	partition coefficient (relative solubility) of molecule "s" for two media
d	membrane thickness
I(t)	electric current
K_m	dielectric constant of membrane
K_w	dielectric constant of water
kT	Boltzmann's constant multiplied by temperature
N_s	number of molecules of type "s"
P_m	permeability of membrane
R_E	resistance of electrolyte pathway
R_p	interior pore resistance
r_{cell}	radius of spherical cell
r_{min}	minimum pore radius (≈1 nm)
U(t)	transmembrane voltage as function of time
U_c	critical transmembrane voltage associated with rupture
U_{SC}	voltage across stratum corneum (SC) of skin
$V_{extracellular}$	total extracellular volume in a cell system
$V_{intracellular}$	total intracellular volume in a cell system
V(t)	voltage applied to electrodes
W_{Born}	Born (electrostatic) energy
z	number of electronic charges on ion or molecule
γ	pore edge energy density
Γ	membrane surface energy
J_s	flux of molecule "s"
ρ_{in}	resistivity of intracellular medium
ρ_e	resistivity of extracellular medium
σ_e	electrical conductivity of extracellular medium ($1/gr_e$)
τ_m	average planar membrane lifetime against rupture

(solubility in the membrane relative to water), D_s the diffusion constant within the membrane and d is the membrane thickness.

If this idea is pursued, the only attractive local rearrangement that significantly increases the solubility of water soluble molecules is creation of an aqueous region, i.e. an aqueous pathway or "pore." According to the view we have presented, electroporation is exactly that, a class of structural rearrangements that amount to aqueous pathways across the membrane. Clearly the process of poration results in enhanced membrane permeability, as pores will increase the diffusive, electrophoretic, and pressure driven flux of water soluble molecules and ions.

But electroporation is consistent with more than electropermeabilization. Specifically, electro-insertion of membrane proteins involves more than a permeability change, as does electrofusion of cells to other cells. Further, the hypothesis of electroporation leads logically to specific predictions that can be tested, some of which are not related directly to steady state transport. Currently these are: (1) the occurrence of rupture or reversible electrical breakdown in the same artificial planar bilayer membrane, depending only on pulse properties, (2) the approximate magnitude of the transmembrane voltage associated with rupture, (3) the magnitude of maximum transmembrane voltage during reversible electroporation (about one volt), (4) the stochastic nature of the membrane lifetime associated with rapture, (5) the maximum fractional aqueous area for a membrane undergoing reversible behavior, and (6) the order of magnitude of transported small charged molecules for a single pulse, including the existence of an approximate plateau in the amount of transport as a function of pulse magnitude. For this reason, at the present stage of our understanding, we regard electropermeabilization as only a part of electroporation, although an extremely important one. The evidence supporting the electroporation hypothesis is, however, indirect. For fundamental reasons relating to their expected size, lifetime and composition, no existing form of microscopy can be expected to image the primary hydrophilic pores.[141] In contrast, secondary pores which arise by enlargement of hydrophilic pores have been visualized in frozen specimens using electron microscopy.[142] Thus, the main support for electroporation comes from the inferences made from electromechanical and molecular transport observations, and the implausibility of other hypotheses.

In conclusion, we also recognize that all hypotheses or models are only guides to our understanding of nature, and can only be disproven when new, surprising experimental results are obtained that cannot be described by the then-prevailing model. At the present time there is considerable research being devoted to the effects associated with large transmembrane voltages, and new insights and surprises can be expected.

ACKNOWLEDGMENTS

We are indebted to a great many colleagues for numerous stimulating discussions. Supported partially by Cygnus Therapeutics Systems, Inc., U.S. Army Research Office Grant No. DAAL03-90-G-0218, NIH Grant ES06010, and a computer equipment grant from Stadtwerke Düsseldorf, Germany.

REFERENCES

1. **Zahn, M.,** *Electromagnetic Field Theory: A Problem Solving Approach,* Wiley & Sons, New York, 1979.
2. **Parsegian, V. A.,** Energy of an ion crossing a low dielectric membrane: solutions to four relevant electrostatic problems, *Nature,* vol. 221, pp. 844–846, 1969.

3. **Stämpfli, R.**, Reversible Electrical Breakdown of the Excitable Membrane of a Ranvier Node, *An. Acad. Brasil. Ciens.*, vol. 30, pp. 57–63, 1958.
4. **Sale, A. and Hamilton J. H.**, Effects of high electric fields on microoranisms: I. killing of bacteria and yeasts, *Biochim. Biophys. Acta,* vol. 148, pp. 781–788, 1967.
5. **Neumann, E. and Rosenheck K.**, Permeability changes induced by electric impulses in vesicular membranes, *J. Membrane Biol.*, vol. 10, pp. 279–290, 1972.
6. **Abidor, I. G., Arakelyan V. B., Chernomordik, L. V., Chizmadzhev Yu. A., Pastushenko, V. F., and Tarasevich, M. R.**, Electric breakdown of bilayer membranes: I. the main experimental facts and their qualitative discussion, Bioelectrochem. Bioenerget, vol. 6, pp. 37–52, 1979.
7. **Benz, R., Beckers, F., and Zimmermann U.**, Reversible Electrical Breakdown of Lipid Bilayer Membranes: A Charge-Pulse Relaxation Study, J. Membrane Bio., vol. 48, pp. 181–204, 1979.
8. **Benz, R. and Zimmermann, U.**, Relaxation studies on cell membranes and lipid bilayers in the high electric field range, *Bioelectrochem, and Bioenerg.*, vol. 7, pp. 723–739, 1980.
9. **Benz, R. and Zimmermann, U.**, Pulse-length dependence of the electrical breakdown in lipid bilayer membranes, *Biochim. Biophys, Acta,* vol. 597, pp. 637–642, 1980.
10. **Benz, R. and Zimmermann, U.**, The resealing process of lipid bilayers after reversible electrical breakdown, *Biochim. Biophys. Acta,* vol. 640, pp. 169–178, 1981.
11. **Benz, R. and Conti F.**, Reversible electrical breakdown of squid giant axon membrane, *Biochim. Biophys. Acta,* vol. 645, pp. 115–123, 1981.
12. **Chernomordik, L. V., Sukharev S. I., Abidor, M. G., and Chizmadzhev, Yu. A.**, The Study of the BLM Reversible Electrical Breakdown Mechanism in the Presence of UO_2^{2+}, *Bioelectrochem. Bioenerget.* vol. 9, pp. 149–155, 1982.
13. **Kinosita, K. Jr. and Tsong, T. Y.**, Formation and resealing of pores of controlled sizes in human erythrocyte membrane, *Nature,* vol. 268, pp. 438–441, 1977.
14. **Kinosita, K. Jr. and Tsong, T. Y.**, Survival of Sucrose-Loaded Erythrocytes in Circulation, *Nature,* vol. 272, pp. 258–260, 1978.
15. **Teissie, J. and Tsong, T. Y.**, Electric field induced transient pores in phospholipid bilayer vesicles, *Biochemistry,* vol. 20, pp. 1548–1554, 1981.
16. **Neumann, E., Sowers, A., and Jordan C., (Eds),** *Electroporation and Electrofusion in Cell Biology,* Plenum, New York, 1989.
17. **Tsong, T. Y.**, Electroporation of Cell Membranes, *Biophys, J.*, vol. 60, pp. 297–306, 1991.
18. **Chang, D. C., Chassy, B. M., Saunders, J. A., and Sowers, A. E., (Eds.),** *Guide to Electroporation and Electrofusion,* Academic Press, 1992.
19. **Blank, M.**, Ed., *Electricity and Magnetism in Biology and Medicine,* San Francisco Press, 1993.
20. **Weaver, J. C.**, Electroporation: a general phenomenon for manipulating cells and tissue, *J. Cellular Biochem.*, vol. 51, pp. 426–435, 1993.
21. **Orlowski, S. and Mir, L. M.**, Cell electropermeabilization: a new tool for biochemical and pharmacological studies, *Biochim. Biophys, Acta,* vol. 1154, pp. 51–63, 1993.
22. **Tien, H. T.**, Planar bilayer lipid membranes, *Progress in Surface Science,* vol. 19, pp. 1–106, Pergamon Press, New York, 1985.
23. **Montal, M. and Mueller P.**, Formation of bimolecular membranes from lipid monolayers and a study of their electrical properties, *PNAS,* vol. 69, pp. 3561–3566, 1972.
24. **Chernomordik, L. V. and Chizmadzhev, Y. A.**, Electrical breakdown of BLM: phenomenology and mechanism, in **Electroporation and Electrofusion in Cell Biology,** ed. E. Neumann, A. Sowers and C. Jordan (Eds), pp. 83–96, Plenum, New York, 1989.
25. **Barnett, A. and Weaver, J. C.**, Electroporation: a unified, quantitative theory of reversible electrical breakdown and rupture, *Bioelectrochem and Bioenerg.*, vol. 25, pp. 163–182, 1991.
26. **Chernomordik, L. V., Sukharev S. I., Abidor, I. G. and Chizmadzhev, Yu. A.**, Breakdown of lipid bilayer membranes in an electric field, *Biochim. Biophys. Acta,* vol. 736, pp. 203–213, 1983.
27. **Schwan, H. P.**, Dielectrophoresis and rotation of cells, in Electroporation and Electrofusion in Cell Biology, ed. E. Neumann, A. E. Sowers and C. A. Jordan, pp. 3–21, Plenum, New York, 1989.
28. **Kinosita, K. Jr., Ashikawa, I., Saita, N., Yoshimura, H., Itoh, H., Nagayama, H., and Ikegami, A.**, Electroporation of cell membrane visualized under a pulsed-laser fluorescence microscope, *Biophys. J.*, vol. 53, pp. 1015–1019, 1988.
29. **Winterhalter, M. and Helfrich, W.**, Deformation of spherical vesicles by electric fields, *J. Colloid and Interfac. Sci.*, vol. 122, pp. 583–586, 1988.
30. **Kinosita, K. Jr., Hibio, M. Itoh, H., Shigemori, M., Hirano, K., and Hayakawa, T.**, Electroporation visualized on the time scale from a submicrosecond to seconds, in Guide to Electroporation and Electrofusion, pp. 29–46, Academic Press, 1992. D. C. Chang, B. M. Chassy, J. A. Saunders and A. E. Sowers (Eds)
31. **Sugar, I. P. and Neumann, E.**, Stochastic model for electric field-induced membrane pores: electroporation, *Biophys. Chemistry,* vol. 19, pp. 211–225, 1984.

32. **Crowley, J. M.,** Electrical breakdown of bimolecular lipid membranes as an electromechanical instability, *Biophys. J.,* vol. 13, pp. 711–724, 1973.
33. **Weaver, J. C. and Mintzer, R. A.,** Decreased bilayer stability due to transmembrane potentials, *Phys. Lett.,* vol. 86A, pp. 57–59, 1981.
34. **Michael, D. H., O'Neill, M. E., and Zuecher, J. C.,** The breakdown of electrical insulation in a planar layer of insulating fluid by electrocapillary instability, *J. Fluid Mech.,* vol. 47, pp. 609–623, 1971.
35. **Maldarelli, C. and Stebe, K.,** An anisotropic, elastomechanical instability theory for electropermeabilization of bilayer-lipid membranes, in *Electrical Trauma: Pathophysiology and Clinical Management,* ed. R.C. Lee, J. F. Burke and E. G. Cravalho (Eds.), pp. 327–360, Cambridge University Press, 1992.
36. **Freeman, S. A., Wang, M. A., and Weaver, J. C.,** Theory of electroporation for a planar bilayer membrane: predictions of the fractional aqueous area, change in capacitance and pore-pore separation. *Biophysical J.,* vol. 67, pp. 42–56, 1994.
37. **Litster, J. D.,** Stability of lipid bilayers and red blood cell membranes, *Phys. Lett.,* vol. 53A, pp. 193–194, 1975.
38. **Taupin, C., Dvolaitzky, M., and Sauterey, C.,** Osmotic pressure induced pores in phospholipid vesicles, *Biochmistry,* vol. 14 pp. 4771–4775, 1975.
39. **Powell, K. T., Derrick, E. G., and Weaver, J. C.,** A quantitative theory of reversible electrical breakdown, *Bioelectrochem. Bioelectroenerg.,* vol. 15, pp. 243–255, 1986
40. **Weaver, J. C. and Barnett, A.,** Progress towards a theoretical model of electroporation mechanism: membrane electrical behavior and molecular transport, in *Guide to Electroporation and Electrofusion,* pp. 91–117, Academic Press, 1992. D. C. Chang, B. M. Chassy, J. A. Saunders and A. E. Sowers (Eds)
41. **Wang, M. A., Freeman, S. A., Bose, V. G., Dyer, S., and Weaver, J. C.,** *Theoretical Modelling of Electroporation: Electrical Behavior and Molecular Transport,* pp. 138–140, 1993. Electricity and Magnetism in Biology and Medicine, M. Blank, Ed.
42. **Zimmermann, U., Pilwat, G., and Riemann, F.,** Preparation of erythrocyte ghosts by dielectric breakdown of the cell membrane, *Biochim. Biophys. Acta,* vol. 375, pp. 209–219, 1975.
43. **Zimmermann, U., Riemann, F., and Pilwat, G.,** Enzyme loading of electrically homogeneous human red blood cell ghosts prepared by dielectric breakdown, *Biochim. Biophys. Acta,* vol. 436, pp. 460–474, 1976.
44. **Mir, L. M., Banoun, H., and Paoletti, C.,** Introduction of definite amounts of nonpermeant molecules into living cells after electropermeabilization: direct access to the cytosol, *Exp. Cell Res.,* vol. 175, pp. 15–25, 1988.
45. **Chakrabarti, R., Wyle, D. E., and Schuster, S. M.,** Transfer of monoclonal antibodies into mammalian cells by electroporation, *J. Biol. Chem.,* vol. 264, pp. 15494–15500, 1989.
46. **Lambert, H., Pankov, R., Gauthier, J., and Hancock, R.,** Electroporation-mediated uptake of proteins into mammalian cells, *Biochem. Cell Biol.,* vol. 68, pp. 729–734, 1990.
47. **Bartoletti, D. C., Harrison, G. I., and Weaver, J. C.,** The number of molecules taken up by electroporated cells: quantitative determination, *FEBS Let.,* vol. 256, pp. 4–10, 1989.
48. **Prausnitz, M. R., Lau, B. S., Milano, C. D., Conner, S., Langer, R., and Weaver, J. C.,** A quantitative study of electroporation showing a plateau in net molecular transport, *Biophysical J.,* vol. 65, pp. 414–422, 1993.
49. **Prausnitz, M. R., Milano, C. D., Gimm, J. A., Langer, R., and Weaver, J. C.,** Quantitative study of molecular transport due to electroporation: uptake of bovine serum albumin by human red blood cell ghosts, *Biophysical J.,* vol. 66, pp. 1522–1530, 1994.
50. **Gift, E. A. and Weaver, J. C.,** Observation of extremely heterogeneous electroporative uptake which changes with electric field pulse amplitude in *saccharomyces cerevisiae. Biochim. Biophys. Acta,* vol. 1234, pp. 52–62, 1995.
51. **Graziadei, L., Burfeind, P., and Bar-Sagi, D.,** Introduction of unlabeled proteins into living cells by electroporation and isolation of viable protein-loaded cells using dextran-fluorescein isothiocyanate as a marker for protein uptake, *Anal. Biochem.,* vol. 194, pp. 198–203, 1991.
52. **Mehrle, W., Zimmermann, U., and Hampp, R.,** Evidence for asymmetrical uptake of fluorescent dyes through electro-permeabilized membranes of Avena Meosphyll protoplasts, *FEBS Lett.,* vol. 185, pp. 89–94, 1985.
53. **Sowers, A. E. and Lieber, M. R.,** Electropore diameters, lifetimes, numbers, and locations in individual erythrocyte ghosts, *FEBS Lett.,* vol. 205, pp. 179–184, 1986.
54. **Liang, H., Purucker, W. J., Stenger, D. A., Kubiniec, R. T., and Huie, S. W.,** Uptake of fluorescence-labeled dextrans by 10T 1/2 fibroblasts following permeation by rectangular and exponential-decay electric field pulses, *BioTechniques,* vol. 6, pp. 550–558, 1988.
55. **Dimitrov, D. S. and Sowers, A. E.,** Membrane electroporation—fast molecular exchange by electroosmosis, *Biochim. Biophys. Acta,* vol. 1022, pp. 381–392, 1990.
56. **Tekle, E., Astumian, R. D., and Chock, P. B.,** Electroporation using bipolar oscillating electric field: an improved method for DNA transfection of NIH3T3 cells, *PNAS,* vol. 88, pp. 4230–4234, 1991.
57. **Sixou, S. and Teissie, J.,** Exogenous uptake and release of molecules by electroloaded cells: a digitized videomicroscopy study, *Bioelectrochem. Bioenerget,* vol. 31, pp. 237–257, 1993.

58. **Hibino, M., Shigemori, M., Itoh, H., Nagyama, K., and Kinosita, K.,** Membrane conductance of an electroporated cell analyzed by submicrosecond imaging of transmembrane potential, *Biophys. J.,* vol. 59, pp. 209–220, 1991.
59. **Shapiro, H. M.,** *Practical Flow Cytometry, Second Edition,* A. R. Liss, New York, 1988.
60. **Melamed, E. R., Lindmo, T. and Mendelsohn, M. L.,** in *Flow Cytometry and Sorting, Second Edition,* Wiley-Liss, New York, 1990.
61. **Givan, A. L.,** in *Flow Cytometry: First Principles,* Wiley-Liss, New York, 1992.
62. **Pawley, J. B., (Ed),** *Handbook of Biological Confocal Microscopy,* Plenum Press, New York, 1990.
63. **Wilson, T.,** *Confocal Microscopy,* Academic Press, London, 1990.
64. **Hui, L., Gift, E. A., and Weaver, J. C.,** Uptake of bovine serum albumin by yeast due to electroporation. (in preparation)
65. **Klenchin, V. A., Sukharev, S. I., Serov, S. M., Chernomordik, L. V., and Chizmadzhev, Yu. A.,** Electrically Induced DNA Uptake by Cells is a Fast Process Involving DNA Electrophoresis, *Biophysical J.,* vol. 60, pp. 804–811, 1991.
66. **Sukharev, S. I., Klencin, V. A., Serov, S. M., Chernomordik, L. V., and Chizmadzhev, Y. A.,** Electroporation and electrophoretic DNA transfer into cells, *Biophysical J.,* vol. 63, pp. 1320–1327, 1992.
67. **Zimmermann, U., Schnettler, R., Klöck, G., and Watzka, H.,** Mechanisms of electrostimulated uptake of macromolecules into living cells, *Naturwissenschaft,* vol. 77, pp. 543–545, 1990.
68. **Kim, D., Lee, Y. J., Rausch, C. M., and Borrelli, M. J.,** Electroporation of extraneous proteins into CHO cells: increased efficiency by utilizing centrifugal force and microsecond electrical pulses, *Exp. Cell Res.,* vol. 197, pp. 207–212, 1991.
69. **Auer, D., Brandner, G., and Bodemer, W.,** Dielectric breakdown of the red blood cell membrane and uptake of SV 40 DNA and mammalian cell RNA, *Naturwiss.,* vol. 63, p. 391, 1976.
70. **Neumann, Schaefer-Ridder, E., M., Wang, Y., and Hofschneider, P. H.,** Gene transfer into mouse lyoma cells by electroporation in high electric fields, *EMBO J.,* vol. 1, pp. 841–845, 1982.
71. **Dekeyser, R. A., Claes, B., De Rycke, R. M. U., Habets, M. E., Van Montagu, M. C., and Caplan, A. B.,** Transient gene expression in intact and organized rice tissues, *Plant Cell,* vol. 2, pp. 591–602, 1990.
72. **Titomirov, A. V., Sukharev, S., and Kistoanova, E.,** *In vivo* electroporation and stable transformation of skin cells of newborn mice by plasmid DNA, *Biochim. Biophys. Acta,* vol. 1088, pp. 131–134, 1991.
73. **Stopper, H., Jones, H., and Zimmermann, U.,** Large scale transfection of mouse L-cells by electropermeabilization, *Biochim. Biophys. Acta,* vol. 900, pp. 38–44, 1987.
74. **Winterbourne, D. J., S. Thomas, J. Hermon-Taylor, I. Hussain, and A. P. Johnstone,** Electric shock-medicated transfection of cells. Characterization and optimization of electrical parameters, *Biochem. J.,* vol. 251, pp. 427–434, 1988.
75. **Andreason, G. L. and Evans, G. A.,** Optimization of electroporation for transfection of mammalian cell lines, *Anal. Biochem.,* vol. 180, pp. 269–275, 1989.
76. **Chernomordic, L. V., Sokolov, A. V., and Budker, V. G.,** Electrostimulated uptake of DNA by liposomes, *Biochim. Biophys. Acta,* vol. 1024, pp. 179–183, 1990.
77. **Pastushenko, V. P. and Chizmadzhev, Y. A.,** Theory of DNA translocation through membranes, *Biol. Mem.,* vol. 6, pp. 287–300, 1992.
78. **Pastushenko, V. F. and Chizmadzhev, Yu. A.,** Stabilization of conducting pores in BLM by electric current, *Gen. Physiol. Biophys.,* vol. 1, pp. 43–52, 1982.
79. **Powell, K. T. and Weaver, J. C.,** Transient aqueous pores in bilayer membranes: a statistical theory, *Bioelectrochem. Bioelectroenerg.,* vol. 15, pp. 211–227, 1986.
80. **Chizmadzhev, Yu. A. and Abidor, I. G.,** Bilayer Lipid membranes in strong electric fields, *Bioelectrochem. Bioenerget.,* vol. 7, pp. 83–100, 1980.
81. **Chernomordik, L. V., Sukharev, S. I., Popov, S. V., Pastushenko, V. F., Sokirko, A. V., Abido, I. G., and Chizmadzhev, Y. A.,** The electrical breakdown of cell and lipid membranes: the similarity of phenomenologies, *Biochim. Biophys. Acta,* vol. 902, pp. 360 373, 1987.
82. **Chernomordik, L. V. and Chizmadzhev, Y. A.,** Electrical breakdown of lipid bilayer membranes: phenomenology and mechanism, in *Electroporation and Electrofusion in Cell Biology,* ed. E. Neumann, A. E. Sowers and C. A. Jordan, pp. 83–95, Plenum, New York, 1989.
83. **Pilwat, G., Zimmermann, U., and Riemann, F.,** Dielectric breakdown measurements of human and bovine erythrocyte membranes using benzyl alcohol as a probe molecule, *Biochim. Biophys. Acta,* vol. 406, pp. 424–432, 1975.
84. **Weaver, J. C., Harrison, G. I., Bliss, J. G., Mourant, J. R., and Powell, K. T.,** Electroporation: high frequency of occurrence of the transient high permeability state in red blood cells and intact yeast, *FEBS Lett.,* vol. 229, pp. 30–34, 1988.
85. **Tsoneva, I., Tomov, T., Panova, I., and Strahilov, D.,** Effective production by electrofusion of hybridomas

secreting monoclonal antibodies against Hc-antigen of *salmonella, Bioelectrochem. Bioenerget.*, vol. 24, pp. 41–49, 1990.

86. **Kwee, S., Nielsen, H. V., and Celis, J. E.,** Electropermeabilization of human cultured cells grown in monolayers: incorporation of monoclonal antibodies, *Bioelectrochem. and Bioenerget.* vol. 23, pp. 65–80, 1990.
87. **Mishra, K. P. and Singh, A. B.,** Temperature effects on resealing of electrically hemolysed rabbit erythrocytes, *Indian J. Exp. Biol.*, vol. 24, pp. 737–741, 1986.
88. **Lee, R. C., River, L. P., Pan, F.-S., Ji, L., and Wollmann, R. L.,** Surfactant induced sealing of electropermeabilized skeletal muscle membranes *in vivo*, PNAS, vol. 89, pp. 4524–4528, 1992.
89. **Michel, M. R., Elgizoli, M., Koblet, H., and Kempf, Ch.,** Diffusion loading conditions determine recovery of protein synthesis in electroporated P3X63 Ag8 cells, *Experientia,* vol. 44, pp. 199–203, 1988.
90. **Weaver, J. C.,** Molecular basis for cell membrane electroporation, *Annals of the New York Academy of Sciences*, vol. 720, pp. 141–152, 1994.
91. **Hamilton, W. A. and Sale, A. J. H.,** Effects of high electric fields on microoranisms: II. killing of bacteria and yeasts, *Biochim. Biophys. Acta,* vol. 148, pp. 789–800, 1967.
92. **Sale, A. J. H. and Hamilton, W. A.,** Effects of high electric fields on microoranisms: III. Lysis of erythrocytes and protoplasts, *Biochim. Biophys. Acta,* vol. 163, pp. 37–43, 1968.
93. **Gift, E. A. and Weaver, J. C.,** Cell survival following electroporation: quantitative assessment using large numbers of microcolonies, pp. 147–150, 1993. Electricity and Magnetism in Biology and Medicine, M. Blank, Ed.
94. **Powell, K. T., Morgenthaler, A. W., and Weaver, J. C.,** Tissue electroporation: observation of reversible electrical breakdown in viable frog skin, *Biophys. J.*, vol. 56, pp. 1163–1171, 1989.
95. **Zimmermann, U.,** Electric field-mediated fusion and related electrical phenomena, *Biochim. Biophys. Acta,* vol. 694, pp. 227–277, 1982.
96. **Kuzmin, P. I., Pastushenko, V. P., Abidor, I. G., Sikharev, S. I., Barbul, A. V., and Chizmadzhev, Y. A.,** Electrofusion of cells: theoretical analysis, *Biol. Membr.*, vol. 5, pp. 600–612, 1988.
97. **Sowers, A. E.,** A long lived fusogenic state is induced in erythrocyte ghosts by electric pulses, *J. Cell Biol.*, vol. 102, pp. 1358–1362, 1986.
98. **Teissie, J. and Rols, M.-P.,** Fusion of mammalian cells in culture is obtained by creating the contact between cells after their electropermeabilization, *Biochem. Biophys. Res. Commun.*, vol. 140, pp. 258–264, 1986.
99. **Grasso, R. J., Heller, R., Cooley, J. C., and Haller, E. M.,** Electrofusion of individual animal cells directly to interaction corneal epithelial tissue, *Biochim. Biophys. Acta,* vol. 980, pp. 9–4, 1989.
100. **Heller, R. and Grasso, R. J.,** Transfer of human membrane surface components by incorporating human cells into interaction animal tissue by cell-tissue electrofusion *in vivo, Biochim. Biophys. Acta,* vol. 1024, pp. 185–188, 1990.
101. **Mouneimne, Y., Tosi, P-F., Gazitt, Y., and Nicolau, C.,** Electro-insertion of xeno-glycophorin into the red blood cell membrane, *Biochem. Biophys. Res. Comm.*, vol. 159, pp. 34–40, 1989.
102. **Mouneimne, Y., Tosi, P-F., and Nicolau, C.,** *Biochim. Biophys. Acta,* vol. 1027, pp. 53–58, 1990.
103. **Mouneimne, Y., Tosi P-F., Barhoumi, R., and Nicolau, C.,** *Biochim. Biophys. Acta,* vol. 1066, pp. 83–89, 1991.
104. **Zeira, M., P-F. Tosi, Y. Mouneimne, J. Lazarte, L. Sneed, D. J. Volsky and C. Nicolau,** Full-length CD4 electro-inserted in the erythrocyte membrane as a long-lived inhibitor of infection by human immunodeficiency virus *PNAS,* vol. 88, pp. 4409–4413, 1991.
105. **Jain, R. K.,** Physiological resistance to the treatment of solid tumors, 1993, *Drug Resistance in Oncology,* B. A. Teicher, Ed., Marcel Dekker, New York, pp. 87–105.
106. **Okino, M. and Mohri, H.,** Effects of a high-voltage electrical impulse and an anticancer drug on *in vivo* growing tumors, *Jpn. J. Cancer Res.*, vol. 78, pp. 1319–1321, 1987.
107. **Mir, L. M., Orlowski, S., Belehradek, Jr., J., and Paoletti, C.,** *In vivo* potentiation of the bleomycin cytotoxicity by local electric pulses, *Eur. J. Cancer,* vol. 27, pp. 68–72, 1991.
108. **Mir, L. M., Orlowski, S., Poddevin, B., and Belehradek, Jr. Jr.,** Electrochemotherapy tumor treatment is improved by interleukin-2 stimulation of the host's defenses, *Eur. Cytokine Netw.*, vol. 3, pp. 331–334, 1992.
109. **Okino, M., Tomie, H., Kanesada, H., Marumoto, M., Esato, K., and Suzuki, H.,** Optimal electrical conditions in electrical impulse chemotherapy, *Jpn. J. Cancer Res.*, vol. 83, pp. 1095–1101, 1992.
110. **Salford, L. G., Persson, R. B. R., Brun, A., Ceberg, C. P., Kongstad, P. C., and Mir, L. M.,** A new brain tumor therapy combining bleomycin with *in vivo* electropermeabilization, *Biochem. Biophys. Res. Comm.*, vol. 194, pp. 938–943, 1993.
111. **Mir, L. M., Belehradek, M., Domenge, C., Orlowski, S., Poddevin, B., Belehradek J. Jr., Schwaab, G., Lkuboinski, B., and Paoletti, C.,** L'ëlectrochimiothërapie un nouveau traitement antiumoreal: premier essai clinique, *C. R. Acad. Sci. Paris,* vol. 313, Sirie III, pp. 613–618, 1993.
112. **Mir, L. M., Orlowski, S., Belehradek, J. Jr., Poddevin, B., Pron, G., Domenge, C., Belehradek, M.,**

Lkuboinski, B., and Paoletti, C., Electrochemotherapy: a new antitumor treatment using electric pulses, 1993. Electricity and Magnetism in Biology and Medicine, M. Blank, Ed., pp. 119–121.

113. **Belehradek, M. and Domenge, C., Orlowski, S., Belehradek, J. Jr., and Mir, L. M.,** *Cancer,* vol. 72, pp. 3694–3700, 1993.
114. **McArdle, F. J., Brown, B. H., and Angel, A.,** Imaging resistivity changes of the adult brain during the cardiac cycle, Proc. 11th IEEE/BME, pp. 480–481, 1989.
115. **Berridge, M. J. and Oschman, J. L.,** *Transporting Epithelia,* Academic Press, New York, 1972.
116. **Elias, P. M. and Menon, G. K.,** Structural and lipid biochemical correlates of the epidermal permeability barrier, *Adv. Lipid Res.,* vol. 24, pp. 1–26, 1991.
117. **Potts, R. O., Bommannan, D. B., and Guy, R. H.,** Percutaneous absorption, In *Pharmacology of the Skin,* H. Mukhtar, Ed., pp. 13–27, 1992 CRC Press
118. **Chandrasekaran, S. K., Bayne, W., and Shaw, J. E.,** Pharmokinetics of drug penetration through human skin, *J. Pharmaceut. Sci.,* vol. 67, pp. 1370–1374, 1978.
119. **Hadgraft, J. and Guy, R. H.,** *Transdermal Drug Delivery: Development Issues and Research Initiatives,* Marcel Dekker, New York, 1989.
120. **Cullander, C. and Guy, R. H.,** Transdermal delivery of peptides and proteins, *Advanced Drug Delivery Rev.,* vol. 8, pp. 291–329, 1992.
121. **Reach, G. and Wilson, G. W.,** Can continuous glucose monitoring be used for the treatment of diabetes?, *Anal. Chem.,* pp. 381A–386A, 1992.
122. **Chizmadzhev, Yu. A., V. Zarnytsin, J. C. Weaver, and R. O. Potts,** Mechanism of electroinduced ionic species transport through a multilamellar lipid system, *Biophysical J.,* vol. 68, pp. 749–765, 1995.
123. **Friend, D. R.,** *In vitro* skin permeation techniques, *J. Controlled Rel.,* vol. 18, pp. 235–248, 1992.
124. **Ussing, H. H.,** *The Alkali Metal Ions in Biology,* Springer-Verlag, Berlin, 1960.
125. **Prausnitz, M. R., Bose, V. G., Langer, R. S., and Weaver, J. C.,** Transdermal Drug Delivery by Electroporation, 1992, Abstract, Proceed. Intern. Symp. Control. Rel. Bioact. Mater. 19, Controlled Release Society, July 26–29, 1992 Orlando, FL, pp. 232–233.
126. **Prausnitz, M. R., Bose, V. G., Langer, R., and Weaver, J. C.,** Transtissue molecular transport due to electroporation of skin, 1993, Electricity and Magnetism in Biology and Medicine, M. Blank, Editor, pp. 122–124.
127. **Prausnitz, M. R., Bose, V. G., Langer, R., and Weaver, J. C.,** Electroporation of mammalian skin: a mechanism to enhance transdermal drug delivery, *PNAS,* vol. 90, pp. 10504–10508, 1993.
128. **Bommannan, D., Leung, L., Tamada, J., Sharifi, J., and Potts, R.,** Transdermal drug delivery of luteinizing hormone releasing hormone: comparison between electroporation and iontophoresis *in vitro*, 1993, Proc. Intl. Symp. Control. Rel. Bioact. Mater., 20, pp. 97–98.
129. **Pliquett, U., R. Langer, and J. C. Weaver,** The change in the passive electrical properties of skin due to electroporation, *Biochim. Biophys. Acta,* (in press), 1995.
130. **Bose, V. G., M. R. Prausnitz, R. Langer and J. C. Weaver,** Electrical changes in human stratum corneum due to electroporation, (submitted).
131. **Pliquett, U. and J. C. Weaver,** Electroporation of human skin: simultaneous measurement of changes in the transport of two flourescent molecules and in the passive electrical properties, *Bioelectrochem Bioenerget.* (accepted), 1995.
132. **Prausnitz, M. R., U. Pliquett, R. Langer, and J. C. Weaver,** Rapid temporal control of transdermal drug delivery by electroporation, *Pharm. Res.* vol. 11, pp. 1834–1837, 1994.
133. **Lee, R. C. and Kolodney, M. S.,** Electrical injury mechanisms: dynamics of the thermal response, *Plast. Reconstr. Surg.,* vol. 80, pp. 663–671, 1987.
134. **Lee, R. C. and Kolodney, M. S.,** Electrical injury mechanisms: electrical breakdown of cell membranes, *Plast. Reconstr. Surg.,* vol. 80, pp. 672–679, 1987.
135. **Gaylor, D. C., Prakah-Asante, K., and Lee, R. C.,** Significance of cell size and tissue structure in electrical trauma, *J. Theor. Biol.,* vol. 133, pp. 223–237, 1988.
136. **Lee, R. C., Gaylor, D. C., Bhatt, D., and Israel, D. A.,** Role of cell membrane rupture in the pathogenesis of electrical trauma, *J. Surg. Res.,* vol. 47, pp. 709–719, 1988.
137. **Bhatt, D. L., D. Gaylor C., and Lee, R. C.,** Rhabdomyolysis due to pulsed electric fields, Plast. Reconstr. Surg., vol. 86, pp. 1–11, 1990.
138. **Gaylor, D. C., Bhatt, D. L., and Lee, R. C.,** Skeletal muscle cell membrane electrical breakdown in electrical trauma, in *Electrical Trauma: Pathophysiology and Clinical Management.* ed. R. C. Lee, J. F. Burke and E. G. Cravalho (Eds.), pp. 327–360, Cambridge University Press, 1992.
139. **O'Neill, J. J. and Tung, L.,** Cell-attached patch clamp study of the electropermeabilization of amphibian cardiac cells, *Biophys. J.,* vol. 59, pp. 1028–1039, 1991.
140. **Tung, L.,** Electrical injury to heart muscle cells, in *Electrical Trauma: Pathophysiology and Clinical Management,* ed. R.C. Lee, J. F. Burke and E. G. Cravalho (Eds.), pp. 361–400, Cambridge University Press, 1992.

141. **Weaver, J. C.,** Electroporation: A dramatic, non-thermal electric field phenomenon, pp. 95–100, 1993. Electricity and Magnetism in Biology and Medicine, M. Blank, Ed.
142. **Chang, D. C.,** Structure and dynamics of electric field-induced membrane pores as revealed by rapid-freezing electron microscopy, in *Guide to Electroporation and Electrofusion,* pp. 9–27, Academic Press, 1992. D. C. Chang, B. M. Chassy, J. A. Saunders and A. E. Sowers (Eds)
143. **Potts, R. O. and Francoeur, M. L.,** Lipid biophysics of water loss through the skin, *PNAS,* vol. 87, pp. 3871–3873, 1990.
144. **Chizmadzhev, Yu. A., Arakelyan, V. B., and Pastushenko, V. F.,** Electric breakdown of bilayer membranes: III. analysis of possible mechanisms of defect origin, *Bioelectrochem. Bioenerget.,* vol. 6, pp. 63–70, 1979.
145. **Bach, D. and Miller, I. R.,** Glyceryl monooleate black lipid membranes obtained from squalene solutions, *Biophys. J.,* vol. 29, pp. 183–188, 1980.

Chapter 7

EPIDEMIOLOGICAL STUDIES OF ELECTROMAGNETIC FIELDS AND HEALTH

Richard G. Stevens

CONTENTS

I. INTRODUCTION

Epidemiological studies to determine whether some aspect of the AC electric and/or magnetic fields (EMF) resulting from electric power is associated with increased risk of cancer have for the most part been crude and indirect. Past EMF exposure cannot be assessed accurately either by questionnaire or by contemporaneous field measurements. This fact has been used to claim, therefore, that the epidemiology does not support a relationship. However, it can equally be used to assert that the epidemiology does not rule out large effects. The epidemiology, so far, is weak both ways.

There are three types of study design used by epidemiologists: 1) cohort, 2) case-control, and 3) other. In a cohort study, a large group of people is identified and characterized as to demographics and exposure to the agent under study (e.g., smoking). This cohort of people then is followed over time and new cases of disease (e.g., lung cancer) are identified. After some specified period (e.g., many years), an analysis is performed. The number of cases in exposed people is divided by the number of person-years of follow-up among exposed to yield a rate in the exposed sub-group. The same rate is calculated for the unexposed subgroup. These two are divided to yield the relative risk. A relative risk of 10 for lung cancer among smokers has high impact on society because both disease and exposure are common. A relative risk of less than 2 for an uncommon exposure and an uncommon disease would have very little impact on society although, of course, any case of cancer has severe impact on the victim and family. A relative risk of less than 2 can have a large impact on society

0-8493-0641-8/96/$0.00+$.50

if the exposure and disease are common. The magnitude of a relative risk should be viewed within the context of the exposure and disease.

There are other commonly used statistics from other types of epidemiologic studies meant to approximate the relative risk. In a case-control study, cases of the disease under study are identified and then a sample of controls chosen from the general population is gathered to compare to the cases for exposure. An odds ratio is calculated, because a direct estimate of disease rates in exposed and unexposed is not possible. Standardized mortality or incidence ratios can be calculated from cohort studies. These take the age structure of the cohort and calculate an expected number of cases based on the experience of some reference population such as that of the entire United States. The ratio of observed cases to that expected from the population at large is the standardized moratlity ratio (SMR). Some occupational studies report a proporationate mortality ratio. (PMR). This is the ratio of the observed proportion of total cancer cases accounted for by the cancer site under study to the expected proportion from some referent population. The PMR is used when there is only information on the numbers of cancer cases and no information on the person years at risk in the occupations studied.

An important limitation of the epidemiology is exposure assessment. To date, attempts have been made to estimate chronic, long-term exposure to EMF expressed as a time weighted average, or TWA. Other potentially biologically active features of EMF such as peaks, or transients, or specific frequency-amplitude combinations, have received much less attention.

II. RESIDENTIAL STUDIES

Wertheimer and Leeper[1] initiated the interest in this area with a case-control study of childhood cancer. They compared 491 residences of children who died of cancer between 1950 and 1973 with 472 residences of controls selected from birth certificates.

The exposure to EMF was characterized by the nearby wiring configurations, with the assumptions that typical current flow past the home could be inferred from the wiring configuration and that this configuration is correlated with long-term, average, magnetic-field exposures in the home. They[1] first defined a high current configuration (HCC) and a low-current configuration (LCC). This was later[2] expanded to include four categories labeled very-high-current-configuration (VHCC), ordinary high-current configuration (OHCC), ordinary low-current configuration (OLCC), and the lowest category called "end-pole." Selected measurements of magnetic fields near homes not in the case-control study tended to corroborate these assumptions, with median measurements ranging from 0.9 to 7 milliGauss (mG) depending on the type of wiring. Other variables potentially related to cancer risk were obtained from the birth certificate (e.g., maternal education, birth order) and from the residence (e.g., urban-suburban residence, traffic flow past the home). The investigators found that high-current configurations were significantly more prevalent among case homes compared to control homes, with calculated odds ratios ranging from 1.5–3.0. This association was consistent for all case sub-groups (males and females, different cancer types, different ages), with some evidence for a dose-response gradient.

Several aspects of the design of Wertheimer and Leeper's study[1] limit the conclusiveness of their study. Their study considered deceased rather than incident cases, but, during the period of case ascertainment (1950–73), most childhood cancers were fatal. Second, several different methods were used to identify controls, with some but not others matched to cases by county of birth, but there is no obvious reason for this to bias their results. Third, the exposure indicator itself (wiring configuration) was not proven to be a valid reflection of EMF exposures in the residence. More importantly, the investigators assigned the exposure status of study homes with prior knowledge of whether the residence was occupied by a case or by a control. Thus, it is plausible that assignment of exposure status contained subtle biases that produced the positive results. The similarity of their findings across all cancer subtypes, all ages, and both sexes would be consistent with either (1) a subtle bias in assignment of exposure or (2) a consistent cancer-promoting effect. Data to counter the possibility of bias were provided by Wertheimer and Leeper,[1] who indicated that blinded coding produced a similar pattern of results to the overall study.

Finally, any attribute that is correlated with living in a high-current configuration home and is also related to childhood cancer (e.g., property value, occupation, nutritional status) could have caused the observed pattern, confounding the EMF/cancer association by mixing the effects of the extraneous variable with the effects of EMF. Because so little is known about the causes of childhood cancer or the correlates of wiring configurations, it is impossible to argue convincingly that such confounding has occurred. Nonetheless, data were not gathered to refute this possibility.

In spite of the limitations in Wertheimer and Leeper's study,[1] it was an important investigation and continues to serve as the prototypical study to which others are compared. None of the above challenges to their results convincingly negate their findings. Given the absence of human studies of cancer in relation to EMF prior to that time, this study raised important questions but could not, as a preliminary study, provide conclusive results.

Wertheimer and Leeper[2] reported a study of adult cancers in relation to residential EMF. Adult cancer cases (N = 1179) in the Denver area were compared to controls identified from death certificates and from a telephone survey. A revised wiring-configuration protocol was used that produced four levels of estimated exposure for above-ground wiring (Very High Current Configuration, Ordinary High Current Configuration, Ordinary Low Current Configuration, and Very Low Current Configuration) rather than the dichotomy of high versus low current. Remarkably similar findings of modest but statistically significant excesses of higher configuration homes were noted among cases compared to controls. For all cases, the odds ratios (with end-pole homes defined as the lowest exposure referent) were 1.46 for Ordinary Low Current Configuration homes, 1.66 for Ordinary High Current Configuration homes, and 2.18 for Very High Current Configuration homes. Although confidence intervals were not calculated, the results were reported as being highly statistically significant.

The same concerns with the validity of the exposure-classification rationale and its objective implementation apply to the study of adult cancers. The failure to consider such well-established risk factors as tobacco use, reproductive history, and occupational exposures to chemicals can be noted more specifically as potential confounders of the apparent association between adult cancer and imputed EMF exposures, in that these are known influences on cancer etiology. If such attributes were related to wiring configurations (an unexplored suggestion), they would produce a spurious association with cancer risk. The lack of specificity as to which cancers are affected might argue against the observed association being causal, but some carcinogens (e.g., ionizing radiation) affect multiple sites and in general, cancer promoters might be expected to produce more widespread effects than cancer initiators.

Tomenius[3] conducted a study of childhood cancer over the years 1958 to 1973 and EMF exposure in Sweden. This case-control study included all 716 children with cancer who were born and diagnosed in Stockholm County. Control children were selected from birth registration records to match each case child on age, gender, and church district at birth. The exposure variable consisted of a measurement of the magnetic-field level present near the front door of the residence, in addition to notation of visible electrical constructions (e.g., substations, overhead transmission lines) within 150 m of the home. Exposure assessments were conducted blindly with respect to case-control status.

Electrical constructions were observed more frequently within 150 m of case homes than control homes, with 200-kV wires more than 2.1 times as frequent near case homes. Furthermore, homes with visible 200-kV lines within 150 m did have higher average exposures than homes without such lines nearby. When the analysis was restricted to homes not within

150 m of electrical constructions, the average magnetic-field values did not differ between case and control homes (means of 0.69 and 0.68 mG, respectively). Among those same homes, however, a cutoff value of 3 mG resulted in an odds ratio of 2.1 comparing cases and controls. This was observed for both birth and diagnosis dwellings with the strongest effect observed for nervous system tumors (odds ratio = 3.7). There was an insignificantly reduced risk of leukemia associated with exposure above 3 mG. Females showed a stronger effect of exposure than did males in this study. It should be noted that there is virtually no contribution to EMF from a transmission line at 150 m.

A potential problem with this study is in the exposure assessment. The one-time magnetic-field measurement is certainly open to criticism as an adequate reflection of long-term levels of EMF within the home, with long-term exposures presumably relevant to cancer development. Another concern is with their decision to dichotomize their exposures at the relatively high value of 3 mG. There was no difference in the mean exposures between case and control homes, so that the choice of this cutoff value may have been critical to their observation of positive results. The pattern of results contained some anomalies, e.g., there was a positive association of magnetic fields and cancer risk only among homes not near visible electrical constructions.

Finally, the role of potential confounding variables remains unexplored in Tomenius's study,[3] as in the studies by Wertheimer and Leeper[1] and Fulton et al.[4] Nonetheless, there is a suggestion in the study reported by Tomenius[3] that inferred and measured magnetic-field exposures are more common among homes occupied by children with nervous system cancers than among homes occupied by controls.

Savitz et al.[5] reported results from a case-control study of childhood cancer in the Denver metropolitan area. Approximately 250 childhood cancer cases diagnosed from 1976 to 1983 were compared to 230 general population controls with respect to wire configuration codes (as developed by Wertheimer and Leeper[2]) and to spot measurements of in-home magnetic fields. Measured magnetic fields showed a tendency to be higher in case homes, with an odds ratio of 1.35 (95 percent confidence interval, 0.63, 2.90) for exposures above 2 mG, with an odds ratio of 1.93 (95 percent confidence interval 0.67, 5.56) for leukemias. Wire codes showed a more consistent and precise relationship. Dichotomizing wire codes into high (OHCC and VHCC) versus low (endpole and OLCC) produced an odds ratios of 1.53 (95 percent confidence interval of 1.04, 2.26) for total cancers and an odds ratio of 1.56 (95 percent confidence interval of 0.91, 2.67) for leukemias. Adjustment for several potential risk factors for childhood cancer and suspected correlates of wire codes and measured fields did not affect these results. The principal limitations in the study are a relatively high non-response rate (especially among controls), residential mobility restrictions on the controls which may have made them different from cases, and the use of a spot measurement of magnetic fields to estimate long-term exposure history. Although the study has limitations which preclude a firm conclusion, this study is consistent with the Wertheimer and Leeper results[1] and supports an association of proximity to power lines and risk of childhood cancer.

Severson et al.[6] reported on a study of residential EMF exposures in relation to adult acute non-lymphocytic leukemia in the Seattle area. Interviews, field measurements, and wire codes were obtained for 114 cases and 133 controls. Magnetic fields were estimated based on wiring configurations and measurements inside the residences. Based on longest-occupied residence, there was no association of magnetic field exposure and leukemia. A very slight increase in risk was noted for residences occupied closest in time to diagnosis date, but the authors interpreted this study as negative with respect to the primary hypothesis. As a negative study, however, the study also suffered from non-response, a small study group size, and the usual difficulties in accurately reconstructing exposure histories.

A study designed to replicate the methods of Savitz et al.[5] was reported by London et al.[7] in 1991. A new data-gathering effort was added to a completed study of childhood leukemia in the Los Angeles area. Cases of childhood leukemia were identified through the Los Angeles County Cancer Surveillance Program over the years 1980 to 1987. A matched-pair design was used in which controls were selected to individually match each case on age, ethnicity, and gender. The first 65 controls were selected as friends of cases and the remaining controls were selected by random-digit dialing. A total of 232 matched-pairs were included for study and there was a successful interview conducted with the mother of each case and each control. There were 24-hr magnetic field measurements collected on 108 matched-pairs, and wiring maps collected on 194 pairs. The Wertheimer-Leeper code was significantly associated with risk (OR = 2.15 for "very high" vs. "very low" , $p < 0.05$). Measured magnetic fields were not significantly associated with risk although the odds ratio for the highest field-level homes was somewhat elevated. This study suffered from the same problems of lack of participation of identified cases and controls as previous studies, and it remains unclear whether some sort of selection-bias due to random digit dialing for controls could account for the results. The authors, however, presented extensive data on how many subjects were eligible and how many were lost to inclusion in the study at each stage of the effort, from phone call, to answering the phone call, to interviewing, to magnetic field measurement.

The study by Feychting and Ahlbom[8,9] has received a great deal of attention. It is seen by many as a convincing replication of the association of proximity to power lines and risk of leukemia in children. The methods of their study could only be accomplished in a country such as Sweden with its extensive record keeping and record linkage on each of its citizens. The study population consisted of all people who had lived within 300 meters of any 220-kV or 400-kV power line in Sweden for at least one year over the period 1960 to 1985. Thus defined, there were 436,503 people available for analysis. There were 142 cases of cancer diagnosed in children (ages 0 to 15) over the study period and 548 cases leukemia and CNS tumor in adults. Four controls were chosen for each case matched on age, gender, and geographic locale within Sweden. Using distance to the power lines and historical load data, a magnetic field was calculated for each subject residence for each year of occupancy. In addition, the contemporary 50-Hz magnetic field was measured in each study residence. For childhood leukemia, the relative risk increased from 1.0 in those whose homes yielded a calculated magnetic field of < 1 mG closest to time of diagnosis, to 2.1 in those with fields between 1 and 2 mG, and to 2.7 in those with fields over 2 mG. It should be noted that there were no cases of leukemia in residences with calculated magnetic fields between 2 and 3 mG., i.e. all cases occuring in the > 2 mG group occured in houses with fields $>$ 3 mG. There was no relationship for brain tumors although the number of brain tumors was comparable (38 leukemias and 33 brain tumors). There was no relationship of either leukemia or brain tumors with measured magnetic fields in the current residence. In adults, there was no relation of calculated or measured magnetic fields and risk of brain tumors or leukemia overall. There was an insignificant elevation of acute myeloid leukemia (AML) and chronic myeloid leukemia (CML) in adults in residences with calculated fields over 2 mG. End-use voltage is higher in Sweden than in the United States (220 vs. 110) and so currents are lower; therefore the contribution from household current to EMF is lower and the contribution from transmission lines might be more accurately estimated by their methods in Sweden than in the U.S.

The Feychting and Ahlbom[8] study is a strong one. It avoided some of the shortcomings of the American studies[1,5,7] yet is consistent with them in its childhood leukemia results. The potential problems with random digit dialing were not an issue in the Swedish study. In addition, the availability of historical load data and precise residence information on each study subject allowed for calculation of magnetic field exposures for the distant as well as immediate past. Supporting the quality of these calculated fields, the calculated contemporary fields were remarkably similar to actual contemporary measured fields; calculated historical fields were not related to contemporary measured fields. However, it is not clear from what has been provided to the public by its authors whether it is, in fact, qualitatively superior to the previous studies or not. The crux of the debate is whether the calculated historic magnetic field means anything more than simply distance to the transmission line. Distance alone is also strongly related to risk of leukemia in children as can be seen in their report. The distinction is important for several reasons including the fact that people have had the opportunity to choose their residence on the basis of distance to power line but not on the basis of magnetic field level.

Ahlbom et al.[10] report a combined analysis of the data from Feychting and Ahlbom[8] and from studies in Denmark and Finland. The combined analysis shows an increased risk of leukemia in children associated with estimated residential magnetic field level, and no relation of brain tumors or lymphoma.

A number of other studies have been reported with mixed results.[11,12,13,14] Two studies from the same research group in England reported on adult haematological malignancies[12] and childhood cancer.[13] Both studies identified cases from the Yorkshire region, and both used calculated magnetic fields based upon current load in nearby overhead power lines. For the childhood study, controls were selected from birth records; for the adult study, hospital controls were chosen excluding diagnoses of cancer. In both studies a very low percentage of study subjects were defined as exposed, and in neither study were there significant associations of exposure with disease. The studies had low power because of the rarity of elevated exposures.

Coleman et al.[14] conducted a case-control study of leukemia and proximity to electricity distribution equipment (100 m) in south-east England. For 771 cases identified from the Thames Cancer Registry over the years 1965 to 1980, 1,432 controls matched on sex, age, and year of diagnosis were identified; these controls were also cancer cases from the same registry diagnosed with a solid tumor excluding lymphoma. There were no interviews conducted, and exposure was estimated on the basis of distance to substations or any overhead power line within 100 meters. In addition, a weight was used based on power company records of peak winter load during winter. Only 0.6% of study subjects lived within 100 m of an overhead power line, whereas 40% lived within 100 m of a substation. An insignificant odds ratio of 1.45 was associated with overhead power line and an odds ratio of 0.99 was associated with substations. There are many limitations of this study; the crude exposure metric would have worked against detecting a true association with magentic field exposure should one actually exist, whereas cancer controls may be inappropriate if other cancers are actually associated with magentic field exposure. McDowall[11] conducted the only cohort study of residential exposures to EMF and cancer mortality in East Anglia, England. A cohort of approximately 8,000 persons was identified in 1971 based on living in the vicinity of electrical transmission facilities (50 m from a substation or 30 m from an overhead power line). The cohort was traced through 1983 and 814 deaths were observed over that period. Mortality from all causes was slightly reduced in the exposed cohort, but mortality from leukemia, other lymphatic cancers, and lung cancers was increased in women (risk ratios of 1.54 to 1.75). Within the cohort, there was also a tendency for the subgroup of people living nearest to the installations to have higher mortality from lung cancer, leukemia, and other lymphatic cancers.

Use of electric blankets has been been examined as a possible risk factor for testicular cancer and leukemia as well as breast cancer. Savitz et al.[15] reported modest increases in risk of leukemia and brain tumors in children whose mothers had used electric blankets during gestation. Verreault et al.[16] reported no assocation at all of electric blanket use and testicular cancer based on a case-control study from Washington state. Cases were identified by the Cancer Surveillance System (a SEER registry) and were 20–69 years old during the period 1981–84. Controls were selected by randon-digit dialing from the same geographic area from which the cases appeared. All men were asked, among other things, if they had used an electric blanket in the past ten years, and if so, how long. Preston-Martin et al.[17] also found no association of electric blanket use and risk of leukemia in a case-control study in Los Angeles. These authors identified 224 leukemia cases (116 AML and 108 CML) and matched neighborhood controls. They then queried all study subjects on electric blanket use as part of their extensive questionnaire on many potential risk factors. Vena et al.[18] examined electric blanket use and risk of female breast cancer in a case-control study and found nonsignificant elevations of risk associated with continuous use throughout the night, although no "dose-response" with exposure duration was seen. The study had low power due to a small number of heavy users, limited information of blanket use restricted to only the previous ten years, and unknown reliability of reported use.

Exposure characterization is the most problematic issue in the studies described above. In the case-control studies, a variety of proxies have been used (wire codes, spot magnetic-field measurements), with the "gold standard" of long-term past exposures unattainable. It is not known what best approximates exposure history, nor is it certain what, if any, parameter of exposure is critical, e.g., total time-weighted exposure, proportion of time above or below a given level. Except for the potential for biased coding in the studies of Wertheimer and Leeper,[1,2] however, the errors in exposure classification would tend to be similar for both cases and controls and thus would produce a bias towards the null (odds ratio of 1.0) rather than produce a spurious elevation in risk.[19] On this basis, these random errors in exposure assignment cannot be invoked to dismiss the positive results that have been obtained.

Selection of a control series is a key concern in all of the case-control studies of residential EMF and cancer. The goal is to select controls who represent the source population from which cases arose with regard to their potential for exposure, so that differences between the cases and the controls can be interpreted as excess exposure to the cases rather than reduced exposure to the controls. Given the limited knowledge of what types of people occupy homes with differing exposures, it is difficult to specify how the selection of control groups might have produced a spuriously elevated exposure level among cases. However, those studies which used random digit dialing to select controls may have suffered a control selection bias[20] which may or may not have biased the results.

In spite of the inconsistencies and uncertainties in the studies of residential EMF and cancer, there are several suggestions of an association of risk with distance to power lines which should not be dismissed. For childhood leukemia, several studies have now provided evidence of elevated risk among children living close to power lines with no obvious common error applicable to all the investigations.[1,3,5,7,8] For studies of childhood brain tumors there is not a consistent result. The data for adult cancers are more limited and inconsistent.[2,6,8]

III. OCCUPATIONAL STUDIES

A number of the early surveys of EMF health were conducted in the Soviet Union as reviewed by Bonnell[21] and Knave et al.[22] For example, high-voltage switchyard electricians reported an excess of such symptoms as headache, irritability, loss of memory, insomnia, sweating, fatigue, general weakness, and sexual impotence,[23] some of which are related to depression. According to Asanova and Rakov,[24] 37 of 45 high-voltage switchyard workers complained of headache, sluggishness, fatigue, or sleepiness, apparently related to the timing of work during the day and to increased duration of employment.[25] As concluded in the reviews cited above, the nature of the symptoms (very common, non-specific, and amenable to reporting biases) and the quality of the study designs (no comparison groups, no verification of self-reported complaints, no documentation of exposure) make the Soviet literature inconclusive.

More recently, several types of study designs have been used to assess an association between cancer and occupational exposures to EMF. None of these studies actually measured such exposures; instead, indirect assessments of exposure were based on job titles. This literature can be divided conveniently into surveys of existing data bases primarily for leukemia patterns, case-control studies of leukemia and brain cancer in which occupational exposures were considered, and cohort studies of industrial populations potentially exposed to EMF.

A series of "Letters to the Editor" were published in The New England Journal of Medicine and The Lancet from 1982 to 1985 concerning men occupationally exposed to electric and magnetic fields. The results of these surveys were aggregated and reviewed by Savitz and Calle.[26] These surveys analyzed data from vital statistics or cancer registries in order to determine whether leukemia diagnoses or deaths were overrepresented among persons engaged in occupations assumed to entail high levels of exposure to EMF ("electrical occupations"). The opportunity to explore inexpensively these existing data bases is appealing, especially because the large populations surveyed result in relatively high statistical precision in risk estimates. Unfortunately, the ability of such investigations to address carcinogenic effects of EMF is limited.

Savitz and Calle[26] included data from the 1950 survey of occupational mortality in the United States,[27] letters to the editor concerning cancer among men in Washington State,[28] Los Angeles,[29] England and Wales,[30] Southeast England,[31] New Zealand,[32,33] and Wisconsin,[34] amateur radio operators;[35] and tabulations of cancer cases or deaths in Massachusetts[36] and California.[37] The designs of these investigations were very similar, with exposure to EMF inferred from the occupation indicated on the death certificate or from membership in an organization of amateur radio operators.[26]

Integration of these data sources provides some evidence of a modest (20 percent) elevation in total leukemia risk across all "electrical occupations," with larger relative risks (ratio of disease risk among exposed divided by risk among unexposed) for acute leukemia of 1.4 and for acute myelogenous leukemia of 1.5. For specific occupational groups, these risks were larger. On the basis of the criteria of magnitude of risk elevation and replication in several data sets, occupation-related increases in leukemia were most suggestive among telegraph, radio, and radar operators (risk ratio for total leukemia of 1.8, 95 percent confidence interval of 1.4, 2.6); power and telephone linemen (risk ratio for total leukemia of 1.3, 95 percent confidence interval of 1.0, 1.6); and electrical and electronic engineers (risk ratio for total leukemia of 1.2, 95 percent confidence interval of 1.0, 1.5). For these groups, the relative risks of acute myelogenous leukemia were 2.6, 2.5, and 1.9, respectively.

There are several methodological concerns that limit these surveys. Many of these reports were based on the relative proportion of deaths[28,35] or cancers[29] due to leukemia rather than the absolute risk (actual incidence rates) of leukemia. Protection from other major causes of death (as is commonly observed among working populations) or other types of cancer would manifest itself as a proportionate excess of leukemia. In many instances, results were based on very small numbers of exposed leukemia cases, reflecting the rarity of this disease. Exposure definition is especially problematic in these studies. Exposure assignment based on job title on the death certificate or in the cancer registry is an inherently error-prone method if one were able to compare the inferred to actual EMF exposures. In general, however, this exposure misclassification would be expected to dilute rather than exaggerate relative risk estimates.

Two case-control studies of leukemia were concerned principally with other exposures, Stern et al.[38] focusing on ionizing radiation and solvents and Flodin et al.[39] on indoor radon. Each found markedly increased risk associated with occupations presumed to involve elevated EMF exposures. Risk of AML according to occupation was studied by Flodin et al.[40] who identified 86 cases for hospital records occurring over the period 1977 to 1985 in Sweden and 172 control subjects matched on age, gender, and county of residence. There had been ~170 cases that had occurred in the study area over the same time interval based upon the regional cancer registry. The study was designed to examine a wide range of possible occupational associations with AML. They found a non-significant association with "electrical work" and an estimated odds ratio of 2.1.

Several cohort studies of cancer among electronics-industry workers provide yet another approach to assessing carcinogenic hazards associated with EMF exposure. Vågerö and Olin[41] examined cancer morbidity in the Swedish electronics industry. Among male workers, small excesses were noted in pharyngeal and respiratory cancers. No evidence for excess leukemia incidence was found. A study of Italian thermoelectric power-plant workers[42] found some evidence of increased cancer risk, but the limited population size precluded evaluation of leukemias and brain tumors in particular.

Juutilainen et al.[43] grouped all male Finnish industrial workers, aged 25 to 64 during 1971–80, into "no exposure," "possible exposure," and "probable exposure" to EMF based on a modification of the scheme used by Lin et al.[44] No measurements of EMFs were presented. Incident cases of leukemia or CNS tumor occurring over the same study period, 1971–80, were identified through the Finnish Cancer Registry. Expected cases in each of the three EMF exposure groups were calculated based on the age- and period-specific incidence rates for the cohort across occupations. They reported relative risks 1.85 ($p < 0.05$) for all leukemia, and 1.47 (NS) for AML in the "probable exposure" group, and 1.42 ($p < 0.05$) and 1.37 (NS) in the "possible exposure" group. For CNS tumors, the relative risks were 1.31 (NS), and 1.29 ($p < 0.05$) for "probable" and "possible" exposures. The authors point out several factors that could militate against observing an effect in their study should it actually exist in the population at large. First, assignment to an exposure category was certainly crude, with many workers in each assigned group actually experiencing exposure apropriate to another group. In addition, occupation was defined on the basis of the 1970 census so that some workers may have left exposed jobs and some may have come from unexposed jobs to exposed jobs over the study period. The authors also point out that socioeconomic status probably does not affect their results because the cohort only included "industrial workers."

Pearce et al.[45] reported on analyses of the New Zealand Cancer Registry designed to determine if there were elevated risks of cancer in workers occupationally exposed to EMF. During the period 1980–84 there were 19,904 cases of cancer in the registry for which there was also occupational information (80% of the total 24,762 cancer cases registered in men aged 20 years or more). The study was done as a series of case-control studies in which each cancer site was compared to the remainder of cancer sites combined. Odds ratios were calculated based on the numbers of cases who had been employed in EMF-exposed jobs compared to the number of "controls" (all other cancer sites combined). Leukemia was significantly associated with EMF occupations; the strongest associations among the EMF occupations were with radio/television repairmen, electricians, linemen, and power station operators. The strongest leukemia subtype association was with chronic lymphatic. In a companion study of the same study population,[46] no increased risk of brain tumor in electrical workers overall was reported, although a non-significant odds ratio of 1.91 was seen for electricians.

Preston-Martin and Peters[47] conducted a case-control study of CML in Los Angeles. They found a strong association risk and occupation as a welder. Of 130 cases included in the study, 22 had worked as welders, whereas only 4 of 130 controls matched individually to the cases on neighborhood, gender, race, and age had been welders. They concluded that the association was so strong that an exposure from the welding occupation probably accounted for an increased risk of CML, although they could not say whether gases, fumes, or radiation (ionizing or non-ionizing) was responsible.

A study by Balli-Antunes et al.[48] provided an examination of risk of death from all leukemia in railway engine drivers in Switzerland, a group presumed to be exposed to EMF. The virtue of the study was the tremendous size of the "exposed" and "unexposed" groups: 70,000 man-years from 1969–1983 in the railway drivers, and 900,000 and 1.56 million in the two control groups of "metal construction and machine building" and "technical personnel" respectively. They reported nonsignificant elevations of PMR (144) and SMR (108) when compared to the builders, and significantly elevated PMR (163) and SMR (171) when compared to the technical personnel. The first control group was chosen because it was believed to entail exposures to chemicals similar to those to which the railway engine drivers are exposed, whereas the second control group was chosen because no exposures were thought to be similar but socioeconomic status was similar. The authors considered potential biases in detail, clearly defining the strengths and weaknesses of their study.

Törnqvist et al.[49] conducted a cohort study of Swedish occupational EMF exposure and risk of leukemia or brain cancer over the years 1961–79. For all "exposed" electrical occupations, SMRs were near unity for both leukemia and brain cancer. With a group of nearly 134,000 defined as "exposed" this study was large, and encompassed the entire working population of Swedish men. Certain subgroups of the all exposed group showed some elevated point estimates of relative risk.

Two large cohort studies designed to survey possible associations of usual occupation with risk of specific cancer types in order to generate hypotheses did not report any remarkable EMF-related risks for leukemia or other cancers.[50,51]

Matanoski et al.[52] reported an elevated leukemia risk of 2.5-fold among telephone employees working in jobs with high EMF exposures. Theirs was a nested case-control study within a cohort of New York State telephone company emplyees. Subjects who died of leukemia (excluding chronic lymphocytic) over the period 1975 to 1980 were matched to controls on gender, age, and year of hire.

Tynes et al.[53] reported on cancer risk among working men in Norway according to possible EMF exposure on the job. They identified 37,945 men who worked in jobs thought to have elevated EMF exposure in 1960 and compared their cancer experience to the total population of working men up to 1985. No measurements of EMF were made in the study. The standardized incidence ratio for leukemia was 1.41 and for brain cancer was 1.14. No other cancer sites were associated with electrical occupations.

Another recently reported study from Sweden was of occupational EMF exposure and risk of leukemia and brain tumors.[54] This was a case-control study based upon all men aged 20–64 living and working in Mid-Sweden or Gothenburg county in 1980. Among these men there were 325 cases of leukemia and 346 of brain cancer that occurred between 1983 and 1987 that were eligible for study. Two controls matched on age were selected for each case from among the base population. The men were therefore representative of the general working population and were not restricted to particular occupations. Exposure was assigned based upon actual field measurements in the working environment of each study subject or a surrogate by use of a personal dosimeter. The job held longest in the ten years preceding diagnosis was chosen, and at least six hours of measurements were taken at 1-second intervals. The authors argued that these contemporary measurements were adequate to at least rank men on exposure if not give a true quantitative assessment of field levels at the time of employment before disease occurrence. Men were grouped according to field levels by four different, but related schemes: mean field level, median, standard deviation, and percentage of time above 2 mG. There was no relationship of AML to assigned exposure category of the men (87 cases). There was a significant relationship of risk of chronic lymphocytic leukemia (CLL) to assigned exposure (104 cases). Results for brain tumors (258 cases) show an insignificant excess in the men with a mean magnetic field in the highest quartile ($> =$ 2.9 mG; OR $=$ 1.38), and a significant excess in men with a median magnetic field in the highest quartile (> = 1.7 mG; OR = 1.48). The results of this study are not consistent with other occupational studies of leukemia in that risk of AML was not related to EMF exposure. The results are consistent with other studies in showing a marginal association of risk of brain cancer with exposure.

Similarly rigorous exposure assessment was done in two other occupational studies by Sahl et al.[55] and Thériault et al.[56] Sahl et al.[55] evaluated mortality from leukemia, lymphoma, and brain cancer in a cohort of 36,221 electric utility workers in California over the period 1960 to 1988. Analyses based on job category were supplemented with anaylses based on exposure scores for specific tasks within job categories. There was no evidence for an association of electrical occupations or EMF exposure scores with risk of these cancers.

Thériault et al.[56] reported increased risk of non-lymphocytic leukemia in workers in jobs expected to have above the median exposure to magnetic fields. Each of ~220,000 workers in three utilities in Canada and France had an estimate of exposure based upon measurements made on 2,066 current wokers in specific job categories. There was no evidence of dose-response and the association was not consistent among the three utilities. No significant associations were observed for any of 29 other cancer sites. In particular, 7 cases of breast cancer occurred whereas 8.25 were expected; the relative risk for prostate cancer was 1.3, but not significant.

Lin et al.[44] presented the results of a case-control study of brain-tumor mortality in relation to estimated occupational exposures to EMF. Occupational histories of 951 brain-tumor decedents in Maryland (1969–1982) were contrasted with the work histories of controls who died of causes other than cancer. Jobs in which EMF exposures were likely to occur were noted more frequently among victims of primary brain tumors (gliomas and astrocytomas) compared to controls. Unlike the preceding studies, this preliminary finding was followed by an effort to improve exposure assessment through a more detailed consideration of job titles. Potential job exposure to EMF was categorized as definite, probable, possible, or no exposure based on a panel of experts consisting of an industrial hygiengist, an occupational physician, and a radiation physicist. In the case-control analysis, an odds ratio of 2.15 (95 percent confidence interval 1.10, 4.06) was found for definite exposure relative to no exposure, an odds ratio of 1.95 for probable exposure (95 percent confidence interval 0.94, 3.91), and an odds ratio of 1.44 (95 percent confidence interval 1.06, 1.95) for possible exposure. The pattern of results was primarily due to the markedly increased risk among electric or electronic engineers and technicians.

The results of Lin et al.[44] provide a stronger suggestion of an association between at least some aspect of occupation and cancer than many of the preceding leukemia reports. The increased strength of association for primary brain tumors as opposed to "brain tumors; type unclassified" is suggestive of an etiologic association. Furthermore, the risk gradient based on the presumed exposure classification suggests that the EMF exposures may have an important role in the findings, though the data are still limited to presumed (rather than measured) exposures. Also, confounding factors were limited to those on the death certificate. Nonetheless, in spite of these real limitations, the study by Lin et al.[44] was of sufficient quality to encourage continued study of brain tumors and electrical occupations.

Some corroborative evidence that EMF might increase brain cancer is derived from Milham's[57] proportionate mortality analysis in which a risk ratio of 1.2 was found for all electrical workers and a risk ratio of 1.5 for electricians. Two-fold increased risks of brain cancer (based on very sparse data) were reported by Coggon et al.[58] for electrical engineers. Thomas and Waxweiler[59] cited data obtained by Guralnick[27] and Petersen and Milham[37] also indicating excessive brain cancer mortality among men who worked as electricians or linemen.

Thomas et al.[60] conducted a large and thorough evaluation of brain cancer mortality through a case-control study covering several areas in the United States. Occupations among 435 men dying of brain cancer were compared to those of controls who died of other causes.

Using Lin et al.'s[44] exposure groups, an odds ratio of 1.6 (95 percent confidence interval 1.0, 2.4) was obtained. The increased risk was concentrated among electrical equipment workers, leading the authors to suggest that solvents are a more likely explanation for their results than EMF. Linemen, however, also showed imprecise elevations in brain cancer mortality. This illustrates the inherent limitations in the ability of job titles to distinguish between chemicals and EMF as the etiologic agent, especially when exposures are estimated rather than measured.

A case-control study in East Texas[61] was very similar in design to that conducted by Lin et al.[44] in Maryland. Men dying of brain cancer (N = 202) were contrasted with 238 controls who died of other causes. Workers presumed to be exposed to EMF (utility workers, electricians, and electronics workers) had a 3.9-fold excess risk of brain cancer. Using Lin et al.'s[44] exposure classification, odds ratios (with 95 percent confidence intervals) of 1.2 (0.7, 1.8), 2.9 (0.8, 10.3), and infinity (6 exposed cases, 0 exposed controls) were found for possible, probable, and definite electromagnetic field exposure, respectively. Though questions remain regarding the accuracy of exposure assessment, there seems to be a relatively firm basis for suggesting that some aspect of "electrical work" increases the occurrence of brain cancer. An editorial accompanying the paper by Speers et al.[61] acknowledges the substantial gaps in the epidemiologic studies but suggests that EMF "must be considered an environmental hazard and dealt with accordingly, until proven otherwise."[62]

Törnqvist et al.[63] assessed cancer incidence over a 19-year period among 3,358 power linemen and 6,703 power station operators, groups very likely to have elevated electric and magnetic field exposures. Relative to the general population cancer rates, linemen had relative risks for leukemia and brain cancers of 1.5 and 1.3 based on 13 and 10 cases, respectively. No excesses were seen among power station operators. Finally, McLaughlin et al.'s[64] brain cancer study found no associations with any of the subgroups of electrical workers considered.

Milham[65] conducted a cohort study of 67,829 amateur radio operators in Washington State and California whose hobby presumably entails elevated exposures to EMF. Modest increases in leukemia (risk ratio of 1.2, 95 percent confidence interval 0.9, 1.7) and brain cancer (risk ratio of 1.4, 95 percent confidence interval 0.9, 2.0) were noted, with a more pronounced excess of AML (risk ratio of 1.8).

In a letter to the editor, Gallagher et al.[66] briefly described a study of death certificates from British Columbia over the years 1950 to 1984. Proportional mortality ratios (PMR) for brain cancer were elevated for 7 of 9 occupations thought to have some exposure to EMF, and in all groups combined, though not significantly so. However, among six occupations with significantly elevated PMRs for brain cancer, the only common thread seemed to be high socioeconomic status, not EMF exposure. Gallagher et al.[67] also reported insignificantly elevated PMRs for leukemia in some "electrical" occupations but not others.

In 1989, Preston-Martin et al.[68] published the results of a case-control study of brain tumors that was designed to identify occupational and other exposures that might be related to risk of brain tumor. They recruited 272 cases of brain cancer (including glioma and meningiomas) from 478 cases reported in Los Angeles county during 1980 to 1984. They also recruited one matched neighborhood control for each case. Their primary finding was that meningiomas, but not gliomas were strongly related to previous history of head injury. Another finding was that astrocytomas, a subset of gliomas, were associated with occupation in a job with probable EMF exposure. In their abstract they state "This risk was greatest for astrocytoma (OR for employment in such jobs for > 5 years = 4.3; CI = 1.2–15.6)." The same group[69] presented another analysis of the same data set with nothing new being contributed.

The case-control study of Loomis and Savitz[70] utilized the mortality data tapes provided by the National Center for Health Statistics (NCHS) on the 16 states that code occupation and industry on death certificates and supply this data to NCHS. They identified 2173 men who died of primary brain cancer during 1985 and 1986, 3400 who died of leukemia, and

10 controls for each case from any death excluding brain cancer or leukemia. They defined EMF exposed occupation on the basis of expected exposure compiled from previous studies. They also attempted to take account of possible chemical exposures that might confound associations with EMF. They reported significant associations of EMF occupational exposure and risk of brain cancer, but not leukemia.

Melanoma[71,72,73] and breast cancer in men[74,75,76] have also been implicated in some studies.

Guénel et al.[77] in Denmark and Vågerö and Olin[41] in Sweden both reported no association of EMF occupation and breast cancer risk in women. However, there may have been extensive exposure misclassification. The definitions of EMF exposure occupations for women were broad. In Vågerö and Olin,[41] about 2.6% of men and 2% of women in Sweden were classified as in exposed occupations. This was broadly defined as the "electronics or electrical manufacturing industry." In Guénel et al.,[77] about 1.7% of men and 0.61% of women were classified as continuously exposed; and 14.5% of men and 12.6% of women were classified as intermittently exposed.

Loomis et al.[78] reported on an occupational study of breast cancer death in women in the United States. There was a significant excess of breast cancer deaths in women working in electrical occupations in contrast to Guénel et al.[77] and Vågerö and Olin.[41] There are differences in the studies. In the U.S. study, women were classified as exposed if they worked in jobs that had been previously used to classify men as being exposed to EMF. In the study population, 2.6% of men were defined as in exposed occupations; similar to Guénel et al.[77] and Vågerö and Olin.[41] For women, however, only 0.18% were in those occupations. It is this group that showed the significant elevation. An additional group of possible exposed occupations made up only 2% of women, and among these there was not a significant elevation of breast cancer risk.

Some methodologic issues merit comment in interpreting the results of all of the occupational studies discussed above. First, the quality of exposure information (duration, intensity, field frequency) that can be derived from occupational titles is not quantified in any of these reports. Although precise methods for defining EMF exposures for occupationally exposed groups remain elusive, sufficient experience has accumulated to show with confidence that either industry of employment or job title alone is incomplete.[22,79] The definition of exposed groups is often arbitrary and judgmental rather than based on empirical evidence.

Second, occupational exposures never occur in isolation. For example, Wright et al.[29] noted that, in addition to electric and magnetic fields, workers were potentially exposed to "metal fumes, solvents (including benzene), fluxes, chlorinated biphenyls, sythetic waxes, epoxy resins, and chlorinated naphthalenes." This same issue can, however, be viewed in the reverse in the case of brain cancer. Presumed chemical exposures in brain cancer etiology were re-interpreted by Lin et al.[44] as having a potential EMF explanation. The problem is that no occupational title unambiguously defines workers who are exposed only to EMF.

Third, other known cancer determinants such as ionizing radiation from medical treatment, tobacco smoking, or asbestos exposure were not examined. Even readily available information on social class was not considered as a possible confounder in most studies. If "electrical occupations" contain a substantial proportion of technically trained, relatively well-paid workers and if leukemia risk is somewhat elevated among the higher social classes, then the observed findings may be attributable to social-class differences.

In spite of these sources of uncertainty, the degree of consistency in these reports should not be ignored. Generally, the misclassification of EMF exposure using job titles as a proxy should apply to both the cancer cases and the controls or the source population of the cases. This phenomenon tends to dilute any real associations with EMF exposure rather than to inflate them. Second, the different approaches used (large population surveys, case-control designs in occupational settings and the general population) have different methodological strengths and weaknesses, yet several of these approaches tend to point in the same direction.

The failure to consider other occupational exposures or other leukemia risk factors allows for the possibility that some other workplace hazard accounts for the observed patterns, but such a factor has yet to be identified.

Occupational studies are often the most feasible studies to perform due to the fact that a cohort of subjects has already been defined and followed. Thus, the marginal cost in both money and effort of examining a possible association between a putative risk factor and disease outcome is lower than for a study in the population at large. It is important to note however that lack of an association of a risk factor and disease in the occupational setting does not mean that there is no problem with exposure to that same risk factor in the population at large. For example, occupational studies set within a society with an already high disease burden (e.g., breast cancer in Sweden) may not detect an increased risk in high exposure job categories because the population exposure level may already be at a saturation level.

IV. REPRODUCTIVE OUTCOME

Studies have reported a possible effect of EMF on six different reproductive endpoints: infertility and sex ratio,[22] congenital malformations,[80] prolonged gestation and fetal loss,[81,82] and paternally mediated childhood cancer.[83] None of the individual findings are convincing that EMF exposures actually caused the observed reproductive problems, although the Juutilainen et al.[82] study is suggestive. Exposures were notably different between the Swedish studies and the American study (power-company occupational exposure versus electric blanket and heated waterbed), and even the sex of the exposed person (father versus mother) varied. There are no corroborations of any of the positive findings across studies. The conclusion from this review of the literature is that evidence of adverse effects of EMF on human reproduction is lacking, although these studies do provide some suggestions which should be addressed in future investigations.

The data on potential adverse reproductive outcomes resulting from work with video-display terminals (VDTs) are given below. This issue has generated a tremendous amount of concern. The exposure characteristics associated with work near VDTs is sufficiently different from a simple, ELF electric or magnetic field that the applicability of VDT work to such fields is questionable.[84]

Goldhaber et al.[85] used records of the Kaiser Permanente clinics and examined pregnancy outcomes among 1,583 women. Women working with VDTs over 20 hours per week were 1.8 times more likely to have a miscarriage compared to other working women and 1.4 times as likely to have an infant with a birth defect. As noted by the authors, the principal concern is that women who have had an adverse outcome may exaggerate their history of VDT exposure given the public attention focused on this potential effect. Furthermore, there was no opportunity to distinguish other aspects of the work situation (e.g., stress) from VDT exposure per se.

Schnorr et al.[86] conducted a retrospective cohort study of telephone operators in two telephone companies serving eight southeastern states. Women were defined as eligible for inclusion if they were between 18 and 33, married during the period 1983 and 1986 and worked as either a directory-assistance operator (VDT user) or a general operator (no VDT use). Of the 5,544 women identified as eligible, 4,246 agreed to participate. Of these, 2,430 were married during the study period and were interviewed in detail. Outcomes of eligible pregnancies yielded 54 spontaneous abortions among 307 pregnancies in the VDT-using directory-assistance operators (14.8%), and 82 spontaneous abortions among 430 pregnancies in the general operators who did not use VDTs at work (15.9%). The authors concluded that work with VDTs did not increase risk of spontaneous abortions. This study has many strengths. The study was designed specifically to examine VDT use and risk of spontaneous abortion. The two groups were comparable with respect to demographics, socio-economic status, and work environment except for the VDT use. Measurements of electric and magnetic fields were made in the typical work environment of 48 directory-assistance operators (VDT users) and 24 general operators. A caution about generalizability concerning EMF, however, is that the non-VDT using general operators were nevertheless working in a rich EMF environment, and were not representative of the general population of women with regard to EMF exposures.

Slesin and Connelly[87] point out that exposure to ELF (extremely low frequency) was similar for the VDT users and the non-users although VLF (very low frequency) was higher for the users. In addition, they point out that the original study was not designed to include field measurements, and such measurements were made years after the subject pregnancies. Schnorr et al.[88] counter that their study successfully examined possible adverse effects of VDT use among office-working women, and that to use a control group of non-working women would have introduced intractable problems of non-comparability of controls with VDT-users. Taken together, these arguments limit the generalizability of the Schnorr et al.[86] study as stated above.

Lindbohm et al.[89] studied spontaneous abortions and VDT use in Finland. Spontaneous abortion differs from EPL (early pregnancy loss) in that EPL is a pregnancy loss that is not realized by the woman whereas a spontaneous abortion is a pregnancy loss that is detected by the woman. In general, therefore, EPL occurs earlier in pregnancy than spontaneous abortion. In their study, Lindbohm et al.[89] identified women who had had a spontaneous abortion over the period 1975–85 from Finnish medical registers; 3 controls from all women with live births were chosen for each case and indivdually matched on age and year of pregnancy. Questionnaires were mailed to 368 identified cases and 1,098 controls. The final study population consisted of 191 cases and 394 controls. There was a low response rate. The questionnaire asked about work history, and VDT use in particular. A subject was classified as a VDT user if she used a VDT during the first trimester of pregnancy and her employer at the time (3 to 13 years prior to the study) confirmed that she used one. There were 87 case users and 172 control users (OR = 1.1, NS). Among the VDT users 17 different models were identified, and these were measured in the laboratory for ELF magnetic fields, VLF magnetic fields, and the time derivitive magnetic flux density in the VLF range (many years after the actual exposure period). These measured magnetic fields were then assigned each women depending on the type of VDT she used. Among users (87 cases and 172 controls) the OR for those using a VDT >20 hours per week compared to those using it between 1 and 10 was 2.0 (NS). Also among users, the OR for use of a VDT with >9 mG compared to those <0.4 mG was 3.4 ($p < 0.05$). However, when users of terminals with >9 mG were compared to nonusers the OR is 1.6 and not significant. The authors discuss several possible interpretations of these data.

The suggestion that VDT use might be associated with miscarriage warrants (and is receiving) the attention of researchers. Regardless of whether it is a physical, psychological, or ergonomic hazard, the widespread use of VDTs argues for a thorough evaluation of its potential health implications.

V. CONCLUSIONS

Among the human health effects examined in relation to EMF, the greatest quantity and most suggestive data are available for cancer. Childhood leukemia, adult leukemias, and adult brain tumors have some epidemiologic support for an association with EMF, as does breast cancer in men. Considerable uncertainty remains, however, regarding the causative relationship between magnetic fields and cancer risk due in part to difficulties in exposure assessment. The latest residential study from Sweden[8] is the first to address long term exposure quantitatively. Indirect measurements of long-term exposures in residences leaves open the possibility that some other correlate of wire configurations (e.g., pesticide use, air pollution) is the causative factor. Similarly, occupational studies of cancer fail to distinguish between EMF and other workplace exposures. However, the weight of evidence supports an association of proximity to power lines and risk of, at least, childhood leukemia, and the weight of evidence supports an association of putative occupational EMF exposure and increased risk of leukemia and brain tumors. It is important to determine if these associations are causal. In addition, other possible cancer associations have begun to be examined that are plausible based on laboratory research on melatonin. These include breast cancer in women, and prostate cancer in men. The hypothesis that prolonged magnetic field exposures might increase the risk of cancer in humans deserves continued serious scrutiny.

Reproductive effects have been examined epidemiologically. The evidence is contradictory and lacks consistency. Although marginal effects are reported in some studies, different effects are reported in others. One of the only studies to attempt to address exposure to ELF magnetic fields directly reported an increased risk of EPL[82] in women living in residences with high fields. The bulk of studies of VDT use show no increased risk of adverse pregnancy outcome associatied with use, although the Lindbohm et al.[89] study does report increased spontaneous abortion among women using VDTs with high fields compared to women using VDTs with low fields.

In the realm of other clinical or subclinical health effects, epidemiologic research has not produced persuasive evidence of widespread adverse effects of occupational EMF exposures. Early suggestions of such effects have been refuted by the more recent methodologically superior research, although a very carefully done study does support an association of residential proximity to a transmission line and depression.[90] There is a biological rationale for a possible association, and this area should not be considered resolved.

In spite of continued questions about the biological rationale and plausibility of EMF harming human health, the suggestive epidemiologic data cannot and should not be dismissed. Even effects which initially seem implausible in the laboratory may prove to be causative, as illustrated by the early difficulties in confirming carcinogenic effects of benzene and inhaled arsenic, known human carcinogens. In the case of EMF, the possibilities are of more than theoretical interest since the exposures are virtually ubiquitous in industrialized societies. Even a small elevation in cancer or other disease among more highly exposed persons would have an important impact due to the pervasiveness of such exposures. For these reasons, research on possible health effects of EMF is an important component of biomedical research.

ACKNOWLEDGMENTS

Battelle, Pacific Northwest Laboratories, Richland, WA 99352. Supported by the U.S. Department of Energy under contract DE-AC06-76RL0 1830 and by the National Cancer Institute grant R01-CA55844.

This chapter is a condensed version of Appendix I of a report that will be published by the National Council on Radiation Protection and Measurement (NCRP) Scientific Committee 89–3 entitled "Extremely Low Frequency Electric and Magnetic Fields" and is included with permission of the NCRP.

REFERENCES

1. **Wertheimer, N. W. and Leeper, E.,** Electrical wiring configurations and childhood cancer, *Am. J. Epidemiol.*, 109, 273, 1979.
2. **Wertheimer, N. W. and Leeper, E.,** Adult cancer related to electrical wires near the home, *Int. J. Epidemiol.*, 11, 345, 1982.

3. **Tomenius, L.**, 50-Hz Electromagnetic environment and the incidence of childhood tumors in Stockholm County, *Bioelectromagnetics,* 7, 191, 1986.
4. **Fulton, J. P., Cobb, S., Preble, L., Leone, L., and Forman, E.**, Electrical wiring configurations and childhood leukemia in Rhode Island, *Am. J. Epidemiol.*, 111, 292, 1980.
5. **Savitz, D. A., Wachtel, H., Barnes, F. A., John, E. M., and Tvrdik, J. G.**, Case-control study of childhood cancer and exposure to 60-hertz magnetic fields, *Am. J. Epidemiol.*, 128, 21, 1988.
6. **Severson, R. K., Stevens, R. G., Kaune, W. T., Thomas, D. B., Hueser, L., Davis, S., and Sever, L. E.**, Acute nonlymphocytic leukemia and residential exposure to power frequency magnetic fields, *Am. J. Epidemiol.*, 128, 10, 1988.
7. **London, S. J., Thomas, D. C., Bowman, J. D., Sobel, E., Cheng, T-C, and Peters, J. M.**, Exposure to residential electric and magnetic fields and risk of childhood leukemia, *Am. J. Epidemiol.*, 134, 923, 1991.
8. **Feychting, M. and Ahlbom, A.**, Magnetic fields and cancer in people residing near Swedish high voltage power lines, *Institutet för miljömedicin, Karolinska institutet,* 1992.
9. **Feychting, M. and Ahlbom, A.**, Magnetic fields and cancer in children residing near Swedish high-voltage power lines, *Am. J. Epidemiol.*, 138, 467, 1993.
10. **Ahlbom, A., Feychting, M., KosKenvuo, M., Olsen, J. H., Pukkala, E., Schulgen, G., and Verkasalo, P.**, Electromagnetic fields and childhood cancer, *Lancet,* 342, 1295, 1993.
11. **McDowall, M. E.**, Mortality of persons resident in the vicinity of electricity transmission facilities, *Br. J. Cancer,* 53, 271, 1986.
12. **Youngson, J. H., Clayden, A. D., Myers, A., and Cartwright, R. A.**, A case-control study of adult haematological malignancies in relation to overhead powerlines, *Br. J. Cancer,* 63, 977, 1991.
13. **Myers, A., Clayden, A. D., Cartwright, R. A., and Cartwright, S. C.**, Childhood cancer and overhead powerlines: a case-control study, *Br. J. Cancer,* 62, 1008, 1990.
14. **Coleman, M. P., Bell, C. M. J., Taylor, H-L., and Primic-Zakelj, M.**, Leukaemia and residence near electricity transmission equipment: a case-control study, *Br. J. Cancer,* 60, 793, 1989.
15. **Savitz, D. A., John, E. M., Kleckner, R. C.**, Magnetic field exposure from electric appliances and childhood cancer, *Am. J. Epidemiol.*, 131, 763, 1990.
16. **Verreault, R., Weiss, N. S., Hollenbach, K. A., Strader, C. H., and Daling, J. R.**, Use of electric blankets and risk of testicular cancer, *Am. J. Epidemiol.*, 131, 759, 1990.
17. **Preston-Martin, S., Peters, J. M., Yu, M. C., Garabrant, D. H., and Bowman, J. D.**, Myelogenous leukemia and electric blanket use, *Bioelectromagnetics,* 9, 207, 1988.
18. **Vena, J. E., Graham, S., Hellmann, R., Swanson, M., and Brasure, J.**, Use of electric blankets and risk of postmenopausal breast cancer, *Am. J. Epidemiol.*, 134, 180, 1991.
19. **Copeland, K. T., Checkoway, H., McMichael, A. J., and Holbrook, R. H.**, Bias due to misclassification in the estimation of relative risk, *Am. J. Epidemiol.*, 105, 488, 1977.
20. **Poole, C. and Trichopoulos, D.**, Extremely low-frequency electric and magnetic fields and cancer, *Cancer Causes Control,* 2, 267, 1991.
21. **Bonnell, J. A.**, Effects of electric fields near power-transmission plant, *J. Royal Soc. Med.*, 75, 933, 1982.
22. **Knave, B., Gamberale, F., Bergstrom, S., Birke, E., Iregren, A., Kolmodin-Hedman, B., and Wennberg, A.**, Long-term exposure to electric fields. A cross-sectional epidemiologic investigation of occupationally exposed workers in high-voltage substations, *Scand. J. Work Environ. Health,* 5, 115, 1979.
23. **Danilin, V. A., Voronin, A. K., and Madorskii, V. A.**, The state of health of personnel working in high-voltage electric fields, *Gig Tr. Prof. Zabol,* 13, 51, 1969.
24. **Asonova, T. P. and Rakov, A. I.**, The state of health of persons working in the electric field of outdoor 400 kV and 500 kV switchyards, Piscataway, NJ: *Institute of Electrical and Electronic Engineers Power Engineering Society,* (Special publ. No. 10), 1972.
25. **Korobkova, V. P., Morozov, Y. A., Stolarov, M. D. and Yakup, Y. A.**, Influence of the electric field in 500 and 750 kV switchyards on maintenance staff and means for its protection, *Conference Internationale desGrands Reseaux Electrique a Haute Tension 23–6,* 1972.
26. **Savitz, D. A. and Calle, E. E.**, Leukemia and occupational exposure to electromagnetic fields: Review of epidemiologic surveys, *J. Occup. Med.*, 29, 47, 1987.
27. **Guralnick L.**, Mortality by Occupation and Cause of Death Among Men 20 to 64 Years of Age: United States, 1950, *Vital Statistics: Special Reports, Vol. 53, No. 3. U.S. Government Printing Office, Washington, D.C.*, 1963.
28. **Milham, S., Jr.**, Mortality from leukemia in workers exposed to electrical and magnetic fields (letter), *New Engl. J. Med.*, 307, 24, 1982.
29. **Wright, W. E., Peters, J. M., and Mack, T. M.**, Leukemia in workers exposed to electrical and magnetic fields (letter), *Lancet,* 2, 1160, 1982.
30. **McDowall, M. E.**, Leukemia mortality in electrical workers in England and Wales (letter), *Lancet,* 1, 246, 1983.
31. **Coleman, M., Bell, J., and Skeet, R.**, Leukemia incidence in electrical workers (letter), *Lancet,* 1, 982, 1983.
32. **Pearce, N. E., Sheppard, R. A., Howard, J. K., Fraser, J., and Lilley, B. M.**, Leukaemia in electrical workers in New Zealand (letter), *Lancet,* 1, 811, 1985.

33. **Pearce, N. E.,** Leukaemia in electrical workers in New Zealand: a correction (letter), *Lancet,* 2, 48, 1988.
34. **Calle, E. E. and Savitz, D. A.,** Leukemia in occupational groups with presumed exposure to electrical and magnetic fields (letter), *N. Engl. J. Med.,* 313, 1476, 1985.
35. **Milham, S., Jr.,** Silent keys: Leukaemia mortality in amateur radio operators (letter), *Lancet,* 1, 812, 1985.
36. **Dubrow, R. and Wegman, D. H.,** Occupational Characteristics of Cancer Victims in Massachusetts 1971–1973, *DHHS (NIOSH) Publ. No. 84-109, (National Institute for Occupational Safety and Health, Cincinnati, Ohio),* 1984.
37. **Peterson, G. R. and Milham S., Jr.,** Occupational Mortality in the State of California 1959–61, *DHEW (NIOSH) Publ. No. 80-104, (National Institute for Occupational Safety and Health, Cincinnati, Ohio),* 1980.
38. **Stern, F. B., Waxweiler, R. J., Beaumont, J. J., Lee, S. T., Rinsky, R. A., Zumwalde, R. D., Halperin, W. E., Bierbaum, P. J., Landrigan, P. J., and Murray, W. E., Jr.,** A case-control study of leukemia at a naval nuclear shipyard, *Am. J. Epidemiol.,* 123, 980, 1986.
39. **Flodin U., Fredriksson M., Axelson, O., Persson, B., and Hardell, L.,** Background radiation, electrical work, and some other exposures associated with acute myeloid leukemia in a case-referent study, *Arch. Environ. Health,* 41, 77, 1986.
40. **Flodin, U., Fredriksson, M., Persson, B., and Axelson, O.,** Acute myeloid leukemia and background radiation in an expanded case-referent study, *Arch. Environ. Health,* 45, 364, 1990.
41. **Vagero, D. and Olin, R.,** Incidence of cancer in the electronics industry: Using the new Swedish Cancer Environment Registry as a screening instrument, *Br. J. Industr. Med.,* 40, 188, 1983.
42. **Cammarano, G., Crosignani, P., Berrino, F., and Berra, G.,** Cancer mortality among workers in a thermoelectric power plant, *Scand. J. Work Environ. Health,* 10, 259, 1984.
43. **Juutilainen, J., Läärä, E., and Pukkala, E.,** Incidence of leukaemia and brain tumours in Finnish workers exposed to ELF magnetic fields, *Int. Arch. Occup. Environ. Health,* 62, 289, 1990.
44. **Lin, R. S., Dischinger, P. C., Conde, J., and Farrell, K. P.,** Occupational exposure to electromagnetic fields and the occurrence of brain tumors. An analysis of possible associations, *J. Occup. Med.,* 27, 413, 1985.
45. **Pearce, N., Reif, J., and Fraser, J.,** Case-control studies of cancer in New Zealand electrical workers, *Int. J. Epidemiol.,* 18, 55, 1989.
46. **Reif, J. S., Pearce, N., and Fraser, J.,** Occupational risks for brain cancer: a New Zealand cancer registry-based study, *JOM,* 31, 863, 1989.
47. **Preston-Martin, S. and Peters, J. M.,** Prior employment as a welder associated with the development of chronic myeloid leukemia, *Br. J. Cancer,* 58, 105, 1988.
48. **Balli-Antunes, M., Pfluger, D. H., and Minder, C. E.,** The mortality from malignancies of haematopoietic and lymphatic systems (MHLS) among railway engine drivers, *Environmetrics,* 1, 121, 1990.
49. **Törnqvist, S., Knave, B., Ahlbom, A., and Persson, T.,** Incidence of leukaemia and brain tumours in some "electrical occupations," *Br. J. Indust. Med.,* 48, 597, 1991.
50. **Blair, A., Walrath, J., and Rogot, E.,** Mortality patterns among U.S. veterans by occupation. I. cancer, *JNCI,* 75, 1039, 1985.
51. **Linet, M. S., Malker, H. S., McLaughlin, J. K., Weiner, J. A., Stone, B. J., Blot, W. J. Ericsson, L. E., and Fraumeni, J. F.,** Leukemias and occupation in Sweden: a registry-based analysis, *Am. J. Indust. Med.,* 14, 319, 1988.
52. **Matanoski, G. M., Elliot, E. A., Breysse, P. N., and Lynberg, M. C.,** Leukemia in telephone linemen, *Am. J. Epidemiol.,* 137, 609, 1993.
53. **Tynes, T., Andersen, A., and Langmark, F.,** Incidence of cancer in Norwegian workers potentially exposed to electromagnetic fields, *Am. J. Epidemiol.,* 136, 81, 1992.
54. **Floderus, B., Persson, T., Stenlund, C., Linder, G., Johansson, C., Kiviranta, J., Parsman, H., Lindblom, M., Wennberg, A., Öst, A., and Knave, B.,** Occupational exposure to electromagnetic fields in relation to leukemia and brain tumors. A case-control study., *National Institute of Occupational Health, Solna, Sweden,* 1992.
55. **Sahl, J. D., Kelsh, M. A., and Greenland, S.,** Cohort and nested case-control studies of hematopoietic cancers and brain cancer among electric utility workers, *Epidemiology,* 4, 104, 1993.
56. **Thérialult, G., Goldberg, M., Miller, A. B., Armstrong, B., Guénel, P., Deadman, J., Imbernon, E., To, T., Chevalier, A., Cyr, D., and Wall, C.,** Cancer risks associated with occupational exposure to magnetic fields among electric utility workers in Ontario and Quebec, Canada, and France: 1970–1989. *Am. J. Epidemiol.,* 139, 550, 1994.
57. **Milham, S., Jr.,** Mortality in workers exposed to electromagnetic fields, *Environ. Health Perspect.,* 62, 297, 1985.
58. **Coggon, D., Pennett, B., Osmond, C. and Acheson, E. D.,** A survey of cancer and occupation in young and middle aged men. II Non-respiratory cancers, *Br. J. Industr. Med.,* 43, 381, 1986.
59. **Thomas, T. L. and Waxweiler, R. J.,** Brain tumors and occupational risk factors. A review, *Scand. J. Work Environ. Health,* 12, 1, 1986.
60. **Thomas, T. L., Stolley, P. D., Stemhagen, A., Fontham, E. T. H., Bleecker, M. L., Stewart, P. A., and**

Hoover, R. N., Brain tumor mortality risk among men with electrical and electronics jobs: A case-control study, *J. Natl. Cancer Inst.*, 79, 233, 1987.

61. **Speers, M. A., Dobbins, J. G., and Miller, V. S.**, Occupational exposures and brain cancer mortality: A preliminary study of East Texas residents, *Am. J. Industr. Med.*, 13, 629, 1988.
62. **Modan, B.**, Exposure to electromagnetic fields and brain malignancy: A newly discovered menace?, *Am. J. Industr. Med.*, 13, 625, 1988.
63. **Törnqvist, S., Norell, S., Ahlbom, A., and Knave, B.**, Cancer in the electric power industry, *Br. J. Industr. Med.*, 43, 212, 1986.
64. **McLaughlin, J. K., Malker, H. S. R., Blot, W. J., Malker, B. K., Stone, B. J., Weiner, J. A., Ericsson, J. L. E., and Fraumeni, J. F., Jr.**, Occupational risks for intracranial gliomas in Sweden, *J. Natl. Cancer Inst.*, 78, 253, 1987.
65. **Milham, S., Jr.**, Increased mortality in amateur radio operators due to lymphatic and hemaopoietic malignancies, *Am. J. Epidemiol.*, 127, 50, 1988.
66. **Gallagher, R. P., McBride, M. L., Band, P. R., Spinelli, J. J., Threlfall, W. J., and Tamaro, S.**, Brain cancer and exposure to electromagnetic fields (letter), *JOM*, 33, 944, 1991.
67. **Gallagher, R. P., McBride, M. L., Band, P. R., Spinelli, J. J., Threlfall, W. J., and Yang, P.**, Occupational electromagnetic field exposure, solvent exposure, and leukemia (letter), *JOM*, 32, 64, 1990.
68. **Preston-Martin, S., Mack, W., and Henderson, B. E.**, Risk factors for gliomas and meningiomas in males in Los Angeles county, *Cancer Res.*, 49, 6137, 1989.
69. **Mack, W., Preston-Martin, S., and Peters, J. M.**, Astrocytoma risk related to job exposure to electric and magnetic fields, *Bioelectromagnetics*, 12, 57, 1991.
70. **Loomis, D. P. and Savitz, D. A.**, Mortality from brain cancer and leukaemia among electrical workers, *Br. J. Indust. Med.*, 47, 633, 1990.
71. **Vagero, D., Ahlbom, A., Olin, R., and Sahlsten, S.**, Cancer morbidity among workers in the telecommunications industry, *Br. J. Industr. Med.*, 42, 191, 1985.
72. **DeGuire, L., Theriault, G., Iturra, H., Provencher, S., Cyr, O., and Case, B. W.**, Increased incidence of malignant melanoma of the skin in workers in a telecommunications industry, *Br. J. Indust. Med.*, 45, 824, 1988.
73. **Swerdlow, A. J.**, Epidemiology of eye cancer in adults in England and Wales, 1962–1977, *Am. J. Epidemiol.*, 118, 294, 1983.
74. **Matanoski, G., Elliott, E., and Breysse, P.**, "Cancer incidence in New York telephone workers," poster presented at the *Annual Department of Energy-Electric Power Research Institute Contractor's Review*, Portland, OR, November 15, 1989, and *Lancet*, 337, 737, 1991.
75. **Tynes, T. and Andersen, A.**, Electromagnetic fields and male breast cancer, *Lancet*, 336, 1596, 1990.
76. **Demers, P. A., Thomas, D. B., Rosenblatt, K. A., Jimenez, L. M., McTiernan, A., Stalsberg, H., Stemhagen, A., Thompson, W. D., Curnen, M. G., Satariano, W., Austin, D. F., Isacson, P., Greenberg, R. S., Key, C., Kolonel, L. N., and West, D. W.**, Occupational exposure to electromagnetic fields and breast cancer in men, *Am. J. Epidemiol.*, 134, 340, 1991.
77. **Guénel, P., Raskmark, P., Andersen, J. B., and Lynge, E.**, Incidence of cancer in persons with occupational exposure to electromagnetic fields in Denmark, *Brit. J. Indus. Med.*, 50, 758, 1993.
78. **Loomis, D. P., Savitz, D. A., and Ananth, C. V.**, Mortality from breast cancer among female electrical workers, *J. Natl. Cancer Inst.*, in press, 1994.
79. **Broadbent, D. E., Broadbent, M. H. P., Male, J. C., and Jones, M. R. L.**, Health of workers exposed to electric fields, *Br. J. Industr. Med.*, 42, 75, 1985.
80. **Nordstrom, S., Birke, E., and Gustavsson, L.**, Reproductive hazards among workers at high voltage substations, *Bioelectromagnetics*, 4, 91, 1983.
81. **Wertheimer, N. W. and Leeper, E.**, Possible *effects of electric blankets and heated waterbeds on fetal development*, Bioelectromagnetics, 7, 13, 1986.
82. **Juutilainen, J., Matilainen, P., Saarikoski, S., Läärä, E., and Suonio, S.**, Early pregnancy loss and exposure to 50-Hz magnetic fields, *Bioelectromagnetics*, 14, 229, 1993.
83. **Spitz, M. R. and Johnson, C. C.**, Neuroblastoma and paternal occupation. A case-control analysis, *Am. J. Epidemiol.*, 121, 924, 1985.
84. **Stuchly, M. A., Lecuyer, D. W., and Mann, R. D.**, Extremely low frequency electromagnetic emissions from video display terminals and other devices, *Health Physics*, 45, 713, 1983.
85. **Goldhaber, M. K., Polen, M. R., and Hiatt, R. A.**, The risk of miscarriage and birth defects among women who use visual display terminals during pregnancy, *Am. J. Industr. Med.*, 13, 695, 1988.
86. **Schnorr, T. M., Grajewski, B. A., Hornung, R. W., Thun, M. J., Egeland, G. M., Murray, W. E., Conover, D. L., and Halperin, W. E.**, Video display terminals and the risk of spontaneous abortion, *N. Engl. J. Med.*, 324, 727, 1991.
87. **Slesin, L. and Connelly, M.**, Video display terminals and spontaneous abortions (letter), *N. Engl. J. Med.*, 325, 811, 1991.

88. **Schnorr, T. M., Grajewski, B. A., and Murray, W. E.,** Video display terminals and spontaneous abortions (letter), *N. Engl. J. Med.,* 325, 812, 1991.
89. **Lindbohm, M-L., Hietanen, M., Kyyrönen, P., Sallmén, M., von Nandelstadh, P., Taskinen, H., Pekkarinen, M., Ylikoski, M., and Hemminki, K.,** Magnétic fields of video display terminals and spontaneous abortion, *Am. J. Epidemiol.,* 136, 1041, 1992.
90. **Poole, C., Kavet, R., Funch, D. P., Donelan, K., Charry, J. M., and Dreyer, N. A.,** Depressive symptoms and headaches in relation to proximity of residence to an alternating-current transmission line right-of-way, *Am. J. Epidemiol.,* 137, 318, 1993.

Chapter 8

EXPERIMENTAL RADIO AND MICROWAVE DOSIMETRY

Maria A. Stuchly and Stanislaw S. Stuchly

CONTENTS

0-8493-0641-8/96/$0.00+$.50

I. INTRODUCTION

Interactions of radio frequency (RF) fields with biological tissues and bodies are complex functions of numerous parameters. Dosimetric studies attempt to quantify these interactions. Radiowaves in free space are characterized by the frequency, intensity of the electric (E) and magnetic (H) fields, their direction, and polarization. However, only the fields inside the tissues and biological bodies can interact with them, and therefore it is necessary to determine these fields for any meaningful and general quantification of biological data obtained experimentally.

Fields inside biological bodies exposed to known external electromagnetic fields can be calculated by solving Maxwell's equations subject to given boundary conditions. A biological body is an inhomogeneous, lossy dielectric, whose macroscopic electrical properties are described by the complex permittivity. The inhomogeneity of the dielectric properties and the complexity of the shape make a solution, or sometimes even a full formulation of the problem, a forbidding task. Only much simplified models can be analyzed. The other approach is an experimental one, which is also subject to considerable limitations.

The intensity of the internal fields depends on the parameters of the external field, the frequency, intensity, and polarization, size, shape, dielectric properties of the exposed body, spatial configuration between the exposure source and the exposed body, and the presence of other objects in the vicinity. With a complex dependence on so many parameters, it is apparent that the internal fields in a mouse and a man exposed to the same external field can be dramatically different, and so will be their biological response, regardless of physiological differences. Conversely, different exposure conditions, e.g., different frequencies, may induce similar fields inside such diverse shapes as a mouse and a man.

A. SPECIFIC ABSORPTION RATE (SAR)

The internal fields can be quantified in various ways. The magnetic permeability of most tissues is practically equal to that of free space, and all known and anticipated interactions at higher RFs occur through mechanisms involving the E field (including the currents induced by the H field). Therefore, the E-field vector, or its distribution throughout the exposed body, fully describes the exposure field-tissue interactions. Some additional information may be needed for full quantification, e.g., frequency characteristics of the exposure field, modulation characteristics, and modulation frequency. A dosimetric measure that has been widely adopted is the SAR defined as "the time derivative of the incremental energy (dW) absorbed by, or dissipated in an incremental mass (dm) contained in a volume element (dV) of a given density (ρ)."[1]

$$\text{SAR} = \frac{d}{dt}\left(\frac{dW}{dm}\right) = \frac{d}{dt}\left[\frac{dW}{\rho(dV)}\right] \tag{1}$$

Using the Poynting vector theorem for sinusoidally varying electromagnetic fields, Equation 1 can be presented as

$$\text{SAR} = \frac{\sigma}{2\rho}|E_i|^2 = \frac{\omega\epsilon_0\epsilon''}{2\rho}|E_i|^2 \tag{2}$$

where σ is the tissue conductivity in siemens per meter, ϵ_0 is the dielectric constant of free space ($\epsilon_0 = 8.85 \times 10^{-12}$F/m), ϵ'' is the loss factor, $\omega = 2\pi f$, f is the frequency in hertz, and E_i is the peak value of the internal E field in volts per meter. The SAR is expressed in watts per kilogram (W/kg) or derived units (e.g., mW/g). The average SAR is defined as a ratio of the total power absorbed in the exposed body to its mass. The local SAR refers to the value within a defined unit volume or unit mass, which can be arbitrarily small.

It is very important to realize that even though Equation 2 expresses the rate at which the electromagnetic energy is converted into heat through well-established interaction mechanisms described in Part I, it provides a valid quantitative measure of all the interaction mechanisms that are dependent on the intensity of the internal E field.[2] At this point, some additional information may be relevant. For instance, since some effects of radiowaves modulated in amplitude at ELF (extremely low frequencies) are dependent on the E field intensity,[3] once specific interaction mechanisms are better understood, they could be expressed in terms of the SAR and modulation characteristics, even though the interaction mechanism may not necessarily be thermal. However, should the direction of the internal E field with respect to the biological structures be of importance for a given interaction mechanism, the SAR would not provide full quantitative information. Similarly, for the interactions directly through the H field, the present SAR concept would not be sufficient. Bearing these limitations in mind, the SAR concept has proven to be a simple and useful tool in quantifying the interactions of RF radiation with living systems. It enables comparison of experimentally observed biological effects for various species under various exposure conditions, and it provides the only means, however imperfect, of extrapolating the animal data into potential hazards to humans exposed to RF radiation. It also facilitates planning and effective execution of therapeutic hyperthermic treatment.

B. EXPERIMENTAL DOSIMETRY AND UNITS

Dosimetry in bioelectromagnetic research has been developing in two parallel but interacting and complementary streams: theoretical and experimental. Theoretical dosimetry deals with the analysis of simplified models as reviewed in Chapter 9, while experimental dosimetry concerns the development of methods and instruments suitable for measurements of the internal E fields, SARs, or related variables. Measurements of the external fields as relevant, for the evaluation of the exposure conditions are usually considered a part of experimental dosimetry (only occasionally but not appropriately named "densitometry"). To place the measurements of the external fields vs. the internal fields in proper perspective, one has to realize that the instruments and methods for measurements of internal fields are not suitable for use on humans except in hyperthermic treatment for cancer therapy. Therefore, to prevent potentially hazardous exposure to man, the external fields are measured. Data on the intensities of external fields that potentially, constitute a hazard to man are obtained through experimental evaluation of the intensities of the internal fields (SARs) that cause damage to animals and determination of external fields which produce similar internal fields (SARs) in man.

Experimental dosimetry includes the following areas: measurements of the total SAR and its distribution in live or dead experimental animals, measurements of the average SAR and

TABLE 1
Quantities, Common Symbols, and Units used in Experimental Dosimetry

Symbol	Quantity	Unit
	External Fields	
$\bar{E}$	E field (vector)	V/m
$\bar{H}$	H Field (vector)	A/m
S	Power density, energy flux density (scalar)	W/m^2 mW/cm^2
$\bar{S}$	Poyinting vector $\bar{S} = \bar{E} \times \bar{H}$	W/m^2
W	Energy density (scalar)	J/m^3
$\bar{k}$	Propagation vector	
Z	Wave impedance, $Z = E/H$ (scalar)	Ω
	Internal Fields	
$\bar{E}_i$	E field (vector)	V/m
SAR	SAR (av and local)	W/kg mW/g
ϵ	Complex relative permittivity, $\epsilon = \epsilon' - j\epsilon''$	—
σ	Conductivity, $\sigma = \omega\epsilon_0\epsilon''$	S/m

SAR distribution in models of animals and humans, and development of exposure devices for animals and in vitro preparations in which quantification of the exposure conditions and internal fields is facilitated.

Experimental dosimetry plays a crucial role in the control of animal exposures and quantification of the data from animal experimentation. It enables comparison of experiments conducted in various laboratories, under different exposure conditions, and for different species. It also plays an important role in verifying the predictions of theoretical dosimetry. While theoretical methods, with appropriate simplifying assumptions, can in a straightforward manner predict the total or average SAR, the SAR distributions in complex models that are close approximations of the actual animal, and in the animal, can usually be more easily determined experimentally.

The fundamental quantities and units in the SI system used in experimental dosimetry are summarized in Table 1.

The dosimetric data can be used to extrapolate the results of animal experiments to human exposures. The physical scaling principle is then employed. For instance, to determine the equivalent human exposure at a frequency f_n (i.e., where roughly similar, but by no means identical SAR distribution exists) for a man of a height 1_n and an animal of a height 1_a exposed at a frequency f_a (with both man and animal having the same orientation in respect to the exposure field), the following relationship can be used:[4]

$$f_n = f_a \frac{1_a}{1_n} \tag{3}$$

It should be emphasized that the SAR distribution is only similar, not identical.[5]

In this chapter the basic knowledge on experimental dosimetry is summarized. The material is divided into three parts. The first deals with measurements of the fields both external and internal, in bodies exposed to RF fields. The second pan covers dosimetric measurements based on temperature measurements. The concept of the SAR is used extensively in these two parts. The last part describes basic principles and some designs of exposure systems that are suitable for exposures with well-defined dosimetry. Effects related to the use of

TABLE 2
Experimental Dosimetry Methods

Av SAR	SAR distribution	E-field vector
Absorbed power measurements	E-field probes	E-field probes
Calorimetry	Nonperturbing thermometers	
	Thermography	

minitoring devices (e.g., electrodes) are also briefly outlined. A listing of dosimetry methods and their applications as described in this chapter are given in Table 2.

II. FIELD MEASUREMENTS

A. EXTERNAL FIELDS

1. Basic Definitions

RFs occupy a relatively wide segment of the electromagnetic spectrum with frequencies ranging from a few hundred kilohertz to about 300 GHz. The exposure fields can be produced intentionally by radiators such as those used in telecommunications and radar, or may be a result of imperfections in the design or manufacture of a given device, resulting in undesirable energy leakage, e.g., in RF dielectric heaters. The nature of these two types of fields and their characteristics that are important from the dosimetry viewpoint, are significantly different. One of the most important descriptors of an exposure field is whether it is the far field or the near field. The far-field region extends from a certain minimum distance from the antenna to infinity. This minimum distance is usually selected as $1 \cong \frac{2D^2}{\lambda}$, where D is the larger dimension of the antenna and λ is the wavelength. This is equivalent to the fact that the power per unit solid angle is almost constant.[1] In this region the field has a predominantly plane-wave character, i.e., the E-field vector is perpendicular to the H-field vector, and they are both perpendicular to the direction of propagation. The ratio of the E-field intensity to the H-field intensity is constant at any location, and for free space is equal to

$$Z = \frac{E}{H} \cong 377\Omega \tag{4}$$

In the far field, the field intensity is therefore fully defined by either the E-field intensity or the H-field intensity. Most frequently, for a plane-wave the power density is used to describe the exposure field.

$$S = \frac{E^2}{Z} = H^2 Z \tag{5}$$

The near-field region close to the radiator (antenna) comprises the radiating near-field region (Fresnel region) and the reactive (non-radiating) near-field regions. The reactive region is adjacent to the antenna and extends to about one wavelength from the antenna. In this region the E and H fields vary rapidly with distance. A portion of the electromagnetic energy is stored in this region. In the radiating near field the energy propagates away from the antenna, and the intensity of the E and H field decreases with distance, but the radiation still lacks the plane-wave character. The power density usually varies with the distance in an oscillatory fashion, rather than inversely proportionally with the square of the distance, as

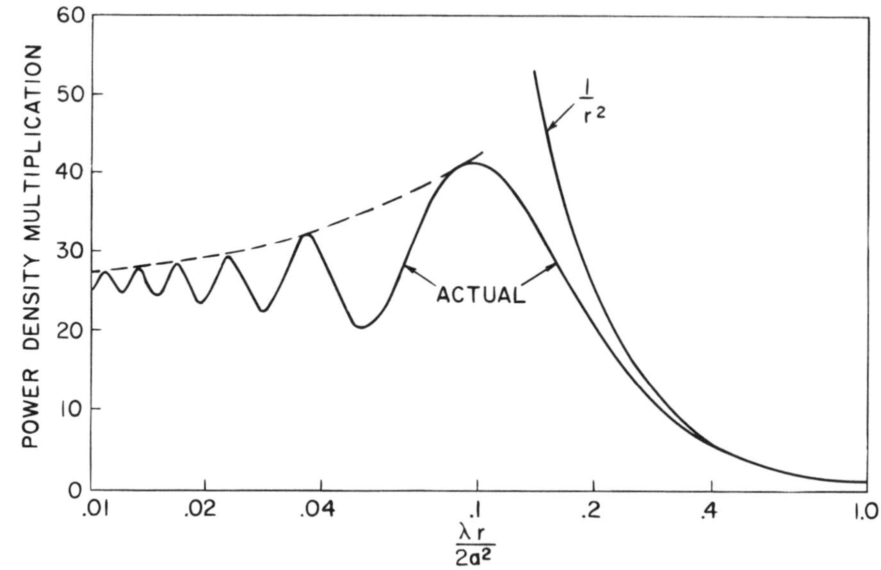

FIGURE 1 Relative power density vs. distance along the axis of an aperture antenna. (r) The distance from the antenna; (a) the diameter of the antenna; (λ) the wavelength. (From Microwave Engineering Handbook, Horizon House.)

TABLE 3
Essential Parameters of the Exposure Field

Frequency	Intensity	Orientation
Carrier frequency	Far field: power density	Far field: polarization, direction with respect to object
Modulation type (e.g., amplitude, sinusoidal, pulse)	Near field: E-field intensity	
Modulation characteristics (e.g., pulse duration, shape, and repetition rate)	H-field intensity	Near-field: directions of $\bar{E}$ and $\bar{H}$ with respect to the object

for the plane-wave (far field). For illustration, the variations of power density vs. distance are given in Figure 1.

From the definitions and character of the fields it is obvious that while in the far field a measurement of only one field component (the E or the H field) is sufficient for a full description of the field, in the near field both fields should be measured. Furthermore, in the reactive zone, a measuring instrument is likely to significantly perturb the field and modify the radiation characteristics. Leakage fields, particularly at longer radiowaves, are in most cases of the near-field type. Essential parameters of an exposure field that have to be defined are summarized in Table 3.

2. Measurement Errors

Measurements of external fields, for evaluation of a potential hazard of exposure or for definition of exposure conditions in animal experimentation, are subject to several possible sources of interference and error.[6,7] Exposure may occur at more than one frequency either because of multiple sources of emission or emissions at more than one frequency from a single source (e.g., harmonics). Exposure fields may have various polarizations which may not be known prior to measurements. Nearby objects may cause partial or total reflection or scattering of the waves, creating interference patterns with maxima and minima interspaced a quarter of the wavelength. The modulation characteristics of the radiation are also important. Pulse shape and duration, repetition rate, and accuracy of measurements. Previously mentioned presence of the measuring instrument or any other object in the reactive field of the antenna can cause considerable change in the exposure field. Finally, other factors such as RF interference and reflecting surfaces in the vicinity may cause errors in measurements of exposure fields.

3. Broadband Instruments—Survey Meters

Two general categories of instruments are suitable, within certain restrictions, for measurements of RF exposure fields. They are broadband and narrowband type instruments. The broadband instruments, which have been developed specifically for hazard field surveys, have frequency-independent sensitivity, measure integrated intensity of the fields, and do not provide any information on the frequency spectrum of the field. The narrowband instruments, on the other hand, are selective and provide information about the field intensity at a selected frequency or rather within a very narrow frequency band, e.g., 1% around the nominal frequency.

The broadband instruments are more suitable for measurements in the near field as they usually cause only minimal perturbation of the field; however, their sensitivity is usually much less than that of the narrowband instruments. The salient characteristics of broadband survey instruments are listed in Table 4.

Various design concepts have been employed in E-field probes and basically one concept in H-field probes. Principles of operation of the broadband probes are outlined below.

One of the simplest E-field sensors consists of a short dipole connected to a diode. Such a sensor responds only to the field that is polarized parallel to the dipole, however when three dipoles are arranged in a mutually orthogonal configuration a nearly isotropic directional radiation characteristic can be obtained. For an electrically short dipole, i.e., $kh < I$, where k is the propagation constant (in free space $k = 2\pi/\lambda$,λ is the wavelength) and 2h is the physical length of the dipole, the effective length h_e, and the input capacitance C_a, are independent of frequency and equal to:[9,10]

TABLE 4
Desirable Characteristics of Broadband Instruments for Measurements of Exposure Fields

Sensor	Meter
Response to one field only (E or H)	Response to modulated fields
Nonperturbing	Accuracy
Suitable for near and far field	Sensitivity
Small-good spatial resolution	Wide dynamic range
Broad frequency range, flat response characteristic	RFI resistance
Isotropic or three directions	Peak/av value option
Multiple signal integration	Short response time
Temperature insensitivity	Portability
Stability	Temperature insensitivity
Overload insensitivity	Stability
Linear response	Overload sensitivity
Low noise	Linear response

$$h_c = \frac{h(\Omega - 1)}{2(\Omega - 2 + 2\ln 2)} \tag{6}$$

$$C_a = \frac{4\pi\epsilon_0 h}{\Omega - 2 - 2\ln 2} \tag{7}$$

where

$$\Omega = 2\ln(2h/a) \tag{8}$$

a is the antenna radius and ϵ_0 is the permittivity of free space. By selecting a proper detector diode and load impedance, a broadband frequency response can be obtained.[10]

The diode is connected to the signal processing circuitry by a pair of high-resistance leads. The high-resistance leads are necessary, because conductive leads act as antennas. The high-resistance leads provide an additional advantage of causing only negligible perturbation to the measured electromagnetic field. A description of an early design of a meter based on this principle is given elsewhere.[8] In some designs the dipole is also made of a resistive strip to achieve broadband operation.[11]

Several commercially available survey meters are based on this principle (e.g., Aeritalia, General Microwave, Holaday, Instruments for Industry). One of the major problems of the sensors with a diode detector is the nonlinearity of the diode characteristic. While for low-intensity E fields the output voltage of the diode is proportional to the square of the field intensity, at higher intensities the relationship becomes almost linear. This limits the dynamic range for accurate measurements.

The E-field intensity can also be measured by a thermocouple-based sensor.[12] The sensor consists of several miniature thermocouples connected in series, forming a thermopile as shown in Figure 2. The output signal of the thermocouple is proportional to the power dissipated in it, which in turn is proportional to the square of the current induced in the thermocouple elements acting as miniature dipoles. As the induced current is proportional to the E-field intensity polarized parallel to the dipole, the thermocouple sensor provides an output signal proportional to the square of the E-field intensity. A wide dynamic range can therefore be obtained. An arrangement of three mutually perpendicular thermocouple elements gives a practically isotropic spatial characteristic. Commercially available instruments based on this principle are manufactured by Loral Narda, and utilize overlapping thin films of

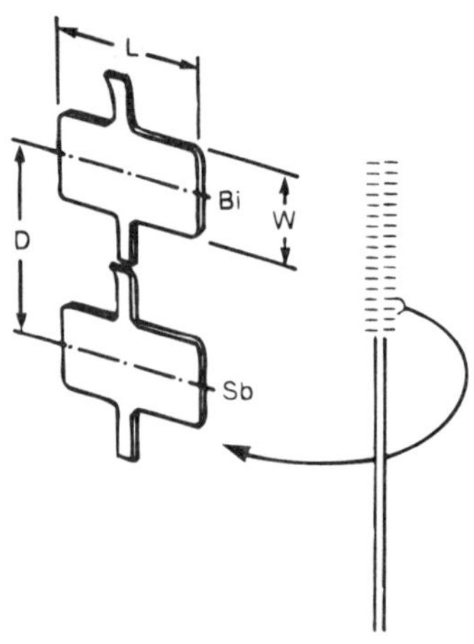

FIGURE 2 Thermocouple E-field probe.[12]

antimony and bismuth (Figure 2).[12] The thermocouple sensor is connected with the external circuitry by high-resistance leads.

Thermocouple probes are true square-law detectors and are inherently accurate indicators of the time-averaged intensity for modulated exposure fields. They are more sensitive to overloading than the diode probes and their dynamic range, although theoretically unlimited, is restricted by the thermocouple sensitivity at one end and by the maximum power dissipated in the junction at the other.

A H-field sensor consists of a small loop with a diode or a thermocouple connected by high-resistance leads to the external circuitry. The diameter of the loop, d, should be small compared with the wavelength, i.e., $d < 0.1\lambda_{min}$, where λ_{min} is the wavelength at the highest frequency of operation of the sensor. Under this condition the effective length h_H of the loop antenna is

$$h_H = \frac{\pi^2 d^2 n}{2\lambda} \tag{9}$$

where n is the number of turns. The induced voltage is

$$V_H = -j\omega\mu_0 HnS \tag{10}$$

where $\omega = 2\pi f$, f is the frequency in hertz, μ_o is the permeability of free space $\mu_o = 4\pi 10^{-7}$ *H/m*), *H* is the intensity of the H field perpendicular to the loop in ampere per meter, and S is the surface area of the loop in square meters. Through a proper resistive loading of the diode it is possible to obtain a relatively flat response of the probe over a relatively wide range of frequencies.[14,15] Three mutually perpendicular loops produce a nearly isotropic response of an H-field probe.

One of the problems that has to be considered for a loop, H-field sensor is its undesirable response to the E field. While a dipole does not respond to the H field, the situation is not that straightforward for the loop. For a plane-wave the ratio of the voltage induced by the E field to that induced by the H field is[16]

$$\frac{V_E}{V_H} \simeq -j2\pi\left(\frac{d}{\lambda}\right) \tag{11}$$

for a circular loop, and

$$\frac{V_E}{V_H} \simeq -j3\pi\left(\frac{W}{\lambda}\right)\frac{\Omega - 4.32}{\Omega - 3.17} \tag{12}$$

for a square loop of side length w and wire radius a, where $\Omega = 2 \ln(4w/a)$ is a wire thickness parameter. Equations 11 and 12 indicate that a circular loop is less sensitive than a square loop, and that the E-field sensitvity can be minimized by decreasing the size of the loop. The other possibility for minimizing the sensitivity of the H-field sensors to E fields is offered by the Moebius design.[17]

A probe equipped with three dipoles and three loops with zero-bias Schottky diode detectors was also developed.[15] Three orthogonal vector components of the H and E field can be measured simultaneously and independently. The antenna elements are arranged to minimize the interactions between individual elements.

All types of nonperturbing field probes can be connected with the external circuitry by an optical-fiber telemetry system. This further minimizes perturbation of the surveyed field

TABLE 5
Comparison between Broadband (Survey Meters) and Narrowband (Spectrum Analyzers, Test Receivers) Instruments

Characteristic	Broadband instruments	Narrowband instruments
Usable frequency range	Broad	Broad
Frequency resolution	None	Excellent
Modulation response	Limited	Excellent
Sensitivity	Moderate	Very high
Dynamic range	Moderate (~30 dB)	Very wide (> 100 dB)
Spatial resolution	Excellent possible (usually ~5 cm)	Usually poor (large antennas)
Near-field response	Accurate (ca. ± 1 dB)	Inaccurate, unless special small sensors used
Directional response	Isotropic	Linear or planar
Field perturbation	Minimal	Considerable, depends on sensor
Antenna size	Small	Comparable with wavelength
Portability	Excellent	Limited or poor

and is particularly useful in near-field measurements. The use of the optical-fibers rather than cables with metal conductors limits perturbation of the test field.[18,19] An additional advantage resulting from the use of optical fibers is the elimination of RF interference (RFI). Single, optical fiber kits, which include a transmitter and a receiver, are commercially available (Hewlett-Packard, Burr-Brown) and a simple end reliable system can be built at low cost.

4. Narrowband Instruments

The second category of instruments for measurment of external fields comprises narrowband, selectively tuned instruments such as spectrum analyzers and field intensity meters or panoramic receivers. Summary comparison between the broadband and narrowband instruments is outlined in Table 5. A typical narrowband instrument consists of two elements, a calibrated, usually broadband or tunable antenna and a measurement system capable of rather sophisticated processing of the received signals. The size of the linearly polarized antennas is usually comparable with the wavelength. An ominidirectional spatial response can be obtained by positioning the antennas along three mutually orthogonal directions and summing up the results, e.g., for the E field in cartesian coordinates

$$E = [E_x^2 + E_y^2 + E_z^2]^{1/2} \tag{13}$$

where $E_X E_Y$, and E_z are the field intensities measured with the antenna positioned along the x, y, and z directions, respectively. Directional antennas can also be assembled to form an isotropic response system. Several types of commercially available antennas, e.g., logperiodic antennas, provide a planar radiation pattern, i.e., giving genuine response for all polarizations in the plane of the antenna.

Special purpose antennas can also be designed and calibrated to provide desired characteristics for a specific task, e.g., a small-diameter, multiple-turn loop to measure the H field emissions from a device.

Calibrated antennas can be used in conjunction with a spectrum analyzer or a field intensity meter. A spectrum analyzer provides visual, quantitative information on the spectral composition of the signal received by the antenna. The display is usually given on a cathode ray tube (CRT). Spectrum analyzers are capable of making absolute frequency and amplitude measurements, operate over a wide-amplitude dynamic range, have high frequency and amplitude resolution, and high sensitivity. They also provide means for observation, storage, and recording the output in a convenient form. More sophisticated signal analyzers capable of digital signal processing, i.e., Fourier analyzers, can be used in place of spectrum analyzers. Field strength meters or test receivers are tuneable, narrowband voltmeters. They do not provide information about modulation characteristics of the signal but give only the rms or peak value. However, information about the modulation characteristics can be retrieved through the analysis of the fequency spectrum.

Attempts have also been made to develop a personal dosimeter. The main goal of such a device is to record, as fully as possible, the exposure conditions of a person occupationally exposed to radiowaves. A fundamental difficulty in developing a useful dosimeter, apart from technical problems, is conceptual. A dosimeter, unless very complex, is only able of recording cumulative exposure, i.e., the total dose equal to the power density integrated over time, and cannot distinguish short-term, high-level from long-term, low-level exposures. However, biological effects and hazards are not a direct function of the total dose but the dose rate, i.e., exposure to 200 mW/cm^2 for 10 min is not equal to exposure to 1 mW/cm^2 for 2000 min. In all attempts to develop a personal dosimeter, this fundamental problem has not been adequately addressed. Additional technical difficulties are posed by the frequency response of the dosimeter, the perturbation and partial shielding of the exposure field by the dosimeter wearer. Efforts were made to design a personal dosimeter whose sensor was to contain isotropic probes for E and H fields at frequencies from 2 to 300 MHz, and E-field probes at frequencies from 300 MHz to 23 GHz,[20] however due to technical difficulties this effort was discontinued. A description of other earlier, unsuccessful attempts is given elsewhere.[7]

5. Instruments for Transient and Pulsed Fields

Some natural and man-made transient and pulsed electromagnetic fields are characterized by fast (sub-nanosecond) rise times and broad (gigahertz) bandwidth. For measurements of such fields transmitting and receiving (sensing) antennas must be inherently broadband and nondispersive. A comprehensive review of the time domain antennas is available.[20A]

Antennas for transients and pulsed fields can be divided into two basic groups: electrically large ($\sim\lambda/2$) passive antennas and electrically small E-field and H-field active antennas.

A new antenna having a potential bandwidth from 20 MHz to 10 GHz has been recently developed in our laboratory.[20B,20C] A monopole (ground plane) version of the new antenna is shown in Figure 3. The sensitive-loaded monopole is 20 cm long, 2 mm thick and has a 15° wedge angle. The Response-Equalizer Circuit (REC) consists of a 100Ω chip resistor and a 10-pF chip capacitor. The probe input resistance is 500Ω.

Figure 4 demonstrates the waveform reproduction fidelity by comparing the received voltage with the incident field of a pulse (approx. gaussian) of the half-magnitude of approximately 500-ps. The received signal is very close to the incident field.

To illustrate wideband performance of electrically small active antennas, the performance of the E & H field sensors is presented.[20D]

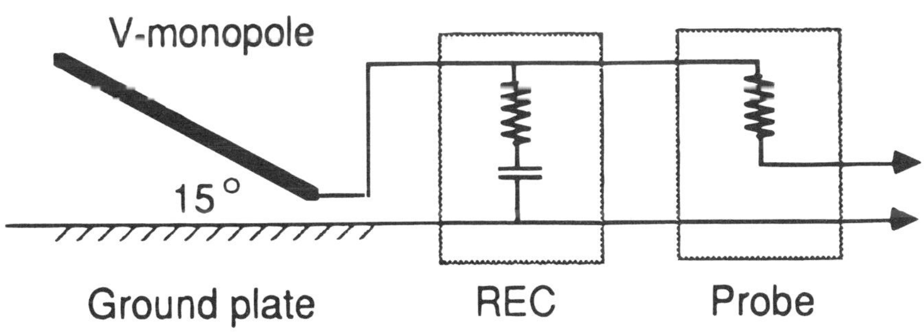

FIGURE 3 The monopole version of the broadband V-antenna.

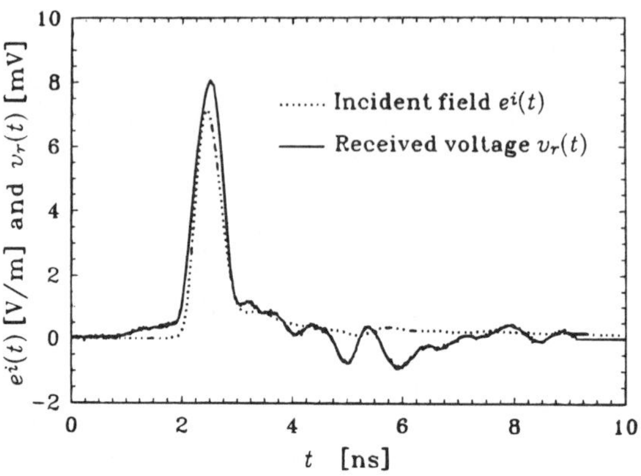

FIGURE 4 The waveform of the electrical field $e^i(t)$ incident in the boreside direction and the response of the V-antenna $v_r(t)$.

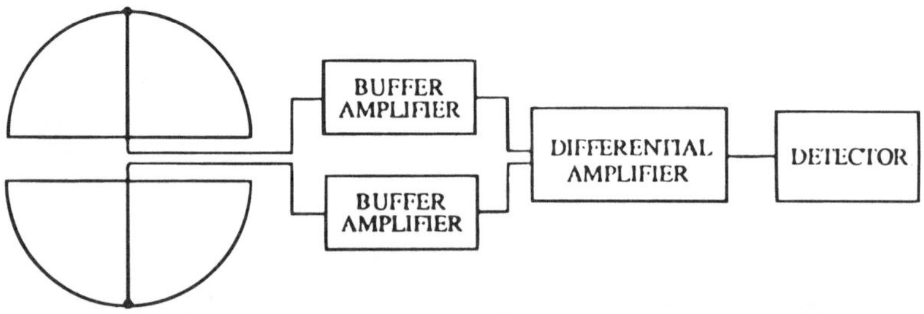

FIGURE 5 Block diagram of an active E-field sensor.

The E-field sensor shown in Figure 5 utilizes a spherical antenna with two high input impedance buffer amplifiers and a differential amplifier located inside the sphere. The antenna consists of two identical hollow metal hemispheres. The apex of each hemisphere is connected to the buffer amplifier using a large-diameter cylindrical conductor.

The sensors were tested and calibrated in a TEM cell in both the frequency domain and time domain. In the frequency domain the transfer function of the sensor was measured using HP3577A and HP8510B network analyzers. Measurements of the sensor responses in the time domain were performed using an HP8082A pulse generator, a PSPL 1000C impulse generator, and a Tektronix 11802 digital sampling oscilloscope.

The experimental results for the E-field sensor are shown in Figure 6. The flat response region extends from 143 Hz to 140 MHz. The sensitivity of the sensor is estimated to be of the order of 0.4 mV/(V/m), or 68.4 db/m in terms of an antenna factor. When tested with a Gaussian-pulse incident field (450-ps width), all sensors provide an output pulse with a rise time of less than 1 ns.

Similarly a magnetic field sensor based on an electrically small loop suitable for measurements of fields in a very wide frequency range and transients has been developed.[20E] Through the use of a current transformer and electronic circuitry, "flat" amplitude and linear phase responses sensitivity 0.18 V/Am^{-1} and dynamic range of 60 dB have been achieved in a broad frequency range. A single sensor covers frequencies from 600 Hz to 200 MHz. The sensor has a minimal response to the electric field (at least -20 dB).

B. INTERNAL FIELDS

Since the present state of knowledge indicates that the thermal and nonthermal interactions at higher frequencies of electromagnetic fields with mammalian tissues are related to the

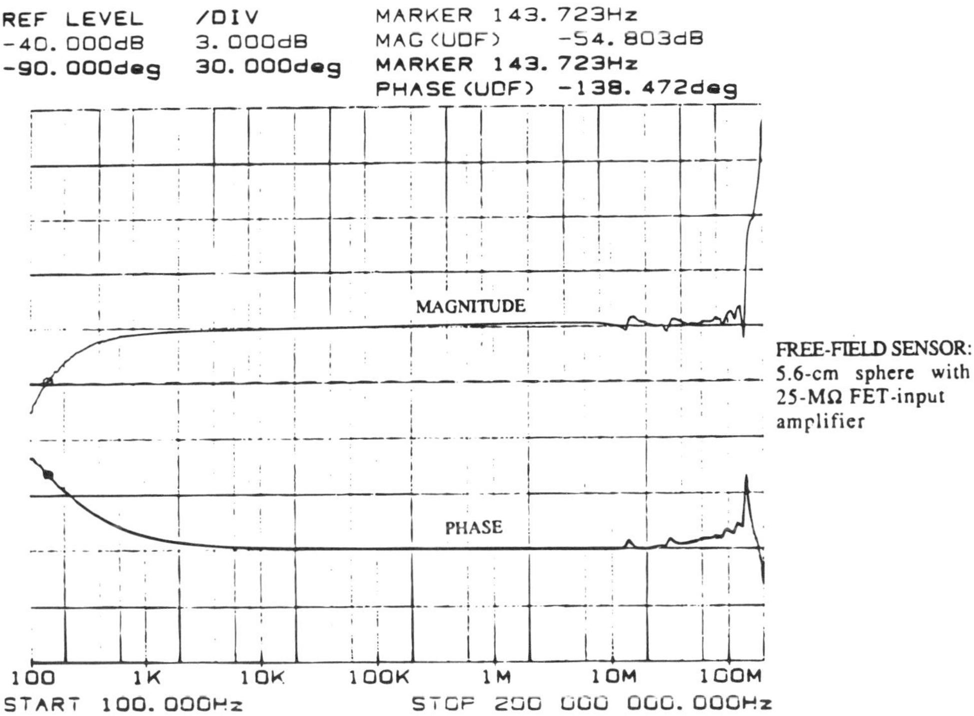

FIGURE 6 Amplitude and phase responses of the E-field sensor.

intensity of the internal E field, only E-field sensors have been developed. There are two possibilities of measuring the intensifies of the E field by direct means. The simplest way is to use a monopole or dipole antenna at the end of a coaxial line which is connected to a conventional detection and measuring system. This technique is effective when the direction of the E-field is known, so that the coaxial line can be placed perpendicularly to the direction of the field, ensuring minimal perturbation to the field. This is frequently the case for applicators used in hyperthermia systems.

A non-perturbing, implantable E-field probe consists of a very short dipole and a miniature Schottky-barrier diobe connected by high-resistance leads with the external circuitry. The dipoles and the high-resistance leads are deposited on a dielectric substrate. Amplitude and frequency responses of such a probe placed in a lossy, dielectric medium have been investigated.[22–26] Two types of probes can be employed, a bare probe and an insulated probe, as illustrated in Figure 7. The bare probe lenght is 2h, and the diameter is 2a. The insulated probe of the same dimensions is surrounded by a cylindrical, dielectric sheath having the dielectric permittivity ϵ_i. The dielectric medium around the probe is assumed homogenous within the probe-sampling volume.[22] For a short probe the sampling volume is approximately equal to a sphere of a radius h centered on the probe.[22] The equivalent circuits of the probes are shown in Figure 8.[22] The input admittance of the bare probe in any dissipative medium is[22]

$$Y = G + j\omega C \frac{\pi h}{\ln(h/a - 1)} (\sigma + j\omega\epsilon) \tag{14}$$

where ϵ is the dielectric permittivity of the biological tissue surrounding the probe and σ is the conductivity of the tissue ($\sigma = \omega\epsilon_0\epsilon''$). Equation 14 is valid for a thin ($a << h$) and electrically short probe ($\alpha h << 1$ and $\beta h << 1$, where α and β are the attenuation and

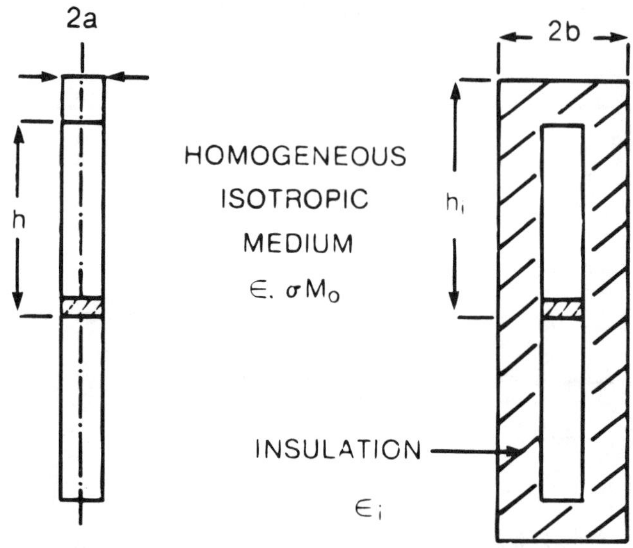

FIGURE 7 Bare and insulated cylindrical, dipole E-field probes.

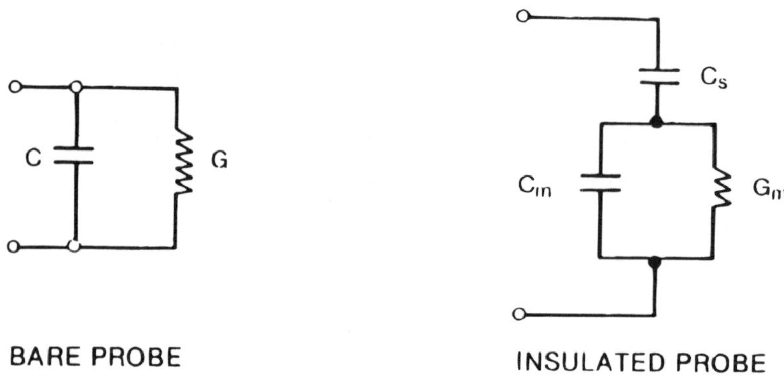

FIGURE 8 Equivalent circuits of electrically short bare and insulated E-field probes.[22]

propagation constants, respectively). The input admittance is therefore dependent on the permittivity of the surrounding medium.

The input admittance of the insulated probe is[22]

$$Y = \frac{\sigma\pi h}{\ln(h/b) - 1} \frac{\gamma^2}{p^2 + (1+\gamma)^2} + j\omega \frac{\epsilon_i \pi h}{\ln(b/a)} \frac{p^2 + (1+\gamma)}{p^2 + (1+\gamma)^2} \tag{15}$$

where

$$\gamma = \frac{\epsilon_i}{\epsilon} \frac{\ln(h/b) - 1}{\ln(b/a)} \tag{16}$$

and

$$p = \frac{\sigma}{\omega\epsilon} = \frac{\epsilon''}{\epsilon'} \tag{17}$$

For ϵ_i small compared with ϵ, and a sufficient thickness of the dielectric sheath, the admittance can be approximated by[22]

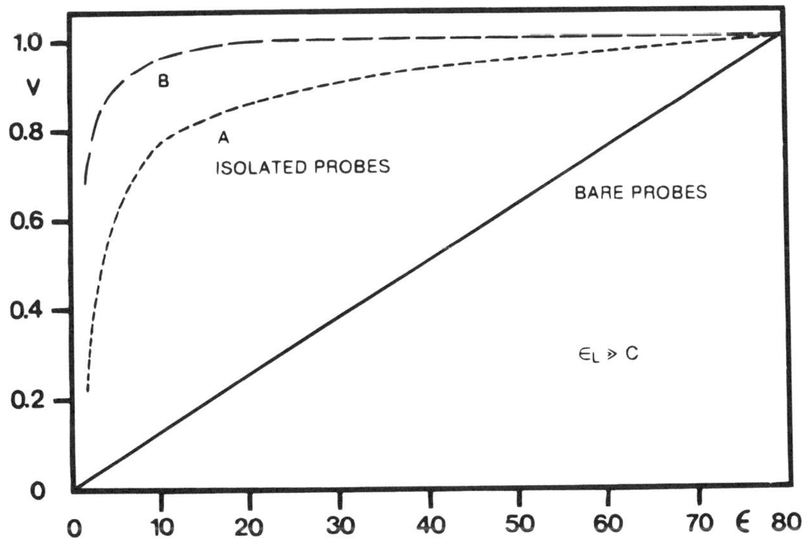

FIGURE 9 Normalized output voltage of bare and insulated probe as a function of relative, dielectric constant of the medium. The probes dimensions are (Probe A) a = 0.10 mm, b = 0.34 mm; (Probe B) a = 0.14 mm, b = 1.34 mm. For both probes h = 5.56 mm, h_i = 7.14 mm, and ϵ_i = 2.1 (Teflon®).[22]

$$Y = j\omega \frac{\epsilon_i \pi h}{\ln(b/a)} \tag{18}$$

and is therefore, independent of the dielectric properties of the medium in which the probe is immersed, but only on the dielectric permittivity ϵ_i of the probe insulation.

The voltage across the load admittance for the bare probe is[22]

$$V = -hE\frac{Y}{Y + Y_L} \tag{19}$$

where E is the intensity of the E-field parallel to the dipole and Y_L is the load admittance. For the voltage to be independent of the medium permittivity the load admittance must be much less than the probe input admittance, i.e.,

$$|Y_L| << |Y| \tag{20}$$

for the permittivities of all measured media. However, the input admittance can be very small,[22] making the condition (Equation 20) difficult to satisfy. Therefore, only a properly insulated probe is capable of providing a signal proportional to the E-field and independent of the dielectric properties of the test medium. Experimental results for a bare probe and the same probe with two dielectric sheaths of various thicknesses are shown in Figure 9.[22]

The normalized voltage increases linearly with the medium, dielectric constant for the bare probe, but is practically independent of the medium, dielectric constant (for values > 10) for the probe in the thick insulation. In all cases the load admittance is much greater than the probe input admittance ($Y_L >> Y$).

Insulation also reduces the error due to proximity of an interface. For instance, the error is <5% for the interface at a distance of h/3 from the probe.[24]

The reception of the incident signal by a parallel wire transmission line formed by the resistive leads was compared with that of the dipole.[26] The leads distort the field pattern somewhat, but for high-resistance leads the distortion is small. High resistance also reduces scattering of the incident wave. The highly resistive transmission line behaves as a low-pass filter, limiting applications of the probe to the fields modulated at frequencies much lower than f_m[26]

$$f_m = \frac{1}{8\pi RCs^2} \tag{21}$$

where C is the capacitance per unit length, R is the resistance per unit length of the thin-film leads, and s is the length of the leads.

Single- and three-dipole (isotropic) probes were designed and tested[18,27] and a few models are commercially available. A three-axis E-field probe with submillimeter dimensions has been recently developed.[27A] The probe can detect signals in the frequency band 0.2–18 GHz ranging from 60 V/m to over 1200 V/m. The overall probe diameter is 1.1 mm individual dipoles are 0.6 mm long and the resistive transmission line is 15 cm long and is deposited on a flexible mylar sheet. The probe is particularly suitable for mapping the fields adjacent to irregularly shaped objects.

A passive, remote-sensing method for measurements of the internal fields has also been proposed.[28] In this method, miniature, insulated diodes are imbedded in the tissue. When irradiated with two signals at different frequencies, diode nonlinearities generate intermodulation products whose amplitudes are related to the intensities of the irradiating fields. This method, the feasibiltiy of which was established,[28] does not seem to have any advantages over implantable E-field probes.

A summary of E-field probes and associated detector circuits for biological applications between 0.2 MHz and 26 GHz and field strengths from < 1 to 1000 V/m has been published recently.[28A]

C. CALIBRATION TECHNIQUES

Probes for measurements of external and internal fields have to be calibrated over the whole frequency and amplitude range of operation. The calibration consists of placing the probe in a well-defined field of known intensity and taking output readings. While external field probes are calibrated in air, internal field probes should be calibrated in the tissueequivalent dielectric, i.e., a dielectric having the complex permittivity close to that of tissues at the frequencies of interest.

To calibrate external field probes, three types of techniques can be used: free space, standard field methods, guided wave methods, and transfer probe methods.[7,29]

1. Free Space Methods

In the free space methods, a plane-wave field of a known power density is produced in free space either in an open range or in an anechoic chamber. A basic block diagram of the arrangement is shown in Figure 11. The power density S at the axis at a distance r from a standard gain horn in the far-field zone is given by

$$S = \frac{P_T G}{4\pi r^2} \tag{22}$$

where P_T is the power transmitted by the standard horn and G is the horn gain. The transmitted power is determined from measuring the forward and reflected powers. The dual directional coupler (Figure 10) has to be calibrated. The standard horn gain can be determined experimentally[30] or calcualted.[31–33] Since the distance between the horn and the calibrated probe is

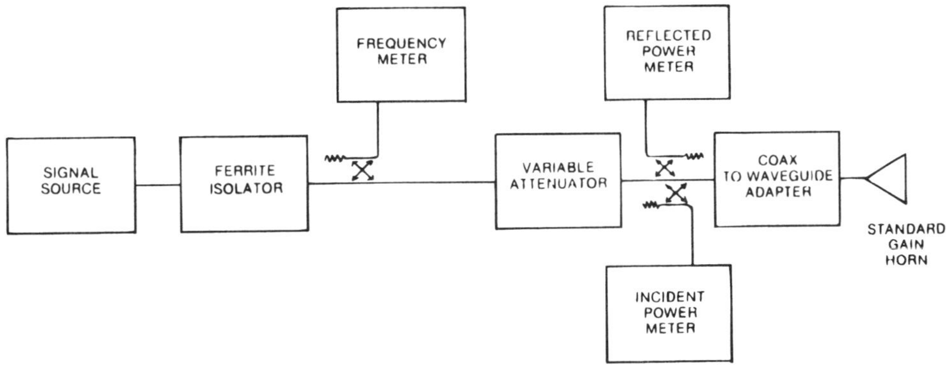

FIGURE 10 Experimental arrangement for the free-space, standard field method of calibration.

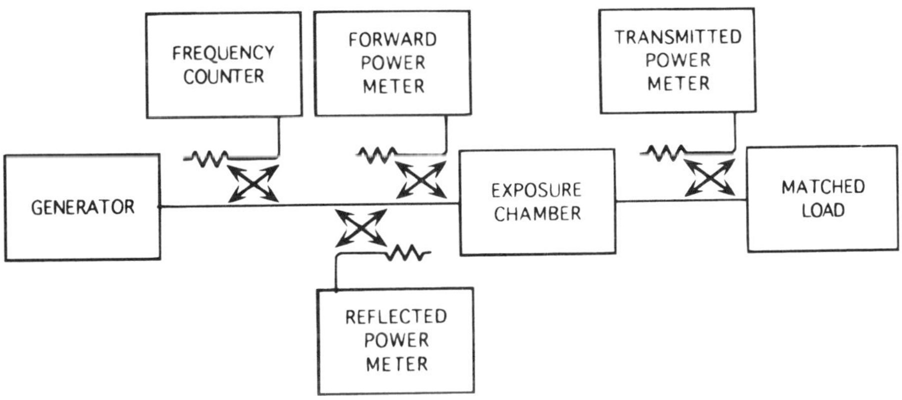

FIGURE 11 Experimental arrangement tor measurement of the average SAR.

frequently too short for far-field operation, near-field corrections have to be included.[32,33] Operation in the far field is frequently precluded because of the limited size of the anechoic chamber or limited power of the signal sources to produce the required power density.

2. Guided Wave Methods

The E-field sensors can be calibrated in guided structures such as rectangular waveguides and transverse electromagnetic (TEM) cells. These methods are only useful when the dimensions of the sensor are small compared with the cross-section of the structure. Furthermore, the calibration guide has to be terminated by a matched load, so that the standing waves are minimized.

In a rectangular waveguide operating in the fundamental TE_{10} mode the E-field intensity (rms value) at the center is equal to

$$E = \left(\frac{2PZ}{ab}\right)^{1/2} \tag{23}$$

where P is the power delivered to the waveguide, a and b are the dimensions of the waveguide, and

$$Z = \frac{Z_0}{[1 - (\lambda/2a)^2]^{1/2}} \tag{24}$$

($Z_0 = 377\Omega$ and λ is the free space wavelength). The power density at the center of the waveguide is equal to

$$S = \frac{E^2}{Z} = \frac{2P}{ab} \tag{25}$$

The E-field intensity varies cosinusoidally along the wider dimension a of the rectangular wavegulde.

A TEM cell can also be conveniently used for calibration of the exteranl field probes. The TEM cell consists of a section of the rectangular, coaxial transmission line tapered on both ends to converge into a standard, coaxial line. The TEM cell can support various propagation modes,[36–38] the lowest being the TEM mode. When only the TEM mode is propagating (for the conditions that have to be satisfied see References 36 to 38), the intensity of the E-field midway between the center and the outer conductor of the cell is equal to[33]

$$E = \frac{(ZP)^{1/2}}{D} \tag{26}$$

where Z is the characteristic impedance of the cell (usually 50 Ω), P is the power transmitted through the cell, and the D is the separation distance between the center conductor and the outer conductor of the cell. A standing wave in the cell can cause considerable calibration errors.

3. The Standard Probe Method

In the standard probe method, a probe to be calibrated is compared with an accurately calibrated "transfer standard."[29] The only requirement is that the field used for calibration be of a desired magnitude, sufficiently uniform over the calibration region and that it remain constant during the time required for calibration.

4. Implantable Probes

Since as indicated earlier, the output voltage of an implantable probe depends on the dielectric properties of the medium surrounding the probe, calibration has to be performed with the test probe in the environment simulating the actual conditions of use. Most implantable probes have a sufficiently thick dielectric insulation around the dipole, so that calibration in one medium having a dielectric constant between 10 and 80 is usually sufficient. Of course, calibration has to be performed at all frequencies of interest, or at a sufficient number of frequencies to obtain the full frequency characteristic of the probe through extrapolation. Various mixtures of electric materials (phantom materials) have been developed to simulate the electrical and thermal properties of biological tissues. These mixtures are used in calibrating, implantable probes. Basically, all three calibration methods that are used for free-space probes can be utilized for implantable probes when free space is replaced by the phantom material.

The simplest method of calibration uses an anechoic chamber and a semi-infinite slab of a tissue phantom material.[39] The E-field intensity in a sufficiently large slab (such that it can be considered infinite in all three dimensions) exposed to a plane-wave of known intensity can be calculated. If E_0 is the tangential E-field intensity at the interface between air and the dielectric slab, the field intensity along the direction of wave propagation is

$$E_Z = E_0 e^{-\alpha z} \tag{27}$$

where z is the distance from the interface and α the attenuation constant.

$$\alpha = \frac{2\pi f}{c}(\epsilon_r/2)^{1/2}\{[1 + (\sigma/2\pi f\epsilon_r\epsilon_0)^2]^{1/2} - 1\}^{1/2}(N/m) \tag{28}$$

where f is the frequency, c is the velocity of light in vacuum, ϵ_r is the relative dielectric constant, σ is the conductivity of the dielectric slab, and ϵ_0 is the permittivity of free space. This method is suitable at relatively high frequencies (above 2 GHz), so that the size of the slab is easily manageable. Also a relatively high power source is needed to provide sufficient power density.

In another calibration technique the probe is moved along the axis of a sphere made of a phantom material and irradiated by a plane-wave of known intensity.[27] The E-field intensity in the sphere can be accurately calculated. To avoid interference from the reflected and scattered fields, the calibration should be performed in an anechoic chamber.

A waveguide method can also be used to calibrate implantable probes.[40] A section of a rectangular waveguide is filled with a liquid phantom material and separated by a very thin ($< 0.01\lambda$) plastic window from the air-filled part of the waveguide. The dielectric constant of the window material should be low, so that the reflections from the window are minimal. The length of the dielectric-filled waveguide should be sufficient to simulate an infinitely long sample. The waveguide operates in the fundamental TE_{10} mode. The probe being calibrated is moved along the center of the waveguide axis in such a way that the dipole is parallel to the E-field direction or remains at a constant known angle (the angle must not be 90°). For the probe in the air section, the output voltage is related to the field intensity as[40]

$$V(\text{air}) = B_A E^2(\text{air}) \tag{29}$$

and in the dielectric

$$V(\text{diel}) = B_E E^2(\text{diel}) \tag{30}$$

At the interface

$$E(\text{diel}) = E(\text{air}) \tag{31}$$

and

$$\frac{B_E}{B_A} = \frac{V(\text{diel})}{V(\text{air})} \tag{32}$$

where B_E is the probe calibration factor in the test dielectric phantom material and B_A the probe calibration factor in air. By measuring V as a function of position on each side of the thin window and extrapolating the results to the interface, the ratio of the calibration coefficients can be determined. The calibration factor in the dielectric material can be then calculated by measuring the calibration factor in air as outlined in Section II.C.2. The dielectric window should be very thin and of low-permittivity material ($\epsilon \simeq 2$ to 3). A window thickness of 1 mm was found to be acceptable at 2.45 GHz.[40]

D. ABSORBED POWER MEASUREMENTS

The average SAR can be determined by measurement of the total absorbed power in an exposed object, e.g., a model or an experimental animal. This technique is suitable only for exposures in electrically enclosed exposure chambers such as a TEM cell, waveguide, or resonant cavity. A simplified block diagram of the measurement system is shown in Figure 11.

The absorbed power can be determined as

$$P_A = P_F - P_R - P_T \tag{33}$$

where P_A is the absorbed power and P_F, P_R, and P_T are the measured forward, reflected, and transmitted powers, respectively. If the exposure chamber is not of a transmission type, e.g., a. one-port cavity, Equation 33 simplifies to

$$P_A = P_F - P_R \tag{34}$$

As the absorption by an exposed animal changes as the animal moves, it is advisable to monitor and record the power levels during the whole exposure period, then to calculate the total absorbed energy from which the average absorbed power and the average SAR can be calculated. Several systems for this purpose have been described elsewhere.[41–46]

Equations 33 and 34 are valid provided that the losses (power absorbed) in the empty exposure chamber are negligible compared with the losses in the exposed object. If this condition is not satisfied, the power absorbed in the empty exposure chamber or the exposure chamber containing the animal cage has to be measured, e.g., using the same method of measuring the forward, reflected, and transmitted powers.

The corrected absorbed power is equal to

$$P_A = P_A - P_E \tag{35}$$

where P_E is the power absorbed in the exposure chamber, animal-restraining devices, etc.

Some errors can be associated with use of animal-restraining devices, since the power absorbed in them may be different in the presence of the animal due to the perturbation of the exposure field.

A correction to Equation 33 may also have to be introduced, if the load is not matched and significant reflections occur[46]

$$P_A = P_F - P_R - (1 - |\Gamma_L|^2)P_T \tag{36}$$

where $|\Gamma_L|$ is the modulus of the reflection coefficient of the load.

III. THERMAL MEASUREMENTS

A. BASIC PRINCIPLES

The rate of temperature change in the subcutaneous tissue *in vivo* exposed to RF or microwave energy is related to the SAR, as[1]

$$\frac{\Delta T}{\Delta t} = (SAR + P_m - P_c - P_b)/C \tag{37}$$

where ΔT is the temperature increase, Δt is the exposure duration, P_m is the metabolic heating rate, P_c is the rate of heat loss per unit volume due to thermal conduction, P_b is the rate of

TABLE 6
Tissue-Specific Heat Capacity and Density[47]

Tissue	Specific Heat Capacity (J/kg°C)	Density (kg/m³)
Skeletal muscle	3470	070
Fat	2260	940
Bone, cortical	1260	1790
Bone, spongy	2970	1250
Blood	3890	1060

heat loss per unit volume due to blood flow, and C is the specific heat. If before the exposure a steady—tate condition exists, i.e.,[1]

$$P_m = P_c + P_b \tag{38}$$

during the initial period of exposure

$$\frac{\Delta T}{\Delta t} = \frac{SAR}{C} \tag{39}$$

and the SAR can be determined from measurements of an increase in the tissue temperature over a short period of time following the exposure. For tissue phantoms and tissues *in vitro,* Equation 39 can be used as long as the thermal conductivity can be neglected, i.e., for short durations of exposure. Several methods of SAR determination are based on thermal measurements and utilization of Equation 39. The specific heat and density of various tissues is summarized in Table 6.[47]

B. CALORIMETRY

All calorimetric methods are suitable for determination of average SAR, but do not allow determination of the SAR distribution (i.e., variation in SAR from position to position in the irradiated body).

1. Analysis of Heating and Cooling Data

Heating and cooling data can be analyzed to estimate the energy absorbed by an irradiated sample.[48,49] For a system of mass m, specific heat C, and initial temperature T_0 heated at a constant rate from time $t = 0$ to $t = \tau$, the temperature at any time reflects a balance between input and loss of heat in the heating period[48]

$$\frac{\partial T}{\partial t} = A - k\Delta T \quad \text{for} \quad 0 \leq t \leq \tau \tag{40}$$

where $\partial T / \partial t$ is the heating rate, A is the rate of temperature increase in the absence of heat losses, and k is the proportionality constant $\Delta T = T - T_0$ In the cooling period

$$\frac{\partial T}{\partial t} = -k\Delta T \qquad t \geq \tau \tag{41}$$

If the system is absorbing energy at a constant rate W then

$$W = mCA \tag{42}$$

Integration of Equations 40 and 41 yields

$$\Delta T = A/k(1 - e^{-kt}) \qquad 0 \leq t \leq \tau \tag{43}$$

and

$$\Delta T = \Delta T_{\tau} e^{-k(t-\tau)} \qquad t \geq \tau \tag{44}$$

where ΔT_{τ} is the temperature difference at the termination of heating. To calculate W (Equation 42), the constant k is first determined from the slope of the cooling curve (Equation 43). This is done by finding the slope of ln ΔT_{τ} vs. time for $t > \tau$ (e.g., by a linear regression analysis). Then A can be found from the slope of ΔT vs. e^{-kt} (Equation 43). Alternatively,[47] if the temperature of the system reaches a steady state before the heating is terminated, the rate of energy input equals the rate of energy loss, and Equation 43 becomes

$$\lim_{t\to\infty} \Delta T = A/k = \Delta T_{SS} \tag{45}$$

where ΔT_{ss} denotes the steady-state temperature increase. From Equations 42 and 45,

$$W = mCk\Delta T_{SS} \tag{46}$$

In this case k and ΔT_{ss} can be obtained directly from the, data, i.e., the temperature vs. time.

Another method of calculation is based on combining Equations 43 and 45 to obtain

$$1n(\Delta T_{SS} - \Delta T) = -Kt + 1n\Delta T_{SS} \qquad t \leq \tau \tag{47}$$

where k is the slope and ΔT_{ss} is the intercept of $1n(\Delta T_{ss} - \Delta T)$ vs. time.

One important advantage of the heating and cooling curve analysis is that this method does not require calibration. The method is particularly suited for *in vitro* measurements. The analysis is difficult to apply to *in vivo* experiments because of the animal thermoregulatory response to electromagnetic heating. A major limitation for both *in vivo* and *in vitro* experiments arises when the energy deposition in the exposed object is non-uniform while the temperature is measured only in few (frequently only one) locations. To limit the amount of power required to achieve a ΔT of a few degrees, the sample may be enclosed in Styrofoam® insulation.[48] In summary, the heating/cooling curve method is simple and can provide accurate results when carefully applied to samples/animals that are exposed under such conditions that energy deposition through the whole volume is relatively uniform and no thermoregulatory response is evoked by the exposure. To obtain reliable temperature measurements, only probes suitable for operation in RF fields should be used.

2. Dewar-Flask Calorimeter

A typical calorimeter contains a vacuum (Dewar) flask containing a measured amount of water and sealed with a cork stopper.[50] The water is stirred by a magnetic stirrer to minimize time to reach thermal equilibrium. Optionally, insulation may be added on the outside of the flask.[51] The temperature of water is measured with a thermistor and recorded with a resolution of 0.01°C.[50]

The SAR (W/kg) is determined as[52]

$$SAR = C_S[T_e(\text{exposed}) - T_e(\text{control})]/t \tag{48}$$

where t is the exposure time, C_S is the specific heat capacity of the exposed sample (e.g., for a rat or mouse $C_S = 3448$ J/kg °C), and

$$T_e = [Z_p + M_eC_e)(T_F - T_i)]/(M_SC_S) + T_F \tag{49}$$

where T_e is the temperature (°C) of the exposed sample upon insertion in the calorimeter, M_s is the mass of the sample (kg), M_e is the mass of water in the calorimeter (kg), C_e is the specific heat capacity of water ($C_e = 4185$ J/kg °C), T_i is the temperature of the calorimeter just before insertion of the sample (°C), T_F is thermal temperature of the calorimeter (°C), and Z_p is the heat capacity of the calorimeter (J/°C).

Several precautions should be followed to obtain accurate results,[52] namely, there should be enough water just to allow total immersion of the exposed sample (animal cadaver) and the water temperature should be adjusted to approximately 0.5°C below the ambient air temperature. Irradiation time should be adjusted so that the resultant temperature following the insertion of the test sample is about 1°C higher. This limits the calorimeter drift. It is assumed that the thermal equilibrium is reached when the change in temperature is not >0.01°C.

The heat capacity of the Dewar flask can be found from Equation 49 by adding a known amount of water at a known temperature.[50] Similarly, the specific heat capacity of an unknown sample can be obtained from Equation 49 when T_e is known.[50]

The Dewar calorimeter is a relatively simple and inexpensive instrument which can easily be calibrated and requires a relatively short time for one measurement (about 1 hr for a rat).[52] When the experiment is carefully performed, accuracy of a few percent can be achieved. Nonuniformity of energy deposition does not affect the accuracy of the average SAR obtained by this method. Liquids other than water can also be used, e.g., dodecane ($C_S = 2100$ J/kg °C).[50]

3. Twin-Well Calorimeter

A twin-well or differential calorimeter[2,53] consists of two identical metal cylinders large enough to hold a particular animal of interest (different sizes are used for different animals). Each cylinder is surrounded by a thermopile, an array of thermocouples connected in series (voltages additive).[2] The thermopiles of the two cylinders are connected in opposite polarities so that the voltages are subtracted. Therefore, when both cylinders are at the same temperature, the resulting output voltage from the two thermopiles is zero. The twin cylindrical wells are surrounded by a cylindrical heat exchanger carrying a circulating fluid. The exchanger is covered by an insulating jacket. The measurement procedure[53] consists of placing a pair of freshly killed animals of the same body weight into the calorimeter and equilibrating their temperatures—this usually takes about 2.5 hr.[2] One animal is then exposed briefly to RF energy, and the second serves as a reference. Immediately after treatment the animals are returned to the calorimeter and the heat added to the animal's body by irradiation is measured. The heat is determined by integrating the output voltage from the thermopiles over time (usually about 10 to 12 hr[2]). The system can be calibrated with a known amount of ice at 0°C or by adding a known amount of heat to one of the two identical thermal bodies in the calorimeter.[2] The use of a microprocessor greatly improves the accuracy of the integration over the long period of time (several hours) required to equilibrate the two wells.

The twin-well calorimeter is more accurate than the Dewar-flask calorimeter, its major drawback being the relatively long time required for one measurement.

TABLE 7
Salient Characteristics of Thermometers

Capability	Description
Sensitivity	Minimum ΔT detectable
Accuracy	Difference between the indicated and true temperature
Stability	Drift of accuracy over time
Response time	Time to reach 67% of the final reading when subjected to a step change in temperature
Size	Size determines the spatial resolution
Indication	Type of temperature indication: direct-indirect (e.g., tables); analog-digital; continuous-intermittent

C. THERMOMETRY

1. Basic Concepts

Various physical principles can form a basis for temperature measurements. One of the major conditions for a thermometer to be suitable for temperature measurements under certain environmental conditions, e.g., RF exposure, is that it should give a true response to temperature but not other environmental conditions, e.g., the intensity of the exposure field. Some of the most popular thermometers, e.g., mercury-in-glass and platinum resistance, are not suitable for use in electromagnetic fields, while others, e.g., thermocouple or thermistor, can be used after special modifications. The salient characteristics of thermometers that should be considered in selecting one for a specific purpose are summarized in Table 7. Various types of popular thermometers were reviewed elsewhere.[54,55]

2. Conventional Thermometer Probes

The difficulty of measuring temperature in electromagnetic fields with many conventional thermometers stems from three types of interaction between the thermometer and the field.[54] They are electromagnetic interference (EMI) (basically pick-up by the leads), direct heating of the temperature sensor, and perturbation of the field by the thermometer. Several techniques can be used to minimize these effects.[54] The interference and pick-up can be minimized by placing the leads of the sensor, e.g., a thermocouple or thermistor, perpendicular to the E-field. Magnetic induction pick-up is reduced when the leads are tightly twisted. Several designs have been proposed for thermocouple and thermistor thermometers for use in electromagnetic fields.[56–61]

A stainless steel hypodermic needle containing a thermistor[57] can be used to measure temperature in saline solutions simulating tissue. All leads and connectors not immersed in the solution should be shielded from microwave radiation. The thermistor probe should be located beneath the surface of the solution. Ultraminiature (125-μm diameter), glass-coated thermistors can also be used for temperature monitoring in vivo, e.g., during heating.[60] The thermistor beads are bound to high-resistance (>164 kΩ/m) carbon-impregnated, flexible Teflon® leads.[60]

In some applications even a carefully designed thermocouple thermometer can provide satisfactory results. A small size of the thermocouple wires (0.05 mm in diameter) is a minimum requirement.[59] The error related to the current flow through the thermocouple junction due to the unequal resistance of the two wires can be minimized by a filter.[58] A 0.1-mm diameter thermocouple in a metal shield and a Teflon® jacket connected through two choke-coils with a shielded digital thermometer was successfully used for monitoring hyperthermia induced by a RF field (at 13.5 MHz).[61] However, use of thermocouples and thermistors in RF fields is subject to interference from many sources and is possible only in certain situations. Unless a sensor is carefully designed and thoroughly tested, serious errors in measurements can arise. As an alternative, thermocouples and thermistors can be used and inserted into the test site while the RF field is turned off for the duration of the temperature measurement. Before the field is turned on the thermometer has to be withdrawn. Usually special insertion guides are employed.[2]

3. Nonperturbing Temperature Probes

Several types of nonperturbing temperature probes have been developed,[62–75] however only few are available commercially.

A small, high-resistance thermistor with high-resistance leads[65] was developed into a commercial product manufactured by Vitek, Inc. The thermistor resistance is sensed by injecting a constant current and measuring the voltage across the thermistor by means of a high-impedance amplifier.[65]

Several optical temperature sensors take advantage of the fact that optical fibers, being low-loss dielectrics, are neither directly heated by an electromagnetic field, nor do they perturb the field. In the liquid-crystal, optical-fiber probe a strong temperature-dependence of the reflectance of red light by a liquid crystal is utilized.[62–64,71] The liquid crystal medium is sealed between a glass lens and a fiberoptic catheter. The catheter contains eight optical fibers in Teflon® tubing. Light (670 nm) emitted by a light-emitting diode (LED) is transmitted by four fibers to the sensor tip, and the remaining fibers receive light scattered by the sensor and carry it to a photodetector. The photodetector converts light into voltage. The actual system is somewhat more complicated as pulse modulation is employed. An increase in the sensor temperature results in an increased reflectance from the crystal and subsequently an increase in the detected voltage. By adjusting the crystal composition its useful range of operation (about 10°C) can be changed. Some problems were encountered with the time stability of the probe.[71]

Reflection of a light beam from a thermodilatable liquid can also be used to measure the temperature.[66–68] The shape and position of the liquid meniscus vary with temperature. The light beam is carried by optical fibers.

In the birefringent crystal thermometer a solid, single crystal of lithium tantalate is used as a sensor.[70] The principle of operation is based on the temperature-dependence of the difference in the indices of refraction along the two optical axes. The optical and electronic systems can then be similar to those used in other optical-fiber systems. To further improve the performance, a reference signal from the LED can be applied to the photodetector. The solid crystal sensor is more time-stable than liquid crystal sensors.

A very small (0.25 mm) and inherently stable sensor is based on the semiconductor band edge shift.[72] In semiconductor materials a forbidden zone separates the allowed electron energy levels with a typical energy spacing that corresponds to the quantum energy of infrared or visible light, e.g., for GaAs $\Delta E = 1.43$ eV, which corresponds to a wavelength of 0.87 μm at room temperature. At temperatures of interest in biological applications, the thermal energy is very small compared with the gap energy, so practically all electrons occupy the lower (valence) band, and the upper (conduction) band is empty. Light of a wavelength corresponding to the gap energy excites the electrons and is absorbed (Planck law). A typical dependence of the light transmission vs. wavelength is shown in Figure 12. It is evident that for a properly selected wavelength the absorption is strongly dependent on the temperature.

Properties of rare earth phosphors and fiber optics are utilized in another temperature monitoring system[73,74] that is manufactured by Luxtron Corp. The principle of operation is based on emissions by the phosphors of well-defined fluorescent lines, whose ratio of intensities depends on the temperature of the phosphor only. The ratio of intensities does not depend on amount of phosphor, level of illumination, or nature of optical coupling, and is not in any way affected by electromagnetic fields. Therefore, the temperature of the phosphor can be determined by measuring relative intensities of two colors from the glow

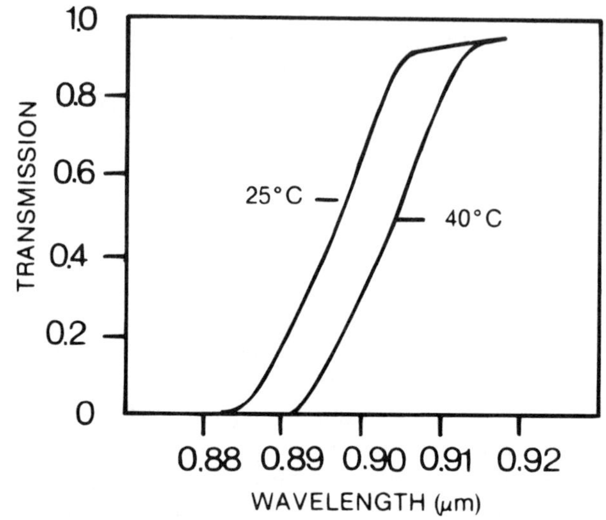

FIGURE 12 Optical transmission through GaAS at two temperatures.[72]

TABLE 8
Characteristics of Nonperturbing Thermometers

Type	Dynamic range (°C)	Accuracy (°C)	Diameter (mm)	Ref.
Thermistor[a]	30	0.1	0.94	65
Thermistor[b]	25	0.15	0.91	76
Liquid crystal	15	0.1	2	62–64,71
Thermodilable liquid	40	0.5	2	66–68
Birefringent crystal	30	0.3	1.5	70
GaAs semiconductor	25	0.1	0.25	72
Phosphor	250	0.1	0.5	73, 74

[a] Available from Vitec Inc.
[b] Available from Narda Corp.

of a very small amount of phosphor. The probes can be very small (0.2 mm) and have a wide range of temperature sensitivity (100°C), and modern electronics utilizing a microprocessor provide for accurate and convenient manual or automatic measurements.

Various methods[71,76,77] can be employed for evaluating the performance of a temperature probe for use in microwave fields, and some probes have been evaluated. Basic characteristics of nonperturbing thermometers are summarized in Table 8.[1,76]

D. THERMOGRAPHY

Infrared thermography technique for determination of temperature and SAR in models of biological bodies and animal cadavers exposed to strong electromagnetic fields was first developed by Guy[78] and has gained wide acceptance.[1,75,77,79–83] Thermographic cameras are commercially available and are used in various fields. A thermographic camera gives temperature distribution on a surface with a capability of spatially resolving the temperature gradients over any selected line within the limits of the camera resolution. A typical camera consists of an infrared radiation detector, usually a liquid-nitrogen-cooled chip of indium antimonide or mercury (cadmium) telluride, an optical scanning system, and a synchronized display unit.[81] Many cameras are also computer-compatible.

The response I(T) of a thermographic camera to the incident radiation can be represented by[81]

$$I(T) = \int_{\lambda_{min}}^{\lambda_{max}} p(\lambda, T)r(\lambda)d \tag{50}$$

where T is the temperature λ A is the light wavelength, p(λ, T) is a function describing the spectral radiation characteristics of the source, and r(λ) is a function characterizing the spectral response of the instrument, including the detector characteristic, attenuation by optical elements, and electronic gain factors. In practice, I(T) is determined from a calibration of the system against black-body sources at a known temperature. For a gray source, the emittance characteristics of the source must be determined experimentally.[81]

To calibrate the camera, the differential characteristic, i.e., dI/dT as a function of temperature is determined by measuring finite differences $[I(T_1) - I(T_2)]/(T_1 - T_2)$ for various temperatures. Then, the average temperature $T_{av} = (T_1 + T_2)/2$ is used as the independent variable in the calibration.[81] Most thermographic cameras have calibration sources at various reference temperatures.

The optical properties of the material on which temperature profile is being measured with a thermographic camera affect the measurement accuracy. These are the emittance, reflectance, transmittance, and penetration depth.[81] The reflectance and emittance are interrelated, and for opaque objects their sum is = 1. The transmittance and penetration depth are similarly related. The penetration depth is defined as the depth in the medium from which 1/e of the radiation emitted by an elemental volume in the medium escapes to the surface. The penetration depth is of importance for materials with nonuniform temperature distribution "in depth." A non-uniformly heated transparent material which is hotter inside than near the surface appears warmer than a similarly heated opaque material. Tissue phantom materials developed for bioelectro-magnetics studies were tested,[81] and for thicknesses >0.4 mm no differences within the noise of 0.2°C were found (it is interesting to note that the muscle equivalent material 0.4 mm thick was visually, but not thermally, translucent). On the other hand, the transmittance of 0.05-mm thick polyethylene is 0.9, but its effect on determining the temperature distribution of the material underneath is negligible.[81]

The emittance of a surface is the ratio of the radiation flux per unit area of the emitter to that of a black-body radiator at the same temperature and under the same conditions. The emittance depends on the material, the surface smoothness, and the presence of oils, moisture, and dust on the surface. The emittance of 2.45-GHz bone equivalent material is about 0.8 to 0.9, and for muscle phantom very close to that of water, i.e., 0.98.[81] the emittance has to be taken into account when information about absolute temperature is needed.

The major limitations in thermographic measurements are the spatial resolution of the camera (the source has to be larger than the minimum surface area resolved) and the oblique angles of viewing (more than 30°).

When a thermographic camera is used to determine the temperature distribution or SAR in tissue equivalent models or animal cadavers exposed to RF radiation fields, the following procedure is employed.[78] A model of an animal or other object is made of tissue-equivalent materials and cut along planes whose two-dimensional SAR (temperature) distribution is to be determined. To avoid water loss from the material, the open surfaces are covered with a very thin (0.05 mm or less) polyethylene film or silk screen. An animal is killed, frozen in dry ice, cast in a Styrofoam® block, bisected, covered with polyethylene film, reassembled, and equilibrated to room temperature.[2] The model or cadaver is then exposed under specific test conditions for a few seconds (preferably <20) to a high-intensity field. The parts are separated and thermographic scans are taken. To obtain temperature or SAR information undistorted by thermal conduction the exposure time and the delay between taking thermographic scans must be minimized. The delay and the recording time should not be longer than 10 sec. High-power RF sources allow exposure times of <20 sec. Large temperature gradients should also be avoided, as they increase thermal conduction. Gradients of 5 to 10°C appear reasonable. If the thermal conduction is neglected, the SAR can be calculated from the measured temperature using Equation 39.

IV. EXPOSURE SYSTEMS

A. GENERAL REQUIREMENTS

Depending on the object and the purpose of the experiment, different characteristics of an exposure system are required. For instance, chronic exposure of a large number of rodents requires different features of the exposure system than exposure of a small *in vitro* preparation. Exposures to investigate potentially hazardous effects of electromagnetic fields usually call for uniform, far-field conditions, while exposures for therapeutic purposes require a specific profile of the energy deposition. Certain requirements are common for all exposure systems. These include (1) a well-defined exposure field; (2) controllable, ambient conditions (temperature, humidity); (3) ability to determine, preferably to measure, the SAR and/or temperature of the specimen.

The exposure characteristics that are important only in some situations are (1) possibility of providing the exposed subject with nutrients; (2) ability to monitor various functions or conditions, electrocardiogram e.g., (ECG), electroencephalogram (EEG), respiration; (3) simultaneous exposure of a number of test subjects or samples. Another underlying factor in selecting an exposure system is its cost. To limit the cost, enclosed systems, such as a TEM cell or cavities, are often used to simulate free space fields.

B. FREE FIELD—ANECHOIC CHAMBERS

This exposure system simulates the far-field, free-space exposure. The exposure takes place in a large room lined with a special material that is only minimally reflective for radiowaves over a frequency range of interest. Such enclosures are called "anechoic chambers" and have been extensively used in antenna research. The anechoic materials are available commercially in the form of pyramids or wedges, and the larger their size the lower the frequency of operation with small reflections. An exposure field is produced by an antenna, and the far-field conditions prevail at distances from the antenna greater than

$$R \geq \frac{2a^2}{\lambda} \tag{51}$$

where a is the greatest dimension of the antenna, and λ is the wavelength. The far-field power density on the antenna axis at a distance r is equal to

$$W = \frac{PG}{4\pi r^2} \tag{52}$$

where P is the power delivered to the antenna and G is the antenna gain. The area within which the power density is relatively constant (e.g., within $\pm$ 0.5 dB) depends on the distance from the antenna and the antenna pattern.

A schematic diagram of an anechoic chamber exposure system is shown in Figure 13. To determine the power density, the system has to be calibrated (see Section II.C.1). Anechoic chambers are used for exposures of experimental animals usually at frequencies above 500

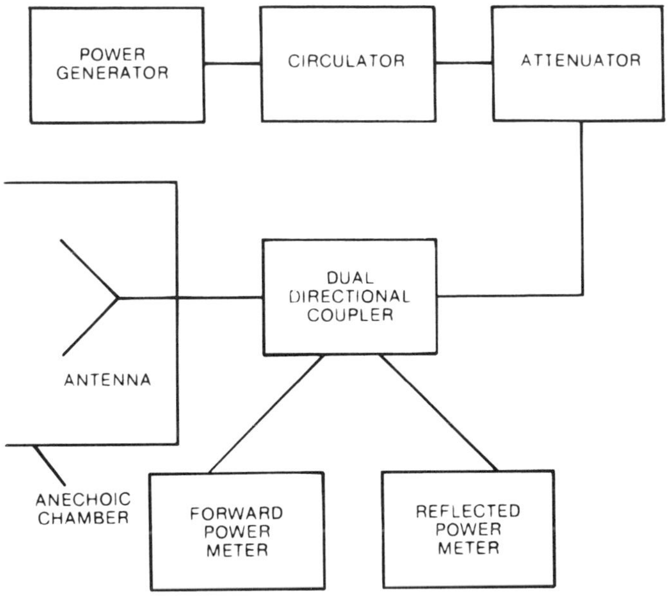

FIGURE 13 Anechoic chamber exposuse system.

MHz. At lower frequencies enclosures become exceedingly large and costly. Temperature and humidity in anechoic chambers can be closely controlled. One of the main disadvantages is that when a large number of animals are exposed simultaneously, a relatively high power is required. To ensure the same exposure conditions of each test animal they have to be separated from each other by a sufficient distance to avoid multiple reflections.

By proper separation of specially designed animal cages, non-uniformities in the exposure field can be minimized.[91] The animal cages should be constructed from materials having electrical properties as close to those of free space as possible. Low-density Styrofoam® is an excellent choice, and low-dielectric, constant plastics (e.g., Plexiglas®, Teflon®, and acrylic) and dry wood can be used when mechanical strength is required. To increase the power density within the exposure area for a given input power a special slot array[92] can be utilized.

An improved space utilization and a simulation of the free-field exposure is provided by "compact ranges." Two structures have been described[93] and used.[94] A prolate-spheroidal reflector having two foci, when illuminated by an open-ended waveguide placed in one focus, produces a uniform wavefront over a diameter of 1.5 wavelength in the other focus.[93] Another design uses absorber-lined horns.[93,94] The absorber-lined horn (rectangular or cylindrical) acts as a small-aperture radiating source with a radiation pattern that is approximately uniform at a plane near to the horn aperture. An absorber-lined horn operating at 2.45 GHz is shown in Figure 14.[93] The absorber is thin and of the resonant type.

C. ENCLOSED EXPOSURE SYSTEMS

1. TEM Cell

Exposure conditions simulating those of free space can be obtained in a TEM cell utilizing a modified, shielded stripline configuration as shown in Figure 15. A simplified sketch of a TEM transmission cell is shown in Figure 16. The tapered sections are designed to match the rectangular, cross-section strip line to standard, 50-Ω coaxial connectors. The characteristic impedance of the shielded strip line (Figure 15) can be calculated from the following equation[95]

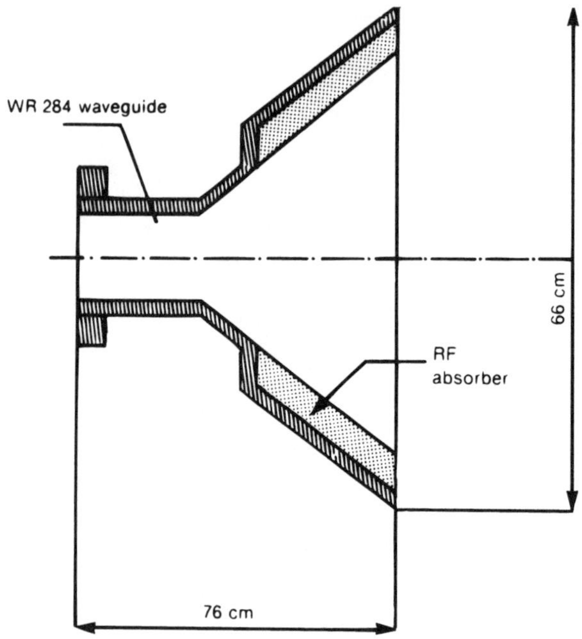

FIGURE 14 Compact range—an absorber-lined horn at 2.45 GHz.[93]

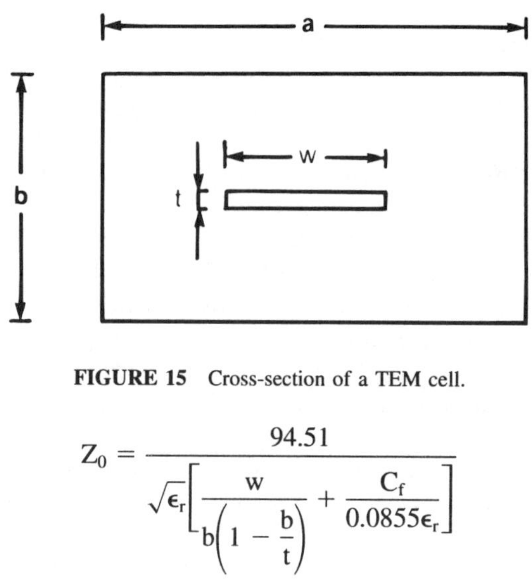

FIGURE 15 Cross-section of a TEM cell.

$$Z_0 = \frac{94.51}{\sqrt{\epsilon_r}\left[\dfrac{w}{b\left(1 - \dfrac{b}{t}\right)} + \dfrac{C_f}{0.0855\epsilon_r}\right]} \tag{53}$$

where ϵ_r is the relative dielectric constant of the medium between the conductors (for air lines $\epsilon_r = 1$), W, b, and t are the line dimensions as shown in Figure 15 and C_f is the fringing capacitance of the line in picofarads per centimeter. The fringing capacitance can be determined experimentally[95] or by various analytical methods.[96] For strip lines having very thin septum ($t/b << 1$ and $w/b \geq 0.5$), the fringing capacitance per unit length can be calculated within an accuracy better than 2% from[96]

$$C_f = \frac{2\epsilon_0}{\pi} \ln\left[1 + \coth \pi\left(\frac{a - w}{2b}\right)\right] \tag{54}$$

where $\epsilon_0 = 8.854\ 10^{-12}$F/m. For thin-septum strip lines with $0.1 \leq w/b < 0.5$, a correction factor ΔC has to be added to C_f.[96]

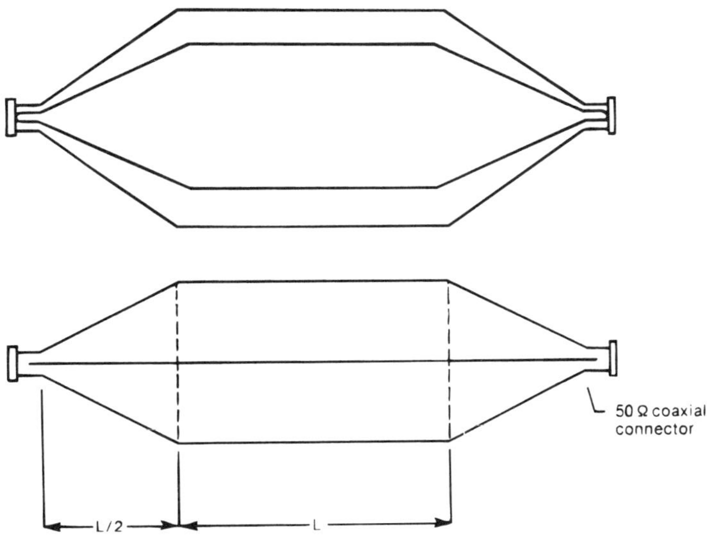

FIGURE 16 Top and side view of a TEM cell.

$$\Delta C \simeq \frac{2\epsilon_0}{\pi}\left[\frac{8}{(1+\sqrt{\lambda})^2(1+\lambda)}\right] \tag{55}$$

where

$$\lambda \simeq [1 - \exp(-2\pi w/b)]^{1/2} \tag{56}$$

For a finite thickness line the fringing field capacitance can be obtained from the following equation[96]

$$C_f = \frac{\epsilon_0}{\pi \ln 2}\left\{\frac{b}{b-t}\ln\left(\frac{2b-t}{6}\right) + \ln\left[\frac{t(2b-t)}{(b-t)^2}\right]\right\}$$
$$\cdot \ln\left[1 + \coth \pi\left(\frac{a-2}{2b}\right)\right] \tag{57}$$

The fields in a TEM cell resemble those of a plane-wave as long as only the fundamental TEM mode propagates in the cell. The approximate cut-off frequencies for the higher order modes are given by[95]

$$(f_C)_{n,m} = \frac{c(n^2a^2 + m^2b^2)^{1/2}}{2ab} \tag{58}$$

where c is the velocity of light ($c \simeq 3 \times 10^8$ m/sec). The E-field distribution for the first few modes is shown in Figure 17. More accurate and detailed data on higher-order modes in TEM cells are given elsewhere.[97] The cut-off frequency of the TE_{10} mode is equal to

$$(f_C)_{1.0} = \frac{c}{2a} \tag{59}$$

TEM transmission cells can be used for exposures even at frequencies higher than that

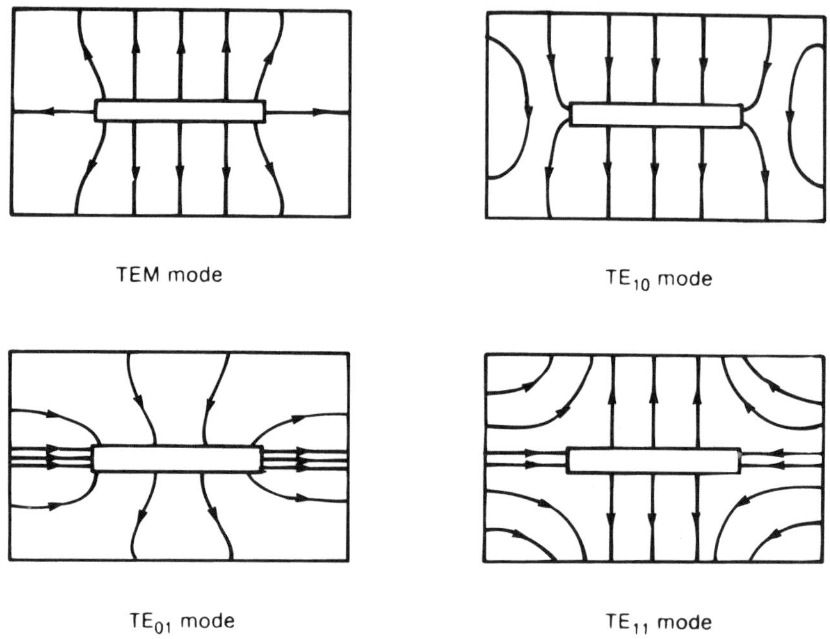

FIGURE 17 The E-field distribution in a TEM cell.

given by Equation 59.[97,98] The exposed object however, should not occupy more than one third of the volume between the septum and outer conductor.[95]

The E-field intensity (when operating in the TEM mode) is[95]

$$E = \sqrt{PZ_0}/d \tag{60}$$

where P is the net power flowing through the cell, Z_0 is the cell characteristic impedance, and d is the distance between the septum and the wall. Sources of error in determination of the field intensity from Equation 60 include[95] (1) the uncertainties in P (uncertainties in coupler calibration and RF power measurements, mismatch between the cell, RF source, and the cell termination); (2) the uncertainties in Z0; (3) the nonuniformity of the E field inside the cell resulting from field perturbation by the test object.

2. Resonant Cavity

Single- and multiple-mode resonant cavities can be used as exposure chambers. One of their major advantages is that the total SAR can be easily determined from measurements of the net input power. Other advantages are low cost and limited space requirements. The main disadvantage is that usually only one object (animal) can be exposed in a well-defined field. However, a system consisting of many cavities connected to a single power source offers a viable alternative to exposures in an anechoic chamber.[99]

A cylindrical cavity can be excited in the TE_{11} mode circularly polarized to provide relatively constant and easily quantifiable coupling of the energy to the test animal. Such cavities have been designed and extensively used in studies of rodents at 918 MHz[99] and 2.45 GHz.[100] The cavities can be made of wire screen to facilitate ventilation and visibility of the test object. The test animal is placed in a plastic cage equipped with a food pellet dispenser and a water bottle. The food pellets are dry and absorb very little microwave energy. The water bottle made of glass is electrically decoupled from the animal by two concentric, one quarter-wavelength coaxial chokes.[99] This is essential for eliminating the marked increases in power absorption in the animal when it comes in contact with the highly conductive water spout.

A rectangular cavity operating in two modes can be employed to produce desired intensities of the E and H fields within an exposure volume occupying a relatively small part of the center of the cavity.[101] This type of a cavity can be used at frequencies between 100 MHz and 1 GHz.[101] A multimode rectangular cavity can also be used for exposures.[102] A mode stirrer is employed to increase the number of modes excited in the cavity and to improve the uniformity of the exposure field.

3. Near-Field Synthesizer

Exposures at lower RFs (a few tens of megahertz) can be conveniently accomplished in a near-field synthesizer.[103,104] The synthesizer consists of two parallel plates. A uniform E-field can be generated between the plates and a loop inductor placed between the plates can produce a H-field of any desired orientation. Since the intensities of the two fields can be controlled independently, any desired exposure conditions simulating actual near-field exposure can be obtained.

D. CELL CULTURE EXPOSURE SYSTEMS

In exposing cell cultures, blood samples, and other solutions containing microorganisms, it is extremely important to ensure that the E-field intensity and temperature are well-defined (and uniform if possible).[104] The situation in which the sample holder may significantly modify the exposure field should be avoided.

Various waveguide[105–107] and cavity[108] exposure systems for cell cultures have been developed. Most of the systems have provisions for temperature control of the culture. A waveguide exposure system[105,106] is shown in Figure 18. The sample is contained in a micropipette immersed in a buffer solution. To couple the energy to the sample effectively, a layer of a dielectric material is used (the thickness of the dielectric is one quarter guide wavelength in the dielectric and the dielectric constant is approximately $= \sqrt{2\epsilon}$, where ϵ is the relative dielectric constant of the buffer). The buffer solution can be circulated to maintain a constant temperature.

A specially designed, coaxial-transmission-line exposure cell[104,109] contains a 5-ml sample, which can be exposed at frequencies up to 1 GHz. A simplified diagram is shown in Figure

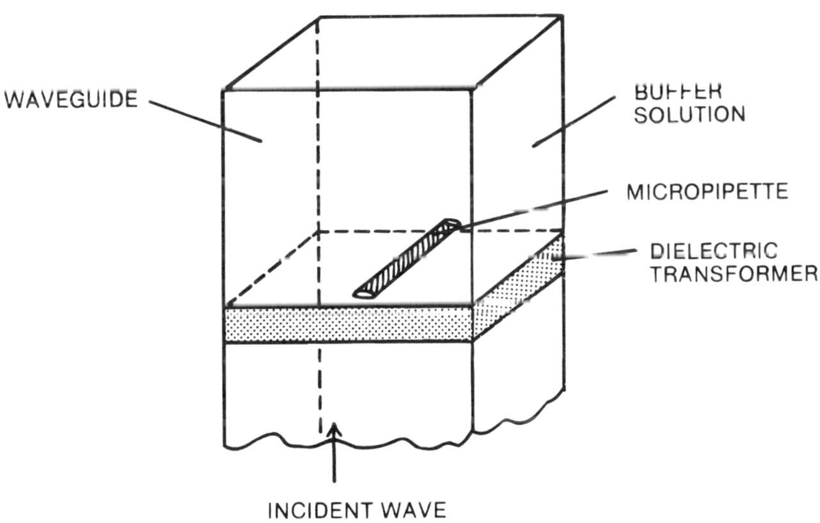

FIGURE 18 Waveguide exposure system for cell cultures.

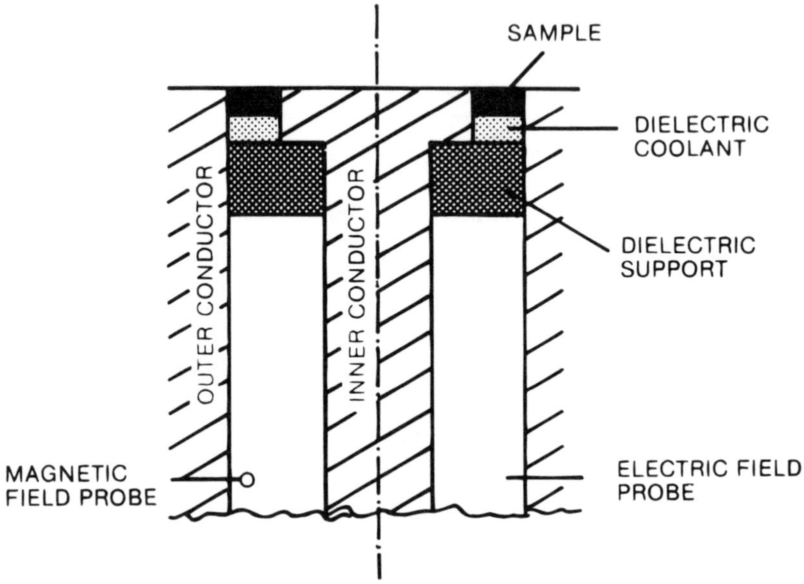

FIGURE 19 Simplified diagram of cell culture transmission line exposure system.[104,109]

19. The temperature of the sample can be controlled, and 37°C or less can be maintained at field strength in excess of 2.5 kV/m. E- and H-field probes are used to monitor the reflection coefficient from the sample. The field intensity in the sample can easily be calculated from an analysis of the equivalent circuit.[104,109]

E. MONITORING ELECTRODES

Metal electrodes and connecting wires conventionally used to record electrophysiological processes when used in electromagnetic fields significantly perturb the exposure field and cause highly localized heating at electrode/tissue boundariest due to the field enhancement.[110] Several types of electromagnetic-field-compatible electrodes have been designed.[110–114] Even nonmetallic electrodes when made of a material of high conductivity can enhance the field.[114]

A non-perturbing electrode system using low conductivity plastics for electrodes and leads[114] is satisfactory for recordings of EEGs and ECGs. The system consists of a pair of electrodes and leads prefabricated as a single unit. The electrode is 10 cm long, 2 mm wide, 0.25 mm thick, and made of carbon-loaded Teflon® (CLT) having a conductivity of 45 S/m (Technical Fluorocarbons Engineering, Warwick, R.I., Polymer Corp., Reading, Pa.) The lead is a strip of 72-S/m CLT of the same width and thickness as the electrode. The lead and the electrode are taped one atop the other with double-sided cellophane adhesive tape. The field enhancement due to these electrodes is very low.[114]

V. SUMMARY OF EXPERIMENTAL RESULTS

The various experimental dosimetry methods described in this chapter have been used to verify theoretical calculations for models and to determine the SAR in experimental animals. A representative summary of the data is given in Table 9.

VI. CONCLUDING REMARKS

Experimental dosimetry is essential in determining the exposure field and the whole-body or the body-part average SARs in experiments with animals exposed to RF fields. Furthermore,

TABLE 9
Representative Results of Experimental Dosimetry Studies

Experiment	Technique	Exposure conditions				SAR	
		Frequency (MHz)	Field	Polarization[a]	Average	Distribution[b]	Ref.
Spheres, prolate spheroids	Near-field synthesizer and anechoic chamber, thermography	3–1600		Various		+	5
Phantom-scale models of man	Anechoic chamber, thermography	80–250 (scaled)	Far	E, H, k	Given	+	117
Prolate spheroidal models	Two-plate stripline, thermography	285–4000	TEM	E, H, k	Approx.	Approx.	116
Mice	Anechoic chamber, flask calorimeter	2600	Far	E	1.15 W/kg/mW/cm^2	—	51
Mice, 24–25 g	Free-space, calorimeter	2450	Far	E	1.2–1.4W/kg/mW/cm^2	—	118
Mice 30–50 g	Anechoic chamber, twin-well calorimeter	2450	Far		0.32–0.42 W/kg/mW/cm^2	—	119
Mice, 27 g	Anechoic chamber, flask calorimeter	2450	Far	E, H, k	0.58–1.21 W/kg/mW/cm^2	—	122
Voles, 22–24 g	Free-space calorimeter	2450	Far	E	1.41–1.5 W/kg/mW/cm^2	—	118
Whiptail lizards, 19–24 g	Free-space, calorimeter	2450	Far	E	1.85–2.68 W/kg/mW/cm^2	—	118
Ring doves, 120–144 g	Free-space calorimeter	2450	Far	E	0.57–0.75 W/kg/mW/cm^2	—	118
Rats, 190–380 g and cylindrical models	Multimode cavity, twin-well calorimeter	2450	Cavity	—	1.4–2.3 W/kg/W	—	53
	Anechoic chamber, twin-cell calorimeter	2880	Far		~0.3–0.39 W/kg/mW/c^2	—	53
Rats, 132–490 g	TE_{11} cavity, differential power	918	Cavity	Variable	1.16–3.15 W/kg/W	—	99
Spheroidal models of rats	TE_{11} cavity, thermography	2450	Cavity	E/H, k	2.64, 1.99 W/kg/W	+	100
Rat, 324 g	TE_{11} cavity, thermography	2450	Cavity	k	2.34–2.47 W/kg/W	+	100
Rats, 96–390 g	Two-plate stripline, thermography	285–4000	TEM	k	Approx.	Approx.	116
Spheroidal models of rats, 300–330g	TE_{11} cavity, thermography	918	Cavity	E/H, k	Given	+	99
Rats, 388 g	TE_{11} cavity, thermography	918	Cavity	E/H	0.98–1.3 W/kg/W	+	99
Rats, 355–550 g	Free-space, calorimeter	2450	Far	E	Enhancement factor given	—	118
Rats, 100–250 g proximity enhancement	Free-space, calorimeter	2450	Far	E	0.15–0.23 W/kg/mW/cm^2	—	119
Rats, 6–440 g	Anechoic chamber, twin-well calorimeter	2450	Far		0.1–1.25 W/kg/mW/cm^2	—	118

TABLE 9
Continued

Experiment	Technique	Exposure conditions				SAR	
		Frequency (MHz)	Field	Polarization[a]	Average	Distribution[b]	Ref.
Rats	Anechoic chamber, calorimeter, and optional fiber probe	2450	Far	E, k	Head and whole-body	—	127
Rabbits, 2–2.25 kg	Free-space, calorimeter	2450	Far	E	0.076–0.085 W/kg/mW/cm^2	—	118
Speroid rhesus monkey phantom, 3.5 kg	TEM cell, differential power	10–50	TEM	E/H, k	Given	—	121
Rhesus monkeys, 3.2–4.2 kg	TEM Cell, differential power	10–50	TEM	E, H, k	Given	—	121
Full-scale model of Rhesus monkey	Anechoic chamber, calorimeter and optical fiber probe	225, 1.29	Far	E	Given	+	129, 130
Rectangular (46 × 79 × 20 cm) human phantom	TEM cell, net power absorbed	10, 20, 30	TEM	k	$10.3 \times 10^{-3}, 5 \times 10^{-3}$ mW/cm^3	—	115
Spheroidal human phantom, 70 kg	TEM cell, differential power	10–50	TEM	E, H, k	Given	—	121
Phantom-scale models of man	Near-field synthesizer anechoic chamber and resonant cavity, thermograph	3–1600		Various	Given	+	2, 80
	TM_{110} and TE_{102} cavity, thermography	24, 31	Cavity	Various	Given	+	5
Men	TEM cell (70 Ω) differential power	8.5, 23.25	TEM	E, H, k	Given	—	123, 128
Sealed spheroidal models of man	Monopole in anechoic chamber, fiber-optic thermometer	75 (scaled)	Far, near	Given	—	124	
Phantom-scale models of man	Anechoic chamber, parallel plate, digital thermometer	70–280 (scaled)	Far	E, H, k	Approx.	Approx.	120, 125, 126
	Anechoic chamber, flash calorimeter	300–470 (scaled)	Far	E, k	Head, whole-body	—	127
Full-scale models of man	Anechoic chamber, implantable E-field probe	350, 915	Far	E, H, k	Given	+	131–136
	Anechoic chamber implantable E-field probe	350, 915	Near (dipole above ground slot)	E, H, k	Given	+	133, 137–140

[a] E designates the E field parallel to the longest axis, H designates the H field parallel to the longest axis, and k designates the waves propagating along the longest axis.

[b] – Distribution not measured, + distribution measured.

carefully performed experiments are crucial in verification of theoretical predictions and delineation of their limitations. For instance, experimentally obtained whole-body, average SAR for humans are three to four times greater than those calculated.[128] The differences in the spatial distribution of the SAR between the theoretical estimations and the measured values are even greater.[133]

The two approaches to dosimetry at RFs, namely the theoretical and the experimental, are complementary, and as both have certain advantages and limitations, these should be considered in selecting one or the other or both approaches to solve a given problem.

ACKNOWLEDGMENTS

Our thanks to the Editors for their constructive and helpful criticism. Special thanks are due to Mrs. Maureen Denning of the Department of Electrical and Computer Engineering, University of Victoria, for the editorial assistance.

REFERENCES

1. Radiofrequency electromagnetic fields. Properties, quantities and units, biophysical interaction, and measurements, National Council on Radiation Protection and Measurements, Report No. 67, Washington, D.C., March 1981.
2. **Guy, A. W.,** Non-ionizing radiation: dosimetry and interaction, *Proc. ACGIH Top. Symp.,* 1979, 75.
3. **Adey, W. R.,** Tissue interactions with nonionizing electromagnetic fields, *Physiol. Rev.,* 61, 435, 1981.
4. **Durney, C. H., Iskander, M. F., Massoudi, H., Allen, S. J., and Mitchell, J. C.,** Radiofrequency Radiation Dosimetry Handbook, 3rd ed., Report SAM-TR-80-32, U.S. Air Force School of Aerospace Medicine, Brooks Air Force Base, Tex., 1980.
5. **Guy, A. W., Webb, M. D., and Sorensen, C. C.,** Determination of power absorption in non exposed to high frequency electromagnetic fields by therrnographic measurements on scale models, *IEEE Trans. Biomed. Eng.,* 23, 361, 1976.
6. **Wacker, P. F. and Bowman, R. R.,** Quantifying hazardous electromagnetic fields: scientific basis and practical considerations, *IEEE Trans. Microwave Theor. Tech., 19,* 178, 1971.
7. **Tell, R. A.,** Instrumentation for measurement of radiofrequency electromagnetic fields: equipment, calibrations and selected applications, in *Proc. Course Adv. Biol. Eff. Dosimetry Low Energy. Electromagnetic Fields,* Grandolfo, M. and Michaelson, S. M., Eds., Plenum Press, New York, 1982.
8. **Bowman, R. R.,** Some recent developments in the characterization and measurement of hazardous electromagnetic fields, in *Proc. Int. Symp. Biol. Eff. Health Hazards Microwave Radiat.,* Polish Medical Publishers, Warsaw, 1974, 217.
9. **King, R. W. P.,** *Theory of Linear Antennas,* Harvard University Press, Cambridge, 1956, 184.
10. **Kanda, M.,** Analytical and numerical techniques for analyzing an electrically short dipole with a nonlinear load, *IEEE Trans. Antennas Propag.,* 28, 71, 1980.
11. **Hopfer, S. and Adler, Z.,** An ultra broad-band (200 kHz–26 GHz) high-sensitivity probe, *IEEE Trans. Instrum. Meas.,* 29, 445, 1980.
12. **Aslan, E.,** Broad-band isotropic electromagnetic radiation monitor, *IEEE Trans. Instrum. Meas.,* 21, 421, 1972.
13. **Fletcher, K. and Woods, D.,** Thin-film spherical bolometer for measurement of hazardous field intensities from 400 MHz to 40 GHz, *Non-ioniz. Radiat.,* 57, September, 1969.
14. **Greene, F. M.,** Development of magnetic near-field probes, Publ. No. 75-127, J.S. Department of Health, Education and Welfare, Cincinnati, 1975, 75.
15. **Babij, T. M. and Bassen, H. I.,** Development of E/H near-field probe, *IEEE Trans. Electromagn. Compat.,* in press.
16. **Whiteside, H. and King, R. W. P.,** The loop antenna as a probe, *IEEE Trans. Antennas Propag.,* 12, 71, 1964.
17. **Duncan, P. H.,** Analysis of the Moebius loop magnetic field sensor, *IEEE Trans. Electromagn. Compat.,* 16, 83, 1974.
18. **Bassen, H., Herman, W., and Hoss, R.,** EM probe with fiberoptic telemetry system, *Microwave J.,* 20, 35, 1977.

19. **Bassen, H. I. and Hoss, R. J.,** An optically linked telemetry system for use with electromagnetic-field measurement probes, *IEEE Trans. Electromagn. Compat.,* 20, 483, 1978.
20. **Postow, E.,** Plans to develop a radiofrequency personnel dosimeter, *Proc. Workshop Prot. Personnel Against Radiofrequency Electromagnetic Radiat.,* Rev. 3-81, Mitchell, J. C., Ed., U.S. Air Force School of Aerospace Medicine, Brooks Air Force Base, Tex., 1981, 93.
20A. **Kanda, M.,** Time-domain sensors and radiators, Time Domain Measurements in Electromagnetics, E. K. Miller (Ed.), Van Nostrand Reinhold Co., New York, 122–174, 1986.
20B. **Esselle, K. P. and Stuchly, S. S.,** An electrically large antenna for transient electromagnetic field measurements, *IEEE Trans. Instrum. Measurement,* 2, 460–464, 1991.
20C. **Esselle, K. P. and Stuchly, S. S.,** A broad-band resistively loaded V-antenna: experimetal results, *IEEE Trans. Antennas and Propagation,* 39, 11, 1587–1591, 1991.
20D. **Thansandote, A., Stuchly, S. S., Stuchly, M. A. and Barski, M.,** Broadband active E-field sensors for measurement of transients, *IEEE Trans. Instrum. Measurement,* 40, 2, 465–468, 1991.
20E. **Stuchly, M. A., LePocher, H., Gibbons, D. T. and Thansandote, A.,** Active magnetic field sensor for measurements of transients, *IEEE Trans. Electromagnetic Compatibility,* 33, 4, 275–280, 1991.
21. **Gajda, G., Stuchly, M. A., and Stuchly, S. S.,** Mapping of the near-field pattern in simulated biological tissues, *Electron. Lett.,* 15, 120, 1979.
22. **Smith, G. S.,** A comparison of electrically short bare and insulated probes for measuring the local radio frequency electric field in biological systems, *IEEE Trans. Biomed. Eng.,* 22, 477, 1975.
23. **Mousavinezhad, S. H., Chen, K.-M., and Nyquist, D. P.,** Response of insulated electric field probes in finite heterogeneous biological bodies, *IEEE Trans. Microwave Theor. Tech.,* 26, 599, 1978.
24. **Smith, G. S.,** The electric-field probe near a material interface with application to the probing of fields in biological bodies, *IEEE Trans. Microwave Theor. Tech.,* 27, 270, 1979.
25. **King, R. W. P., Prasad, S., and Sandler, B. H.,** Transponder antenna in and near a three-layered body, *IEEE Trans. Microwave Theor. Tech.,* 28, 586, 1980.
26. **Smith, G. S.,** Analysis of miniature electric field probes with resistive transmission lines, *IEEE Trans. Microwave Theor. Tech.,* 29, 1213, 1981.
27. **Stuchly, M. A., Kraszewski, A., and Stuchly, S. S.,** Implantable electric field probes—some performance characteristics, *IEEE Trans. Biomed. Eng.,* 31, 526, 1984.
27A. **Howerton, P. H. and Batchman, T. E.,** Design of a three-axis microwave E-field probe having submillimeter dimensions, *IEEE Trans. Instrum. Measurement,* 36, 2, 567–571, 1987.
28. **Brodwin, M. E., Taflove, A., and Matz, J. E.,** A passive electrodeless method for determining the interior field of biological materials, *IEEE Trans. Microwave Theor. Tech.,* 24, 514, 1976.
28A. **Bassen, H. I. and Smith, G. S.,** Electric field probes—a review, *IEEE Trans. Antennas Propag.,* 31(5), 710, 1983.
29. **Baird, R. C.,** Methods of calibrating microwave hazard meters, in *Proc. Int. Symp. Biol. Health Hazards Microwave Radiat.,* Polish Medical Publishers, Warsaw, 1974, 228.
30. **Hickman, T. G. and Heaton, R. A.,** Measurement of gain, in *Microwave Antenna Measurements,* Hollis, J. S., Lyon, T. J., and Clayton, L., Eds., Scientific Atlanta, 1970, chap. 8.
31. **Jull, E. V.,** Finite-range gain of sectoral and pyramidal horns, *Electron. Let.,* 6, 680, 1970.
32. **Hunter, J. D. and Morgan, I. G.,** Near-field gain correction for transmission between horn antennas, *IEEE Trans. Instrum. Meas.,* 26, 58, 1977.
33. **Larsen, E. B.,** Techniques for producing standard EM fields from 10 kHz to 10 GHz for evaluating radiation monitors, in *Proc. Symp. Electromagnetic Fields Biol. Syst.,* Stuchly, S. S., Ed., Intern. Microwave Power Inst., Edmonton, Alberta, Canada, 1979, 96.
34. **Bowman, R. R.,** Field strength above 1 GHz: measurement procedures for standard antennas, *Proc. IEEE,* 55, 981, 1967.
35. **Bassen, H. I. and Herman, W. A.,** Precise calibration of plane-wave microwave power density using power equation technique, *IEEE Trans. Microwave Theor. Tech.,* 25, 701, 1977.
36. **Crawford, M. L.,** Generation of standard EM fields using TEM transmission cells, *IEEE Trans. Eleclromagn. Compat.,* 16, 189, 1974.
37. **Weil, C. M., Joines, W. T., and Kinn, J. M.,** Frequency range of large-scale TEM mode rectangular strip lines, *Microwave J.,* 24, 93, 1981.
38. **Hill, D. A.,** Bandwidth, limitations of TEM cells due to resonances, *J. Microwave Power,* 18, 181, 1983.
39. **Cheung, A. Y.,** Electric field measurements within biological media, in *Proc. Symp. Biol. Eff. Meas. Radio Publ. 77-8026, Frequency/Microwaves,* Food and Drug Administration, U.S. Department of Health, Education and Welfare, Rockville, Md., 1977, 117.
40. **Hill, D. A.,** Waveguide technique for the calibration of miniature implantable electric-field probes for use in microwave-bioeffects studies, *IEEE Trans. Microwave Theor. Tech.,* 30, 92, 1982.

41. **Christman, C. L., Ho, H. S., and Yarrow, S.,** A microwave dosimetry system for measured sampled integral-dose rate, *IEEE Trans. Microwave Theor. Tech.,* 22, 1267, 1974.
42. **Greene, F. M.,** Measurement of RF power-absorption in biological specimens (10 to 100 MHz), Publ. 77-146, U.S. Department of Health, Education and Welfare, Washington, D.C., 1977.
43. **Edwards, W. P. and Ho, H. S.,** RF cavity irradiation dosimetry, *IEEE Trans. Microwave Theor. Tech.,* 23, 311, 1975.
44. **Allen, S. J.,** Measurement of power absorption by human phantoms immersed in radiofrequency fields, *Ann. N.Y. Acad. Sci.,* 247, 494, 1975.
45. **Hill, D. A.,** Human microwave absorption from 7 to 40 mHz, in Proc. Workshop Prot. Personnel Against Radiofrequency Electromagnetic Radiat., Rev. 3-81, Mitchell, J. C., Ed., U.S. Air Force School of Aerospace Medicine, Brooks Air Force Base, Tex., 1981, 170.
46. **Hill, D. A.,** Human whole-body radiofrequency absorption studies using a TEM-cell exposure system, *IEEE Trans. Microwave Theor. Tech.,* 30, 1847, 1982.
47. **Guy, A. W., Lehmann, J. F., and Stonebridge, J. B.,** Therapeutic applications of electromagnetic power, *Power IEEE,* 62, 55, 1974.
48. **Allis, J. W., Blackman, C. F., Fromme, M. L., and Benane, S. G.,** Measurement of microwave radiation absorbed by biological systems. I. Analysis of heating and cooling data, *Radio Sci.,* 12(6S), 1977.
49. **McRee, D. J.,** Determination of energy absorption of microwave radiation using the cooling curve technique, *J. Microwave Power,* 9, 263, 1974.
50. **Blackman, C. F. and Black, J. A.,** Measurement of microwave radiation absorbed by biological systems, 2 analysis by Dewar-flask calorimetry, *Radio Sci.,* 12(6S), 9, 1977.
51. **Allen, S. J. and Hunt, W. D.,** Calorimetric measurements of microwave energy absorption by mice after simultaneous exposure of 18 animals, *Radio Sci.,* 14(6S), 1, 1979.
52. **Durney, C. H., Iskander, M. F., Massoudi, H. Allen, S. J., and Mitchell, J. C.,** Radiofrequency Radiation Dosimetry Handbook, 3rd ed., Report SAM-TR-80-32, U.S. Air Force School of Aerospace Medicine, Brooks Air Force Base. Tex., 1980, 29.
53. **Phillips, R. D., Hunt, E. L., and King, N. W.,** Field measurement, absorbed dose, and biologic dosimetry of microwaves, Ann. IV.Y. *Acad. Sei.,* 247, 499, 1975.
54. **Cetas, T. C.,** Thermometry in strong electromagnetic fields, Publ. 77-8055, *U.S. Department of Health, Education and Welfare, Washington, D.C.,* 1977, 261.
55. **Cetas, T. C.,** Invasive thermometry, in *Proc. AAPM Summer Sch.,* Phys. Aspects Hyperthermia, Dartmouth College, Hanover, N.H., 1981.
56. **McRee, D. J. and Pendergrass, F. T.,** Interaction of 2450-MHz microwave field with thermocouples and thermistors, *Health Phys.,* 25, 180, 1973.
57. **McAfee, R. D., Cazenavette, L. L., and Shubert, H. A.,** Thermistor probe error in an X-band microwave field, *J. Microwave Power,* 9, 177, 1974.
58. **Chakraborty, D. P. and Brezovich, I. A.,** A source of themrocouple error in radiofrequency electric fields, *Electron. Lett.,* 16, 853, 1980.
59. **Szwarnowski, S., Sheppard, R. J., Grant, E. H., and Bleehen, N. M.,** A thermocouple for measuring temperature in biological material heated by microwaves at 2.45 GHz, *Br. J. Radiol.,* 53, 711, 1980.
60. **Samaras, G. M., Harrison, G. H., and Blaumanis, O. R.,** *In vivo* temperature/heat flow sensing in microwave fields, in Proc. 3rd Int. Sym. Cancer Ther. Hyperthermia, Fort Collins, Colo., 1980, 192.
61. **Eno, K., Kato, H., Nishida, T., Kano, E., Sugahara, T., Tanaka, H., and Ishida, T.,** Physical basis of RF hyperthermia for cancer therapy, a non-perturbed and non-perturbing thermometer at RF heating, *J. Radiat. Res. (Tokyo),* 22, 265, 1981.
62. **Johnson, C. C. and Rozzell, T. C.,** Liquid crystal fiberoptic RF probes, temperature probe for microwave fields, *Microwave J.,* 18, 55, 1975.
63. **Johnson, C. C., Durney, C. H., Lords, J. L., Rozzell, T. C., and Livingston, G. K.,** Fiberoptic liquid crystal probe for absorbed radio-frequency power and temperature measurement in tissue during irradiation, *Ann. N.Y. Acad. Sci.,* 247, 527, 1975.
64. **Rozzell, T. C., Johnson, C. C., Durney, C. H., Lords, J. L., and Olsen, R. G.,** A non-perturbing temperature sensor for measurements in electromagnetic fields, *J. Microwave Power,* 9, 241, 1974.
65. **Bowman, R. R.,** A probe for measuring temperature in radiofrequency-heated material, *IEEE Trans. Microwave Theor. Tech.,* 24, 43, 1976.
66. **Deficis, A. and Priou, A.,** Non-perturbing microprobe for measurement in electromagentic fields, *Microwave* J., 20, 55, 1977.
67. **Deficis, A.,** Fiberoptic microprobes for microwave electromagnetic field measurements, Digest Microwave Theor. Tech. Int. Symp., Palo Alto, Calif., 1975, 300.
68. **Deficis, A. and Priou, A.,** Non-perturbing microprobe for measurements in electromagnetic fields, Publ. 77-8055, Food and Drug Administration, U.S. Department of Health, Education and Welfare, 1977, 283.

69. **Cain, C. A. and Chen, M. M.,** The viscometric thermometer, Publ. 77-8055, Food and Drug Administration, Department of Health, Education and Welfare, 1977, 295.
70. **Cetas, T. C.,** A birefringent crystal optical thermometer for measurements of electromagnetically induced heating, Symp. Proc. Biol. Eff. Meas. Radio-frequency/Microwaves, Publ. 77-8026, Food and Drug Administration, Department of Health, Education and Welfare, Rockville, Md., 1977, 338.
71. **Livingstone, G. K.,** Thermometry and dosimetry of heat with specific reference to the liquid crystal optical fiber temperature probe, *Radiat. Environ. Biophys.,* 17, 233, 1980.
72. **Christensen, D. A.,** A new non-perturbing temperature probe using semiconductor band edge shift, in Dig. IEEE/MTTS Int. Microwave Symp., San Diego, Calif., 1977.
73. **Wickesheim, K. A. and Alves, R. V.,** Ratioing fluoroptic temperature sensor for induced hyperthermia, presented at the 3rd Int. Symp. Cancer Ther. Hyperthermia, Drugs, Radiat., Fort Collins, Colo., 1980.
74. **Wickersheim, K. A. and Alves, R. B.,** Recent advances in optical temperature measurement, *Ind. Res. Dev.,* December, 1979.
75. **Cetas, T. C.,** Temperature measurement in microwave diathermy fields: principles and probes, Proc. Int. Syrup. Cancer Ther. Hyperthermia Radiat., Washington, D.C., 1975, 193.
76. **Hochuli, C. U.,** Procedure for evaluating non-perturbing temperature probes in microwave field, Publ. 818143, Food and Drug Administration, U.S. Department of Health and Human Services, 1981.
77. **Hochuli, C. U. and Kantor, G.,** An analysis of minimally perturbing temperature probe and thermographic measurements in microwave diathermy, *IEEE Trans. Microwave Theor. Tech.,* 29, 1285, 1981.
78. **Guy, A. W.,** Analyses of electromagnetic fields induced in biological tissues by thermographic studies on equivalent phantom, *IEEE Trans. Microwave Theor. Tech.,* 19, 205, 1971.
79. **Guy, A. W.,** History and state of the art on the quantitation of the interaction of electromagnetic fields with biological structures, NATO ASI Interaction Electromagnetic Fields with Structures, Norwich, England, 1979.
80. **Guy, A. W., Webb, M. D., and McDougall, J. A.,** RF radiation absorption patterns: human and animal modeling data, Publ. 77-183, U.S. Department of Health, Education and Welfare, Washington, D.C., 1977.
81. **Cetas, T. C.,** Practical thermometry with a thermographic camera—calibration, transmittance, and emittance, measurements, *Rev. Sci. Instrum.,* 49, 245, 1978.
82. **Ho, H. S. and Faden, J.,** Experimental and theoretical determination of absorbed microwave dose rate distributions in phantom heads irradiated by and aperture source, *Health Phys.,* 33, 13, 1977.
83. **Cristensen, D. A.,** Current techniques for non-invasive thermometry, *Proc. AAPM Summer Sch.,* Phys. Aspects Hyperthermia, Dartmouth College, Hanover, N.H., 1981.
84. **Christensen, D. A.,** Thermal dosimetry and temperature measurements, *Cancer Res.,* 39, 2325, 1979.
85. **Bolzano, Q., Garay, O., and Steel, F. R.,** An attempt to evaluate exposure of operators of portable radios at 30 MHz, Conf. Proc. 29th IEEE Vehicular Technol. Soc., Arlington Heights, Fla., 1979, 187.
86. **Cheung, A. Y. and Koopman, D. W.,** Experimental development of simulated biomaterials for dosimetry studies of hazardous microwave radiation, *IEEE Trans. Microwave Theor. Tech.,* 24, 669, 1976.
87. **Thansandoe, A.,** Monitoring Time-varying Biological Impedance at Microwave Frequencies, Ph.D. thesis, Carleton University, Ottawa, Canada, 1981.
88. **Paulsson, L. E.,** Measurements of 0.915, 2.45 and 9 GHz absorption in the human eye, Proc. 6th Eur. Microwave Conf., Rome, 1976.
89. **Toler, J. and Seals, J.,** RF dielectric properties measurement system: human and animal data, Tech. Rep. A-1862, Biomedical Research Group, Engineering Experimental Station, Georgia Institute of Technology, Atlanta, 1977.
90. **Hand, J. W., Robinson, J. E., Szwarnowski, S., Sheppard, R. J., and Grant, E. H.,** A physiologically compatible tissue-equivalent liquid bolus for microwave heating of tissues, *Phys. Med. Biol.,* 24, 426, 1979.
91. **Oliva, S. A. and Catravas, G. N.,** A multiple-animal array for equal power density microwave irradiation, *IEEE Trans. Microwave Theor. Tech.,* 25, 433, 1977.
92. **Hagmann, M. J. and Gandhi, O. P.,** Radiators for microwave biological effects research—waveguide slot array with constant radiation intensity, *Radio Sci.,* 12(6S), 97, 1977.
93. **Bassett, H. L., Ecker, H. A., Johnson, R. C., and Sheppard, A. P.,** New techniques for implementing microwave biological-exposure systems, **IEEE Trans. Microwave Theor. Tech.,** 19, 197, 1971.
94. **McRee, D. and Walsh, P.,** Microwave exposure system for biological specimens, *Rev. Sci. Instrum.,* 42, 1860, 1971.
95. **Crawford, M. L.,** Generation of standard EM fields using TEM transmission cells, *IEEE Trans. Electromagn. Compat.,* 16, 189, 1974.
96. **Weil, C. M.,** The characteristic impedence of rectangular transmission lines with thin center conductor and air dielectric, *IEEE Trans. Microwave Theor. Tech.,* 26, 238, 1978.
97. **Weil, C. M., Joines, W. T., and Kinn, J. B.,** Frequency range of large-scale TEM mode rectangular strip lines, *Microwave J.,* 24, 93, 1981.
98. **Hill, D. A.,** Bandwidth limitations of TEM cells due to resonances, J. *Microwave Power,* 18, 1983.
99. **Guy, A. W. and Chou, C. K.,** System for quantitative chronic exposure of a population of rodents to UHF

fields, Biological Effects of Electromagnetic Waves, Publ. 77-8011, Bureau of Radiological Health, U.S. Department of Health, Education and Welfare, Rockville, Md., 1976, 389.

100. **Guy, A. W., Wallace, J., and McDougall, J. A.,** Circularly polarized 2450-MHz waveguide system for chronic exposure of small animals to microwaves, *Radio Sci.,* 14(6S), 63, 1979.
101. **Spiegel, R. J., Oakey, W. E., and Bronaugh, E. L.,** A variable-volume cavity electromagnetic nearfield simulator, IEEE Trans. Electromagn. Compat., 22, 289, 1980.
102. **Heynick, L. N., Polson, P., and Karp, A.,** A microwave exposure system for primates. *Radio Sci.,* 12(6S), 103, 1977.
103. **Greene, F. M.,** Development of an RF near-field exposure synthesizer (10 to 40 MHz), Publ. 76-160, National Institute for Occupational Safety and Health, U.S. Department of Health, Education and Welfare, Cincinnati, 1976.
104. **Guy, A. W., Webb, M. D., and McDougall, J. A.,** RF radiation absorption patterns: human and animal modeling data, Publ. 77-183, National Institute for Occupational Safety and Health, U.S. Department of Health, Education and Welfare, Cincinnati, 1977.
105. **Chen, K. C. and Lin, C. J.,** A system for studying effects of microwaves on cells in culture, *J. Microwave Power,* 13, 251, 1978.
106. **Lin, J. C.,** A new system for investigating nonthermal effect of microwaves on cells, in Biological Effects of Electromagnetic Waves, Publ. 77-8011, Bureau of Radiological Health, Department of Health, Education and Welfare, Rockville, Md., 1976, 350.
107. **Partlow, L. M., Bush, L. G., Stensaas, L. J., Hill, D. W., Riazi, A., and Gandhi, O. P.,** Effects of millimeter-wave radiation on monolayer cell cultures. I. Design and validation of a novel exposure system, *Bioelectromagnetics,* 2, 123, 1981.
108. **McEwen, A., Tanabe, E., Vaguine, V., Williams, N., Li, G. C., and Hahn, G. M.,** Temperature controlled cavity applicator and radiofrequency wave irradiation of mammalian cells *in vitro,* Proc. Symp. Electromagnetic Fields Biol. Syst., Stuchly, S. S., Ed., International Microwave Power Institute, Edmonton, Alberta, Canada, 1979, 113.
109. **Guy, A. W.,** A method of exposing cell cultures to electromagnetic fields under controlled conditions of temperature and field strengths, *Radio Sci.,* 12(6S), 87, 1977.
110. **Johnson, C. C. and Guy, A. W.,** Nonionizing electromagnetic wave effects in biological materials and systems, *Proc. IEEE,* 60, 692, 1972.
111. **Larsen, L. E., Moore, R. A., and Acevedo, J.,** A microwave decoupled brain-temperature transducer, *IEEE Tralls. Micrwave Theory. Tech.,* 22, 438, 1974.
112. **Lords, J. L., Durney, C. H., Borg, A. M., and Tinney, C. E.,** Rate effects in isolated hearts inducted by microwave irradiation, *IEEE Trans. Microwave Theor. Tech.,* 21, 834, 1973.
113. **Tyazhelov, V. V., Tigranian, R. E., and Khizhniak, E. P.,** New artifact-free electrodes for recording of biological potentials in strong electromagnetic fields, *Radio Sci.,* 12(6S), 121, 1977.
114. **Flanigan, W. F., Bowman, R. R., and Lowell, W. R.,** Nonmetallic electrode system for recording EEG and ECG in electromagnetic fields, *Physiol. Behav.,* 18, 531, 1977.
115. **Allen, S. J.,** Measurements of power absorption by human phantoms immersed in radiofrequency fields, *Ann. N.Y. Acad. Sci.,* 247, 494, 1975.
116. **Gandhi, O. P.,** Strong dependance of whole animal absorption on polarization and frequency of radiofrequency energy, *Ann. N.Y. Acad. Sci.,* 247, 532, 1975.
117. **Chou, C. K. and Guy, A. W.,** Quantitation of microwave biological effects, In Proc. Symp. Biol. Eff. Meas. Radio Frequency/Microwaves, Hazzard, D. G., Ed., Publ. 77-8026, Bureau of Radiological Health, U.S. Dcpartment of Health, Education and Welfare, Rockville, Md., 1977, 81.
118. **Gandhi, O. P., Hagmann, M. J., and D'Andrea, J. A.,** Part-body and multi-body effects on absorption of radio-frequency electromagnetic energy by animals and models of man, *Radio Sci.,* 14(6S), 15, 1979.
119. **Kinn, J. B.,** Whole-body dosimetry of microwave radiation in small animals: the effects of body mass and exposure geometry, *Radio Sci.,* 12(6S), 61, 1977.
120. **Gandhi, O. P., Sedigh, K., Beck, G. S., and Hunt, E. L.,** Distribution of electromagnetic energy deposition in models of man with frequencies near resonance, in Biological Effects of Electromagnetic Waves, Publ. 77-8011, Bureau of Radiological Health, Food and Drug Administration, U.S. Department of Health, Education and Welfare, Rockville, Md., 1976, 44.
121. **Allen, S. J., Hurt, W. D., Krupp, J. H., Ratliff, J. A., Durney, C. H., and Johnson, C. C.,** Measurement of radiofrequency power absorption in monkeys, monkey phantoms, and human phantoms exposed to 10–50 MHz fields, in Biological Effects of Electromagnetic Waves, Publ. 77-8011, Bureau of Radiological Health, Food and Drug Administration, U.S. Department of Health, Education and Welfare, Rockville, Md., 1976, 83.
122. **Carnie, A. B., Hill, D. A., and Assenheim, H. M.,** Dosimetry for a study of effects of 2.45-GHz microwaves on mouse testis, *Bioelectromagnetics,* 1, 325, 1980.

123. **Hill, D. A.,** Human whole-body radiofrequency absorption studies using a TEM-cell exposure system, *IEEE Trans. Microwave Theor. Tech.,* 30, 1847, 1982.
124. **Iskander, M. F., Massoudi, H., Durney, C. H., and Allen, S. J.,** Measurements of the RF power absorption in speroidal human and animal phantoms exposed to the near field of a dipole source, *IEEE Trans. Biomed. Eng.,* 28, 258, 1981.
125. **Gandhi, O. P.,** Conditions of strongest electromagnetic power in man and animals, *IEEE Trans. Microwave Theor. Tech.,* 233, 1021, 1975.
126. **Gandhi, O. P.,** State of the knowledge for electromagnetic absorbed dose in man and animals, *Proc. IEEE,* 68, 24, 1980.
127. **Hagmann, M. J., Gandhi, O. P., D'Andrea, J. A., and Chatterjee, I.,** Head resonance: numerical solutions and experimental results, *IEEE Trans. Microwave Theor. Tech.,* 27, 809, 1979.
128. **Hill, D. A.,** The effect of frequency and grounding on whole-body absorption of humans in E-polarized radiofrequency field, *Bioelectromagnetics,* 5, 131, 1984.
129. **Olsen, R. G.,** Distribution of absorbed energy in primate models, in Proc. Workshop Prot. Personnel Against Radiofrequency Electromagnetic Radiat., Rev. 3-81, Mitchell, J. C., Ed., U.S.A.F School of Aerospace Medicine, Brooks Air Force Base, Tex., 1981, 149.
130. **Olsen, R. G., Griner, T. A., and Pettyman, G. D.,** Far-field dosimetry in a Rhesus monkey model, *Bioelectromagnetics,* 1, 149, 1980.
131. **Stuchly, S. S., Barski, M., Tam, B., Hartsgrove, G., and Symons, S.,** Computer-based scanning system for electromagnetic dosimetry, *Rev. Sci. Instrum.,* 54, 1547, 1983.
132. **Kraszewski, A., Stuchly, M. A., Stuchly, S. S., Hartsgrove, G., and Adamski, D.,** Specific absorption rate distribution in a full-scale model of man at 350 MHz, *IEEE Trans. Microwave Theor. Tech.,* 32, 779, 1984.
133. **Stuchly, M. A., Kraszewski, A., and Stuchly, S. S.,** Exposure of human models in the near and far field—a comparison, *IEEE Trans. Biomed. Eng.,* 32, 609, 1985.
134. **Stuchly, S. S., Stuchly, M. A., Kraszewski, A., and Hartsgrove, G.,** Energy deposition in a model of man: Frequency effects, *IEEE Trans. Biomed. Eng.,* 33, 7, 1986.
135. **Stuchly, S. S., Kraszewski, A., Stuchly, M. A., Hartsgrove, G. W. and Spiegel, R. J.,** RF energy deposition in a heterogeneous model of man: Far-field exposures, *IEEE Trans. on Biomed. Eng.,* 34, 12, 1987.
136. **Spiegel, R. J., Fatmi, M. B. A., Stuchly, S. S. and Stuchly, M. A.,** Comparison of finite-difference time-domain SAR calculations with measurements in a heterogeneous model of man, *IEEE Trans. Biomed. Eng.,* 36, 8, 1989.
137. **Stuchly, S. S., Kraszewski, A., Stuchly, M. A., Hartsgrove, G., and Adamski, A.,** Energy deposition in a model of man in the near-field, *Bioelectromagnetics,* 6, 1985.
138. **Stuchly, M. A., Kraszewski, A., Stuchly, S. S., Hartsgrove, G. W. and Spiegel, R. J.,** RF energy deposition in a heterogeneous model of man: Near-field exposures, *IEEE Trans. Biomed. Eng.,* 34, 12, 1987.
139. **Cleveland, R. F. and Athey, T. W.,** Specific absorption rate (SAR) in models of human head exposed to hand-held UHF portable radios, *Bioelectromagnetics,* 10, 173, 1989.
140. **Kuster, N and Bolzano, Q.,** Energy absorption mechanism by biological bodies in the near field of dipole antennas above 300 MHz, *IEEE Trans. Vehic. Tech.,* 41, 17, 1992.

Chapter 9

COMPUTATIONAL METHODS FOR PREDICTING FIELD INTENSITY

James C. Lin and Om P. Gandhi

CONTENTS

0-8493-0641-8/96/$0.00+$.50

I. INTRODUCTION

The distribution of electromagnetic fields in biological materials is governed by dielectric constant, conductivity, source configuration, and the geometrical factors that describe the tissue structure. These parameters also determine the quantity of energy a given biological body extracts from the field. When the radius of curvature of the body surface is large compared to the wavelength and beam width of the impinging radiation, planar tissue models may be used to estimate the absorbed energy and its distribution inside the body. Otherwise, the absorbed energy will be dictated by the size of the body, curvature of its surface, the ratio of body size to wavelength, and the source characteristics.

The purpose of this chapter is to present a selective account of the distribution of electromagnetic fields in biological media, with special emphasis on the energy coupling and distribution characteristics in models of biological structures. Such information is essential for analyzing the interrelationships among various observed biological effects, for separating, known and substantiated effects from those that are speculative and unsubstantiated, for assessing therapeutic effectiveness of electromagnetic waves, and for extracting diagnostic information from field effects.

There exists a wide variety of methods for quantifying fields in biological bodies. The extent of computer usage varies, depending on specific information sought and complexity of tissue geometry. The whole-body absorption of electromagnetic energy by humans and laboratory animals is of interest because it is related to the energy required to alter the thermoregulatory system of the exposed subject, and because it may serve as an index for extrapolating experimental results to human exposures. The distribution of absorbed electromagnetic energy within an exposed body is important because (1) it relates to specific responses of the body, (2) it facilitates understanding of phenomena, and (3) it contributes to definition of mechanisms of interaction.

This chapter outlines a number of techniques that have been successfully employed to analyze the propagation and absorption characteristics of electromagnetic energy in tissue structures. There are two general approaches: one involves extensive use of analytical development and the other relies more heavily on numerical formulation. Analytical computations are best suited for calculation of distribution of absorbed energy in simplified tissue geometries such as plane slabs, cylinders, and spheroids, whereas numerical methods offer the prospect of analyzing the coupling of electromagnetic energy to animal and human bodies which is difficult, if not impossible, to approach analytically. The advantages and limitations of various methods for field computation, along with representative results, are considered in the following sections.

II. PLANAR TISSUE GEOMETRIES

A. SEMI-INFINITE LAYERS

The reflection and transmission of a plane wave at a planar tissue interface depend on the frequency, polarization and angle of incidence of the wave, and on the dielectric constant and conductivity of the tissue. For a linearly polarized plane wave impinging normally on a boundary separating two semi-infinite media, the reflection and transmission coefficients are respectively given by

$$R = \frac{\eta_2 - \eta_1}{\eta_2 + \eta_1} \qquad (1)$$

and

$$T = \frac{2\eta_2}{\eta_2 + \eta_1} \tag{2}$$

where η_1 and η_2 are the intrinsic impedance of the media. When $\eta_1 \cong \eta_2$, e.g., the electromagnetic properties of the media are approximately equal, there is no reflection and the transmission is maximum. In contrast, there is complete reflection when T in Equation (2) is zero. Note that these equations are sufficiently simple to be easily programmed on a hand-held calculator to assess electromagnetic reflection and transmission behavior at planar tissue interfaces.

Table 1 summarizes the magnitude of the reflection coefficient at the boundary separating various tissues. The fraction of normally incident power reflected by the discontinuity is given by R^2. Clearly, about one half of the incident power is reflected at these boundaries. Further, the reflection coefficient for tissue-tissue interfaces generally is smaller than air-tissue interfaccs. The percent reflected power for tissue-tissue interfaces range from a low of 5 for muscle-blood to a high of 50 for bone-biological fluid interfaces. This suggests that the closer are the dielectric properties on both sides of the interface, the smaller is the power reflection.

The fraction of power transmitted is related to the power transmission coefficient ($1 - R^2$). It is readily apparent from Table 1 that the power transmitted at air-tissue interfaces is

TABLE 1
Reflection Coefficient (Magnitude in Percent) Between Biological Tissues at 37°C

	Frequency (MHz)	Air	Fat (bone)	Lung	Muscle (skin)	Blood	Saline
Air	433	0	46	76	82	81	83
	915	0	43	73	78	79	80
	2,450	0	41	71	76	77	79
	5,800	0	39	70	75	76	78
	10,000	0	37	70	74	76	78
Fat (bone)	433		0	46	56	56	60
	915		0	43	52	54	57
	2,450		0	42	50	53	57
	5,800		0	42	50	53	56
	10,000		0	45	52	54	58
Lung	433			0	14	13	19
	915			0	12	14	18
	2,450			0	10	15	19
	5,800			0	10	14	19
	10,000			0	10	13	18
Muscle (skin)	433				0	4	6
	915				0	4	7
	2,450				0	5	10
	5,800				0	4	9
	10,000				0	3	9
Blood	433					0	6
	915					0	4
	2,450					0	5
	5,800					0	5
	10,000					0	6
Saline	433						0
	915						0
	2,450						0
	5,800						0
	10,000						0

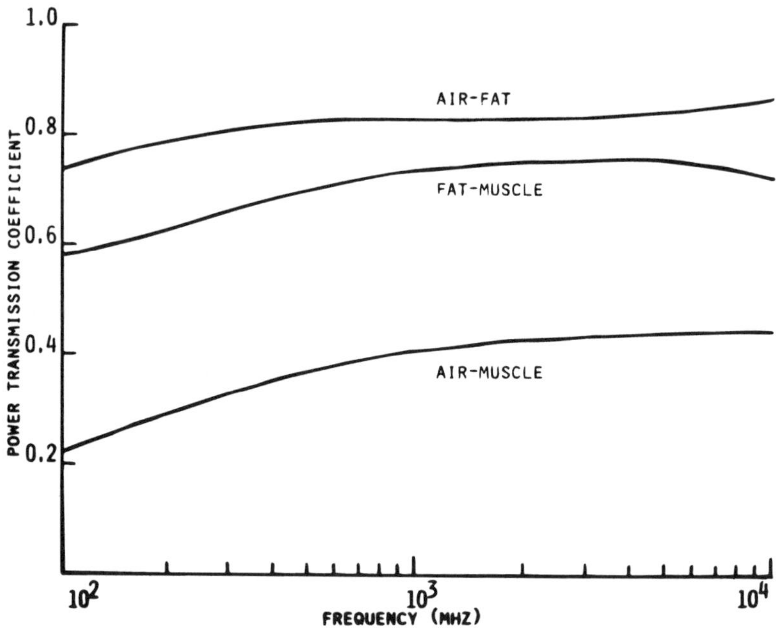

FIGURE 1 Power transmission coefficients at three tissues interfaces as functions of frequency.

quite substantial at radio and microwave frequencies. Moreover, Figure 1 shows that the power transmission coefficient is highly frequency dependent, especially at lower frequencies.

As the transmitted wave propagates in the tissue medium, energy is extracted from the wave and absorbed by the medium, resulting in a progressive reduction of the power density of the wave as it advances in the tissue. This reduction is quantified by the depth of penetration, which is the distance in which the power density decreases by a factor of e^{-2}. Table 2 presents the calculated depth of penetration in selected tissues using typical dielectric constants and conductivities. A graphical representation of penetration depth vs. frequency for blood, muscle, and fat is given in Figure 2. It is seen that the penetration depth is frequency dependent and takes on different values for different tissues. In particular, the penetration depth for fat and bone is nearly five times greater than for higher-water-content tissues.

A wave of general polarization usually is decomposed into its orthogonal linearly polarized components whose electric or magnetic field parallels the interface. These components can be treated separately and combined afterward. Figures 3 and 4 illustrate the magnitude and phase of the reflection coefficients of representative tissue interfaces at a temperature of

TABLE 2
Depth of Penetration of an Electromagnetic Wave in Biological Tissues as a Function of Frequency

Frequency (MHz)	Tissue: Saline	Blood	Muscle (skin)	Lung	Fat (bone)
	Depth of Penetration (cm)				
433	2.8	3.7	3.0	4.7	16.3
915	2.5	3.0	2.5	4.5	12.8
2,450	1.3	1.9	1.7	2.3	7.9
5,800	0.7	0.7	0.8	0.7	4.7
10,000	0.2	0.3	0.3	0.3	2.5

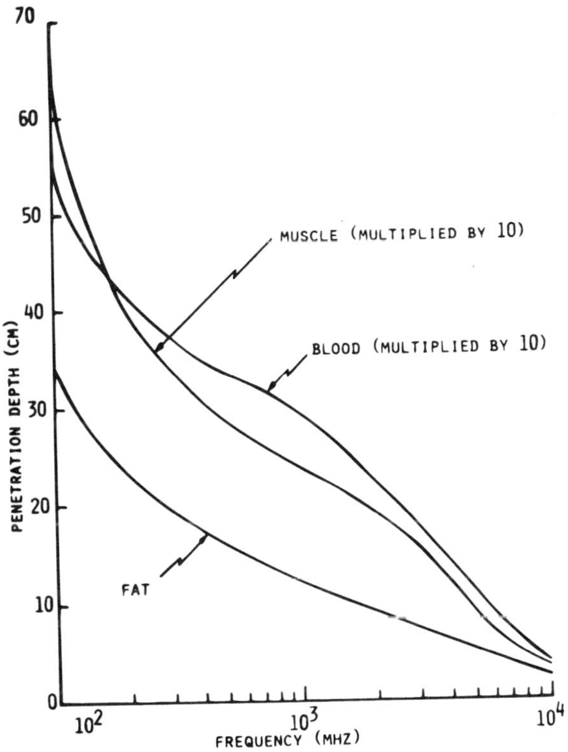

FIGURE 2 Depth of penetrations for blood, muscle, and fat as functions of frequency.

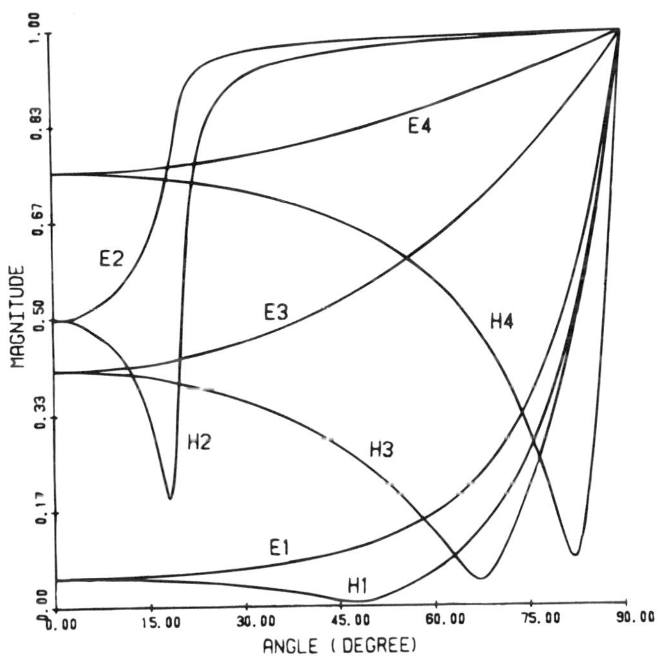

FIGURE 3 Magnitudes of reflection coefficients for E and H polarized plane waves at 2450 MHz.

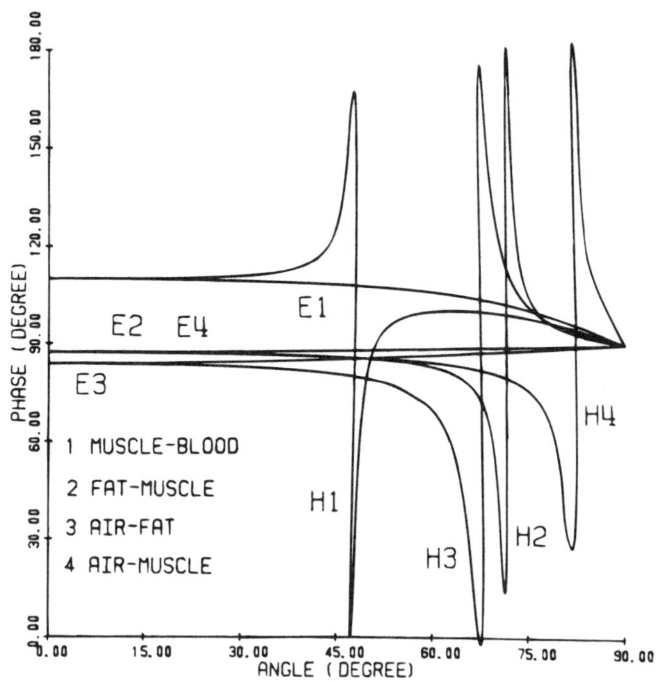

FIGURE 4 Phase of reflection coefficients for E and H polarized 2450 MHz plane waves.

37°C for irradiation at 2450 MHz. The figures clearly show the difference between E and H polarization. E polarization, also called perpendicular polarization and H polarization, also referred to as parallel polarization, are defined in Chapter 1 of this book. For E polarization, there is only a slight variation in magnitude and phase of the reflection coefficient with incidence angle. For H polarization, however, there is a pronounced dependence on incidence angle. The reflection coefficient reaches a minimum magnitude and has a phase angle of 90° at Brewster's angle. Thus, the H polarized wave is totally transmitted into the muscle medium at Brewster's angle.

B. MULTIPLE LAYERS

When there are several layers of different tissues, the reflection and transmission characteristics become more complicated. Multiple reflections can occur between the skin and subcutaneous tissue boundaries, with a resulting modification of the reflection and transmission coefficients.[1,2] In general, the transmitted wave will combine with the reflected wave to form standing waves in each layer. This phenomenon becomes especially pronounced if the thickness of each layer is less than the penetration depth for that tissue. Plane waves impinging on the human body considered as consisting of parallel layers of subcutaneous fat and more deeply lying muscle have been studied in detail by Schwan and Li.[3]

For the situation depicted in Figure 5, the electric field strength in the fat layer is given by

$$E_f = F_l E_0 [e^{-(\alpha_2 + j\beta_2)} + R_{32} e^{(\alpha_2 + j\beta_2)z}] \tag{3}$$

and the electric field in the underlying muscular tissue is given by

$$E_m = F_t E_0 e^{-(\alpha_3 + j\beta_3)z} \tag{4}$$

where α_2, β_2 and α_3, β_3 are the attenuation and propagation coefficients in fat and muscle, respectively. The layer function F_l and the transmission function F_t, are given by

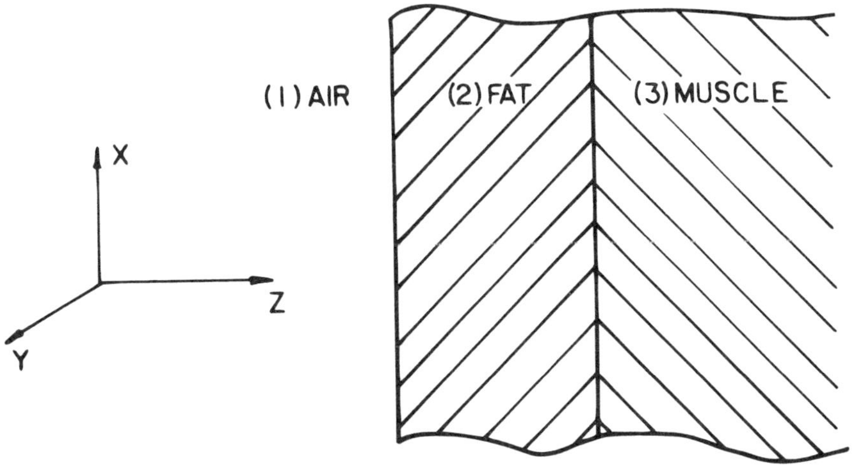

FIGURE 5 Plane wave impinging on a composite fat-muscle layer

$$F_l = T_{12}/[e^{(\alpha_2+j\beta_2)l} + R_{21}R_{32}e^{-(\alpha_2+j\beta_2)l}] \tag{5}$$

$$F_t = (T_{12}T_{23})/[e^{(\alpha_2+j\beta_2)l} + R_{21}R_{32}e^{-(\alpha_2+j\beta_2)l}] \tag{6}$$

where T_{12} and T_{23} are the transmission coefficients at the air-fat and fat-muscle boundaries, respectively. R_{21} and R_{32} denote, respectively, the reflection coefficients at these boundaries; l is the thickness of the fat layer. The power absorption in a given layer can be obtained from the equation

$$W_a = \frac{1}{2}\sigma E_f^2 \tag{7}$$

where σ is the electrical conductivity.

Figure 6 shows the numerical results obtained using the above equations. The values are normalized to the specific absorption rate (SAR) in muscle at the fat-muscle boundary. Note the absorbed energy is much lower in fat than in muscle. The standing wave maximum becomes bigger in fat and the penetration into muscle becomes less as the frequency increases.

The electromagnetic energy absorbed in models composed of planar layers of skin, fat, and muscle can be analyzed in a similar manner,[3,4] except the distribution of absorbed energy becomes more complex. Figure 7 shows that in addition to frequency dependence, the peak SAR exhibits considerable fluctuation with thickness of the subcutaneous fatty layer. The incident power density is, in this case, 10 W/m^2 and the skin layer is 0.2 cm thick. Note that the peak SAR is always higher in the skin layer for planar models at microwave frequencies. The depth of penetration for 10 GHz radiation in skin is less than 0.05 cm; thus, the transmitted energy is almost completely absorbed in the skin and the SAR is rather unaffected by changing fatty layer thickness. The fact that the SAR is highest in skin is significant, since skin is populated with thermosensitive free nerve endings which may be excited along with cutaneous pain receptors when the absorbed energy exceeds the normal range that can be handled by thermoregulation.

Figure 8 shows the distribution of induced electric field strength in a layer of muscle beneath layers of fat, muscle, and bone for two frequencies.[2] It is seen that in addition to frequency dependence, the electric fields exhibit considerable fluctuation within each tissue layer. While the standing-wave oscillations are larger at 2450 than at 915 MHz, microwave

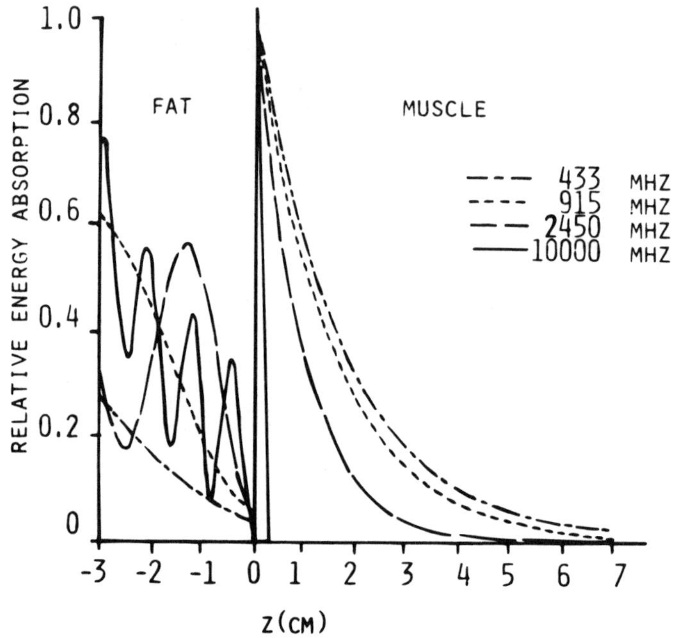

FIGURE 6 SAR (absorbed power density) in plane fat-muscle layers.

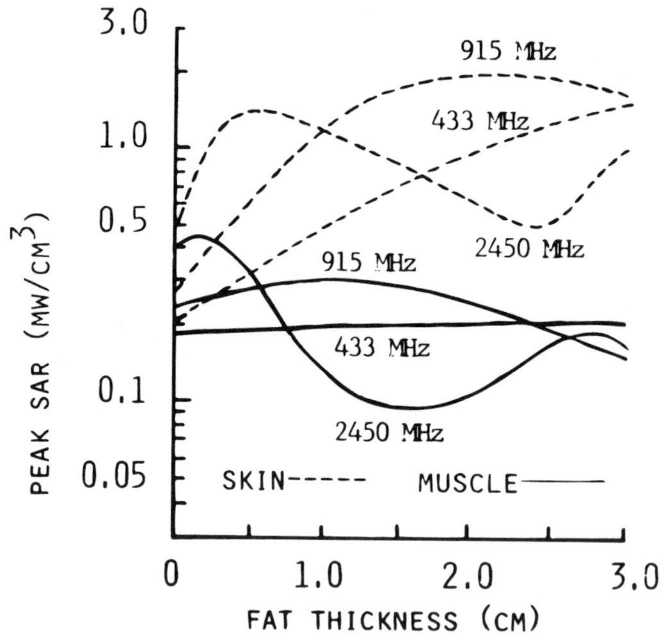

FIGURE 7 Peak SAR (absorbed power density) in models composed of fat-skin-muscle layers.

energy at both frequencies penetrates to deeper tissues. This result, together with Figures 6 and 7, implies that at frequencies between 300 and 3000 MHz, sufficient energy may be transmitted and reflected to allow sensing of organs within the body. Furthermore, at these frequencies, electromagnetic energy can penetrate into more deeply situated tissues, making it especially desirable for therapeutic applications.

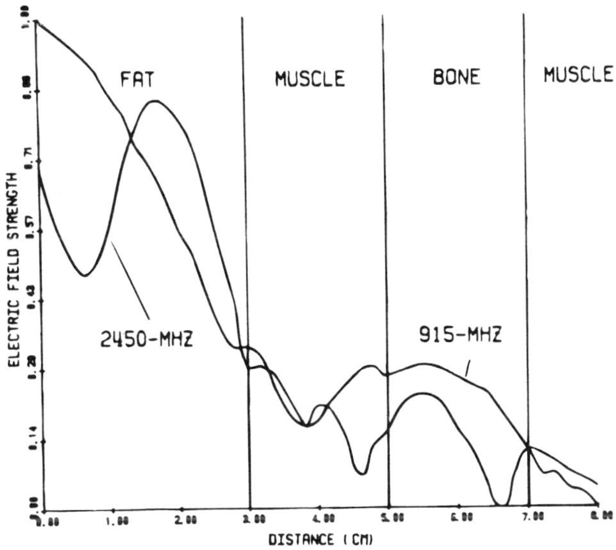

FIGURE 8 Distribution of electric field strength in planar layer of fat-muscle-bone-muscle tissue model.

III. BODIES OF REVOLUTION

Although depth of penetration, and reflection and transmission characteristics in planar tissue provide considerable physical insight into coupling and distribution of electromagnetic energy, biological structures generally are more complex in form and exhibit substantial curvature that can modify electromagnetic energy transmission and reflection. For bodies with complex shapes, the propagation characteristics depend critically on polarization and on orientation of the incident wave with respect to the body, as well as on the ratio of body size to wavelength. These complications place severe limitations on calculations of reflected and transmitted energy for bodies of arbitrary shape and complex permittivity. This section presents a summary of analytic/numeric results that have been obtained for homogeneous and multilayered models based on bodies of revolution that approximate mammalian tissue structures.

A. SPHERICAl MODELS

The distribution of the electric field induced inside a homogeneous dielectric sphere by a plane wave propagating in the positive z direction and polarized along the x axis is given by

$$\mathbf{E} = \mathbf{E}_0 \, e^{j\omega t} \sum_{n=1}^{\infty} (-j)^n \frac{2n + 1}{n(n + 1)} [a_n^t \underline{m}_{oln}^{(1)} + jb_n^t \underline{n}_{eln}^{(1)}] \tag{8}$$

where a_n^t and b_n^t denote the magnetic and electric modes of oscillations inside the sphere, respectively, and $\underline{m}_{oln}$ and $\underline{n}_{eln}$ are two independent solutions called vector spherical wave functions. They are all functions of the wavelength, dielectric constant, conductivity, and size of the sphere. The infinite series of Equation 8 is readily adapted for computer calculation using tabulated numerical values of dielectric permittivity and conductivity. The power absorption is once again given by Equation 7 and the total time rate of energy absorption is obtained from

$$W_a = W_t - W_s \tag{9}$$

with the time rate of total energy derived from the incident waves and the energy scatterd by the sphere given by

$$W_t = -\frac{E_0^2}{120k_0^2} R_e \sum_{n=1}^{\infty} (2n + 1)(a_n^r + b_n^r) \tag{10}$$

$$W_s = \frac{E_0^2}{120k_0^2} \sum_{n=1}^{\infty} (2n + 1)(|a_n^r|^2 + |b_n^r|^2) \tag{11}$$

and

$$a_n^r = a_n^t \left[\frac{j_n(Nk_0a)}{h_n(k_0a)} - \frac{j_n(k_0a)}{h_n](k_oa)}\right] \tag{12}$$

$$b_n^r = \left[b_n^t \frac{Nj_n(Nk_0a)}{h_n(k_0a)} - \frac{j_n(k_0a)}{h_n(k_0a)}\right] \tag{13}$$

where $N^2 = \epsilon$ is the relative dielectric permittivity, $k_0 = 2\pi/\lambda_0$, and λ_0 is the wavelength in vacuum; j_n and h_n are the spherical Bessel functions and a is the radius of the sphere.

Some representative calculations of the SAR are shown in Figure 9 for two models at 918 and 2450 MHz.[2,6] The 6-cm-diameter sphere approximates a cat or rhesus monkey brain,

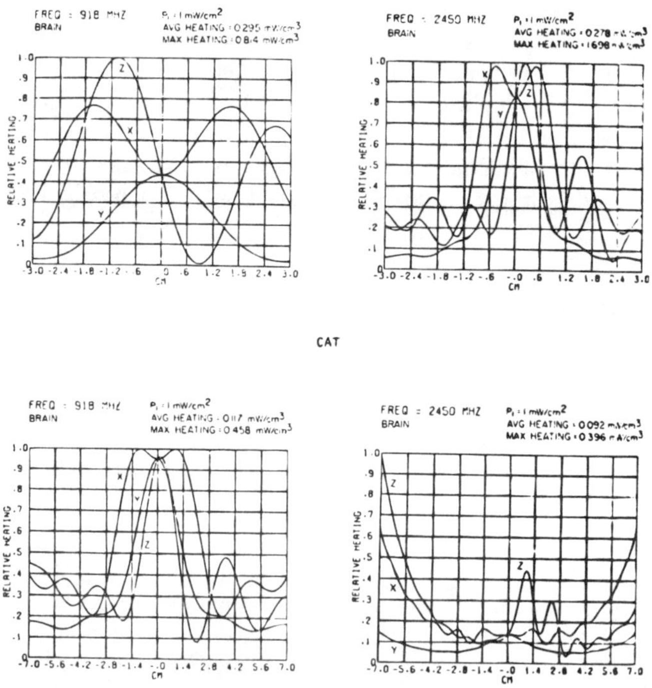

FIGURE 9 Predicted SAR distribution (heating pattern) along the three rectangular axes of spherical models of brain exposed to plane waves.

while the 14-cm-diameter sphere is more typical of a human adult brain. The figures illustrate the SAR distributions along the three perpendicular axes whose origin coincides with the center of the sphere. An incident plane wave power density of 10 W/m^2 is assumed. It is seen that for 918 MHz, maximum absorption occurs near the center of both the cat- and human-size brain sphere. When the frequency is increased to 2450 MHz, the location of peak SAR for the cat-size brain sphere remains near the center, whereas that for a human-size brain sphere is moved to an anterior location.

In general, standing wave patterns with many oscillations are observed. Note that while peak and average SARs in the cat brain are larger by a factor of two than in the human brain at 918 MHz; at 2450 MHz the peak absorption is four times and the average absorption is three times greater in the cat brain than in the human brain. Other studies[6–9] indicate that the peak absorption may be as much as five times greater than the average, and the enhanced absorptions near the center of these brain models may be two to three orders of magnitude greater than that expected from the planar tissue models. The increased absorptions are due to a combination of high dielectric constant and curvature of the model, which produces a strong focusing of energy toward the interior of the sphere that more than compensates for the transmission losses through the tissue.[10]

The peak absorption per unit volume, average absorption per unit volume, and average absorption per unit surface area as functions of frequency and radius of spherical brain model are illustrated in Figure 10. It can be seen that the absorbed energy varies widely with sphere size and frequency. In general, the absorption increases rapidly with increasing radius and is then followed by some resonant behavior. The peaks of these resonant oscillations are related to the maxima, or hot spots, in the distribution of absorbed energy inside the head model, as shown in Figure 9. Therefore, for $(2\pi a/\lambda_0) < 0.4$, where a is the sphere radius and λ_0 is the wavelength in vacuum, hot spots do not occur inside the sphere. However, for some combinations of irradiation frequency and radius hot spots will occur, e.g., in spheres with radii between 2 and 8 cm at 918 MHz and between 0.9 and 5 cm at 2450 MHz. For spheres whose radii exceed the size ranges mentioned above, the maximum absorption appears at the anterior portion (exposed surface) of the sphere, and the penetration depth at the surface becomes a dominating factor for exposures at frequencies in this range. The planar model discussed previously may be applied to obtain a theoretical estimation of the absorbed energy in this case.

The frequency dependence of energy absorption is illustrated in the upper graphs in Figure 10 for the head of a small animal, such as a cat or rhesus monkey, and a human-head-size sphere. Besides the occurrence of resonant peaks, with increasing frequency, energy is absorbed in a decreasingly smaller volume as result of shortened penetration depth.

The effects of skin, fat, bone, dura, and cerebrospinal fluid on the absorption of radio frequency radiation by the brain have been investigated in several laboratories[7,11–14] using more complex spherical structures where the spherical core of brain is surrounded by five concentric shells of tissues (see Table 3). It is interesting to note that if brain sizes remain unchanged, but the overall sphere diameter is increased to account for the outer tissue layers, absorption in brain tissues may be increased by 25% for human- and cat-size heads at 918 MHz or decreased by 70% or more in the case of 2450 MHz (see Figures 11 and 12). Moreover, surface absorption is greatly increased in the case of layered models, while fat and bone always absorb the least amount of energy.

If the outer diameter of the sphere remains the same, while the tissue layers are allowed to be either layered or homogeneous, the peak and average SARs show very little change except when the radius of the spherical head is between 0.1 and 1.0 times the wavelength in air. The peak and average SARs for layered models may be several times greater than for homogeneous models. Enhancement is apparently the result of resonant coupling of energy into the sphere by the outer tissue layers.

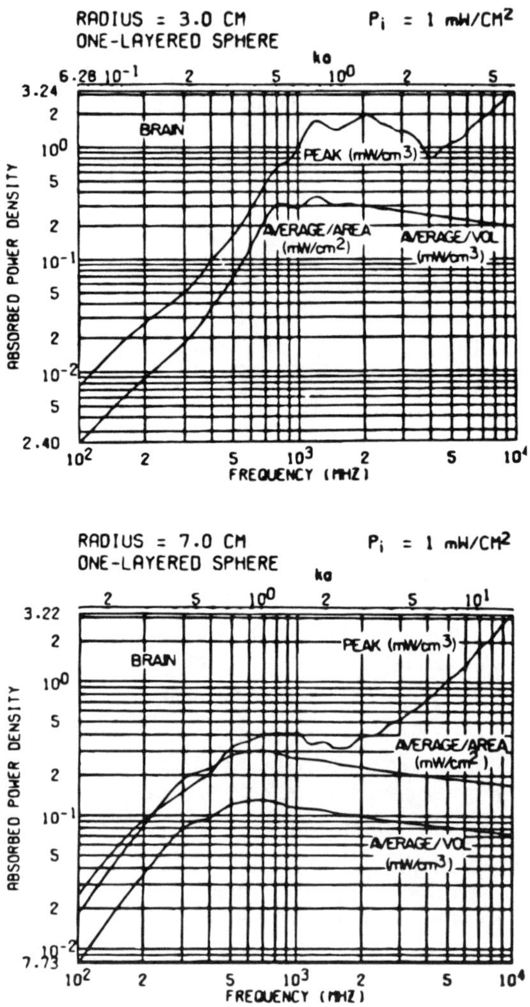

FIGURE 10 Electromagnetic energy absorption characteristics for spherical models of brain.

TABLE 3
Composition of a Six-layer Model of a Mammalian Head

	Dielectric constant (MHz)		Thickness (cm)	
	918	**2450**	**3.5**	**10**
Brain	34.42-j15.49	32.78-j15.37	2.88	6.98
CSF	80.85-j14.05	77.0-j13.94	0.20	1.10
Dura	51.40-j25.08	47.52-j11.42	0.05	0.8
Bone	5.56-j0.856	5.0-j0.857	0.20	0.7
Fat	5.56-j0.856	5.0-j0.857	0.07	0.27
Skin	51.40-j25.08	47.52-j11.42	0.10	0.15

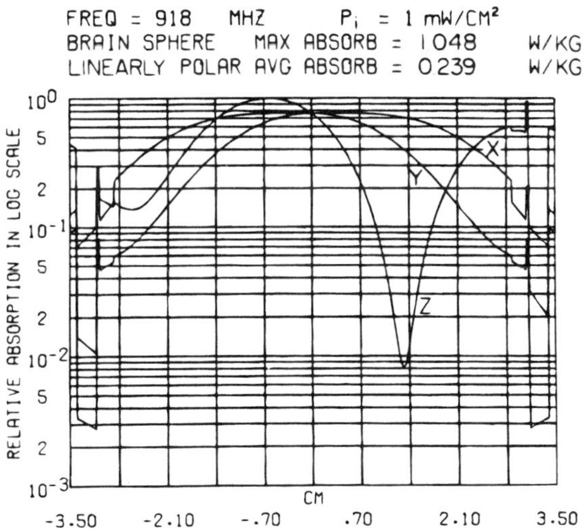

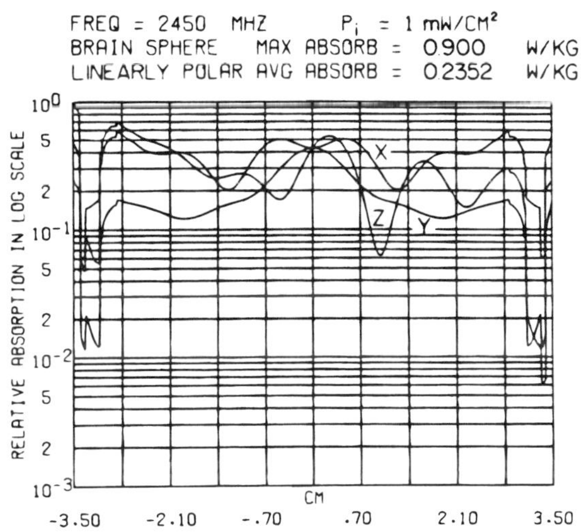

FIGURE 11 Distribution of absorbed energy in a six-layered spherical model of the cat head.

It should be mentioned that the computed absorption characteristics are highly sensitive to the dielectric constant and conductivity assumed as a function of frequency. There are two different sets of brain dielectric constant and conductivity in the literature.[2,8,13,15,16] Depending on the particular values chosen for these quantities, the interpretation of the relative energy absoption at 2450 MHz can vary from low to quite high.[7,11,13,14] The conclusion that 2450 MHz could be a hazardous frequency for humans most likely stems from the inordinately high (by a factor of 2) dielectric constant used in these computations[17–19] (compare Table 6 in chapt. 1).

A study also has been made of the interaction of circularly polarized plane electromagnetic waves with six-layered spherical models of the mammalian head.[14] The approach is a classic one; Mie equations as given in Equations 8 to 13 are modified to account for the two polarizations that are orthogonal in space. Two sets of representative calculations are shown

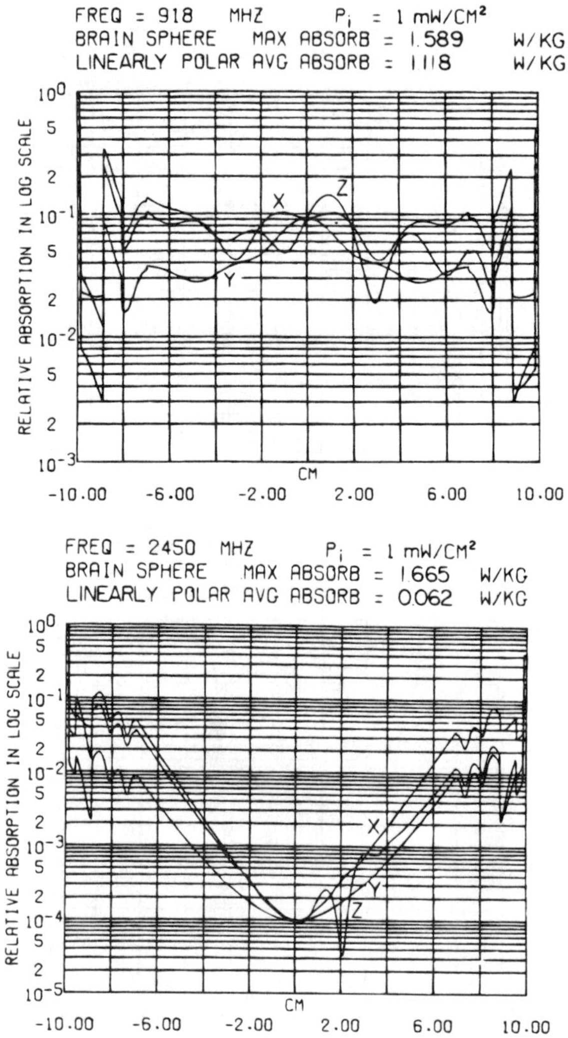

FIGURE 12 SAR distribution (absorbed energy) in a six-layered spherical model of an adult human head.

in Figures 13 and 14 for 918 and 2450 MHz, respectively. The 7-cm-diameter sphere represents a cat or monkey head, while the 20-cm-diameter sphere is typical of a human adult head. In each case, the absorbed power distribution is normalized to the maximum along the z axis. An incident power density of 10 mW/cm^2 is again assumed.

As for the case of linearly polarized plane waves, the maximum absorption for 918 MHz occurs near the center of a cat-size head, whereas the maximum absorption for a human-size head is at the surface. When the frequency is increased to 2450 MHz the location of maximum absorption for both the smaller and larger spheres shifts to the leading surface. The distribution of absorbed energy for circularly polarized waves is more uniform compared with the linearly polarized case. In fact, the absorbed energy distribution in the planes transverse to the direction of propagation is rotationally symmetric, i.e. it is independent of angular variation (see the B curves). Note also that the maximum energy absorbed in the spherical head models varies only slightly between the two frequencies studied. However, a greater quantity of energy is deposited in the inner sphere (representing the brain of a human head) for 918 than 2450 MHz radiation.

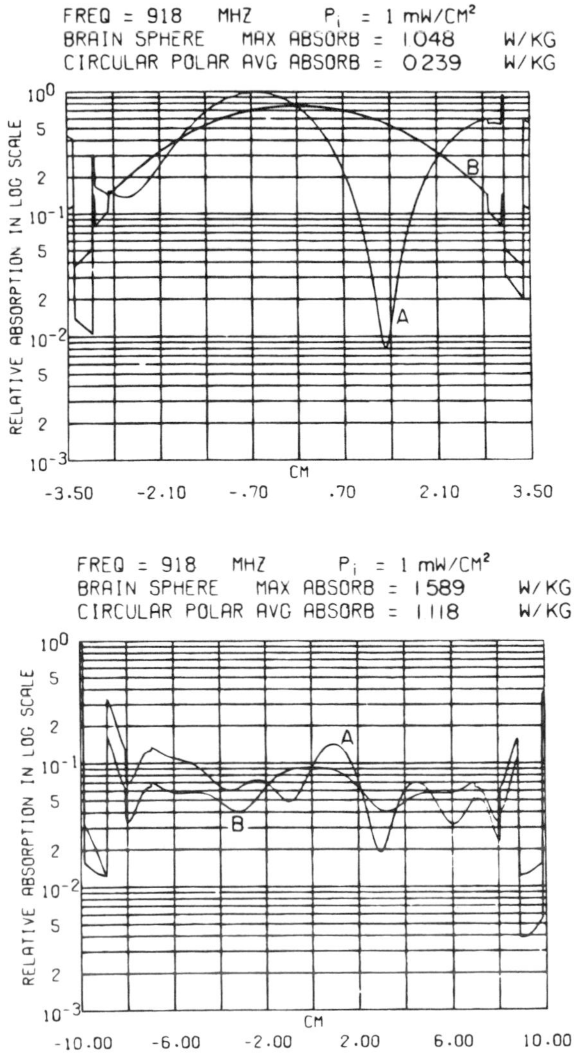

FIGURE 13 Relative absorption of circularly polarized 918 MHz radiation in spherical models of mammalian head.

A number of investigations[2,20–22] have used muscle spheres as a first-order approximation for the extrapolation to human beings of results obtained from laboratory animals and as an index of whole-body absorption of electromagnetic energy as a function of frequency. The spherical model is attractive since exact solutions for absorbed energy can be obtained for all frequencies and body sizes. The absorptions for homogeneous muscle spheres, whose volumes correspond to small animals, such as a rat, and standard man, computed as functions of frequency, are shown in Figure 15. While in this case the peak absorption is of very limited utility, the average absorption per unit surface area is related to the time and power required to overload the thermoregulatory capacity of an exposed subject. Note that the average absorption for the rat model is at least ten times higher than for a muscle sphere representing a human body at frequencies greater than 500 MHz. Absorption increases, in both cases, rapidly with frequency until the free space wavelength of the impinging radiation approaches the diameter of the sphere. A number of resonant oscillations appear which tend to increase the amount and nonuniformity of absorbed energy. Above this range the absorption falls off slowly, indicating that details of body surface curvature are of little significance.

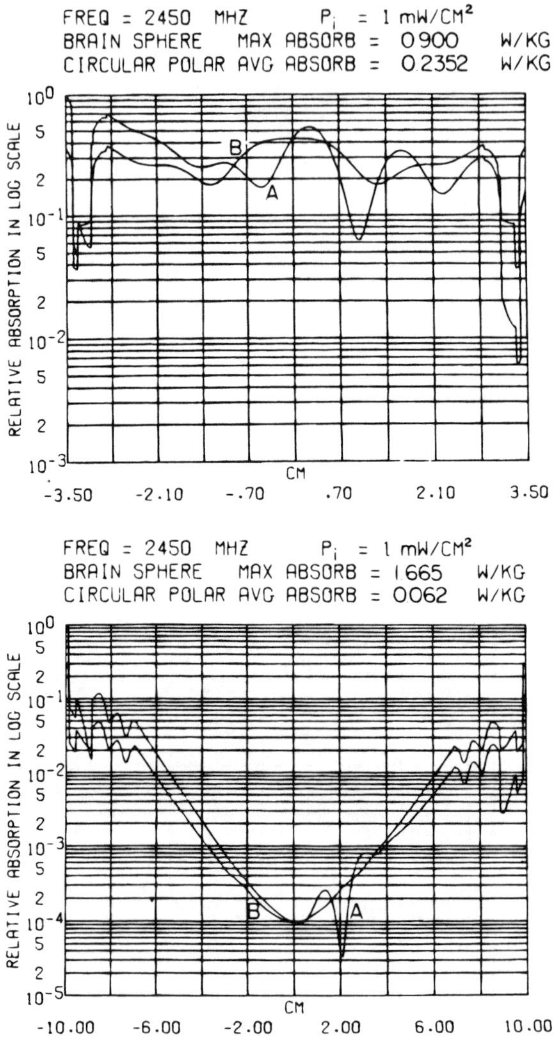

FIGURE 14 Circularly polarized 2450 MHz microwave radiation-induced SAR distribution in spherical models of mammalian head.

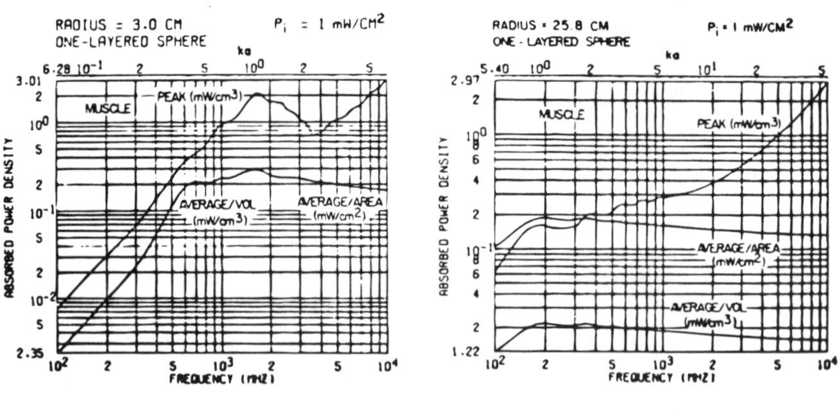

FIGURE 15 Absorption characteristics of spherical body models.

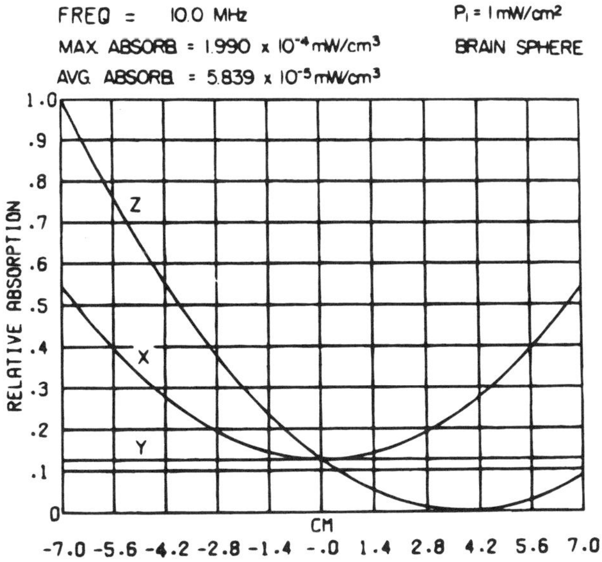

FIGURE 16 SAR distribution (relative absorption) along the X, Y, and Z axes of a small brain sphere.

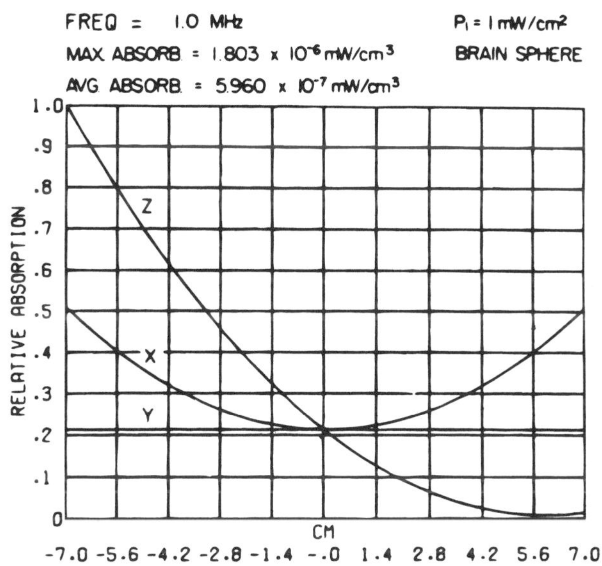

FIGURE 17 Predicted SAR distribution (relative absorption) in a spherical brain whose size is extremely small compared to the wavelength, along the X, Y, and Z axes.

We have, thus far, dealt mainly with situations where the diameter of the sphere is comparable to or larger than the wavelength in air. When the dielectric sphere is small compared with a wavelength, the absorbed energy distribution varies almost as the square of the radius or distance (x) from the magnetic axis (y, the axis parallel to the direction of magnetic field vector) as shown in Figure 16. If the sphere is extremely small compared with wavelength, the absorbed energy distribution becomes more uniform in the x direction, but decreases continuously with distance (z) from the exposed surface (see Figure 17). This behavior can be explained by a quasi-static field theory.[21] Accordingly, for an incident field polarized in the x direction which advances in the z direction, the induced electric field inside a dielectric sphere is given by

$$\mathbf{E}_t = E_0 e^{j\omega t}\left[\frac{3}{\epsilon}\hat{x} - j\frac{kr}{2}(\cos\phi\,\hat{\theta} - \cos\theta\,\sin\phi\,\hat{\phi})\right] \tag{14}$$

where E_0 is the strength of the incident electric field and r is the radial variable in spherical coordinates. The whole-body absorption rate is given by

$$P_a = \frac{1}{2}\sigma E_0^2 V\left[\frac{9}{|\epsilon|^2} + \frac{1}{10}(ka)^2\right] \tag{15}$$

where V and a are respectively, the volume and radius of the sphere.

The electric component of the incident field couples to the object in the same fashion as an electrostatic field. This gives rise to a constant induced electric field inside the sphere which has the same direction, but is reduced by $3/\varepsilon$ from the applied electric field for biological materials and is independent of sphere size. Similarly, the magnetically induced electric field inside the body is identical to the quasi-static solution whose magnitude is given by $E = \pi f \mu r H$, where f is the frequency, μ is the permeability, r is the radius, and H is the magnetic field component. Thus, the magnetic component of the incident field produces an internal electric field that varies directly with distance away from the axis and in proportion to the frequency.

A sketch depicting how the induced electric fields combine inside the sphere is shown in Figure 18. It is seen that the magnetically induced electric field encircles the y axis (magnetic axis) and gives rise to an eddy current whose magnitude increases with distance from the y axis. This indicates that while the H-induced energy absorption in a mouse or larger animal is much greater than the E-induced component, electrically and magnetically induced absorption may be equally significant in even smaller animals at lower frequencies (below 30 MHz). Moreover, for a small insect or pupae the electric field will be the predominant factor.

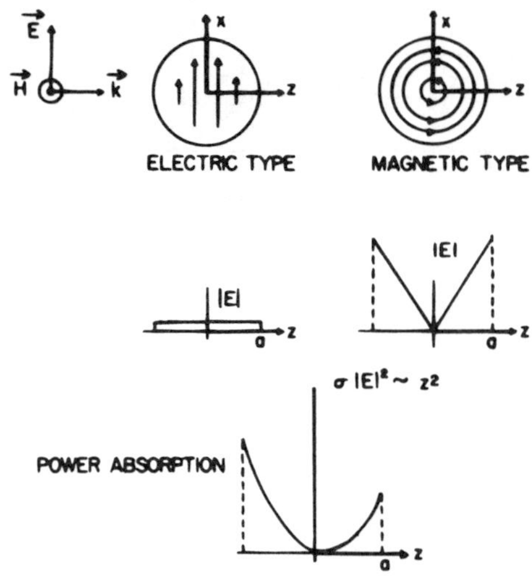

FIGURE 18 Illustration of the behavior of electric and magnetic fields under quasi-static conditions of exposure (applied magnetic field H in y-direction).

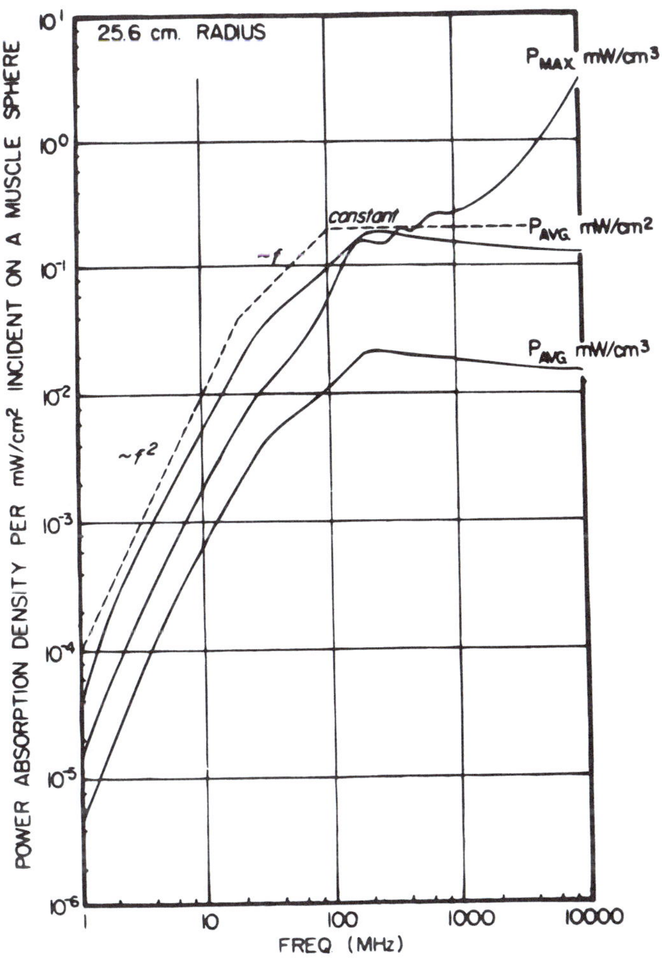

FIGURE 19 Frequency dependence of absorption in a spherical model of the human body. Maximum and average per unit volume, average per unit area.

The variation of average and maximum energy absorption with frequency for a human-size sphere is illustrated in Figure 19. In the frequency range from 1 to 20 MHz, the maximum absorption rate is only 10^{-6} to 10^{-3} W/kg/W/m^2 of incident power. Inspection of the maximum absorption rate induced by a plane wave, a quasi-static electric field, and a quasi-static magnetic field shows that absorption at frequencies below 20 MHz is primarily due to the magnetic induction and is characterized by a square-of-frequency dependence. The approximate frequency dependence of average or total energy absorption throughout the frequency range from 1 MHz to 10 GHz is indicated by the dashed line. For frequencies below 20 MHz the average absorption varies as the square of the frequency. In the frequency range of 20 to 200 MHz, the average absorption increases directly in proportion to frequency and attains a maximum of about 2×10^{-3} W/kg per W/m^2 of incident power at 200 MHz. The average absorption rate remains fairly constant with increasing frequency. (It actually is inversely proportional to frequency for higher frequencies.) There is thus little doubt that electromagnetic energy absorption varies both with frequency and body size, and in a predictable manner.

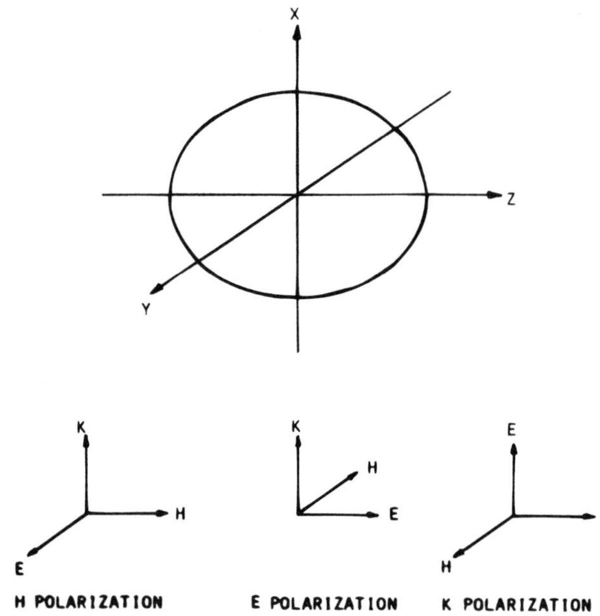

FIGURE 20 Polarization states of a plane wave impinging on a spherical body model.

B. PROLATE SPHERIODAL MODELS

Since the bodies of humans and experimental animals are seldom spherical in shape, we need a better model to analytically and numerically describe the induced fields and absorbed energy inside experimental subjects. A prolate spheroid emulates more closely the shape of mammalian bodies, but most analyses have been restricted to homogeneous models for humans and experimental animals.[23–29] The basic approach[30] is to expand the incident, scattered, and transmitted electric fields in terms of vector wave functions in spheroidal coordinates. The expansion coefficients are determined from the boundary conditions demanding that the tangential components of the fields must be continuous across the surface of the spheroid. The absorbed power is obtained from Equation 7 and the average absorptions are found from relationships similar to those of Equations 9 to 13.

In addition, for frequencies below resonance (resonance is defined as the condition of maximum absorption), long-wavelength formulations[24,25] and quasi-static approximations[31] have been used to obtain absorption information. Also, geometric-optics approximations have been developed for computation of absorption characteristics of prolate spheroidal models of humans at frequencies whose wavelengths are short compared with body size.[19]

Three orientations of the impinging plane wave with respect to the body must be distinguished (see Figure 20): E-polarization in which the electric field is parallel to the major axis of the spheroid, H-polarization in which the magnetic field vector is parallel to the major axis, and K-polarization in which both electric and magnetic field vectors are perpendicular to the major axis of the spheroid. In general, E-polarization produces the highest energy absorption for frequencies up to and slightly beyond the resonance region.

The induced electric fields inside a dielectric prolate spheroid (semi-major axis. a: semi-minor axis, b) in a plane wave electromagnetic field with long wavelength ($\lambda > a$) may be represented by

$$\mathbf{E}_e = \frac{E_0}{C_1}\,\hat{z} - j\frac{1}{2}E_0\,\frac{\omega\mu_0}{377}\,(z\cos\phi\,\hat{\rho} - z\sin\phi\,\hat{\phi} - \rho\cos\phi\,\hat{z}) \tag{16}$$

for E-polarization and

$$\mathbf{E}_h = \frac{-E_0\hat{y}}{C_2} + \frac{1}{2} E_0 \frac{\omega\mu_0\rho\hat{\phi}}{377} \tag{17}$$

for H-polarization. The whole-body energy absorption rate is generally given by

$$P_a = \frac{1}{2}\sigma \int_{\text{Volume}} \mathbf{E} \cdot \mathbf{E}^* \, dv \tag{18}$$

where $\sigma = \omega\epsilon_0\epsilon''$ is the electrical conductivity and $\mathbf{E}^*$ denotes the complex conjugate of the induced field $\mathbf{E}$ inside the body. The whole-body rates of absorbed energy, by substituting Equation 16 and 17 into 18, are

$$P_{ae} = \frac{1}{2}\sigma E_0^2\left(\frac{4}{3}\pi ab^2\right)\left[\frac{1}{C_1^2} + \frac{1}{20}\left(\frac{\omega\mu_0}{377}\right)^2(a^2 + b^2)\right] \tag{19}$$

for E-polarization and

$$P_{ah} = \frac{1}{2}\sigma E_0^2\left(\frac{4}{3}\pi ab^2\right)\left[\frac{1}{C_2^2} + \frac{1}{10}\left(\frac{\omega\mu_0}{377}\right)^2 b^2\right] \tag{20}$$

for H-polarization.[31] The constants C_1 and C_2 are given by

$$C_1 = (1 - \epsilon)\eta_0[(1 - \eta_0^2)\coth^{-1}\eta_0 + \eta_0] + \epsilon \tag{20a}$$

$$C_2 = 1 + 1/2(\epsilon - 1)\eta_0[(1 - \eta_0^2)\coth^{-1}\eta_0 + \eta_0] \tag{20b}$$

where $\epsilon = \epsilon' - j\epsilon''$ is the complex relative permittivity and η_0 is the eccentricity given by

$$\eta_0 = \frac{a}{(a^2 - b^2)^{1/2}} \tag{20c}$$

Since C_1 and C_2 do not depend on spatial variables, the electrically induced fields within the spheroid are uniform and independent of size when the external field is uniform. For $\epsilon > 1$, the field inside the spheroid is weaker than the applied field. Moreover, the whole-body energy absorption depends not only on the strength of impressed fields, but also on the orientation of the field with respect to the major axis of the body.

As in the case of spherical models, the absorption is produced by an electrically induced current in the direction of the applied E-field vector, combined with a circulating current induced by the incident H field. One would therefore expect the electrically induced absorption to be uniform, whereas the absorption due to the circulating eddy current would be zero at the center and increase as the square of the distance from the center.

Computations of the absorbed energy as a function of frequency and body size have been made using homogeneous muscle material to serve as an index of SARs in mammals and to serve as a guide for extrapolating data from experimental animals to human beings,

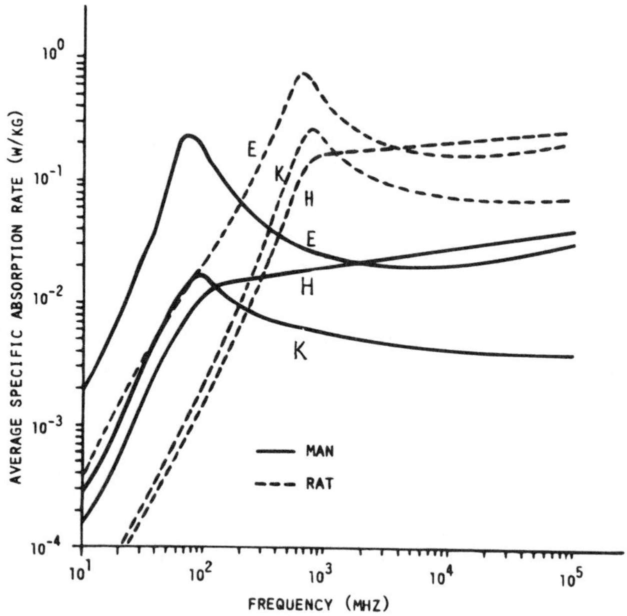

FIGURE 21 Predicted frequency dependence of absorbed energy in spherical models of biological bodies.

particularly with regard to averages of absorbed energy.[32] Figure 21 shows the theoretically projected frequency dependence of absorbed energy for humans and for laboratory rats. Note that for a given incident field orientation, the average SAR for humans may either be higher or lower than for rats, depending on the frequency. For example, at 70 MHz, the average SAR is the highest for humans, having a value of 0.25 W/kg for an incident power density of 10 W/m^2; the average SAR for a rat is only 0.0125 W/kg. In contrast, the average SAR of 0.8 W/kg at 700 MHz is the highest for rats; the corresponding value for humans is less than 1/25. It is thus extremely important to take into account the body size and operating frequency to establish any relationship between the biological effects that arise in the laboratory and corresponding effects that might occur in humans at a given incident power density.

The frequency for maximal absorption (resonance frequency) depends on the subject and its orientation with respect to the incident field. In general, the shorter the subject, the higher the resonance frequency and vice versa. Further, the frequency dependence of whole-body or average absorption may be partitioned into three regions. This may be illustrated using the orientation that is most efficient in energy coupling, E-polarization. Figure 22 gives the whole-body absorption data for a prolate spheroidal model of the rat exposed to uniform fields in free space. For frequencies well below resonance such that the ratio of the longest body dimension (L) to free space wavelength (λ) is less than 0.2, the average SAR is characterized by a f^2 dependence. The average absorption goes through a resonance in the region where $0.2 < L/\lambda < 1.0$. In this case, the average SAR rapidly increases to a maximum near $L/\lambda = 0.4$ and then falls off as 1/f. At frequencies for which $L/\lambda > 1.0$, the whole-body absorption decreases slightly but approaches asymptotically the geometrical optics limit of about one half of the incident power (1 − power reflection coefficient).

It should be noted that the resonant absorption length of 0.4 λ is in good agreement with results from antenna analysis. In addition, whole-body absorptions for H- and K-polarizations are totally different. The resonances are not nearly as well defined as for E-polarization. In fact, the whole-body absorption curve for H-polarization gradually reaches a plateau and stays at that plateau for higher frequencies.

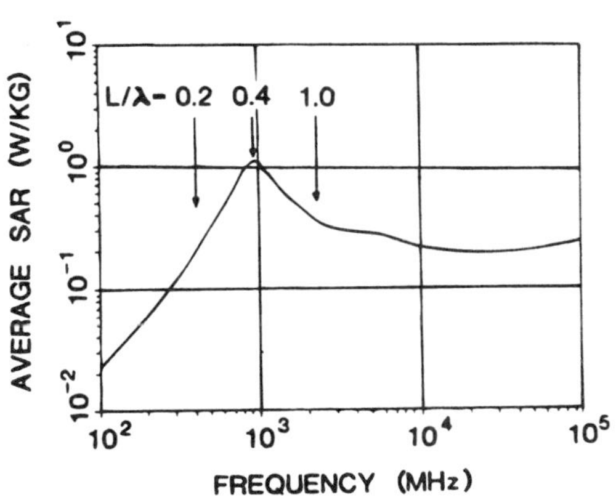

FIGURE 22 The three distinct regions of absorption as a function of frequency.

The distribution of absorbed energy in a prolate spheroidal model whose size is small compared with wavelength is shown in Figure 23 for the three distinct polarizations. The height of the prolate spheroid equals 1.75 m with a major-to-minor axis ratio of 6.34 and a 70-kg mass corresponding to a human body. The dielectric constant and conductivity are those for muscular tissue at 10 MHz. These graphs are qualitatively similar to those for spherical models. As expected, the absorbed energy is highest for E-polarization. In fact, there is approximately an order of magnitude difference.

For E-polarization the eddy current is zero on the x axis, thus a low relative absorption is seen along the x axis which is only due to the electrically induced current. The absorption rate is elevated considerably, however, indicating a strong coupling of the impressed electric field into the interior of the prolate spheroid. The slender spheroid forces the internal electric field to more closely correspond to the incident field because of the boundary condition that requires continuity of tangential fields. The relatively flat energy distribution along the z axis again shows the strong coupling of impressed electric field and relatively weak eddy-current contribution due to a smaller cross-section for intercepting the magnetic flux. The distribution, along the y axis, indicates that the electrically and magnetically induced field components are nearly equal. The electric polarization field and the circulating eddy current add at the front side and subtract on the back side of the spheroid to render an absorption pattern that peaks at the front surface and is reduced to almost zero deeper inside the spheroid.

For H-polarization, the electrically induced current flows along the x axis (direction of incident E field) and the eddy current field encircles the z axis (direction of incident H field). The relatively low power on the z axis comes solely from the incident electric field. The combination of E- and H-induced components generates a displaced parabolic energy absorption pattern along both the x and y axes. Clearly, the magnetically induced eddy current predominates in this case and the absorption is highest along the transverse circumference at the middle of the prolate spheroid. For K-polarization, both the electric and magnetic components of the incident field are along the minor axes of the spheroid. The electrically induced current flows along the x axis. The incident magnetic field induces an eddy current electric field that encircles the y axis and the absorption is lowest at the axis. Whereas in both E- and H-polarization cases, the peak absorption occurs at the front surface of the spheroid (the surface of the spheroid initially encountered by the incident field), this is not the case for K-polarization. Maximum absorption appears at $x = \pm b$ and the absorbed energy varies parabolically along the x axis. This results from the large quantity of magnetic flux intercepted by the broad cross-section (and resulting concentration of eddy current). It should be noted that the results illustrated in Figure 23 match very well with experimental measure-

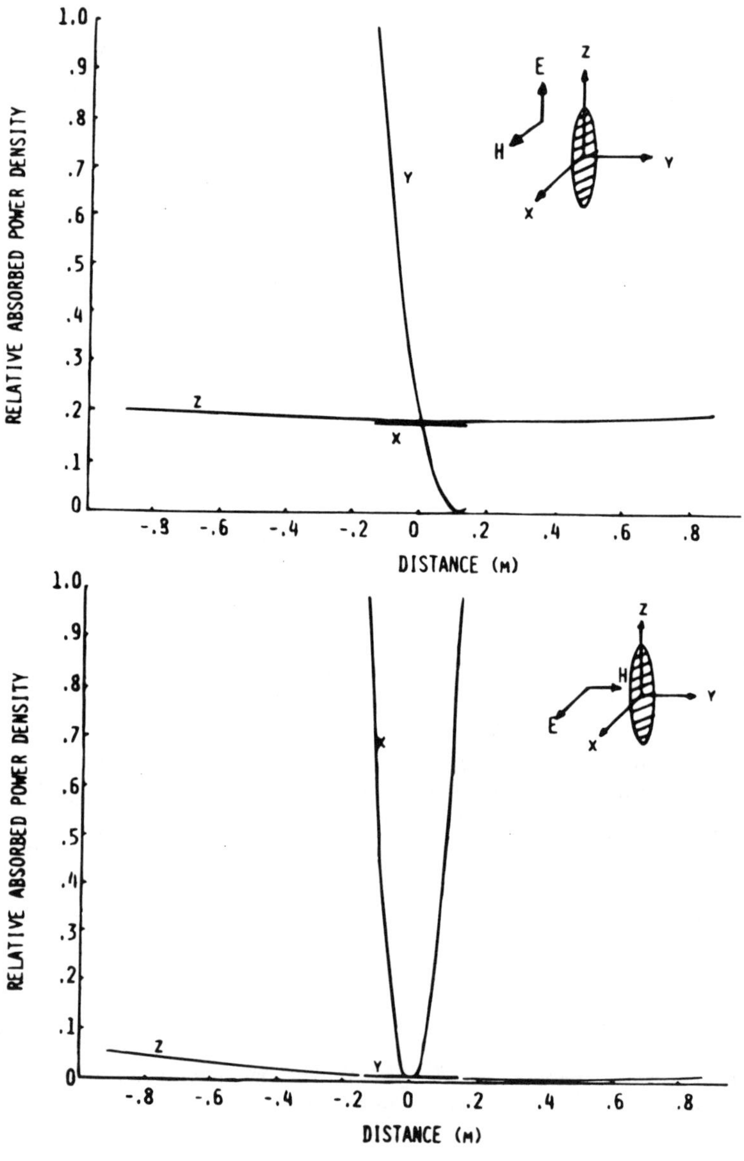

FIGURE 23 SAR distributions along the X, Y, and Z axes of a prolate spheroidal model of humans. a: E-polarization, b: K-polarization, c: H-polarization.

ments.[33] Moreover, the peak absorptions may be two orders of magnitude higher than those for dielectric spheres of equal mass.

IV. MODELS WITH SMALL CELL SIZES

We have summarized above some of the computational approaches to quantify the absorbed energy in simple models of biological objects. It should be recognized that while spheres and spheroids are good models of some animal bodies and certain body parts, they may not always be adequate for humans and experimental animals under a variety of exposure situations. More realistic models, such as models of human bodies formed from cubes, have

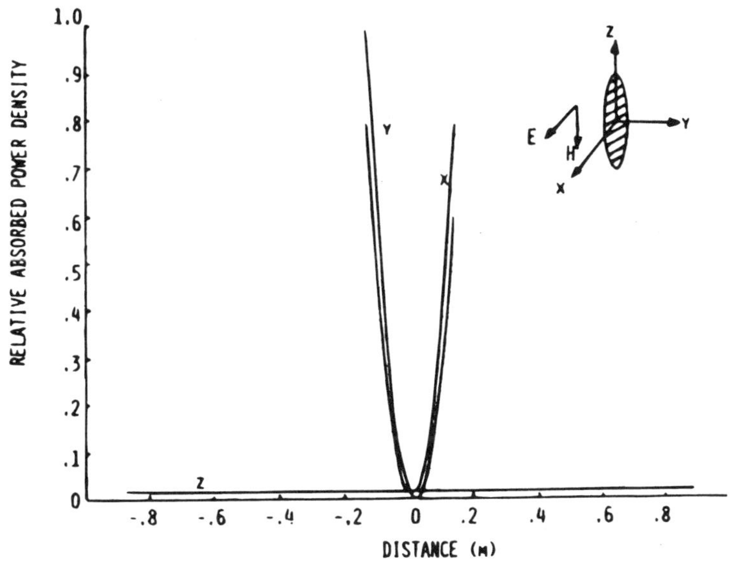

FIGURE 23 Continued

been developed to account for the irregular shapes.[34–40] These models, based on numerical techniques, hold great promise for accurately predicting energy absorption and its distribution in biological objects exposed to electromagnetic fields (EMF) and radio frequency radiation (RFR). In what follows, we shall summarize a number of computer techniques that have been applied with some success in solving electromagnetic energy absorption and SAR distribution problems. We shall also describe some results obtained using these methods.

A. A BRIEF SURVEY OF NUMERICAL METHODS

The numerical methods used to predict induced fields in biological bodies of realistic shape and composition include the quasi-static impedance method, the method of moments, the finite element method, and the finite difference time domain method.

1. The Quasi-static Impedance Method

For low frequency situations, where the dimensions of the biological body are small compared to the wavelength, the impedance method has been found to be highly efficient as a numerical procedure for calculating internal current densities and induced electric fields.[41–45] In this method, the biological body or the exposed part thereof is represented by a three-dimensional (3-D) network of impedances whose individual values are obtained from the complex conductivities $\sigma + j\omega\epsilon$ for the various locations of the body. The impedances for various directions for the three-dimensional network can be written as:

$$Z_m^{i,j,k} = \frac{\delta_m}{\delta_n \delta_p (\sigma_m^{i,j,k} + j\omega\epsilon_m^{i,j,k})} \tag{21}$$

where i, j, k indicate the cell index; m is the direction which can be x, y, or z for which the impedance is calculated; and $\sigma_m^{i,j,k}$ and $\epsilon_m^{i,j,k}$ are the conductivities and the electrical permittivities for the cell i, j, k. δ_m is the thickness of the cell in the m-th direction, and δ_n and δ_p are the widths of the cell in directions at right angles to the m-th direction.

In the impedance method formulation, it can be seen that the cells need not be identical so that fairly thin features of the body can be modeled as well as the interfaces between the various tissues and organs. Also, the conductivity for a given cell can be directionally dependent. This feature will be useful in allowing for the highly anisotropic conductivities of the tissues that have been reported for low frequencies including power-line frequencies.[46,47]

Using anatomically based models of the human body, the impedance method has, in the past, been used for the following applications:

a. Calculation of SAR distributions for operator exposure to spatially variable fields of induction heaters.[43]
b. SAR distributions for linearly or circularly polarized RF magnetic fields representative of magnetic resonance imagers.[44]
c. SAR distributions due to capacitive-type electrodes used for hyperthermia.[45]
d. SAR distributions for interstitial RF needle applicators for hyperthermia.[48]

Some recent calculations using the impedance method are listed below:

a. Internal electric fields and current densities induced in the human body for exposure to magnetic fields of high-voltage power transmission lines.[49]
b. Electric fields and current densities induced in the human head by magnetic fields of a hair dryer.[50]
c. Current densities induced in the arm and the body by magnetic fields of an electric hand drill.[51]
d. Currents induced in the anatomically based model of the human body by the electric and magnetic fields of electric blankets.[52]

In the section that follows the use of the impedance method is described, in detail, for calculating currents in models of the human body exposed to electric and magnetic fields of both the conventional (pre-1990) electric blanket and the new Low-Magnetic-Field electric blanket.

2. The Volume Integral Equation MoM Method

The method of moments (MoM)[53] is used either in conjunction with the volume integral equation method or the surface integral equation method for finding solutions to the unknown fields inside the body. The approaches differ, however, in specifics, in that the surface integral equation MoM (SMoM) finds the unknown currents on the body surface and calculates the interior fields from the surface currents, the reciprocity theorem, and a "measurement matrix." In contrast, the volume integral equation MoM (VMoM) requires determination of unknown fields throughout the volume of the body using the volume equivalence principle and the method of moments.

The numerical technique that has been adopted for most of the early field intensity computations is the VMoM, employing the volume equivalence principle.[36,38,54,55] The method of moments is used to transform the integral equation into a matrix equation by subdividing the body into N simply shaped cells. This is accomplished with the aid of an appropriate set of expansion functions, chosen to satisfy the boundary conditions, and a set of weighting (testing) functions to reduce the matrix fill-in time. The total electric field in each of the N cells is given by matrix inversion. A more detailed description of the volume integral equation method is included in the next section.

However, it should be noted that a fundamental limitation of this method is the use of full or nearly full matrices and therefore, the requirement of extensive computer storage and long computation time. Even with the availability of larger and faster computers, this difficulty is not completely resolved. The need for excessively large numbers of mathematical cells to render a more accurate representation of the body will give rise to an equally large and full matrix. The inversion of large, full matrices often leads to numerical instabilities in the solution. Nevertheless, the method does allow the use of inhomogeneous models with up to 1000 cells. In fact, this method has been employed, successfully, to calculate whole-body averaged absorption and to obtain regional distribution of absorbed radio frequency energy using inhomogeneous block models composed of rectangular cells.[36–38,40,56] This method also has been used to study the interaction of the near-zone field of an antenna with biological bodies.[37,57,58]

3. The Surface Integral Equation MoM Method

Another approach for predicting the distribution of absorbed electromagnetic energy is the SMoM method.[59–62] This method makes use of two coupled integral equations, i.e., the electric field and magnetic field integral equations for the tangential components of the field on the surface separating the biological body from air. The unknown surface currents are found by Fourier decomposition and the moment method. The fields inside the biological body are calculated using the previously computed surface currents, the reciprocity theorem, and the concept of measurement matrix.[60,63–65]

The method begins with the matrix representation of the coupled integral equations. If the body is assumed to be rotationally symmetric, the incident wave and the induced current could then be expanded in a Fourier series expansion in the angle of rotation. This reduces the problem to that of solving a system of orthogonal modes. The method further expands the surface components in terms of triangular expansion or basis functions and allows the testing functions to be the complex conjugate of the basis functions taking advantage of the orthogonality property. Thus, the major advantage of introducing the Fourier series is to enable each mode to be treated completely independently of all other modes. This results in a much smaller size, manageable matrix equation to be evaluated for the unknown expansion coefficients which determine the surface currents. It should be noted that for biological bodies, triangular expansion and testing functions are preferred over flat pulse expansion functions.[60] In fact, an expansion function with a continuous first derivative may constitute an even better choice for the expansion basis function. In any event, once the surface currents are obtained, the fields everywhere, or SAR at each point inside the body, can be calculated using the reciprocity theorem.[64,65] The total absorption can be found by integrating the surface Poynting vector.

The validity of SMoM has been substantiated by using a dielectric sphere.[59] Calculations for a human torso modeled by a homogeneous muscle body of revolution with a height of 1.78 m at 30, 80, and 300 MHz showed enhanced absorption in the neck region for all three frequencies and both vertical and horizontal polarizations.[60] In this calculation the vertical direction is aligned with the long dimension of the torso and serves as the axis of symmetry. The strongest absorption in the torso model was found to occur with vertical polarization and near the first resonance frequency of the torso (80 MHz). In general, the surface integral equation method is applicable to any arbitrarily shaped homogeneous body of revolution. The method can be used not only with incident plane waves, but also with a wide variety of other field exposure conditions, including direct contact situations and near-zone sources.

Since both the surface and volume integral equation methods for field intensity prediction rely on the MoM method for implementation, it is instructive to compare the relative advantages of these two techniques. For simplicity, consider a homogeneous cube with N samples on each side: the computer storage requirements are N^2 and N^3 for the surface and volume integral methods, respectively.[60] For sufficient sampling to ensure accurate description of field variations, N is usually a large number. Thus, the surface integral equation method requires significantly fewer unknowns for homogeneous models. Moreover, in cases where permittivity and conductivity values are large, such as in biological bodies, the wavelength becomes contracted inside the body, and a much-larger number of cells than that indicated above may actually be needed. If the model is inhomogeneous, then the volume integral equation would prove to be more suitable. It is possible, however, to generalize the surface integral equation technique to account for inhomogeneities by employing the invariant imbedding procedure.[66]

4. The Fast Fourier Transform Method

The fast Fourier Transform (FFT) method makes use of the fact that the integral equation for the electric field is in the form of a convolution integral. This integral, when Fourier transformed, becomes a simple algebraic equation in the frequency domain which can be solved, iteratively, to obtain the unknown electric field. Specifically, the integral is first replaced by a double discrete summation to preserve the convolutional nature of the equation. A FFT algorithm is used to efficiently evaluate the kernel and current distributions, which are then algebraically multiplied to give the transform of the internal electric field. The inverse FFT then yields the internal electric field in the spatial domain. Since the FFT method computes a circular rather than a linear convolution, the FFT lengths must be chosen to ensure that the convolution is linear inside the biological body. This choice of FFT lengths decouples the fields due to the periodic scattering, introduced by the discrete Fourier transform. This also obviates the need for large intervals which would otherwise have restricted the number of cells that could be used.

It is important for the success of the FFT approach that a fast convergence is achieved so that the solution is reached within a few iterations. There are several iteractive techniques which may be employed with the FFT method. These include the stationary iterative scheme associated with the K-space formulation,[67] the steepest descent scheme,[68] and the conjugate gradient method.[69,70] In general, the number of iterations required for convergence could decrease by as much as an order of magnitude for these three schemes, in the same order as the presentation.

Clearly, the numerical implementation of the FFT method is iterative. Therefore, the computer storage requirement is proportional to N and the computation time is proportional to 4N log 8N per iteration, where N is the number of cells used to represent the body. This may allow use of 10,000 to 100,000 homogeneous cells to model three-dimensional bodies at higher frequencies, as well as with greater anatomic details, for distributive SAR calculations. Results obtained for two-dimensional problems demonstrate accuracy comparable to known analytic solutions for homogeneous and layered circular cylindrical bodies.[71]

5. The Finite Difference Time Domain Method

The finite difference time domain (FDTD) approach is an attempt to solve Maxwell's curl equations by directly modeling propagation of waves into a volume of space containing the biological body. By repeatedly implementing a finite difference representation of the curl equations at each cell of the corresponding space lattice, the incident wave is tracked as it first propagates to the body and then interacts with it through surface current excitation, transmission, and diffraction. This wave-tracking process is completed when the steady-state behavior is observed at each lattice cell. Considerable simplification is achieved by analyzing the interaction of the wavefront with a part of the body surface at a time, rather than attempting a simultaneous solution of the entire problem.

Time-stepping for the FDTD method is accomplished by an explicit finite difference procedure.[72,73] For a cubic cell lattice space, this procedure involves positioning the electric and magnetic field components about a unit cell of the lattice and then evaluating the components at alternate half-time steps. In this manner, centered difference expressions can be used for both the space and time increments without solving simultaneous equations to compute the fields at the latest time step.

The body of interest is mapped into the lattice space by first choosing the lattice increment and then assigning values of permittivity and conductivity to each cell. The boundary conditions at media interfaces are naturally generated by the curl equations. Thus, once a computer program is developed, the basic routines need not be changed for different model geometries. In fact, inhomogeneities and fine structural details could be modeled with a maximum resolution of one unit cell.

The explicit formulation of the FDTD method is particularly suited for execution with minimum computer storage and run-time using current array-processing computers. The required computer storage and run-time increase only linearly with N, the number of cells. In fact, it has been shown that the FDTD method is capable of solving for more than 1 million unknown field components within a few minutes on an array-processing computer. Field intensities have been predicted to within 2.5% accuracy relative to known analytical and experimental bench marks. Recently, this FDTD technique has been improved to allow solutions for field penetration and absorption in large, complex, inhomogeneous, and irregularly shaped biological bodies in three dimensions, with millimeter range spatial resolution. With the exception of a few early attempts with lossy biological objects,[74–77] a majority of early efforts have been directed toward application of the FDTD method to electromagnetic interaction in time-varying inhomogeneous media,[78] and with metallic bodies of revolution.[79] However, during the last few years, the FDTD method has become the most extensively used numerical procedure for bioelectromagnetic computations. Further discussions of methodological details and specific applications of the method to computation of induced electromagnetic fields in biological bodies are given in the section that follows.

6. The Finite Element Method

The finite element method (FEM) has been a preferred numerical algorithm in many fields of application. However, its use and popularity in predicting field intensities in biological systems have been modest until recent progress in mesh generation, boundary conditioning, and large matrix solvers. The FEM method is a near-neighbor, volume method for solving Maxwell's differential equations and is associated with a sparce system of equations.[80,81] Aside from the low memory requirement (on the order of N), an inherent attraction of FEM is its adaptability in modeling inhomogeneities and complex geometries. The feature of comforming and variable-sized cell elements of the computational volume are extremely important in bioelectromagnetics.

The basic approach of the FEM method for predicting electromagnetic field distributions inside biological bodies starts by subdividing the physical space and biological body of interest into meshes of small volumes or cells of tetrahedral elements. This step is very important since the manner in which the volume is subdivided will dictate the computational resources required, and the speed of the computation and accuracy of the results. Each cell element and node location have to be systematically numbered and described. Once the volume has been subdivided, labeled, and appropriate property values have been ascribed, the unknown field within each element is then approximated using linear extrapolation. A major step in FEM is the formulation of the system of linear equations using either the Ritz or Galerkin algorithm with proper boundary conditions. There are two approaches to solving the system of linear algebraic equations: the direct method of Gaussian elimination, or the iterative method that starts with an initial guess. In practice, either method can produce an approximate solution to the unknown field intensity with a prescribed accuracy.

It should be noted that a large region exterior to the biological body is often encountered in bioelectromagnetic situations, where the biological body or portion of it, is part of a region into which electromagnetic energy is radiated and scattered. The region of space exterior to the biological body and applicator must be truncated with an artificial boundary to limit the volume elements and the number of unknowns. Consequently, an appropriate boundary condition needs to be established at this artificial boundary for a unique finite element determination of the induced fields inside the body. The most common boundary condition selected for this purpose is the absorbing boundary conditions which minimizes the nonphysical reflections from the artificial boundaries by making boundaries transparent to the scattered field.

Fairly large scale calculations, on the order of 200,000 elements, have been conducted effectively in the workstation computing environment. Specifically, detailed power deposition pattterns have been simulated in full and partial models of the human body undergoing electromagnetic hyperthermia treatment for cancer.[82] In this case, the cell elements were generated from computerized tomographic data obtained on human patients.

While the quasi-static impedance method is restricted to lower frequencies (<30–40 MHz for the human body), the FEM and FDTD methods may be used for any frequency of interest. It should be noted, in addition, that both FEM and FDTD involve solving Maxwell's equations in the differential form for the computation of induced fields.

B. HUMAN BODIES EXPOSED TO ELECTROMAGNETIC FIELDS

1. Realistic Models of the Human Body

Several models have been proposed as better representations of the complex geometry and composition of the human body. They include constructions using cubic cells of small volumes, assemblies based on anatomical cross-sectional diagrams, and cell meshes generated from computerized tomographic and magnetic resonance image data.

Cubic Cell Models

Models of the human body consisting of 200 to 1000 cubic cells that account more realistically for the gross anatomic and biometric characteristics of human bodies have been used by several investigators.[34–40] The models are 1.75 m tall and can be made either homogeneous or inhomogeneous by choosing an equivalent or a volume-weighted complex permittivity for each cell. The cubic cell model has been employed, successfully, to calculate whole-body averaged absorption. It is important to note that for subdivision with less than three cells per wavelength, the magnitude and phase resolutions would be such that, even with convergence, the reliability of the MoM computed SAR would be questionable. Therefore, if the interest is primarily in whole-body SAR, this model may provide quite adequate results for frequencies lower than 30 MHz. To achieve more accurate structural representation of the human body, anatomically based models have been offered in recent years. They are described briefly in the following paragraphs.

Models Based on Anatomical Cross-Sectional Diagrams

These models are based on the cross-sectional diagrams of the human body given in the book by Eycleshymer and Schoemaker.[83] Cross-sectional diagrams of the human body were obtained by making cross-sectional cuts at spacings of about one inch in human cadavers. The process for creating the data base of the man model was the following: a quarter-inch grid was taken for each single cross-sectional diagram and each cell on the grid was assigned a number corresponding to one of the sixteen tissue types (muscle, fat, bone, blood, intestine, cartilage, liver, kidney, pancreas, spleen, lung, heart, nerve, brain, skin, eye) or air. Thus the data associated with a particular layer consisted of three numbers for each square cell: x and y positions relative to some anatomical reference point in this layer, usually the center of the spinal cord; and an integer indicating which tissue that cell contained.[84] Since the cross-sectional diagrams available are for somewhat variable separations, typically 2.3–2.7 cm, a new set of equispaced layers was defined at 1/4-in (0.635 cm) intervals by interpolating the data onto these layers. Because the 1/4-in cell size is too small for the memory space of readily accessible computers, the proportion of each tissue type was calculated next for somewhat larger cells of size 1/2 in (1.27 cm) combining the data for $2 \times 2 \times 2 = 8$ cells of the smaller dimension. Without changes in the anatomy, this process allows some variability in the height and weight of the body. A final cell size of 1.31 cm (rather than 1/2 in) was used to obtain a total height and body weight of 175.5 cm and 69.6 kg, respectively. With a resolution of 1.31 cm, the model consists of a total of 45,024 cells that are either totally or partially within the human body. If the shoe-wearing condition of the model is needed, this is simulated by using a separation layer of rubber ($e_r = 4.0$) that is 2 layers or 2.62 cm thick between the feet and the ground plane.

Millimeter-Resolution Model Based on MRI Scans of the Human Body

A new millimeter-resolution model of the human body has been developed from the magnetic resonance imaging (MRI) scans of a male volunteer of height 176.4 cm and weight 64 kg.[85,86] The MRI scans were taken with a resolution of 3 mm along the height of the body and 1.875 mm for the orthogonal axes in the cross-sectional planes. Even though the height of the volunteer was quite appropriate for an average adult male, the weight was somewhat lower than an average of 71 Kg, which is generally assumed for an average male. This problem can, to some extent, be ameliorated by assuming that the cell dimensions for the cross sections are larger than 1.875 mm by the ratio of $(71/64)^{1/2} = 1.053$. By taking the larger cell dimensions of $1.053 \times 1.875 = 1.974$ mm for the cross-sectional axes, the volume of the model can be increased by $(1.053)^2 = 1.109$, i.e., by about 10.9 percent which results in an increase of its weight by approximately the same percentage, i.e., to a new weight of approximately 71 Kg. The MRI sections were converted into images involving 29 tissue types whose electrical properties can then be prescribed at the exposure frequency. The tissue types are fat, muscle, bone, cartilage, skin, brain, nerve, cerebrospinal fluid (CSF), intestine, spleen, pancreas, heart, blood, eye, eye humor, eye sclera, eye lens, ear, liver, kidney, lung, bladder, stomach, ligament, compact bone, testicle, spermatic cord, prostate gland, and erectile tissue. This model has, to date, been used to calculate the electromagnetic absorption in the human head, neck, and shoulders for cellular telephones operating at frequencies of 800–900 MHz. Because of the localized nature of electromagnetic fields, it was possible to use the model corresponding to the top 42 cms of the body for SAR calculations.

2. Currents Induced in the Human Body by the EMFs of Electric Blankets–Impedance Method

To illustrate the use of the impedance method, currents induced in the human body by the EMFs of electric blankets have been calculated.[87] Two blankets used for the calculations are: a Low-Magnetic-Field blanket, and a conventional (pre-1990) electric blanket. Typical wiring configurations of the two blankets are shown in Figure 24 parts a and b, respectively. An equivalent circuit of the positive temperature coefficient (PTC) twin-lead construction used for the Low-Magnetic-Field blanket is shown in Figure 24c. As seen from Figures 24a and 24c, the Low-Magnetic-Field blanket uses two parallel leads carrying equal and opposite currents to reduce the net magnetic field around the conductors. The two leads are separated typically by 1.5 mm and are embedded in a PTC conductive polymer and insulated by PVC. As shown in Figure 24c, the PTC conductive polymer surrounding the two leads may be represented by a set of distributed resistors which would result in linearly decreasing equal but opposite currents flowing through the two leads over the length of the wiring used for the blanket. By comparison, the conventional electric blanket uses a resistive alloy wire wrapped on a nylon cord and insulated with PVC. Because of the distributed resistance of the wire, this blanket would therefore have a linearly diminishing voltage and identical magnitude of current over the length of the wiring used for the blanket.

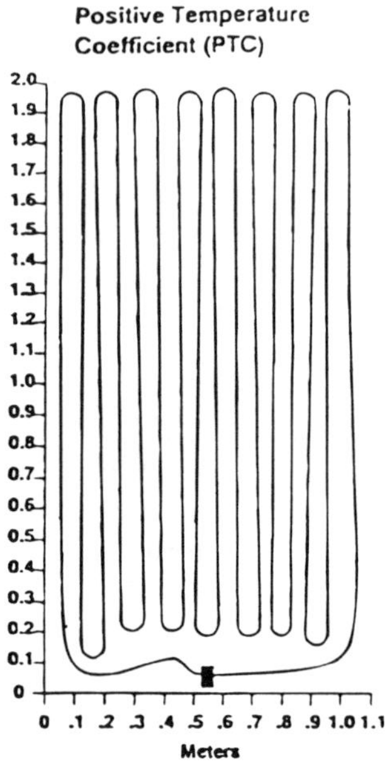

a. Low-Magnetic-Field Blanket.

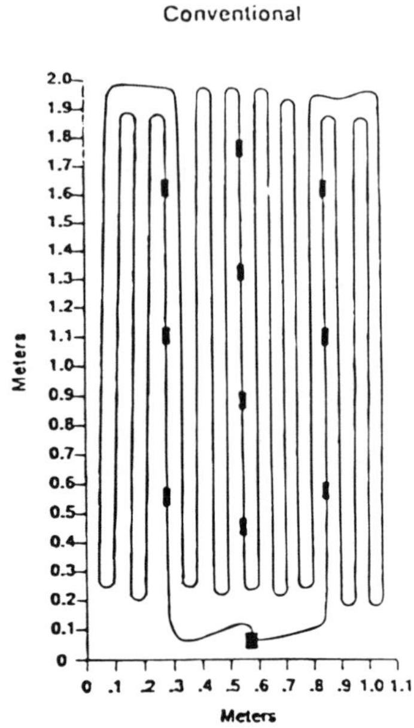

b. The conventional electric blanket.

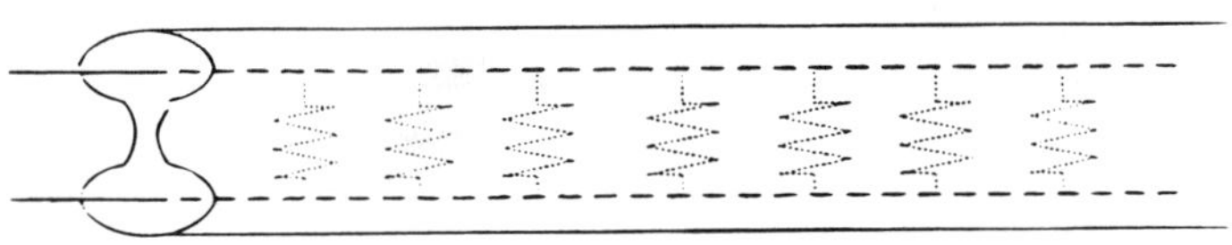

c. The twin-lead positive temperature coefficient (PTC) construction used for the low-magnetic-field blanket. Two parallel wires separated by 1.5 mm embedded in a PTC conductive polymer and insulated with PVC is used for the wiring.

FIGURE 24 Arrangements for the heating elements of electric blankets.

Florig et al.[88,89] and Hayashi et al.[90] have calculated the electric and magnetic fields produced by the conventional electric blankets. Simplified ellipsoidal and cylindrical models were used to represent the human body. The results given below are obtained using the 1.31-cm resolution, anatomically based model of the human body. It is recognized that the conductivities of various tissues, e.g., skeletal muscle, heart, bone, etc., are anisotropic for power-line frequencies.[46,47,91] This has, however, been neglected in the first instance and average values of conductivities given in Table 4 have been taken for the various tissues for the calculations in the first instance. Also, since the term $j \omega \epsilon$ can be neglected as compared to σ for the various tissues at power frequencies (50/60 Hz), the impedances for the various cells of the model given by Eq. 21 can be replaced by resistances $R_m^{i,j,k}$. Currents are induced in the body due to the following sources:

a. Time- and spatially varying magnetic fields of the blanket which result in the induced voltages or EMFs $-\frac{d}{dt} \int \mathbf{B} \cdot ds$ for the various resistance loops that represent the equivalent impedance model of the body.

TABLE 4
Tissue Conductivities used for Calculations at the Power-line Frequency of 60 Hz.

Tissue Type	σ S/m
Air	0
Muscle	0.52 or 0.11*
Fat, Bone	0.04
Blood	0.6
Intestine	0.11
Cartilage	0.04
Liver	0.13
Kidney	0.16
Pancreas	0.11
Spleen	0.18
Lung**	0.04
Heart	0.11
Nerve, Brain	0.12
Skin	0.11
Eye	0.11

* See section N.B.5. (calculations using the Anatomically Based Models).
** We have used 33 percent lung tissue and 67 percent air for the dielectric properties of the lung.

b. Currents launched into different subareas at the body surface by means of the capacitively coupled currents from the various conductors of the blanket.

The Calculational Steps

The spatial variations of the magnetic fields were calculated from the Biot-Savart's law of electromagnetics using the configurations of the conductors shown in Figure 24. From Biot-Savart's law, the magnetic field due to a small element $d\ell$ of a conductor carrying current I at a distance R is given by

$$d\mathbf{H} = \frac{Idl \times \hat{R}}{4\pi R^2} \tag{22}$$

where $\hat{R}$ is a unit vector from the element dl to a point P where the magnetic field is to be calculated. By integrating Equation 22 over the entire length of the current-carrying conductors, one can obtain the vector magnetic fields at the centers of the cells representing the model of the human body or any of the other points in space.

From the vector magnetic fields thus calculated for each of the cell centers for the impedance model of the human body, the induced voltages or EMFs for each of the faces of the cells can be written.[44] This information is then used to calculate the induced currents for the various impedances (resistances at power frequencies). The average current densities J_x, J_y, and J_z for each of the cell centers can be obtained by taking the average of the currents through the resistances representing the four edges each of the cell in the respective directions and dividing them by the cross-sectional area δ^2 (= 1.31 × 1.31 cm^2).

To calculate the electric-field distribution in air, a three-dimensional impedance model consisting of capacitors representing the space between the various faces of the cells was used. For cubic cells of dimension $\delta = 1.31$ cm, the capacitances used are $\epsilon_0\,\delta^2/\delta = 0.116$ pF.

In order to calculate the currents induced in the human body due to electric fields, it should be recognized that the energized conductors of the blanket are capacitively coupled to the body. The capacitance between a given conductor and the highly conducting human body can be obtained by using an expression similar to that for a conductor at a distance S from the ground plane. Capacitance per unit length C_l of a wire of diameter d parallel to but separated a distance S from the ground plane is given by

$$C_l = \frac{2.73\epsilon_{eff}}{\log_{10}\left(\frac{4S}{d}\right)} \text{ pF/m} \tag{23}$$

For a spacing S = 5 mm and diameter d = 0.8 mm for the wire, and for $\epsilon_{eff} = 2.5\ \epsilon_0$, which is a value intermediate between the permittivity ϵ_0 for air and 4 ϵ_0 for the material of the blanket, we can calculate C_l = 43.1 pF/m. For a cell length $\delta = 1.31 \times 10^{-2}$ m, the coupling capacitance C_c between the wire and the cell can be calculated to be 0.565 pF. Since the interconductor spacing of 1.5 mm for a PTC blanket is fairly small as compared to the cell size, a proportionately smaller resistance is taken for the tissue-equivalent cells immediately underneath the conductors for the direction parallel to the interconductor spacing. Capacitances of 0.565 pF are taken from each of the conductors of the PTC blanket to the appropriate points on the impedance model of the human body. In the presence of an electrical grounding surface, the space underneath the model is represented by a 3D network of capacitors each of value 0.116 pF representing the air space between the various faces of the cubic cells of dimension δ = 1.31 cm for each of the sides.

For the PTC Low-Magnetic-Field blanket, a constant voltage of 110 V ac is taken between the conductors of the twin-lead wiring for calculation of currents induced/injected into the human body due to electric fields. For calculating the magnetic fields, an input current of 1 A is taken at the lower right-hand side of the blanket (see Figure 24a) close to the feet of the model. On account of the conductive polymer surrounding the parallel wires, this current diminishes linearly to zero at the end of the PTC wiring, which corresponds to the lower left-hand side of the blanket. This assumes a blanket input power of 110 W during operating conditions. If magnetic fields or induced current densities due to larger input powers are desired, the numbers calculated for 1 A input current may then be multiplied by the appropriate factor.

The conventional blanket, on the other hand, uses a resistive conductor (see Figure 24b) for which the voltage diminishes linearly from 110 V to zero over the length of the wiring. For this blanket the current throughout the length of the wiring is the same as at the input, i.e., 1 A that is assumed for the calculation of magnetic fields.

To verify the accuracy of the code, several test runs were made and are briefly described in the following.[87]

a. A configuration of periodically placed current-carrying conductors was used. The conductors were spaced 7 cm apart, transversing a cylindrical model of the human body assumed previously by Hayashi et al.[90]
   The calculated periodically varying components B_z, B_r of the field are in excellent agreement with the data given by Hayashi et al.[90]
b. For electric-field calculations, an arrangement of the current-carrying conductors of a conventional blanket used by Florig et al.[88] was adopted. Here, too, the variations of the calculated electric fields at the surface of a high-conductivity ($\sigma \geq 20$ S/m) grounded cylindrical model of the human body are in good agreement with Florig et al.'s paper.[89] Similar to these authors, it was found that the most intense surface fields (~6.5 KV/m)

TABLE 5
Comparison of the Calculated and Measured Magnetic and Electric Fields Close to a Flat Blanket. A Distance of 5 cm is Taken from the Conductor Layouts Sketched in Figs 24a and 24b.

	Low Magnetic Field Blanket		
	For this blanket, the magnetic-field results are normalized for a blanket input current of 1.227 A, i.e., a power input of 135 W.		
	Magnetic Field (μT)		
	Grid Size cm	Calculated	Measured
Average	1.31 × 1.31	0.056	—
	10.5 × 10.5	0.072	0.09
Peak	N.A.	0.20	0.26
	Electric Field (V/m)		
Average	1.31 × 1.31	103.7	—
	10.5 × 10.5	144.6	111.2
Peak	N.A.	159.7	176.0
	Conventional Electric Blanket		
	For this blanket, an input current of 1 A is assumed.		
	Magnetic Field (μT)		
	Grid Size cm	Calculated	Measured
Average	1.31 × 1.31	2.16	—
	10.5 × 10.5	2.45	2.18
Peak	N.A.	3.52	3.94
	Electric Field (V/m)		
Average	1.31 × 1.31	57.3	—
	10.5 × 10.5	70.1	95.4
Peak	N.A.	176.1	167.2

occur when the blanket user is grounded, the current is off, and all sections of the heating element are at the potential of 115 V.

Magnetic and Electric Fields Close to A Flat Blanket: Comparison With Experimental Data

To compare with measured data from Casamento et al.,[92] the magnetic (**H**) and electric (**E**) fields were calculated, both for the Low-Magnetic-Field blanket (Figure 24a) and the conventional blanket (Figure 24b) for an xy plane at a distance of 5 cm parallel to the surface of the blanket when these blankets were assumed to be flat. For numerical calculations, a grid size of 1.31 × 1.31 cm, and also a coarser grid that was 8 times larger and of dimensions 10.48 × 10.48 cm were used. The coarser grid was taken to compare the calculated data with the measured values that were obtained by Casamento et al.[92] for a grid of dimensions 10 × 10 cm. Since the magnetic-field probe diameter of about 3.2 cm used by Casamento et al. is not really negligible as compared to the blanket-to-grid separation of 5 cm, magnetic fields were calculated for xy planes at distances of 3.4 and 6.6 cm (5 ± 1.6 cm), respectively, and the averaged data are given in Table 5.

For electric-field measurements by Casamento et al.,[92] a metallic grounded plate was used 1 m above the blanket. Therefore, a grounded plate was also used for calculation of the electric fields. The space immediately above the blanket conductors was divided into cubic cells of dimension 1.31 cm to about half of the volume between the blanket conductors and the grounded plate. As mentioned, this region was represented by means of an impedance model consisting of capacitors of value $\epsilon_0\delta^2/\delta = 0.116$ pF. The remainder of the volume up to the ground plate was represented by the equivalent capacitances from the individual nodes of this network. These capacitances are given by the equation $\epsilon_0 \delta^2/w$, where w is the thickness for the remainder of the volume from the boundary of the impedance network to the metal plate. Since modeling of the region close to the blanket conductors is of primary significance, it was found that the results were nearly the same for different values of w as long as w did not exceed 0.5 m, i.e., half of the space between the blanket and the ground plane.

The average and peak H- and E-fields are given in Table 5 for both of these blankets. The calculated E- and H- fields are in good agreement with the values that have been determined experimentally.[92]

Induced Current Densities for Magnetic Fields of the Blankets

The 45,024–cell, 1.31-cm resolution anatomically based model of the human body described in Section IV.B,1. was used to calculate the induced current densities due to the magnetic fields of both the Low-Magnetic-Field blanket (Figure 24a) and the conventional blanket (Figure 26b). For these calculations, the wiring of the blanket was taken to be 0.5 cm from the surface of the body. By prescribing the loop EMFs for the various cells of the impedance model due to the time-varying magnetic fields at the cell centers, the currents for the various elements could be calculated, and hence the current densities for the various regions of the model. Induced current density vectors **J** have thus been calculated for all 45,024 cells of the body. Due to lack of space, only the section-averaged magnitudes of the total current densities $J_T = (J_x^2 + J_y^2 + J_z^2)^{1/2}$ are given for the various sections of the body, from head to feet for the two types of blankets in Figures 25a and 25b, respectively. Nearly identical current densities were also obtained for a grounding plate underneath the body at distances of 0.25, 0.5, and 1.0 m, respectively. It is interesting to note that the induced current densities are larger by a factor of about 500 for the conventional blanket *vis à vis* those for the Low-Magnetic-Field blanket.

Induced Current Densities for Electric Fields of the Blankets

The calculated section-averaged magnitudes of the total current densities due to electric fields of the blankets in the absence of a grounded plane are shown in Figures 26a and 26b, respectively. It should be noted that while the current densities induced by the electric fields of a Low-Magnetic-Field blanket (Figure 26a) are considerably higher than those due to magnetic fields (Figure 25a), the converse is true for a conventional blanket. For this blanket, the current densities induced by the magnetic fields (Figure 25b) are higher than those due to electric fields (Figure 26b). In fact, while the current densities due to magnetic fields are fairly small for a Low-Magnetic Field blanket as compared to those for a conventional blanket, the current densities due to electric fields of a Low-Magnetic Field blanket are even higher than those for a conventional blanket (see Figures 26a and 26b). The reasons for these observations can be seen from the values of magnetic and electric fields given in Table 5 for the two types of blankets. While fairly small magnetic fields are calculated for the Low-Magnetic-Field blanket as compared to those for the conventional blanket, the converse is true for the electric fields created by these blankets. As seen in Table 5, somewhat higher electric fields are created by the Low-Magnetic-Field blanket as compared to those for the conventional blanket. This is likely due to the higher potential difference between the twin-lead conductors that are used for the Low-Magnetic-Field blankets.

Different spacings d_g of 0.25, 0.5 and 1.0 m from the bottom of the body to a grounded metallic plate were also used to understand the effect of a grounded plate underneath the model of the body.

3. SARs From Cubic Cell Models–MoM Method

In this section we shall briefly summarize the volume integral equation MoM method for field computation and present some results obtained for plane wave exposures.

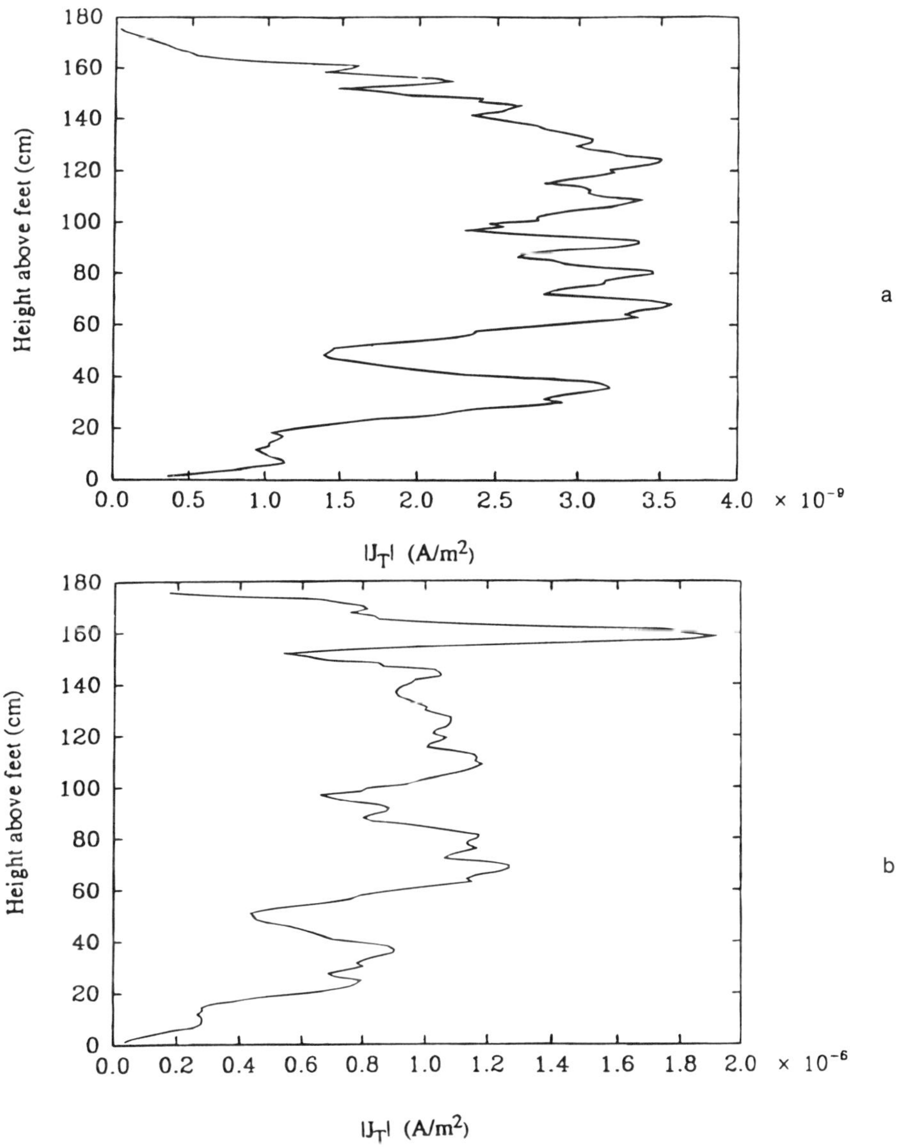

FIGURE 25 Section-averaged magnitudes of total current densities for the various sections of the body for magnetic fields of the blankets. Nearly identical current densities were obtained also for a grounding plate at distances of 0.25, 0.5, and 1.0 m underneath the body. Input current = 1 A. a—low magnetic field blanket, b—conventional electric blanket.

The basic equation for the total induced electric field $\mathbf{E}(r)$ inside a biological body of volume V with conductivity $\tau(r')$, and absolute dielectric permittivity $\epsilon(r')$, and an incident electric field $\mathbf{E}^i(r)$ is given by the equation

$$\left[1 + \frac{\tau(r)}{3j\omega_0\epsilon_0}\right]\mathbf{E}(r) - PV\int_V \tau(r')\mathbf{E}(r')\cdot\mathbf{G}(r, r')dv' = \mathbf{E}^i(r) \quad (24)$$

with the Green's function,

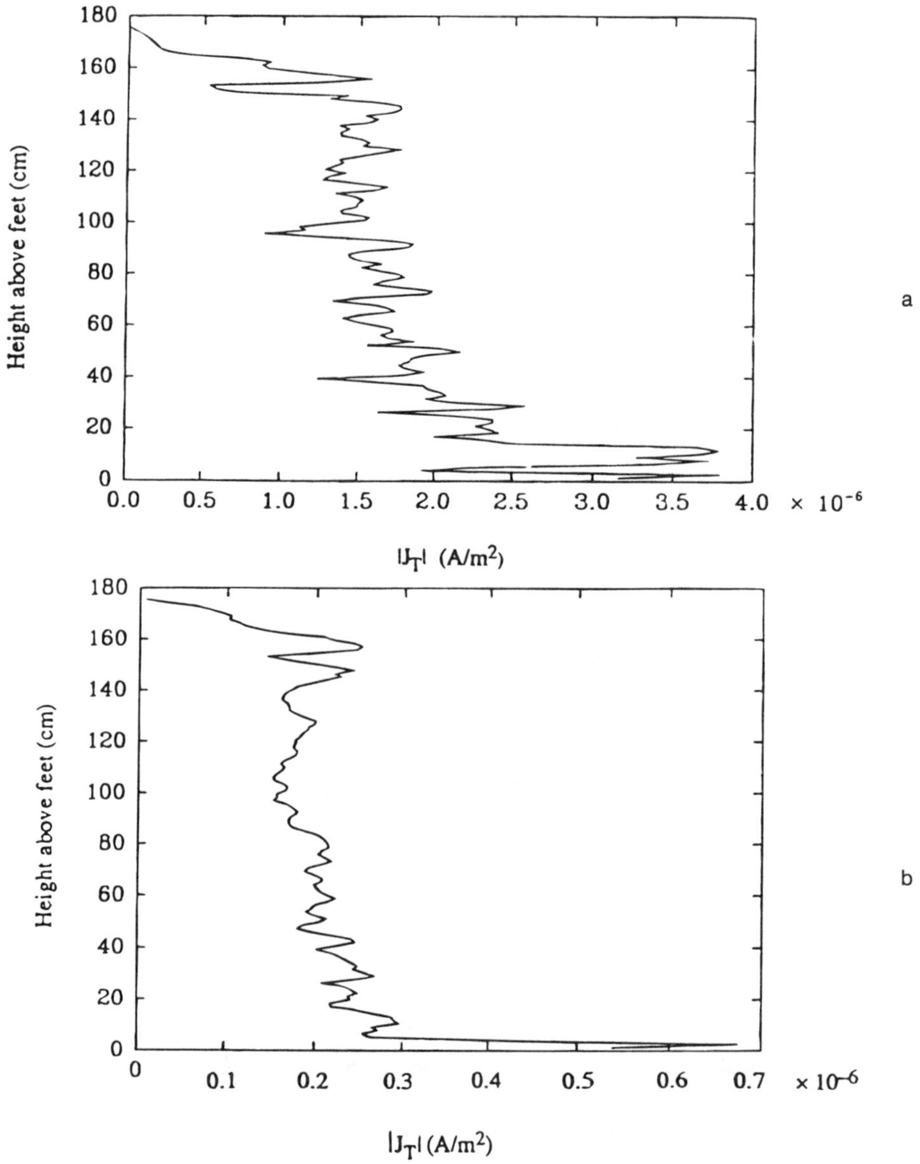

FIGURE 26 Section-averaged magnitudes of total current densities for the various sections of the body for electric fields of the blankets. No ground plane underneath the body. a—Low magnetic field blanket, b—Conventional electric blanket.

$$\mathbf{G}(r, r') = j\omega\eta_0\left(\mathbf{I} + \frac{1}{k_0^2}\nabla\nabla\right)\frac{\exp(jkl_0|r - r'|)}{4\pi|r - r'|} \tag{25}$$

where $\tau(r) = \sigma(r) + j\omega[\epsilon(r) - \epsilon_0]$, r is the vector from the origin of coordinates to a given point, r′ is the vector from the origin to the differential volume element dv′, **I** is the unity dyad, k_0 is the free space wave number, ∇ signifies the gradient operation, and PV symbolizes the principal value of the integral.

According to the MoM method, the body may be partitioned into N cubic sub-volumes or cells which are sufficiently small for the electric field and dielectric permittivity to be constant within each cell. The integral equation is then transformed into a system of 3N simultaneous linear equations for the three orthogonal components of the electric field at the center of each cell. The simultaneous equations may be written in matrix form as

$$[G][E] = -[E^i] \tag{26}$$

where [G] is a 3N × 3N matrix and [Ei] and [E] are column matrices representing incident and induced electric fields at the center of each cell. The elements of [G] can be evaluated as shown in References 34 and 37. In particular, the diagonal elements of the [G] matrix may be evaluated exactly by approximating each subvolume with a sphere of equal volume centered at the position of an interior point. If the actual shape of the cell differs appreciably from that of a sphere, this approximation may lead to unsatisfactory numerical results.[37] In such cases, a small cylindrical volume may be created around an interior point. It may also be necessary to evaluate these terms by numerical integration throughout the cubic sub volume for increased accuracy. The evaluation of off-diagonal elements of the [G] matrix is considerably simplified since it does not involve principal value operations. Therefore, for a given applied field configuration, the induced electric fields inside the body are obtained by matrix inversion. That is,

$$[E] = -[G]^{-1}\,[E^i] \tag{27}$$

Several computer programs have been developed to implement this numerical procedure. Factors that influence the computational accuracy include frequency, body size, cell dimensions, and computer memory. It has been found that reliable numerical results can be obtained if the linear dimensions of the cell do not exceed a quarter free-space wavelength.[34] For a computer with sufficient capacity to invert a 120 × 120 matrix, the maximum number of cells is limited to 40. If we assume, for simplicity, symmetries between the right and the left half and the front and the back of a 1.7-m-tall adult human body, this computer would handle approximately a cell size around $10^{-3}\ -\text{m}^3$. Once the $10^{-3}\ -\text{m}^3$ cell size is adopted, 750 MHz would be the highest frequency that can be considered for field intensity calculation without violating the criterion that the linear dimension of the cell not exceed a quarter free-space wavelength.

The computational resources necessary to obtain even a regional SAR using this MoM approach are quite extensive. A relatively full complex matrix, 3N × 3N in dimensions, is required for a model with N cells. The computation time required for a noniterative solution of the matrix equation is therefore proportional to between N^2 and N^3, which increases rapidly as N increases. The faithfulness with which a cubic cell model approximates the detailed structure of a biological body and the maximum usable frequency increases with the number of cells. In fact, substantial errors will occur if

$$N \lesssim \frac{L}{\lambda'}\frac{2\pi}{\sqrt{6}} \tag{28}$$

where $\frac{L}{\lambda'}$ is the ratio between the linear dimension of the body and the wavelength in the body.

The accuracy of the numerical method can be verified by comparison with known results from exact analytic solutions based on well-characterized geometric bodies, such as spheres. It should be noted that perfect agreement between the exact solution, based on Mie theory, and the numerical method, based on the volume integral equation, is not expected unless a

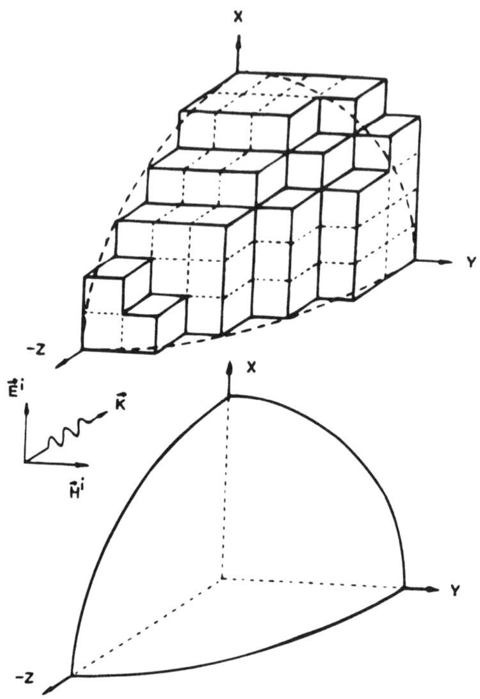

FIGURE 27 The approximation of one-eighth of a sphere by an equivalent cubic-cell-formed structure.

large number of cubic cells are used to simulate the sphere. Figure 27 shows one eighth of a sphere approximated by one eighth of a "cubic model of a sphere," which is constructed from 73 cubic cells. Clearly, a better approximation can be achieved by a larger number of smaller cubic cells. Nevertheless, for a brain sphere constructed from 40 cubic cells at a frequency of 918 MHz, the computed maximum field intensity deviated from the exact solution by less than 9%.[93]

A model of a human body consisting of 180 cubic cells that accounts for the anatomic and biometric characteristics of human beings is shown in Figure 28. The model is 1.75 m tall and can be made either homogeneous or inhomogeneous by using an equivalent or a volume-weighted complex permittivity for each cell.[94] The distributions of absorbed energy inside the model in free space at 80 MHz (near resonance) for homogeneous and inhomogeneous tissue properties are illustrated in Figures 29 and 30, respectively. The electric field vector is oriented along the height of the model and the plane wave propagates from front to back of the body. It is interesting to observe that the absorbed energy in the upper torso is higher in the middle and back layers than the front layer on which the plane wave impinges; the highest absorption in the transverse dimension occurs generally near the center of the torso. Energy deposition along the height of the body shows several peaks which appear in the neck, the thoracic region, the pelvic/thigh region, and the calf, for the homogeneous model. The SARs for 10 W/m^2 of incident power density are 86 W/kg for the pelvic/thigh region; 79 W/kg for the calf; 47 W/kg for the thorax; and 34 W/kg for the neck. It should be noted that the distribution of absorbed energy in the homogeneous model is close to that measured using scaled models.[37,38]

The effect of inhomogeneity produced by volume-weighted cell complex permittivity is mostly revealed in the relative strength of absorbed energy in different regions of the body. The general distribution of the absorbed energy shows very little difference between homogeneous and inhomogeneous models. That is, enhanced absorptions occur in the same region,

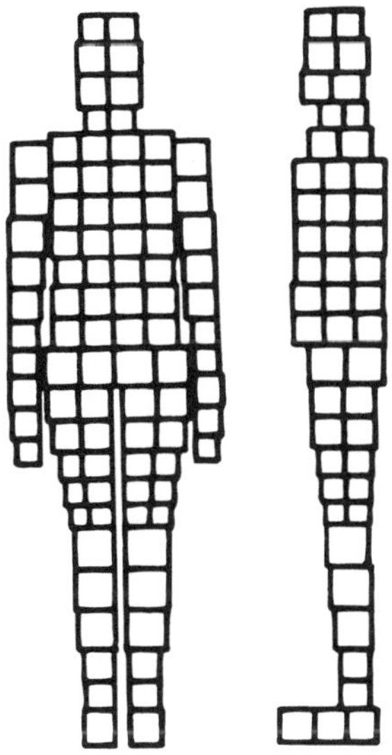

FIGURE 28 A cubic-cell representation of human body.

but the SARs in the pelvic/thigh and thoracic regions are now, respectively, 12 and 5.1 W/kg per W/m^2 incident power density. The SARs in the neck and the calf, however, are reduced to 2.8 and 6.8 W/kg, respectively. Since the heart and the pelvic/thigh regions are characterized by tissues with high water content, whereas the neck and the calf are composed largely of tissues with low water content, this suggests that energy absorption in the body is increased in regions with larger quantity of muscular tissue and is decreased in regions that consist mainly of bony tissues. Changes in coupling characteristics that stem from the inhomogeneity, however, may also have contributed to the difference in absorbed energy distribution.

The average absorption or whole-body SAR for the model of the human body shown in Figure 28 as a function of frequency is illustrated in Figure 31. Again, the electric field vector is along the height of the body and the plane wave propagates from front to back of the model with an incident power density of 10 W/m^2. A homogeneous complex permittivity approximately two thirds of that for muscle is used in the calculations. Note that the whole-body SAR increases with frequency until it reaches a maximum of about 0.23 W/kg at 77 MHz (resonance frequency) and then decreases as l/frequency. The experimental data shown in Figure 32 are obtained from a saline-filled scale model of the human body. It can be seen that the calculated absorption is in good agreement with that found experimentally,[38,95] except for the resonant frequency which is somewhat lower (70 MHz) in the experimental case. It should be mentioned that whole body SAR given in Figure 31 is typically within 10% of that estimated from prolate spheroidal models of the same height and dielectric property. Further, when' inhomogeneous complex permittivities are used with the model, the whole-body SAR changes less than 2% from that depicted in Figure 31. Thus, if one is primarily concerned with average absorption over the body, a homogeneous prolate spheroidal model may be quite adequate.

While MoM methods based on the volume integral equation have been a very useful numerical procedure for computation of SAR distribution in complex tissue geometries, the requirement of a full 3N × 3N matrix presents severe limitations. The computation times required to provide even regional SAR distribution of sufficient resolution to delineate the resonant frequency for the head region are enormous. A minimum of 340 cells was needed, increasing the computation time by a factor of four over the 180-cell models.[95,96]

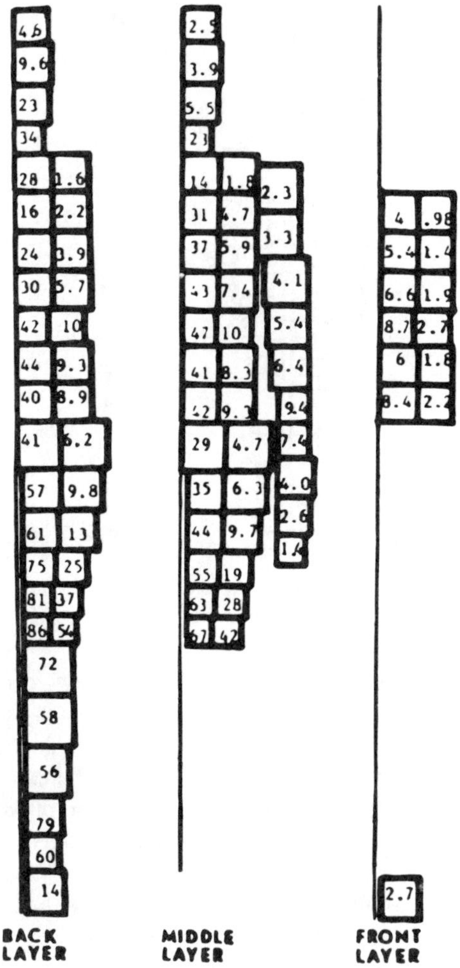

FIGURE 29 SARs for a homogeneous 180-cell model of the human body exposed to vertically polarized, 80 MHz plane wave. The values are expressed as W/kg/mW/cm^2 × 100.

Matrix inversion operations consume the largest block of time in moment-method solutions for the cubic cell models. The computer time required is proportional to the cube of the number of cells. However, the matrix generated is usually diagonally dominant and well conditioned. For a human-size body, iterative procedures for matrix inversion are practical at frequencies below about 60 MHz. The convergence rate decreases with increasing frequency and fails above 90 MHz. This is most likely caused by the decrease in degree of diagonal dominance with increasing frequency. A number of approaches have been investigated to alleviate this difficulty. A semi-iterative procedure called Band Approximation Method appears to be an efficient algorithm that can be profitably used to invert large matrices generated by the number of cells for human models at arbitrarily high frequencies and converges significantly faster than standard iterative algorithms.[96]

4. Field Intensity Computations Using the FDTD Method

One of the most successful methods for SAR calculations is the FDTD method. This method was first proposed by Yee[72] and later developed by Taflove and colleagues;[74,78,79] Holland,[97] and Kunz and Lee.[98] A recent book by Kunz and Luebbers[73] is devoted to the

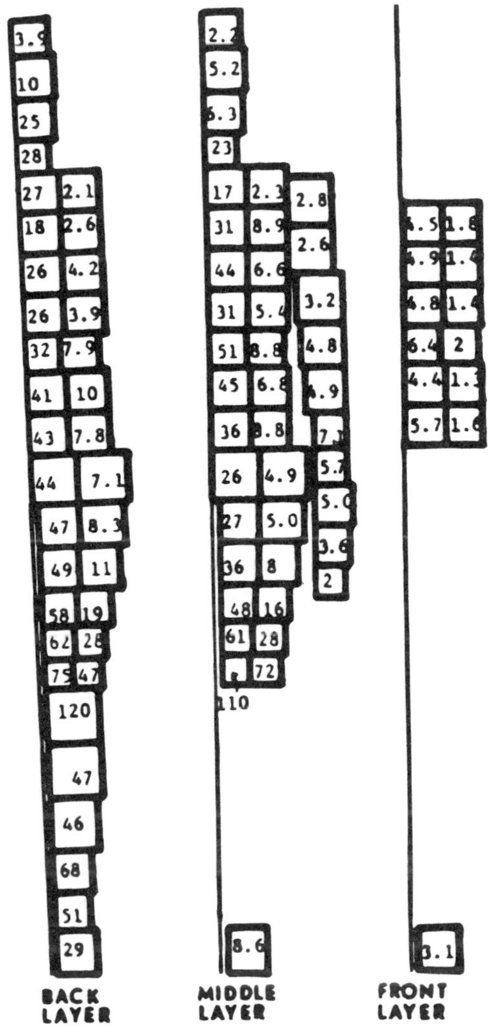

FIGURE 30 SARs for an inhomogeneous 180-cell model of the human body exposed to vertically polarized, 80-MHz plane wave. The values are expressed as W/kg/mW/cm^2 × 100.

FDTD method and some of its applications. For bioelectromagnetic applications the FDTD method has been found to be extremely versatile and has been used for whole-body or partial-body exposures due to spatially uniform or nonuniform fields (far- or near- fields), sinusoidally varying electromagnetic fields, and for transient fields such as those of an electromagnetic pulse (EMP).[76,77,84,99,100–104]

The Traditional FDTD Method

In this method, the time-dependent Maxwell curl equations

$$\nabla \times \mathbf{E} = -\mu \frac{\partial \mathbf{H}}{\partial t} \tag{29}$$

$$\nabla \times \mathbf{H} = \sigma \mathbf{E} + \epsilon \frac{\partial \mathbf{E}}{\partial t} \tag{30}$$

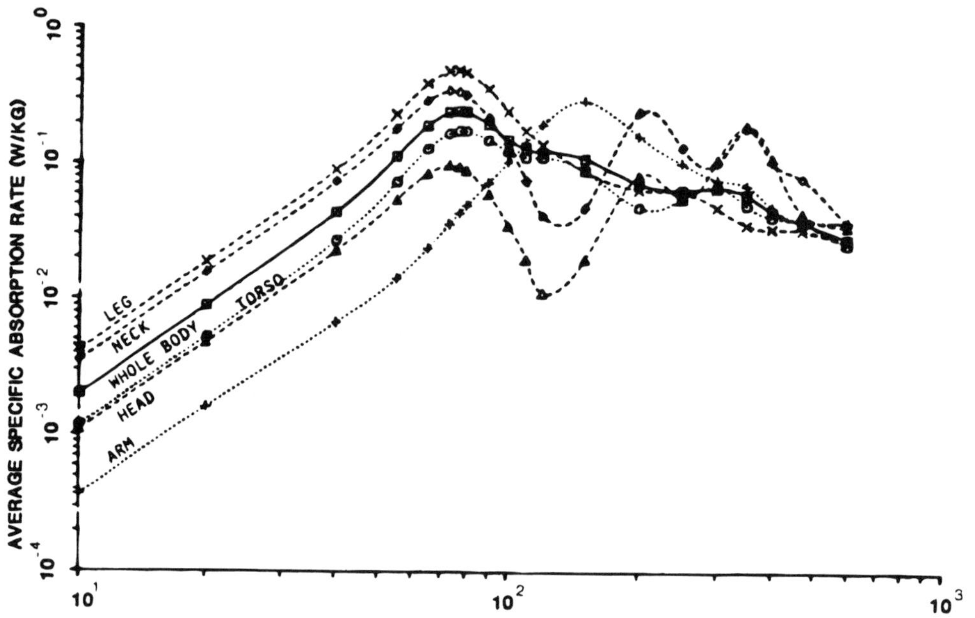

FIGURE 31 Average SARs for a homogeneous 180-cell model of the human body exposed to vertically polarized, 80-MHz plane wave. The incident power density is 10 mW/cm^2.

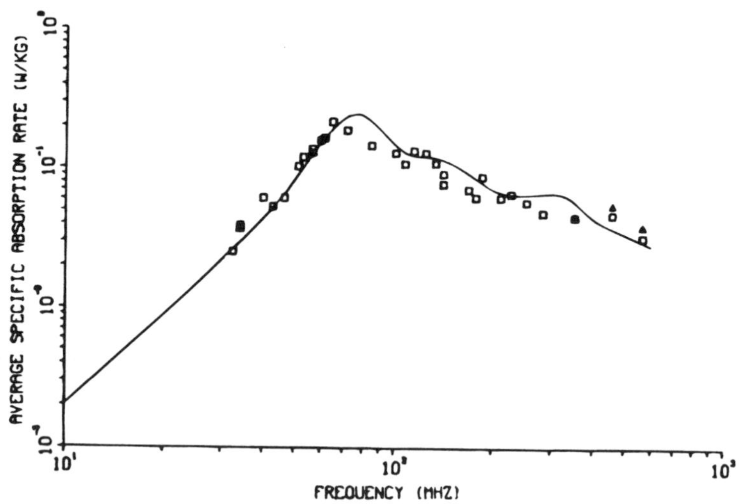

FIGURE 32 Whole-body averaged absorption for a homogeneous cubic-cell model of humans exposed to vertically polarized plane in free space. The incident power density is 10 mW/cm^2.

are implemented for a lattice of subvolumes or Yee "cells" that may be cubical or parallelepiped with different dimensions δ_x, δ_y, and δ_z in x-, y-, or z-directions, respectively. The components of **E** and **H** are positioned about each of the cells as shown in Figure 33 and calculated alternately with half-time steps where the time step $\delta t = \delta/2c$. Here δ is the smallest of the dimensions used for any of the cells and c is the maximum phase velocity of the fields in the modeled space. Since some of the modeled volume is air, c corresponds to the velocity of EM waves in air.

In the FDTD method, it is necessary to represent not only the scatterer/absorber such as the human body or a part thereof, but also the EM sources, including their shapes, excitations,

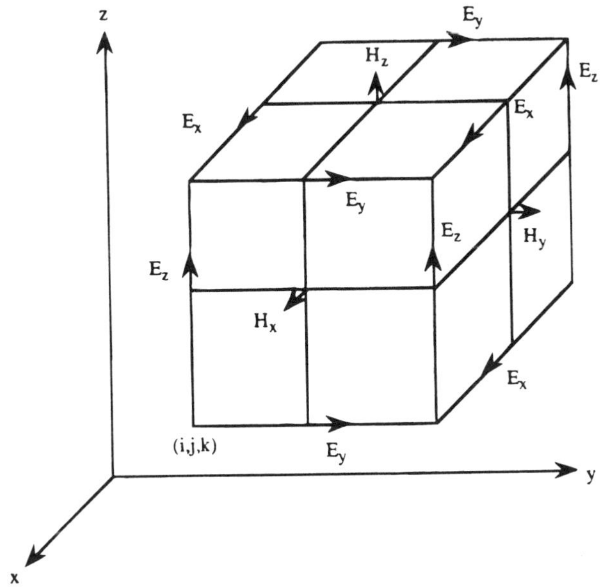

FIGURE 33 Unit cell of Yee lattice indicating positions for various field components.

etc., if these sources are in the near-field region.[102,103] The far-field sources, on the other hand, are described by means of incident plane-wave fields prescribed for a "source" plane,[100,101,76,84,104] typically 6–10 cells away from the exposed body. The source-body interaction volume is subdivided into Yee cells of the type shown in Figure 33. The interaction space consisting of several hundred thousand to a few million cells is truncated by means of absorbing boundaries. The prescribed incident fields are tracked in time for all cells of the interaction space. The solution is considered completed when either the fields have died off or, for sinusoidal excitation, when a sinusoidal steady-state behavior for **E** and **H** is observed for the interaction space.

The Frequency-Dependent FDTD Method

The frequency-dependent FDTD or $(FD)^2TD$ method is needed for short pulses where wide bandwidths are generally involved. Two general approaches have been used for the $(FD)^2TD$ method. One approach is to convert the complex permittivity from the frequency domain to the time domain and convolve this with the time-domain electric fields to obtain time-domain fields for the dispersive material. This discrete time-domain method may be updated recursively for some rational forms of complex permittivity, which removes the need to store the time history of the fields and makes the method feasible. This method has been applied to materials such as water for which the permittivity may be described by a first-order Debye relaxation equation[105–107] or more complex materials with dielectric properties given by a second-order Lorentz equation with multiple poles.[108]

A second approach is to add a differential equation relating the electric flux density **D** to the electric field **E** and solve this new equation simultaneously with the standard FDTD equations. This method has been applied to 1D and 2D examples with materials described by a first-order Debye equation or second-order single-pole Lorentz equation,[109] and to 3D sphere and homogeneous two-thirds muscle-equivalent man models with properties described by a second-order Debye equation.[110,111] In the following we describe this differential equation approach, which has now been used for induced current and SAR calculations for a heterogeneous model of the human body.[112]

The time-dependent Maxwell's curl equations used for the FDTD method have already been given as Equations 29 and 30. The curl **H** can also be written as follows:

$$\nabla \times \mathbf{H} = \frac{\partial \mathbf{D}}{\partial t} \tag{31}$$

where the flux density vector **D** is related to the electric field through the complex permittivity $\epsilon^*(\omega)$ of the local tissue by the following equation:

$$\mathbf{D} = \epsilon^*(\omega)\mathbf{E} \tag{32}$$

Since Equations 29 and 31 are to be solved iteratively in the time domain, Equation 32 must also be expressed in the time domain. This may be done by choosing a rational function for $\epsilon^*(\omega)$ such as the Debye equation with two relaxation constants (see Chapter 1):

$$\epsilon^*(\omega) = \epsilon_0\left[\epsilon_\infty + \frac{\epsilon_{s1} - \epsilon_\infty}{1 + j\omega\tau_1} + \frac{\epsilon_{s2} - \epsilon_\infty}{1 + j\omega\tau_2}\right] \tag{33}$$

Rearranging Equation 33 and substituting in Eq. 32 gives

$$\mathbf{D}(\omega) = \epsilon^*(\omega)\mathbf{E}(\omega) = \epsilon_0 \frac{\epsilon_s + j\omega(\epsilon_{s1}\tau_2 + \epsilon_{s2}\tau_1) - \omega^2\tau_1\tau_2\epsilon_\infty}{1 + j\omega(\tau_1 + \tau_2) - \omega^2\tau_1\tau_2}\mathbf{E}(\omega) \tag{34}$$

where the dc (zero frequency) dielectric constant is given by

$$\epsilon_s = \epsilon_{s1} + \epsilon_{s2} - \epsilon_\infty \tag{35}$$

Assuming $e^{j\omega t}$ time dependence Equation 34 can be written as a differential equation in the time domain

$$\tau_1\tau_2\frac{\partial^2\mathbf{D}}{\partial t^2} + (\tau_1 + \tau_2)\frac{\partial \mathbf{D}}{\partial t} + \mathbf{D} = \epsilon_0\left[\epsilon_s\mathbf{E} + (\epsilon_{s1}\tau_2 + \epsilon_{s2}\tau_1)\frac{\partial \mathbf{E}}{\partial t} + \epsilon_\infty\tau_1\tau_2\frac{\partial^2\mathbf{E}}{\partial t^2}\right] \tag{36}$$

For the $(FD)^2TD$ method, Eqs. 29 and 31 need to be solved subject to Equation 36. These equations can be written in the difference form,[110,111] and solved to find **E**, **H**, and **D** at each cell location. The **E** → **H** → **D** loop is then repeated until the pulse has died off.

The FDTD algorithm has recently been used for the following applications.

a. For ultrashort pulses where extremely wide bandwidths (on the order of 1 GHz) are involved, the code has been modified to include frequency-dependent properties of the tissues using the best-fit two relaxation Debye equations.[111,112] This $(FD)^2TD$ method can also be used to determine SAR distributions at multiple frequencies based on a single numerical run with an impulse using the Fourier components at various frequencies.
b. The FDTD code has also been used to calculate SARs in the exposed parts of the human body due to RF magnetic fields that are used in magnetic resonance imaging (MRI). Newer techniques for MRI are leading to the use of higher static magnetic fields and correspondingly higher RF magnetic fields. Based on the solution of a complete set of Maxwell's equations, the FDTD method is usable at any frequency, including RF magnetic-field frequencies as high as about 200 MHz that are proposed for the MRIs of the future.[113]

c. In one recent application, the FDTD code has been adapted for calculations of internal **E** and **H** fields and induced current densities for human exposure to electromagnetic fields (EMFs) at power-line frequencies.[49] For these calculations, both the magnetic and electric incident fields have been considered and use is made of a scaled higher quasi-static frequency[114] on the order of 5–10 MHz to help reduce the computation time by over five orders of magnitude.
d. The FDTD code has also been used for calculations of SAR distributions in the human head, neck, and shoulders for exposure to electromagnetic fields of cellular telephones.[85] For this application, the new millimeter-resolution MRI-based model of the human body described in section IV B has been used.

5. Currents Induced in the Human Body by EMFs of Power Transmission Lines–FDTD Method

The FDTD method has been adapted for calculations of internal **E** and **H** fields and induced current densities for exposure to purely electric, purely magnetic, or combined electric and magnetic fields (EMFs) at power-line frequencies.[49] While recognizing that the conductivities of several tissues (skeletal muscle, bone, etc.) are highly anisotropic for power-line frequencies, however, this has been neglected in the first instance. They could be included in the future when models based on separate identification of these tissues are available.

Both sinusoidal and prescribed time-varying incident fields can be used with the FDTD procedure—hence the method is well-suited also for transient exposures that are often of interest at power-line-related frequencies. For sinusoidally varying fields, the solution is completed when a sinusoidal steady-state behavior for **E** and **H** fields is observed for each of the cells. For lossy biological bodies this typically takes a stepped time on the order of 3 to 4 time periods of oscillation. Since δt is fixed for a given cell size, a larger number of iterations is therefore needed at lower frequencies. Because of the very large number of iterations, the FDTD procedure would be clearly inapplicable for calculations at power-line frequencies were it not for the quasi-static nature of the coupling as previously pointed out by Kaune and Gillis[115] and Guy et al.[114] Using a logic similar to these authors, the fields outside the body depend not on the internal tissue properties, but only on the shape of the body as long as the quasi-static approximation is valid, i.e., the size of the body is a factor of 10 or more smaller than the wavelength, and $|\sigma + j\omega| >> \omega\epsilon_0$ where σ and ϵ are the conductivity and the permittivity of the tissues, respectively, $\omega = 2\pi f$ is the radian frequency, and ϵ_0 is the permittivity of the free space outside the body. Under these conditions, the electric fields in air are normal to the body surface and the internal tissue electric fields are given from the boundary conditions in terms of the fields outside

$$j\omega\epsilon_0\hat{n}\cdot\mathbf{E}_{air} = (\sigma + j\omega\epsilon)\hat{n}\cdot\mathbf{E}_{tissue} \tag{37}$$

A higher quasi-static frequency f′, of say, 5 to 10 MHz, may therefore be used for irradiation of the model and the internal fields E thus calculated may be scaled back to frequency f of interest, e.g., 60 Hz. Since in the FDTD method, one needs to calculate in the time domain until convergence is obtained (typically 3–4 time periods), this frequency scaling to 5–10 MHz for f reduces the needed number of iterations by over 5 orders of magnitude. From Equation 37 we can write

$$\begin{aligned}\mathbf{E}_{tissue}(f) &= \frac{\omega}{\omega'}\frac{(\sigma' + j\omega\epsilon')}{(\sigma + j\omega\epsilon)}\mathbf{E}'_{tissue}(f') \\ &\simeq \frac{f\sigma'}{f'\sigma}\mathbf{E}'_{tissue}(f')\end{aligned} \tag{38}$$

assuming that $\sigma + j\omega\epsilon \simeq \sigma$ at both f′ and f.

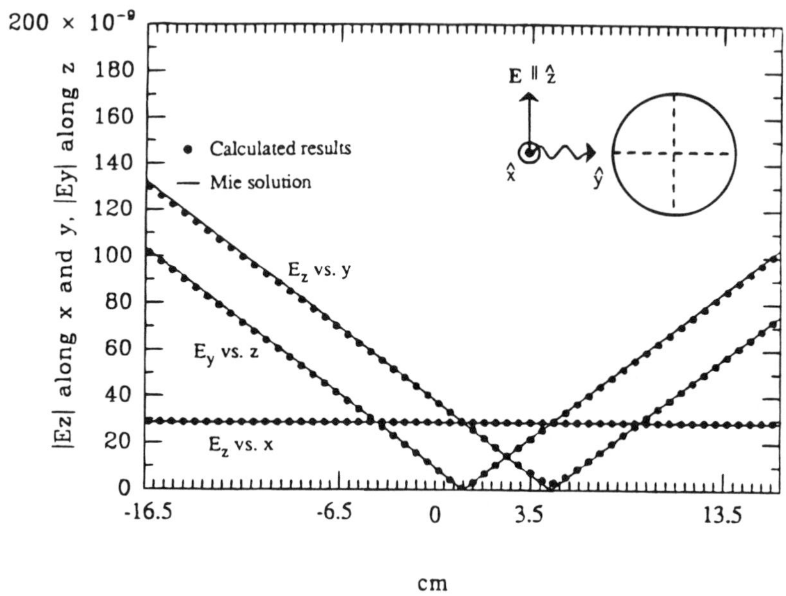

FIGURE 34 Comparison of the calculated internal E fields in V/m along the various central axes for a homogeneous sphere. Also shown are the values obtained from the Mie series solution. Assumed for the properties of the sphere are $\sigma = 0.35$ S/m, $\epsilon_r = 1$. A plane wave of $E_{inc} = 1$ V/m and $H_{inc} = 1/377$ A/m is assumed.

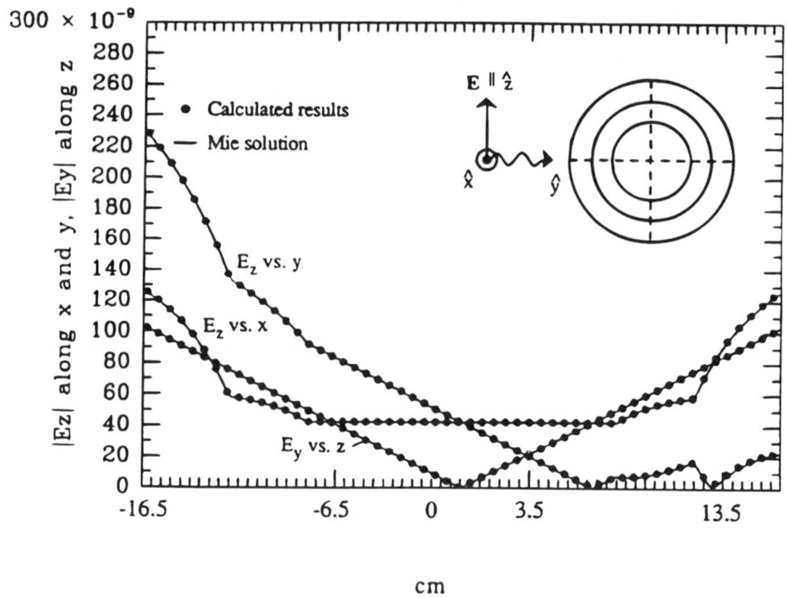

FIGURE 35 Comparison of the calculated E fields in V/m for a three-layer concentric sphere along the various central axes. Also shown for comparison are the values obtained from the Mie series solution. Properties assumed are $\sigma = 0.52$ S/m for the innermost sphere $0 < r < 8$ cm, $\sigma = 0.2$ S/m for the intermediate sphere $8 < r < 12$ cm, and $\sigma = 0.04$ S/m for the outermost sphere $12 < r < 16.5$ cm.

Test Cases

To verify the validity of using a higher frequency f′ to obtain induced E fields at ELF frequencies, test cases involving homogeneous and layered spheres have been used. The results for two representative test cases are given in Figures 34 and 35 for homogeneous and layered spheres, respectively. In Figure 34 the calculated results scaled to 60 Hz are shown for E_z vs. x and y and E_y along z for the central axes of the sphere. These results shown by dots were obtained using an irradiation frequency f′ of 20 MHz followed by scaling to 60 Hz using Equation 38. A homogeneous sphere of radius 16.5 cm and electrical conductivity $\sigma = 0.35$ S/m and dielectric constant $\epsilon_r = 1$ was assumed for the calculations. Since $\sigma >> \omega\epsilon$ at both 20 MHz and 60 Hz, choice of ϵ_r is relatively immaterial for the calculations. Assumed for the calculations is a plane wave of electric field $E_{inc} = 1$ V/m and magnetic field $H_{inc} = 1/377$ A/m. Also shown for comparison in Figure 34 are the results obtained using the Mie series solution.[116] The agreement can be seen to be excellent, lending support to the validity of using the scaled-frequency FDTD method to obtain induced E fields at power-line and other ELF frequencies.

The reason for a high degree of variation in the induced E-fields from the front to back of the sphere is that significant circular E-fields (in the yz plane) are also induced due to the incident magnetic fields. These fields add to the internal z-directed E-fields induced by the incident electric field ($E_z = 3\ E_{inc}/(2 + \sigma/j\omega\epsilon_0) \simeq 3\ j\omega\epsilon_0\ E_{inc}/\sigma$) for the front half of the sphere ($-16.5 < y < 0$ cm) and subtract from these fields for the back half of the sphere ($0 < y < 16.5$ cm). The net induced E-fields for a part of the back half of the sphere are, in fact, 180° out of phase (negative) because of the preponderance of the oppositely directed induced E field due to the incident magnetic field. In Figure 34, however, only the magnitudes of E_z and E_y are shown.

A second representative test case is that of a three-layer concentric sphere for which the assumed properties are as follows: $\sigma_1 = 0.52$ S/m corresponding to that of the skeletal muscle for the inner most sphere $0 < r < 8$ cm, a medium value $\sigma_2 = 0.2$ S/m for the intermediate sphere $8 < r < 12$ cm, and $\sigma_3 = 0.04$ S/m (corresponding to fat and bone) for the outer most sphere $12 < r < 16.5$ cm. For each of the spheres $\epsilon_r = 1$ since the choice of ϵ_r results in a negligible difference in calculated results ($\sigma >> \omega\epsilon_0$ for each of the layers). The results for the variations of the different field components along the three central axes of the layered sphere are shown in Figure 35 where the agreement with the results obtained using the Mie series solution can be seen again to be excellent. As for Figure 34, here too an irradiation frequency f′ of 20 MHz was used and the calculated results for E′ scaled back to obtain **E** at f = 60 Hz using Eq. 38.

Excellent agreement of the numerical results with the analytical results lends support to the validity of the FDTD method for calculating internal E-fields and current densities at power-line related frequencies. It should be noted that incident E- and H-fields of any orientation and relative magnitudes can be prescribed in the FDTD method allowing the possibility of calculations for realistic exposure conditions. Also the choice of a considerably higher frequency f′ of 5–20 MHz reduces the number of iterations needed to obtain converged results by 5–6 orders of magnitude as compared to those that would be needed at ELF frequencies of, say, 10–1000 Hz.

Calculations Using the Anatomically Based Models

As aforementioned, the FDTD method is versatile and will allow incident E and H fields of any orientation and relative magnitudes that need to be prescribed for source plane or planes surrounding the exposed body. For the results shown, the sixteen-tissue, 1.31-cm-resolution, anatomically based model of the human body was used (see section IV.B1) and a frequency f′ of 5–10 MHz to reduce the computation time. At the higher irradiation frequency f′, $\sigma' = \sigma$ was assumed, i.e., conductivities of the various tissues at 60 Hz. Furthermore, we have taken the incident E-field $\mathbf{E}_i(f') = 60\ \mathbf{E}_i(f)/f'$ to obtain $\mathbf{E}_{tissue}(f)$ at say $\mathbf{E}_i(f) = 10$ kV/m. The incident magnetic field $\mathbf{H}_i(f')$ has similarly been taken to be considerably lower ($= 60\ \mathbf{H}_i(f)/f'$) to account for the fact that the induced current densities and internal

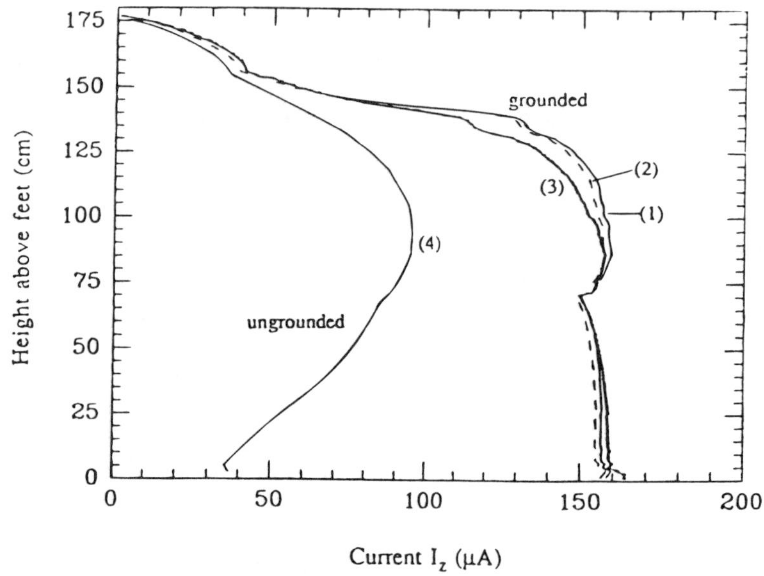

FIGURE 36 Calculated layer currents for anatomically based grounded and ungrounded models exposed to EMFs at 60 Hz. For curves (1) and (2) σ = 0.52 S/m for skeletal muscle and σ = 0.11 S/m for the interior muscle. For curves (3) and (4) σ = 0.11 S/m for all of the muscle. E = 10 kV/m (vertical), H = 26.5 A/m from side to side for all of the curves except for (2) for which only E-field exposure is assumed.

electric fields are proportional to the frequency of the incident fields and would therefore be higher at the assumed frequency f′.

In addition to the test cases using homogeneous and layered spheres, the results can be checked against the experimental data on a mannequin given by Deno[117] and the variations of the induced currents calculated along the height of the body by DiPlacido et al.[118] The agreement with the results of these two authors who had used a vertical electric field such as that under a high-voltage power line was found to be very good.[49]

Shown in Figure 36 are the calculated results using the anatomically based model where the conductivities used for the various tissues are as given in Table 4. Recognizing the anisotropy in the conductivity of the skeletal muscle, two different values of muscle conductivities are taken for curves (1) and (2). For these curves a higher conductivity of 0.52 S/m is taken for the skeletal muscle and an average value of 0.11 S/m is taken for the muscle in the interior of the body. For curves (3) and (4), however, a lower conductivity of 0.11 S/m is taken for all of the muscle, interior or skeletal. The results shown in Figure 36 are for exposures directly underneath a power line. Curves (1), (3), and (4) are for E_{inc} = 10 kV/m (vertical) and H_{inc} = 26.5 A/m (B_{inc} = 33.3 μT) from side to side of the model. To point out the preponderance of the induced currents due to the incident electric field, H_{inc} = 0 is assumed for the calculations shown in curve (2). It is interesting to note that the layer currents due to E-field exposure alone are almost 98–99 percent of the currents calculated for the combined electric and magnetic fields. It is also interesting to note that the calculated foot currents of 155–160 μA are in excellent agreement with 165 μA that would be projected from the measurements of Deno[117] for the human.

For the combined electric- and magnetic-field exposure condition of curve (1), the current density distributions for three representative sections through the brain, the heart, and the ankles are shown Figures 37a–c, respectively. For each of the sections the letter F denotes the front of the body. For comparison the current density distributions for the same sections due to incident magnetic field alone (H_{inc} = 26.5 A/m or B_{inc} = 33.3 μT) are shown in

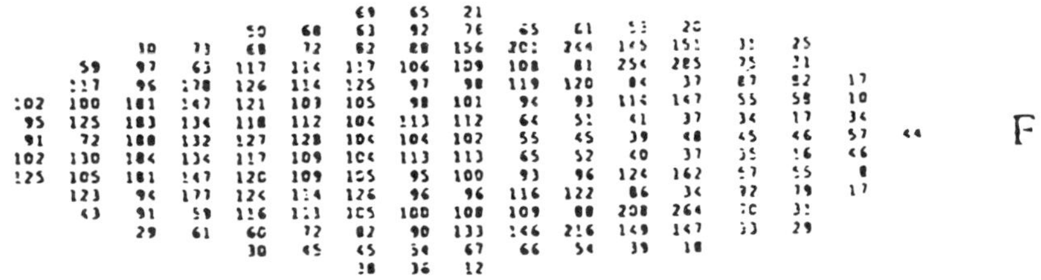

a. Section through the brain (height above ground = 167 cm).

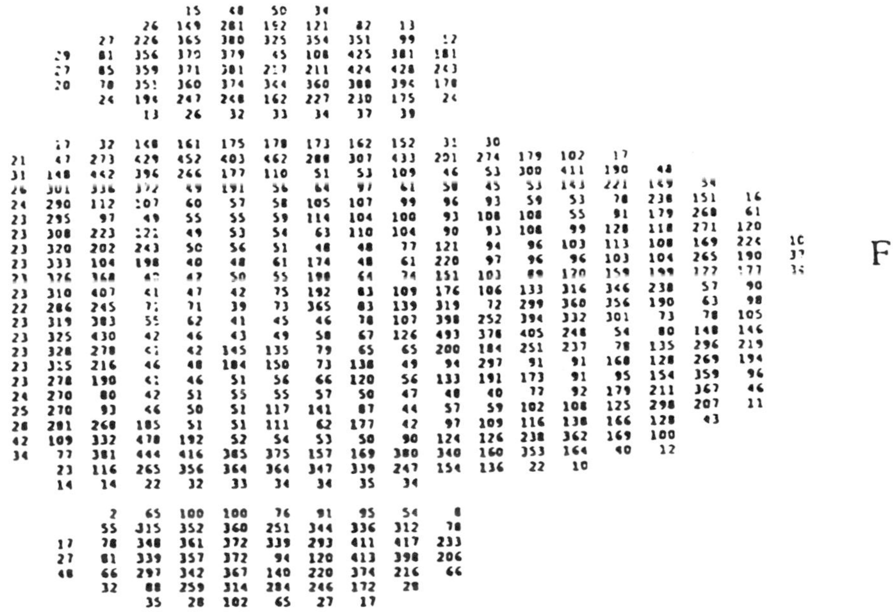

b. Section through the heart (height above ground = 133 cm).

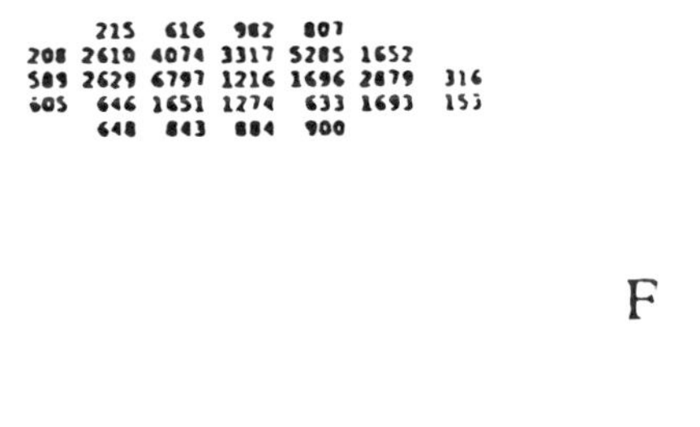

c. Section through the ankles (height above ground = 15.1 cm).

FIGURE 37 Calculated total current densities in nA/cm^2 for the various cells (δ = 1.31 cm) of a grounded anatomically based human model. Figures a–c are for combined E and H fields, E = 10 kV/m (vertical), H = 26.5 A/m (from side to side of the body), while Figs. d–f (next page) are for magnetic field alone.

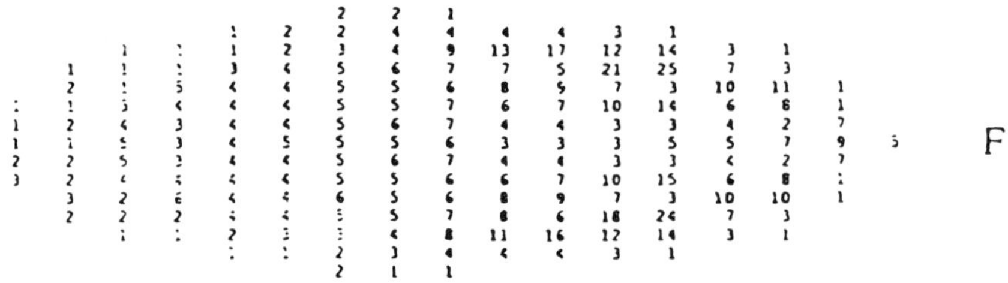

d. Section through the brain (height above ground = 167 cm).

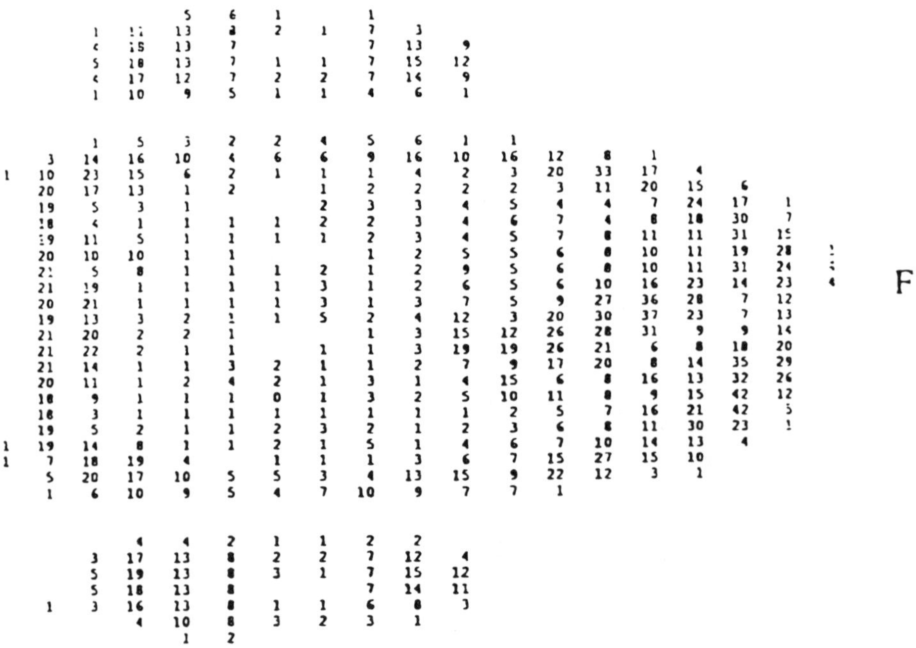

e. Section through the heart (height above ground = 133 cm).

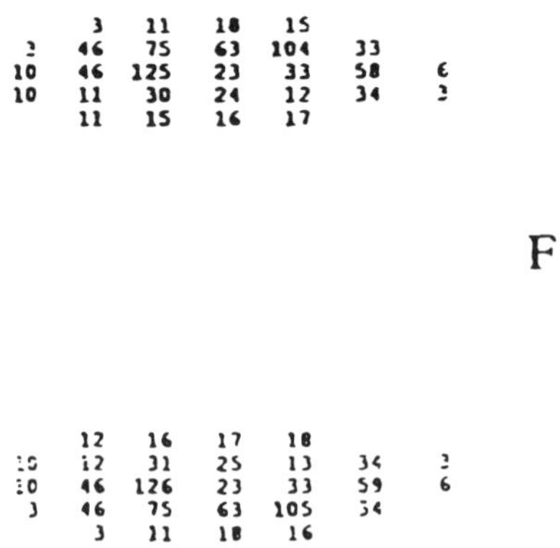

f. Section through the ankles (height above ground = 15.1 cm).

FIGURE 37 Continued

Figures 37d–f. It is interesting to note that considerably smaller current densities are obtained due to magnetic fields alone, as compared to the combined EMFs under the specified exposure conditions.

6. Coupling of Ultrawide Band Pulse to The Human Body—FDTD Method

Modeling of Tissue Properties with the Debye Equation

For ultrawide band calculations using the $(FD)^2TD$ method, the measured properties for the various tissues have been fitted to Equation 33 (the Debye equation) with two relaxation constants.[110–112] The measured properties of biological tissues (muscle, fat, bone, blood, intestine, cartilage, lung, kidney, pancreas, spleen, lung, heart, brain/nerve, skin, and eye) were obtained from reference 32. Optimum values for ϵ_{s1}, ϵ_{s2}, e_∞, τ_1, and τ_2 in Equation 33 were obtained by nonlinear least squares matching to the measured data for fat and muscle. All other tissues have properties falling roughly between these two types of tissue. Optimum values of ϵ_{s1}, ϵ_{s2}, and ϵ_∞ shown in Table 6 for all tissues were then obtained with τ_1 and τ_2 being the average of the optimized values for fat and muscle. This was done to facilitate volume averaging of the tissue properties in cells of the inhomogeneous man model. Having τ_1 and τ_2 constant for all tissues, allowed linear (volume) averaging of the ϵ values for each tissue in a given cell to calculate ϵ values for that cell. The measured tissue properties and those computed from the Debye equation with τ_1 and τ_2 being the average of fat and muscle are shown in Fig. 38 for fat and muscle. Similar comparisons can be also obtained for the other tissue types.

Induced Currents and Specific Absorptions

The $(FD)^2TD$ method has been used to calculate coupling of an ultrashort pulse to the heterogeneous model of the human body. From the calculated internal fields, the vertical currents passing through the various layers of the body are calculated, by using the following equation:

$$I_z(t) = \delta^2 \sum_{i,j} \frac{\partial D_z}{\partial t} \tag{39}$$

where δ is the cell size (= 1.31 cm), and the summation is carried out for all cells in a given

TABLE 6
Debye Constants for Tissues

$\tau_1 = 46.2 \times 10^{-9}$ s
$\tau_2 = 0.91 \times 10^{-10}$ s
(average of optimum for fat and muscle)

Tissue	ϵ_∞	ϵ_{s1}	ϵ_{s2}
Muscle	40.0	3948.	59.09
Bone/Cartilage	3.4	312.8	7.11
Blood	35.0	3563.	66.43
Intestine	39.0	4724.	66.09
Liver	36.3	2864.	57.12
Kidney	35.0	3332.	67.12
Pancreas/Spleen	10.0	3793.	73.91
1/3 Lung	10.0	1224.	13.06
Heart	38.5	4309.	54.58
Brain/Nerve	32.5	2064.	56.86
Skin	23.0	3399.	55.59
Eye	40.0	2191.	56.99

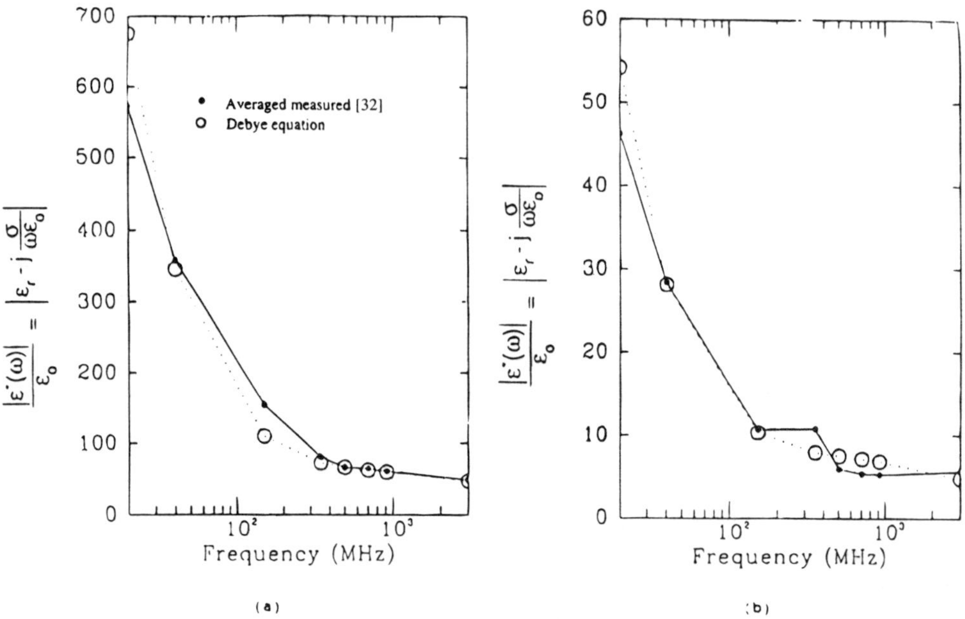

FIGURE 38 Fit of Debye equations with two relaxation constants (6) to measured tissue properties of (a) muscle and (b) fat.

layer. The layer-averaged absorbed energy density or specific absorption (SA) and the total energy W absorbed by the whole body can be calculated using the following relationships:

$$SA|_{\text{layer k}} = \frac{\delta t}{N_k} \sum_{i,j,t} \frac{\mathbf{E}(i, j, k, t)}{\rho(i, j, k)} \cdot \frac{\partial \mathbf{D}(i, j, k, t)}{\partial t} \tag{40}$$

$$W = \delta t \cdot \delta^3 \sum_{i,j,k,t} \mathbf{E}(i, j, k, t) \cdot \frac{\partial \mathbf{D}(i, j, k, t)}{\partial t} \tag{41}$$

In Equations 40 and 41, δt is the time step ($= \delta/2c = 0.02813$ ns) used for the time-domain calculations, N_k is the number of cells in layer k of the body, and $\rho(i, j, k)$ is the mass density in kg/m^3 for each of the cells in the corresponding layers.

A typical ultrawide band pulse with a prescribed peak amplitude of 1.1 V/m is shown in Figure 39 in the time domain. It is interesting to note that the pulse has a rise time of about 0.2 ns and a total duration of about 7–8 ns. The Fourier spectrum of the prescribed pulse is shown in Figure 40. Most of the energy in the pulse is concentrated in the 200–900 MHz band with the peak of the energy being at about 500 MHz.

For purposes of illustration, the results to follow, assume the incident fields to be vertically polarized, since this polarization is known to result in the strongest coupling for standing individuals. Also, a uniform plane wave illumination of the whole body is assumed by the incident fields. The (FD)^{2}TD procedure is used to calculate the temporal variations of total vertical currents for the various sections of the body, for both the shoe-wearing grounded, and ungrounded exposure conditions of the model, respectively. The current variations for two representative sections such as those through the eyes and the bladder are given in Figures 41a and 41b, respectively. The calculated peak currents for the various sections are on the order of 1.1 to 3.2 mA/(V/m). It is interesting to note that there is very little difference

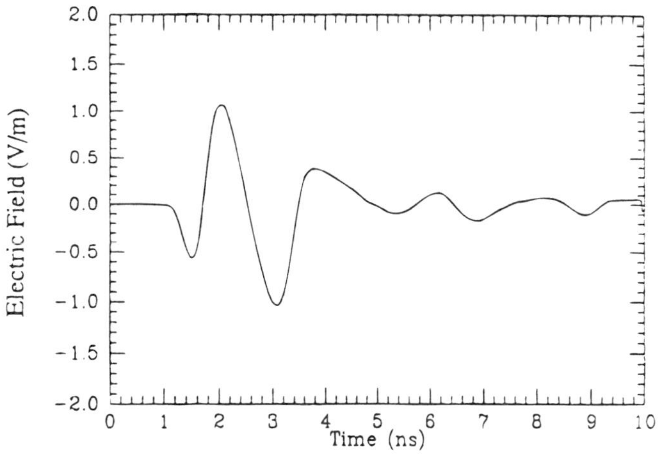

FIGURE 39 The prescribed electromagnetic pulse. Peak incident field = 1.1 V/m.

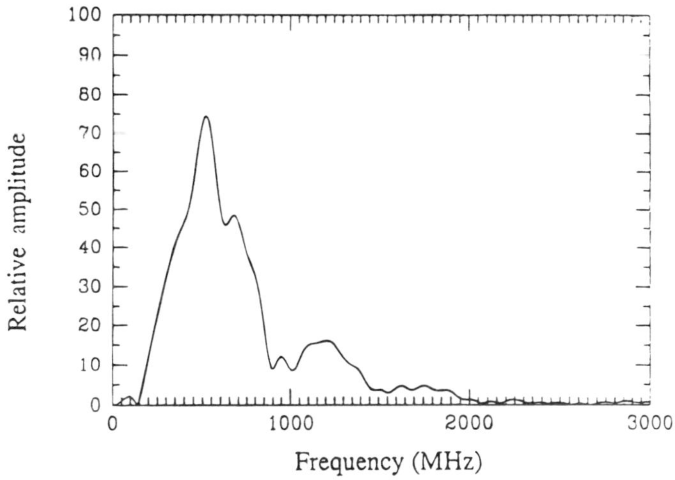

FIGURE 40 Fourier spectrum of the electromagnetic pulse of Fig. 39.

in the induced currents whether the model is grounded or not. This is because most of the energy in the pulse is at frequencies in excess of 300 MHz, where the effect of the ground plane on the induced currents or the SARs is minimal.[104]

In Figure 42, the peak current for each section of the body is plotted with a resolution of 1.31 cm. The maximum peak current of 3.5 mA, which is 3.2 mA/(V/m) occurs at a height of 96.3 cm above the bottom of the feet. A very similar result also had been observed for calculations using isolated and grounded models of the human body for plane-wave exposures at frequencies of 350–700 MHz, where the highest induced currents on the order of 3.0–3.2 mA/(V/m) were calculated for sections of the body that are at heights of 85–100 cm relative to the feet.[104]

The specific absorption (SA) and the total absorbed energy for exposure to the ultrawide band pulse can be calculated using Equations 40 and 41. The total energy absorbed by the body exposed to a single pulse of Figure 39 is 2.0 and 1.91 pico joules for isolated and shoe-wearing grounded conditions, respectively.

a. Section through the eyes (height above the bottom of the feet = 168.3 cm).

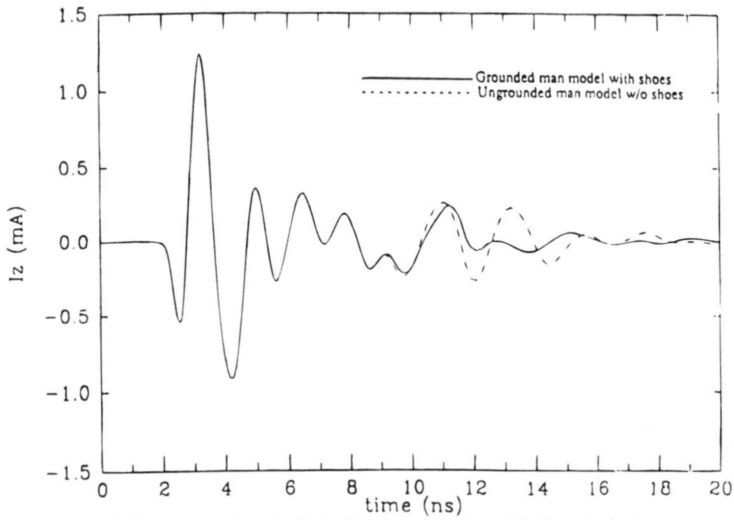

b. Section through the bladder (height above the bottom of the feet = 91.0 cm).

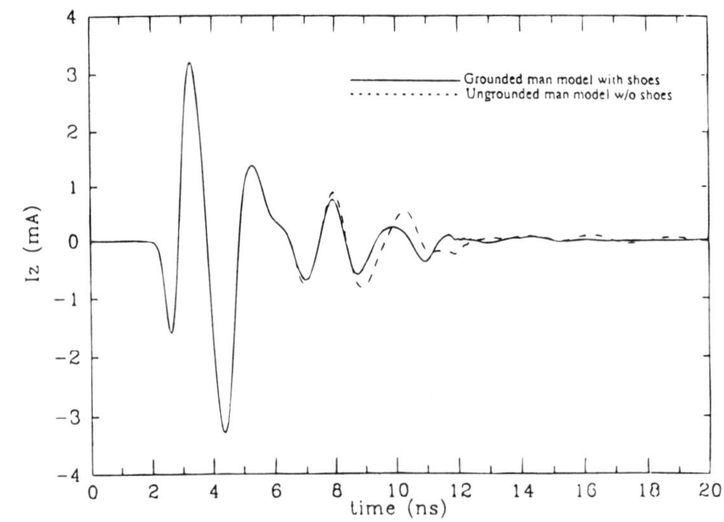

FIGURE 41 Currents induced for the various sections of the body for shoe-wearing grounded and ungrounded conditions of exposure. E_{peak} = 1.1 V/m.

7. EM Absorption in The Human Head for Cellular Telephones–FDTD Method

The millimeter-resolution model of the human body described in section IV B1 has been used to calculate the SAR distributions due to three 2-element Yagi antennas of a proposed 6-GHz mobile communication system[85] and for a number of cellular telephones operating in the frequency band 820–850 MHz.[86] The FDTD method was applied to calculate the electromagnetic fields for the coupled region consisting of the shape and size of the antenna and the handset of the cellular telephone and the heterogeneous MRI-based model of the head, neck, shoulder, and upper torso for which anatomically based tissue properties have been prescribed for subvolumes or "cells" of dimensions 1.974 × 1.974 × 3 mm. The weight of the individual cells is approximately 11.7 mg. The dielectric properties used for the various

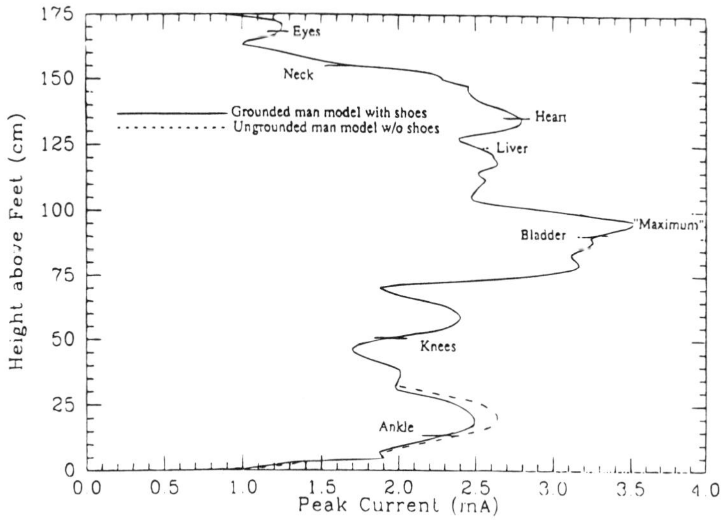

FIGURE 42 Peak currents induced for the various sections of the body for shoe-wearing grounded and ungrounded conditions of the model. E_{peak} = 1.1 V/m.

tissues at the midband cellular telephone frequency of 835 or 900 MHz are given in Table 7. The reason for truncating the human body model at a depth of 42 cm from the top of the head rather than using the whole body is that negligible SARs are obtained for the lower sections of the body. To simulate a handset that is typically tilted forward by about 33° for a vertically erect head, the MRI-based model was modified so that it is tilted forward by 33°. With this forward tilt, the handset may then be held in the vertical position for SAR calculations. This 33° forward tilt in the model of the head was accomplished by displacing each successive layer of the MRI scans separated by 3 mm back by 1 cell, i.e., by 1.974

TABLE 7
Dielectric Properties of the Various Tissues Assumed for the Model of the Head, Neck, and Upper Torso at the Midband Cellular Telephone Frequency of 835 MHZ or 900 MHZ [taken from unpublished data by C. Gabriel—personal communication].

Tissue Type⁺	Tissue	ϵ_r	σ(S/m)
0	air	1.0	0.0
1	muscle	50.0	1.08
2	fat	11.0	0.17
3	bone	21.0	0.33
4	cartilage	37.0	0.8
5	skin	35.0	0.60
6	brain	41.0	0.86
11	blood	55.0	1.86
12	eye	67.0	1.97
15	lung	12.0	0.24
17	cerebrospinal fluid (CSF)	78.0	1.97
18	vitreous humor	67.0	1.68
19	sclera/cornea	51.0	1.13
20	lens	45.0	0.75

⁺ These numbers correspond to the tissue types used for the model of the whole body.

mm. Because of the vertical orientation of the handset and the antenna, it is easier to model their exact shapes and dimensions. Due to the proximity of the hand to the telephone, it is essential to also model the hand for numerical calculations. For the present calculations, the hand was modeled by a region of 2/3 muscle-equivalent material (ϵ_r = 33.3, σ = 0.72 S/m) of thickness 1.974 cm (10 δ_y), wrapped around the handset on three sides, with the exception of the side facing the head up to two-thirds of the height of the handset, and connected to the shoulder of the model.

To normalize the calculated SAR distributions for a peak operating power of 600 mW for the cellular telephone, the procedure was as follows. A sinusoidal driving voltage of 3 V rms was assumed across the gap for the cell representing the driving point. Starting with this prescribed uniform z-directed E-field of 1000 V/m rms, the various converged components of E and H were calculated for all the cells of the modeled region. From the calculated tangential magnetic fields for the cells representing the antenna, the z-directed current distributions along the lengths of the various antennas were obtained from the expression

$$I = -\oint H \cdot dl \tag{42}$$

The z-directed current calculated for the cell corresponding to the driving point was then used to calculate the complex feedpoint impedance and, hence, the power fed into the antenna. The calculated SARs were subsequently scaled for a maximum antenna power of 0.6 W.

The effect of the antenna length (3 λ/8 vs. λ/4) on the power absorbed by the head and neck, and the peak 1-g SARs anywhere in the head, including the ear, and the tissues in the brain have been examined. The salient features of the results obtained for a few representative cellular telephones are given in Table 8. It is interesting to note that the powers absorbed and the peak SARs are lower for 3 λ/8 antennas as compared to λ/4 antennas. This is due to the fact that the peak current region for the longer antennas is higher up on the antenna. Unlike the high-current region for λ/4 antennas, which is very close to the ear, this high-current region is further from the head, which reduces the coupling and, hence, the SARs. A similar argument can also be given for 33° tilted antennas, which result in lower absorbed

TABLE 8
Comparison of the Powers Absorbed and Peak SARs for λ/4 and 3λ/8 Antennas†

Antenna length	Tilt	Frequency MHz	Box size cm	Peak 1-g SAR W/kg	Brain Peak 1-g SAR W/kg	% power absorbed by "hand"	% power absorbed by head and neck
3λ/8	33°	835	15 × 6 × 3	0.61	0.58	—	26.9
3λ/8	33°	835	15 × 6 × 2.5	0.74	0.68	12.5	34.4
				(795 mg)*	(877 mg)*		
λ/4	33°	900	15 × 6 × 2.5	1.37	1.13	14.9	54.1
				(830 mg)*	(877 mg)*		
λ/4	0°	900	15 × 6 × 2.5	2.27	1.18	14.8	60.4
				(526 mg)*	(877 mg)*		
0.22λ	33°	835	13.7 × 5.5 × 2.1	1.48	1.10	—	43.3

†All of the calculations were done using the new values of σ, ε given in Table 7 [C. Gabriel—personal communication]. The calculations also assume that the ear is pressed against the head.
*Actual weight of 1 cm^3 of volume.

powers and SARs *vis à vis* a vertically held antenna, which, because of its proximity to the head throughout its length couples more strongly (see Table 8).

V. A NEW CONVOLUTION PROCEDURE FOR NUMERICAL DOSIMETRY

Detailed knowledge of SAR distribution is required for safety assessment, biological effects and medical applications of electromagnetic energy. Efficient models and computational techniques are needed to alleviate the problem of having to run computer-memory-intensive anatomically based models of the human body repeatedly. More recently, there have been several efforts to develop simple and efficient techniques other than the FEM and FDTD methods to predict field intensity. One such technique is based on the convolution theory, where the impulse response of the heterogeneous model would be obtained and stored using an impulse in the time domain or space domain (spatially localized incident fields) or both, and convolved with the prescribed variation of the incident fields in time and space domains. This procedure is illustrated using an example of the response of the anatomically based model for exposure to pulses or transients of prescribed shapes.[119] For impulse response calculations, a narrow unit magnitude flat-top "impulse" of time duration $T' = 5$ to $10\ \delta t$ is assumed. The impulse-induced current response $H_i(t)$ for the various sections of the body (such as sections through the eyes, neck, heart, liver, etc.) is calculated and stored. Induced currents $I_i(t)$ for these sections for a prescribed incident pulse $E(t)$ can then be calculated from the convolution integral

$$I_i(t) = \int_0^T E(\tau)H_i(t - \tau)\, d\tau \tag{43}$$

where the integration time T is a period in excess of the time duration of the incident pulse. An alternate procedure that has also been found to be equally applicable is to work with Equation 43 in terms of the Fourier transforms of the induced currents $H_i(t)$, the broadband initial impulse $E^{im}(t)$, and the prescribed pulse $E(t)$ for which the induced currents $I_i(t)$ are desired. In terms of the Fourier transforms Equation 43 can be written as

$$I_i(t) = F^{-1}\left\{\frac{F\{E(t)\}F\{H_i(t)\}}{F\{E^{im}(t)\}}\right\} \tag{44}$$

where the discrete Fourier transforms needed can be efficiently calculated using the Fast Fourier Transform (FFT) algorithm. Both of the above approaches are relatively simple and do not need a large computer memory, making it possible to use a small computer or PC for calculating the response of the human body to any desired incident pulse. Figures 43a and 43b show the comparison of the vertical currents induced for a couple of representative sections of the body for the ultrawide band pulse of Figure 39. In both Figures 43a and 43b the time-domain variations of the currents calculated with the simpler impulse response and the convolution method are in excellent agreement with the results obtained from the exact simulation with the conventional $(FD)^2TD$ method.

A similar procedure may be also usable for prescribed spatial variations of the incident fields.

VI. CONCLUDING REMARKS

Knowledge of internal electric and magnetic fields, induced current densities, and SARs is of basic interest in the assessment of biological effects and medical applications of electro-

a. Section through the liver (height above the bottom of the feet = 123.8 cm).

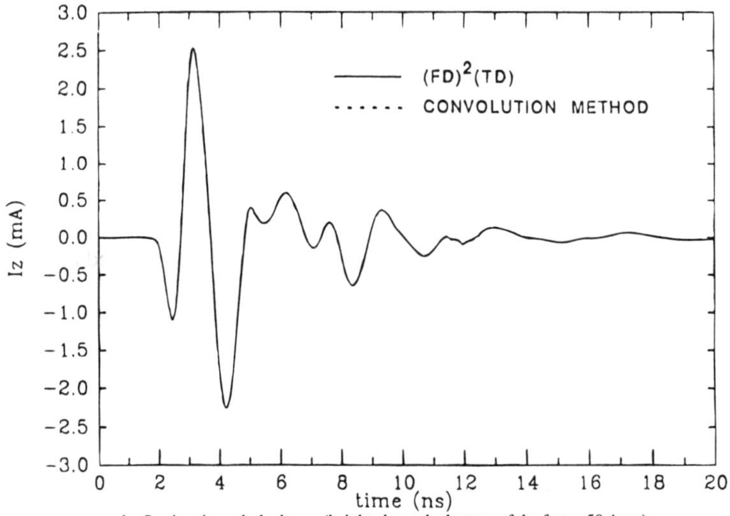

b. Section through the knees (height above the bottom of the feet = 50.4 cm).

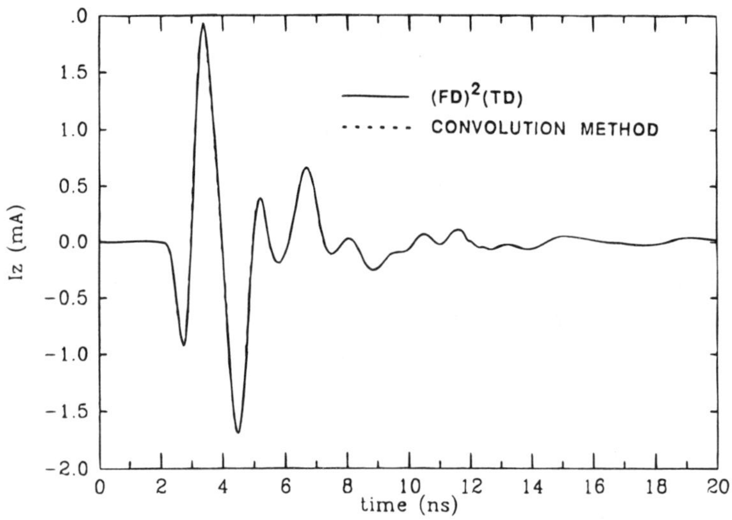

FIGURE 43 Comparison of the induced currents calculated using the convolution method with the conventional $(FD)^2TD$ method for two representative sections of the body. Shoe-wearing grounded model of the body was used for the calculations.

magnetic fields. Because of the inhomogeneities of the tissues and the nonuniformity of the incident fields, closed-form analytical solutions are impossible and computer methods are needed to obtain the internally coupled fields. Some of the earlier approaches involved the method of moments[34,36,38,120] and its variations involving use of fast Fourier transforms.[71] These methods are described in detail in the above references and in some review articles.[39,99,121,122] A handicap of the MoM method is the need to invert a $3N \times 3N$ matrix involving coupling between the N subvolumes or cells into which the exposed body or a part thereof is divided. This takes on the order of $(3N)^2$ for storage, and $(3N)^2$ to $(3N)^3$ of computation steps. Since the computational requirements expand rapidly with N, it has not been possible to apply the traditional MOM to cubic cell model representations of the human body with N larger than 1132.[96] Even though the FFT-based MOM[71] alleviates the problem somewhat by allowing an iterative solution of the electric-field integral equation in the Fourier domain, here too the computational resource requirements are on the order of $N^{3/2}$ log N, which increases rapidly as N is increased. Using the MOM/FFT algorithm with the conjugate gradient method, it has been possible to calculate SAR distributions for a 5607-cell model of the human body for exposure to plane-wave EM fields.[122]

Great strides have been made in the last few years in the area of numerical dosimetry using anatomically based models of the human body. Among the most promising of the numerical methods for induced currents and SAR calculations are the impedance method for use at lower frequencies where quasi-static approximations may be made[41–45]($\leq$40 MHz for the human body) and the FEM and FDTD methods which may be used at any frequency of interest.[72,74,75,82,97,98,100,101] For numerical calculations the FDTD method requires a computer memory and computation time that is proportional to N. This is a considerable advantage over computing methods such as MOM or MOM/FFT. In this chapter we have described the salient features of these methods and the many bioelectromagnetic exposure conditions for which they have been used. Because of the limitations on the length of the chapter, only a few of the important recent applications of some of these methods were presented in detail.

Note that numerical methods have matured to a level that they are being increasingly used by researchers in many laboratories for dosimetric calculations for important and meaningful bioelectromagnetic problems. Some of the future developments will involve improving the efficiencies of the various codes by techniques such as use of the expanding grid[123] rather than the regular grid, elimination of the relatively shielded interior regions of the modeled space at higher frequencies, and use of truncated models of the body at microwave frequencies where there is a lack of coupling between the various regions of the body. Because of accurate modeling of the tissue heterogeneities and shapes, these models are likely to play an increasing role in emerging technologies with bioelectromagnetic concerns.[82,124] Some of the likely applications are: personal wireless and mobile communications systems, automotive devices such as electric automobiles and collision avoidance systems, medical devices such as MRIs, implantable pacemakers and defibrillators, applicators for hyperthermia, hypothermia, and other minimally invasive therapeutic procedures. Another area of likely application of numerical design is emerging technologies such as electroporation for drug delivery, bone regeneration and growth, etc.

ACKNOWLEDGMENTS

This work was supported in part by the U.S. National Science Foundation under Grant BES-9110184 (JCL) and the ROC National Science Council under Grant 83-0102-C033-001 (JCL).

REFERENCES

1. **Schwan, H. P. and Piersol, G. M.,** Absorption of electromagnetic energy in body tissues. *Am. J. Phys Med.,* 33, 371, 1954.
2. **Johnson, C. C. and Guy, A. W.,** Nonionizing electromagnetic wave effects in biological materials and systems. *Proc. IEEE,* 60, 692, 197
3. **Schwan, H. P. and Li, K.,** Hazards due to total body irradiation by radar. *Proc. IRE,* 44, 1572, 1956.

4. **Schwan, H. P. and Li, K.,** The mechanism of absorption of ultrahigh frequency electromagnetic energyin tissue as related to the problem of tolerance dosage, *IRE Trans. Med. Electron.,* 4, 45, 1956.
5. **Stratton, J. A.,** *Electromagnetic Theory,* McGraw-Hill, New York, 1941.
6. **Lin, J. C., Guy, A. W., and Kraft, G. H.,** Microwave selective brain heating, *J. Microwave Power,* 8, 275, 1973.
7. **Shapiro, A. R., Lutomirski, R. F., and Yura, H. T.,** Induced fields and heating within a cranial structureirradiated by an electromagnetic plane wave, *IEEE Trans. Microwave Theory Tech.,* 19, 187, 1971.
8. **Kritikos, H. N. and Schwan, H. P.,** Hot spot generated in conduction spheres by EM waves and biological implications, *IEEE Trans. Biomed. Eng.,* 19, 53, 1972.
9. **Ho, H. S. and Guy, A. W.,** Development of dosimetry for RF and microwave radiation, *Health Phys.,* 29, 317, 1975.
10. **Lin, J. C.,** Microwave biophysics, *in Microwave Bioeffects and Radiation Safety,* Stuchly, M. A., Ed., IMPI, Edmonton, Alberta, Canada, 1978, 15.
11. **Weil, C. M.,** Absorption characteristics of multi-layered sphere models exposed to UHF/microwave radiation, *IEEE Trans. Biomed. Eng.,* 22, 468, 1975.
12. **Guy, A. W., Lin, J. C., and Chou, C. K.,** Electrophysiologic effects of electromagnetic fields on animals, in *Fundamental and Applied Aspects of Nonionizing Radiation,* Michaelson, S. M. et al., Eds., Plenum Press New York. 1975, 167,
13. **Joines, W. T. and Spiegel, R. J.,** Resonance absorption of microwaves by the human skull. *IEEE Trans. Biomed. Eng.,* 21, 46, 1975.
14. **Lin, J. C.,** Interaction of two cross-polarized electromagnetic waves with mammalian cranial structures, *IEEE Trans. Biomed. Eng.,* 23, 371, 1976.
15. **Schwan, H. P.,** Electrical properties of tissues and cell suspensions, *Adv. Biol. Med. Phys.,* 4, 147, 1957.
16. **Schwan, H. P.,** Survey of microwave absorption characteristics of body tissues. in *Proc. 2nd Tri-Service Conf. Biol. Effects of Microwave Energy,* Pattishall, E.G. and Banghart F. W., Eds., 1958, 126.
17. **Lin, J. C.,** Microwave properties of fresh mammalian brain tissues at body temperature, *IEEE Trans. Biomed. Eng.,* 22, 74, 1975.
18. **Lin, J. C., Grove, H. M., and Sharp, J. C.,** Comparative measurement of dielectric properties of fresh mammalian tissues, in *Proc. Conf. Precision Electromagnetic Meas.,* 1975, 246.
19. **Schwan, H. P. snd Foster, K. R.,** RF-field interaction with biological systems: electrical properties and biophysical mechanisms, *Proc. IEEE,* 68, 104, 1980.
20. **Schwan, H. P.,** Microwave biophysics, in *Microwave Power Engineering,* Okress, E. C., Ed., Academic Press, New York, 1968, 213.
21. **Lin, J. C., Guy, A. W., and Johnson, C. C.,** Power deposition in a spherical model of man exposed to 1–20-MHz electromagnetic fields, *IEEE Trans. Microwave Theory Tech.,* 21,791, 1973
22. **Kritikos, H. N. and Schwan, H. P.,** The distribution of heating potential inside lossy spheres, *IEEE Trans. Biomed. Eng.,* 22, 457, 1975.
23. **Johnson, C. C., Durney, C. H., and Massoudi, H.,** Long-wavelength electromagnetic power absorption in prolate spheroidal models of man and animals, *IEEE Trans. Microwave Theory Tech.,* 23, 739, 1975.
24. **Durney, C. H., Johnson, C. C., and Massaudi, A.,** Long wave-length analysis of plane wave irradiation of a prolate spheroidal model of man, *IEEE Trans. Microwave Theory Tech.,* 23, 246, 1975.
25. **Massoudi, H., Durney, C. H., and Johnson, C. C.,** Long wavelength electromagnetic power absorption in ellipsoidal models of man and animals, *IEEE Trans. Microwave Theory Tech.,* 24, 41, 1977.
26. **Durney, C. H., Iskander, M. F., Massoudi, H., and Johnson, C. C.,** An empirical formula for broadband SAR calculations of prolate spheroidal models of humans and animals, *IEEE Trans. Microwave Theory Tech.,* 27, 758, 1979.
27. **Wu, C. L. and Lin, J. C.,** Absorption and scattering of EM waves by prolate spheroidal models of biological structures, in *IEEE AP-S Int. Symp.,* Stanford. Calif., 1977, 142.
28. **Barber, P.,** Electromagnetic power absorption in prolate spheroidal models of man and animals at resonance, *IEEE Trans. Biomed. Eng.,* 24, 513, 1977.
29. **Rowlandson, G. I. and Barber, P. W.,** Absorption of high frequency RF energy by biological models:calculations based on geometrical optics, *Radio. Sci.,* 14, 43S, 1979.
30. **Asano, A. and Yamamoto, G.,** Light scatteriny by spheroidal particles, *Appl. Opt,* 14, 29, 1975.
31. **Lin, J. C.,** Whole-body exposure in the near zone of HF electromagnetic fields, in *URSI Int. Electromagnetic Waves and Biology Syrup.,* Jouy en Josas, France, 1980.
32. **Durney, C. H., Iskander, M. F., Massoudi, H., Allen, S. J. and Mitchell, J. C.,** *Radio Freguency Radiation Dosimetry Handbook,.* 3rd ed., Brooks AFB, Tex., 1980.
33. **Guy, A. W., Webb, M. D., and Sorenson, C. C.,** Determination of power absorption in man exposed to HF electromagnetic fields by thermographic measurements on scale models, *IEEE Trans. Biomed. Eng.,* 23, 361, 1976.
34. **Livesay, D. E. and Chen, K. M.,** Electromagnetic fields induced inside arbitrary shaped biological bodies, *IEEE Trans. Microwave Theory Tech.,* 22, 1273, 1974.

35. **Guru, B. S. and Chen, K. M.,** Experimental and theoretical studies in electromagnetic field induced inside finite biological bodies, *IEEE Trans. Microwave Theory Tech.,* 24, 433, 1976.
36. **Chen, K. M. and Guru, B. S.,** Internal EM field and absorbed power density in human torsos induced by 1–500 MHz EM waves, *IEEE Trans. Microwave Theory Tech.,* 25, 746, 1977.
37. **Chen, K. M.,** Interaction of electromagnetic fields with biological bodies, in *Research Topics in Electromagnetic Theory,* Kong J. A. Ed., John Wiley & Sons, New York, 1980, 290.
38. **Hagman, M. J., Gandhi, O. P., and Durney, C. H.,** Numerical calculation of electromagnetic energy deposition for a realistic model of man, *IEEE Trans. Microwave Theory Tech.,* 27, 804, 1979.
39. **Durney, C. H.,** Electromagnetic dosimetry for models of humans and animals: a review of theoretical and numerical techniques, *Proc. IEEE,* 68, 33, 1980.
40. **Gandhi, O. P.,** Electromagnetic absorption in inhomogeneous model of man for realistic exposure conditions, *Bioelectromagnetics,* 3, 81, 1982.
41. **Armitage, D. W., Leveen, H. H., and Pethig, R.,** Radio-frequency induced hyperthermia: Computer simulation of specific absorption rate distributions using realistic anatomical models, *Physics in Medicine and Biology,* 28, 31, 1983.
42. **Gandhi, O. P., DeFord, J. F., and Kanai, H.,** Impedance method for calculation of power deposition patterns in magnetically induced hyperthermia, *IEEE Transactions on Biomedical Engineering,* 31, 644, 1984.
43. **Gandhi, O. P. and DeFord, J. F.,** Calculation of EM power deposition for operator exposure to RF induction heaters, *IEEE Transactions on Electromagnetic Compatibility,* 30, 63, 1988.
44. **Orcutt, N. and Gandhi, O. P.,** A 3-D impedance method to calculate power deposition in biological bodies subjected to time-varying magnetic fields, *IEEE Transactions on Biomedical Engineering,* 35, 577, 1988.
45. **Orcutt, N. and Gandhi, O. P.,** Use of the impedance method to calculate 3-D power deposition patterns for hyperthermia with capacitive plate electrodes, *IEEE Transactions on Biomedical Engineering,* 37, 36, 1990.
46. **Epstein, B. R. and Foster, K. R.,** Anisotropy in dielectric properties of skeletal muscle, *Medical and Biological Engineering and Computing,* 21, 51, 1983.
47. **Zheng, E., Shao, S., and Webster, J. G.,** Impedance of skeletal muscle from 1 Hz to 1 MHz, *IEEE Transactions on Biomedical Engineering,* 31,477, 1984.
48. **Zhu, X. L. and Gandhi, O. P.,** Design of RF needle applicators for optimum SAR distributions in irregularly-shaped tumors, *IEEE Transactions on Biomedical Engineering,* 35, 382, 1988.
49. **Gandhi, O. P. and Chen, J. Y.,** Numerical dosimetry at power-line frequencies using anatomically based models, *Bioelectromagnetics,* Supplement 1, 43, 1992.
50. **Tofani, S., Ossola, P., d'Amore, G., and Gandhi, O. P.,** Electric fields and current density distributions induced in an anatomically based model of the human head by magnetic fields from a hair dryer, *Health Physics,* 68, 71, 1995.
51. **Wu, D., Ren, J., and Gandhi, O. P.,** Currents induced in the arm and the body by the magnetic fields of an electric hand drill, *Abstracts of the Fifteenth Annual Meeting of the Bioelectromagnetics Society,* Los Angeles, California, 31, 1993.
52. **Ren, J., Gandhi, O. P., Casamento, J. P., and Bassen, H. I.,** Currents induced in the human body by electric blankets, *Abstracts of the Fifteenth Annual Meeting of the Bioelectromagnetics Society,* Los Angeles, California, 32, 1993.
53. **Harrington, R. F.,** *Field Computation by Moment Methods,* McGraw-Hill, New York, 1968.
54. **Schelkunoff, S. A.,** Field equivalence theorems, *Comm. Pure Appl. Math.,* 4, 43, 1951.
55. **Lin, J. C. and Wu, C. L.,** Scattering of microwaves by dielectric materials used in laboratory animal restrainers, *IEEE Trans. Microwave Theory Tech.,* 24, 219, 1976.
56. **Gandhi, O. P., Hagman, M. J., and D'Andrea, J. A.,** Part-body and multi-body effects on absorption of radio frequency electromagnetic energy by animals and by models of man, *Radio Sci.,* 14, 155, 1979.
57. **Karimullah, K., Chen, K. M., and Nyquist, D. P.,** Electromagnetic coupling between a thin-wire antenna and a neighboring biological body, *IEEE Trans. Microwave Theory Tech.,* 28, 1218, 1980.
58. **Chatterjee, I., Hagman, M. J., and Gandhi, O. P.,** Electromagnetic energy deposition in an inhomogeneous block model for near-field irradiation conditions, *IEEE Trans. Microwave Theory Tech.,* 28, 1452, 1980.
59. **Wu, T. K. and Tsai, L. L.,** Electromagnetic fields induced inside arbitrary cylinders of biological tissue *IEEE Trans. Microwave Theory Tech.,* 25, 61, 1977.
60. **Wu, T. K. and Tsai, L. L.,** Scattering from arbitrary-shaped lossy dielectric bodies of revolution, *Radio Sci,* 12, 709, 1977.
61. **Poggio, A. J. and Miller, E. K.,** Integral equation solutions of three-dimensional scatteriny problems, in *Computer Techniques for Electromagnetics,* Mittra, R., Ed. Pergamon Press, Elmsford, N.Y., 1973, 159.
62. **Massoudi, H., Durney, C. H., Barber, P. W., and Iskander, M. F.,** Post resonance EM absorption by man and animals, *Bioelectromagnetics,* 3, 333, 1982.
63. **Wu, T. K.,** Electromagnetic fields and power deposition in body of revolution models of man, *IEEE Trans. Microwave Theory Tech.,* 27, 279, 1979

64. **Mautz, J. R. and Harrington, R. F.,** Radiation and scattering from bodies of revolution, *Appl. Sci. Res.,* 20, 405, 1969.
65. **Harrington, R. F. and Mautz, J. R.,** Green's functions for surfaces of revolution, *Radio Sci.,* 7, 603, 1972.
66. **Pogorzelski, R. J. and Wu, T. K.,** Computations of scattering from inhomogeneous penetrable elliptic cylinders by means of invariant imbedding, in *URSI Symp. Electromagnetic Wave Theory.* Stanford, 1977, 323.
67. **Bojarski, N. N.,** N-Dimensional fast Fourier transform tomography for incomplete information and its application to electromagnetic inverse scattering theory in *Mathematical Methods and Applifations of Scattering Theory,* Desanto, J. A. et al., Eds., Springer-Verlag. Berlin, 1980, 277.
68. **Sarkar, T. and Rao, S. M.,** An iterative method for solving electrostatic problems, *IEEE Trans. Ant. Propagation,* 30, 611, 1982.
69. **Hestenes, M. and Stiefel, E.,** Method of conjugate gradients for solving linear systems, *J. Res. Natl. Bur Standards,* 49, 409, 1952.
70. **Sarker, T, K., Siarkiewicz, K, R., and Stratton, R. F.,** Survey of numerical methods for solution of large systems of linear equations for electromagnetic field problems, *IEEE Trans. Ant. Propagation.,* 29, 847, 1981.
71. **Borup, D. T. and Gandhi, O. P.,** Calculation of high-resolution SAR distributions in biological bodies using the FFT algorithm and conjugate gradient method, *IEEE Trans. Microwave Theory and Tech.* 33, 417, 1985.
72. **Yee, K. S.,** Numerical solutions of initial boundary value problems involving Maxwell's equations in isotropic media, *IEEE Trans. Ant. Propagation,* 14, 303, 1966.
73. **Kunz, K. S. and Luebbers, R. J.,** *The Finite-Difference Time-Domain Method for Electromagnetics,* CRC Press, Inc., Boca Raton, Florida, 1993.
74. **Taflove, A. and Brodwin, M. E.,** Computation of the electromagnetic fields and induced temperatures within a model of the microwave-irradiated human eye, *IEEE Trans. Microwave Theory Tech.,* 23, 888, 1975.
75. **Taflove, A. and Lin, J. C.,** Finite difference time domain computation of microwave absorption in models of biological bodies, in *Abstr. Bioelectromagnetic Soc. Annu Meet.,* Washington. D.C., 1981, 62.
76. **Spiegel, R. J., Fatmi, M. B. E, and Kunz, K. S.,** Application of a finite-difference technique to the human radio-frequency dosimetry problem, *Journal of Microwave Power,* 20, 241, 1985.
77. **Spiegel, R. J.., Fatmi, M. B. A., Stuchly, S. S., and Stuchly, M. A.,** Comparison of finite-difference time-domain SAR calculations with measurements in a heterogeneous model of man, *IEEE Trans. Biomedical Engineering,* 36, 849, 1989.
78. **Taflove, A. and Brodwin, M. E.,** Numerical solution of steady-state EM scattering problems using the time dependent Maxwell's equation, *IEEE Trans. Microwave Theory Tech.,* 23, 623, 1975.
79. **Taflove, A.,** Application of the finite-difference time domain method to sinusoidal steady-state electromagnetic-penetration problems, *IEEE Trans. EM Compatibility,* 22, 191, 1980.
80. **Jin, J.,** *The Finite Element Method in Electromagnetics,* Wiley, New York, 1993.
81. **Boyes, W. E., Lynch, D. R., Paulsen, K. D., and Minerbo, G. N.,** Nodal-based finite element modelling of Maxwell's equations in three dimensions, *IEEE Trans. Antennas Propagat.,* 40, 642, 1992.
82. **Paulsen, K. D., Jia, X., and Sullivan, J. M., Jr.,** Finite element computations of specific absorption rates in anatomically conforming full-body models for hyperthermia treatment analysis, *IEEE Trans. Biomedical Engineering,* 40, 933, 1993
83. **Eycleshymer, A. C. and Schoemaker, D. M.,** *A Cross-Section Anatomy,* Appleton-Century-Crofts, New York, 1970.
84. **Chen, J. Y. and Gandhi, O. P.,** Currents induced in an anatomically based model of a human for exposure to vertically polarized electromagnetic pulses, *IEEE Trans. Microwave Theory and Tech.,* 39, 31, 1991.
85. **Gandhi, O. P. and Chen, J. Y.,** Electromagnetic absorption in the human head for a proposed 6 GHz mobile communication system, to appear in *IEEE Transactions on Electromagnetic Compatibility.*
86. **Gandhi, O. P.,** Some numerical methods for dosimetry: ELF to microwave frequencies, *Radio Science,* 30, 161, 1995.
87. **Gandhi, O. P., Ren, J., and Chen, J. Y.,** Currents induced in the human body by electric blankets, Final Report on Contract No. FDA 0961340091, submitted to DHHS/PHS,FDA/CDRH, Rockville, Maryland, 20857, July 31, 1992.
88. **Florig, H. K. and Hoburg, J. F.,** Power frequency magnetic fields from electric blankets, *Health Physics,* 58, 493, 1990.
89. **Florig, H. K., Hoburg, J. F., and Morgan, M. G.,** Electric-field exposure from electric blankets, *IEEE Transactions on Power Delivery,* 2, 527, 1987.
90. **Hayashi, N., Isaka, K., and Yokoi, Y.,** Analysis of magnetic-field profiles in electric blanket users, *IEEE Transactions on Power Delivery,* 4, 1897, 1989.
91. **Rush, S., Abildskov, J. A., and McFee, R.,** Resistivity of body tissues at low frequencies, *Circulation Research,* 12, 40, 1963.
92. **Casamento, J., Crowl, B., and Bassen, H.,** Spatially mapped magnetic and electric fields from electric

blankets, paper presented at the First World Congress for Electricity and Magnetism in Biology and Medicine, Lake Buena Vista, Florida, June 14–19, 1992 (unpublished data kindly provided by the authors for comparison).

93. **Rukspollmuang, S. and Chen, K. M.,** Heating of spherical vs. realistic models of human and infrahuman heads by electromagnetic waves, *Radio Sci.,* 14, 51, 1979.
94. **Hagmann, M. J., Gandhi, O. P., D'Andrea, J. A., and Chatterjee, I.,** Head resonance: numerical solutions and experimental results, *IEEE Trans. Microwave Theory Tech.,* 27, 809, 1979.
95. **Gandhi, O. P., Hunt, E. L., and D'Andrea, J. A.,** Deposition of EM energy in animals and in models of man with and without grounding and reflector effects, *Radio Sci.,* 12, 39S, 1977.
96. **DeFord, J. F., Gandhi, O. P., and Hagmann, M. J.,** Moment-Method solutions and SAR calculations for inhomogeneous models of man with large number of cells, *IEEE Trans. Microwave Theory Tech.,* 31, 848, 1983.
97. **Holland, R.,** THREDE:A free field EMP coupling and scattering code, *IEEE Transactions on Nuclear Science,* 24, 2416, 1977.
98. **Kunz, K. S. and Lee, K. M.,** A three-dimensional finite-difference solution of the external response of an aircraft to a complex transient EM environment: Part 1—The method and its implementation, *IEEE Transactions on Electromagnetic Compatibility,* 20, 328, 1978.
99. **Spiegel, R. J.,** A review of numerical models for predicting the energy deposition and resultant thermal response of humans exposed to electromagnetic fields, *IEEE Trans. on Microwave Theory and Tech.,* 32, 730, 1984.
100. **Sullivan, D. M., Borup, D. T. and Gandhi, O. P.,** Use of the finite-difference time-domain method in calculating EM absorption in human tissues, *IEEE Transactions on Biomedical Engineering,* 34, 148, 1987.
101. **Sullivan, D. M., Gandhi, O. P., and Taflove, A,** Use of the finite-difference time-domain method in calculating EM absorption in man models, *IEEE Transactions on Biomedical Engineering,* 35, 179, 1988.
102. **Chen, J. Y. and Gandhi, O. P.,** Numerical simulation of annular-phased arrays of dipoles for hyperthermia of deep-seated tumors, *IEEE Transactions on Biomedical Engineering,* 39, 209, 1992.
103. **Chen, J. Y., Gandhi, O. P., and Conover, D. L.,** SAR and induced current distributions for operator exposure to RF dielectric sealers, *IEEE Transactions on Electromagnetic Compatibility,* 33, 252, 1991.
104. **Gandhi, O. P., Gu, Y. G., Chen, J. Y., and Bassen, H. I.,** Specific absorption rates and induced current distributions in an anatomically based human model for plane-wave exposures, *Health Physics,* 63, 281, 1992.
105. **Luebbers, R., Hunsberger, F. P., Kunz, K. S., Standler, R. B., and Schneider, M.,** A frequency-dependent finite-difference time-domain formulation for dispersive materials, *IEEE Transactions on Electromagnetic Compatibility,* 32, 222, 1990.
106. **Bui, M. D., Stuchly, S. S., and Costache, G. I.,** Propagation of transients in dispersive dielectric media, *IEEE Trans. Microwave Theory and Tech.,* 39, 1165, 1991.
107. **Sullivan, D. M.,** A frequency-dependent FDTD method for biological applications, *IEEE Trans. Microwave Theory and Tech.,* 40, 532, 1992.
108. **Luebbers, R. J., Hunsberger, F., and Kunz, K. S.,** FDTD for Nth order dispersive media, *IEEE Transactions on Antennas and Propagation,* 40, 1297, 1992.
109. **Joseph, R. M., Hagness, S. C., and Taflove, A.,** Direct time integration of Maxwell's equations in linear dispersive media with absorption for scattering and propagation of femtosecond electromagnetic impulses, *Optics Letters,* 16, 1412, 1991.
110. **Gandhi, O. P., Gao, B. Q., and Chen, J. Y.,** A frequency-dependent finite-difference time-domain formulation for induced current calculations in human beings, *Bioelectromagnetics,* 13, 543, 1992.
111. **Gandhi, O. P., Gao, B. Q., and Chen, J. Y.,** A frequency-dependent finite-difference time-domain formulation for general dispersive media, *IEEE Trans. Microwave Theory and Tech.,* 41, 658, 1993.
112. **Furse, C. M., Chen, J. Y., and Gandhi, O. P.,** A frequency-dependent finite-difference time-domain method for induced current and SAR calculations for a heterogeneous model of the human body, *IEEE Transactions on Electromagnetic Compatibility,* 36, 128, 1994.
113. **Gandhi, O. P. and Chen, J. Y.,** Absorption and distribution patterns of RF magnetic fields used or planned for MRI with frequencies up to 200 MHz, *Abstracts of the Fifteenth Annual Meeting of the Bioelectromagnetics Society,* Los Angeles, California, 23, 1993.
114. **Guy, A. W., Davidow, S., Yang, G. Y., and Chou, C. K.,** Determination of electric current distributions in animals and humans exposed to a uniform 60-Hz high-intensity electric field, *Bioelectromagnetics,* 3, 47, 1982.
115. **Kaune, W. T. and Gillis, M. F.,** General properties of the interaction between animals and ELF electric fields, *Bioelectromagnetics,* 2, 1, 1981.
116. **Mautz, J. R.,** Mie series solution for a sphere, *IEEE Transactions on Microwave Theory and Techniques,* 26, 375, 1978.
117. **Deno, D. W.,** Currents induced in the human body by high voltage transmission line electric field—measurement and calculation of distribution and dose, *IEEE Transactions on Power Apparatus and Systems,* 96, 1517, 1977.

118. **DiPlacido, J., Shih, C. H., and Ware, B. J.,** Analysis of the proximity effects in electric field measurements, *IEEE Transactions on Power Apparatus and Systems,* 97, 2167, 1978.
119. **Chen, J. Y., Furse, C. M., and Gandhi, O. P.,** A simple convolution procedure for calculating currents induced in the human body for exposure to electromagnetic pulses, *IEEE Transactions on Microwave Theory and Techniques,* 42, 1172, 1994.
120. **Hagmann, M. J. and Gandhi, O. P.,** Numerical calculation of electromagnetic energy deposition in models of man with grounding and reflector effects, *Radio Science,* 14(6s), 23, 1979.
121. **Lin, J. C.,** Computer methods for field intensity predictions, in *Handbook of Biological Effects of Electromagnetic Fields,* Polk, C. and Postow, E., Eds., CRC Press, Boca Raton, Florida, 1988, part III, chap. 2.
122. **Gandhi, O. P.,** Numerical methods for specific absorption rate calculations, in *Biological Effects and Medical Applications of Electromagnetic Energy,* Gandhi, O. P., Ed., Prentice Hall, Inc., Englewood Cliffs, New Jersey, 1990, chap. 6.
123. **Gao, B. Q. and Gandhi, O. P.,** An expanding-grid algorithm for the finite-difference time-domain method, *IEEE Transactions on Electromagnetic Compatibility,* 34, 277, 1992.
124. **Lin, J. C.,** Microwave technology for minimally invasive interventional procedures, *Chinese J. Med. Biol. Engineering,* 13, 293, 1993.

Chapter 10

THERMOREGULATION IN THE PRESENCE OF MICROWAVE FIELDS

Eleanor R. Adair

CONTENTS

0-8493-0641-8/96/$0.00+$.50

I. INTRODUCTION

Thermoregulation is the maintenance of the body temperature within a prescribed range under conditions in which the thermal load on the body may vary. Thermal loads result from changes in metabolic heat production and from alterations in ambient conditions (temperature, ambient vapor pressure, air velocity, insulation, and other environmental variables that may alter the temperature of the skin). The deposition of thermalizing energy in the body by exposure to radiofrequency electromagnetic (RF/EM) radiation is a unique exception to the energy flows normally encountered by earthly organisms, although metabolic activity in skeletal muscle can also deposit large amounts of thermal energy directly into deeper tissues.

All animals exhibit substantial response to changes in body temperature. Each species is characterized by a preferred level of body temperature at which its functioning and well being are optimal. While cold-blooded animals (ectotherms) regulate their body temperatures largely through behavioral selection of a preferred microclimate, warm-blooded animals (endotherms) also have available an involuntary physiological system that functions to produce and dissipate body heat. In general, endotherms also rely largely on behavioral thermoregulation even though physiological responses can maintain a narrower range of body temperatures. When behavioral responses, which include the thermostatic control of the immediate microenvironment, are rendered difficult or inoperative, the body temperature of endotherms is regulated by the physiological mechanisms that control heat production, the distribution of heat in the body, and the avenues and rate of heat loss from the body to the environment.[1–5]

The goal of much research into the biological consequences of exposure to RF/EM radiation is the understanding of how such exposure may compromise the normal biological functioning of *human beings.* Since, for ethical reasons, many experimental maneuvers cannot be performed on human subjects, studies of animal subjects must be substituted. The investigation of thermoregulation in the presence of microwave fields is no exception in this regard. While humans have been exposed inadvertently to ultralow, "background" levels of both naturally occurring and man-made RF/EM sources, few individuals have voluntarily exposed themselves to significant levels for the scientific assessment of any responses, thermoregulatory or otherwise. Most of the extant data[6] quantify the thermal sensations of human subjects derived from stimulation of restricted body areas at a few discrete microwave frequencies as compared with similar sensations derived from stimulation at infrared frequencies. Since thermal sensation is the initial event that often underlies changes in thermoregulatory responses, these data are extremely valuable.

The physiological mechanisms in the human body that function to maintain a nearly constant internal temperature are well described and well understood quantitatively, far better, indeed, than the responses of assorted laboratory animals. Therefore, the following discussion of thermoregulation in the presence of RF/EM fields will be based on our knowledge of how human thermoregulatory responses change (1) in the presence of conventional thermal stimuli in the environment, (2) during exercise-induced hyperthermia, (3) during RF energy deposition concomitant with MRI imaging or hyperthermia in the clinic, and (4) on predictions made by sophisticated computer models of the human thermoregulatory system. The best quantitative data derived from animal subjects are cited whenever applicable.

II. FUNDAMENTALS OF THERMOREGULATION

Most of the vital internal organs of human beings function most efficiently when maintained at a relatively constant temperature near 37 °C. While the temperature of individual body parts may vary characteristically from this norm, significant departures are associated with disease states or possibly lethal conditions. The usual range of "normal" body temperatures

(35.5 to 40 °C) encompasses circadian-variation, vigorous exercise, variations in ambient temperature, sequelae of food intake, age factors, cyclical variation in females, and emotional factors. Temperatures outside of this range must be related to disease states, unusual activity, or extraordinary environmental conditions.

Body tissues are extremely vulnerable to changes in temperature, particularly to overheating. The intricate system of mechanisms developed by humans for regulating internal body temperature is therefore not surprising. In humans, as in all endotherms, two distinct control systems are available to accomplish thermoregulation: a behavioral system involving conscious, voluntary acts that adjust the physical characteristics of the air-skin interface, and a physiological system involving the involuntary responses of the body that generate and dissipate body heat. It is the physiological system to which we first turn our attention.

A. BODY HEAT BALANCE

The most important principle involved in the study of physiological thermoregulation is the first law of thermodynamics or the law of conservation of energy.[7] In the steady state, the heat produced in the body is balanced by the heat lost to the environment such that storage of heat is minimal. This can be expressed by a generalized heat balance equation:

$$M \pm W = \pm R \pm C \pm E \pm S \qquad [W/m^2] \tag{1}$$

in which M = the rate at which thermal energy is produced through metabolic processes (always +), W = rate of work produced by or on the body (+ for work against external forces), R = rate of radiant heat exchange with the environment (+ for gain), C = rate of convective heat exchange with the environment (+ for gain), E = rate of heat loss due to the evaporation of body water (− for net loss), and S = rate of heat storage in the body (+ for net gain).

As the equation is written, negative values of R, C, and E may all cause a rise in the body temperature; positive values may cause a fall. A strict definition of work (W) is used in Equation 1, i.e., a force operating through a certain distance in the direction of the force. A person walking on the level is not performing any work because of the absence of force in the direction of motion. On the other hand, W is positive when accomplished by the body on the environment (e.g., climbing stairs), and this energy must be subtracted from metabolic energy (M) to find the net heat developed within the body. When W is negative (e.g., walking downhill), this heat is added to the metabolic heat. Usually, E, evaporative heat loss, is positive; when E is negative, condensation occurs and thermal injury is possible.

Because heat exchange by radiation (R), convection (C), or evaporation (E) is always related in some way to the surface area of the body, it useful to describe each term in Equation 1 in terms of power per unit surface area, e.g., W/m^2. The most commonly used measure of body surface area in humans is that proposed by DuBois,[8]

$$A_D = 0.202\ \underline{w}^{0.425}\underline{h}^{0.725} \tag{2}$$

where A_D = DuBois surface [m^2], $\underline{w}$ = body mass [kg], and $\underline{h}$ = height [m].

As noted by Kleiber,[9] $\underline{h}$ for similar body shapes is proportional to a mean linear dimension equal to $\underline{w}^{1/3}$. Therefore, Equation 2 may be generalized to compare humans of different sizes:

$$\begin{aligned} A_D &= k_1\underline{w}^{0.425}k_2\underline{w}^{1/3\times 0.725} \\ &= k_1k_2\underline{w}^{(0.425+0.242)} \\ &= k_1k_2\underline{w}^{2/3} \end{aligned} \tag{3}$$

The surface area-to-body mass ratio varies between species so that it is difficult to give a general rule for the determination of surface area. Indeed, many methods have been devised for the direct measurement of the surface area of experimental animals, a discussion of which is beyond the scope of this chapter. In most cases, the surface area is some function of $\underline{w}^{2/3}$ and the resting M of most species varies as $\underline{w}^{3/4}$. Since the latter relationship is most generally accurate, metabolic heat production is often expressed in units of W/kg.

No term appears in Equation 1 for heat transfer through *conduction* which is usually insignificant in human beings under normal conditions. However, conduction combined with mass transfer forms the mode of heat transfer called *convection,* a significant form of heat loss in humans. Convective heat transfer in air (C) is a linear function of body surface area (A_D) and the convective heat transfer coefficient (h_c) is a function of ambient air motion to the 0.6 power ($V^{0.6}$). The amount of heat lost by the body through convection depends upon the difference between the surface temperature of the skin (T_{sk}) and the air temperature, usually taken as the dry bulb temperature (T_{db}). The expression for heat loss via convection is

$$C = kV^{0.6}\,A_D(T_{sk} - T_{db}) \qquad [W] \tag{4}$$

where the heat transfer coefficient (h_c) equals $kV^{0.6}$ and k depends upon certain properties of the surrounding medium such as density, viscosity, etc., as well as a shape/dimension factor for the body. Clothing complicates the analysis and is often evaluated in terms of insulation (clo) units.[10–12] This variable is discussed in more detail below (cf. Section III.A).

Heat transfer by *radiation* (*R*) is independent of T_{db}. The wavelengths of the radiation exchanged between two objects are related to their respective surface temperatures and the net heat transfer by radiation is proportional to the difference between their absolute temperatures to the fourth power and to the relative absorptive and reflective properties of the two surfaces. In general, the net radiant heat exchange (where h_r = radiant heat transfer coefficient) between a nude person and the environment involves estimation of the mean radiant temperature (MRT). MRT (alternate symbol T_r) can be derived from the temperature (T_g) of a blackened hollow sphere of thin copper (usually 0.15 m diameter) having similar heat transfer characteristics to those for the human body (e.g., $h_c/h_r = 0.178$).[13] Thus,

$$MRT = (1 + 0.222V^{0.5})(T_g - T_{db}) + T_{db} \qquad [°C] \tag{5}$$

Again, clothing complicates the analysis as it does for heat transfer by other modes. Solar heat load poses other special problems.[14] Heating by RF/EM radiation may provide further complications to the analysis of radiant heat exchange between a human and the environment, although Berglund[15] has demonstrated that the latter can be accomplished using conventional methods.

The fourth avenue of heat loss available to human beings is that due to the *evaporation* of water. The latent heat of vaporization of water at normal body temperature is 0.58 kcal/g; thus, the body loses this amount of heat when water is evaporated from its surfaces. Water in the expired air is continually being lost from the respiratory surfaces as is water that continuously diffuses through the skin (insensible water loss). These two avenues make equal contribution to a heat loss totalling ~25% of the resting metabolic heat production (*M*) in a thermoneutral environment. However, the major avenue of evaporative heat loss in humans is sweating. Evaporation depends on the vapor pressures of the ambient air and the evaporating surface and is thus a direct function of both dry bulb (T_{db}) and wet bulb (T_{wb}) temperatures. For a body under conditions of $T_{db} = T_{wb}$, the air is at 100% relative humidity (RH) and thus no water can be evaporated from the skin surface. When the ambient air is at less than 100% RH, evaporation can take place. The interrelationships between these variables, as well as tolerance limits for heat and cold by humans, can be determined from a standard psychrometric chart.

Evaporative heat transfer (E) for a human body can be quantified as:

$$E = 2.2\ h_c A(P_{sk} \cdot Ø_a P_a)F_{pcl} \qquad [W] \tag{6}$$

where the evaporative area (A) is traditionally taken as 1.83 m^2 for an average or "standard" man, P_{sk} is the vapor pressure of water at skin temperature, $Ø_aP_a$ is the ambient vapor pressure at RH = $Ø_a$, h_c is the convective heat transfer coefficient, and F_{pcl} is a dimensionless permeability factor. The reader is referred to the *ASHRAE Handbook*[16] for additional details related to Equation 6. This relationship applies to a nude human and will be confounded by the insulation value and permeability of clothing. It should also be noted that E represents the evaporative cooling allowed by the environment (E_{max}) and is in no way related to the evaporative cooling required (E_{req}) by the individual.

B. ENDOGENOUS HEAT PRODUCTION (M)

The basal metabolic rate (BMR) is defined as the heat production of a human in a thermoneutral environment (33 °C), at rest mentally and physically, at a time exceeding 12 hr from the last meal. The standard BMR for man is ~250 ml O_2 per minute, or 84 W or 0.8 MET (where 1 MET = 58.2 W/m^2). The BMR also corresponds to ~1.2 W/kg for a 70-kg "standard" man. The BMR is altered by changes in active body mass, diet, and endocrine levels, but probably not by living in the heat.[17] In a resting human, most of the heat is generated in the core of the body—the trunk, viscera, and brain—despite the fact that these regions represent only about one third of the total body mass. This heat is conducted to the other body tissues and its elimination from the body is controlled by the peripheral vasomotor system. The range of M for humans, considering work performed and assorted physiological variables such as age, sex, and size, is roughly 40 to 800 W/m^2 (1 to 21 W/kg for "standard" man) depending on physical fitness and level of activity. If deep body temperature is altered, either from heat storage in warm environments or febrile disease, there is a comparable change in M.[18] Similar changes occur when deep body temperature rises during exposure to RF/EM radiation.[19]

In cold environments, shivering is usually preceded by a generalized increase in muscle tone and piloerection which can increase M by about 35%.[20] Active shivering begins locally, recruits more and more of the body's skeletal musculature, and eventually increases heat production to as much as four to five times the resting level (i.e., 160 to 200 W/m^2). However, if cold exposure lasts many hours, a doubling of the resting M is all that can be expected.[21] Further increases in M must be produced by exercise.

C. AVENUES OF HEAT LOSS

Both changes in vasomotor tonus and evaporation of body water through active sweating are mechanisms of heat loss. In general, vasomotor control normally operates to achieve regulation of the body temperature when humans are in thermoneutral environments; sweating is activated in warmer environments and during exercise.

1. Vasomotor Control

Convective heat transfer via the circulatory system is under the control of the sympathetic nervous system. In the cold, vasoconstriction of the peripheral vasculature in arm, leg, and trunk skin minimizes heat loss from the body core to the skin, leaving a residual conductive heat flow of 5 to 9 W/m^2 per °C temperature difference between body core and skin. In thermoneutral environments, when the peripheral vessels are vasodilated, each liter of blood at 37 °C that flows to the skin and returns 1 °C cooler allows the body to lose ~1 kcal or

1.16 W·hr of heat.[22] During vigorous exercise in the heat, peripheral blood flow can increase almost tenfold; this increase is essential to eliminate the increased metabolic heat produced in the working muscles.

Tissue "conductance" represents the combined effect of two channels of heat transfer in the body, conduction through layers of muscle and fat and convective heat transfer by the blood. Tissue thermal conductance is defined as the rate of heat transfer per unit area in the steady-state when a temperature difference of 1 °C is maintained across a layer of tissue [W/m²/°C].[7] Although this variable is not amenable to direct measurement in the living person, tissue conductance can be estimated for resting humans if we assume that all the heat is produced in the core of the body and is transferred to the skin and thence to the environment. Thus we may write

$$H_L = K(T_{re} - T_{sk}) \qquad [W/m^2] \qquad (7)$$

in which H_L represents the heat loss (neglecting that lost through respiration), K is conductance, T_{re} is rectal (or core) temperature, and T_{sk} is the average skin temperature. In the cold (22 to 28 °C), conductance is minimal for both men and women, ranging between 6 and 9 $W/m^2/°C$. In warm environments, conductance increases markedly, with women showing a faster increase than men.[23] In general, measurements of heat transfer made at the skin surface yield only minimal rates of blood flow, since countercurrent heat exchange between arteries and veins in the peripheral tissues affects the observed heat flow.

2. Evaporative Water Loss through Sweating

Evaporation of sweat from the skin is an efficient way of losing heat even in environments warmer than the skin. Thus, evaporative heat loss must take care of both metabolic heat and that absorbed from the environment by radiation and convection. Thermalizing energy from RF/EM fields will be handled similarly to heat produced by normal metabolic processes in resting humans or absorbed by exposure to warm environments.

Normal secretory functioning of the ~2.5 million sweat glands in the skin of a human being is essential to prevent dangerous hyperthermia. Secretion is under the control of the sympathetic nervous system and occurs when the ambient temperature rises above 30 to 31 °C and/or the internal body temperature rises above 37 °C. Local sweating rate has also been found to depend on the local skin temperature.[24] Individuals who are physically fit and/or acclimated to warm environments sweat at a lower internal body temperature and show an increased responsiveness of the sweating mechanism when challenged by exercise.[25] Dehydration or increased salt intake will alter plasma volume and decrease the efficiency of sweating during exercise.[26]

It is important to recognize that evaporative water loss through sweating occurs in humans, the great apes, certain other primates (to a greater or lesser extent), and a few other species (e.g., horses). Many animals (e.g., dogs and cats) use panting (increased respiratory frequency coupled with copious saliva production) to lose heat by evaporation. Other species (e.g., rodents) have no such physiological mechanisms and must rely on behavioral responses that include seeking shade, burrows, or aqueous environments and/or grooming their bodies with fluid (water, urine, saliva) to aid evaporative cooling.[27]

D. NEUROPHYSIOLOGICAL CONTROL OF THERMOREGULATION

The operative characteristics of the thermoregulatory system have much in common with those of an automatic control system involving negative feedback. The body temperature appears to be regulated at a set or reference level. Temperature sensors located in various parts of the body detect thermal perturbations and feed this information to a central integrator and controller. The controller integrates sensory information from various parts of the body and controls the effector systems for heat production and heat loss. If a thermal disturbance shifts the body temperature away from the set or reference level, this change will be detected by the controller through signals from the temperature sensors, and appropriate effector responses will be mobilized to return the body temperature back to the set level.

1. Thermosensitive Neural Tissue

The type and vigor of thermoregulatory effector activity depends on the temperature of the skin; thus, specialized receptors must exist in the outer layers that detect changes in their own temperature. Two populations of cutaneous free nerve endings increase their activity to warming (warm receptors) or to cooling (cold receptors), respectively.[28] Hensel and Kenshalo[29] heated and cooled the facial skin of cats, while recording from single fibers in the infraorbital nerve, and determined response functions for both types of receptor. An increase in skin temperature (T_{sk}) above thermoneutrality (~33 °C) produced a sharp increase in impulse frequency of warm fibers, while cold fibers decreased theirs. Warm fiber response ceased at a T_{sk} above 45 °C indicating thermal death of the fibers. Skin temperatures below 33 °C produced an increase in impulse frequency of cold fibers while suppressing activity in the warm fibers.

Extensive neurophysiological research, beginning more than a century ago with studies of lesioned animals,[30,31] has shown that certain central nervous system (CNS) tissue is either essential for or involved in the control of autonomic mechanisms of heat production and heat loss in endotherms. Barbour,[32] in 1912, showed that a primitive thermode implanted in the anterior brainstem of rabbits could be used to trigger many thermoregulatory responses. When this device was warmed, heat loss responses were initiated; when cooled, heat production and conservation responses were stimulated. These primitive experiments provided the impetus for more than 80 years of neurophysiological research, much of which has involved the search for temperature sensors in those areas of the CNS where lesions or thermal stimulation influence thermoregulatory effector responses.

Many CNS sites have been explored with microelectrodes in experimental animals to reveal the existence of single units whose firing rates depend upon the temperature of the tissue in which they are located.[33–35] In these experiments, local tissue temperature was controlled with implanted thermodes.[36] The activity patterns of CNS thermal sensors resemble those of the skin sensors in many respects. CNS sites that exhibit such thermosensitivity include the medial preoptic area of the hypothalamus, lateral hypothalamus, midbrain reticular formation, medulla, motor cortex, thalamus, spinal cord, and deep viscera.[37–40] Many preoptic cells respond to temperature changes in the skin, the spinal cord, and other CNS thermosensitive sites as well as in their own vicinity.[41–43] Neurophysiologists hold this to be evidence that the preoptic area of the hypothalamus is the site of the "central thermostat" as well as the locus of the "integrator-controller." However, there is still no direct evidence that these thermally responsive neurons are the actual sensors involved with thermoregulation. A further assumption is that the CNS of humans contains thermal sensors in sites analogous to those that have been found in experimental animals. The strongest evidence marshaled for this assumption is the uniformity of the neurophysiological findings between ectotherms and endotherms.[44]

Temperature changes localized in several thermosensitive sites trigger behavioral as well as physiological thermoregulatory responses.[45–49] Any means by which thermosensitive CNS structures may be thermally stimulated has the potential for altering the normal thermoregulatory responses of endotherms. RF/EM radiation is such a potential stimulus. Under appropriate conditions of frequency, intensity, and duration, RF/EM radiation can be absorbed by biological targets in such a way that tissues below the skin surface may be heated. Often, the patterns of absorbed energy are extremely complex, depending not only on the parameters of the radiation, but also on the size, shape, and electrical characteristics of the exposed tissue.[50] Complex absorption may give rise to nonuniform heating of tissue, a potentially unique challenge to the thermoregulatory system.[51]

2. Characteristics of the Thermoregulatory "Integrator-Controller"

The integration of neural signals from the various thermal sensors in the body by a "controller" located (perhaps) in the hypothalamus, the comparison of the integrated signal with a reference level, and the generation of neural effector "commands" are all hypothetical constructs that aid our understanding of thermoregulatory processes. There are nearly as many models of the thermoregulatory "controller" as there are researchers in the field. Although many early models featured two centers for thermoregulation (one controlling heat loss, the other heat production[31,52]), we now believe that one center can accomplish both types of regulation.[36] All models have other features in common: the controlling center resides in the brain and contains a reference against which error signals of various types are compared.

Models have been of many types: verbal descriptions,[31,52,53] pictorial models of the neurophysiological system,[54–57] physical analogs of the thermoregulatory system or portions thereof,[58] electrical analogs,[59] mathematical models,[60–65] neuronal models,[66–68] and chemical or neutransmitter models.[69,70] Hardy[4] gives a comprehensive account of the development of modeling in thermal physiology. The use of such simulation models to predict the human response to RF/EM radiation is very useful and is discussed in a later section of this chapter.

III. LIMITS OF HUMAN HEAT TOLERANCE TO RF RADIATION

The human thermoregulatory response to the deposition of significant amounts of thermalizing energy from RF/EM radiation (i.e., sufficient to elevate the temperature of tissues in all or part of the body) will not differ substantially from the response to heat generated in the body tissues by other means. The basic problem posed by excessive heating of the body, whatever the source, is whether the heat-loss capability of the thermoregulatory system is sufficient to prevent heat storage in the body. Thus for a human being exposed to a RF/EM field, the generalized heat balance Equation 1 may be rewritten:

$$(M \pm W) + A_{RF} = \pm R \pm C \pm E \pm S \qquad [W/m^2] \tag{8}$$

where A_{RF} represents the rate of energy absorption from the RF/EM field. Neglecting the work factor, the sum of the heat production, heat exchange by radiation and convection, and the absorbed RF energy will provide a useful estimate of the evaporative cooling required (E_{req}):

$$E_{req} = M \pm C \pm R \pm A_{RF} \qquad [W/m^2] \tag{9}$$

If the maximum available evaporative cooling (E_{max}), calculated by Equation 6, is less than E_{req}, S will be positive and body temperature will rise.

In general, the degree of heat stress placed on the body can be predicted by the simple ratio of E_{req}/E_{max} which yields a measure of the percentage of the skin surface that is wet with sweat.[71] Values of $E_{req}/E_{max} < 20\%$ yield a state of thermal comfort, while higher percentages indicate tolerance limits. This ratio has also been called the Heat Stress Index (HSI).[72] HSI values in excess of 30% are judged to be uncomfortable and may interfere with concentration and fine motor performance, but are judged to be tolerable; values up to 60% result in finite tolerance times and interfere with work; values between 60 and 100% represent severe, intolerable conditions.

TABLE 1
Comfort-Health-Index—Human Responses

C-H-I	Sensations	Physiological responses	Health hazard
(T_{db} at 50% RH) (°C)			
45	Unbearable	Rapid body heating; failure of body temperature regulation	Circulatory collapse
40	Very uncomfortable, hot	100% Wetted area	Temporary exposure dangerous
35	Uncomfortable, hot	50% Wetted area; physiological strain	"Danger line" for heat stroke
30	Slightly uncomfortable		
25	Warm; comfort zone; cool	25% Wetted area; zone of vasomotor regulation	No health hazard
20	Slightly uncomfortable; cold	Vasoconstriction of hands and feet	Muscular pains; peripheral circulation impaired

The amount of heat storage (S) in the body, calculated from Equation 8, can also be used to predict heat tolerance times (i.e., when $E < E_{req}$).[73] Thus, values of S between 0 and 25 kcal/hr may not even be noticed; $S > 80$ kcal/hr represents the usual voluntary tolerance limit, double this value (160 kcal/hr) incurs a 50% risk of collapse, and triple this value (240 kcal/hr) is intolerable.

A. ENVIRONMENTAL VARIABLES: TEMPERATURE, VAPOR PRESSURE, AND CLOTHING

Tolerance limits for human beings can be determined from a standard psychrometric chart which diagrams the vapor pressure of the ambient air as a function of both T_{wb} and T_{db}. A comfort-health-index (C-H-I) for human responses, extracted from such a chart, is shown in Table 1 (adapted from Hardy[74]). The table gives thermal sensations, physiological thermoregulatory responses, and associated health hazards as related to T_{db} at RH = 50%. This level of RH prevails in much of the U.S. (except for certain high-altitude and desert regions); therefore, the T_{db} given will be appropriate for a high percentage of the U.S. population and can be applied to sedentary humans in the indoor environment. No data exist related to exposure to RF/EM radiation that can be added to this table, but it is most probable that intensities of exposure that provide significant energy deposition in the body core will alter the C-H-I in much the same way as would an equivalent metabolic load. This conclusion was reached by Nielsen and Nielsen[75] who demonstrated the equivalence of passive heating (through diathermy) and active heating (through exercise) on thermoregulation in two male subjects. Stolwijk[76] has predicted, using a mathematical model of the human thermoregulatory system, that at T_{db} = 30 °C and RH = 50% deposition of 500 W of RF power for 30 min into the trunk core of a sedentary man with intact thermoregulation would elevate the temperature of the central blood by ~1.5 °C. This rate of energy deposition is equivalent to moderate exercise (5 METS) such as pedaling a bicycle ergometer at 50 rpm with a load of 600 kpm/min.[77] (The kilopondmeter per minute, kpm/min, is used as the unit of work, *W*, by exercise physiologists. A kilopondmeter is a kilogram mass under normal gravity raised through a distance of 1 m.) If the sweating response were disabled (as when the body is immersed in water or RH = 100%), deposition of 500 W for 30 min into the trunk core would elevate the temperature of the central blood by > 4 °C. Stolwijk's model has also been used to predict shifts in many other thermoregulatory parameters (body temperatures, heat storage, metabolic rate, evaporative water loss, skin blood flow, cardiac output, and effective thermal conductance) that may occur when significant amounts of RF/EM energy are depositied in the head core or leg.[76]

A somewhat different approach has been taken in the *Radiofrequency Radiation Dosimetry Handbook, 4th ed.*.[78] Using a rectal temperature limit of 39.2 °C as a criterion of heat-stress tolerance and an arbitrary 60-min period of exposure to electromagnetic radiation, the radiation conditions (SAR_{60}) were determined that would fulfill these criteria under different environmental conditions (i.e., T_{db} from 30 to 68 °C and RH = 80, 50, and 20%). Equation 8 was used to predict the maximum whole-body specific absorption rate (SAR) that a healthy man could tolerate for 1 hr in a specified environment. One such calculation yielded an SAR_{60} of 3.11 W/kg for T_{db} = 40 °C and RH = 80%. Then the SAR curves for an average man[78] were used to calculate the power density for any given frequency and polarization that will produce this SAR_{60}. These SAR curves, derived from a homogeneous prolate spheroidal model of man at 10 W/m^2, are found in Figure 6.3 of the work cited. Thus, under the assumption of uniform power absorption over the entire body, an SAR_{60} of 3.11 W/kg would represent a power density of roughly 130 W/m^2 for-irradiation by an electromagnetic plane wave (E-polarization) at the resonant frequency of 70 MHz. SAR_{60} values calculated in this manner are not to be interpreted as safe exposure limits, however; they are simply estimates of the upper tolerance levels for a healthy man with normal heat loss capabilities.

Such calculations assume that the subject is a healthy nude male in a resting state. The clothed human body presents a resistance to the flow of heat from the skin to the environment that is a direct function of the thickness of the air layer trapped by the clothing. The standard insulation unit, called the clo[10], was derived empirically and equals 0.18 °C·m^2·hr/kcal or 0.155 °C·m^2/W. The insulation value of a standard business suit is defined as 1 clo. Typical indoor clothing worn in offices today ranges from 0.4 clo in summer (short, light dress; light slacks and short sleeved shirt) to ~1.0 clo in winter (heavy slacks, blouse, light sweater, and jacket; heavy trousers, shirt, sweater, and jacket).[79] A rule of thumb[80] states that for each 0.1-clo deviation from the usual 0.6-clo baseline insulation for sedentary persons doing light office work (1 MET), the air temperature for comfort can be offset by 1 °F (0.55 °C). This temperature offset can be doubled when the workload is increased to 4–5 METS.

The limiting factor to lowering office temperatures for energy conservation appears to be related to cold hands and feet. It is difficult to design practical mittens that warm the hands and still allow the manual dexterity required for most jobs. A proposal[81] to provide RF/EM comfort heating of human beings in otherwise cool surroundings is an interesting alternative. On the other hand, avoidance of thermal discomfort in hot environments that may include a source of RF/EM radiation will be accomplished primarily by minimizing the percentage of the body surface area that is wet with sweat. This can be accomplished by (1) increasing air movement, (2) reducing the ambient vapor pressure, and (3) removing as much clothing as possible consistent with acceptable social norms.

B. EXERCISE

During physical exercise, the human thermoregulatory system functions in the presence of an internal disturbance. When a proportional, negative-feedback-control system (e.g., thermoregulatory system) encounters such a disturbance, it usually maintains an offset in its regulated variable (the regulated internal body temperature, in this case). This occurs during muscular exercise. At the initiation of activity, heat production in the skeletal muscles rises rapidly, producing a temporary large increase in heat storage within the body. The rising body temperature generates an error signal relative to the "set" body temperature, which drives the mechanisms of heat loss at an ever-increasing rate until heat loss equals heat production. When a balance is attained, the body temperature will equilibrate at a hyperthermic level that is characteristic of the level of exercise being performed.[82] Exercise represents the single condition that provides a maximal strain to the thermoregulatory system under normal circumstances and allows measurement of the various modes by which the body adjusts to this strain.[83]

Exercise changes neither the "set point" for body temperature nor the normal operating characteristics of the thermoregulatory system. It seems reasonable to postulate that whole-body absorption of thermalizing RF/EM energy may result in the same thermoregulatory consequences as the energy generated through exercise. In a landmark study that tested this hypothesis, Nielsen and Nielsen[75] used short-wave diathermy to deposit heat directly into the deep tissues of the trunk in human subjects. In other experiments, the same subjects exercised on a bicycle ergometer at a work rate that was adjusted so that the heat load during cycling and diathermic heating was the same (approximately 5 METS). Experiments of both types (passive and active heating) were conducted at four discrete ambient temperatures ranging from ~10 to 27 °C. In the steady state (after 30–60 minutes of diathermy or exercise), the rectal temperature increased by the same amount during the two procedures and appeared to be independent of the ambient temperature. At the same mean skin temperature, thermal conductance (as assessed by changes in skin blood flow) and sweating rates were much higher with passive and active heating than during normal rest. But under the two heating regimes, both skin blood flow and sweating rate were the same. Thus, passive heating of the trunk by diathermy and the heat generated during active exercise produce the same kind of thermal disturbance in the body as a whole, although the distribution of heat in individual tissue compartments of the body may have been different in the two cases. It is evident that convective heat transfer via blood flow provides the major avenue by which thermal averaging over the body's tissues occurs during both active and passive heating.

Nielsen and Nielsen's study indicated that the major stimulus for increased heat dissipation during exercise, caused by the work itself, is the increased internal body temperature. On the other hand, the ambient temperature, within wide limits, controls heat loss via changes in skin temperature. These important findings also demonstrate that the thermoregulatory consequences of whole-body energy deposition by RF/EM radiation may be predicted by the consequences of equivalent heat loads produced by exercise. As discussed in greater detail below, this rationale underlies predictions of changes in thermoregulatory responses to microwave power deposition made using mathematical models of the human thermoregulatory system.

C. FEBRILE STATES

It is important to differentiate between the hyperthermia produced by exercise and that occurring during febrile disease. Exercise produces an elevation in body temperature to a level above the normal, regulated (or "set") level. The magnitude of the heat dissipation response that is stimulated during exercise is directly related to the magnitude of this temperature offset.[84] On the other hand, during fever the increase in the regulated body temperature is defended in the same way as is the normal body temperature during euthermia. During the generation and maintenance of fever the thermoregulatory system behaves in a manner consistent with an elevation of the "set" temperature.[85] This was demonstrated in a human subject during daily bouts of exercise and rest.[86] When the subject contracted a febrile disease, his rectal temperature rose and fell during the periods of exercise and rest, but at an overall level about 2 °C higher than the other (healthy) subjects. An elevated internal temperture at the threshold for vasodilation also occurs during fever.[83]

Fever is generated differently in different thermal environments. In warm environments, heat loss will be curtailed and vasoconstriction will occur; if these responses are inadequate to increase heat storage, heat production will rise. In cold environments where heat loss responses are already minimal, greatly increased heat production (vigorous shivering) is the only way to raise the body temperature. Once a pyrogenic substance introduced into the body displaces the set temperature, the controller for thermoregulation will mobilize any appropriate effector mechanism to increase the storage of body heat.[87] Behavioral responses that increase skin temperature (by selection of a warmer environment or by increasing body insulation) generate an identical fever to that produced by changes in physiological responses when a given dose of pyrogenic substance is administered.[88]

The evolution of a febrile episode is well understood, although the actual mechanism by which the set point is elevated by the pyrogen remains a mystery. When the set temperature is elevated by a circulating pyrogen, a large load error is generated because the body temperature is below the new set point. Heat loss is curtailed, vasoconstriction occurs, and metabolic heat production increases in an attempt to raise the body temperature to the new set level. This "chill" phase gives way to a "plateau" phase once the body temperature has risen to the febrile level. At this point, the error signal no longer exists and the effector responses return to the magnitudes existing before fever onset, regulating in normal fashion around the new elevated set point. When the disease runs its course or an antipyretic is taken, the normal set point is restored, giving rise to a new load error of opposite sign to that occurring at fever onset. During this "crisis" phase, vigorous heat loss responses are stimulated (vasodilation and sweating) which rapidly return the body temperature to normal. Thus, it is clear that the hyperthermia that may result from exposure to RF/EM radiation differs greatly from that occurring during febrile episodes, resembling much more closely the hyperthermia of exercise.

D. HUMAN DATA FROM CLINIC AND LABORATORY

No mammalian species is as well equipped, both physiologically and behaviorally, as human beings to withstand heat generated in the body by exogenous and endogenous sources. Thermalizing energy deposited in the body during exposure to RF/EM fields is no exception. A comprehensive study[89] estimates that more than 99% of the U.S. population is exposed to background RF radiation at less than 1 $\mu W/cm^2$. At the resonant frequency for humans, this value represents a whole-body SAR of 0.03% of the resting M, completely insignificant for thermoregulation. Even the whole-body SAR = 0.4 W/kg, adopted as the basis for the ANSI/IEEE C95.1-1992 guideline,[90] represents only 35% of the resting M (equivalent to donning a light sweater) and provides little more than noise during most daily activity.

Thermoregulatory processes are ongoing in each of us all the time; small perturbations in endogenous heat production or in the thermal characteristics of the environment result in finely-tuned thermoregulatory adjustments and maintenance of a stable internal body temperature. Most humans have the capacity to cope with exercise or work loads, often undertaken in thermally stressful environments, that are the equivalent of 15 times the resting M. Exceptional rates of human sweat production, plus efficient behavioral thermoregulation, ensure a minimal rise in deep body temperature. It is not surprising, then, that the few reports of humans exposed to modest levels of RF/EM energy in laboratory or clinic show no adverse health effects.

In addition to the extensive use of microwave energy for diathermic heating of body tissues[91] and as an adjunct to cancer treatment,[92] techniques have been developed that use RF/EM energy for rewarming the hypothermic body[93] or individual body parts.[94,95] RF coils, energized at frequencies of 13–30 MHz, heat deep tissues more rapidly and efficiently than warm-water immersion, insulated sacks, or nichrome wire coils energized by DC. Skin areas remote from the RF coils are warmed convectively by heated blood while no adverse thermal effects have been observed.

One study,[96] designed to induce temporary sterilization in adult males by high intensity RF/EM exposure of the testes, provided results of long-term, intermittent, partial-body exposure. Thirteen volunteers underwent weekly, 30-min localized exposures (915 or 2450 MHz) at power levels (20–30 watts) sufficient to raise scrotal skin temperature by 10 °C. Seven subjects underwent 100 sessions, the remainder somewhat fewer. Biopsies of testicular tissue, conducted 6 months after the last exposure session, showed considerable cellular damage but no significant morphological abnormalities. Although the RF treatment reduced spermatogenesis, sterilization was not reliably achieved; two men fathered children during the course of treatment.

A considerable amount of literature is now available that describes thermal responses of human subjects exposed to RF fields during MRI procedures. Various skin, corneal, and deep body temperatures, skin blood flow, and other physiological variables have been assessed during MRI with a 1.5 T head coil at 64 MHz,[97,98] before, during and after MRI in a 1.5 T body coil,[99] and during scrotal MRI at 1.5 T.[100] Scan durations ranged from 8 to 30 min and estimated peak SARs ranged from 0.56 to 4.0 W/kg. Because surface heating predominates during MRI, observed increases in deep body temperature (generally ~0.1 °C) were judged to be physiologically inconsequential; also, modest increases in skin temperatures and skin blood flow were judged to be trivial. No tissue temperature changes occurred when subjects were exposed to 1.5 T static magnetic fields only,[101,102] indicating that the RF component of MRI was correlated with temperature rise in body tissues.

IV. THERMAL SENSATION AND THERMOREGULATORY BEHAVIOR

Since thermalization can occur in body tissues during RF/EM exposure, such radiation must be considered as part of the thermal environment to which humans and animals may potentially be exposed.[81] Although physiological responses (e.g., vasodilation, sweating) may be triggered automatically by thermal stimuli, the sensation of tissue heating is necessary for the initiation of appropriate behavioral action. Excitation of temperature-sensitive neural structures residing within the outermost 0.6 mm of mammalian skin[103] is believed to underlie the sensation of changes in skin temperature. Whether or not an exposure to RF/EM radiation produces a sensation of warmth depends on many parameters of the signal (e.g., frequency, modulation, intensity, duration, and the exposed surface area of the body). Many of these same parameters influence the magnitude of thermal sensations derived from exposure to IR radiation.[104–111] IR energy is absorbed in the uppermost layers of the skin in close proximity to the temperature-sensitive structures; a similar absorption profile will be obtained for the highest microwave frequencies (10 GHz or higher). However, lower RF/EM frequencies will be absorbed in complex patterns at other depths, rendering prediction of thermal sensations difficult. It has been suggested that psychophysical studies of thermal sensations aroused by microwaves should be guided by past experience with more conventional means of thermal stimulation, such as IR radiation.[6]

A. THERMAL SENSATIONS PRODUCED BY MICROWAVES

Absolute thresholds for the detection of RF irradiation by human observers have been determined in several studies,[112–118] all of which have involved brief exposures (10 sec or less) and restricted areas of forehead or forearm skin. In general, the shorter the wavelength of the irradiation, the less energy is required to provoke a just-detectible sensation of warmth.[119] When a 37 cm^2 area of the forehead was irradiated for 4 sec, the mean absolute threshold of warmth was 335 W/m^2 for 3 GHz irradiation, 126 W/m^2 for 10 GHz, and 42 W/m^2 for frequencies > 1000 GHz (far-IR).[115] Irradiation of small areas of skin by 3 or 10 GHz pulsed (2500 pps, 0.4 μsec pulse width) microwaves has to last at least 5 sec in order for the minimal intensity to evoke a thermal sensation, the exact intensity depending upon the area stimulated.[113,115] At shorter durations, stimulus intensity has to be greatly increased in order to evoke comparable sensations of warmth. An unusual aspect of the sensation derived from RF/EM irradiation is its persistence following cessation of the stimulus.[112,115] This property, together with longer-than-average latencies to detection, probably derives from the thermal inertia of the affected tissues[116,118] and the greater volume of tissue involved when deeply penetrating radiation is concerned.[50]

One study incorporated indirect assessment of the energy absorbed during 10 sec exposures of the human forearm (average area = 107 cm^2) to 2450 MHz CW microwaves.[118] Sensations of warmth were reported by the subjects when the measured energy dose rate was ~900 J/m^2, as opposed to ~200 J/m^2 when the same skin area was exposed to IR radiation. These threshold values correspond to incident powers of 270 and 17 W/m^2, respectively, and are similar to the results reported above.

B. CHANGES IN THERMOREGULATORY BEHAVIOR IN THE PRESENCE OF MICROWAVES

Thermoregulatory behavior refers to voluntary actions by an organism that control the thermal characteristics of the air-skin interface, thereby facilitating regulation of the body temperature at a stable level. Few experimental data exist that relate changes in thermoregulatory behavior of human subjects to exposed RF/EM fields. However, experimental results from animal subjects afford meaningful insights with good predictive value.

Low intensity microwave fields influence thermoregulatory behavior like other sources of heat. Lizards bask in proximity to a 900 W/m^2 microwave source and regulate their body temperatures thereby.[120] Rats, trained to press a lever for IR heat in the cold, suppress lever pressing in inverse proportion to the strength of an imposed 2450 MHz microwave field.[121] Mice select a cooler part of a thermal gradient as the imposed RF field inside a waveguide becomes more intense.[122] A discrete threshold power density governs such behavioral changes[123] and the behavior, once initiated, persists until the field is extinguished.[124] Rhesus monkeys learn to press a lever for intermittent pulses of 6.4 GHz microwaves when they are cold.[125] Similarly, squirrel monkeys can be trained to turn an RF source on and off; the chosen duration of RF field exposure is inversely proportional to the field strength.[126] Even at whole-body SARs that are the equivalent of about two times the resting M (expressed in W/kg), changes in thermoregulatory behavior can regulate the body temperature at a level close to normal.[127]

If only part of the body (such as head or trunk) is exposed to RF/EM radiation, the change in selected ambient temperature is governed by an integrated energy deposition over the whole body, not by energy deposited in some specific locus such as the brain. Indeed, the hypothalamic "thermostat" appears to play no more vital a role in behavioral thermoregulation during RF exposure than do other thermosensitive regions of the body.[128]

Behavioral thermoregulatory responses are mobilized in orderly fashion when squirrel monkeys undergo whole-body exposure at the resonant frequency (450 MHz, CW, E-polarization).[129] The threshold SAR for behavioral alteration is about 40% higher than the threshold SAR determined in identical experiments when exposure is at a supra-resonant frequency of 2450 MHz.[123] During exposure at resonance, behavioral responses regulate the skin temperature at the normally-preferred level, but because of the deep penetration of RF energy at resonance, a stable hyperthermic offset or bias occurs in the deep body temperature. As noted above, this situation is identical to that occurring during exercise. Thermophysiological responses of peripheral vasodilation and sweating, manifested on the skin surface, are stimulated at SARs similar to the behavioral threshold, indicating the possibility that such responses could serve as auxiliary cues to behavior.

Extensive training of the experimental animal in thermoregulatory behavior[123] is certainly advantageous and may be essential to survival when RF/EM fields are present.[130] The presence of visible or auditory cues correlated with RF/EM irradiation facilitates the learning of suitable responses leading to escape from potentially lethal fields,[131,132] while careful shaping of appropriate responses accelerates such learning dramatically.[133] Previous experience with similar stimulus-response contingencies has also been demonstrated to be helpful.[126] Extrapolation of these general findings to the human condition, especially in industrial settings that incorporate high-power sources of RF fields, has led to the conclusion that education of the worker, together with environmental signals (visual and auditory) that are associated with equipment function and malfunction, could aid recognition of the potential for over-exposure and insure timely retreat from potentially-injurious fields.[133]

V. ADDITIONAL SUPPORTING DATA FROM ANIMAL STUDIES

In addition to the behavioral data summarized above, several studies have assessed individual physiological responses of heat production and heat loss in the presence of RF/EM fields with varying degrees of success. Experimental animals must be restrained to afford the opportunity to monitor thermoregulatory variables, and restraint in and of itself often imposes a stress that interferes with normal thermoregulation. On the other hand, it is difficult to assess changes in body temperatures and other thermoregulatory responses in unrestrained, freely-moving animals, particularly if a RF/EM field is present. An added complication is that many of the animal studies have involved exposure solely to 2450 MHz radiation. There is a continuing, clear need for definitive experiments of this type, using both higher primates and human subjects, and a wider range of frequencies.

A. THRESHOLD EFFECTS

For any given species, under any given environmental conditions, an intensity of imposed RF/EM radiation can be determined that will reliably initiate or alter whatever thermoregulatory response is appropriate to those environmental conditions. Such determination requires, for comparison, adequate baseline or control data collected under identical conditions but with the RF/EM field absent. The thermoregulatory characteristics of the species in question[78,134] should be invoked as a guide to correct response selection, and appropriate consideration should be made for differences in humidity and air movement, as well as T_{db}. The RF/EM intensity so determined can be designated a threshold for response mobilization. Some[135,136] prefer a statistical definition of "threshold" derived from a "hockey-stick" analysis[137] of experimental data collected across a wide range of RF intensities at a specific T_{db}. For either method, coherent data are essential for accurate determination of a threshold. By definition, sub-threshold RF/EM intensities will not produce response alteration or mobilization.

In general, RF/EM intensities above an experimentally-determined threshold will also alter the response in question, usually to a degree that is intensity dependent.[19,123,138–140] If the field strength is great enough, the response under observation will be altered maximally and the next response in the thermoregulatory hierarchy will be mobilized. Adair[141] provides a detailed discussion of these concepts and their implications for human thermoregulation in the presence of RF/EM fields.

1. Metabolic Responses

The metabolic heat production, M, of endothermic mammals equilibrated to cool environments will be elevated by an amount that is directly proportional to T_{db}. During acute, far-field exposure of the whole body to RF/EM fields, the elevated M of nonhuman primates (Macaca mulatta or Saimiri sciureus) will be reduced by an amount proportional to the field strength or SAR.[127,138–140,142–144] As a result of this response adjustment, the internal body temperature is usually regulated within the limits normal for the species. Similar results have been demonstrated in rodents after a period of whole-body irradiation in either a multimodal cavity[145] or a waveguide.[146,147] Furthermore, chronic low-level RF irradiation produces no measureable alteration in the normal metabolism of infant rats,[148] of rats irradiated throughout their lifetimes,[149] or of squirrel monkeys irradiated for 15 weeks.[150]

A threshold SAR must be surpassed before a reliable reduction in M occurs; this threshold is between 0.5 and 1.5 W/kg in nonhuman primates[139,140,143] but has not been systematically explored in other species. It should vary with T_{db} in accordance with the thermoregulatory characteristics of each species. Abundant evidence indicates that both the threshold and the magnitude of the M reduction depend, in regular fashion, on the magnitude of the cold stress.[19,127,139,140,142,147] If only part of the body is exposed to RF/EM energy, the magnitude of the change in *M* reflects the total absorbed energy, as though it were integrated over the whole body.[128]

The maximum whole-body absorption of RF/EM energy occurs when the long axis of the body is parallel to the electric field vector (E-polarization) and is about 0.4 of the free-space wavelength (resonant frequency).[78] RF/EM exposure of nonhuman primates at their resonant frequency yields somewhat less efficient thermoregulation than does exposure to other frequencies.[151,152] Although the threshold for M reduction may be lower at resonance,[140] the magnitude of the response change may be less for a given SAR than at non-resonance and the body temperature may rise. The hyperthermia is modest, however, and well regulated even at whole-body SARs similar (in W/kg) to the level of resting M.[143] The situation is identical to that occurring in humans during exercise[82,83] and the experimental data from the exercise literature, as well as the RF/EM literature, reinforced by Nielsen and Nielsen's results,[75] suggest that human exposure at resonance may pose no greater hazard than exposure at other frequencies.

2. Vasomotor Responses

Vasomotor tonus in the cutaneous blood vessels of the extremities can be indexed by changes in local skin temperature during exposure of the whole body to RF/EM fields. At a T_{db} just below that at which the vessels normally vasodilate, the stage is set to initiate peripheral vasodilation as soon as *M* has been reduced to the resting level. In laboratory animals, the vessels of the tail and the ears usually vasodilate before those of the feet. Generally, once the threshold field strength has been achieved, the response occurs rapidly at field onset and the magnitude or degree of vasodilation is a direct function of whole-body SAR[138,139,153] or the total heat load.[154] As T_{db} increases to higher levels, the SAR at threshold is reduced.[138,139,141,142] Extinction of the RF/EM field induces rapid vasoconstriction. In nonhuman primates, both the threshold and the degree of vasodilation depend on the imposed frequency as well as the T_{db}, although in a species-dependent manner. In rhesus monkeys, the closer the exposure frequency to whole-body resonance, the less energy is required to induce vasodilation at a given T_{db} and the greater the response magnitude at a given SAR.[140] However, for squirrel monkeys exposed to RF/EM fields in cold T_{db}, higher SARs are required to induce vasodilation during exposure at resonance than at a frequency above resonance.[143] This finding relates to the different thermoregulatory characteristics of the species in question[134] and the peripheral control of M reduction in the cold.

Peripheral vasodilation can also occur spontaneously during the course of a prolonged RF/EM exposure carried out in cool environments.[138,139,141,142,155] In this case, vasodilation is mobilized because the internal body temperature slowly rises during the exposure, eventually surpassing a threshold for initiation of the response. It is clear from these data that vasomotor control is exerted by a combination of central and peripheral thermal inputs. In addition, changes in the caliber of blood vessels deep in the body, as well as in the periphery, accompany RF/EM exposure and the rate of local blood flow increases dramatically whenever the temperature of the heated tissue exceeds 41 to 42 °C.[23,156] This phenomenon forms the basis of the treatment of localized malignancies by microwave hyperthermia.[92]

On the other hand, attempts to demonstrate "thresholds" for vasodilation in laboratory animals held at thermoneutral or warm T_{db} are doomed to failure because the vasomotor response is already partial or complete when the exposure begins. For example, very low SAR thresholds (at 600 MHz) for an increase in the temperature of the ear skin have been reported for rabbits held at T_{db}s of 20 and 30 °C inside a waveguide. Associated measures of the SAR threshold for an increase in body temperature (T_{co}) in the same animals were low at T_{db} = 20 °C and virtually indeterminate at T_{db} = 30 °C.[135] An evaluation of these results in terms of the thermoregulatory capability of the rabbit reveals that the zone of thermoneutrality ranges from 15 to 25 °C and the upper survival limit is 32 °C.[157] Thus, at T_{db} = 20 °C the animal subjects were already partially vasodilated and at T_{db} = 30°C they were fully vasodilated and close to the upper limit of survival. Since the rabbit pants when heat stressed, a measure of respiratory rate developed for use in RF/EM-exposed rodents[158,159] could have been used to advantage in the study cited.[135]

3. Evaporative Responses

When the peripheral vasculature is fully vasodilated, dry heat loss from the body nears its maximum. To prevent significant heat storage and a resulting rise in body temperature, heat loss by evaporation must be initiated. Any attempt to predict the evaporative capability of a particular endothermic species must consider its thermoregulatory characteristics,[134] the avenue of evaporative heat loss available (panting or sweating), or whether such capability may not exist (as is the case with rodents). Fortunately, humans have an extraordinary capability to lose body heat through sweating when T_{db} rises above 30–31 °C or the deep body temperature exceeds 37 °C.[160]

Sweating from the foot of squirrel monkeys equilibrated to a T_{db} of 34 °C can be reliably initiated by RF/EM exposure (2450 MHz) at a threshold SAR equivalent to 20% of the animals' resting *M*.[142,161] Below this T_{db}, the threshold SAR is linearly related to the exposure T_{db}. Like heat production, the magnitude of the sweating response depends on the integration of absorbed RF/EM energy over the whole body, not energy deposited in some locus such as the "central thermostat" of the hypothalamus. Indeed, thermoregulatory sweating occurs in the presence of a RF/EM field even when the hypothalamus is artificially cooled to prevent its temperature from rising.[128] During partitional calorimetric studies at several T_{db}, cool (20 °C), neutral (26 °C), and warm (32 °C), steady-state tolerance limits have been determined for squirrel monkeys, exposed to 2450 MHz CW microwaves, in terms of both power density and SAR.[127] The limiting response in all cases was sweating, which is restricted to the extremities in this species.[162] Above the sweating threshold, the magnitude of the change in sweating rate at a given T_{db} was directly related to the power density of the imposed field. These responses are qualitatively similar to those of human subjects.[163] The local sweat rate of exercising humans increases linearly as the temperature of the body core rises during exercise at a given ambient temperature. But a higher core temperature is necessary to initiate sweating at lower skin temperatures and the rate of increase in sweating during exercise becomes lower as the environment becomes cooler. This comparison indicates that, in the monkey, the rate of thermoregulatory sweating initiated by RF/EM exposure depends not only on the ambient (skin) temperature, but also upon the temperature of the body core as it is directly increased by absorbed RF/EM energy.

Sweating from the calf of rhesus monkeys, during RF/EM exposure at the resonant frequency in neutral T_{db} of 26 and 32 °C, is reported to occur at somewhat higher SARs, equivalent to 80% of the animals' resting M.[164] At T_{db} = 26 °C, peripheral vasodilation preceded the onset of sweating as would be predicted. However, sweating in neutral T_{db}, like reduced M in cooler environments,[139] failed to prevent a substantial rise in the deep body temperature of the rhesus monkey during RF/EM exposure at resonance. Lotz concluded that careful control over ambient conditions (temperature and air movement) is essential during such exposure to ensure that heat loss from the skin is not impeded.

Gordon[122,165] describes a threshold SAR of 29 W/kg for initiation by 2450 MHz microwaves of an increase in the evaporative water loss of mice held at T_{db} = 22 °C inside a waveguide. However, mice and other rodents neither sweat nor pant when hyperthermic; thermoregulation is restricted to behavioral maneuvers such as spreading saliva or urine on the pelt to aid evaporative cooling. None of these responses could be observed nor could M or body temperatures be measured to provide evidence that the animals were thermoregulating normally. Since the T_{db} in the waveguide was well below thermoneutrality for the mouse, changes in M and conductance, not changes in E, would first be anticipated during RF/EM exposure. The latter should occur only after M was reduced to resting level and the animal was fully vasodilated. With this perspective, the high evaporative threshold for the mouse may be more easily understood.[166,167]

B. INTENSE OR PROLONGED EXPOSURE

The thermoregulatory consequences of exposure to intense microwave fields have been explored by Michaelson and co-workers.[168,169] Whole-body exposure of dogs to 2880 MHz pulsed (360 pps, 2 to 3 μsec pulse width) microwave fields at an average power density of 1 kW/m² (SAR = 3.7 W/kg) for 6 hr or 1.65 kW/m² (SAR = 6.1 W/kg) for 2 to 3 hr produces a characteristic triphasic change in internal body temperature. These phases are (1) an initial increase in core temperature, (2) a plateau phase at the hyperthermic level, and (3) thermoregulatory failure. Presumeably, specific heat loss mechanisms (e.g., panting) mobilized by these exposures are able to counterbalance partially the thermalizing energy absorbed by the animal, but only temporarily. The strain placed upon the thermoregulatory system ultimately exhausts the heat loss capabilities of the dog and death ensues due to hyperpyrexia unless the animal is removed from the field.

The environmental temperature at which the exposure occurs is also very important.[170] Thus, dogs can tolerate exposures at SARs of 3.7 and 6.1 W/kg when T_{db} = 11 °C and exposures at 3.7 W/kg when T_{db} = 22 °C without becoming hyperthermic. However, at T_{db} = 40.5 °C, dangerous hyperthermia can occur within 20 min at SAR = 6.1 W/kg (a value roughly twice the resting M of the dog). Hydration during microwave exposure permits an extended tolerance at high SARs, presumeably through an increased capacity for respiratory evaporative heat loss (panting). One of Michaelson's notable findings was the development of tolerance to RF/EM exposure at SAR = 6.1 W/kg as the number of such exposures increased; e.g., it took 60 min on the first day, but 220 min on the 34th day, for dogs to generate a rise in rectal temperature of 1.5 °C. This phenomenon resembles acclimatization to hot environments (cf. Goldman[171]).

Thermographic studies on tissue-equivalent models of humans and animals[49,172] have demonstrated regions of high local SAR during exposure of the whole body to planewave RF/EM fields. Wrists, ankles, and neck (also animal tails) are predicted to be foci of enhanced energy absorption where excessive elevations of tissue temperature may occur. Studies by Krupp[152] of anesthetized rhesus monkeys exposed to planewave 210-MHz RF/EM radiation at power densities from 50 to 270 W/m² demonstrate substantial increases in core temperature, but no evidence for tissue "hot spots" in wrist, ankle, thigh, or biceps. Increased convective heat transfer via blood flow during such exposure clearly results in the protection of individual tissues.[76,173] Similarly, mathematical models[174–177] have predicted regions of high local SAR within the brain of animals exposed to RF/EM fields. Should such "hot spots" occur in the vicinity of the hypothalamic thermoregulatory center, this would have great import for normal thermoregulation. Burr and Krupp[178] mapped the distribution of RF energy absorption in the heads of live rhesus monkeys and concluded that normal thermoregulatory processes (especially increased blood flow) prevent inordinate increases in local temperature until a threshold SAR of 5 W/kg is reached or exceeded. Significant elevation of the brain temperature of anesthetized rats was measured following exposure to 1600 MHz microwaves at 200 to 800 W/m².[179] The deep brain temperature of conscious rats can rise by as much as 1.0 °C during 60-min exposure to 2800 MHz pulsed fields at low power densities.[180] Studies of squirrel monkeys[181,182] featured hypothalamic heating and cooling by an implanted thermode (tube perfused with temperature-controlled fluid) during microwave exposure to a head-resonant frequency. The experiments not only failed to find evidence for brain hot spots, but also led to the conclusion that the hypothalamic thermoregulatory center is no more important than other thermally-sensitive sites in the body for the mobilization of autonomic thermoregulatory responses. Other reports[183,184] indicate that the anterior hypothalamic temperature of squirrel monkeys engaged in behavioral thermoregulation seldom rises more than 0.2 to 0.3 °C above normal baseline levels when the animals are exposed to 2450-MHz CW fields; indeed, efficient thermoregulatory behavior acts to prevent such a temperature rise by increasing the skin-to-ambient temperature gradient, thereby facilitating heat loss from the body.

It is essential that quantitative studies of thermoregulatory function always be conducted at the same time of day, because a circadian rhythm of regulated changes in body temperature is characteristic of all endothermic species. Lotz[185] has demonstrated an influence of this circadian rhythm on the elevation of rectal temperature in rhesus monkeys exposed or sham-exposed to 1.29 GHz pulsed radiation at a whole-body SAR of 4.1 W/kg. The 8-hr exposures occurred either during the day or during the night. The change in body temperature resulting from these exposures was nearly identical (~1.6 °C), but the peak temperature was ~1.0 °C lower at night, a value equal to the normal nocturnal fall in regulated body temperature in this species. Plasma cortisol was elevated only during the day exposure. Lotz suggested that this response may depend simply on the absolute level of body temperature rather than the magnitude of temperature change during RF/EM exposure.

Studies too numerous to mention corroborate the general finding in many species that absorbed doses of microwave energy that are equivalent to the resting metabolic heat production will elevate the deep body temperature of an experimental animal by 1 °C or more. Of special note is the work of deLorge.[186–189] who has correlated the rise in deep body temperature with the cessation of food-reinforced operant behavior (work stoppage). Estimates of whole-body SAR may not be the best predictors of work stoppage from one frequency of RF/EM radiation to another; however, deLorge suggests that energy deposition in the head or brain may be a more reliable predictor than is the increase in body temperature. This suggestion runs counter to results described above,[183,184] which indicate that changes in hypothalamic temperature are minimal and may be irrelevant in this context. Since the animals in deLorge's experiments were often tested in restraining devices, it should be possible to monitor local body (including brain) temperatures, as well as individual thermoregulatory responses, in order to derive a more complete assessment of the effects of RF/EM radiation on changes in thermoregulatory processes during ongoing behavior.

C. DRUG—MICROWAVE INTERACTIONS

More than 20 years ago, it had been recognized that experimental animals treated with certain drugs were especially vulnerable in the presence of microwave fields. Michaelson and colleagues reported that dogs display a greater susceptibility to microwave heating after administration of pentobarbital sodium, morphine sulphate, or chlorpromazine, indicating that mechanisms of heat loss may be compromised by treatment with these drugs.[168] A similar effect was demonstrated[190] in rats rendered hypothermic by cortisone injections; following this treatment, RF/EM exposure at 40 W/kg produced greater increases in core temperature than occurred in nontreated control animals. However, increased vulnerability in rabbits and rats was also shown to take the form of a delayed rise in deep body temperature, reflecting the hypothermic action of the anesthetic.[168] On the other hand, the duration of hexobarbital anesthesia in mice[191] and pentobarbital anesthesia in rabbits[192] was found to be directly related to the field strength of a 2450 MHz microwave exposure at thermalizing levels. This shortening of anesthesia apparently results from a higher rate of redistribution of the drug through increased blood flow, a thermoregulatory response. Still other studies[193,194] have reported an increased survival of rodents, treated with general anesthetics (including chloroform, halothane, ether, chloral hydrate and barbiturates) prior to RF/EM exposure at high power densities.

The picture is further complicated, however, by other factors. Lai, et al.[195] have reported that both the orientation of the animal during exposure (anterior vs posterior irradiation) and the order of presentation of anesthetic and microwaves can influence the magnitude and duration of anesthesia. In these studies, rats were exposed for 45 min to 2.45 GHz circularly polarized, pulsed microwaves (2 μsec pulse, 500 pps) at SAR = 0.6 W/kg. When microwave exposure preceded injection of the anesthetic, recovery from hypothermia was lengthened. When anesthesia preceded irradiation, posterior-exposed animals recovered from hypothermia significantly faster than anterior- or sham-exposed animals. Polarization is also important. When Ketamine-anesthetized rats were exposed to 1.2 GHz CW microwaves at an SAR of about 8 W/kg, E-polarization generated greater tissue heating (colonic, tympanic, subcutaneous and tail surface) than H-polarization.[196] Similar results were found for 2.45 GHz, although tissue heating was more superficial at this higher frequency.[197]

Smialowicz[198–200] has used drug-microwave interactions to demonstrate subtle heating produced by low intensity ($\leq$ 10 W/m^2) exposure to 2.45 GHz CW microwaves in both mice and rats. Mice injected I.P. with 5-HT creatinine sulphate complex (20 mg/kg) and rats injected I.V. with *Salmonella typhimurium* LPS (100 μg/kg) became hypothermic when held at T_{db} = 22 °C. Postinjection exposure to RF/EM radiation generated an increased colonic temperature that was a direct function of the power density of the imposed field. Saline-injected control animals, similarly exposed, did not show comparable elevations of core temperature. Thus, chemical impairment of thermoregulatory mechanisms can permit detection of significant changes in the body temperature of experimental animals exposed to low-intensity RF/EM fields that were formerly believed to be nonthermogenic.

Many other classes of drugs interact with RF/EM exposure to alter body temperatures in interesting ways. Although an exhaustive summary of drug/microwave interactions is beyond the scope of this chapter, it is appropriate to mention a few studies that have demonstrated clear thermoregulatory effects in experimental animals. One study[128] assessed the potential for disruption of autonomic thermoregulatory responses by an altered metabolic state (induced by injections of isoproterenol) in squirrel monkeys exposed to 2.45 GHz CW microwaves at thermoneutrality (T_{db} = 33 °C). Unlike shivering thermogenesis, chemically-mediated nonshivering thermogenesis was unaffected by RF exposure at SARs from 1.5 to 4.0 W/kg. Instead, the elevated energy production/absorption in the body, coupled with partially disabled heat loss responses, produced an exaggerated rise in body temperature. These results confirm and extend a conclusion drawn by others that an endothermic organism whose thermoregulatory system is compromized by drugs or other agents may be at a disadvantage, in terms of its ability to regulate the body temperature, during exposure to thermalizing RF/EM fields.

The psychotropic drug chlorpromazine has been rather widely investigated for possible synergistic action with RF/EM exposure.[201] Early studies[169,194] indicated a greater susceptibility to tissue thermalization and a decreased survival time in unanesthetized animals that had been treated with chlorpromazine prior to RF/EM exposure at high power densities. More recent studies[202,203] have involved microwave irradiation of anesthetized rats injected with either saline or chlorpromazine (5 mg/kg) and held at T_{db} = 24 °C. Two RF conditions were studied: 2.8 GHz PW (2μsec pulse, 500 pps) at SAR = 14 W/kg and 5.6 GHz CW at SAR = 12 W/kg. Under both exposure conditions, chlorpromazine tended to enhance thermoregulatory efficiency when the rat's colonic temperature was not allowed to rise above 39.5 °C but the drug increased susceptibility to RF when the exposure was continued until death from hyperpyrexia occurred. The effects of similar psychotropic drugs were also tested in one study;[202] amytryptaline HCl (10 mg/kg) and haloperidol (0.1 mg/kg) did not alter the thermal responses of anesthetized rats to 2.8 GHz PW exposure at SAR = 14 W/kg. While important, these studies have limited generality because *ranges* of SAR, drug dose and T_{db} have yet to be explored.

The importance of parametric studies[141] becomes clear from exploration of the synergy between RF/EM exposure and other substances whose physiological action tends to alter the body temperature. A classic example is ethanol. It is known that the magnitude of ethanol-induced hypothermia in the rat is highly dependent on both the dose administered and the T_{db} to which the animal is exposed.[204] Hyperthermia can even be observed in ethanol-treated rats if the prevailing T_{db} is well above thermoneutrality (e.g., 36 °C), a response that has been interpreted to reflect alteration of heat loss mechanisms.[205] However, careful parametric studies[204] have demonstrated, for T_{db} from 17 to 32 °C, that acute adminstration of ethanol at a commonly-used dose of 3 g/kg interferes with metabolic heat production rather than with heat loss. More important was the finding that rats equilibrated to T_{db} = 26 °C prior to ethanol dosing and testing at T_{db} = 17 °C exhibited an initial depression of metabolic heat production that slowed the developing hypothermia.

Such information is essential to the proper interpretation of colonic temperature changes measured in ethanol-treated rats that had previously been either sham exposed or irradiated for 45 min by 2.45 GHz, circularly polarized, pulsed microwaves (2 μs pulse, 500 pps) at SAR = 0.6 W/Kg.[206] In this study, the mean T_{db} was 22 ± 0.1°C, but the range was 21–24°C. All rats were injected I.P. with ethanol at 3 g/kg; no control injections of the vehicle were given. Most importantly, since no dependent variable other than colonic temperature was measured, it was impossible to determine whether alterations in heat production or heat loss contributed to the observed attenuation of hypothermia in the microwave-exposed animals relative to the sham exposed. Microwave exposure alone was reported to produce no change in colonic temperature. However, abundant evidence exists for the maintenance of normothermia (constant deep body temperature) during RF/EM exposure, even at moderately high SARs[127,138,140,141]; careful titration of heat production and/or heat loss is the underlying cause. Thus, the speculation by Lai, et al.[206] that their results might reflect stress-related interference with heat loss, perhaps mediated by endogenous opioids, is certainly open to question. Clearly, a complete understanding of the subtle actions of RF/EM exposure requires measurement of many more thermoregulatory responses than the changes in "colonic temperature" measured at arbitrary intervals in an experimental subject.

Lai and colleagues at the University of Washington have undertaken an extensive research program to determine how endogenous opioids may be involved in the interaction between certain psychoactive drugs administered to rats and prior acute irradiation with circularly-polarized 2.45 GHz pulsed microwaves (SAR = 0.6 W/kg).[207] The early hypothesis-generating experiments usually followed the same protocol: a 45-min sham or microwave exposure in a circular waveguide was followed sequentially by colonic temperature measurement, injection of a drug at a single dose, and measurement of colonic temperature every 15 min for periods ranging from 45 min to 2 hr. Most studies suffered from the limitations noted above, including a narrow T_{db} range, a single drug dose, infrequent assessment of the effects of vehicle injection and no evaluation of specific contributing thermoregulatory mechanisms. Apomorphine-induced hypothermia was found to be enhanced by RF/EM irradiation while amphetamine-induced hyperthermia was attenuated.[208] A more elaborate study determined that injection of 1 or 2 ml/kg of distilled water produced a greater post-irradiation hyperthermia in microwave exposed animals than in sham exposed; this hyperthermia was attenuated by injections of the narcotic antagonist naltrexone or one of the serotonin antagonists cinanserin, cyproheptadine or metergoline.[209] But, the hyperthermia was not altered by the peripheral serotonin antagonist xylamidine nor the dopamine antagonist haloperidol. The previously reported[208] attenuation of amphetamine-induced hyperthermia by prior microwave irradiation was blocked by pretreatment with naloxone, a narcotic antagonist, suggesting that microwave exposure may activate endogenous opioids which may, in turn, alter drug action.[210] The magnitude of amphetamine-induced hyperthermia was found, in a separate experiment, to depend on the dose of morphine (1, 5, 10, 15 and 20 mg/kg, S.C.) injected before the amphetamine dose (5 mg/kg, I.P.). The attenuation of amphetamine-induced hyperthermia by low-level pulsed microwaves could also be classically conditioned by repeated (10 days) microwave exposures prior to the attenuation test. Sensory cues in the waveguide, adaptation to stress produced by microwave exposure and involvement of endogenous opioids were all considered as mediators in these effects. Direct thermal effects on metabolic heat production, thermal conductance, and convective or evaporative heat loss are also prime candidates for investigation in this regard.

VI. APPLICATIONS OF SIMULATION MODELS OF HUMAN THERMOREGULATORY SYSTEM TO DEPOSITION OF RF/EM POWER WITHIN THE BODY

Since the thermal characteristics of the human body are reasonably well known, it is possible to develop mathematical simulation models of physiological thermoregulation in considerable detail and validate them against experimental data.[59–62,64,65,211] Thermophysiological modeling becomes especially important when its purpose is to extrapolate variables that are not experimentally attainable and to simulate experiments that cannot be performed (e.g., those involving RF/EM exposure of human beings).

The basis of thermophysiological modeling is the energy (or heat) balance equation (Equation 1). Simulation models incorporate the physical characteristics of the body, heat production and heat loss responses, and all relevant characteristics of the environment.[15] As Equation 8 shows, absorbed RF/EM energy is added to M and must be balanced by appropriate heat loss from the body to avoid changes in the body temperature.

As in all modeling efforts, thermophysiological models are based on simplifications of the actual system. It is useful to separate the total physiological system into two major components: the regulated or passive system, and the regulatory or controlling system. Both systems are simplified to a degree necessitated by knowledge or required by the application.[212]

The passive system consists of a simplified representation of the thermal characteristics of all body tissues including metabolic heat production, heat capacitance, local tissue temperature, heat transfer by conduction and convection inside the body, and by evaporation, conduction, convection and radiation between skin and environment. In the case of RF/EM exposure of the whole body, the absorbed energy will be added to the metabolic energy generated in the body's core. Partial-body exposure or allowance for distributive dosimetry requires the absorbed energy to be proportioned appropriately in various body compartments.

The controlling system consists of the response characteristics of the structures which sense body temperatures, the central neural integration, neural effector pathways, and the effector mechanisms themselves (e.g., shivering muscles, secreting sweat glands, and the tone of peripheral blood vessels in the skin and elsewhere which control the convective transfer between different parts of the body).

In order for such a model to have maximum utility, it must contain all the changes in physiological responses that may occur within the ranges of imposed environmental thermal stress, metabolic heat production, and body temperatures that encompass its anticipated uses. The model of Stolwijk and Hardy[64] accurately predicts these response changes to both sudden changes in environmental conditions[213] and to bouts of exercise.[65] It seems well suited to predict changes in thermoregulatory responses that may result from deposition of RF/EM energy into the body and has provided the basis for several such analyses.[173,214,215]

In recent years, considerable progress has been made in the quantification of the absorption of RF/EM energy by animals and humans.[49,172,177,216,217] Many studies have pointed out nonuniformities of energy deposition within exposed tissue volumes and the potential for hot spots, especially in the cranial cavity.[218–222] Mathematical models embodying considerable detail[174–176,216] are now available to simulate the deposition of RF/EM energy in the human body. Thus, it is logical to combine the two types of models for the evaluation of physiological thermoregulatory responses to microwave power deposition.[174,213] However, the interweaving of two simulation models, each of which was developed for different purposes using different rationales and objectives, poses particular problems and should be approached with caution. Prior assumptions and simplifications of different types that may have been incorporated into the separate models may lead to internal incompatibilities which may not be readily apparent.

Stolwijk[213] has used a combined simulation model to demonstrate that during the localized deposition of RF/EM power the rise in local temperatures in the human body, especially in the brain, may be very much less than anticipated. Similar conclusions have been drawn by Way et al.,[173] who used a modified Stolwijk model that provided for focusing of RF/EM energy in the hypothalamus. Greatly enhanced evaporative heat loss, skin blood flow, and core-to-skin conductance serve to protect individual tissues of the body during deposition of RF/EM energy. Thus, regions of high tissue temperature (i.e., thermal "hot spots") are rather unlikely, because the convective heat transfer via blood flow will dissipate deposited energy over the heat capacitance of the whole body.

It is possible that poorly perfused tissues such as the ocular lens, gut or resting muscle may sustain substantial temperature elevation if high local SARs are created during RF/EM exposure. However, most body tissues are well vascularized and very large increases in local blood flow will occur as soon as a tissue temperature of 42 to 43 °C is reached.[23,223] Thus, cases where a combined simulation model predicts local tissue temperatures in excess of this level[215] should be viewed with great caution until the presence of an increased blood perfusion response has been verified.

A considerably simpler model of physiological thermoregulation has recently been adapted to predict the thermophysiological consequences of exposure to RF/EM fields in the MRI environment.[224] The model is similar to the Stolwijk and Hardy formulation but has only two nodes, core and skin. Based on equation 8, the model predicts physiological heat loss responses in real time as a function of selected T_{db}, air movement (v), and rate of whole-body energy deposition (SAR). Assuming a criterion elevation in deep body temperature (ΔT_{co}) of 0.6 °C, T_{db} = 20 °C, and v = 0.8 m/sec, the model predicts that a 70 kg patient could undergo an MRI scan of infinite duration at SAR $\leq$ 5 W/kg. Lowering T_{db} or increasing v permits a rise in permissible SAR for a given ΔT_{∞}. More restrictive ΔT_{∞} criteria result in lower permissible SARs and shorter exposure durations. The limiting response is usually the rate of peripheral blood flow, although sweating can play a role in limiting ΔT_{∞}.

Restrictions on the rate of skin blood flow (SkBF), ranging from 0 to 89% of normal, have also been studied with this model.[225] Under conditions that are desireable in the clinic (T_{db} = 20 °C, 50% RH, still air), moderate restrictions (up to 67%) of SkBF yield tolerable ΔT_{co} ($\leq$ 1 °C) during MRI scans (SAR $\leq$ 4 W/kg) of 40 min or less. Increased T_{db} and RH exacerbate the thermal stress imposed by absorbed RF/EM energy; severely impaired SkBF encourages short MRI exposures (e.g., 20 min or less) at SARs $\leq$ 3 W/kg.

At a given T_{db}, a person can absorb some level (SAR) of RF/EM energy indefinitely (i.e., thermal equilibrium with the prevailing conditions can be reached). At low SARs, T_{co} will rise initially and then stabilize at some elevated level. If the cardiovascular system is impaired, the maximum SAR for thermal equilibrium will be lower. The two-node model adapted by Adair and Berglund[225] has been used to predict zones of stable and rising body temperature as a function of T_{db}, RH, clothing insulation (clo), and cardiovascular impairment. These predictions have been used[226] to calculate the effect of T_{db} on the average rate of increase in T_{co} under conditions of 40% impairment in SkBF, clo = 0.2, and a 60-min exposure duration. Under these conditions, when SAR = 4 W/kg and T_{db} = 19 °C, then ΔT_{co} can be held within 1.0 °C. Further, if T_{db} = 25 °C, an SAR of 3 W/kg is appropriate. A 1-hr RF/

EM exposure to 1 W/kg, even at T_{db} = 27 °C, should result in no body temperature rise. These predictions do not conflict with data collected in the clinical setting and, indeed, are most reassuring for the generality of whole-body exposure to RF/EM fields.

Precooling of patients to the prevailing T_{db} prior to MRI exposure has little value in terms of preventing a rise in body temperature and use of added insulation (i.e., a blanket) should be discouraged in the normal clinical setting.[227] Clinical tests of two normal male subjects have provided limited confirmation of the predictions of the two-node model. During 20-min MRI scans at a whole-body SAR of 1.2 W/kg, core and skin temperatures, sweat rate, and judgments of thermal sensation and discomfort were very similar to the predicted values. However, the authors hasten to point out the limitations of this model because the simulations are based on RF/EM exposure of the whole body, an atypical situation for the MRI clinic. A model having many more nodes[213] would be much more useful for general predictions of the thermophysiological responses to RF/EM exposure (partial- and whole-body), as well as specific predictions for clinical use.

REFERENCES

1. **Adair, E. R.,** Autonomic thermoregulation in squirrel monkey when behavioral regulation is limited, *J. Appl., Physiol.,* 40, 694, 1976.
2. **Adair, E. R. and Wright, B. A.,** Behavioral thermoregulation in the squirrel monkey when response effort is varied, *J. Comp. Physiol. Psychol.,* 90, 179, 1976.
3. **Corbit, J. D.,** Behavioral regulation of body temperature, in *Physiological and Behavioral Temperature Regulation,* Hardy, J. D., Gagge, A. P., and Stolwijk, J. A. J., Eds., Charles C Thomas, Springfield, Ill., 1970, 777.
4. **Hardy, J. D.,** Models of temperature regulation–a review, in *Essays on Temperature Regulation,* Bligh, J. and Moore, R., Eds., North-Holland, Amsterdam, 1972, 163.
5. **Stitt, J. T., Adair, E. R., Nadel, E. R., and Stolwijk, J. A. J.,** The relation between behavior and physiology in the thermoregulatory response of the squirrel monkey, *J. Physiol. (Paris),* 63, 424, 1971.
6. **Stevens, J. C.,** Thermal sensation, infrared and microwaves, in *Microwaves and Thermoregulation,* Adair, E. R., Ed., Academic Press, New York, 1983, 191.
7. **Bligh, J. and Johnson, K. G.,** Glossary of terms for thermal physiology, *J. Appl. Physiol.,* 35, 941, 1973.
8. **DuBois, D. and DuBois, E. F.,** A formula to estimate approximate surface area, if height and weight are known, *Arch. Int. Med.,* 17, 863, 1916.
9. **Kleiber, M.,** *The Fire of Life,* R. E. Kreiger, Huntington, N.Y., 1975.
10. **Gagge, A. P., Burton, A. C., and Bazett, H. C.,** A practical system of units for the description of the heat exchange of man with his environment, *Science,* 94, 428, 1941.
11. **Winslow, C.-E. A., Gagge, A. P., and Herrington, L. P.,** Heat exchange and regulation in radiant environments above and below air temperature, *Am. J. Physiol.,* 131, 79, 1940.
12. **Sprague, C. H. and Munson, D. M.,** A composite ensemble method for estimating thermal insulation values of clothing, *ASHRAE Trans.,* 80(1), 120, 1974.
13. **Woodcock, A. H., Pratt, R. L., and Breckenridge, J. R.,** Theory of the globe thermometer, Res. Study Rep. BP-7, Quartermaster R & E Command, U.S. Army Natick Laboratories, Natick, Mass, 1960.
14. **Roller, W. L. and Goldman, R. F.,** Estimation of solar radiation environment, *Int. J. Biometeorol.,* 11, 329, 1967.
15. **Berglund, L. G.,** Characterizing the thermal environment, in *Microwaves and Thermoregulation,* Adair, E. R., Ed., Academic Press, New York, 1983, 15.
16. American Society of Heating, Refrigerating and Air-Conditioning Engineers, Inc., Physiological principles, comfort, and health, in *ASHRAE Handbook, 1981 Fundamentals,* ASHRAE, Atlanta, Ga., 1981, 8.1.
17. **Goldman, R. F.,** Effect of environment on metabolism, in *Proc. 1st Ross Conf. on Metabolism, 26–28 June, 1978, Jackson Point, Maine,* Ross Laboratories, Columbus, Ohio, 1980, 117.
18. **Stitt, J. T., Hardy, J. D., and Stolwijk, J. A. J.,** PGE_1 fever: its effect on thermoregulation at different low ambient temperatures, *Am. J. Physiol.,* 227, 622, 1974.
19. **Adair, E. R. and Adams, B. W.,** Adjustments in metabolic heat production by squirrel monkeys exposed to microwaves, *J. Appl. Physiol. Respir. Environ. Exercise Physiol.,* 52, 1049, 1982.

20. **Swift, R. J.,** The effects of low environmental temperature upon metabolism. II. The influence of shivering, subcutaneous fat, and skin temperature on heat production, *J. Nutr.,* 5, 227, 1932.
21. **Iampietro, P. F., Vaughn, J. A., Goldman, R. F., Kreider, M. B., Masucci, F., and Bass, D. E.,** Heat production from shivering, *J. Appl. Physiol.,* 15, 632, 1960.
22. **Hardy, J. D.,** Regulation of body temperature in man–an overview, in *Energy Conservation Strategies in Buildings,* Stolwijk, J. A. J., Ed., Yale University Printing Service, New Haven, Conn., 1978, 14.
23. **Cunningham, D. J.,** An evaluation of heat transfer through the skin in the human extremity, in *Physiological and Behavioral Temperature Regulation,* Hardy, J. D., Gagge, A. P., and Stolwijk, J. A. J., Eds., Charles C Thomas, Springfield, Ill., 1970, 302.
24. **Nadel, E. R., Mitchell, J. W., Saltin, B., and Stolwijk, J. A. J.,** Peripheral modifications to the central drive for sweating, *J. Appl. Physiol.,* 31, 828, 1971.
25. **Nadel, E. R., Pandolf, K. B., Roberts, M. F., and Stolwijk, J. A. J.,** Mechanisms of thermal acclimation to exercise and heat, *J. Appl. Physiol.,* 37, 515, 1974.
26. **Greenleaf, J. E.,** Blood electrolytes and exercise in relation to temperature regulation in man, in *The Pharmacology of Thermoregulation,* Schoenbaum, E. and Lomax, P., Eds., S. Karger, Basel, 1973, 72.
27. **Cabanac, M.,** Thermoregulatory behavioral responses, in *Microwaves and Thermoregulation,* Adair, E. R., Ed., Academic Press, New York, 1983, 307.
28. **Hensel, H., Iggo, I., and Witt, I.,** A quantitative study of sensitive cutaneous thermoreceptors with C afferent fibers, *J. Physiol. (London),* 153, 113, 1960.
29. **Hensel, H. and Kenshalo, D. R.,** Warm receptors in the nasal region of cats, *J. Physiol. (London),* 204, 99, 1969.
30. **Aronsohn, E. and Sachs, J.,** Die Beziehungen des Gehirns zur Korperwarme und zum Fieber, *Pflugers Arch. Ges. Physiol.,* 37, 232, 1885.
31. **Ott, I.,** The heat center in the brain, *J. Nerv. Ment. Dis.,* 14, 152, 1887.
32. **Barbour, H. G.,** Die Wirkung unmittelbarer Erwarmung und Abkuhlung der Warmezentra auf die Korpertemperatur, *Arch. Exp. Pathol. Pharmakol.,* 70, 1, 1912.
33. **Nakayama, T., Eisenman, J. S., and Hardy J. D.,** Single unit activity of anterior hypothalamus during local heating, *Science,* 134, 560, 1961.
34. **Nakayama, T., Hammel, H. T., Hardy, J. D., and Eisenman, J. S.,** Thermal stimulation of electrical activity of single units of the preoptic region, *Am. J. Physiol.,* 204, 1122, 1963.
35. **Guieu, J. D. and Hardy, J. D.,** Integrative activity of preoptic units. I. Response to local and peripheral temperature changes, *J. Physiol. (Paris),* 63, 253, 1971.
36. **Hammel, H. T.,** Regulation of internal body temperature, *Annu. Rev. Physiol.,* 30, 641, 1968.
37. **Hellon, R. F.,** Thermal stimulation of hypothalamic neurones in unanaesthetized rabbits, *J. Physiol. (London),* 193, 381, 1967.
38. **Cabanac, M., Stolwijk, J. A. J., and Hardy, J. D.,** Effect of temperature and pyrogens on single-unit activity in the rabbit's brain stem, *J. Appl. Physiol.,* 24, 645, 1968.
39. **Simon, E.,** Temperature regulation: the spinal cord as a site of extra-hypothalamic thermoregulatory functions, *Rev. Physiol. Biochem. Pharmacol.,* 71, 1, 1974.
40. **Rawson, R. O. and Quick, K. P.,** Evidence of deep-body thermoreceptor response to intra-abdominal heating of the ewe, *J. Appl. Physiol.,* 28, 813, 1970.
41. **Wit, A. and Wang, S. C.,** Temperature-sensitive neurones in preoptic/anterior hypothalamic region: effects of increasing ambient temperature, *Am. J. Physiol.,* 215, 1151, 1968.
42. **Hellon, R. F.,** Environmental temperature and firing rate of hypothalamic neurones, *Experientia,* 25, 610, 1969.
43. **Wunnenberg, W. and Hardy, J. D.,** Response of single units of the posterior hypothalamus to thermal stimulation, *J. Appl. Physiol.,* 33, 547, 1972.
44. **Cabanac, M., Hammel, H. T., and Hardy, J. D.,** *Tiliqua scincoides:* temperature sensitive units in lizard brain, *Science,* 158, 1050, 1967.
45. **Satinoff, E.,** Behavioral thermoregulation in response to local cooling of the rat brain, *Am J. Physiol.,* 206, 1389, 1964.
46. **Adair, E. R.,** Hypothalamic control of thermoregulatory behavior: preoptic posterior hypothalamic interaction, in *Recent Studies of Hypothalamic Function,* Lederis, K. and Cooper, K. E., Eds., S. Karger, Basel, 1974, 341.
47. **Lipton, J. M.,** Behavioral temperature regulation in the rat: effects of thermal stimulation of the medulla, *J. Physiol. (Paris),* 63, 325, 1971.
48. **Cabanac, M.,** Thermoregulatory behavior, in *Essays on Temperature Regulation,* Bligh, J. and Moore, R., Eds., North-Holland, Amsterdam, 1972, 19.
49. **Carlisle, H. J. and Ingram, D. L.,** The effects of heating and cooling the spinal cord and hypothalamus on thermoregulatory behaviour in the pig, *J. Physiol (London),* 231, 353, 1973.
50. **Guy, A. W.,** Analysis of electromagnetic fields induced in biological tissues by thermographic studies on equivalent phantom models, *IEEE Trans. Microwave Theory Tech.,* MTT-19, 205, 1971.

51. **Adair, E. R.**, Sensation, subtleties, and standards: synopsis of a panel discussion, in *Microwaves and Thermoregulation*, Adair, E. R., Ed., Academic Press, New York, 1983, 231.
52. **Meyer, H. H.**, Theorie des Fiebers und seiner Behandlung, *Verh. Dtsch. Bes. Inn. Med.*, 30, 15, 1913.
53. **Rubner, M.**, *Die Gesetze des Energieverbrauchs bie der ernahrung*, Franz Deutiche, Leipzig, 1902.
54. **DuBois, E. F.**, *Fever and the Regulation of Body Temperature*, Charles C Thomas, Springfield, Ill., 1948.
55. **Benzinger, T. H.**, On physical heat regulation and the sense of temperature in man, *Proc. Natl. Acad. Sci. U. S. A.*, 45, 645, 1959.
56. **Hensel, H.**, Physiologie der thermoreception, *Ergeb. Physiol.*, 47, 166, 1952.
57. **Chatonnet, J. and Cabanac, M.**, The perception of thermal comfort, *Int. J. Biometerol.*, 9, 183, 1965.
58. **Aschoff, J. and Wever, R.**, Modellversuche zum gegenstrome-Warmeaustausch in der Extremitat, *Z. Gesamte Exp. Med.*, 130, 385, 1958.
59. **MacDonald, D. K. C. and Wyndham, C. H.**, Heat transfer in man, *J. Appl. Physiol.*, 3, 342, 1950.
60. **Burton, A. C.**, The application of the theory of heat flow to the study of energy metabolism, *J. Nutr.*, 7, 497, 1934.
61. **Wissler, E. H.**, Steady state temperature distrubution in man, *J. Appl. Physiol.*, 16, 734, 1961.
62. **Wissler, E. H.**, A mathematical model of the human thermal system, *Bull. Math. Biophys.*, 26, 147, 1964.
63. **Smith, P. E. and James, E. W.**, Human responses to heat stress, *Arch. Environ. Health*, 9, 332, 1964.
64. **Stolwijk, J. A. J. and Hardy, J. D.**, Temperature regulation in man–a theoretical study, *Pflugers Arch. Gesamte Physiol.*, 291, 129, 1966.
65. **Stolwijk, J. A. J. and Hardy, J. D.**, Control of body temperature, in *Handbook of Physiology, Sec. 9*, Lee, D. H. K., Ed., American Physiological Society, Bethesda, Md., 1977, 45.
66. **Hammel, H. T.**, Neurons and temperature regulation, in *Physiological Controls and Regulation*, Yamamoto, W. S. and Brobeck, J. R., Eds., W. B. Saunders, Philadelphia, 1965, 71.
67. **Hardy, J. D. and Guieu, J. D.**, Integrative activitiy of preoptic units. II. Hypothetical network, *J. Physiol. (Paris)*, 63, 264, 1971.
68. **Bligh, J.**, Neuronal models of mammalian temperature regulation, in *Essays on Temperature Regulation*, Bligh, J. and Moore, R., Eds., North-Holland, Amsterdam, 1972, 105.
69. **Feldberg, W. and Myers, R. D.**, Effects on temperature of amines injected into the cerebral ventricles. A new concept of temperature regulation, *J. Physiol. (London)*, 173, 226, 1964.
70. **Myers, R. D.**, Hypothalamic mechanisms of pyrogen action in the cat and monkey, in *CIBA Symp. on Pyrogens and Fever*, Churchill, London, 1970.
71. **Gagge, A. P.**, A new physiological variable associated with sensible and insensible perspiration, *Am. J. Physiol.*, 120, 277, 1937.
72. **Belding, H. S. and Hatch, T. F.**, Index for evaluating heat stress in terms of resulting physiological strain, *Heat./Piping/Air Cond.*, 27, 129, 1955.
73. **Goldman, R. F.**, Prediction of human heat tolerance, in *Environmental Stress*, Follinsbee, L. J. et al., Eds., Academic Press, New York, 1978. 53.
74. **Hardy, J. D.**, Thermal comfort and health, *ASHRAE J.*, 77, 43, 1970.
75. **Nielsen, B. and Nielsen, M.**, Influence of passive and active heating on the temperture regulation of man, *Acta Physiol. Scand.*, 64, 323, 1965.
76. **Stolwijk, J. A. J.**, Whole body heating–thermoregulation and modeling, in *Physical Aspects of Hyperthermia*, Nussbaum, G. H., Ed., American Institute of Physics, New York, 1982, 565.
77. **Gagge, A. P., Stolwijk, J. A. J., and Nishi, Y.**, An effective temperature scale based on a simple model of human physiological regulatory response, *ASHRAE Trans.*, 77, 247, 1971.
78. **Durney, C. H., Massoudi, H. and Iskander, M. F.**, *Radiofrequency Radiation Dosimetry Handbook, 4th ed.*, Report USAFSAM-TR-85-73, prepared for U. S. Air Force School of Aerospace Medicine, Aerospace Medical Division (AFSC), Brooks Air Force Base, TX., 1986.
79. **Goldman, R. F.**, The role of clothing in achieving acceptability of environmental temperatures between 65F and 85F (18C and 30C), in *Energy Conservation Strategies in Buildings. Comfort, Acceptibility and Health*, Stolwijk, J. A. J., Ed., John B. Pierce Foundation of Connecticut, New Haven, Conn., 1978.
80. ASHRAE monographs for practical application of ASHRAE research: thermal comfort conditions, *ASHRAE J.*, 16, 90, 1974.
81. **Pound, R. V.**, Radiant heat for energy conservation, *Science*, 208, 494, 1980.
82. **Nielsen, M.**, Die regulation der Korpertemperatur bei Muskelarbeit, *Skand. Arch. Physiol.*, 79, 193, 1938.
83. **Nadel, E. R.**, Ed., *Problems with Temperature Regulation During Exercise*, Academic Press, New York, 1975.
84. **Robinson, S.**, Physiological adjustments to heat, in *Physiology of Heat Regulation and the Science of Clothing*, Newburg, C. H., Ed., W. B. Saunders, Philadelphia, 1949, 193.
85. **von Liebermeister, C.**, *Handbuch der Pathologie und Therapie des Fiebers*, Vogel, Leipzig, 1875.
86. **Macpherson, R. K.**, The effect of fever on temperature regulation in man, *Clin. Sci.*, 18, 281, 1959.
87. **Stitt, J. T.**, Fever versus hyperthermia, *Fed. Proc., Fed. Am. Soc. Exp. Biol.*, 38, 39, 1979.

88. **Crawshaw, L. I. and Stitt, J. T.,** Behavioural and autonomic induction of prostaglandin E_1 fever in squirrel monkeys, *J. Physiol. (London),* 244, 197, 1975.
89. **Tell, R. A. and Mantiply, E. D.,** Population exposure to VHF and UHF broadcast radiation in the United States, *Proc. IEEE,* 68, 6, 1980.
90. IEEE C95.1-1991, *IEEE Standard for Safety Levels with Respect to Human Exposure to Radio Frequency Electromagnetic Fields, 3 kHz to 300 GHz,* IEEE, New York, 1991.
91. **Lehmann, J. F.,** (ed.) *Therapeutic Heat and Cold,* Williams and Wilkins, Baltimore, 1982.
92. **Hornback, N. B.,** *Hyperthermia and Cancer: Human Clinical Trial Experience,* Vols. I and II, CRC Press, Boca Raton, 1984.
93. **Hesslink, R. L., Pepper, S., Olsen, R. G., Lewis, S. B., and Homer, L. D.,** Radio-frequency (13.56 MHz) energy enhances recovery from mild hypothermia, *J. Appl. Physiol.,* 67, 1208, 1989.
94. **Olsen, R. G.,** RF energy for warming divers' hands and feet, in *Emerging Electromagnetic Medicine,* O'Connor, M. E., Bentall, R. H. C., and Monahan, J. C., Eds., Springer-Verlag, New York, 1990, 135.
95. **Lloyd, J. R. and Olsen, R. G.,** Radiofrequency energy for rewarming of cold extremities, *Undersea Biomed. Res.,* 19, 199, 1992.
96. **Liu, Y.-H., Li, X.-M., Zou, R.-P., and Li, F.-B.,** Biopsies of human testes receiving multiple microwave irradiation: An histological and ultra-microscopical study, *J. Bioelectricity,* 10, 213, 1991.
97. **Shellock, F. G. and Crues, J. V.,** Corneal temperature changes induced by high field-strength MR imaging with a head coil, *Radiology,* 167, 809, 1988.
98. **Shellock, F. G. and Crues, J. V.,** Temperature changes caused by MR imaging of the brain with a head coil, *Am. J. Neuroradiol.,* 9, 287, 1988.
99. **Shellock, F. G., Schaefer, D. J., and Crues, J. V.,** Alterations in body and skin temperatures caused by magnetic resonance imaging: is the recommended exposure for radiofrequency radiation too conservative? *Brit. J. Radiol.,* 62, 904, 1989.
100. **Shellock, F. G., Rothman, B., and Sarti, D.,** Heating of the scrotum by high-field-strength MR imaging, *Am J. Roentgenology,* 154, 1229, 1990.
101. **Shellock, F. G.,** Thermal responses in human subjects exposed to magnetic resonance imaging, *Ann. N.Y. Acad. Sci.,* 649, 260, 1992.
102. **Shellock, F. G., Schaefer, D. J., and Crues, J. V.,** Exposure to a 1.5T static magnetic field does not alter body and skin temperatures in man, *Mag. Res. Med.,* 11, 371, 1989.
103. **Hensel, H.,** Neural processes in thermoregulation, *Physiol. Rev.,* 53, 948, 1973.
104. **Cain, W. S.,** Spatial discrimination of cutaneous warmth, *Am. J. Psychol.,* 86, 169, 1973.
105. **Hardy, J. D. and Oppel, T. W.,** Studies in temperature sensation. III. The sensitivity of the body to heat and the spatial summation of the end organ responses, *J. Clin. Invest.,* 16, 533, 1937.
106. **Herget, C. M., Granath, L. P., and Hardy, J. D.,** Warmth sense in relation to the area of skin stimulated, *Am. J. Physiol.,* 135, 20, 1941.
107. **Kenshalo, D. R., Decker, T., and Hamilton, A.,** Spatial summation on the forehead, forearm, and back produced by radiant and conducted heat, *J. Comp. Physiol. Psychol.,* 63, 510, 1967.
108. **Marks, L. E. and Stevens, J. C.,** Temporal summation related to the nature of the proximal stimulus for the warmth sense, *Percept. Psychophys.,* 14, 570, 1973.
109. **Stevens, J. C. and Marks, L. E.,** Spatial summation and the dynamics of warmth sensation, *Percept. Psychophys.,* 9, 391, 1971.
110. **Stevens, J. C., Marks, L. E., and Simonson, D. C.,** Regional sensitivity and spatial summation in the warmth sense, *Physiol. Behav.,* 13, 825, 1974.
111. **Stevens, J. C., Oculicz, W. C., and Marks, L. E.,** Temporal summation at the warmth threshold, *Percept. Psychophys.,* 14, 307, 1973.
112. **Schwan, H. P., Anne, A., and Sher, L.,** *Heating of Living Tissues,* Report NAEC-ACEL-534, U. S. Naval Air Engineering Center, Philadelphia, 1966.
113. **Eijkman, E. and Vendrik, A. J. H.,** Dynamic behavior of the warmth sense organ, *J. Exp. Psychol.,* 62, 403, 1961.
114. **Vendrik, A. J. H. and Vos, J. J.,** Comparison of the stimulation of the warmth sense organ by microwave and infrared, *J. Appl. Phys.,* 13, 435, 1958.
115. **Hendler, E., Hardy, J. D., and Murgatroyd, D.,** Skin heating and temperature sensation produced by infrared and microwave irradiation, in *Temperature: Its Measurement and Control in Science and Industry,* Vol. 3, Hardy, J. D., Ed., Reinhold, New York, 1963, 211.
116. **Hendler, E. and Hardy, J. D.,** Infrared and microwave effects on skin heating and temperture sensation, *IRE Trans. Med. Electron.,* ME-7, 143, 1960.
117. **Hendler, E.,** Cutaneous receptor response to microwave irradiation, in *Thermal Problems in Aerospace Medicine,* Hardy, J. D., Ed., Unwin, Surrey, 1968, 211.
118. **Justesen, D. R., Adair, E. R., Stevens, J. C., and Bruce-Wolfe, V.,** A comparative study of human sensory thresholds: 2450-MHz microwaves vs. far-infrared radiation, *Bioelectromagnetics,* 3, 117, 1982.

119. **Michaelson, S. M.,** Cutaneous perception of microwaves, *J. Microwave Power,* 7, 67, 1972.
120. **D'Andrea, J. A., Cuellar, O., Gandhi, O. P., Lords, J. L., and Nielson, H. C.,** Behavioral thermoregulation in the whiptail lizard (*Cnemidophorus tigris*) under 2450 MHz CW microwaves, in *Biological Effects of Electromagnetic Waves,* Abstr. URSI General Assembly, Helsinki, 1978, 88.
121. **Stern, S., Margolin, L., Weiss, B., Lu, S.-T., and Michaelson, S. M.,** Microwaves: effect on thermoregulatory behavior of rats, *Science,* 206, 1198, 1979.
122. **Gordon, C. J.,** Behavioral and autonomic thermoregulation in mice exposed to microwave radiation, *J. Appl. Physiol.: Resp. Environ. Exercise Physiol.,* 55, 1242, 1983.
123. **Adair, E. R. and Adams, B. W.,** Microwaves modify thermoregulatory behavior in squirrel monkey, *Bioelectromagnetics,* 1, 1, 1980.
124. **Adair, E. R. and Adams, B. W.,** Behavioral thermoregulation in the squirrel monkey: adaptation processes during prolonged microwave exposure, *Behav. Neurosci.,* 97, 49, 1983.
125. **Marr, M. J., deLorge, J. O., Olsen, R. G., and Stanford, M.,** Microwaves as reinforcing events in a cold environment, in *Electromagnetic Fields and Neurobehavioral Function,* O'Connor, M. E. and Lovely, R. H., Eds., Alan R. Liss, New York, 1988, 219.
126. **Bruce-Wolfe, V. and Adair, E. R.,** Operant control of convective cooling and microwave irradiation by squirrel monkeys, *Bioelectromagnetics,* 6, 365, 1985.
127. **Adair, E. R.,** Microwave challenges to the thermoregulatory system. Technical Report No. USAFSAM-TR-87-7, USAF School of Aerospace Medicine, Brooks AFB, TX, August, 1987.
128. **Adair, E. R.,** Microwave challenges to the thermoregulatory system, in *Electromagnetic Waves and Neurobehavioral Function,* O'Connor, M. E. and Lovely, R. H., Eds., Alan R. Liss, New York, 1988, 179.
129. **Adair, E. R.,** Thermoregulatory Consequences of Resonant Microwave Exposure. Report No. USAFSAM-TR-90-7, USAF School of Aerospace Medicine, Human Systems Division (AFSC), Brooks, AFB, TX, June, 1990.
130. **Carroll, D. R., Levinson, D. M., Justesen, D. R., and Clarke, R. L.,** Failure of rats to escape from a potentially lethal microwave field, *Bioelectromagnetics,* 1, 101, 1980.
131. **Levinson, D. M., Riffle, D. W., Justesen, D. R., Carroll, D. R., Bruce-Wolfe, V., and Clarke, R. L.,** Escape behavior by mice and rats in 2450- and 918-MHz fields: evaluation of sonic stimulation as a cue, in *Abstr. 3rd Annu. Meet. Bioelectromagnetics Society,* BEMS, Gaithersburg, Md., 1981, 40.
132. **Levinson, D. M., Grove, A. M., Clarke, R. L., and Justesen, D. R.,** Photic cuing of escape by rats from an intense microwave field, Bioelectromagnetics, 3, 105, 1982.
133. **Justesen, D. R.,** Sensory dynamics of intense microwave irradiation: a comparative study of aversive behaviors by mice and rats, in *Microwaves and Thermoregulation,* Adair, E. R., Ed., Academic Press, New York, 1983, 203.
134. **Adair, E. R.,** Thermophysiological effects of electromagnetic radiation, in *Biological Effects and Medical Applications of Electromagnetic Energy,* Gandhi, O. P., Ed., Prentice Hall, Englewood Cliffs, N.J., 1990, 256.
135. **Gordon, C. J., Long, M. D., and Fehlner, K. S.,** Temperature regulation in the unrestrained rabbit during exposure to 600 MHz radiofrequency radiation, *Int. J. Radiat. Biol.,* 49(6), 987, 1986.
136. **Gordon, C. J., Long, M. D., Fehlner, K. S., and Stead, A. G.,** Temperature regulation in the mouse and hamster exposed to microwaves in hot environments, *Health Physics,* 50(6), 781, 1986.
137. **Hasselblad, V., Creason, J. P. and Nelson, W. C.,** Regression using "hockey stick" functions, *Environ. Health Effects Research Series, (EPA-600/1-76-024),* U.S. Environmental Protection Agency, Research Triangle Park, NC, 1976.
138. **Candas, V., Adair, E. R., and Adams, B. W.,** Thermoregulatory adjustments in squirrel monkeys exposed to microwaves at high power densities, *Bioelectromagnetics,* 6, 221, 1985.
139. **Lotz, W. G. and Saxton, J. L.,** Metabolic and vasomotor responses of rhesus monkeys exposed to 225 MHz radiofrequency energy, *Bioelectromagnetics, 8,* 73, 1987.
140. **Lotz, W. G. and Saxton, J. L.,** Thermoregulatory responses in the rhesus monkey during exposure at a frequency (225-MHz) near whole-body resonance, in *Electromagnetic Fields and Neurobehavioral Function,* O'Connor, M. E. and Lovely, R. H., Eds., Alan R. Liss, New York, 1988, 203.
141. **Adair, E. R.,** Thermophysiological effects of electromagnetic radiation, *IEEE Eng. Med. Biol.,* 6, 37, 1987.
142. **Adair, E. R.,** *Microwave radiation and thermoregulation,* Technical Report No. USAFSAM-TR-85-3, USAF School of Aerospace Medicine, Brooks AFB, TX, May, 1985.
143. **Adair, E. R., Adams, B. W. and Hartman, S. K.,** Physiological interaction processes and radiofrequency energy absorption, *Bioelectromagnetics,* 13, 497, 1992.
144. **Lotz, W. G.,** Hyperthermia in radio-frequency exposed rhesus monkeys: A comparison of frequency and orientation effects, *Rad. Research,* 102, 59, 1985.
145. **Phillips, R. D., Hunt, E. L., Castro, R. D., and King, N. W.,** Thermoregulatory, metabolic, and cardiovascular responses of rats to microwaves, *J. Appl. Physiol.,* 38, 630, 1975.
146. **Ho, H. S. and Edwards, W. P.,** Oxygen-consumption rate of mice under differing dose rates of microwave radiation, *Radio Sci.,* 12 (6S), 131, 1977.

147. **Ho, H. S. and Edwards, W. P.,** The effect of environmental temperature and average dose rate of microwave radiation on the oxygen-consumption rate of mice, *Radiat. Environ. Biophys.,* 16, 325, 1979.
148. **Spiers, D. E. and Adair, E. R.,** Thermoregulatory responses of the immature rat following repeated postnatal exposure to 2,450 MHz microwaves, *Bioelectromagnetics,* 8, 283, 1987.
149. **Chou, C.-K., Guy, A. W., Kunz, L. L., Johnson, R. B., Crowley, J. J., and Krupp, J. H.,** Long-term, low-level microwave irradiation of rats, *Bioelectromagnetics,* 13, 469, 1992.
150. **Adair, E. R., Spiers, D. E., Rawson, R. D., Adams, B. W., Sheldon, D. K., Pivirotto, P. J., and Akel, G. M.,** Thermoregulatory consequences of long-term microwave exposure at controlled ambient temperatures, *Bioelectromagnetics,* 6, 339, 1985.
151. **Krupp, J. H.,** *In vivo* measurement of radio-frequency radiation absorption, in *Proceedings of a Workshop on the Protection of Personnel Against Radio-frequency Electromagnetic Radiation,* Mitchell, J. C., Ed., USAF Aeromedical Review SAM-3-81, Brooks AFB, TX, School of Aerospace Medicine, 1981.
152. **Krupp, J. H.,** *In vivo* temperature measurements during whole body exposure of *Macaca mulatta* to resonant and non-resonant frequencies, in *Microwaves and Thermoregulation,* Adair, E. R., Ed., Academic Press, New York, 1983, 95.
153. **Adair, E. R. and Adams, B. W.,** Microwaves induce peripheral vasodilation in squirrel monkey, *Science,* 207, 1381, 1980.
154. **Gordon, C. J.,** Influence of heating rate on control of heat loss from the tail in mice, *Am. J. Physiol.,* 244, R778, 1983.
155. **Ward, T. R. and Spiegel, R. J.,** Thresholds of microwave-induced vasodilation in the squirrel monkey, *BEMS Sixth Annual Meeting Abstracts,* Gaithersburg, MD: BEMS, 37, 1984.
156. **Lehmann, J. F., Guy, A. W., Stonebridge, J. B., and deLateur, B. J.,** Evaluation of a therapeutic direct-contact 915-MHz microwave applicator for effective deep-tissue heating in humans, *IEEE Trans. Microwave Theory Tech.,* MTT-26, 556, 1978.
157. **Gonzalez, R. R., Kluger, M. J., and Hardy, J. D.,** Partitional calorimetry of the New Zealand white rabbit at temperatures of 5–35°C, *J. Appl. Physiol.,* 31, 728, 1971.
158. **Gordon, C. J. and Ali, J. A.,** Measurement of ventilatory frequency in unrestrained rodents using microwave radiation, *Resp. Physiol.,* 56, 73, 1984.
159. **Gordon, C. J. and Long, M. D.,** Ventilatory frequency of mouse and hamster during microwave-induced heat exposure, *Resp. Physiol.,* 56, 81, 1984.
160. **Wenger, C. B.,** Circulatory and sweating responses during exercise and heat stress, in *Microwaves and Thermoregulation,* Adair, E. R., Ed., Academic Press, New York, 1983, 251.
161. **Adair, E. R.,** Microwaves and thermoregulation, in *USAF Radiofrequency Radiation Bioeffects Research Program–A Review,* Mitchell, J. C., Ed., USAFSAM Report No. SAM-TR-81-30, Brooks AFB, TX, 1981, 145.
162. **Stitt, J. T. and Hardy, J. D.,** Thermoregulation in the squirrel monkey *(Saimiri sciureus), J. Appl. Physiol.,* 31, 48, 1971.
163. **Nadel, E. R., Bullard, R. W., and Stolwijk, J. A. J.,** Importance of skin temperture in the regulation of sweating, *J. Appl. Physiol.,* 31, 80, 1971.
164. **Lotz, W. G. and Saxton, J. L.,** Influence of radiofrequency exposure on sweating and body temperatures in rhesus monkeys, *BEMS Eighth Annual Meeting Abstracts,* Gaithersburg, MD, BEMS, 79, 1986.
165. **Gordon, C. J.,** Effects of ambient temperature and exposure to 2450-MHz microwave radiation on evaporative heat loss in the mouse, *J. Microwave Power,* 17, 145, 1982.
166. **Adair, E. R., Spiers, D. E., Stolwijk, J. A. J., and Wenger, C. B.,** Technical Note: On changes in evaporative heat loss that result from exposure to nonionizing electromagnetic radiation, *J. Microwave Power,* 18, 209, 1983.
167. **Adair, E. R., Wenger, C. B., and Spiers, D. E.,** Technical Note: Beyond allometry, *J. Microwave Power,* 19, 145, 1984.
168. **Michaelson, S. M.,** Thermal effects of single and repeated exposures to microwaves—a review, in *Biological Effects and Health Hazards of Microwave Radiation,* Czerski, P., et al., Eds., Polish Medical Publishers, Warsaw, 1974, 1.
169. **Michaelson, S. M., Thomson, R. A. E., and Howland, J. W.,** Physiologic aspects of microwave irradiation of mammals, *Am. J., Physiol.,* 201, 351, 1961.
170. **Michaelson, S. M.,** Thermoregulation in intense microwave fields, in *Microwaves and Thermoregulation,* Adair, E. R., Ed., Academic Press, New York, 1983, 283.
171. **Goldman, R. F.,** Acclimation to heat and suggestions, by inference, for microwave radiation, in *Microwaves and Thermoregulation,* Adair, E. R., Ed., Academic Press, New York, 1983, 275.
172. **Guy, A. W., Webb, M. D., and Sorensen, C. C.,** Determination of power absorption in man exposed to high frequency electromagnetic fields by thermographic measurements in scale models, *IEEE Trans. Biomed. Eng.,* 23, 361, 1974.
173. **Way, W. I., Kritikos, H., and Schwan, H. P.,** Thermoregulatory physiologic responses in the human body exposed to microwave radiation, *Bioelectromagnetics,* 2, 341, 1981.

174. **Kritikos, H. and Schwan, H. P.,** Hot spots generated by EM waves in lossy spheres and biological implications, *IEEE Trans. Biomed, Eng.,* 19, 53, 1972.
175. **Kritikos, H. and Schwan, H. P.,** The distribution of heating potential inside lossy spheres, *IEEE Trans. Biomed. Eng.,* 22, 457, 1975.
176. **Kritikos, H. and Schwan, H. P.,** Formation of hot spots in multilayer spheres, *IEEE Trans. Biomed. Eng.,* 23, 168, 1976.
177. **Kritikos, H. and Schwan, H. P.,** Potential temperature rise induced by elecromagnetic field in brain tissues, *IEEE Trans. Biomed. Eng.,* 26, 29, 1979.
178. **Burr, J. G. and Krupp, J. H.,** Real-time measurement of RFR energy distribution in the *Macaca mulatta* head, *Bioelectromagnetics,* 1, 21, 1980.
179. **Merritt, J. H., Chamness, A. F., Hartzell, R. H., and Allen, S. J.,** Orientation effects on microwave-induced hyperthermia and neurochemical correlates, *J. Microwave Power,* 12, 167, 1977.
180. **Brainard, G., Albert, E., DeSantis, M., Postow, E., and Parker, J.,** The effects of low power density microwaves on rat hypothalamic temperatures, *Proc. 1983 Symp. on Electromagnetic Fields in Biological Systems,* IMPI, Edmonton, Alberta, Canada, 1979, 14.
181. **Adair, E. R., Adams, B. W., and Rawson, R. O.,** Role of the hypothalamus in thermoregulatory responses to microwaves (Abstr. Bull.), in 6th Annu. Meet. Bioelectromagnetics Soc., Atlanta, 1984, 37.
182. **Adair, E. R. and Adams, B. W.,** Brain 'thermostat' plays minor role in metabolic response to microwaves (Abstr.), in *Int. Symp. Interaction of Electromagnetic Fields with Biological Systems,* International Union of Radio Science (U.R.S.I.), Florence, Italy, 1984, 56.
183. **Adair, E. R.,** Microwave irradiation and thermoregulatory behavior, in *Behavioral Effects of Microwave Radiation,* Monahan, J. C. and D'Andrea, J. A., Eds., U.S. Department of Health and Human Services Publ. FDA 858238, PHS-FDS, Rockville, Md., 1985, 84.
184. **Adair, E. R., Adams, B. W., and Akel, G. M.,** Minimal changes in hypothalamic temperature accompany microwave-induced alteration of thermoregulatory behavior, *Bioelectromagnetics,* 5, 13, 1984.
185. **Lotz, W. G.,** Influence of the circadian rhythm of body temperature on the physiological response to microwaves: day vs night exposure, in *Microwaves and Thermoregulation,* Adair, E. R., Ed., Academic Press, New York, 1983, 445.
186. **deLorge, J. O.,** Operant behavior and colonic temperature of squirrel monkeys *(Saimiri sciureus)* during microwave irradiation, *NAMRL-1222,* Naval Aerospace Medical Research Laboratory, Pensacola Fla., 1976.
187. **deLorge, J. O.,** The effects of microwave radiation on behavior and temperature regulation in rhesus monkeys, in *Biological Effects of Electromagnetic Waves,* Selected Papers of the USNC/URSI Annual Meeting, Boulder, Colo., October 20 to 23, 1975, Johnson, C. C. and Shore, M. L., Eds., HEW Publ. (FDA) 77-8010, U. S. Government Printing Office, Washington, D.C., 1977, 158.
188. **deLorge, J. O. and Ezell, C. S.,** Observing-responses of rats exposed to 1.28 and 5.62-GHz microwaves, *Bioelectromagnetics,* 1, 183, 1980.
189. **deLorge, J. O.,** The thermal basis for disruption of operant behavior by microwaves in three animal species, in *Microwaves and Thermoregulation,* Adair, E. R., Ed., Academic Press, New York, 1983, 379.
190. **Putthoff, D. L., Justesen, D. R., Ward, L. B., and Levinson, D. M.,** Drug-induced ectothermia in small mammals, the quest for a biological microwave dosimeter, *Radio Sci.,* 12(6S), 73, 1977.
191. **Blackwell, R. P.,** Effects of microwave exposure on anesthesia in the mouse, in *Proceedings of the International Symposium on the Biological Effects of Electromagnetic Waves,* Jouy en Josas, France, URSI, CNFRS, 1980, 71.
192. **Cleary, S. F. and Wangemann, R. T.,** Effect of microwave radiation on pentobarbital-induced sleeping time, in *Biological Effects of Electromagnetic Waves,* HEW Publication (FDA) 77-8010, 1975, 311.
193. **Koldaev, V. M.,** Effect of general anesthetics on mice after microwave irradiation, *Bull. exp. Biol. Med. USSR,* 87, 425, 1979.
194. **Lobanova, Y. A.,** Investigations on the susceptibility of animals to microwave (MW) irradiation following treatment with pharmacologic agents, in *Biological Effects of Radiofrequency Electromagnetic Fields,* Gordon, Z.V., ed., Joint Publications Reprint Service, Washington, DC, 1974, 201.
195. **Lai, H., Horita, A., Chou, C. K. and Guy, A. W.,** Effects of acute low-level microwaves on pentobarbital-induced hypothermia depend on exposure orientation, *Bioelectromagnetics,* 5, 203, 1984.
196. **Jauchem, J. R., Frei, M. R. and Padilla, J. M.,** Thermal and physiologic responses to 1200-MHz radiofrequency radiation: Differences between exposure in E and H orientation, *Proc. Soc. Exp. Biol. Med.,* 194, 358, 1990.
197. **Frei, M. R., Jauchem, J. M. and Merritt, J. H.,** Thermoregulatory responses of rats exposed to 2.45 GHz radiofrequency radiation: A comparison of E and H orientation, *Radiat. Environ. Biophys.,* 28, 235, 1989.
198. **Smialowicz, R. J., Compton, K. L., Riddle, M. M., Rodgers, R. R., and Brugnolotti, P. L.,** Microwave radiation (2450 MHz) alters the endotoxin-induced hypothermic response of rats, *Bioelectromagnetics,* 1, 353, 1980.
199. **Smialowicz, R. J., Riddle, M. M., Brugnolotti, P. L., Rodgers, R. R., and Compton, K. L.,** Detection of microwave heating in 5-hydroxytryptomine-induced hypothermic mice, *Radiat. Res.,* 88, 108, 1981.

200. **Smialowicz, R. J.**, Detection of radiofrequency radiation-induced whole body heating following chemical impairment of thermoregulation, in *Microwaves and Thermoregulation,* Adair, E. R., Ed., Academic Press, New York, 1983, 431.
201. **Jauchem, J. R.**, Effects of drugs on thermal responses to microwaves, Gen. *Pharmacol.,* 16, 307, 1985.
202. **Jauchem, J. R., Frei, M. R., and Heinmets, F.**, Effects of psychotropic drugs on thermal responses to radiofrequency radiation, *Aviat. Space Environ. Med.,* 56, 1183, 1985.
203. **Jauchem, J. R., Frei, M. R. and Heinmets, F.**, Thermal responses to 5.6-GHz radiofrequency radiation in anesthetized rats: effect of chlorpromazine, *Physiol. Chem. Phys. Med. NMR,* 20, 135, 1988.
204. **Spiers, D. E., Threatte, R. M. and Fregly, M. J.**, Response to thermal stress in the rat following acute administration of ethanol, *Pharmacology,* 28, 155, 1984.
205. **Myers, R. D.**, Alcohol's effect on body temperature: hypothermia, hyperthermia or poikilothermia?, *Brain Res. Bull.,* 7, 209, 1981.
206. **Lai, H., Horita, A., Chou, C. K. and Guy, A. W.**, Ethanol-induced hypothermia and ethanol consumption in the rat are affected by low-level microwave irradiation, *Bioelectromagnetics,* 5, 213, 1984.
207. **Lai, H., Horita, A., Chou, C. K. and Guy, A. W.**, A review of microwave irradiation and actions of psychoactive drugs, *IEEE Eng. Med. Biol. Mag.,* March, 31, 1987.
208. **Lai, H., Horita, A., Chou, C. K. and Guy, A. W.**, Psychoactive drug response is affected by acute low-level microwave irradiation, *Bioelectromagnetics,* 5, 213, 1984.
209. **Lai, H., Horita, A., Chou, C. K. and Guy, A. W.**, Microwave-induced post-exposure hyperthermia: involvement of endogenous opioids and serotonin, *IEEE Trans. Microwave Theory Tech.,* MTT-32, 882, 1984.
210. **Lai, H., Horita, A., Chou, C. K. and Guy, A. W.**, Effects of low-level microwave irradiation on amphetamine hyperthermia are blockable by naloxone and classically conditionable, *Psychopharmacology,* 88, 354, 1986.
211. **Crosbie, R. J., Hardy, J. D., and Fessenden, E.**, Electrical analog simulation of temperature regulation in man, *IEEE Trans. Biol. Med. Electron.,* 8, 245, 1961.
212. **Stolwijk, J. A. J.**, Mathematical models of thermal regulation, *Ann. N.Y. Acad. Sci.,* 335, 98, 1980.
213. **Stolwijk, J. A. J.**, A Mathematical Model of Physiological Temperature Regulation in Man, *Report CR-1855 NASA,* Washington, D.C., 1971.
214. **Emery, A. F., Short, R. E., Guy, A. W., Kraning, K. K., and Lin, J. C.**, The numerical thermal simulation of the human body when undergoing exercise or nonionizing electromagnetic irradiation, *Trans. Am. Soc. Mech. Eng.,* 284, 1976.
215. **Spiegel, R. J., Deffenbaugh, D. M., and Mann, J. E.**, A thermal model of the human body exposed to an electromagnetic field, *Bioelectromagnetics,* 1, 253, 1980.
216. **Gandhi, O. P.**, Electromagnetic absorption in an inhomogeneous model of man for realistic exposure conditions, *Bioelectromagnetics,* 3, 81, 1982.
217. **Guy, A. W. and Chou, C.-K.**, Electromagnetic heating for therapy, in *Microwaves and Thermoregulation,* Adair, E. R., Ed., Academic Press, New York, 1983, 57.
218. **Burr, J. G., Cohoon, D. K., Bell, E. L., and Penn, J. W.**, Thermal response model of a simulated cranial structure exposed to radiofrequency radiation, *IEEE Trans. Biomed. Eng.,* 27, 452, 1980.
219. **Johnson, C. C. and Guy, A. W.**, Nonionizing EM effects in biological materials and systems, *Proc. IEEE,* 60, 692, 1972.
220. **Lin, J. C., Guy, A. W., and Johnson, C. C.**, Power deposition in a spherical model of man exposed to 1–20 MHz electromagnetic fields, *IEEE Trans. Microwave Theory Tech.,* 21, 791, 1973.
221. **Shapiro, A. R., Lutomirski, R. F., and Yura, H. T.**, Induced fields and heating within the cranial structure irradiated by an electromagnetic wave, *IEEE Trans. Microwave Theory Tech.,* 19, 187, 1971.
222. **Weil, C. M.**, Absorption characteristics of multilayered sphere models exposed to UHF/microwave radiation, *IEEE Trans. Biomed. Eng.,* 22, 468, 1975.
223. **Sekins, K. M.**, Microwave Hyperthermia in Human Muscle: An Experimental and Numerical Investigation of the Temperature and Blood Flow Fields Occurring During 9.5-MHz Diathermy, Ph.D. Dissertation, University of Washington, Seattle, 1981.
224. **Adair, E. R. and Berglund, L. G.**, On the thermoregulatory consequences of NMR imaging, *Magn. Res. Imaging,* 4, 321, 1986.
225. **Adair, E. R. and Berglund, L. G.**, Thermoregulatory consequences of cardiovascular impairment during NMR imaging in warm/humid environments, *Magn. Res. Imaging,* 7, 25, 1989.
226. **Schaefer, D. J.**, Dosimetry and effects of MR exposure to RF and switched magnetic fields, *Ann. N.Y. Acad. Sci.,* 649, 225, 1992.
227. **Adair, E. R. and Berglund, L. G.**, Predicted thermophysiological responses of humans to MRI fields, *Ann. N.Y. Acad. Sci.,* 649, 188, 1992.

Chapter 11

INTERACTION OF NONMODULATED AND PULSE MODULATED RADIOFREQUENCY FIELDS WITH LIVING MATTER: EXPERIMENTAL RESULTS

Sol M. Michaelson* and Edward C. Elson†

CONTENTS

* Deceased.
† Author of revision.

0-8493-0641-8/96/$0.00+$.50

I. INTRODUCTION

The motivations for research in this field are scientific curiosity and public health considerations. The latter primarily determine the amount and type of research supported and performed. Manmade sources of radiofrequency energy continue to increase in the environment and so do environmental and health concerns. The rate of output of radiofrequency-effects research, however, has decreased since the publication of the last edition of the Handbook as judged from a survey of leading research journals from 1980 to 1993. From the mid 1980s Extremely Low Frequency (ELF) research has increasingly engaged the imagination and interest of the scientific community stemming from increased societal concerns in the area of "power" frequency, effects of power transmission lines and appliances.

Other factors have affected the enthusiasm with which new investigators have ventured into the field and have affected the motivation of research administrators who influence public policy on resource allocation. The field reached its own organic "middle age" in the 1980s marked by some self doubt[1] and questioning of direction. The concerns expressed by Pickard and Foster produced a voluble exchange of views. While their concerns over non-reproducibility of results and non-robustness of effects apply in many branches of science the problems in bioelectromagnetics seem especially vexatious. These issues will be addressed at the end of this chapter.

The problems were brought into a sharp focus by the fact that the data-base of this field, its scholarly output, is conferred directly into the public health policy forum and used to establish permissible exposure limits, directly affecting diverse economic interests.[2] This subjects the research to a second, critical review, following scholarly peer review, but in a more pragmatic context. The veracity of the work is again scrutinized but, in addition, its relevance for formulating safe limits for human exposure is analyzed. This exercise requires extrapolation from conditions of experiments, often with animals, to human exposure. Inevitably the process requires the imposition of interpretations and judgments as to whether an effect might or might not constitute a "hazard" to a human. The data base is seldom clear on that point as is often the case in toxicology generally. Utilization of the data base in standards setting has been described by Peterson.[3] The role of culture and philosophy in the divergence of eastern European and western safety standards has been the subject of commentary by Sliney.[4]

The fundamental interaction of radiofrequency (RF) radiation, defined by the Institute of Electrical and Electronics Engineers (IEEE) as extending from 3 kiloHertz to 300 GigaHertz,[2] with biological systems, considered as water-dominated dielectrics richly endowed with electrolytes and intricately packaged polar and non-polar molecules, is set forth in detail in other chapters of this book. At sufficiently high RF intensities bulk thermal energy is imparted to the system causing an increase in random kinetic energy and temperature. The thermal gradients differ from those of ambient heating, a fact which is martialed occasionally to explain differences in effects between RF and ambient heating.

Above certain intensities RF energy can quickly produce morbidity, and, after thermoregulatory mechanisms are overwhelmed, mortality. In recent years work has focused on possible effects at "non-thermal" levels or under conditions in which physiologic temperature can be maintained in the presence of normally thermalizing specific absorption rates.

The human perception of microwave heating has been studied.[5] From these studies it has been concluded that the greatest cutaneous sensitivity to microwave heating is at frequencies with wavelengths comparable to or smaller than the thickness of skin, the millimeter wave range. In this range most of the energy is absorbed in the superficial dermis containing thermal sensors. At lower frequencies, 1 to 10 GHz, with wavelengths equal to or longer than the human body, much of the energy is absorbed below the superficial dermis. In this range the threshold temperature for cellular injury, about 42°C, seconds to tens of seconds, is below the threshold for pain, about 45°C. Consequently cutaneous perception of RF energy may not be a reliable response which protects against potentially harmful levels of RF radiation at the lower microwave frequencies.[5]

A. THE PROBLEM OF THERMAL VERSUS ATHERMAL OR NON-THERMAL EFFECTS

From an experimental point of view one can specify an effect as athermal or non-thermal if during exposure of a system, using a non-perturbing temperature probe, the real time temperature excursion is below an arbitrarily finite ΔT deemed non-thermal. A typical excursion could be $\Delta T = 1°C$, but, in any event, represents a judgment by an investigator carrying a set of presumptions into the experiment which he may or may not have made explicit. There will always be some temperature change because of the finite heat capacity of any material system. An effect can also be identified as non-thermal, even in the presence of a temperature excursion exceeding an arbitrary threshold if an RF-induced effect is found and the same effect is not elicited with ambient heating producing an identical ΔT.

Theoretical definitions are also arbitrary. The introduction of RF energy into a system is inherently diffuse on a scale large compared with macromolecules or whole cells. In thermodynamic terms one can posit a target structure, possessing a mode (e.g. vibration, rotational degree of freedom) in equilibrium with a surrounding heat bath (tissue). RF energy couples into the structure and heat bath simultaneously but unequally which, perturbed from equilibrium, proceed to interact with each other on the order of time of radiative or thermal transfers of energy, to achieve equilibrium. In this construct a rigorously non-thermal interaction would be one resulting from direct electromagnetic coupling with the target structure independent of thermal interactions with the heat bath. In any experimental setting such a distinction would be difficult to make. The "displacive phase transition" in a macromolecule produced by pumping RF energy into "low frequency" vibrational modes of DNA leading to unwinding and strand separation was postulated by Prohofsky, et al.[6] This process, if indeed it happens, could be considered "non-thermal." One also finds in the literature the term "field" effect, again the concept of direct transduction of RF energy into a specific kinetic or metabolic change unrelated to a change in random kinetic energy mediating the effect.

A more empirical way of approaching the thermal versus non-thermal argument would be to identify an experimental effect that occurs or is observed at "low" RF energy and look for the presence or absence of the effect following the insertion of energy by other means such as conductive heating from a water bath. Changing the ambient temperature of a biological system, by whatever means, produces thermal gradients which are different from those produced by RF energy. If such differences can be assumed not significant (an analysis is required for each experiment) then one may look for the presence or absence of a biological endpoint in each exposure condition and identify the effect as "non-thermal" if it occurs following RF exposure but not ambient heating to the equivalent small ΔT.

In evaluating the claims for thermal versus non-thermal effects one must remember how elusive the distinction can be. Ease of application varies, depending on the experimental problem. Early claims that RF energy could produce a change in the intrinsic permeability of the blood-brain barrier were later challenged, both as to the reality of the altered permeability and the need for generalized heating. "Window" effects with regard to calcium exchanges between compartments in the brain are viewed as inherently non-thermal effects by many investigators, but the validity of such effects has been challenged. The topic of window effects is treated in Part III, Chapter 5.

B. VAGARIES AND PITFALLS IN THE LITERATURE

A critical analysis of the literature is a daunting task. Many interesting reports have not been replicated. Efforts to replicate some have shown a negative or antithetical result. Discussions then arise over whether an attempt to reproduce a result was the same or, in some way, a significantly different experiment. The following presentation will identify still unresolved issues, and, where possible, describe a prevailing concensus.

Tables 1 and 2 identify factors which must be addressed in performing research in the field. In fact, since some of the parameters are not addressed in research reports appearing in journals known for meticulous peer review, perhaps because of practical limitations of space, the reader does not know if they were considered. It is even harder to establish whether a particular laboratory is imbued with a "culture of rigorous science." Does it have a systematic approach to measurement assurance and quality control? The literature alone does not answer such questions. Because of the large number of potential sources of error and sources of variation from one laboratory to another, it is not surprising that the literature contains contradictions and produces frustration, especially in the risk assessment community.

C. PHYSICAL AND PHYSIOLOGIC SCALING

The cost-benefit analyses used to justify exposure of human beings to drugs of unknown hazard, but potential benefit, do not, in general allow, systematic experimentation on human

TABLE 1
Factors that affect Microwave/RF Absorption

Physical Parameters	Biological Parameters	Artifacts	Environmental Factors
Frequency	Tissue dielectric properties	Ground or conductor plate	Temperature
Polarization	Size; geometry	Container (material, size)	Humidity
Modulation (AM, FM, Pulse, CW)	Animal orientation relative to polarization	Metal implants	
Power density (peak and average)	Spatial relations among animals	Shielding materials	
Field pattern (near or far)		Metal or nonmetallic objects in the field	
Field uniformity			
Type of transmitting and radiating equipment			
Chamber material			
Chamber dimensions			

TABLE 2
Factors that Influence Biological Responses to the Same SAR

Subject Variable	Concomitant Variables	Environmental Variables	Experimental Variables
Species: sex; age; weight	Genetic predisposition	Temperature; humidity	Acclimation procedures
Sensitivity	Base line of the response	Air flow	Duration of exposure
Number of subjects	Functional and metabolic disorders	Lighting	Number and schedule of exposures
Interventions (anesthetics; drugs; electrodes; lesions)		Noise	Mode of exposure (partial- or whole-body)
Animal husbandry		Odor	Sampling technique
			Time between exposure and sampling
			Time of day of exposure
			Restraint devices
			Investigator-animal interaction

beings for RF effects. A noteworthy exception has existed in medical oncology and, more recently, in the treatment of musculoskeletal, degenerative disorders where treatment efficacy has been established. Research in these areas focuses on maximizing benefit while minimizing adverse consequences. The subject is briefly considered in this chapter Section, IID.

Although the human is the ultimate "test object" the bulk of research has been on animals. This requires an understanding and appreciation of biophysical principles and "comparative medicine." Such studies require interspecies "scaling,"[7,8,9,10,11,12,13,14] the selection of biomedical parameters that reflect basic physiological functions, and differentiation of adaptational or compensatory changes from pathological manifestations. In comparing results of experiments performed in the same or different laboratories, standardization of conditions is important, and unfortunately all too often not attained.

Even by using approaches where absorbed energy patterns in a test animal are set to approximate as closely as possible the patterns that may exist in man under certain exposure conditions, the intrinsic physical and physiological dissimilarities between species further confound the problem of extrapolating between animals and to man. In addition to the obvious external geometric differences, the differences in internal vascular anatomy and mechanisms of heat dissipation in fur-bearing animals compared to man must be taken into consideration.

Because of the use of animals as a surrogate for humans in hazard analysis, one must create a set of experimental conditions which are as relevant as possible for the purpose of the study. Many factors, such as methods of animal care, the role of seasonal and circadian rhythms, temperature and humidity, etc., as well as psychosocial interactions, must be considered in experimental design and analysis of results.

One should not extrapolate to larger animals, or man, results obtained in small laboratory animals without consideration of size and energy distributions, as well as metabolic and physiological differences.

The possible difference between an effect and a deleterious change in function is an important question in hazard assessment. Remarkable changes following exposure of cells or tissues *in vitro* do not necessarily have any significance *in vivo*. Most exposures are necessarily short-term and do not provide information on possible long-term consequences. Even short-term exposures can produce interpretive problems. For example, could the human central nervous system tolerate specific absorptions up to 30 kJ/kg delivered in periods from 360 msec to 1 μsec with no long term sequelae as do rats?[15] Although rats are stunned in such an exposure they appear to make a complete recovery within minutes. No responsible investigator would be willing to assert that a human would make a similar recovery.

Johnson[16] has described factors that affect absorption of electromagnetic energy in animals. This absorption is dependent on the size and geometry of the animal relative to the wavelength and polarization. The wavelength-to-animal size relationship (λ-to-a, where a is the longest axis dimension of the body and the electric field vector is parallel to the longest axis) is a critical factor in the relative absorption cross-section, the ratio of the absorbed energy per second to the power incident on the geometrical cross-sectional area of the animal.[17] This produces the immediate result that, at a given frequency and power density, the specific absorption rate is vastly different in animals depending on size. To produce an identical whole body average specific absorption rate one must scale from one frequency to another. It is of practical importance to realize that experiments on biological effects at 2.45 GHz on small animals, like mice and rats, do not scale to man at 2.45 GHz, but rather approximately to effects on man at approximately 100 MHz. Much information on absorption of RF energy has been collected in the Air Force RF Handbook.[18] In recent years much research has been done on the variation of localized absorption in animals and man, both experimental measurements and modelling.[19]

The process of scaling in RF bioeffects research is not restricted to extrapolation of the physical absorption of energy, but includes the problems of quantitatively comparing the relative efficiency of various biological processes in different animals to each other and to those processes in man. The same biological processes in different animals are not necessarily affected by the same specific absorption rate. It is also possible that the absorbed energy in a particular area or tissue may not be the most useful indicator of disturbances of some biological functions, if those functions are perturbed by a systemic stimulus instead of a localized stimulus. The need for proper dosimetry in experimental procedures and the importance of using appropriate scaling factors for extrapolation of data obtained with small laboratory animals to man are thus clearly indicated.

D. SAR-SPECIFIC ABSORPTION RATE

RF energy absorption is converted to thermal energy. This thermal energy is then rapidly redistributed by conduction, convection (blood flow) within and to a lesser extent, radiation from the biological target. The tissue heat capacity and heat transfer processes influence the dose and dose distribution within the body. Present modelling techniques use the internal

TABLE 3
Ratio of SAR to Basal Metabolic Rate for an Average Man Exposed to Far-field Incident Power Densities of 10 and 50 W/m²

	Av SAR/BMR (%)	
Frequency (GHz)	**10 W/m²**	**50 W/m²**
0.01	0.13	0.65
0.02	0.60	3.00
0.05	5.80	29.00
0.06	10.00	50.00
0.08	16.00	80.00
0.10	12.00	60.00
0.20	5.20	26.00
0.50	3.70	18.00
1.00	2.90	14.50
2.00	2.50	12.50
5.00	2.50	12.50
10.00	2.50	12.50
20.00	2.50	12.50

Adapted from Stuchly, M.A., Health Aspects of Radiofrequency and Microwave Radiation Exposure, Part 2, Department of National Health and Welfare, Ottawa, 1978.

geometry and dielectric properties of different tissues to develop a profile of energy absorption and current flow as a function of position as described in Part III, Chapter 1. These calculations produce excellent agreement with phantoms (anatomically realistic models of the body with similar dielectric properties), but are not yet capable of simulating the complex redistribution of absorbed energy in living systems. Of course, they have only limited predictive value for actual physiologic or pathologic response. There is no substitute for animal experimentation.

There is a concensus that an important dosimetric measure of RF exposure is the specific absorption rate (SAR). This is the unit-mass, time-average rate of RF energy absorption specified in SI units of watts per kilogram (W/kg).[20] The amount of energy absorbed by a given mass of material, which is termed specific absorption (SA) is given in Joules per kilogram (J/kg) and is the integral of the instantaneous SAR over the duration of exposure in seconds. Thus the SAR is the rate at which RF electromagnetic energy is imparted to unit mass of a biological body. The SAR is applicable to any tissue or organ of interest, or is expressed as a whole-body average. As the science of dosimetry has evolved localized SARs are more frequently reported, ie specified for a particular location or tissue, such as the eye.

Whole-body SAR is maximal when the long axis of a body is parallel to the E-field vector and is four tenths the wavelength of the incident field. At 2.45 GHz (λ = 12.5 cm), for example, Standard Man (long axis is 175 cm) will absorb about half of the incident energy. If the human whole-body SAR is divided by the basal metabolic rate (BMR) for man, a ratio is obtained that provides a measure of the thermal load incurred due to a known incident power density.[21] Table 3 illustrates the variation of this ratio with frequency at two incident power densities. In the region of human whole-body resonance (0.06 to 0.08 GHz), this ratio reaches a maximum value (about 16 for an incident far-field power density of 10 W/m²). The ratio drops off rapidly on either side of this peak.

II. CELLULAR AND MOLECULAR BIOLOGY

A. CHROMOSOME-GENETIC EFFECTS

Some investigators have reported chromosome changes in various plant and animal cells in tissue culture.[22,23,24,25,26] Huang et al.[27] have reported no changes. Reported chromosomal changes include translocations, structural aberrations,[24,25] polyploidy,[28] and stickiness.[29] Exposures ranged from 70 W/m^2 to more than 2 kW/m^2. In a study of meiotic cells removed from exposed male mice,[29] translocations were observed at meiosis I at 10 and 50 W/m^2. The problems inherent in the use of chromosomal aberrations as indicators of genetic damage have been noted by Savage.[30]

Some interest remains in the interaction of millimeter waves with biological media in order to understand the mechanisms of biological effects of RF energies. A number of investigators reported sharp, distinct resonances in the absorption of millimeter wave RF energy by various biochemical and biological preparations[31,32,33,34,35] or that exposure to millimeter waves produces biological effects which exhibit a sharp frequency dependence.[34,36,37,38,39,40] These studies suggest sharp millimeter wave frequency-dependent lethal and mutagenic effects on microorganisms, on metabolic control of growth of cells, on oncogenic viruses, viability properties of cells, protective effects on X-irradiated and cytotoxic-drug-treated animals, etc. Because of poor documentation of a number of these studies and failure to replicate the observation of resonances in studies which were adequately documented[41] there is little confidence in the validity of such resonances. The original observations have not been followed up. Additional information is presented in Part III, Chapter 5.

Absorption resonances were similarly reported for DNA in solution for the range 8 to 12 GHz.[42] These observations were later retracted by the authors who identified artifacts of the experiment.[43] An attempt to identify such resonances in another laboratory was not successful.[44]

Hamrick,[446] studying the response of mammalian lymphocytes exposed to 2.45-GHz MW (CW), reported that power densities from 67 to 160 W/kg had an effect identical to that of heating on the secondary structure of DNA, as determined by comparison of thermal denaturation curves.

Circular DNA plasmids, exposed to 2.55 GHz microwaves at SARs ranging up to 85 Watts/kg for 20 minutes exhibited a decrease in supercoiled DNA with an increase of relaxed and linear DNA indicating single and double strand breaks. The authors indicated that this could be a result of excitation of internal vibrational modes of the DNA but that non-specific or thermal mechanisms could not be excluded.[45]

The mutagenic potential of microwave energy has been evaluated by various techniques including point mutations in bacterial assays,[46] the dominant lethal test in mammalian systems,[47] or genetic transmission in *Drosophila,*[48,49] with inconsistent results. These reports are summarized in Table 4.

Anderstam et al.[50] found no change in mutation induction in *Eschericia coli* or *Salmonella typhimorium* exposure to 0.027 and 2.45 GHz continuous wave (CW) and 3.07 pulsed wave (PW) at SARs from 4 to 100 W/kg. Baranski et al.[51] were not able to attribute mutagenic effects or metabolic changes in *Physarum polycephalum* or *Aspergillus nidulans* to 2.45 GHz, 100-W/m^2 CW or PW microwaves. Correlli et al.[52] also reported no mutagenesis after exposure to RF at 2.6 to 4.0 GHz, 20 W/kg. They investigated the effects of RF on colony-forming ability (CFA) and molecular structure (determined by IR spectroscopy) of *E. coli B* bacterial cells in aqueous suspension. Cells were exposed for 10 hr at SARs of 20 W/kg (equivalent to 5 kW/m^2). No RF-induced effects on either CFA or molecular structure were observed.

Other studies of mutagenic effects of RF in bacteria and yeasts have given negative results. A study by Blackman et al.[53] involved exposure of *E. coli WWU* to 1.70- or 2.45-GHz RF at 20 to 500 W/m^2 for 3 to 4 hr. Dutta et al.[54] exposed *Saccharomyces cerevisiae D4* to 2.45-GHz CW RF at 400 W/m^2 or to 8.5- to 9.6-GHz pulsed RF at 10 to 450 W/m^2 for 120 min with negative results.

TABLE 4
Genetic and Mutagenic Effects of RF Exposure

Effects	Source of Test Specimen	Frequency (GHz)	Exposure Conditions: Intensity (W/m^2)	Exposure Conditions: SAR (W/kg)	Duration (days × min)	Ref.
Change in thermal denaturation profile and hyperchromicity of DNA extracted from testes following exposure	Mouse	1.7 (CW)	≤500	≤2.4 (est)	1 × 80	62
		0.985 (CW)	100	≤0.26 (testis)		
No chromosome aberrations in white blood cells	Chinese hamster	2.45 (CW)	50–450	21	5 × 15	27
No sister chromatid exchange in bone marrow cells	Mouse	2.45 (CW)	200	15	28–480	497
No chromosome aberrations in CHO-K1 cell line if temperature maintained	Chinese hamster	2.45 (CW)	≤2,000	≤360 (est)	1 × 30	498
No chromosome aberrations or change in mitotic activity in regenerating liver cells in rat	Rat	0.013 (CW)	(4.45 kV_{p-p}/m)	1.3 (est)	1 × 1,680	499
			(44.1 kV_{p-p}/m)		2,640	
No mutation induction	Escherichia coli	2.45 (CW)	100 or 500	15 or 70	1 × 180–240	53
	E. coli	1.7 (CW)	(250 kV_{p-p}/m)	3	1 × 120	
No mutation induction observed in Ames tester strains	Sallmonella typhimurium	2.45 (CW)	200	40	1 × 90	488
		8.6–9.6 (PW)	100, 450	18, 80 (est)	1 × 90	
Reduction in survival concomitant with rise in sample temperature	E. coli	8.6–9.0 (PW)	10–200	≥50	1 × 90	500
	S. typhimurium	8.6–9.0 (PW)	≤450	≤80	1 × 90	
	Saccharomyces cerevisiae	8.6–9.0 (PW)	≤450	≤80	1 × 120	
No reduction in survival or mutational events	S. cerevisiae	70 (CW)	≤600	≤17 (est)	1 × 180	501
		73 (CW)				
	S. cerevisiae	9.4 (CW)		≤28 (est)	1 × 300	502
		17 (CW)			1 × 1,440	
No detectable lethal events as measurable change in CFUs	E. coli	2.6–4.0 (CW)		19	1 × 480	52
No observable change in molecular structure as measurable change in infrared spectrum	E. coli	3.2 (CW)		21 or 16	1 × 600–720	52
No repairable DNA damage	E. coli	8.6 (CW)		12	1 × 60–720	488
No change in growth pattern; enhanced colony-forming activity	E. coli	2.45 (CW)	0.05–500	0.008–75	1 × 240	489

TABLE 4
Continued

Effects	Source of Test Specimen	Frequency (GHz)	Exposure Conditions: Intensity (W/m^2)	Exposure Conditions: SAR (W/kg)	Duration (days × min)	Ref.
No change in mutation frequencies at either of the two loci controlling requirements for adenine and tryptophan	S. cerevisiae	2.45 (CW)	200	40	1 × 120	488
		8.5–9.6 (PW)	10–450	≤80 (est)	1 × 120	
No mutagenic effects in exposed embryos	Drosophila melanogaster	2.45 (CW)		100	1 × 360	490
No change in generation time, sex ratio, or sex-linked lethal mutations in offspring	D. melanogaster	2.45 (CW)	46,000–65,000	150–210 (est)	1 × 45	49
No mutations in adult males as evidenced by chromosome loss; Nondisjunction; or sex-linked recessive lethals	D. melangaster	0.029 (CW)	(600 V_{nm}/m)	0.024 (est)	1 × 720	48
		0.146 (CW)	(62.5 V_{nm}/m)	0.015 (est)	1 × 720	
No change in mutation induction	E. coli	0.027 (CW)		≤4	1 × 60–400	491
		2.45 (CW)		35–100	1 × 60–400	
		3.07 (PW)		35–100	1 × 60–400	
	S. typhimurium	0.027 (CW)		≤4	1 × 60–400	
		2.45 (CW)		35–100	1 × 60–400	
		0.027 (PW)		35–100	1 × 60–400	
No mutagenic changes (recessive lethals) in adult females	D. melanogaster	0.098 (CW)	(0.3 V_{nm}/m, [FM] at radio)	0.0004 (est)	224 × 1,140	492
No significant germ-cell mutagenesis in weekly breedings	Rat	2.45 (CW)	50	4.7–0.9	106 × 240	64
	Rat	2.45 (CW)	100	2	5 × 300	64
Same, except decrease in pregnancies, indicating temporary sterility caused by elevated testicular temperatures	Rat	2.45 (CW)	280	5.6	20 × 240	64
Induction of repressed protein, colicin, indicating a change in the genetic processes	E. coli	37 (CW)	0.01–10		1 × 30–120	493
Change in growth rate that was very frequency specific, indicating an alteration in the process of the cell	S. cerevisiae	41.650–41.825 (CW)	10–30		1 × 180	39
			≤100			

TABLE 4
Continued

Effects	Source of Test Specimen	Frequency (GHz)	Exposure Conditions Intensity (W/m^2)	SAR (W/kg)	Duration (days × min)	Ref.
Chromosome aberrations in lung cells in vitro in two frequencies, but not at two closely related frequencies, 0.015 or 0.025 GHz	Chinese hamster	0.019 (PW)	(Up to 300 kV$_{p\text{-}p}$/m)		1 × 30	22
		0.021 (PW)				494
Increase in chromosome translocations in sperm cells	Mouse	9.4 (PW)	1–100	0.05–5 (est)	10 × 60	495
Increased mutations and lethality	S. typhimurium	2.45 (CW)	51,000		1 × 0.03–0.48	496
No change in dominant lethality	Mouse	2.45 (CW)		43	1 × 30	63
No effect on colony survival and chromosome damage	E. coli	9.4	<600		1 × 30	503
		70				
Frequency-specific inhibition of cell growth	E. coli	73				38
No effect on mutation or meiosis efficiency	S. cerevisiae	9.4	<600	1 × 30		503
		70		1 × 120		
Chromatid breaks, rings and exchanges	Chinese hamster	2.45	1,000		1 × 5 or 30	26

Note: est = Estimated; av = average.

Adapted and expanded from U.S. Environmental Protection Agency, Biological Effects of Radiofrequency Radiation, Elder, J. A. and Cahill, D. F., Eds., Rep. No. EPA-600/8-83-026F, Research Triangle Park, N.C., 1984.

Two studies report effects of RF on sister chromatid exchange.[55,56] In the first study, Chinese hamster ovary cells were exposed in vitro to 2.45-GHz RF at unstated power density levels and durations. Sister chromatid exchanges were observed in RF-exposed cells; however, the same level of exchanges was produced in control cells by heating them to the same temperature as that produced by RF exposure. The authors concluded that the production of sister chromatid exchanges is not related to RF. In the second study, mice were exposed to 2.45 GHz RF at 200 W/m^2 (SAR 21 W/kg) for 8 hr/day for 28 days. Incidences of sister chromatid exchange in bone marrow cells of irradiated mice, sham-irradiated control mice and standard control mice were compared. No statistically significant differences were detected.

Saunders et al.,[57] exposed mice to 100 W/m^2 at 2.45 GHz CW for 6 hr/day for an 8 week period. Exposed male mice were mated to an unexposed group of female mice and followed. There was no significant reduction in pregnancy rate and cytogenetic analysis at the end of the mating period showed no significant variation from the controls.

Meltz and Walker[58,59] investigated whether there were any RF-induced alterations in DNA repair in normal human fibroblasts maintained in vitro after the DNA was damaged by a selected dose of UV light. Power densities of 10 or 100 W/m^2 (0.350 and 1.2 GHz) caused no perturbation of the DNA repair process. Brown et al.[60] treated mice with streptozocin, a mutagenic/carcinogenic agent known to damage DNA in the rodent liver, and exposed the mice to 400-MHz RF to determine if excision repair of the DNA would be inhibited. Power densities of 16 and 160 W/m^2 (SAR 0.29 and 2.9 W/kg) did not alter the level of excision repair.

Although Mickey[61] reported increased mutagenesis in *Drosophila* exposed to a 0.02-GHz field, no effects on mutagenesis were observed by Pay et al.[49] using 2.45 GHz or by Mittler[48] with 0.003- and 0.150-GHz RF energy.

Varma and co-workers[62] reported increased mutagenesis, using the dominant lethal test with male mice exposed to 1.7 GHz continuous wave at 100 and 500 W/m^2 for 90 and 30 min, respectively. Mice exposed to 2.45 GHz CW at 1 kW/m^2 for 10 min and at 500 W/m^2 three times, 10 min each, within 1 day also showed increased mutagenesis. Mice subjected to four 10-min exposures at 500 W/m^2, over a period of 2 weeks, showed no increase in dominant lethality above control levels. Saunders et al.[63] found no evidence of change in dominant lethality in mice exposed to 2.45 GHz CW, 43 W/kg.

Berman et al.[64] exposed male rats daily to 2.45 GHz CW (day 12 of gestation to 90 days of age, 100 W/m^2, 4 hr/day) or 2.45 GHz CW at 50, 100, or 280 W/m^2 from day 6 of gestation to 90 days of age, 4 or 5 hr/day. No significant evidence of germ cell mutagenesis or alteration in reproductive efficiency was detected.

It is known that the rate of induction of mutations will increase with increasing temperature. It is possible that artifacts or thermal stress could be factors in some of the reported studies. Baranski and Czerski[65] note that there is no satisfactory evidence of microwave-induced genetic effects at low to modest power densities. Recent experimental studies of primitive organisms and rodents[62,66,67] have confirmed earlier findings that microwave exposure at power densities below 100 W/m^2 is not mutagenic in these organisms. Other studies[68] have failed to find effects that differ from those resulting from RF heating. Janiak and Szmigielski[69] likewise reported no significant differences in the sequence and time-course of cell membrane injury between cells treated in a water bath and those heated with 2.45-GHz microwaves.

Ciaravino et al.[70] exposed Chinese Hamster Ovary (CHO) cells simultaneously to Adriamycin and pulsed 2.45 GHz microwaves at a SAR of 33.8 W/kg. for two hours. Temperatures did not not exceed 39.7°C. Radiofrequency radiation did not affect changes in cell progression produced by Adriamycin and did not change the number of sister-chromatid exchanges induced by Adriamycin, an agent which is known to damage DNA. Similar results were obtained for Mitomycin C.[71] Lloyd[72] et al. examined human peripheral blood lymphocytes *in vitro* exposed up to 200 W/kg of 2.45 GHz radiation for 20 minutes with temperatures ranging from 37 to 40°C. Using standard techniques for evaluating unstable chromosome and chromatid-type breaks, exchange aberrations or sister chromatid exchanges, no significant variations from appropriate control cultures were found. Hamnerius et al,[73] exposed *S. typhimurium* in exponential phase growth to a number of different frequencies in pulse and continuous wave modes producing SARs up to 130 W/kg over exposure periods ranging from 2.5 to 6 hours. Temperatures were closely controlled between exposed and control samples. Ames' assays on bacteria showed no elevated mutation frequency. The same group, utilizing similar exposure conditions, studied mutations in a gene controlling eye pigmentation in Drosophila. No elevated mutation frequency was found.

In summary, there is no solid evidence that exposure to RF radiation induces mutations in bacteria, yeasts, or fruit flies. The results of some studies suggest that RF radiation induces mutations in mammals. Critical review has cast doubt on these findings. Other studies have shown no mutagenic effects of RF on mammals. Evidence for cytogenetic effects of RF is mixed. The lowest power density at which cytogenetic effects have been reported is 200 W/m^2,[74] but these results are contradicted by Chen et al.[25] who failed to find cytogenetic effects at 2 to 5 kW/m^2. Genetic and mutagenic effects of RF exposure are summarized in Table 4.

B. KINETIC OR FUNCTIONAL CHANGES, AND MEMBRANE EFFECTS

Kinetic or functional changes in cells exposed to RF radiation have been reported. Dutta et al.[75] exposed neuroblastoma cells in culture to varying powers of RF at carrier frequencies of 915 and 147 MHz, amplitude modulated at 16 Hz. Enhanced release of calcium ions from the cells was found at two different power densities between which which no enhancement was found. There was enhanced acetycholinesterase activity under the same exposure conditions. Saffer et al.[76,77,78] found an increase in beta-galactosidase expression exposed to RF between 2 and 4 GHz at 10 W/kg for 4.5 hours with temperature changes not exceeding 0.1°C. Control cultures were adjusted to identical temperatures. The authors concluded that the effect was not thermal but possibly caused by thermal gradients due to RF radiation different from conductive heating produced by ambient conditions.

Parker et al.[79] exposed rodent cell lines to 2.45 GHz-microwave radiation at SARs up to 103 W/kg for varying lengths of time at temperatures from 37 to 45°C, and, using DNA probes for oncogenes, heat protein and long terminal repeat sequences, hybridized to mRNA in exposed and control cells. No significant differences in mRNA expression were observed after microwave exposure.

D'Inzeo, et al.[80] studied acetyl-choline-activated single channel openings of cultured myotubes from chick embryos by patch-clamp techniques. During exposure to RF at 10.75 GHz, 0.5 mW/m^2 at the surface of a bathing solution, the frequency of channel openings decreased but channel opening time and conductance were not affected. The authors speculated that exposure to microwave energy decreased channel opening probability via an alteration of intra-cellular enzymatic processes through a non-thermal mechanism.

Rotkovska et al.[81] studied stem cell colony forming units (CFUs) derived from bone marrow cells of female mice utilizing a spleen exocolony method. At 2.45 GHz CW at a range of temperatures there was a significant enhancement of the number of CFUs in RF-exposed systems versus temperature matched controls.

Rotkovska et al.[82] attempted to assess the role of a membrane receptor in the enhancement effect of microwave energy described above. Bone marrow preparations were incubated with a beta receptor blocker, Trimepranol (TMP), during microwave exposure and controlled with preparations without TMP; in addition, comparable samples were exposed only to ambient heat. The microwave stimulatory effect appeared to be blocked in the presence of TMP. The authors stated that activation of the beta receptors results in enhancement of cAMP synthesis and an increase of Ca^{++} ions in the cell caused by increased membrane permeability. This is followed by reactions of the adenylate cyclase system, resulting the enhancement described above.

Cleary et al.[83] exposed cultured glioma cells to 0.027 and 2.45 GHz CW RF radiation for 2 hours at maintained temperatures of 37°C. At SARs up to 50 W/kg there was significantly more incorporation of tritiated thymidine and uridine into exposed versus control cultures. These changes persisted for up to five days postexposure. Kinetic changes of cell populations exposed to RF energy are also described in the section titled "Effects on Hematopoiesis and Hematology" in this chapter.

Brown and Marshall[84] reported that microwave energy produces no perturbation in cultures of murine erythroleukemic cells undergoing erythroid differentiation in response to hexamethylene bisacetamide (HMBA). Exposures were for 48 hr at 1.180 GHz at SARs up to 73 W/kg with temperature maintained at 37.4°C.

Effects of RF on cell membrane dynamics have been reported. Liburdy and Vanek[84] exposed rabbit erythrocytes to 2.45 GHz microwaves under a number of conditions and observed a change in membrane permeability to sodium ions within the temperature range 17.7 to 19.5°C. This change was described as linear with the internal electric field strength inside the sample, saturating at 600 V/m. The same investigators studied the effects of microwave energy on liposomes carrying cytosine arabinofuranoside (ARA-C).[85] Below the membrane phase transition temperature of 41°C liposome membranes are normally not leaky. In buffered saline exposure to 2.45 GHz-microwaves at a SAR of 60 W/kg ARA-C is released at 33°C. In plasma release occurs at 27°C. Release is enhanced by oxygen and attenuated by antioxidants.

Phelan et al.[86] studied the effect of microwave energy on membrane fluidity using a B16 melanoma cell line exposed to 2.45 GHz pulsed waves 10 μs wide at 100 pps for one hour at 100 W/m^2. An increase in membrane ordering as measured by an electron paramagnetic resonance (EPR) technique, reflecting a shift from a more fluid-like phase to a more solid-like phase, was found. A similar result with artificially prepared liposomes constituted with melanin was discovered. Melanotic cells rendered amelanotic did not show the change. The ordering process was inhibited by superoxide dismutase implicating oxygen radicals as a cause of the membrane changes. The authors proposed that microwaves stimulate melanin to free radical production and that reactive oxygen species may subsequently produce membrane alterations via lipid peroxidation resulting in the ordering effect.

Sandblom and Theander[87] studied the effects of microwaves on the kinetics of gramicidin A channels in lipid bilayer membranes. Artificial membranes into which the deca-pentapeptide, gramicidin-A, was inserted were exposed to 10 GHz pulsed microwaves, pulse width 1 microsecond at a pulse repetition rate of 1000/s in waveguide for one minute at an average output power of 300 W, peak power 300 kW. Temperature excursions were up to 5.1°C and estimated average SAR was 350 W/kg. Although microwave energy did not affect single-channel conductance or channel lifetime compared to matched temperature controls, the rate of channel formation was decreased during exposure. The authors stated that this is opposite of the effect expected for a purely thermal effect and offer theories for a non-thermal mechanism.

Philippova et al.[88] observed that microwave radiation decreased specific ^{13}H-camphor binding to rat olfactory membrane. Exposures were achieved in a waveguide using 0.9 GHz fields pulsed-modulated at frequencies at 1–100 Hz. SAR was varied from 0.5 to 18 W/kg and steady state temperatures of 4°C maintained. The inhibition of binding appeared to decrease with increasing SAR. Inhibition did not depend on pulse-modulation rate. The shedding of protein during microwave exposure of erythrocytes and lymphocytes had been previously reported by Liburdy.[89]

Allis and Sinha-Robinson[90] investigated the ATPase activity in human erythrocyte membranes *in vitro* exposed to 2.45 GHz CW microwave radiation for periods of twenty minutes at a SAR of 6 W/kg and stabilized temperatures from 17 to 31°C. At 25°C, but not at other temperatures, in microwave-exposed samples, there was a 35% decrease in the activity of the enzyme. The authors proposed that a transition state of the protein-lipid interaction within the membrane is available at the transition temperature and not at other temperatures.

The activity of a specific enzyme, ornithine decarboxylase, is increased in certain cell lines following exposure to amplitude modulated microwave fields.[91] This work and other studies of amplitude modulated RF are taken up in Chapter 12.

C. CARCINOGENESIS

There is a concensus among students of the subject that radiofrequency radiation does not initiate carcinogenesis by inflicting direct damage on the genome by any mechanism resembling the effect of ionizing radiation. This is evident from a consideration of the low photon energies involved and the failure to demonstrate genomic damage despite a few suggestive reports to the contrary.[22–26] There remains, at least in principle, the possibility that RF energy could function as a promoter or co-promoter for neoplastic change based on a variety of effects or could function indirectly, as through the endocrine system, or merely as a non-specific stressor in some such way as fibrosis and chronic inflammation function as the "setting" for the origin of some malignancies. Epidemiologic studies have offered conflicting evidence. More recently the focus of the scientific community has rightly or wrongly shifted to the relationship of extremely low frequency (ELF) energy and carcinogenesis. In the related area of teratology, despite observed kinetic effects on cells *in vitro* no developmental abnormalities have been documented. This statement must be modified in light of recent evidence of developmental abnormalities induced by ELF amplitude-modulated microwave energy, but because available evidence relates effects to the ELF frequency and not the microwave carrier frequency this topic is treated in Part III, Chapter 5. U.S. safety standards have not been influenced by any formulation of an increased cancer risk.

RF as a sole stressor and as part of a compound stress on intact animals has been studied. In 1962, Prausnitz et al.[92] reported the induction of leukemia in mice exposed to 9.27 GHz RF at 1000 W/m^2, 4.5 min/day, 5 days a week for 59 weeks. This report is now viewed as a flawed study, principally with regard to interpretation of the histology of the hematopoietic abnormality. Szmigielski et al.[93,94] described accelerated development of spontaneous and benzopyrene-induced skin cancer in mice exposed to 2.45 GHz, 150 W/m^2 (6 to 8 W/kg). At 50 W/m^2 (2 to 3 W/kg) the cancer induction was comparable to that which occurs with chronic stress resulting from confinement. Preskorn et al.,[95] however, reported decreased tumor growth and greater longevity in mice exposed to intense levels of 2.45 GHz energy, ie sufficient to make the mice hyperthermic.

In a study of long-term, low-level microwave exposure of rats Chou et al.[96] surveyed the effects of irradiation of rats on general health and longevity. One hundred exposed and 100 control animals were maintained individually in circularly polarized waveguides. During this period, test animals were exposed to 2.45 GHz pulsed microwaves, 10 μsec pulse width, at 800 pulses per second. The pulsed microwaves were square-wave modulated at 8 Hz. Power was adjusted to an average SAR ranging from 0.15 to 0.4 W/kg. and exposures lasted 21.5 hours/day for 25 months. The study concluded that overall there were no definitive biological effects in the chronically exposed rats. The incidence of benign tumors in exposed animals was not elevated. However a fourfold increase of primary malignancies in the exposed animals was found and considered statistically significant. The biological significance of this finding was questioned because of the need to collapse sparse data without regard for the specific type of malignancy. No single type of malignancy appeared with increased frequency. The authors reiterate the view of the U.S. Environmental Protection Agency that such a pattern is minimal evidence of "carcinogenic action."

In another long-term, low-level microwave exposure study on rats, but at 435 MHz, 10 W/m^2, pulsed-wave (1.0 μs pulse width, 1 kHz pulse rate, est. SAR 0.3 to 0.35 W/kg, 22 hr daily, 7 days/week, for 6 months) Toler et al.[622] found no significant adverse effects. They reported that there was no evidence to identify the 435 MHz pulsed-wave environment as a source of subtle stress. The issue of tumor incidence was not addressed in this study but is the subject of a follow-up study, near completion.

With regard to tissue culture work Balcer-Kubiczek and Harrison[97] exposed fibroblasts in culture to 2.45 GHz microwaves, SAR 4.4 W/kg with and without benzpyrene and the promoter, 12–O–tetradecanoylphorbol. They reported a several-fold increase in transformation frequency with microwave exposure versus chemical exposure alone. In later work Balcer-Kubiczek and Harrison[98] exposed fibroblasts for 24 hr to 2.45-GHz microwaves pulsed at repetition rate of 120/sec. The cells were assayed for neoplastic transformation with or without post-treatment administration of 12–O–tetradecanoylphorbol–13–acetate (TPA). RF fields alone had no effect on cell survival or induction of neoplastic transformation. In the presence of TPA transformation was reported as highly significant and dependent on the level of RF exposure.

In summary radiofrequency energy is not a carcinogen by itself. Although a number of studies have not been able to demonstrate a role as a promoter or co-promoter Balcer-Kubiczek and Harrison[98] have reported significant transformation with a phorbol ester. Phorbol esters are usually thought of as promoters, so that the role of RF radiation in such a situation is unclear. Further work is needed to confirm the findings in this case.

D. HYPERTHERMIA IN MEDICAL THERAPEUTICS

The major application of microwave hyperthermia has been for treatment of malignant tumors. The application of microwave energy alone for control or ablation of tumors has been infeasible because the heat required to produce ablation inevitably produces unacceptable damage to surrounding normal tissue. The practical application of microwave hyperthermia has been as an adjunct to ionizing radiation or chemotherapy in which an enhancement of the partial or complete response rate has been demonstrated.[99] The technical problems, study designs, and evaluation of responses can be found in the clinical oncology and therapeutic radiology literature. The issues of therapeutic efficacy and impact on survival are treated in that literature.

Within the last few years the application of microwave hyperthermia to the treatment of rheumatoid joint disease has been studied. Such treatments may be of benefit.[100]

III. BIOCHEMICAL CHANGES

Biochemical alterations have been reported to result from exposure to RF energies. Such effects generally appear to be reversible and no well-defined characteristic response pattern has been determined, nor is it known whether the changes are direct or indirect effects of exposure.

Effects on mitochondria isolated from exposed animals have been reported,[101] but there was no effect on rat liver mitochondria exposed in vitro to 2.4 GHz, 1 to 4 W/Kg, or 10 to 12 GHz, at a maximum of 20 W/m^2.[102] No effect of microwave exposure has been found on a number of enzymes and proteins irradiated *in vitro.*[102,103,104,105] Albert et al.[106] exposed Chinese hamsters at 2.45, 500 W/m^2, for 0.5 to 4.5 hr over a period of 1 to 21 days and found no change in liver ATP.

The limited number of studies on oxidative enzyme systems has yielded mixed results. Exposure of suspension of the membrane-bound enzyme cytochrome oxidase to sinusoidally modulated 2.45-GHz MW energy at an SAR of 26 W/kg did not significantly affect its activity during exposure. On the other hand Sanders et al.,[107] utilizing rat brain *in vivo,* and exposing the brain surface to 591 MHz CW RF at 50 or 138 W/m^2 under conditions of negligible systemic temperature change, reported that increases in NADH and decreases in

ATP concentration occurred supporting an hypothesis of RFR inhibition of electron transport in brain mitochondria. These observations were later extended to effects of CW, pulsed, and sinusoidal-amplitude-modulated microwaves on brain energy metabolism where similar changes were found.[108]

Dumansky, et al.[109] reported a decrease in liver glycogen content together with increased lactic acid levels and phosphorylase activity which was interpreted as evidence of impairment of glycogen synthesis in the livers of rats chronically exposed to 2.45- or 10-GHz MWs at 500 to 2000 W/m^2. Exposure to 2.45- or 10-GHz MWs at 0.25 to 10 W/m^2 decreased the proteolytic activity of the mucous membrane of the small intestines of experimental animals, where the invertase and adenosine triphosphatase activity increased. Altered proteolytic activity was suggested to be due to alteration in the structure and physicochemical characteristics of the mucous membrane. Reduced synthesis of macroglobulin and macroglobulin antibodies resulted from exposure of experimental animals to 0.5 W/m^2.[110]

Dose-dependent transient elevations in serum glucose, blood urea nitrogen, and uric acid were noted following far-field exposure of rabbits to 2.45 GHz for 2 hr at intensities of 50, 100, and 250 W/m^2.[111] There were detectable differences between CW and PW exposures of equivalent average power density. There was an increase in colonic temperature of 1.7 to 3.0°C at 100 and 250 W/m^2. Exposure of rats for 15 min to pulsed 2.86 GHz at 50, 100, 500, or 1000 W/m^2 resulted in statistically significant changes in serum albumin and phosphorous levels only at 1000 W/m^2, and there was no change in serum glucose levels in rats exposed to 2.86- and 0.43-GHz pulse-modulated fields at an average power density of 50 W/m^2. Single or repeated exposures of rabbits to 3 or 10 GHz at 500 to 2500 W/m^2 resulted in alterations in serum albumin/globulin ratio, which was attributed to effects on the liver or adrenals.[112]

Baranski[113] found inconsistent changes in cholinesterase activity in both rabbit and guinea pig brains following 3 months of exposure, 1 hr/day, with pulsed fields of 250 W/m^2. He also found a decrease in cholinesterase activity in the brains of guinea pigs after a single 3-hr exposure to 35 W/m^2 of pulsed 2.45-GHz MWs. Increase to 250 W/m^2 caused a further decrease in activity. Pulsed energy was found to produce a more severe effect than CW exposures of the same average power density, suggesting that these effects are due to peak fields. Nikogosyan[114] found an increase in blood cholinesterase activity after a single 90-min exposure to 3 GHz waves at 400 W/m^2. Revutsky and Edelman[115] also reported an increase in specific cholinesterase activity in rabbit blood exposed in vitro to 2.45-GHz MWs. It should be noted that what is measured in blood is "pseudocholinesterase" which has no neural-related activity. On the other hand, Olcerst and Rabinowitz[116] found no effect on aqueous cholinesterase exposed to 2.45-GHz CW up to 1.25 kW/m^2 for 1/2 hr or 250 W/m^2 for 3 hr. No effect was found on cholinesterase activity in defibrinated rabbit blood exposed for 3 hr to 210, 350, or 640 W/m^2, 2.45 GHz, CW or pulsed. Under similar exposure conditions, there was no effect on release of bound calcium or magnesium from rabbit red blood cells.

Mitchell et al.[117] exposed or sham-exposed rats to 100 W/m^2 CW radiation at 2.45 GHz for a period of 7 hours. Animals were subjected to behavioral, biochemical and electrophysiological measures during and immediately after exposure. In this joint US/USSR study, the US group found Na^+, K^+, and ATPase activity to be significantly lower in microwave exposed animals than in the shams. No significant differences were found by the Soviet group. On review the authors concluded that the US result was spurious. This and other studies done emphasized the value of multi-laboratory collaboration in promoting rigor in scientific design.

Ho and Edwards[118] studied oxygen consumption in mice exposed to 2.45 GHz radiation at various SARs in a waveguide exposure system and found that the animals compensated homeostatically to SARs of 10.4 W/kg or greater by a decrease in metabolic rate to compensate for thermal loading. Normal metabolic activity was resumed following cessation of exposure.

Some of the reports on effects of RF on subcellular systems and biochemical changes are summarized in Tables 5 to 7.

IV. RESPONSE TO LOCAL RF EXPOSURE

Some consideration has been given in animal and human experiments to the temperature rise in various organs and tissues as functions of incident energy wavelength, thickness of the subcutaneous fatty layer, blood-circulation rate, and other factors. The increase in temperature of tissues during local exposure is linear for short periods (1 to 3 min) and proportional to the magnitude of the MW energy absorbed. With exposures in excess of 3 min, the extent of the thermal effect and distribution of heat in tissues is determined by heat-regulating mechanisms[119,120] (Chapter 10). The thermal effect depends on exposure duration. Deep-lying muscles are heated to a greater extent only during the first 20 min of exposure. When the thigh region is exposed to microwaves, there is a greater temperature rise in the muscles than in the skin and subcutaneous (s.c.) fatty layer.[121,122,123,124] The distribution of temperatures among the various tissues depends on the rate of blood circulation. Thus, rapid heating of all tissues in rabbit thighs was observed during MW exposure when the local blood supply was occluded. When the paws of rabbits were exposed to 2.45 GHz radiation, the temperature rise in the deep tissues of the leg could vary substantially with repeated local exposures. While the temperature in the thigh muscles rose 4°C after the first 10-min exposure at 1.2 to 1.5 kW/m^2, it rose only about 2°C after the sixth or seventh treatment. suggesting that adaptive reactions had come into play. Adaptive reactions were also noted in the opposite unexposed leg after daily conditioning exposures to MWs. After denervation of the rear extremity, the temperature rise in the thigh muscles remained the same regardless of the number of exposures.[125]

Engle et al.,[126] Gersten et al.,[127] and Rae et al.[124] exposed the hind legs of dogs and forearms of human volunteers and found that increments in temperature of muscle were slightly higher than those of s.c. fat. Lehmann et al.[128,129] note that variations in the thickness of s.c. tissues involved may explain apparent differences in results of various investigations. Examination of the hindleg of a dog reveals that the s.c. layer is usually only a few millimeters thick and in many cases contains little fat. Similarly, the human forearm has a comparatively thin layer of s.c. fat in most cases. Thus, one explanation for the greater increases in temperature of muscle than of fat in some studies might be that, under the particular conditions of the experiments, relatively small amounts of energy were absorbed in the thin layer of s.c. fat. For a quantitative discussion of such effects, see the "Introduction" and Part III, Chapters 1 and 2.

According to some investigators,[130,131,132,133] hyperthermia develops more efficiently with MWs than with IR exposure. With 2.45 GHz, 500 W/m^2, 30 min exposure of the spinal region in rabbits, a 1 to 1.5°C colonic temperature rise was observed, while 3.5 kW/m^2 was necessary to produce the same effect with IR.

Several studies, notably those of Lehmann and associates,[134,135,136] have compared the heating patterns and temperature distributions resulting from 2.45- and 0.915-GHz MW exposure to those resulting from IR exposure, using human volunteers. The highest temperature distributions with infrared application were obtained in the most superficial tissues, while the highest temperatures in MW application were obtained close to the s.c. fat-muscle interface. MW application produced somewhat higher temperatures in all cases in the deep tissues. A marked difference was found between temperature distributions produced by the two MW frequencies prior to the blood flow changes resulting from temperature increases in the tissues. The 0.9-GHz exposure produced a greater increase in the temperature of the deeper musculature. This difference was much less pronounced after blood flow changed, about 10 min after start of exposure, since the increased flow cooled the skin and s.c. tissues

TABLE 5
Biochemical Effects of RF

End Point Measured or Effects	Experimental System or Model	Frequency (GHz)	Exposure Conditions Duration (min)	SAR (W/kg)	Modality	Ref
No change in activity of membrane-bound enzymes measured spectrophotometrically	Red blood cell membrane; mitochondrial inner membrane	2.45 (SW)[a]	10	26	Waveguide	504
	Endoplasmic reticulum	2.45 (CW)	5	42	Waveguide	102
No difference in respiratory activity	Mitochondria	2.45 (CW)	30–120	17.5, 87.5	Far field	505
	Mitochondria	2–4 (Swept)	10	1.6–2.3	Coaxial airline	506
		3.1 (CW)		41		
No change in formation of microtubules	Tubulin (rabbit brain)	3.4 (CW)	15	112–430	Far field	507
No change in migration of proteins within axonal membrane	Vagus nerve cell	3.1 (PW)	24 hr	~10–100	Far field	508
No change in infrared spectra of proteins and necleic acids in Escherichia coli exposed before drying	Dried file of E. coli cells	3.2 (CW)	8, 10, 11 hr	20	Waveguide	52
No change in acetylcholinesterase activity	Ray electric organ	2.45 (CW)		4		509
Decrease in adenosine triphosphatase activity	Human RBC	2.45 (CW)		6	Waveguide	90
	Rat liver cell membrane	2.45 (CW)	30 × 4 days	80 W/m^2		510
No change in cytochrome C oxidase activity	Cytochrome C oxidase	2.45 (16, 30, 90, 120 Hz AM)		26	Waveguide	511
Decrease in succine dehydrogenase activity	C57-BL mice	3 (PW)	300	0.1–5 W/m^2		513
Decrease Na^+-, K^+-ATPase activity	Rat cerebral cortex synaptic membrane	2.45 (CW)	420	10 W/m^2	Far field	117

Note: SW = Sine-wave modulated. AM = Amplitude modulated.

Adapted and expanded from U.S. Environmental Protection Agency, Biological Effects of Radiofrequency Radiation, Elder, J. A. and Cahill, D. F., Eds., Rep. No. EPA-600/8-83-026F, Research Triangle Park, N.C., 1984.

TABLE 6
Effects of RF on Molecular Systems

End Point Measured or Effects	Experimental System or Model	Frequency (GHz)	Exposure Conditions Duration (min)	SAR (W/kg)	Modality	Ref.
No change in UV difference spectra measured over pH range 25–55	Bovine serum albumin	1.70 (CW) 2.45 (CW)	30	30–100	Waveguide	103
UV spectra and binding constants for mononucleotides showed no difference from controls	Ribonuclease	1.70 (CW) 2.45 (CW)	30	39	Waveguide	514
No change in enzyme activity	Glucose-6-phosphate dehydrogenase; adenylate kinase; NADPH cytochrome c reductase	2.45 (CW)	5	42	Waveguide	102
No difference in melting curves	DNA	2.45 (CW)	60, 960	67, 160	Far field	446
Inactivation of enzyme; probably temperature inhomogeneity effect at very high doses	Horseradish peroxidase	2.45 (CW)	5, 10, 20, 30	62,500–375,000	Waveguide	515
Heat inactivation of enzymes found at higher SAR (T = 50°C) corresponded closely to heat-treated controls	Glucose-6-phosphate dehydrogenase; lactate dehydrogenase; acid phosphatase; alkaline phosphatase	2.8 (PW)	4.5, 18.5	~200–500	Waveguide	104
Heat inactivation of enzyme found at SARs > 165 W/kg	Lactate dehydrogenase	3.0 (CW)	20	33–960	Waveguide	105

Adapted from U.S. Environmental Protection Agency, Biological Effects of Radiofrequency Radiation, Elder, J. A. and Cahill, D. F., Eds., Rep. No. EPA-600/8-83-026F, Research Triangle Park, N.C., 1984

TABLE 7
Effects of RF on Clinical Chemistry and Metabolism

Effects	Test Specimen	Exposure Conditions: Frequency (GHz)	Intensity (W/m²)	SAR (W/kg)	Duration (days × min)	Δt (°C)	Ref
No effect on serum chemistry values	Rat	0.918 (CW)	25	1.0	91 × 600	0	516
Increase in serum glucose	Rabbit	2.45 (CW and PW)	50, 100, 250	0.8, 1.6, 4.0 (est)	1 × 120	0, 0, 1.7 (PW) 0, 0, 2.9 (CW)	111
Increase in blood urea nitrogen	Rabbit	2.45 (CW)	250	4.0 (est)	1 × 120	2.9	
No increase in blood urea nitrogen	Rabbit	2.45 (CW)	50, 100	0.8, 1.6 (est)	1 × 120	0, 0	
	Rabbit	2.45 (PW)	50, 100, 250	0.8, 1.6, 4.0 (est)	1 × 120	0, 0, 1.7	
Increase in uric acid values	Rabbit	2.45 (CW and PW)	100, 250	1.6, 4.0 (est)	1 × 120	0, 1.7 (PW) 0, 2.9	
No change in uric acid values	Rabbit	2.45 (CW and PW)	50	0.8 (est)	1 × 120	0	
No effect on other serum chemistry values	Rabbit	2.45 (CW and PW)	50, 100, 250	0.8, 1.6, 4.0 (est)	1 × 120	0, 0, 1.7 (PW) 0, 0, 2.9 (CW)	
Increase iron and manganese levels in brain	Rat	1.6 (CW)	800	48 (est)	1 × 10	4.5	517
Decrease in specific metabolic rate (ambient temperature = 24°C)	Mouse	2.45 (CW)		10.4	1 × 30		118

TABLE 7
Continued

Effects	Experimental System of Model	Frequency (GHz)	Exposure Conditions Duration (min)	SAR (W/kg)	Modality	Ref
Increase in red blood electrophoretic mobility 30 min postexposure (SAR ≥ 10 W/kg)	Red blood cell	1.0 (CW)	4, 8, 15, 30	5–45	Stripline	518
Increase in K^+ and Na^+ influx	Red blood cell	1.0 (CW)	30	45	Stripline	519
K^+ transport no different from heat-treated controls; no change in osmotic fragility	Red blood cell	2.45 (CW)	60, 120, 180, 240	3–57	Far field	520
K^+ transport not different from controls at corresponding temperature; no difference in hemoglobin release	Red blood cell	2.45 (CW)	45	200	Far field	521
Passive transport of Na^+ and Rb^+ increased at transition temperature	Red blood cell	2.45 (CW)	60	100, 190, 390	Waveguide	522
No significant changes in K^+ efflux, hemoglobin release, or osmotic fragility	Red blood cell	2.45 (CW) 3.00 (CW) 3.95 (CW)	20, 180	22–200	Waveguide	68
Increase in β-galatosidase activity	E. coli containing plasmid pUC8	2–4 (CW)	270	10	Waveguide	76
Decreased frequency of opening of ACh-activated channels	Chicken embryo myoblasts	10.75 (CW)	0.5–3	0.5 W/m^2	Far field	79
Increase in Na^+ permeability	Rabbit red blood cells	2.45 (CW)	≤ 60	100	Waveguide	84
Change in membrane fluidity	B16 melanoma cells	2.45 (PW)	60	100 W/m^2	Far field	86
Decrease in membrane ATPase activity	Human red blood cells	2.45 (CW)	20	6	Waveguide	90

TABLE 7
Continued

Effects	Test Specimen	Frequency (GHz)	Exposure Conditions Intensity (W/m^2)	SAR (W/kg)	Duration (days × min)	Δt (°C)	Ref
Increase in specific metabolic rate (ambient temperature = 35°C)	Mouse	2.45 (CW)		8.6	1 × 30	0	523
Increase NADH fluorescence	Rat (exposed brain)	0.591 (CW)	138	0.36–2.2	1 × 0.5	0	524
		0.591 (Sine wave)	100, 200	1.81, 3.62	1 × 5	0	525
		0.591 (PW, 250 pps)	45–132.5	0.82–2.38	1 × 5	0	
		0.591 (PW, 500 pps)	30–138	0.54–2.50	1 × 5	0	
No change in NADH fluorescence	Rat (exposed brain)	0.591 (PW, 250 pps)	10, 15, 30	0.18, 0.27, 0.54	1 × 5	0	
		0.591 (PW, 500 pps)	10, 15	0.18, 0.27	1 × 5	0	
Decreased ATP	Rat (exposed brain)	0.591 (CW)	50	0.13–0.8	1 × 0.5	0	524
		0.591 (CW)	138	2.5	1 × 0.5, 1, 5	0, 0, 0	525
		0.591 (16 Hz sine)	138	2.5	1 × 0.5, 1, 5	0, 0, 0	
		0.591 (PW, 500 pps)	138	2.5	1 × 0.5, 1	0, 0	
No change in ATP		0.591 (PW, 500 pps)	138	2.5	1 × 5	0	
Decreased CP	Rat (exposed brain)	0.591 (CW)	50	0.13–0.8	1 × 0.5	0	524
		0.591 (CW)	138	2.5	1 × 0.5, 1, 5	0, 0, 0	525
		0.591 (16 Hz sine)	138	2.5	1 × 0.5, 1, 5	0, 0, 0	
		0.591 (PW, 500 pps)	138	2.5	1 × 0.5, 1, 5	0, 0, 0	
Increase in oxygen consumption	Rat	2.45 (120 Hz PW)		6.5, 11.1	1 × 30	0.9, 1.8	526
Decrease in metabolic heat production	Monkey	2.45 (CW)	60	0.9	1 × 10	0	527
No effect on blood coagulation	Human plasma	2.45 (CW)	100–2800	1.3–38 (est)	1 × 30	Not reported	528

Note: Δt = Colonic temperature increase.

Adapted and expnaded from U.S. Environmental Protection Agency, Biological Effects of Radiofrequency Radiation, Elder, J. A. and Cahill, D. F., Rep. No. EPA-600/8-83-026F, Research Triangle Park, N.C., 1984.

specifically. Thus after blood flow increased, the difference in the deep heating properties between the two frequencies diminished. A more detailed discussion of thermoregulation in the presence of RF fields is found in Chapter 10.

V. REPRODUCTION, GROWTH, AND DEVELOPMENT

A. THE GONADS

The effect of MWs on the testes has been studied fairly extensively.[137,138,139] Exposure of the scrotal area at high power densities (>500 W/m^2) results in varying degrees of testicular damage such as edema, enlargement of the testis, atrophy, fibrosis, and coagulation necrosis of seminiferous tubules in rats and rabbits, exposed to 2.45, 3, and 10 GHz radiation. Exposure to 3 GHz radiation, 80 W/m^2 did not affect mating of mice or rats.[140] Pituitary gonadotropic function was preserved in female mice exposed to 3 GHz radiation, 100 W/m^2, twice daily for 5 months.[141]

Saunders and Kowalczuk[142] exposed the caudal area of anesthesized mature male mice to 2.45-GHz RF in a waveguide system for 30 min. Half-body SARs ranging from 18 to 75 W/kg were estimated from measurements of forward, reflected, and transmitted powers. The corresponding colonic temperatures at the end of exposure ranged from 35.3 to 42.2°C. Other anesthetized mice were sham-exposed. Their mean colonic temperature was 32.6°C (4 to 5°C lower than unanesthetized mice). For comparison, the caudal areas of other anesthetized mice were inserted for 30 min in a copper well heated by a water bath to 37, 41, 43, or 45°C, which resulted in colonic temperature from 36.4 to 40.7°C. Six days after treatment, sections of testes were scored for cell damage and enumeration of sperm. Extensive degeneration of the spermatogenic epithelium was evident for RF exposure at 75 W/kg and for direct heating to 45°C. At SARs of 57 and 46 W/kg (43 and 41°C) marked depletion of spermatids and spermatocytes, but not spermatogonia, was observed. At the lower SARs (37, 30, and 0 W/kg) or a temperature of 37°C, no effects were seen.

Temperature-sensing probes were also implanted in the testes of other groups of mice, and testicular temperatures were related to SAR values. Such measurements of testicular temperature indicated the existence of a threshold of about 39°C for depletion of spermatocytes and of about 41°C for 50% cell death after 6 days of RF exposure or direct heating. The corresponding SARs for these two thresholds were 20 and 30 W/kg, respectively.

Lebovitz and Johnson[143] exposed rats for 8 hours to 1.3 GHz CW radiation at a whole-body-averaged SAR of 9 W/kg. This produced a deep rectal temperature elevation of 4.5°C. Subsequent sequential histologic examination, measurements of daily sperm production, circulating follicle-stimulating hormone (FSH) and leutinizing hormone (LH) were all normal, suggesting that the level of acute hyperthermia was not sufficient to produce disruption for any of the indicated measures of testicular function.

Lebovitz et al,[144] compared pulse-modulated 1.3 GHz MW, 1 μsec pulse width, 600 pulses/second, to conventional heating in rats. Anesthetized animals were sacrificed over a 13 day cycle of seminiferous epithelium after exposure. At 7.7 W/kg, whole-body-average SAR for 90 minutes, there was a modest decline in daily sperm production, primarily from effects on primary spermatocytes. Exposure at 4.2 W/kg produced no change. Temperatures in the first exposure reached 40°C and in the second 38°C. Above 7.7 W/kg all germ cell types were destroyed. Conventional heating in excess of 39°C for 60 minutes produced a significant decrease in daily sperm production. The authors stated all damage could be explained by heating.

There is general agreement that high-power density exposure can affect the testes and ovaries. There are reports, however, that chronic "low-level" exposure can result in impairment of spermatogenesis and reproductive function without measurable temperature increase of the testes.[109,142] Cleary et al.,[145] exposed sperm suspensions *in vitro* to RF radiation for 1 hour at 0.027 or 2.45 GHz CW following which an *in vitro* fertilization assay was performed. Exposures in the range 0 to 80 W/kg but with temperature control within 0.1°C at 37°C were studied. The authors reported a statistically significant reduction in the ability of irradiated sperm to fertilize (or penetrate) mouse ova at SARs of 50 W/kg or greater. They asserted that, because of excellent temperature control, it could be concluded that the effect was not thermal but "direct," although a mechanism was not advanced. Effects of RF on the gonads are tabulated in Table 8.

B. EMBRYONIC DEVELOPMENT

There are a few reports that suggest particular combinations of frequency, duration, and power density produce effects on embryonic development and postnatal growth. Alterations in development have been reported in insects,[146] chick embryos,[147,148] and rodents.[141,149,150,151]

Olsen[152] has provided a thermal basis for explanation of the MW-induced teratogenic effects in the meal worm *Tenebrio molitor* reported by Carpenter and Livstone.[146] Lindauer et al.,[153] and Liu et al.,[154] Olsen,[155] and Pickard and Olsen[156] noted the threshold of teratogenesis in *Tenebrio* to be associated with a rise of pupal temperature in excess of 10°C.

Van Ummersen[147,148] reported inhibition of growth and development of chick embryos exposed to 2.45-GHz (CW), 200 to 400 W/m^2, 4.5 min to 5 hr. Since all embryos in which effects were produced experienced a significant temperature increase, the observed deleterious effects were ascribed to heating.

Quail embryos exposed to 300 W/m^2, 2.45 GHz (CW) for 4 hr on each of the first 5 days of incubation did not develop gross deformities or changes in hatchability that could be correlated with exposure.[157] Microwave-induced temperature changes within the eggs varied from 34 to 37°C, which is within the normal range of incubation temperatures.

In another study Japanese quail embryos were exposed during the first 12 days of development to 2.45-GHz MWs at an incident power density of 50 W/m^2 and SAR of 4.03 W/kg. No gross deformities were observed in the exposed quail when examined and sacrificed at 24 to 36 hr after hatch. No significant changes in the total body weight or weights of the heart, liver, gizzard, adrenals, and pancreas were found in the treated birds. Hematological parameters were also measured in the study. The results showed a statistically significant increase in hemoglobin and statistically significant decrease in monocytes in birds exposed to MWs. No statistically significant changes in hematocrit, red blood cells, total white blood cells, lymphocytes, neutrophils, basophils, or eosinophils were detected.[158]

Spiers[159] exposed Japanese quail eggs from day 1 through 15 of incubation to sham or 2.45-GHz microwave radiation, 8 hours/day at SARs from 3.3 to 15.2 W/kg at a number of ambient temperatures. It was concluded that microwave radiation can be used to increase egg temperature and embryonic growth rate at ambient temperatures below the normal incubation level without altering basic metabolic processes of the embryo.

A number of studies, at various MW frequencies, have been concerned with effects of exposure on mammalian embryonic and fetal development. In most of these studies a single acute exposure was administered at a power density that causes an increase in body temperature.[150,160,161,162] Some investigators, however, have reported studies in which protracted exposures have been given at power density levels which apparently do not cause a significant increase in colonic temperature.[163,164,165,166] These studies and results are described below.

Rugh et al.[150,151] reported abnormalities in mouse fetuses exposed in an environmentally controlled waveguide on gestation days 7 to 13 to 2.45-GHz (CW) MWs in the range of 84 to 112 W/kg (10.3 to 33.5 J/g, 2 to 5 min), equivalent to 1.23 kW/m^2 incident power. Hemorrrhages, resorptions, exencephaly, stunting, and fetal death were observed. In a further study, Rugh and McManaway[167,168] demonstrated that lowering the body temperature of the dam with pentobarbital anesthesia could prevent the teratogenic effects of thermal loading with MWs, thus demonstrating the thermal influence in the earlier studies.

TABLE 8
Effects of RF on Testes

Effects	Test Animal	Frequency (GHz)	Exposure Conditions Intensity (W/m²)	SAR (W/kg)	Duration (days × min)	Ref.
No change	Mouse	1.7	100	15	1 × <100	62, 66
Abnormal germinal cells, normal interstitial cells			100	15	1 × 100	
All tissue necrotic, altered spermatogenesis			500	75	1 × 30–40	
Scrotal skin burns			2000	300	1 × 20	
"Minimal" injury	Mouse	3.0	500	50	1 × 20	
No change in tissue, sperm	Mouse	2.45	<370	<8	Many × 16 hr	529
Abnormal spermatogenic tissue	Rat	2.45	800	16	1 × 10–73	530
				16	5 × 10–73	
No change	Rat	2.45	50	0.9–4.5	many ×240	64
No change			100	2	5 × 360	
Temporary sterility			280	5.6	20 × 240	
Testicular degeneration	Mouse	9.27 (pulsed)	1000		295 × 4.5	531
Testicular lesions	Mouse	10.0 (pulsed)	4000		1 × 5	138
No change	Mouse	10.0 (pulsed)	3.44		20–50 × 30	532
Testicular damage	Rat	24	250		1 × 5–15	533
No change in sperm production, circulatory FSH or LH	Rat	1.3 (CW)	—	9	1 × 480	143
Destruction of germinal cells	Rat	1.3 (pulsed)	—	7.7	13 × 90	144

Adapted and expanded from U.S. Environmental Protection Agency, Biological Effects of Radiofrequency Radiatopm, Elder, J. A. and Cahill, D. F., Eds., Rep. No. EPA-600/8-83-026F, Research Triangle Park, N.C., 1984.

Lin et al.[169] studied the effects of repeated exposure of C3H mice to 0.048 GHz. The animals were exposed to 5 W/m^2 (63.25 V/m peak E field) in a TEM exposure chamber for 1 hr/day, 5 days/week, beginning on the 4th to 7th days post partum, for 10 weeks. The formed elements in the blood were not affected by the exposure. The means of body mass of the irradiated and control animals were comparable. No significant differences in lesion onset, incidence, prevalence, extent or type were observed when animals repeatedly exposed to RF were compared with sham-control groups. During the period of the study, no cataracts were noted. Fertility differences among irradiated and control animals were not detected. Body growth patterns did not differ among sham and RF-exposed animals.

Employing a multimodal cavity at 2.45 GHz, Chernovetz et al.[161] irradiated pregnant mice with a single "intense" dose of 38 W/kg for 10 min (22.8J/g) on gestation day (GD) 11, 12, 13, or 14. This dose resulted in 10% maternal lethality. The exposure during late organogenesis or early fetal stage caused no change in fetal mortality or morbidity when compared to shams. In another study, Chernovetz et al.[162] noted an increased rate of resorptions in rats after an absorbed dose of 30 W/kg for 20 min once on GDs 10 to 16.

Berman et al.[163] exposed CD-1 pregnant mice to 2.45 GHz at 34, 136, or 280 W/m^2, GD 6 through 17 for 100 min daily. Estimates of mean dose rate ranged from 2.0 to 22.2 W/kg. No significant increase in colonic temperature was reported. Another group was similarly sham-exposed. The mice in half of each group were examined on GD 18. The incidence of pregnancy, the number of live, dead, and resorbed fetuses, and the total number of fetuses were found to be similar for the exposed and sham-exposed mice. The mean body weight of the live fetuses in the RF-exposed group, however, was significantly smaller (by 10%) than those in the sham-exposed group, a finding consonant with their previous results. In addition, ossification of sternal centers was significantly delayed in the RF-exposed mice. The mice in the other half of each group were permitted to come to term. Significant fetal growth retardation occurred at the highest power density level. A significant increase in central nervous system (CNS) anomalies occurred, but only if the data from three irradiated groups were combined. At 7 days of age, the mean body weight of the suckling mice of the RF-exposed group was also significantly smaller (by 10%) than that for the sham-exposed group. The survival rate was not affected. The authors concluded that MW radiation at a frequency of 2.45 GHz was embryopathic at a power density of 280 W/m^2 (22.2 W/kg).

Bereznitskaya[141,149] exposed mice to 3 GHz, 100 W/m^2 which resulted in increased fetal wastage. No definite abnormalities or inborn genetic defects were found. Neonatal mice exposed to 0.105, 0.01927, or 0.0266 GHz pulsed in a magnetic field of 55 A/m and an electric field of 8000 V/m, 40 min/day for 5 days did not show any evidence of alteration in growth and development.

Nelson et al.[170] exposed C3H mice in a rectangular coaxial exposure system at 0.148 GHz at 5 W/m^2 for 1 hr daily from GD 2 through day GD 19, corresponding to a SAR of 0.013 W/kg. The experiment was conducted as three separate replications. In each of the experiments some of the dams were allowed to come to term and the fetuses were assessed for weight at birth and again at 60 days of age, while in others the uterus was extirpated on GD 19 of gestation. No differences in percentage of resorbed, stillborn, or abnormal fetuses were observed in any of the three experiments.

The embryofetal toxicity and teratogenicity of 2.45-GHz (CW) MWs at different intensities were investigated in the CD-1 mouse by Nawrot et al.[165] Mice were exposed on GD 1 to 15 at an incident power density of 50 W/m^2 (SAR of 5 W/kg), and either on GDs 1 to 6 or 6 to 15 to 210 W/m^2 (SAR of 28.14 W/kg) or to 300 W/m^2 (SAR of 40.2 W/kg) for 8 hr daily. Exposure either on GD 1 to 6 or 6 to 15 to a power density of 210 or 300 W/m^2 caused an increase in colonic temperature of exposed dams of 1 and 2.3°C, respectively. To distinguish between "thermal" and "nonthermal" effects of 210 or 300 W/m^2, groups of mice were also exposed to elevated ambient temperature to raise their body temperature to the level of the animals exposed to MW. Ambient temperatures of 30 and 31°C increased the deep colonic temperature to that obtained with the 210- and 300 W/m^2 MW exposure, respectively. The mice exposed to higher ambient temperature were handled in exactly the same manner as the MW-exposed mice. A significant decrease in implantation sites per litter and reduction in fetal weight was noted in the group exposed at 300 W/m^2 during GDs 1 to 6 compared to that of mice exposed to elevated ambient temperature. Exposure of mice at 300 W/m^2 (GDs 6 to 15) resulted in a slight but significant increase in the percentage of malformed fetuses, predominantly with cleft palate, when compared to all other groups.

Dietzel et al.[171,172,173] exposed pregnant rats using a 27.12-MHz diathermy unit at 55, 70, or 100 W which was sufficient to raise the animals' colonic temperatures to 39, 40, or 42°C. The rats were exposed once between GDs 1 and 16. The fetuses were examined near term. The peak incidence of anomalies was found to occur on gestation days 13 to 14, when 16% of the fetuses were abnormal. This fetal wastage, as in the studies by Rugh et al.,[150,151] is clearly associated with a general body temperature increase in the dam.

Brown-Woodman et al.[174] exposed pregnant rats at GD 9 to thermalizing levels of RF at 27.12 MHz and found increasing developmental abnormalities with increasing core temperature. Maximum temperature elevation was 5°C. Only a few seconds at this temperature produced abnormalities. Rats heated to comparable levels and for comparable durations in a water bath demonstrated similar levels of abnormalities, suggesting that the teratogenicity of RF radiation is related to hyperthermia.

In studies of effects of *in utero* exposure of Long-Evans rats to 2.45-GHz CW microwaves, Michaelson et al.[175] found no adverse effects on the dam or offspring when gestation length or litter size were examined. The rats were exposed at 100 or 400 W/m^2 for 1 hr on GD 9 (organogenesis) or day 16 of gestation. Enhanced maturation, as indicated by adrenocortical response, was suggested.

Laskey et al.[160] observed an increase in resorption rate and a decrease in term fetal weight in offspring of rats exposed to 2.45 GHz for 8 to 13 min on GD 2, 2 and 5, 8, or 15 at 1 kW/m^2. Maternal body temperature increased 3 to 4 deg C during the exposure. Smialowicz et al.[176] exposed rats to 2.45-GHz MWs daily for 4 hr/day at a power density of 40 W/m^2 *in utero* from GD 6 day through 40 days of age. The SAR was determined by twin-well calorimetry for several ages of the animals. Pregnant dams weighing 300 to 350 g had a mean SAR of 0.7 W/kg; animals that were 1 to 5 days of age and weighed 6 to 10 g absorbed approximately 4.7 W/kg. There was no significant difference between the mean body weights of the males (female offspring were not used in this experiment) in the 12 sham-irradiated litters when compared to the mean body weights of the males in the 12 MW-irradiated litters.

Jensh et al.[177] reported on the effects of protracted prenatal exposure of Wistar-derived rats to 2.45-GHz MWs for 8 hr daily throughout gestation. The mean exposure time to incident energy at 200 W/m^2 was 115 hr. No statistically significant differences were observed between control and exposed animals for maternal mass, embryonic and fetal resorption rate, abnormality rate, and the term fetal and placental masses. These authors[178,179] also reported similar experiments in which 0.915-GHz MW were employed. They exposed Wistar rats at 100 W/m^2 for 8 hr daily throughout the period of gestation for an average of 110 hr of exposure. Postexposure colonic temperatures of dams showed no increase over baseline rectal temperatures. No differences were observed in embryonic or fetal death, abnormalities, fetal mass, litter size, placental mass, fetal sex ratio, maternal mass, or maternal gain of mass.

In another study,[180] gravid Wistar rats were exposed 8 hr daily throughout the pregnancy to 0.915-GHz MWs at 100 W/m^2. The average exposure time was 109 hr. In this study, dams were allowed to deliver their offspring. Within 3 days after birth the pups were given three tests for reflexes. After the pups were weaned, a series of performance tests was given. The postweaning tests began on post partum day 60 and were completed by the 90th day. No detrimental effects were observed on performance in a water T-maze, avoidance behavior, open-field behavior, forelimb hanging, or performance in a 24-hr activity wheel. Jensh et al.[181] also reported no observable teratology in rats exposed throughout gestation to 350-W/m^2 MWs at 6 GHz.

A series of experiments was conducted on rats to assess the teratologic effects of 0.027 GHz because of the prevalence of this frequency in industries using RF heat sealers.[182,183,184,185,186] At various days of gestation the investigators used energy levels sufficient to cause and sustain high body temperatures (up to 43°C, 20 to 40 min). These experiments showed that GD 9 was the most sensitive day for the rat. A variety of malformations, including decreased body weight, was found. Thermal (maternal) threshold was 41.5°C. Gross microscopic teratogenic effects were equivalent when increased maternal temperature was caused by MW or hot-water immersion. The threshold was approximately 12 W/kg for 30 min.

Conover et al.[187] reported on single, 20- and 30-min exposures of fetal rats on GD 10, 12, or 14 to 0.02712 GHz at SARs of 17 to 35 W/kg. Preliminary results suggested that fetal wastage occurred after exposure on the 10th or 14th day of gestation. Grossly observable malformations occurred in animals exposed during the 10th day for 30 min, while visceral abnormalities occurred after exposure on the 10th and 14th days. Fetal mass as well as crown-rump length was smaller numerically in the fetuses exposed to MWs. The same investigators reported on 8 groups of 16 to 28 gravid rats exposed to 0.02712 GHz. The facility for exposure was an RF near-field synthesizer operating in the dominant magnetic field mode at a field strength of 55 A/m. The dams were exposed on GDs, 2, 4, 8, 10, 14, or 16 for 20 to 40 min at an average dose rate of 125 W/kg until their colonic temperature reached 43°C. Eight groups were sham-irradiated for 30 min and one group of 29 rats served as cage controls. Rats irradiated on GDs 8 to 16 (organogenesis) had a significant increase in gross malformations and a significant decrease in fetal weight and fetal crown-rump length.

Boak et al.[188] administered "shortwave" radiation (0.01 GHz) to rabbits from the 29th day of life through several matings and pregnancies. The total exposure time ranged between 30 and 75 hr, during which the temperatures of the animals were raised to 41 to 42°C. There was no interference with mating, fertilization, or development of the young *in utero*. Litter sizes were not significantly different from those of the control animals.

Roux et al.[189] reported a study of rats exposed to 434-MHz radiation and ionizing radiation, indicating that the two forms of radiation potentiate teratogenic effects in the rat.

In a study designed to examine possible effects of chronic RF exposure on mother-offspring behavioral patterns and the EEG, Kaplan et al.[190,191] exposed 33 female squirrel monkeys near the beginning of the second trimester of pregnancy to 2.45-GHz RF in multimode, mode-stirred MW cavities at whole-body SARs of 0.034, 0.34, or 3.4 W/kg (the last value equivalent to about 100 W/m^2 of plane wave RF) for 3 hr/day, 5 days/week, until parturition. Eight pregnant monkeys were sham-exposed for the same periods. After parturition, 18 of the RF-exposed dams and their offspring were exposed to RF for an additional 6 months; then the offspring were exposed without the dams for another 6 months. No differences were found between RF- and sham-exposed dams in the numbers of live births or in the growth rates of the offspring. The major difference between RF- and sham-exposed offspring was that four of the five exposed at 3.4 W/kg both prenatally and after birth unexpectedly died before 6 months of age. They apparently had developed a form of pneumonia. Although the numbers of animals used in the behavioral and EEG studies were adequate, the mortality values were too small to place much confidence in statistical inferences.

O'Connor[192] noted that with respect to basic design, procedure, and variables assessed, the teratogenic studies reported were more diverse than decisive. Wide variation in exposure parameters makes it difficult to compare the results; additional difficultly is generated because many of the reports do not contain information on critical variables such as the manner in which the day of gestation was timed. The day on which the animal is sperm- or sperm-plug positive can be timed as day 0, although a more common procedure is to consider this day 1. The manner in which control animals were treated is often not given. Many kinds of controls have been employed including passive cage controls, sham-exposed controls, heat (IR irradiated) controls, and historic controls. In some reports, probability statements have been substituted for data from concurrently studied control animals. Multiple control procedures in single experiments have not been used extensively.

Many of the investigators who have reported defects have employed acute, intense irradiation which has placed a thermal burden on the exposed subject. Some studies attempt to control for heat by including controls heated by means of IR. Of importance in this regard are the comparisons with the literature on heat stress.

Many of the teratogenic studies of RF exposure have been performed without a sufficient number of animals, and others lack sophisticated design and thus do not allow for observation of low probability events. Statistical analyses are usually not given in enough detail to permit evaluation.[192]

If attention is focused, not on procedural questions, but only on similarities in the results of the studies, several trends are apparent. The most common result from fetal exposure to MWs appears to be a nonspecific, general response, that of reduced or retarded gain of body mass. Without further study it is impossible to know if this decrease is maintained after birth and when, if ever, the exposed stunted fetuses catch up. It is important to note that this general suppression of body mass is the only effect that appears to be common across the range of species that have been studied.

A more general deleterious response seen in the mammal is increased rate of fetal resorption. The increase may be indicative of malformed fetuses, but the resorbed nature of the fetal material precludes a more fine-grained analysis and thus identification of which, if any, specific structure was damaged. The increased rate of resorption and the range of exposures within which it occurs is remarkably similar to the effects of heat stress. Particularly in the rat, the resorption rate appears to increase within a rather narrow thermal window, the other side of which is febrile death. The majority of defects have been observed following high level, acute exposures with obvious thermal effect.[192]

It does appear that abnormalities in small animal fetuses can be produced with MWs in conjunction, however, with systemic hyperthermia of 2.5 to 5°C above the normal temperature for the species for some time period during specific critical developmental stages. From a survey of the literature, it appears that it is the temperature rise in the fetus, irrespective of the manner in which it is produced, that causes damage. It is important to realize that in all species there is a dynamic pattern of maturation during gestation, and in the rat this continues during the first 3 weeks of postnatal life. In interpreting teratological effects of MWs, as with any agent, it is important to realize that many fetal defects such as hemorrhage, resorption, stillbirth, and exencephaly occur spontaneously in mice.[161]

Thermal stress appears to be the primary mechanism by which RF energy absorption exerts a teratogenic action. Chernovetz et al.[161] and others have pointed out evidence that indicates increases in mortality and resorption are probably related to peak body temperature and its duration regardless of the method by which the temperature elevation is elicited. The teratogenic effects of hyperthermia, regardless of the source, have been well documented.[193,194,195,196,197,198] Rugh and McManaway[167,168] were able to prevent the increase in incidence of teratogenic activity, which they had previously reported, by lowering the maternal body temperature through controlled use of pentobarbital anesthesia.

Concepts of thermoregulation and thermal stress have been discussed by Michaelson[199] and Way et al.[200] The absence of a core temperature change does not necessarily indicate that physiologic adjustments induced by exposure to RF energy are not heat related.[199] The organism has numerous thermoregulatory mechanisms by which it can maintain homeostasis such as modification of skin blood flow, alterations in peripheral conductance, changes in blood temperature, and alterations in evaporative loss mechanisms.[200] This does not imply that the organism is not stressed. The stress induced by maintaining thermal homeostasis may result in abnormal growth and development in the offspring.

In man, infections such as rubella, influenza, and smallpox, occurring during early pregnancy, are known to cause abortions and fetal malformations.[201,202] In general, it appears that any infection giving rise to fever in early stages of pregnancy in man or animals is capable of producing fetal malformation or abortion. It is well known that induction of fever can lead to the early termination of pregnancy.[203] There are numerous reports of abnormalities from the induction of systemic hyperthermia of 2.5 to 5°C above the normal temperature for the species for 1 hr or longer during specific critical developmental stages of the fetus by exposure of the pregnant animals to elevated temperatures and humidity in environmental chambers. Fetal resorption, growth retardation, microphthalmia, and malformations affecting the CNS, musculo-skeletal system and other organs have been observed in mammalian species, e.g., the guinea pig[197,204] and the rat.[193,196,205] In experiments on rats, hyperthermia of 4 to 4.5°C for 40 to 60 min during specific developmental stages produces increased fetal resorptions, retardation of growth, microphthalmia, anencephaly, and defects of tails, limbs, toes, palate, and body, depending upon the gestational stage at which hyperthermia occurred.[196] These results indicate that the occurrence of fetal malformations in mammals in early pregnancy is probably related, not as much to the viral or bacterial toxemia, but to the fever, hyperthermia occurring at a particular critical stage of organogenesis. The threshold appears to be an elevation of 2.5 to 5.0°C above the normal temperature of the species sustained for an hour or more.

As noted by Marston and Voronina[206] from the standpoint of public health, one must consider the difficulty of extrapolating data from experimental teratology to the human fetus. Such an extrapolation becomes feasible only after detailed analysis of the fine mechanisms of teratogenesis. Also, of great importance is the need for appropriate scaling factors to permit extrapolation of experimental data obtained on small animals, to the human. The concepts of scaling are discussed in Chapters 8 and 9.

The reports of effects of MW exposure on early development have been reviewed by Baranski and Czerski,[65] who concluded that no serious effects are to be expected at power densities below 100 W/m^2 under usual exposure conditions. They further note that defects, when observed, are the result of hyperthermia. There are numerous reports of abnormalities from the induction of systemic hyperthermia of 2.5 to 5°C above the normal temperature for the species, by exposure of the pregnant animal to elevated temperatures at specific critical developmental stages of the fetus.[196,197] It would thus appear that in the reports of MW-induced developmental abnormalities, it is the temperature rise in the fetus, irrespective of the manner in which it was produced, that caused the damage.

RF energies must be applied at high SAR (>15 W/kg) to rodents, approaching lethal levels, for teratogenesis to result. High maternal body temperatures are known to be associated with birth defects. There appears to be a threshold for the induction of experimental birth defects when a maternal colonic temperature of 41 to 42°C is reached. Any agent capable of producing elevated internal temperatures in this range is a potential teratogen.

Most studies involving prenatal exposures have not shown effects on growth and development. Temperature in the testes of >45°C induced by any modality can cause permanent sterility; from 37 to 42°C mature sperm may be killed with a temporary loss of spermatogenic epithelium. Changes in reproductive efficiency have not been directly associated with RF exposure. Studies concerning the effects of RF exposure on neonatal development have been summarized[207] and presented in Table 9.

TABLE 9
Effects of RF Exposure on Growth and Development

Effects	Animal	Exposure Conditions: Frequency (GHz)	Intensity (W/m²)	SAR (W/kg)	Duration (days × min)	Ref.
No effect on weight gain	Rat	2.45 (CW)	50	0.7–4.7	55 × 240	176
No effect on growth, neurological and immunological development, or mutagenicity	Rat	0.1 (CW)	460	2.8	112 × 240	534
No change in infant mortality	Monkey	2.45 (CW)		3.4	285 × 180	190, 191
No effect on growth	Mouse	2.45 (CW)	100	6–8 (est)	24 × 48	535
No effect on body weight	Infant rat	2.45 (CW)	400	20–60 (est)	6 × 5	536
No effect on growth	Mouse	0.01, 0.019, 0.026 (CW)	8.9×10^4	0.9, 1.8, 3.6 (est)	1 × 20	537
	Mouse	0.019	1.7×10^5 11.4×10^5	6.3 (est)	5 × 40	
No effect on weight gain	Mouse	0.148 (CW)	5	0.013	50 × 60	169
Embryonic LD_{50}	Chicken egg	2.45	2000 2800 4000	70 98 140	1 × 12 1 × 7 1 × 4	147, 148
Decreased postnatal survival	Mouse	2.45		104	1 × 4	167, 168
Teratogenesis	Mouse	2.45		85–112	1 × 2–5	150, 151
No change in teratogenesis	Mouse	2.45		38	1 × 10	161
Increase postnatal survival	Mouse	2.45		38	1 × 10	161
Maternal lethality, resorptions, decreased fetal weight	Rat	2.45		31	1 × 20	162
Decreased fetal weight	Mouse	2.45	280	22	12 × 100	163
No change in hatchability, posthatching hemogram, body or organ weight	Japanese quail egg	2.45		14	1 × 1440	538
No change	Rat	2.45		14	1 × 20	539
	Rat	2.45	400	10	1 × 120	175
	Mouse	2.45	34–140	2–8	12 × 100	163
Teratogenesis	Japanese quail egg			4	12 × 1440	540
No change	Rat	2.45	50	0.7–4.7	Many × 240	176
	Rat	2.45	280	4.2	12 × 100	541
	Rat	0.915 2.45 7	100–350	1–3.5	Many	166 180
	Rat	0.918	50	2.5	19 × 1200	542
Decreased body and brain weight	Rat	2.45	100	2.2	16 × 300	543

Adapted from U.S. Environmental Protection Agency, Biological Effects of Radiofrequency Radiation, Elder J. A., and Cahill, D. F., Eds., Rep. No. EPA-600/8-83-026F, Research Triangle Park, N.C., 1984.

C. POST-NATAL DEVELOPMENT AND BEHAVIOR

Few studies have been reported in which postnatal psychophysiologic parameters have been examined in animals exposed prenatally to MWs. Johnson et al.[164] observed behavioral alterations in rats subsequent to protracted (20 hr daily for 19 days during gestation) prenatal exposure to 0.918 GHz at a power density level of 50 W/m^2. Such exposure resulted in a significantly lower level of achievement in the Conditioned Avoidance Response test. Jensh et al.[166] were not able to detect significant behavioral changes associated with protracted 0.915-GHz MWs at 100 W/m^2 in Wistar rats, a power density level which approached that required to increase maternal colonic temperature. Chernovetz et al.[161] subjected mice to a single 10-min exposure to 2.45-GHz radiation at an SAR of 38 W/kg, without observing any postnatal functional alterations.

Japanese quail, irradiated continuously in ovo during the first 12 days of embryogenesis at 50 W/m^2, SAR 4.03 W/kg, core temperature 37.5°C for exposed and controls, were followed to reproductive maturity. The reproductive histories of the exposed hatchlings were no different from the controls.[208]

VI. EFFECTS ON THE NERVOUS SYSTEM

Transient changes in CNS function have been reported following MW exposure. Although some reports describe the thermal nature of MW energy absorption, others implicate non-thermal or "specific" MW effects at the molecular and cellular level. The first report of the effect of MW energy in the centimeter range on the conditional response activity of experimental animals was made by Gordon et al.[209] In subsequent years, study of "non-thermal" effects of MWs gradually occupied the central role in electrophysiological studies in the Soviet Union.[210]

Yakovleva et al.[211] reported that single and repeated exposures of rats to MWs, 50 to 150 W/m^2, weakened the "excitation process" and decreased the "functional mobility" of cells in the cerebral cortex. Edematous changes were most often noted throughout the cortex. The greatest number of altered cells was noted with repeated exposures at 150 W/m^2.

Seizure response to noise was transiently suppressed in audiogenic-seizure susceptible mice and rats after exposure to 3-GHz pulsed MWs at an average power density of 100 W/m^2.[212,213] Changes in olfactory threshold in humans following occupational exposures have been reported.[214,215,216]

Tolgskaya et al.[217] studied the effects of CW and PW 3 and 10-GHz MWs at various intensities on rats. More pronounced morphological changes in the CNS were reported following 3 than 10 GHz at 10 to 100 W/m^2. Pulsed waves were considered more effective than CW, a phenomenon also reported by Marha.[218] Additional studies comparing pulsed and CW MWs can be found in the monograph by Tolgskaya and Gordon,[219] who noted that the disturbances MWs can cause in conditioned reflex activity and which characterize functional changes in CNS activity are functional in nature, are reversible, and disappear at the same time as the conditional reflex activity of the animals is being restored upon cessation of exposure to MWs. A proliferative reaction of glial cells was also described.

Some investigators suggest that MW energy absorption may affect hypothalamic and midbrain function and also affect cerebral, cortical, and reticular system function.[220,221] Acording to Gvozdikova et al.[222] the greatest cortical sensitivity occurs in the meter range, less in the decimeter, and least in the centimeter MW band.

Guy and Chou[15,223] studied response thresholds in rats for short, high-intensity microwave pulses. The energies were substantially in excess of those producing auditory responses (500 microseconds wide or less, 5 to 180 mJ/kg in brain). See Section XII, Auditory Responses. Rats were subjected to local exposures to the head and especially the brain of a single 0.915 GHz pulsed magnetic field sufficient to produce a local SAR as high as 4 × 10^5 W/kg and with pulse widths from 1 μsec to 360 ms. Animals displayed no reaction other than that due to the hearing effect until the specific absorption exceeded 28 kJ/kg. This correlated with a temperature rise in the brain of 8°C. Seizures lasted for one minute after exposure followed by a 4 to 5 minute unconscious state. Rats then recovered without apparent effect from the exposure. Brain histology revealed some demyelination of neurons one day after exposure and focal gliosis one month after exposure. Similar studies were performed by Petin, et al.[224]

By contrast, Brown et al.,[225] also looking at high-power pulses in intervals from 25 msec to 3.2 sec, found effects at lower energies. Utilizing a 1.25 GHz source, comparing pulses at a repetition rate of 80 per second to equivalent "CW" for each episodes of exposure, varying peak powers and total duration for different episode of exposure, a threshold was found for an "evoked body movement," an involuntary movement measured by a motion sensor attached to the tail of the rat. No differences were observed between PW and CW (or wide pulses) at any given SAR and SA. When the SAR was held constant at 7.3 kW/kg under either mode, and the SA was allowed to increase, by varying the total exposure time from 0 to 3.2 seconds the percentage of animals displaying the body movement increased proportionally up to 0.9 kJ/kg and then leveled off at 84% of the animals reflecting a dependency on SA. On the other hand, holding SA constant at 0.9 kJ/kg and varying SAR revealed that above 900 W/kg most animals responded. However, lowering SAR (by lengthening the inter-pulse interval at constant SA) produced a proportionate decrease in percentage of responding animals out to 3.2 seconds, reflecting the ability of the thermoregulatory mechanisms to adjust to these lower SARs. At the highest energies brain temperature rose by 0.4°C. Skin and subcutaneous temperatures facing the energy source at the higher energies were measured and found to be at levels known to produce a non-painful thermal sensation. The authors were uncertain as to the mechanism underlying the evoked movement. In further work of Brown et al. (unpublished) skin and subcutaneous regions of the snout and surrounding areas were anesthetized and animals subjected to pulsed energy as before. Thresholds and nature of responses were identical to those of the previous exposures, suggesting that cutaneous mechanisms did not mediate the "evoked movement." Wachtel, reflecting on his earlier work with *Aplysia ganglia*[226] has proposed that the evoked movement reported by Brown et al., may result from the rate of temperature rise rather than the temperature change per se in a direct coupling with nervous tissue (personal communication). The issue is, however, unresolved. Akyel et al.[227] demonstrated that the evoked body movement produced by PW microwaves is differently affected by a number of neuropharmacologic drugs from that of the classic startle response. Additional work on high-peak-low-average power acute and short-term effects is summarized in Section VIII, Behavioral Effects.

A. ELECTROENCEPHALOGRAPHIC CHANGES

Several investigators have reported that MW exposure produces alterations in the electroencephalogram (EEG).[228–236] Chizhenkova[237] exposed unanesthetized rabbits to 2.4 GHz for 1 min at 400 W/m^2 and observed an increase in the number of slow waves in the EEG, and a change in the discharge frequency of neurons in the visual cortex. Evoked response of visual cortex neurons to light was enhanced and was regarded as the most sensitive indicator of an MW effect. Baldwin et al.[228] exposed the heads of rhesus monkeys to 0.225 to 0.400 GHz CW in a resonant cavity and noted a progressively generalized slowing and some increase in amplitude of EEG patterns accompanied by signs of agitation, drowsiness, akinesia, and nystagmus, as well as autonomic sensory and motor abnormalities. There were signs of diencephalic and mesencephalic disturbances: alternation of arousal and drowsiness, together with confirming EEG signs. The response depended on orientation of the head in the field and reflections from the surrounding enclosure. In this report from 1960 the animals were described as being exposed in a copper mesh resonant cavity into which a quarter wave antenna was inserted at the top. No direct dosimetry is described, but instead temperature measurements were performed in a vessel filled with water the size of an average monkey head. Temperature changes were merely described as trivial. Monkeys were fitted with metallic electrodes driven into the outer table of the skull for EEG measurements. It is a reasonable speculation that greater amounts of energy than suspected produced localized heating around electrodes affecting the results.

In a joint US/USSR study[116] Mitchell et al., studied rats exposed for 7 hours at 2.45 GHz and 100 W/m^2; in comparing results between US and USSR investigators, he found statistically significant effects in the power spectral analysis of EEG frequency, but the two groups found such effects at different frequencies. EEG tracings in rabbits exposed to 3-GHz (PW) 50 W/m^2 showed slight desynchronization from the motor region; at 200 W/m^2 variations in the amplitude were observed; 0.300 GHz had a greater effect than 3 GHz. It was suggested that PW MWs produced a greater effect than CW MWs.

Baranski and Edelwejn[238] reported that rabbits exposed to 10 GHz (pulsed), 40 W/m^2 (single exposure) showed no changes in EEG tracings, but exposure to 3 GHz, 70 W/m^2, 3 hr/day for 60 days, produced functional and morphological changes. Changes in conditional reflexes were also reported in rats exposed to 0.07 GHz, 150 V/m, 60 min/day for 4 months.

A review of the literature on EEG effects requires awareness of certain deficiencies in methodology and interpretation. The EEG is difficult to quantitate due to its time-varying waveform. The use of metallic electrodes either implanted in the brain or attached to the scalp also makes many of the reports on EEG or evoked responses (ER) questionable. Johnson and Guy[239] have pointed out that such metallic electrodes grossly perturb the field and produce greatly enhanced absorption of the energy in the vicinity of the electrodes. Such enhancement produces artifacts in the biological preparation under investigation. Recording artifacts also result from pickup of fields by the electrodes and leads during the recording of EEGs while the animal is being exposed. For example, Tyazhelov et al.[240] have pointed out that, even for the coaxial electrode described by Frey,[241] diffraction of EM waves is still a major source of error because of the metallic nature and large dimensions of the electrode.

Bawin et al.[242] reported that electromagnetic energy of 0.147 GHz, amplitude modulated at brain wave frequencies, influenced spontaneous and conditioned EEG patterns in the cat at 10 W/m^2. These amplitude-modulated 0.147-GHz fields induced changes only when the amplitude modulation frequency approached that of physiological bioelectric function rhythms; no effects were seen at modulation frequencies either below 8 Hz or above 16 Hz.

B. CALCIUM EFFLUX

Extremely low frequency (ELF) amplitude-modulated RF fields have been used to study calcium efflux changes in chick and rat brains.[222,243,244] The entire subject of the calcium efflux studies is discussed in Chapter 12.

C. HISTOPATHOLOGY

According to some authors, cellular changes have been found in the nervous system of small animals following MW exposure at 100 W/m^2.[217,245,246] Albert and Sherif[247] exposed one-day and six-day old rats in the far field to 2.45 GHz, CW microwaves at 100 W/m^2 for 5 consecutive days, 7 hours per day (SAR 2W/kg). Animals were sacrificed one day after exposure cessation, and histology revealed evidence of widespread cell death with associated mononuclear cell infiltration.

Degeneration of neurons in the cerebral cortex and tissue changes in the kidney and myocardium of rabbits have been produced by exposure to 0.2 GHz. Accinni et al.[248] found evidence for a condition resembling immune complex glomerulonephritis in rabbits exposed to 27.12 MHz with a diathermy apparatus sufficient to elevate rectal temperatures 2–3°C per episode of exposure. The exposures occurred daily for three weeks up to 20 min/day.

Exposure of cats for 1 hr to 10 GHz, 4 kW/m^2 resulted in injury to cerebral and spinal cord nerve cells; changes occurred in tigroid substance (Nissl bodies) and other components of nerve cells.[249] On the other hand, rabbits exposed to 10 GHz (pulsed), 40 W/m^2 showed no evidence of morphological damage to the brain, but comparable exposure to 3 GHz did produce such changes.[233]

Tolgskaya and Gordon[217] have investigated the influence of PW and CW 3 and 10 GHz on the morphology of nervous tissue in rats and rabbits. With exposure to 3 GHz (1.1 kW/m^2 and 400 W/m^2), severe symptoms of overheating were observed, often leading to death. Severe vascular disorders such as edema and hemorrhages in the brain and internal organs were prominent. In repeated but less prolonged exposure, vascular disorders and degenerative changes in internal organs and the nervous system were less severe. With repeated exposures, the animals were better able to withstand successive exposures; they continued to gain weight; body temperature after irradiation quickly recovered, and temperature increase was not evident. Such a response resembles thermal adaptation on a physiological level.

At high field intensities, when death is a result of hyperthermia, the vascular changes are those of hyperemia, hemorrhage, and acute dystrophic manifestations.[250,251,252] At low field intensities, the changes are of a more general dystrophic character, and proliferation of the glia and vascular changes are not as prominent.

Albert and DeSantis[253] reported morphological changes in the brains of Chinese hamsters following exposure to 2.45-GHz CW MWs at power densities of 250 and 500 W/m^2. Exposure durations vaired from 30 min to 14 hr/day for 22 days. Both light and electron microscopic findings varied from 30 min to 14 hr/day for 22 days, revealing alterations in the hypothalamus and subthalamic structures of exposed animals, whereas other regions of the brain appeared unaltered. It should be noted that extremely high local SAR could be present under these conditions. Peak SAR could reach 40 to 200 W/kg in selected brain regions.

In subsequent studies at 1.76 and 2.45 GHz, 100 and 200 W/m^2, Albert [254,255] described cytoplasmic vacuolization of neurons, irregular swelling of axons, and decrease in dendritic spines of cortical neurons. The axonal swelling and spine changes were seen only in chronic exposures, whereas neuronal changes were observed in acute exposure. In all studies no signs of permanent degenerative changes were recorded and reversibility was noted 2 hr after exposure.[254] The author concluded that while it was possible that higher exposure levels (250 and 500 W/m^2) could have resulted in thermal effects, it was unlikely that 100 W/m^2 would result in significant thermalization of the whole brain. He did not, however, rule out the possibility of hot spots. Exposure of rats to 100 W/m^2 of 2.45- or 2.8-GHz RF resulted in average hypothalamic temperature increases of 0.4°C or less. This increase is lower than hypothalamic temperature increases observed during the normal activity of an animal.[256,257]

Albert et al.[258] looked at the effects of RF exposure pre-and postnatally on the Purkinje cells of the rat cerebellum. In one experiment, Sprague-Dawley rats were exposed *in utero* to 100-MHz CW RF at 460 W/m^2 (SAR 2.81 W/kg) for 6 hr/day on GDs 16 to 21, and then for 4 hr/day for 97 days after birth. Four exposed and four sham-exposed animals were sacrificed 14 months after cessation of irradiation. Quantitative assessment of the cerebella showed that the relative number of Purkinje cells was significantly smaller (12.7%) in experimental animals than in control animals. In another experiment, Sprague-Dawley rats were exposed *in utero* to 2.45-GHz CW RF at 100 W/m^2 (SAR 2 W/kg) for 21 hr/day on GD 17 to 21. Power density measured was variable from 40 to 300 W/m^2 because of group exposure conditions. Half of the litters were used shortly after delivery and the other half 40 days after cessation of irradiation to assess effects of the exposure on cerebellar Purkinje cells. Because of the immaturity of the neonates, the Purkinje cell layer was not clearly displayed and quantitative results could not be obtained. Experimental animals sacrificed 40 days postexposure showed significantly fewer (25.8%) cells than did the controls. In a final experiment, rat pups were exposed to 2.45-GHz RF at 100 W/m^2 for 7 hr/day on postnatal days 6 to 10. Half the pups were sacrificed at the end of exposure and half at 40 days postexposure. Only those sacrificed immediately showed a significant decrease in the relative number of Purkinje cells as compared with sham-exposed controls. Thus, exposure to two frequencies of RF at similar SAR values (2.8 and 2 W/kg) yielded a reduction in the relative number of Purkinje cells for fetuses and newborn rats. Although the change appeared permanent for rats exposed *in utero* it appeared to be reversible for those exposed postnatally.

In a related study, Albert et al.[259] examined the effects of RF exposure on the Purkinje cells of squirrel monkey cerebella. Pregnant squirrel monkeys were exposed to 2.45-GHz pulsed RF at an equivalent power density of 100 W/m^2 (SAR 3.4 W/kg) 3 hr/day starting in the first trimester of pregnancy. The offspring were exposed similarly for the first 9.5 months after birth. At the end of the irradiation period, seven exposed and seven sham-exposed animals were sacrificed and their cerebella examined. There were no statistically significant differences between control and exposed animals in any of the Purkinje cell parameters examined. Several factors were suggested by Albert et al.[258] to explain the discrepancy between these results and those with the rat. Factors that might have contributed were differences in geometrical configurations of the head, exposure methods, and daily exposure durations, as well as variations in gestational periods and species differences. It is quite possible that high local SAR values in the rat brain, but not in the squirrel monkey brain, at comparable whole-body SARs may be the important factor in explaining these differences.

D. EFFECTS ON THE BLOOD-BRAIN BARRIER

The existence of a blood-brain barrier (BBB) in most regions of the brain has been established experimentally, although its specific morphology is still conjectural. This barrier normally provides resistance to movements of large-molecular-weight, fat-insoluble substances (e.g., proteins or polypeptides) from the blood vessels into the surrounding cerebral extracellular fluid, presumably to protect the brain from invasion by various blood-borne pathogens and toxic substances. Several investigators have reported that RF can increase the permeability of the BBB to certain substances of large molecular weight, as described below. However, others were unable to confirm such effects.

The concept of the BBB is replete with definitional problems even preceding the interpretation of experiments. It is known that normal physiologic mechanisms of substance transport can occur by simple diffusion, facilitated diffusion, active transport or pinocytosis and vesicular transport. Different substances are subject to different mechanisms, making comparisons of differing substances as markers an uncertain process. The concept of an intrinsic permeability which can be separated from other factors such as local cerebral blood flow has proved to be especially difficult to characterize. Changes in temperature, which are inevitable concomitants of RF experiments, alter several processes simultaneously making it difficult to identify changes in an intrinsic permeability.

Sutton et al.[260] used 2.45 GHz to produce selective hyperthermia of the brain in rats. He then studied the integrity of the BBB with horseradish peroxidase (HRP), a protein tracer that can be detected both morphologically and quantitatively. Brains were heated to 40, 42, and 45°C. Barrier integrity was disrupted after heating for more than 45 min at 40°C. Animals with brains heated to 45°C survived for only 8 to 15 min. The most common site of vascular leakage was the white matter adjacent to the granular cell layer of the cerebellum. Sutton concluded that to prevent BBB disruption, brain temperatures must not exceed 40°C in the absence of body-core hypothermia.

Albert et al.[255] also used HRP as a tracer and reported regions of leakage in the microvasculature of the brains of Chinese hamsters exposed to 2.45-GHz MWs at 100 W/m^2 for 2 to 8 hr. In control animals, extravascular reaction product was found only in brain regions normally lacking a BBB. In a later paper, Albert[261] reported that continuation of these earlier studies indicated that a partial restoration of the BBB's impermeability may have occurred within 1 hr after exposure ceased, and that restoration was virtually complete within 2 hr. Albert believed that the transient changes may be clinically subacute and probably cause no lasting ill effects. It is important to note, however, that such leakage of the microvasculature of the brain occurs irregularly; this was observed in approximately 50% of exposed and 20% of control animals studied by Albert.[253,260]

In another study, Albert[261] exposed 52 animals (34 Chinese hamsters and 18 rats) to 2.8 GHz RF for 2 hr at 100 W/m^2. Of these, 30 were euthanized immediately, 11 at 1 hr after exposure, and 11 at 2 hr after exposure. Another 20 animals (12 hamsters and 8 rats) were sham-exposed. Leakage of HRP in some brain regions was reported for 17 of the 30 animals euthanized immediately after RF exposure and for 4 of the 20 sham-exposed animals. Fewer areas of increased BBB permeability were evident for animals euthanized 1 hr after RF exposure, and except for one rat, virtually no leakage of HRP was seen for the animals euthanized 2 hr after RF exposure. These results indicate that increased BBB permeability due to RF exposure at levels insufficient to denature brain tissue is a reversible effect. (Albert suggested that such BBB changes may be clinically subacute and would probably cause no lasting ill effects.) However, the increased BBB permeability seen in 4 of 20 sham-exposed animals may indicate that factors other than RFR in the experimental procedure could alter the BBB. One possible confounding point in the use of injected HRP as a tracer is the existence of endogenous peroxidase, the detection of which could yield false-positive results. Also, no positive (BBB-altering) control agent was used in these studies for comparative purposes.

Comparable results were reported by Albert and Kerns[262] on 51 Chinese hamsters exposed to 2.45-GHz CW RF for 2 hr at 100 W/m^2. There were 39 original sham-irradiated controls. Of the RF-exposed animals, 12 were allowed to recover for either 1 or 2 hr prior to HRP injection and subsequent fixation. This study appears to be an extension (for Chinese hamsters) of the work reported by Albert[261] for 2.8-GHz RF exposure, with the same conclusions as that paper.

Neubauer et al.[263] reported that 2.45 GHz microwave radiation at 100 W/m^2, SAR approximately 2 W/kg, pulse width 10 μsec at 100 pps, resulted in enhanced uptake of rhodamine-ferritin complex by capillary endothelial cells of the cerebral cortex in rats, and was related to exposure duration and power density. Brain temperatures were not reported.

Oscar and Hawkins[264] reported that a single exposure of rats to 1.3-GHz pulsed MWs for 20 min (1 to 20 W/m^2) induces a temporary change in the permeability of small inert polar molecules across the BBB of rats. Increases in permeability were observed for mannitol and inulin, but not for dextran, both immediately and 4 hr after exposure, but not 24 hr after exposure. They found statistically significant changes in the brain uptake index (BUI), also known as the Oldendorf technique, at average power densities less than 30 W/m^2. They also found that, depending on the specific pulse characteristics used, pulsed RF could be more or less effective in altering BBB permeability than CW RF of the same average power density. For pulses of long duration and high pulse power density, but only a few pulses per second, mannitol permeation could be induced at average power densities as low as 0.3 W/m^2. The authors considered the possibility of local heating due to hot spots since the greatest BBB alteration occurs in the cerebellum and medulla or close to the neck region of rats.

The Oldendorf technique involves injecting a measured amount of a ^{14}C labeled test substance and ^{3}H water arterially, allowing one pass of the bolus through the brain, following by decapitation and measurement of radioactivities in specific parts of the brain. A measurement of the ratio $^{14}C/^{3}H$ in brain tissue is normalized to the identical ratio in the injectate. The technique assumes that tritiated water freely diffuses between the brain and its vascular system. In fact, the passage of tritiated water is a function of local cerebral blood flow, which itself varies, and therefore distorts the interpretation of the overall ratio.

Oscar, et al.[265,266] used a technique employing ^{14}C iodoantipyrine to measure local cerebral blood flow (LCBF) in rats. Male rats weighing 250 to 300 g were individually sham-exposed for 5, 30, and 60 min, or individually exposed for 5, 15, 30, 45 or 60 min to pulsed RF at 150 W/m^2 average power density. Carrier frequency was 2.8 GHz, pulse rate was 500 pps, and pulse width was 2 μsec. Within 5 min after sham or RF exposure, the previously venous-catheterized, but conscious, animals were infused with isotonic saline containing 5 μCI/ml of ^{14}C iodoantipyrine. The rats were decapitated 50 sec after the start of the infusion. Brain regions were dissected out and assayed for radioactivity by routine liquid scintillation counting. Local cerebral blood flow was then calculated by established procedures. The results show that MW exposure caused a significant increase in LCBF (minimum 39% to >100%) in all 17 brain regions sampled. Because of these findings, the authors indicated that their earlier reported ratio measurements[264] may be an overestimate.

Preston et al.,[267] using methods similar to those of Oscar and Hawkins,[264] attempted to determine whether exposure to 2.45-GHz CW RF increased BBB permeability to ^{14}C labeled D-mannitol. They exposed rats to 1, 5, 10, or 100 W/m^2 Compared with sham-exposed rats for controls, they found no evidence that RFR exposure increased the permeability of the BBB for mannitol. In a second series, rats were exposed to 3, 10, 30, 100, and 300 W/m^2. Again, there were no differences between results from exposed and sham-exposed animals. Like Oscar et al.,[266] Preston et al.[267] believed changes in LCBF confounded the results of earlier studies.

Chang et al.[268] used a technique involving ^{131}I-labeled albumin to investigate alterations of the BBB in dogs. The heads of dogs were exposed at various average power densities between 20 and 2000 W/m^2. In general, no statistically significant differences were found between exposed and sham-exposed animals, but the number of animals used in this study was too small to ascribe a high level of statistical confidence.

Attempts to duplicate the findings of Oscar and Hawkins[264] have yielded equivocal results[268] or failure.[269–273] Merritt et al.[274] exposed rats for 30 min to 1.2-GHz pulsed RF at peak power densities in the range of 20 to 750 W/m^2 and 0.5 duty cycle, corresponding to average power densities of 10 to 380 W/m^2, or for 35 min to 1.3-GHz pulsed or CW RF at average power densities in the range of 1 to 200 W/m^2. They examined brain slices under UV light for transfer of fluorescein and under white light for transfer of Evans blue dye (a visual tracer) across the BBB, and chemically analyzed various brain regions for fluorescein content. They also measured the brain uptake of ^{14}C-labeled D-mannitol and determined the BUI values. To validate these detection methods, they used hypertonic urea, known to alter the BBB, as an alternative agent to RF. Last, sham-exposed rats were heated for 30 min in a 43°C oven to approximate the hyperthermia obained at 380 W/m^2. In their examination of brain slices, Merritt et al.[274] found no evidence of enhanced fluorescein or Evans blue dye transfer across the BBB of RF-exposed rats, whereas penetration of the BBB was apparent for rats treated with urea instread of RF. The analyses of fluorescein content corroborated these findings. However, fluorescein uptake was high for the sham-exposed rats that were heated in the oven, an indication that hyperthermia of the brain is necessary to alter BBB permeability. In the ^{14}C-mannitol study of the various brain regions, there were no significant differences in BUI between RF- and sham-exposed rats, whereas BUI changes were evident for rats treated with urea. Also, the results showed no evidence of the power density window reported by Oscar and Hawkins.[264] No change in the BBB occurred unless brain temperature was increased by 4°C.

By using a small, dielectrically loaded coaxial applicator, Lin and Lin[275–276] were able to irradiate only the heads of anesthetized adult male Wistar rats with pulsed 2.45-GHz RF (10 microsec, 500 pps) for 20 min at average power densities of 5 W/m^2 to 30 kW/m^2. The distribution of absorbed MW energy inside the head was determined by thermographic procedures, and average SARs were found to range from 0.4 to 240 W/kg. Evans blue dye was injected into a catheterized femoral vein following sham or RFR exposure, and 5 min later the animal was perfused via the left ventricle with normal saline. The brain was removed, examined and scored for degree of tissue staining by the tracer. For average power densities up to and including 26 kW/m^2 (200 W/kg), staining was not significantly different between exposed and control animals. For exposures at 30 kW/m^2 (240 W/kg), extravasation of Evans blue dye could be seen in the cortex, hippocampus and midbrain. The degree of staining decreased with increasing time to euthanasia postexposure, indicating that the effect was reversible. No alteration in the BBB was evident unless brain temperature was increased by 4°C. Neilly and Lin[277,302] found that ethanol inhibits microwave-induced permeation of the blood-brain barrier through reduced heating of the brain.

It is important to realize that the methods used to investigate BBB permeability are still controversial. Permeability changes in cerebral blood vessels occur under various condition, including those that produce heat necrosis.[278] Most techniques used to measure BBB permeability in fact measure the net influence of several variables on brain uptake, and do not differentiate among the effects of changes in the vascular space, alterations of blood flow, and membrane permeability.[264]

The uncertainty in most earlier research on the influence of RF on the BBB is no doubt related to significant artifacts introduced by the kinds of biological techniques used. Several investigators have indicated that exposure to RF may alter the size of vascular and extravascular volumes and cerebral blood flow rate, thereby yielding changes in the BUI that are not necessarily related to BBB permeability alterations. Blasberg[279] reviewed many of the methods previously used for investigating BBB changes and the problems associated with these methods. Rapoport et al.[280] developed a method for measuring cerebrovascular permeability to ^{14}C-labeled sucrose that yields results independent of cerebral blood flow rate. As already noted, Oscar et al.[266] confirmed experimentally that LCBF is increased in the rat brain by exposure to pulsed RFR at 150 W/m^2 average power density.

In repeating a number of earlier experiments Williams et al.[273] concluded that studies reporting increased permeability of the BBB in many cases were subject to technically derived artifact invalidating the conclusion. The authors found decreased entry of horseradish peroxidase (HRP) and ^{14}C sucrose into the microvessel endothelium of hyperthermic rats at cerebral temperatures up to 43°C. They stated that it is possible that temperatures in excess of 43°C might disrupt the barrier by overt thermal injury as has been reported in clinical studies. They concluded that hyperthermia produces disruption of membrane functions including pinocytosis, but pointed out that other modes of transport, such as active and facilitated transport mechanisms have not been evaluated. There is still not a complete concensus among students of the BBB question. There is a widespread view that little confidence can be placed in the results of early experiments on RF-induced BBB alterations. The concensus is that hyperthermic levels of RFR, producing temperatures >43°C, can alter the permeability of the BBB.

Following the original suggestions that RF might increase the permeability of the BBB clinicians considered that the delivery of hydrophilic chemotherapeutic drugs might be facilitated by RF exposure. The inability to deliver such drugs to tumor sites in the brain has long been a problem for oncologists. After several years of investigation it now appears that the level of a drug entering the brain might actually be reduced at "moderate" levels of hyperthermia.[272]

E. COMBINED EFFECT OF RF AND DRUGS, AND RF EFFECTS ON NEUROTRANSMITTERS

A pharmacodynamic approach has been taken by some investigators in the study of MW exposure effects on the CNS and behavior.[281–283] Following exposure to 3 GHz PW MWs altered sensitivity to neurotropic drugs was noted. Decreased tolerance of rabbits to pentylenetetrazole and increased tolerance to strychnine were observed after a single exposure to 200 W/m^2. Repeated exposures at 70 W/m^2 produced a decreased tolerance to pentyletetrazole, strychnine, and acetophenetidin.[276]

Servantie et al.[284] reported that exposure of rats to 3 GHz at 50 W/m^2 for several days resulted in an altered reaction to pentylenetetrazole. Using curare-like compounds, a neuromuscular site of action for this MW effect was implicated. Edelwejn[282] observed alterations in the effects of chlorpromazine and/or D-tubocurarine on EEG recordings in rabbits repeatedly exposed at 70 W/m^2. The author concluded that synaptic structures at the level of the brain stem are affected by MWs. It should be noted that drugs such as pentylenetetrazole and chlorpromazine influence physiologic thermoregulation.

In some cases drugs and microwave irradiation can have parallel effects, suggesting effects via a common mechanism. Lai, for example, has suggested that the parallel effects of MW and endogenous opioids on choline metabolism may be due to activation of endogenous opioids by MWs. Lai et al.,[285] studying the effect of RF on central cholinergic activity, measured sodium-dependent, high-affinity choline uptake in different parts of the rat brain *in vivo* exposed to 2.450 GHz microwave radiation, 100 W/m^2, CW and PW (100 W/m^2, 2 μsec pulses, 500 pulses/sec), whole body average SAR 0.6 W/kg. PW was found to decrease choline uptake in the hippocampus and frontal cortex. There was no effect on the hypothalamus, striatum and inferior colliculus. Pretreatment with a narcotic antagonist blocked the effect of PW on hippocampal choline uptake but did not alter the effect on the frontal cortex. CW decreased uptake in the frontal cortex but not in the hippocampus, striatum or hypothalamus. These effects have been further characterized.[286–291] The effect on central acetylcholine metabolism of RF is therefore probably a secondary effect of endogenous opioids which may result from the character of RF as a stressor. The auditory effect of pulsed RF could act as such a stressor. According to Lai microwave-induced changes in central high-affinity choline uptake may also be mediated by corticotropin-releasing factor.[292]

Microwave energy can affect the action of psychoactive drugs. Modulation of the action of barbiturates and related effects on the EEG have been reported. It has been reported that warming an animal with microwaves during anesthesia can attenuate the effects of pentobarbital.[293] Microwave energy can prolong the narcolepsy and hypothermia induced by pentobarbital. The role of local absorption patterns was found to be important.

A number of drugs can modify the thermal responses to microwaves. For example, chlorpromazine can attenuate the thermal effect of microwaves in ketamine-anesthetized rats.[294] On the other hand chlorpromazine was reported to cause an increase in susceptibility to microwave heating in the dog.[295]

Anticonvulsant drugs such as chloral hydrate can decrease susceptibility of rats to MW-induced convulsions.[296] The subject of drug modifications of thermal responses to microwaves has been reviewed by Jauchem.[297]

Pulsed microwaves have been reported to enhance the hypothermic effects of acetycholinesterase inhibitors in the central nervous system.[298] Quock, et al.[299] reported that microwave exposure facilitates the antagonism of Domperidone to Apomorphine-induced stereotypic climbing in mice. Quock et al.[300] reported that microwave exposure facilitates the antagonism of Methylatropine to central cholinomimetic drug effects.

Lai et al.[301] found that microwave radiation enhances ethanol consumption but attenuates ethanol hypothermia at low, non-thermalizing levels in the rat.

Neilly and Lin[302] studied the possible impact of ethanol on the blood brain barrier in rats. Varying amounts of ethanol were administered followed by high intensity (30kW/m^2 for 15 min) microwaves. Rectal temperature remained constant at 37°C. and brain temperatures were monitored. After radiation Evans blue dye in saline was injected and degree of staining in brain parenchyma examined. The steady state brain temperatures were found to be highest (48°C) in animals receiving saline or the lowest doses of alcohol. The degree of staining decreased with increase in the amount of alcohol administered. The authors suggested that ethanol inhibits microwave-induced permeation of the blood-brain-barrier through reduced heating of the brain. See Section IV. D, this chapter.

Hjeresen et al.[303,304] further characterized the ethanol effect. These investigators found, utilizing low-intensity exposures (0.3 W/kg, 45 min exposures, 2.45 GHz), that attenuation of ethanol-induced hypothermia by microwaves is more effective at lower doses of ethanol than at higher doses. They argued for a non-thermal effect of microwaves, possibly an effect on neurotransmitter systems based on similarities to pentobarbital-induced hypothermia.

The direct effect of microwave energy on neurotransmitters is presently under study.[291] In one investigation a decrease in brain concentration of acetylcholine was seen in mice exposed to pulsed microwaves which produced a brain temperature increase of 2 to 4°C. This was believed to be due to an increased release of the transmitter.[305]

Gandhi and Ross investigated effects of nonionizing radiation on neurotransmitters.[306] Rats were subject to 0.700 GHz radiation at 150 W/m^2 raising core temperature by 2.5°C. Six brain regions were investigated, but only the hypothalamus showed significant changes in receptor states. The study concluded that both norepinephrine and acetylcholine are released in response to heat. A case was made that norepinephrine suppresses efferent impulses for heat production, whereas the cholinergic system initiates the heat-loss or heat-dissipation mechanism.

Inaba et al.[307] investigated the effect on biogenic amines in the rat brain exposed to 2.45-GHz microwaves at 50 and 100 W/m^2 for 1 hr at ambient temperatures of 21 to 23°C. At the lower intensity rectal temperatures stabilized after rising 2.3°C and at the upper intensity 3.4°C. Under these conditions norepinephrine content was reported to be significantly reduced. 5-hydroxyindoleacetic acid was increased and serotonin was reported to be unchanged. These and other results suggested that microwaves can affect the function of some neurotransmitters. Browning and Haycock investigated the effect of low-level MWs on a neuron-specific phosphoprotein, synapsin I, and found no effect on neurotransmitter levels or brain function at the intensity of microwaves studied.[308] Gandhi and Ross[309] investigated the effect of low-level MWs on rat brain synaptosomes and found that MWs alter the metabolism of inositol phospholipids by enhancing their turnover. It was suggested that this may affect transmembrane signalling in the nerve endings.

Further information on combined nonionizing radiation and drugs effects can be found in the review of Lai et al.[310]

F. NEUROPHYSIOLOGIC EFFECTS *IN VITRO*

Although this section is concerned with effects on the central nervous system, it seems appropriate to summarize work on the direct effect of RF on transmission of the nerve impulse, although the bulk of this work has been done on peripheral nerves and ganglia *in vitro.*

Chou and Guy[311] studied the effects of CW and PW 2.45 GHz MW on frog-sciatic, cat-saphenous, and rabbit superior cervical ganglia and vagus nerves. These investigations revealed that there is no direct production of an action potential by CW or pulsed microwaves (up to pulsed RF at 220 kW/kg peak or CW RF at 1500 W/kg). Compound action potentials were excited electrically in the presence and absence of RF. No changes in conduction velocity or amplitude was detected except with temperature elevations of at least 1°C in the solution. Identical changes could be produced by ambient changes in temperature.

Wachtel et al.[312,313] studying the effect of 1.5- or 2.45-GHz CW RF on abdominal ganglia from Aplysia, reported that patterns of pacemaker activity could be altered by MW and that comparable temperature elevations by ambient heating did not produce such changes. Studies by McRee and Wachtel[314] on frog sciatic nerves suggested that MWs could produce diminished vitality of nerve function independent of temperature changes.

Khramov et al.[315] investigated the effects of 34–78 GHz MWs on the spontaneous firing rate of the stretch-receptor neurons of the crayfish under thermally controlled conditions. The firing rates slowed in the presence of RF and temperature increases from 0.1 to 0.2°C., but not in the absence of temperature elevations. An analysis of the data suggested a thermal rather than direct RF effect.

On the other hand Bolshakov and Alekseev[316] studying neurons of *Lymnea stagnalis* compared the effects of CW and PW at 900 MHz in cells exhibiting a burst-like irregularity in firing. Abovc 0.5 W/kg PW (0.5 to 110 pps), but not CW, enhanced the probability of the burst-like irregularity. This was identified as possibly a MW-pulse specific effect which could have arisen from mechanical oscillations in the exposure system.

Pakhomov et al.[317] studied the effect of CW RF exposure (6.5 GHz, 0.1–2 kW/kg, 0.1–20 min) on isolated frog spinal cord preparations. No effect was found in the absence of heating and, in the presence of RF-induced heating, effects were identical to conventional heating. Pakhomov[318] then looked at the effects of 915-MHz PWs (peak SAR of 5–70 kW/kg) on isolated frog nerves under a variety of conditions. In the exposed groups compound action potential amplitudes decreased significantly over the controls. At the highest SARs microwave heating produced temperature increases of 2.7°C, but comparable conventional heating did not produce the effect. Nerve velocities and refractory characteristics did not change. It was concluded that a particular type of exposure can produce MW-specific changes.

Ginsburg et al.[319–324] investigated the effects of MW (CW 2.45 GHz, SAR from 12.5 to 125 W/kg on input resistance and action potential in snail neurons. Although changes occurred they were not consistent. On the other hand conventional temperature changes caused consistent changes in input resistance and action potentials. The authors suggested that CW MWs may enhance degenerative effects in neuron metabolism but do not indicate a specific mechanism for MW interaction with neurons.

Effects of RF on the nervous system have been tabulated in Table 10.

VII. BEHAVIORAL EFFECTS

Studies have been conducted on the effects of RF on performance of trained tasks or operant behavior by rats, rhesus, and squirrel monkeys.[325–337] All of the studies indicated that the exposure would result in suppressed performance of the trained task, and that an energy-power density/dose threshold for achieving the suppression existed. Depending on duration and other parameters of exposure, the threshold power density for affecting trained behavior ranged between 50 and 500 W/m^2.

A. EXPERIMENTAL BEHAVIORAL STUDIES

Justesen and King[331] used a 2.45-GHz (CW) multimodal resonating cavity to investigate conditioned operant behavior in rats. They used a recurrent cycle of exposure, 5 min on and 5 min off, over a 60-min period at average absorbed energy rates of 3.0, 6.2, and 9.2 W/kg. The performance of the animal usually stopped near the end of the 60-min test period during exposure with an energy absorption rate of 6.2 W/kg; at 9.2 W/kg, this effect occurred much earlier in the test period. Hunt et al.,[332] also using a multimodal resonating cavity to expose rats to 2.45 GHz PW, found effects on exploratory activity, swimming, and discrimination performance of vigilance task, after a 30-min exposure at about 6 W/kg.

In an unusual test animal, Liddle et al.[338] found no effect on the web-spinning ability of the cross spider exposed to 9.6 GHz pulsed RF for 16 hr at power densities of 0.01, 0.1, and 1 W/cm^2 (estimated SARs of 0.4, 4,0, and 40 W/kg).

Lobanova[333] exposed rats to 3-GHz pulsed MWs, after which the rats were tested for swimming time. A decrease in endurance was noted after exposure to power-time combinations ranging from 1 kW/m^2 for 5 min to 100 W/m^2 for 90 min.

TABLE 10
Effects of RF on the Nervous System

Effects	Species	Frequency (GHz)	Exposure Conditions Intensity (W/m^2)	SAR (W/kg)	Duration (days × min)	Ref.
Desynchronized EEG	Rabbits	3 (PW)	200 (av)	3.0 (est)	1 × 20	281
Greater effect of CNS stimulating drugs	Rabbits	3 (PW)	70 (av)	1.0 (est)	24–26 × 180	281
Biphasic effect of latency to a convulsive drug effect	Mice	3 (PW)	50	5	8–36 × Unknown	284
Decreased effect of paralyzing drugs	Rats	3 (PW)	50	1	10–15 × Unknown	284
Changes in EEG patterns of unanesthetized animals	Rabbits	9.3 (CW)	7–28	0.1–0.3 (est)	1 × 5	544
Potentiation of drug response	Male rats	2.45 (PW)	10 (av)	0.2 (est)	1 × 30	545, 546
Decreased hypothalamic NE, DA, and hippocampal serotonin in hyperthermic animals	Rats	1.6 (CW)	800	24 (est)	1 × 10	547
Decreased hypothalamic NE, DA	Rats	1.6 (CW)	200, 800	6, 24 (est)	1 × 10	548
No effect on neurotransmitter levels	Rats	1.6 (CW)	100	3.0 (est)	1 × 10	548
No effect on GABA content	Rats	2.86 (PW)	800	16.0 (est)	1 × 5	549
			400	8.0 (est)	1 × 20	
			100	2.0 (est)	5 × 480	
			100	2.0 (est)	40 × 240	
Swollen neurons in hypothalamus and subthalamus	Chinese hamster	2.45 (CW)	500	15 (est)	1 × 30	253
			250	7.5 (est)	22 × 840	
Swollen neurons in hypothalamus and suthalamus	Chinese hamster	1.7 (CW)	100	3 (est)	1 × 30–120	550
Myelin figures in dendrites 6 weeks postexposure	Female rats	2.45 (CW; multimodal cavity)	100	2.3 (est)	110 × 300	551
Increased permeability of BBB to fluorescein	Rats	1.2 (CW)	24	1.0 (est)	1 × 30	552
		1.2 (PW)	2 (av)	0.08 (est)	1 × 30	
Myelin degeneration and metabolic alterations; glial cell proliferation	Guinea pigs, rabbits	3 (CW; PW)	35	0.5 (est)	90 × 180	553
		3 (CW; PW)	250	3.5	1 × 180	
		3 (CW; PW)	50	0.4 (est)	90 × 180	
Focal areas of increased BBB permeability to peroxidase	Chinese hamsters	2.8 (CW)	100	1.9	1 × 120	261
	Rats	2.8 (CW)	100	0.9	1 × 120	
Increased peroxidase in brain, absent after recovery period	Chinese hamsters	2.45 (CW)	100	2.5	1 × 120	262
Increased peroxidase in brain	Chinese hamsters	2.45 (CW)	100	2.5	1 × 120	254

TABLE 10
Continued

Effects	Species	Frequency (GHz)	Exposure conditions: Intensity (W/m^2)	Exposure conditions: SAR (W/kg)	Duration (days × min)	Ref.
Brain temperature elevation (40–45°C); increased permeability of BBB	Rats	2.45 (CW)	(80 W)	—	1 × 10–30	554
Increased permeability of BBB (mannitol and inulin)	Rats	1.3 (CW)	10	0.4	1 × 20	264
		1.3 (PW)	3 (av)	0.1	1 × 20	
EEG effects seen after chronic, but not acute, exposures	Rabbits	0.001–0.01 (AM)	(60–500) (V_{rms}/m)	10^{-5}–10^{-4}	1 × 120–180	555
		0.001–0.01 (AM)	(90–500) (V_{rms}/m)	10^{-4}–10^{-3}	20–30 × 120–180	
Change of predominant EEG frequencies	Cat	0.147 (AM)	10	0.015 (est)	Varying	556
Reversible neuronal morphology alterations	Rat	3	100	2 (est)	35 × 30	557
	Rat	3	100	2 (est)	35 × 30	219
Mild pyknosis of hyppocampal neurons, increased brain, and colonic temperature	Rat	2.45	(60–90 W)	Head only	1 × 2.5–7	558
Increased brain serotonin turnover rate	Rat	3	400	8.0 (est)	1 × 60	559
	Rat	3	100	2.0 (est)	7 × 480	
No decrease in cerebellar Purkinje cells in offspring	Squirrel monkey	2.45	100	3.4	368 × 180	259
Decreased cerebellar Purkinje cells after perinatal exposure	Rat	2.45	100	2.0	5 × 1260	258
		0.100	460	2.7	110 × 240	
Involuntary tail flick	Rat	1.25 (PW)	—	≤7,300	≤3.2 sec	225
Inflammation, necrosis of brain tissue	Rat	2.45 (CW)	100	2.0	5 × 420	247
Enhanced uptake of rhodamine-ferritin complex by cerebral endothelial cells	Rat	2.45 (PW)	100	2.0		263
Ethanol inhibits microwave-induced permeation of blood-brain-barrier	Rat	3.15	30,000		1 × 15	277
Decreased in choline uptake in frontal cortex	Rat	2.45 (CW and PW)		0.6	1 × 45	287

Note: AM = amplitude modulation, NE = norepinephrine, DA = dopamine, BBB = blood-brain barrier, EEG = electroencelphalogram, CNA = central nervous system, GABA = gamma-aminobutyric acid.

Adapted and expanded from U.S. Environmental Protection Agency, Biological Effects of Radiofrequency Radiation, Elder, J. A., and Cahill, D. F., Eds., Rep. No. EPA-600/8-83-026F, Research Triangle Park, N.C., 1984.

Lin et al.[339] exposed rats to 0.918 GHz CW at levels of 100, 200, or 400 W/m^2 for 30 min. No effects on response rates were noted at the two lower levels, but at 400 W/m^2 the performance of the animal decreased after 5 min of exposure and ceased after about 15 min of exposure. The average energy absorption rate measured thermographically was 2.1 W/kg/W/m^2 incident or 8.4 W/kg absorbed at 400 W/m^2.

Diachenko and Milroy[340] studied the effects of pulsed and CW MWs on an operant behavior in rats trained to perform a lever-pressing response on a DRL (Differential Reinforcement of Low Rate) schedule and tested immediately after 1 hr daily exposure to 10, 50, 100, and 150 W/m^2, 2.45-GHz MWs. No behavioral effects were found at these levels. The subjects exposed to 100 W/m^2, while showing no significant decrement in performance, did show signs of heat stress.

Thomas et al.[341] reported response-rate changes in rats exposed between 50 and 200 W/m^2 to 2.86-GHz (CW) and 9.6-GHz PW MWs. Response rates increased in five of ten tests. Decrement in timing of responses were suggested as indicating that "low-level" MWs produce effects on the CNS.

Galloway[330] studied the performance of four monkeys when using a 2.45-GHz (CW) waveguide applicator (total absorbed power of 10, 15, or 25 W) applied to the head. The duration of the irradiation was 2 min or until convulsions began (20 and 25 W produced convulsions). Because of skin burns, only two subjects completed this series of experiments. There were no performance decrements in a discrimination task that the subjects performed immediately after exposure. Acquisition of a new task during the first ten trials of training was impaired at 25 W, which resulted in convulsions.

Roberti et al.[342] measured the running time of rats in an electrifiable runway in which each subject was trained to peak performance. Exposure for 185 hr at 10.7 GHz (CW), 3 GHz (CW), and 3 GHz pulsed, to 10 W/m^2, caused no performance decrements. No change in baseline performance was noted when rats were irradiated with 3 GHz pulsed for 17 days at a power density of 250 W/m^2.

Thomas and Maitland[283] investigated the effects of pulsed 2.45-GHz MWs, 1 W/m^2 in combination with dextroamphetamine on behavior in rats. Both acute and repeated exposures modified the normal dose-effect function. The maximum drug effect was obtained at lower MW exposures as the drug dose was increased.

Galloway and Waxler[343] employed a serotonin-depleting drug (fenfluramine) to investigate the effect of 2.45-GHz CW RF applied to the head of 7- to 9-kg rhesus monkeys at dose rates of 2 to 15 W. Combinations of the drug and RF at a dose rate of 15 W resulted in behavioral deficits, whereas the drug or RF alone up to 15 W failed to produce this effect. In respect to drugs such as dextroamphetamine and fenfluramine, their influence on thermal regulation may be significant in these results.

D'Andrea et al.[344] observed significant differences in shuttlebox performances and schedule-controlled lever pressing for food pellets in rats exposed to 2.45-GHz CW microwaves at an average power density of 0.05 W/m^2, 7 hr/day, 7 days/week, up to 630 hr. Similar findings occurred at 2.5 W/m^2[345] DeWitt et al.[346] exposed rats 7 hr per day for 90 days to CW 2.45-GHz microwaves at an average power density of 0.05 W/m^2 followed by operant testing. In general no behavioral alterations were found except for a variable effect on a time-related operant task. Galvin et al.[347] studied long term exposures of rats prenatally and perinatally (4.45 GHz, 100 W/m^2, 3 hr/day, GDs 5–20 and days 2–20 postnatally) and found altered endurance and changes in gross motor activity in exposed versus control animals.

Mitchell et al.[348] exposed rats to 2.45-GHz (CW) microwave radiation at 100 W/m^2 for 7 hr and then performed a number of behavioral tests. Exposed mice exhibited reduced locomotor activity and diminished responsiveness to acoustic stimuli. Average SAR of the freely moving animals was estimated to be 2.7 W/kg.

Lebovitz[349] compared CW and PW radiation at levels known to affect behavior. The pulsed source utilized could produce 1 μsec wide pulses at 600 pulses per second with a peak pulse power of 750 kW versus the CW source which could produce a maximum power of 100 Watts. Rats, trained in an operant task, were exposed during the behavioral sessions for 3 hr per day to CW or PW (1.3 GHz) at 3.7, 5.8 and 6.7 W/kg. At the two higher SARs core temperature was elevated from 0.5 to 1.0°C. The two modes of energy delivery were equally effective in reducing response rates of the operant behavior. From the dosimetry the auditory effect was probably present.

Akyel et al.[350] utilizing a high-peak-power 1.25 GHz source which could produce a peak SAR of 0.21 MW/kg and SA of 2.1 J/kg per pulse, exposed rats for 10 min and then subjected them to an operant schedule. By varying the pulse repetition rate a range of whole-body-average SARs from 0.84 to 213 W/kg could be produced. The highest SAR produced a colonic temperature elevation of 2.5°C. No interference with operant behavior was found at the lower dose levels. At the highest dose animals failed to respond to the operant schedule. No specific effect of pulse power could be found and the interference with behavior appeared to be of thermal origin.

In contrast to the previous studies in which postexposure behavior was measured, deLorge[351] exposed rats, squirrel monkeys, and rhesus monkeys to 2.45-GHz MWs under far-field conditions while the animals were performing on operant schedules for food reinforcement. Exposure sessions lasted 60 min and were repeated on a daily basis. Stable performance on the operant schedules was disrupted in all three species at power densities positively correlated with the body mass of the animals. The behavior of the squirrel monkeys was disrupted at the middle level of power densities. When the averages of these power densities (280, 450, and 670 W/m^2) are plotted as a function of body mass (0.3, 0.7, and 5 kg) a semilog relationship becomes evident. Extrapolation along the resulting curve could permit prediction of the power densities needed to disrupt ongoing operant behavior in larger animals. The power densities associated with behavioral disruption approximated those power densities that produced an increase in colonic temperature of at least 1°C above control levels in the corresponding animals. These data support the need for scaling factors to extrapolate from small animals to larger animals.

With new industrial and military applications the issue of high-peak, low repetition rate (and low average power) exposure has been studied. The safety standards of the 1980s[352,353] did not limit peak power, setting standards only for temporal average powers. Thomas et al.[354] reported disruption of operant behavior in rats at whole-body average SARs of 0.2–3.6 W/kg and peak SARs of 0.2 to 3.6 kW/kg, 500 pps, 2 μsec pulse widths, 30 minutes/week exposure just before operant testing. However, Lebovitz[349] found no effect on operant behaviors of pulsed exposures with whole-body-average SAR of 6.7 W/kg and peak SAR 11.0 kW/kg, 600 pps, 1 μsec pulse width, 3-hr exposure daily during behavior. D'Andrea et al.[355] also found no effects on monkeys, utilizing operant behavioral studies. In these studies in which tested behaviors were concurrent with exposures, whole-body-average SARS ranged from 0.05 to 0.80 W/kg and peak whole-body-average SAR was 8.3 W/kg.

In a later study D'Andrea, et al.[223] exposed rhesus monkeys to high-peak-power 5.62-GHz microwave pulses during the performance of an operant task. Two different peak powers at each of three different SARs were evaluated. At a whole-body-average SAR of 2 W/kg and peak powers of 560 W/m^2 or 5.18 kW/m^2 no alterations in behavior were found. At 4 W/kg with 1.28 kW/m^2 and 12.7 kW/m^2 peaks, and also at 6 W/kg with 2.77 kW/m^2 and 25.2 kW/m^2 peaks, decrements in performance were found. The differences in peak power at the two higher SARs did not produce any differences in behavior. Performance returned to baseline levels following exposure and the animals exhibited no discomfort during the exposure.

Hjeresen et al.[356] utilizing a source capable of producing peak power densities up 1.8 kW/cm^2, 10 μsec pulse widths, 10 pulses per second, at 1.3 GHz with resultant whole-body-average SARS from 1.8 to 26.2 W/kg, studied a number of behaviors in rats and concluded that disrupted behaviors correlated only with increased colonic temperature and therefore average SAR. Hjeresen and Umbarger[357] utilizing a high-power-microwave source with frequency range of 2.01 to 2.57 GHz producing peak-power densities up to 241 MW/m^2 in pulses 85 nsec wide, with 50 such pulses delivered over a 5-min period, produced no interference in behaviors in rats. No aversive properties were found and the authors suggested that peak power is apparently not as important as average power in causing deleterious biological effects.

Raslear and Akyel et al.[358–361] also studied high-peak-power pulsed exposures with short pulse widths and slow repetition rates (irregularly shaped pulse with a 2 MW/m^2 peak and an average of 500 MW/m^2 over the pulse period, 2 W/m^2 average, 80 nsec pulse-width, repetition rate = 0.125, 3 GHz, 200 pulses over 26 min, average SAR = 0.07 W/kg, peak SAR = 7 MW/kg. These were the highest peak powers and peak SARs found in the literature, and degraded performance of rats was found in a number of test protocols including Y maze negotiation and treadmill running. The authors concluded that at such high peak powers impairments can be found despite the low average SAR. In all of the pulse-power studies described the dosimetry suggests production of the auditory effect. What role it might have played in studies revealing effects is not clear.

Most of the research on the nervous system and behavior has been carried out in rodents and other lower animals. Behavior among animal species reflects adaptive brain-behavior patterns. Behavioral thermal regulation is seen as an attempt to maintain a nearly constant internal thermal environment. Changes in body temperature bring about not only autonomic drives, but also behavioral drives.[362] That MWs can influence behavioral thermoregulation has been shown by Stern et al.[363] and Adair.[364] This subject is treated in greater detail in Chapter 10.

Behavioral responses are not necessarily manifestations of specific changes in the CNS and may be a function of direct or indirect action of MWs on other body systems. Extrapolation of brain-behavior functions from lower animals to man is thus subject to many difficulties.

In assessing the significance of the reported behavioral changes, it is important to recognize certain fundamental factors. The resting metabolic rate for rats is approximately 7 W/kg. When this level is exceeded, disruption of behavior could be elicited. In most of the reports, alterations in the behavior of rats were observed with exposures at average energy absorption rates of 5 to 8 W/kg or greater, i.e., at similar exposure levels to those that produce increases in circulating corticosterone (CS) concentrations in rats.[365] Behavioral changes may be related to more subtle heat alterations with the body. Heat may produce a general debilitation effect or a decreased motivation for food, since it has been shown that rats maintained in hot environments eat less food[366] and rats show decreased response and food reinforcement frequency on an operant schedule when environmental temperature is 35°C, but not 25°C. Behavioral responses may be influenced by the interaction of the organism with the environment.

B. BEHAVIORAL THERMOREGULATION

The regulation of body temperature can be accomplished by complex thermoregulatory patterns of responses of the skeletal musculature to heat and cold which modify the rates of heat production and/or heat loss (e.g., by exercise, change in body conformation, the thermal insulation of bedding and [in man] of clothing, and by the selection of an environment which reduces thermal stress)[367] (see also Chapter 10).

Thermoregulation is part of a complex control system involving circulation, metabolism, and respiration, as well as neural structures. Temperature signals from cutaneous thermoreceptors reach the somatosensory region of the cerebral cortex. The main processing of thermal signals and generation of a controlling signal for the effector part of thermoregulation takes place in the hypothalamus. Adair et al.[368] found minimal changes in hypothalamic temperature in conjunction with alteration of thermoregulatory behavior in squirrel monkeys exposed to 2.45-GHz, RF, 40 to 200 W/m^2 (SAR 0.2 to 1 W/kg).

Successful strategies or mechanisms to maintain body temperatures in a narrow and desirable range in a complex and varying thermal environment are termed thermoregulatory. Although such strategies or mechanisms are found in great variety, they fall into two main categories: voluntary and behavioral adjustments, and involuntary physiological adjustments. The limits of effectiveness of involuntary physiological thermoregulation are rather narrow and we must rely on behavioral methods of thermoregulation over most of the range of environmental temperatures to which we are often exposed. Changes in body temperatures bring about not only autonomic drives, but also behavioral drives.[362]

Thermal motivation arises in situations of thermal stress. On the warm side, it is the uncomfortable feeling of excessive warmth and the desire for temperature reduction; on the cold side, it involves the unpleasant feeling of being too cold and the desire for temperature increase. The biological significance of thermal motivation is that, by acting in such a way as to minimize thermal discomfort and maximize thermal comfort, the organism tends to escape from situations of thermal stress and to locate itself in a physiologically neutral thermal environment, thereby solving the problem of physiological temperature regulation.[369]

Smialowicz et al.[370] made unrestrained, unanesthetized mice hypothermic by injecting them with 5-hydroxytryptamine (5-HT) in a controlled environment of 22°C and 50% relative humidity. Mice injected with saline were used as controls. Colonic temperatures were measured prior to injection. Following injection, groups of mice were exposed to 2.45-GHz CW RF for 15 min at 100, 50, or 10 W/m^2 (equivalent SARs 7.2, 3.6, and 0.7 W/kg) or were sham exposed, after which their colonic temperatures were measured again. The experiments were performed with BALB/c and CBA/J mice. For saline-injected mice of either type, there were no significant colonic-temperature differences between mice exposed at 100 W/m^2 and sham-exposed mice. For BALB/c mice made hypothermic with 5-HT, colonic temperatures were significantly higher for those exposed at all three power densities than those sham exposed, and the differences increased monotonically with power density. The results for the CBA/J mice were similar, but the increases were statistically significant only at 50 and 100 W/m^2. The investigators conclude that subtle heating by RF can alter the thermoregulatory capacity of hypothermic mice, whereas the colonic temperature of normal (saline-injected) mice was not significantly altered by exposure at 100 W/m^2.

Adair and Adams[364] exposed squirrel monkeys to 2.45-GHz CW RF for 10 or 90 min in ambient temperatures of 15, 20, or 25°C. The power densities ranged from 25 to 100 W/m^2 (SARs 0.4 to 1.5 W/kg). Skin and colonic temperatures were monitored continuously during exposure. The metabolic heat production was calculated from the oxygen deficit in the expired air of each monkey. At all three ambient temperatures, 10-min exposure of two monkeys to a threshold power density of 40 W/m^2 and one monkey to 20 W/m^2 reliably initiated a reduction of their metabolic heat production, and the magnitudes of the reduction were a linear function of the power density above the threshold values. At exposure termination, the metabolic heat production often rebounded sharply and overshot normal levels. For the 90-min exposures at 20°C, the initially large reduction of metabolic heat production gradually diminished towards normal levels, apparently to ensure precise regulation of internal body temperature at the normal value.

Adair[371] also exposed four squirrel monkeys to 2.45-GHz CW RF in warm ambient temperatures ranging from 32 to 35°C. After an initially 90-min or longer equilibration period, each monkey was exposed for 10 periods to power densities in an increasing sequence from 25 to 200 W/m^2, with sufficient time between exposures for reestablishment of equilibrium. The colonic temperature and the skin temperature at the abdomen, tail, leg, and foot were monitored continuously. As in the previous investigation, the metabolic heat production was determined from the oxygen deficit in the expired air. In addition, thermoregulatory sweating from the foot was determined by sensing the dewpoint of the air in a special boot over the foot. The results indicate that at ambient temperatures below about 36°C, at which sweating in a sedentary monkey may occur spontaneously, the threshold power density (or SAR) for initiating thermoregulatory sweating from the foot decreased with decreasing ambient temperature.

Adair and Adams[372] equilibrated squirrel monkeys for a minimum of 2 hr to constant environmental temperatures (22 to 26.5°C), cool enough to ensure that the cutaneous blood vessels in the tail and extremities were fully vasoconstricted (an effect produced by the thermoregulatory system to minimize heat loss). The monkeys then underwent 5-min exposures to 2.45-GHz RF at successively higher power densities, starting at 25 to 40 W/m^2, until vasodilation in the tail occurred, as evidenced by an abrupt and rapid temperature increase in the tail skin. For example, a monkey equilibrated to 25°C exhibited tail vasodilation when exposed to RF at 100 W/m^2 (whole-body SAR 1.5 W/kg), whereas it did not when exposed to IR radiation at the equivalent power density, an indication that the effect resulted from stimulation of thermosensitive elements of the thermoregulatory system by the RF rather than from heating of the tail skin. RF exposure at higher power densities was required to cause tail vasodilation in monkeys equilibrated to lower environmental temperatures. Specifically, an increase of 30 to 40 W/m^2 was found necessary for every 1°C reduction in environmental temperature.

Behavioral reactions to RF exposure are summarized in Table 11.

VIII. NEUROENDOCRINE EFFECTS

A. MECHANISMS OF INTERACTION

Many of the studies of the endocrine system have involved whole-body RF exposures of the rat. The issue has been the level of organization of the system at which thermal interactions exert an effect and the nature of the effect. Some investigators believe that endocrine changes result from stimulation of the hypothalamic-hypophysial system at the hypothalamic or adjacent level of organization; others have advanced evidence that effects are mediated at the hypophysis or the particular endocrine gland or end-organ under study. The effects detected are not necessarily pathologic, because the function of the neuroendocrine system is to maintain homeostasis and hormone levels will fluctuate to maintain such organismic stability.

Effects of RF exposure on endocrine function are generally consistent with both immediate and long-term responses to thermal input and to nonspecific stress, which can also arise from thermal loading. Changes found in plasma levels of CS and growth hormone are typical reactions of animals to nonspecific stress; indeed, great care is required in performing experiments to ensure that the changes in hormone level do not result from stress caused by handling of the animals or the novelty of the experimental situation. The topic has been reviewed by Lu et al.[373]

B. HYPOTHALAMIC-HYPOPHYSIAL-ADRENAL RESPONSE

Several investigators have reported biochemical and physiological changes as a result of MW exposure which suggest an adrenal effect. According to Petrov and Syngayevskaya,[374] 3 and 24 hr after dogs were exposed to 3 GHz, 100 W/m^2, the serum corticosteroid content had increased by 100 to 150% above the original level. Serum potassium was decreased by 5 to 10% and sodium was increased by the same amount. They also noted that the susceptibility of rats to MW exposure was sharply increased 1 week after bilateral adrenalectomy. Chronic

TABLE 11
Behavioral Reactions to RF Exposure

Effects	Species	Frequency (GHz)	Exposure Conditions: Intensity (W/m^2)	SAR (W/kg)	Duration (days × min)	Ref.
Decreased exploratory activity and swimming speed, ΔT increase 2.5°C	Rat	2.45 (PW, multi-modal cavity, 120 Hz AM)		6.3	1 × 30	332
No effect on spontaneous activity or forced running	Rat	10.7 (CW)	6–9	0.2	7.7 × 1440	342
		3 (CW)	5–10	0.3	7.7 × 1440	
		3 (PW)	15–20	0.6	7.7 × 1440	
		3 (PW)	240–260 (av)	8.3	17 × 1440	
Increased locomotor activity	Rat	2.45 (CW, multi-modal cavity)	100	2.3	110 × 300	560
Decreased spontaneous activity and food intake	Rat	0.918 (CW)	100	3.6	21 × 600	561
No effect on spontaneous activity or food intake	Rat	0.918 (CW)	25	1.0	91 × 600	516
Decreased activity on stabilimetric platform, no significant increase in wheel running	Rat	2.45 (CW)	50	1.2	80 × 480	349
Increased activity on stabilimetric platform and in wheel running	Rat	0.915 (CW)	50	2.5	80 × 480	563
Decreased time on treadmil and inclined rod, decreased exploratory activity, increased then decreased shock sensitivity; decreased activity and shock sensitivity persisted 90 days after exposure	Rat	2.375 (CW)	5	0.1	30 × 420	564
Colonic temperature rise = 0.37°C before start of test; ΔT = 1.5°C with microwaves	Rat	2.45 (Multimodal cavity, AM)		420	10 × 0.17	565
				220	10 × 0.5	
Response decreased during exposure on random interval schedule (lowest intensity for effect. ΔT = 1.8°C	Rat	0.500a: (CW)	250	10	1 × 11	566
Response decreased during exposure (maximum effect) on random interval schedule, ΔT = 1.8°C	Rat	0.600 (CW)	100	7.5	1 × 55	327
Decreased observing response on vigilance task, ΔT = 2°C	Rhesus monkey	2.45 (120 Hz, AM)	720	5.0	1 × 60	567
No effect on observing responses			160	1.1	1 × 20	

TABLE 11
Continued

Effects	Species	Frequency (GHz)	Exposure Conditions Intensity (W/m^2)	SAR (W/kg)	Duration (days × min)	Ref.
Decreased observing responses on vigilance task	Squirrel monkey	2.45 (120 Hz, AM)	500	2.8	1 × 30	568
No effect on observing responses				0.6–1.7	1 × 60	
Decreased observing responses on vigilance task	Rat	1.28 (PW)	100	2.5	1 × 40	570
		5.62 (PW)	260	4.9	1 × 4	
Response rate decreased on fixed interval schedule in rats with high base-line rates, spending time away from level	Rat	2.45 (120 Hz, AM)	375	7.5	1 × 60	325
No effect on response rate			88–184	1.8–3.7	1 × 60	
No effect on visual tracking task	Rhesus monkey	1.2 (CW)	200	1.6	1 × 120	571
Response rate decreased on FR and increased on DRL schedules	Rat (120 days)	2.45 (CW)	50	1.4	1 × 30	341
		2.86 (PW)	50	1.4	1 × 30	
		9.60 (PW)	50	1.5	1 × 30	
Decreased length of runs and fewer reinforcers on FCN schedule	Rat	2.45 (CW)	50	?	1 × 30	572
Decreased response rate on FR operant schedule	Rat	2.45 (CW)	100	2.7	1 × 900	329
Increased rate of missing observing responses on vigilance task	Young rat	2.45 (PW, AM, Multi-modal cavity)		6.5	1 × 30	332
Decreased rate of responding on repeated acquisition task	Rat	2.80 (PW)	50	0.7	1 × 30	573
Increased response rates in extinction, decreased stimulus control, no effect on Sidman avoidance	Rat	2.45 (CW, multi-modal cavity)	100	2.3	110 × 300	560
No effect on flavor aversion test	Rat	0.918 (CW)	100	3.9	21 × 600	561
	Rat	0.918 (CW)	25	1.0	91 × 600	516
Microwave detected as stimulus	Rat	2.45 (PW, 120 Hz, AM, multi-modal cavity		0.6–2.4	1 × 1	574
	Rat	0.918 (PW)	150	7.5	1 × 0.5	575

TABLE 11
Continued

Effects	Species	Frequency (GHz)	Exposure Conditions: Intensity (W/m^2)	Exposure Conditions: SAR (W/kg)	Duration (days × min)	Ref.
Spending more time in shielded vs. unshielded compartment	Rat	1.20 (CW)	2	0.2	4 × 30	576
Spending equal time in shielded vs. unshielded compartment			24	2.2	3 × 30	
Spending more time in shielded vs. unshielded compartment (occurred in first of seven sessions)	Rat	1.20 (PW)	4	0.4	1 × 90	552
Spending more time in unirradiated compartment	Rat	2.80 (PW)	95	2.1	9 × 60	557
Decrease in SAR at 24°C	Mouse	2.45 (CW)		28	1 × 15	578
Decrease in SAR when ambient temperature increased from 20 to 35°C	Mouse	2.45 (CW)		43.6–0.6	1 × 20	579
No preferential orientation of rats or mice in far field of plane wave	Rat (200–360)	2.45 (CW)	150	3.3	1 × 50	580
	Mouse	2.45 (CW)	150	6.2–12.3 (depending on orientation)	1 × 60	
Cannot take specific action to reduce intensity of irradiation	Rat	0.918 (PW, 60 Hz, AM, multi-modal cavity)		60	5 × 2	581
Augmentation of increased response rates produced by chlordiazepoxide	Rat	2.45 (PW)	10	0.2	1 × 30	545
Shift to left on dose response curve for d-amphetamine in DRL schedule	Rat	2.45 (PW)	10	0.2	1 × 30	283
			10	0.2	4 × 30	
No effect on dose response curve for chlorpromazine or diazepam	Rat	3.8 (PW)	10	0.2	1 × 30	546
Chlordiazepoxide reduced responses, decreased avoidance responses, and increased escape responses to microwaves	Mouse	2.45 (CW)		46	1 × 30	361
Response rate decreases were augmented after exposures at higher ambient temperatures	Rat	2.45 (CW)	100	2.0	1 × 930	582
Reduced responding for heat lamp in a cold room	Rat	2.45 (CW)	50	1.0	1 × 15	363

TABLE 11 (continued)
Behavioral Reactions to RF Exposure

Effects	Species	Frequency (GHz)	Exposure Conditions Intensity (W/m^2)	SAR (W/kg)	Duration (days × min)	Ref.
Selection of a lower ambient air temperature	Squirrel monkey	2.45 (CW)	60	1.0	1 × 10	583
No effect on web-spinning ability	Spider	9.6 (PW)		0.4–4.0	1 × 960	338
Effects on shuttle box performance and schedule controlled lever pressing	Rat	2.45 (CW)	5	0.14	90 × 630	344
			25	0.7	98 × 420	345
No effect on several tasks, variable effect on time-related operant task	Rat	2.45 (CW)	—	0.05	90 × 420	346
Reduced locomotor activity, reduced response to acoustic stimuli	Rat	2.45 (CW)	100	2.7 (est)	1 × 420	348
Interference with operant behavior above colonic ΔT of 2.5°C only	Rat	1.25 (PW)	—	0.84–23	1 × 10	350
No effect on behavior	Rhesus monkey		1,310 (peak)	0.05–0.80	1 × 60	355
No behavioral effect related to PW, effects only at elevated colonic temperature	Rat	1.3 (PW)	≤10,800 (peak)	1.8–26.2	1 × 10	356
No effect on behavior	Rat	2.0–2.6 (PW)	241,000 (peak)	—	1 × 5	357
Interference with time-interval perception	Rat	3.0 (PW)	500×10^6 (peak)	0.7	1 × 26	359
Reduction of time on treadmill	Rat	3.0 (PW)	500×10^6 (peak)	0.7	1 × 26	358
Disruption of Y maze performance	Rat	3.0 (PW)	500×10^6 (peak)	0.7	1 × 26	360

Note: If measured SAR was not reported, SAR was estimated when possible.

Adapted and expanded from U.S. Environmental Protection Agency, Biological of Radiofrequency Radiation, Elder, J.A., and Cahill, D.F., Eds., Rep. No. EPA-600/8-83-026F, Research Triangle Park, N.C., 1984.

exposure of animals to MWs (CW or PW) was accompanied by reduced cholinesterase activity and an increased amount of 17-ketosteroids in the urine, reduced ascorbic acid, and reduced weight of the adrenal glands.

Demokidova[375] reported increased adrenal and pituitary gland weight in rats exposed 1.5 months to 0.0697 GHz, 12 V/m, 1 hr/day and increased adrenal weight in infant rats exposed to 48 V/m, 4 hr/day. The same author, however, reported decreased weight of the adrenal glands in infant rats exposed to 0.01488 GHz at 70 V/m.

The pituitary gland in female mice retained its gonadotropic function when exposed to 3 GHz (100 W/m^2) twice daily for 5 months, although its activity was reduced in comparison with that of nonexposed animals.[141] Tolgskaya and Gordon[219] noted the reversibility of changes in the neurosecretory function of the hypothalamus when exposure was terminated. Rabbits exposed to 3 GHz (500 to 600 W/m^2) 4 hr daily for 20 days tended to show a decline in the amount of urinary 17-hydroxycorticosteroids (17-OHCS) at the beginning of exposure, followed by a gradual return to normal.[376] No change was evident in the excretion of 17-ketosteroids in the urine.

In rats exposed to MWs of varying intensity, no quantitative changes in CS were found in the adrenals and blood plasma.[377] Prepubescent hypophysectomized rats displayed no differences in adrenal growth rate when treated with pituitary homogenates collected either from rats exposed to MWs or from control rats.

Rats exposed to 2.45 GHz (CW), 100 W/m^2 for 4 hr showed no change in adrenal weight, phenylethanolamine-N-methyl transferase (PNMT) activity, or epinephrine levels.[378] After 16 hr of exposure, however, decrease in adrenal epinephrine (32%) was significant and PNMT activity was elevated (25%). There were no statistically significant differences ($p > 0.1$) in adrenal or plasma CS levels between exposed and sham-exposed animals. It should be noted, however, that similar alterations in epinephrine levels can occur in rats subjected to a stressor such as immobilization or acute exposure to cold.

Increased adrenal function due to MW exposure has been correlated with colonic temperature increase in rats.[365] Plasma CS levels in hypophysectomized rats exposed at 600 W/m^2 for 60 min were below control levels. When rats were pretreated with dexamethasone before being exposed at 500 W/m^2 for 60 min the CS response was suppressed. These results suggest that the MW-induced CS response observed in intact rats is dependent on adrenocorticotropic hormone secretion by the pituitary, i.e., the adrenal gland is not the primary endocrine gland stimulated by MW. The evidence obtained in these experiments is consistent with the hypothesis that the stimulation of the adrenal axis in MW-exposed rats is a systemic, integrative process due to a general hyperthermia.[365] The subject of effects of MW on the adrenal cortex has been reviewed by Lu, et al.[379]

C. HYPOTHALAMIC-HYPOPHYSIAL-THYROID RESPONSE

The literature offers comparatively few experimental studies of the effect of RF or MWs on the thyroid. No alterations in thyroid structure or function attributable to MW exposure were noted in rats subjected to 2.45 GHz, CW, 10 W/m^2 continuously for 8 weeks or 100 W/m^2, 8 hr/day for 8 weeks.[380] On the other hand, a stimulatory influence of 50 W/m^2 on the trapping and secretory function of the thyroid gland of rabbits has been reported.[381] These functional changes were in agreement with altered histology of the thyroid.

In rats exposed for 16 hr to 2.45 GHz (CW) at 100 to 250 W/m^2, tests of thyroid function in general showed no statistically significant deviations from the norm, except that in animals with a 1.0 to 1.7°C increase in colonic temperature there was a reduction in the ability of the thyroid to concentrate iodide.[378] Decreased thyroid gland weight was noted in infant rats exposed to 0.0697 GHz, 48 V/m, 4 hr/day, 1.5 months.

Increased thyroid hormone secretion has been correlated with MW-induced thyroid temperature increase in dogs.[382,383] Vetter[384] found that serum protein levels increased as a function

TABLE 12
Neuroendocrine Response to RF Exposure

Effects	Species	Frequency (GHz)	Exposure Conditions Intensity (W/m^2)	SAR (W/kg)	Duration (days × min)	Δt (°C)	Ref.
Increased thyroxine and triidothyronine	Dog	2.45 (CW)	720–2360	58–190	1 × 120	2–8 (Thyroid temp)	382
							383
No effect on thyroid gland or thyroid hormone	Rat	2.45 (CW)	10, 100, 1000	0.25–25	1 × 10–45	≤100 W/m^2 = 0	380
			10, 100	0.25–2.5	56 × 480	1000 W/m^2 ≥ 5	
No effect on thyroid function	Rat	2.45 (CW)	100, 200, 250	2, 5, 6.5 (est)	1 × 240–960	≤200 W/m^2 = 0	378
						250 W/m^2 = 1.7	
Decrease in serum protein-bound iodide, thyroxine, and thyroxine/serum ratio	Rat	2.45 (CW)	150	3.8 (est)	2.5 × 1440	0	378
Increase in thyroid hormone	Rabbit	3 (PW)	50	0.25–0.75 (est)	48 × 180	Not reported	381
Decrease in serum thyroxine levels	Rat	2.45 (CW)	200	5	1 × 240–480	0–0.6	385
			(Neg 10–100)	0.25–2.5	1 × 60–480	0.6–1.4	
Increase in CS levels	Rat	2.45 (120 Hz AM)	400–700	8.4–14	1 × 60	1.3–3.0	584
			(Neg 10–200)	(Neg 0.21–4.2)		0–0.6	
Decrease in TSH levels			100, 400–700	2.1–8.4		0–3.0	
			(Neg 10–50, 200)	(Neg 0.21–4.2)		0–0.6	
Increase in CS levels			100–400	2.1–8.4	1 × 240		
			(Neg 10–50)	(Neg 0.2–1.0)			
Decrease in TSH levels			250, 400	0.6–2.1			
			(Neg 10–200)	0–1			
No effect on thyroid, pituitary, or adrenal gland weights or growth hormone levels	Rat	2.45 (CW)	10–200	0.25–2.5	1 × 60–480	0–1.4	385
No effect on thyroid, anterior pituitary gland, adrenal, prostate, or testes weights; no change in follicle-stimulating hormone or gonadotropic hormone levels	Rat	2.86–2.88 (CW)	100	1–2 (est)	36 × 360	Not reported	585

TABLE 12
Continued

Effects	Species	Frequency (GHz)	Exposure Conditions: Intensity (W/m^2)	Exposure Conditions: SAR (W/kg)	Duration (days × min)	Δt (°C)	Ref.
Increase in leutinizing hormone	Rat	2.86–2.88 (CW)	100	1–2 (est)	36 × 360	Not reported	585
Increased adrenal weights and significant adrenal response	Infant rat	2.45 (CW)	400	20–60 (est)	6 × 5	1.5–2.5	536
Increased plasma CS levels	Rat	2.45 (CW)	500, 600	11.5, 13.8 (est)	1 × 30–60	130 W/m^2 = 0.5	365
			(Neg 130–400)	(Neg 3.0–9.2)		200 W/m^2 = 0.7	
			200–400	4.6–9.2 (est)	1 × 120	300 W/m^2 = 0.9–1.4	
			(Neg 130)	(Neg 3) (est)		400 W/m^2 = 1.3–1.4	
						500 W/m^2 = 1.6–2.4	
						600 W/m^2 = 2.5–2.9	
Increase in CS	Rat	2.45 (120 Hz AM)	500, 600	8.3, 9.6	1 × 60	Not reported	586
No effect on serum CS levels	Rat	0.918 (CW)	25	1.0	91 × 600		516

Note: Δt (°C) = Colonic temperature increase; Neg = effect not found at value indicated.

Adapted from U.S. Environmental Protection Agency, Biological Effects of Radiofrequency Radiation, Elder, J. A., and Cahill, D. F., Rep. No. EPA-600/8-83-026F, Research Triangle Park, N.C., 1984.

of power density, indicating an alteration of protein synthesis or catabolism; levels of thyroid hormone decreased as power density of 1.45-GHz CW was increased from 50 to 250 W/m^2. In agreement is the finding of Lu et al.,[385] who reported that serum thyroxine levels were transiently elevated after exposure of rats at 10 W/m^2 (2.45 GHz) and were depressed after exposure at 200 W/m^2. None of the reported alterations was irreversible or resulted in morbidity.

Perturbation of the thyroid gland may be the result of an indirect effect, the thermal stress on the body producing a hypothalamic-hypophysial response. This is consistent with MW-induced thermal stimulation of hypothalamic-hypophysial-thyroid (HHT) activity.[386] McLees and Finch[387] pointed out that temperature elevation and heat stress have been associated with alterations in radioactive iodine (RAI) turnover rate. The HHT axis has been shown to be sensitive to environmental temperature.[388] Differences in rate of temperature change or alteration in thermal gradients could also result in qualitative differences in endocrine response.

In a study described as a delineation of acute neuroendocrine responses to MW exposure, Lu et al.[389] exposed "gentled" rats to the same frequency at power densities ranging from 10 to 700 W/m^2, up to 8 hr, at an environmental temperature maintained at 24°C. Sham-exposed rats were used as controls. After treatment, the rats were decapitated, colonic temperatures were taken, and blood was collected for assays of thyroxine (T-4), thyrotropin (TSH), growth hormone (GH), and CS. For exposure of 1 hr, colonic temperature increased with power density at 200 W/m^2 and higher, but consistent elevation of serum CS did not occur below 500 W/m^2. Lower serum TSH and GH levels also occurred at this and higher power densities. Significant serum T-4 elevations were noted at 400 and 700 W/m^2, but they were not consistently related to power density. For sham exposures and exposures at 10 to 200 W/m^2 for longer durations (2 to 8 hr), the results were equivocal, presumably, because such exposures encompassed significant portions of the circadian cycle. Specifically, in the sham-exposed rats, the level of T-4 did not change significantly under sham conditions with exposure duration, and significant increases of CS and decreases of TSH and GH were seen, so it was difficult to discern consistent differences in these hormones ascribable to RF exposure. The investigators suggest that the divergent responses may be due to two different mechanisms that are dependent on RF intensity and the time of the exposures relative to circadian rhythms.

In a review article, Lu et al.[373] discuss evidence for the existence of threshold intensities for various RF-induced neuroendocrine effects. They indicate that such thresholds are dependent on the intensity and duration of exposure and can be different for each endocrine parameter.

Abhold et al.[390] exposed rats to 2.45-GHz CW RF for 8 hr (continuously) at 20 or 100 W/m^2 (SAR 0.44 or 2.2 W/kg with the long body axis of the rat parallel to the electric component of the RF). Other rats were sham exposed for the same period, and still others served as untreated controls. Within 15 min after treatment, the rats were euthanized and their blood was assayed for serum T-4, triiodothyronine (T-3), and T-3 uptake. In addition, serum CS concentrations were measured. The results indicated that concentrations of T-4, T-3, and T-3 uptake were not altered by the treatments. However, the rats that were sham-exposed or exposed to RF at 20 W/m^2 had higher levels of CW than the cage-control rats, whereas the rats exposed to 100 W/m^2 had values similar to those of the untreated rats. The changes observed were more likely ascribable to stress and other factors in the experimental protocols than the RF exposure.

In summary, although some effects of RF exposure on the endocrine system appear to be relatively straightforward and predictable from physiological considerations, other more subtle effects require further study, notably those related to the interactions among the pituitary, adrenal, thyroid, and hypothalamus and/or their secretions. Part of the the problem in interpreting results appears to arise from uncertainties regarding stressors inadvertently introduced into the experimental design and the response of the animal to such stressors. Animals placed in novel situations are much more prone to exhibit stress responses than animals that have been adapted to the situation. However, there may be large variations in adaptation among animals in a given situation or among experimental situations in different laboratories. Moreover, the use of sham-treated controls may not always reduce the problem. A review of the effects of RF on physiologic regulation has been published by Michaelson.[391] Neuroendocrine effects of RF exposure are summarized in Table 12.

IX. CARDIOVASCULAR EFFECTS

Low intensity exposures at "non-thermalizing" levels and frank thermal levels of MW have been studied. The role of heating and the possibility of unique MW-induced effects even at thermalizing levels of exposure have been addressed. The data-base is anything but straightforward and not easily distilled into a brief summary.

Clinicians in the former Soviet Union identified a so-called "microwave neurasthenia," or "microwave radiation sickness" the consequence of chronic occupational exposure to low levels of MW.[392] This included bradycardia, tachycardia, hypertension, hypotension, T-wave changes, and other neurologic and behavioral changes. Such a syndrome is not considered adequately documented or well defined in western medicine. The possible hemodynamic changes in the cardiovascular system have been studied in animals ever since. The chronotropic effects (bradycardia or tachycardia) have been of special interest. At non-thermal levels both brady- and tachycardia have been found in different studies of different animals and with inconsistent results in the same animal. Some investigators have proposed that the heart rate response of rabbits depended on exposure technique, distribution of energy as a function of location, and the possibility of differing "regional reflexes" producing differing outcomes.[393] This has not been verified and it is just as likely that biologic variability among animals or the numerous sources of error could account for the findings. Even the distinction between thermal and non-thermal is not a tidy one and minor changes in temperature, perhaps less than one 1°C, at the right anatomic location could affect hemodynamics by a neural mechanism even though the exposure is reported as non-thermal.

Hyperthermia of non-RF origin produces tachycardia and a decrease in total peripheral resistance caused by vasodilation, a heat dissipating response to the thermal burden. This is true whether heat is imparted by environmental heating, water bath or even a general whole-body hyperthermia of RF origin.[394]

Evidence that hemodynamic changes produced by MWs might be very sensitive to a precise dosimetric configuration was presented by Lu et al.[395] Rats were subjected to ventral head-neck exposures in a waveguide at 1.25 GHz, PW and CW at 2.0 and 6.4 W average powers, 10 μsec pulses, 400 kW peak power, brain average SAR 9.5 and 30.4 W/kg, neck SAR 34.3 and 110.0 W/kg, for 5 minutes. Both tachycardia and bradycardia were observed but a decreased stroke volume was also found with consistency. This produced a consistent subnormal cardiac output. Mean arterial pressure was unaltered. From the data it was possible to conclude that total peripheral resistance had increased. This would be an anomalous or unphysiologic response to the insertion of thermal energy into the body. This kind of a response is not observed in experimental animals subjected to whole-body microwave exposure.[396] The investigators concluded that the targeting of the head and neck, even in the presence of generalized hyperthermia, may have elicited specific signals at receptor or effector sites in that area to contravene the normal physiologic response. The baroreptors were identified as a possible locus of stimulation of efferents which raised blood pressure by increasing peripheral resistance. An hypothesis that the baroreptors might respond to pressure waves induced by

TABLE 13
Effects of RF Exposure on Cardiac Function

Effects	Species	Exposure Conditions Frequency (GHz)	Intensity (W/m²)	SAR (W/kg)	Duration (days × min)	Ref.
Bradycardia developes after whole-body exposure, along with hyperthermia	Rat	2.45 (PW)	280, 480	6.5, 11.1	1 × 30	526
Exposure to head promotes tachycardia; exposure to back raises respiratory rate, but not heart rate	Rabbit	2.4 (CW and PW)	200	3	1 × 60	587
Increased respiration	Rabbit	2.4	400–1000	8–20 (est)	1 × 20	588
Increased heart rate from dorsal exposure of the head	Rabbit	2.4	1000	20 (est)	1 × 20	588
Alterations in ECG (shortening of QT interval increased height of T-wave, appearance of U-wave	72-Hr chick heart	2.4	740		1 × 30	589
No effect on heart rate that cannot be attributed to MW heating	Quail embryo	2.45 (CW and PW)		0.3–30	1 × 5–10	590
Pulses synchronized with each R-wave do not affect heart rate	Frog	1.42–10 (PW)	0.32		(100-μs pulses)	591
Synchronized pulses with QRS complex cause increase in heart rate with some arrhythmias	Frog	1.425 (PW)	0.006		(10-μs pulses)	592
Increased heart rate	Rabbit	2.45 (CW)	800	12	10 × 20	593
No effect on heart rate			50	0.3 and 0.093	10 × 20	
Low power levels cause bradycardia in the isolated heart	Turtle	0.960 (CW)		2–10	1 × 60	594
Causes slight rate decrease in the isolated heart	Rat	0.960 (CW)		1.3 and 2.1	1 × 5–10	595
Synchronized exposures with ECG have no effect on heart rate	Frog	1.42–3 (PW)	0.006		(2-, 10-, 150-μs pulses)	596
Bradycardia caused by head and neck exposure	Rat	1.25 (CW and PW)		9.5 and 30.4 (brain) 34.3 and 109.8 (neck)		395

Adapted and expanded from U.S. Environmental Protection Agency, Biological Effects of Radiofrequency Radiation, Elder, J. A. and Cahill, D. F., Eds., Rep. EPA-600/8-83-026F, Research Triangle Park, N.C., 1984.

pulsed MW was not supported, in that both pulse and CW produced the anomalous effect at thermalizing levels.

Rasbury and Seaman[397] also studied exposures of the head and neck in the rat under different conditions from those of Lu et al. The carotid sinus was surgically exposed and rats exposed to the radiation of a coaxial probe under anesthesia for brief periods, 10 sec and local SARs up to 70 kW/kg. At the highest SARs where effects were found approximately half of the animals exhibited decreased blood pressure and heart rate and half the animals increased blood pressure and heart rate. No changes in colonic or tail temperature occurred for any exposures. The authors could not account for the divergent results, but noted that the inconsistent findings were consistent with other studies on hemodynamic changes associated with MW exposure.

Cooper et al.[398,399] and Pinakatt et al.[400,401] studied the influence of various drugs in pentobarbital-anesthetized rats exposed to 2.45 GHz, 800 W/m^2 for 10 min which resulted in a 40.5°C colonic temperature. Pyridoxine and digitoxin did not alter blood pressure or heart rate. Reserpine, vagotomy, and pharmacological ganglioplegia diminished the reaction to MW-induced hyperthermia. See Table 13 for a tabulation of reported effects on the cardiovascular system.

X. EFFECTS ON HEMATOPOIESIS AND HEMATOLOGY

Effects on the hematopoietic system have been reported.[402–415] A summary of studies on hematologic effects of RF exposure is presented in Table 14.

Hyde and Friedman[416] exposed anesthetized female CF-1 mice to 3 GHz, 200 W/m^2 and 10 GHz, 170, 400, or 600 W/m^2 up to 15 min. No significant effect on total or differential leukocyte count or hemoglobin concentration was noted immediately, 3, 7, or 20 days after exposure. There were no changes in femoral bone marrow other than a variable, but slight, increase in the eosinophil series of the exposed animals which was not reflected in peripheral blood counts.

Miro et al.[417] exposed Swiss albino mice to pulsed 3.105 GHz at 200 W/m^2 average power density (4 kW/m^2 peak) for 145 hr. Stimulation of splenic lymphopoiesis and increased ^{35}S methionine incorporation in spleen, thymus, and liver were found. The authors interpret these results as a sign of stimulation of cells belonging to the reticuloendothelial system.

Kitsovskaya[407] subjected rats to 3 GHz, 100 W/m^2, 60 min/day for 216 days; 400 W/m^2, 15 min/day for 20 days; 1 kW/m^2, 5 min/day for 6 days. At 400 W/m^2 and 1 kW/m^2, total RBC, WBC, and absolute lymphocytes were decreased; granulocytes and reticulocytes were elevated. At 100 W/m^2, total WBC and absolute lymphocytes decreased, and granulocytes increased. Bone marrow examination revealed erythroid hyperplasia at the higher power densities. Decreased leukocyte count and phagocytic activity had been reported in rats and rabbits exposed to 0.05 GHz, 0.5 to 6 V/m, 10 to 12 hr/day, 180 days.[418]

Baranski[409,410] exposed guinea pigs and rabbits to 3 GHz (PW or CW), 35 W/m^2 for 3 months, 3 hr daily. Increases in absolute lymphocyte counts, abnormalities in nuclear structure, and mitosis in the erythroblastic cell series in the bone marrow and in lymphoid cells in lymph nodes and spleen were observed. No alteration in the granulocyte series was noted. Baranski suggests that extrathermal complex interactions seem to be the underlying mechanism for the changes.

Spalding et al.[419] exposed mice to 0.8 GHz, 2 hr daily for 120 days in a waveguide at an incident power density of 430 W/m^2. Red and white blood cell count, hematocrit, hemoglobin, growth, voluntary activity, and life span remained normal.

In dogs exposed whole-body to 2.8 GHz pulsed, 1 kW/m^2 for 6 hr, there was a marked decrease in lymphocytes and eosinophils.[406] The neutrophils remained slightly increased at 24 hr postexposure, while eosinophil and lymphocyte values returned to normal levels.

TABLE 14
Hematologic and Hematopoietic Effects of RF

Effects	Species	Exposure Conditions: Frequency (GHz)	Intensity (W/m^2)	SAR (W/kg)	Duration (days × min)	Ref.
Increased: WBC, lymphs, PMN, RBC, Hct, and Hgb	Rat	24 (PW)	100	1.5 (est)	1 × 180	507
			200	3.0 (est)	1 × 420	
No change	Dog	24 (PW)	240	1 (est)	400 × 400–900	597
No change	Rat	3 (PW)	100	2 (est)	216 × 60	407
Decreased: RBC, WBC, and lymphs			400	8 (est)	20 × 15	
Increased: PMN			1000	20 (est)	6 × 5	
Decreased: lymphs and eos	Dog	2.8 (PW)	1000	4 (est)	1 × 360	404
Decreased: WBC, PMN, and eos		2.8 (PW)	1650	6 (est)	1 × 120	
Decreased: WBC, lymphs, and eos; Increased: PMN		1.28 (PW)	1000	4.5 (est)	1 × 360	
Decreased: lymphs		0.200 (CW)	1650	25 (est)	1 × 360	
No change	Mouse	0.800	430	12.9 (est)	175 × 120	419
Increased: lymphs and mitotic index of lymphoid cells	Guinea pig	3 (CW or PW)	35	0.5	120 × 180	409
Increased: RBC, Hct, and Hgb	Rat	2.4 (CW)	100	2 (est)	30 × 120	598
No change	Rat	2.4 (CW)	50	1 (est)	90 × 60	599
Increased: eos	Rabbit	2.45 (CW)	100	1.5	180 × 1380	600
Increased: WBC, CFU; Decreased: ^{59}Fe uptake	Mouse	2.45 (CW)	1000	70 (est)	1 × 5	601
Accelerated recovery following X-irradiation; increased erythropoiesis and myelopoiesis	Mouse	2.45 (CW)	1000	70 (est)	1 × 5	602
Accelerated recovery from X-irradiation	Dog	2.8 (CW)	1000	4 (est)	1 × 3600	603, 604
Increased: PMN and RBC; Decreased: lymphs; Accelerated recovery from X-irradiation	Chinese hamster	2.45 (CW)	600	28 (est)	1 × 30	605
Increased: lymphs; Decreased: PMN (not reproduced consistently)	Rat (perinatal exposure)	0.425 (CW)	100	3–7	47 × 240	606

TABLE 14
Continued

Effects	Species	Frequency (GHz)	Exposure Conditions Intensity (W/m^2)	SAR (W/kg)	Duration (days × min)	Ref.
No change	Rat (perinatal exposure)	2.45 (CW)	50	1–5	57 × 240	176
	Rat (perinatal exposure)	0.1 (CW)	460	2–3	57 × 240	534
	Quail egg	2.45	300	14	1 × 1440	538
Decreased: Hct, WBC, and lymphs	Rat (young)	2.736 (PW)	244	5–25 (est)	35 × 240	607
No change	Mouse	2.45 (CW)	300	22	22 × 30	421
Decreased: lymphs; Increased: PMN	Mouse	0.026 (CW)	86, 100	13 (est)		422
No effect on oxidative activity of mononuclear cells	Human	2.45 (CW)	—	103.5	1 × 30	440
Increase in blast transformation, PW over CW	Human	2.45 (PW and CW)	—	≤12.3	5 × 720	438
Increase in ^{3}H-thymidine uptake in lymphocytes at ≤50 W/kg	Human	0.027 (CW) 2.45 (CW)	—	≤196	1 × 120 1 × 120	439

Note: WBC = white blood cell PMN = polymorphonuclear leukocytes, RBC = red blood cell, Hct = hematocrit, Hgb = hemoglobin, CFU = colony-forming unit, lymphs = lymphocytes, and eos = eosinophils.

Adaped and expanded from U.S. Environmental Protection Agency, Biological Effects of Radiofrequency Radiation, Elder, J. A. and Cahill, D. F., Eds., Rep. No. EPA-600/8-83-026F, Research Triangle Park, N.C., 1984.

Following 2 hr of exposure at 1.65 kW/m^2, there was a slight leukopenia and decrease in neutrophils. When the exposure was of 3-hr duration, leukocytosis was evident immediately after exposure and was more marked at 24 hr, reflecting the neutrophil response. After exposure to 1.285 GHz pulsed 1 kW/m^2 for 6 hr, there was an increase in leukocytes and neutrophils. At 24 hr. the neutrophil level was still noticeably increased. Lymphocyte and eosinophil values were moderately depressed initially, but at 24 hr slightly exceeded their initial value. An exposure of 6 hr to 0.2 GHz (CW), 1.65 kW/m^2 resulted in a marked increase in neutrophils and a mild decrease in lymphocytes. On the following day, the leukocyte count was further increased, and the lymphocytes markedly increased. Such shifts in the white blood cell picture are consistent with focal thermal lesions to be expected under these exposure conditions.

Exposure of mice to 2.45 GHz, 1 kW/m^2 for 5 min resulted in a decrease followed by an increase in ^{59}Fe uptake in the spleen and bone marrow.[412] Alteration in ferrokinetics was also found in rabbits and guinea pigs exposed to 3 GHz at 10 or 30 W/m^2, 2 to 4 hr daily for 14 to 79 days.[413,414]

Possible alteration in the circadian rhythm of bone-marrow mitoses in guinea pigs and mice was noted. No effects were seen on precursors of granulocytes and only minimal effects were found on the erythroid series, but pronounced phase shifts were noted in the pool of stem cells. In inbred Swiss albino mice exposed once for 4 hr at 5 W/m^2 to pulsed 3-GHz MWs, the diurnal rate of proliferation of the stem-cell population was amplified and the phase shifted from that of controls.[414] A comparable phase shift in the circadian rhythm of body temperature was observed in rats exposed to 2.45 GHz, CW, 10 W/m^2, 1 to 8 hr, suggesting an interrelated example of physiologic regulation.[389]

In evaluating reports of hematological changes (Table 14) one must be cognizant of the relative distributions of blood cells in a population of animals or humans and the thermal influence on these alterations. Early and sustained leukocytosis in animals to thermogenic levels of MWs may be related to stimulation of the hematopoietic system, leukocytic mobilization, or recirculation of sequestered cells. Eosinopenia and transient lymphocytopenia with rebound or overcompensation, when accompanied by neutrophilia, may be indicative of increased hypothalamic-hypophysial adrenal function as a result of thermal stress.[420] Hematologic and hematopoietic effects of RF are summarized in Table 14.

XI. EFFECTS ON THE IMMUNE RESPONSE

Some interest has continued in the effect of MW exposure on the immune system. A number of articles published in the early 1980s[421–437] were motivated by the possibility of enhancing the immune response with MW exposure with therapeutic goals in mind. By the late 1980s and early 1990s, with a re-direction of interest to ELF investigation, the rate of output of studies on immune system effects had declined. Reports of lymphoblastoid transformation by MW were confirmed by some studies and not others (see below). The hope of exploiting such a phenomenon to augment beneficial immune responses in certain disorders was tempered by the concern that there might be a hazard involved if such processes led to autonomous and unregulated lymphoid proliferations. There is no evidence that the latter concern has any validity. On the other hand no consistent and robust evidence has been elicited to date that MW-induced lymphoblastoid responses can be used to augment immunotherapy of infectious disease or cancer.

A. LYMPHOCYTE KINETICS

The many reports on effects of MW on lymphocytes have not produced a clear and consistent picture amenable to concise summary. Although lymphocytes can be straightforwardly identified with standard hematologic stains, and mature lymphocytes give the appearance of small, uniform, metabolically inert cells, lymphocytes are composed of immunologically distinct subpopulations and, in response to antigenic stimulation or exposure to many chemical substances, can quickly enlarge, exhibiting a diameter 2 to 5 times that of the resting cell, with a striking increase in amount of cytoplasm and transformation of the nucleus from a dark, dense mass of chromatin material to fine, lacy, or decondensed structure. This process, known as lymphoblastoid transformation, has been intensively studied immunologically and biochemically. The lymphocyte may or may not proceed to mitosis depending on the nature of the challenge or stimulus. A number of plant lectins have been used to stimulate lymphoblastoid transformation, protein, RNA and DNA synthesis *in vitro* to study immune status. In immunologically competent animals unequivocal, vigorous responses can be elicited. Responses to RF have been reported but of a much less vigorous nature.

Lymphoblastoid transformation, *in vitro,* after free field exposure to 3-GHz pulsed MWs, 70 W/m^2, for 4 hr daily and 200 W/m^2 for 15 min daily, 3 to 5 days, has been reported.[425] At this power density, the temperature of the media increased by 0.5°C after 15 min, and after 20 min the increase was 1°C. Changes in the mitotic index were dependent on the exposure time. Although a 5-min exposure did not influence the proportion of dividing cells, slight differences compared with controls were observed after 10- and 15-min exposures and significant differences were seen following 3- or 4-hr exposures at 70 W/m^2.

Czerska et al.[438] studied human peripheral blood lymphocytes *in vitro,* comparing conventional heating to CW and to pulsed (PW) exposures at 2.45 GHz, each type of exposure being evaluated at temperature elevations up to 2°C. Average SARS ranged up to 12.3 W/kg for exposures continuous over 5 days. At non-heating levels CW had no effect on transformation. At heating levels both conventional and CW heating produced a small but significant increase in blast transformation; at non-heating levels PW enhanced transformation significantly and increasingly so at heating levels. The authors speculated that pulse power might act via the larger field amplitudes or by microacoustic shock effects.

Cleary et al.[439] presented evidence that lymphocytes are stimulated to take up ^{3}H-thymidine following exposure at 27 MHz or 2.45 GHz *in vitro* under isothermal conditions, with and without PHA. Maximal uptake occurs at SARs in the range of 25 to 40 W/kg depending on frequency. The authors argue in favor of a non-thermal mechanism.

Kiel et al.[440] investigated the effects of 2.45 MHz CW MW, SAR = 103.5 W/kg from the point of view of impact on oxidative activity in mononuclear cells, primarily lymphocytes, using a chemiluminescence assay for level of nonphosphorylating oxidative activity. Exposure to MWs at 37°C does not appear to affect oxidative metabolic activity of mononuclear cells.

Huang et al.[441] reported that lymphocytes from Chinese hamsters exposed to 2.45 GHz CW, from 50 to 450 W/m^2, 15 min/day for 5 days showed changes in blast transformation and mitosis. No chromosomal aberrations were evident. These studies noted increased but reversible transformation of lymphocytes, without mitogenic stimulation. These authors, noting temperature changes of less than 2°C, were among the first to suggest that hyperthermia could be exploited to manipulate the immune response, assuming the changes were thermally mediated. Zafra et al.[442] reported that MW at 2.45 GHz and power density of 320 W/m^2 for 15 min enhanced a number of activities of murine macrophages in vitro normally elicited in the immune response.

MWs have been reported to induce an increase in the frequency of complement receptor-bearing lymphoid spleen cells in mice,[423] which may represent a maturation of B lymphocytes. Wiktor-Jedrzejczak et al.[423–424] exposed adult male mice to 2.45-GHz CW at an absorbed dose of 12 to 15 W/kg for 30 min in an environmentally controlled waveguide facility and then measured the function of different classes of lymphocytes in vitro. Such exposure failed to produce any detectable changes in function of T lymphocytes or increase in DNA, RNA, or protein synthesis, as measured by incorporation of tritiated-thymidine,

-uridine, and -leucine by spleen, bone marrow, and peripheral blood lymphocytes in vitro. However, the maturation of B lymphocytes from the spleen of exposed mice was stimulated. Consistent with this effect on B lymphocytes are the results reported by Czerski and associates.[413,414,443]

Smialowicz[425] examined the proliferative capacity of lymphocytes that are responsible for cellular immune responses (T cells) and humoral immune responses (B cells) following 2.45-GHz exposure *in vitro.* The ability of mouse spleen lymphocytes to undergo blast transformation in response to mitogens that selectively stimulate either T or B cells was measured by the incorporation of ^{3}H-thymidine into DNA. No consistent difference was found between the blastogenic response of exposed (100 W/m^2, 19 W/Kg, 1 to 4 hr) and control cells.

Roberts et al.[444] exposed influenza virus-infected human mononuclear leukocytes to 2.45 GHz RF, CW or PW at 16 or 60Hz at an SAR of 4 mW/ml. No alteration by RF of mitogen-stimulated DNA synthesis occurred either in virus-infected leukocytes or uninfected leukocytes.

Mice exposed to 2.45-GHz CW, 50 to 350 W/m^2 (SAR = 4 to 25 W/kg) for 1 to 22 consecutive days (15 to 30 min/day) showed no consistent significant alterations in several parameters such as mitogen-stimulated response of T- and B-splenic lymphocytes, and the primary antibody response of mice to sheep erythrocytes.[421] Subsequently, Schlagel et al.[427] reported that RF-induced complement-receptor-positive (CR$^+$) lymphocyte increase was under genetic control. Smialowicz et al.[428] showed, in addition, the age and strain of the mouse, the RF exposure characteristics (waveguide vs. far field), and the environmental conditions are all sources of variation that affect the CR$^+$ cell appearance.

Smialowicz et al.[176] exposed rats *in utero* and neonatally through 40 days of age (4 hr/day, 7 days/week) in a controlled environment to either 2.45 GHz (CW, 50 W/m^2, SAR = 1 to 5 W/kg) or to 0.425 GHz (CW, 100 W/m^2, SAR 3 to 7 W/kg). At 40 days of age significant increases in the response of lymphocytes from exposed rats to *in vitro* stimulation with several mitogens was observed in several experiments. While these results have not been consistently reproduced, the trend in the results suggests that chronic exposure during fetal and neonatal development may change either the frequency or responsiveness of lymphocyte subpopulations. The biological significance of these observed changes is unknown. The mechanism by which these changes are initiated may be related to a thermally induced stress response. Similar responses have been observed in animals following prolonged exposure to nonspecific stressors.

Gildersleeve et al.[445] exposed Japanese quail eggs to 2.45-GHz MW at 50 W/m^2 for 12 days. After hatching an immunologic challenge did not elicit any modification of a cell-mediated or humoral immune response.

Hamrick[446] examined the response of mammalian lymphocytes exposed to 2.45-GHz (CW) in cultures at 200 W/m^2 (7 W/kg) for 48 hr. Changes in the stimulation caused by phytohemagglutinin were not affected by MW exposure.

Decrease in natural killer (NK) activity in lymphocytes of hamsters after exposure to 13 W/kg, 2.45-GHz CW for 1 hr was reported by Yang et al.[429] Roberts et al.[430] examined the effects of exposure to 2.45-GHz continuous wave RF on human mononuclear synthesis after exposure to RF at SARs up to 4 W/kg. They also measured effects on synthesis of specific host defense proteins, namely, interferons, and examined for morphological lymphoblastoid transformation as well as changes indicated by incorporation of radiolabeled precursors. Leukocytes were exposed in 37°C chambers without attempts to counteract RF-induced heating, and final culture temperatures were approximately 0.9°C higher than those of sham-exposed cultures, similar to changes induced by Stodolnik-Baranska.[425] Such exposures resulted in no detectable effects on viability or on unstimulated or mitogen-stimulated DNA, RNA, total protein, or interferon synthesis by the human mononuclear leukocytes.[429]

The difficulties of comparing reports, analyzing the literature and extrapolating from *in vitro* to *in vivo* effects, together with a presentation of additional studies can be found in a review by Budd and Czerski.[447]

B. ADAPTATION

Czerski and associates[413,414] have reported that inbred Swiss mice, immunized with sheep red blood cells and exposed 2 hr/day for 6 or 12 weeks to 5 W/m^2 at 2.95 GHz, showed increased serum hemagglutinin titers and antibody-producing cells in lymph node homogenates. The increase was greater in mice exposed 6 weeks than in those exposed 12 weeks. Similar results were obtained in rabbits exposed 2 hr/day to 30 W/m^2 for 6 months with the maximum increase occurring after exposure of 1 or 2 months, and then returning to control values. According to Czerski,[448] this may indicate that after a period of response, the animals become adapted to the MWs. The phenomenon of physiologic adaptation or decreased reaction as a result of repeated exposure to microwaves has also been reported by others.[65,408,420,449,450,451]

Baranski[410] exposed adult guinea pigs 3 hr/day for 3 months to 35 W/m^2 at 3 GHz and found an increase in lymphopoiesis over controls, as indicated by increased incorporation of ^3H-thymidine and increased mitotic indices. No differences could be detected between pulsed or CW MWs of the same average power level. Twofold increases in lymphocyte numbers were found in the spleen and lymph nodes.

C. INFLUENCE OF HYPERTHERMIA

RF-induced hyperthermia in mice has been associated with transient lymphopenia and neutrophilia, with a relative increase in splenic T and B lymphocytes, and with decreased *in vivo* local delayed hypersensitivity.[422,452,431] The latter was not affected by a comparable increase in core temperature produced by warm air. Reduced thymic mass and cell density,[421,452] suppressed inflammatory response,[421,451] and suppressed allograft transplant rejection[453] have been reported. Such alterations in lymphocyte distribution and function are concomitant with a state of immunosuppression. Qualitatively, similar changes can be induced by administration of synthetic glucocorticoids or CS.[421,430] Liburdy[430] reported elevated plasma CS levels in mice following exposure to RF energy sufficient to cause hyperthermia. A similar response has been reported by Lotz and Michaelson,[365] who showed a correlation between MW-induced body heating and CS levels in the blood of rats. These studies suggest that exposure to MWs of sufficient intensity results in stimulation of the hypophysial-adrenal axis which could affect the immune system.[431]

It would thus appear that RF exposure initially causes a general stimulation of the immune system, but if the exposure continues, the stimulatory effect disappears, suggesting a phase of adaptation to continued RF exposure. There is a body of literature on the influence of heat per se on immunity. Significant influences of microwaves on immune responsiveness would be expected on the basis of the known effects of hyperthermia. Although there has been some uncertainty whether fever enhances host resistance to infection,[454,455] recent evidence suggests that fever may enhance survival after infection in an animal model.[456] Cell-mediated immunity plays a role in defense against facilitative intracellular bacteria, viruses, and certain other infectious agents.[457,458] Roberts and Steigbigel[459] have shown that increased temperature (38.5°C) enhances human lymphocyte response to mitogen (PHA) and antigen (streptokinase-streptodornase) and enhances, but does not accelerate, certain bactericidal functions of human phagocytic leukocytes. One, therefore, has to be circumspect in assessing the mechanisms of MW exposure related to alterations in immune processes. It may be that if MWs do, in fact, increase the proportion of lymphocytes undergoing transformation, it may not in itself be harmful, but actually beneficial. The immune system has a

TABLE 15
Immunologic Effects of *In Vivo* RF Exposure

Effects	Species	Frequency (GHz)	Intensity (W/m^2)	Exposure Conditions SAR (W/kg)	Duration (days × min)	Ref.
Increase in lymphoblasts in lymph nodes and increased response to SRBC	Mouse	2.95 (PW)	5	0.5 (est)	42 × 120	448
Increase in "spontaneous" lymphoblast transformation of cultured lymphocytes	Rabbit	2.95 (PW)	50	0.8 (est)	24–48 × 120	448
Increase in lymphoblasts in spleen and lymphoid tissue	Mouse	3.105 (PW)	20	2 (est)	6–8700	417
Increased transformation of unstimulated cultured lymphocytes and decreased mitosis in PHA-stimulated lymphocyte cultures	Chinese hamster	2.45 (CW)	50, 150, 300, or 450	2.3, 6.9, 13.8, or 20.7	5 × 15	27
Transient decrease and increased response of cultured lymphocytes to PHA, Con A, and LPS	Mouse	2.45 (CW)	50 or 150	3.6 or 10	1–17 × 30	432
Increase in Cr^+, Fc^+, and Ig^+ spleen cells; increased response to B-cell mitogens; decrease in primary response to SRBC	Mouse	2.45 (CW)	—	14	1 or 3 × 30	423, 424 433
Increase in Cr^+ and Fc^+ spleen cells	Mouse	2.45 (CW)	—	11.8 5	1 × 15 1 × 30	434
Increase in response of cultured lymphocytes to T- and B-cell mitogen	Rat	2.45 (CW)	50	1–5	57 × 240	176
		0.425 (CW)	100	3–7	47 × 240 (perinatal exp)	606
No change	Mouse	2.45 (CW)	50–350	4–25	1–22 × 15 or 30	421
	Rat	0.1 (CW)	460	2–3		534
Increase in T and B lymphocytes in spleen; decrease in DTH	Mouse	0.026 (CW)	8,000	5.6 (est)	1 × 15 or 10 × 15	431
Reduction of lymphocyte traffic from lung to spleen	Mouse	2.6 (CW)	50 or 250	3.8 or 19	1 × 60	435
Decreased response to PWM	Rabbit	2.45 (CW)	100	1.5	180 × 1,380	600
No change	Quail	2.45 (CW)	50	4.03	12 × 1,440	456

TABLE 15
Continued

		Exposure Conditions				
Effects	Species	Frequency (GHz)	Intensity (W/m^2)	SAR (W/kg)	Duration (days × min)	Ref.
Decrease in tumor development	Mouse	2.45 (CW)	(Near-field application)	35	11–14 Day of gestation or 11–14 and 19–45 × 20	95
Decreased granulocytic response	Rabbit	3 (CW)	30	0.5	42–84 × 360	437
Tumor regression and increase in antitumor antibodies and anti-BSA	Rabbit	1.356	(Near-field application)	(Local hyperthermia)	1 × 10–15	608
Tumor inhibition and immune stimulation	Rat	2.45 (CW)	(200 W)	(Local hyperthermia)	3 or 6 × 45	609
Increased tumoricidal activity in lymphocytes and macrophages	Mouse	1.356	6,000–9,000	(Local hyperthermia)	1 × 5	610
Tumor regression	Mouse	3 (CW)	400	28 (est)	1 × 14–120	611
Increase in lung cancer colonies and inhibition of contact sensitivity to oxazolone	Mouse	2.45 (CW)	500	36 (est)	4, 7, 10 or 14 × 120	612
Decrease in response to BSA	Rabbit	1.356	(Near-field application)	(Local hyperthermia)	3 × 60	613
Decrease in CFU for erythroid and granulocyte-macrophage series	Mouse	2.45 (CW)	150	10	9 × 30	432
Reduction in CFU granulocyte-macrophage precursors exposed *in vitro*	Mouse	2.45 (CW)	600–10,000	120–10,000	1 × 15	614
No change in cell-mediated or humoral immune response	Quail	2.45 (CW)	50	—	12 × 1440	446

Note: SRBC = sheep red blood cells, PHA = phytohemagglutinine, Con A = concanavaline A, LPS = lipopolysaccharide, CR^+ = complement-receptor positive, Ig^+ = immunoglobulin positive, Fc^+ = Fc portion of immunoglobulin, DTH = delayed-type hypersensitivity, PWM = pokeweed mitogen, BSA = bovine serum albumin, and CFU = colony-forming unit.

Adapted and expanded from U.S. Environmental Protection Agency, Biological Effects of Radiofrequency Radiation, Elder, J. A., and Cahill, D. F., Eds., Rep. No. EPA-600/8-83-026F, Research Triangle Park, N.C., 1984.

TABLE 16
Immunologic Effects on *In Vitro* RF Exposure

Effects	Species	Frequency (GHz)	Exposure Conditions Intensity (W/m^2)	SAR (W/kg)	Duration (days × min)	Ref.
Increased blastogenesis of exposed lymphocytes *in vitro*	Human lymphocytes	3 (PW)	70 140	[a]	3–5 × 240 3–5 × 15	615
Increased blastogenesis	Human lymphocytes	10	50–150	[a]	Observed effect only when culture temperature approached 38°C	65
No effect on viability, DNA, RNA, total protein, interferon synthesis	Human lymphocytes	2.45 (CW)	(Waveguide)	0.5–4 W/l	1 × 120	430
No change in mitogen response to PHA, Con A, or LPS	Mouse spleen cells	2.45 (CW)	100	19	1 × 60, 120 or 240	425
No change in viability or growth	Rat blood lymphocytes	2.45 (CW)	50, 100, or 200	0.7, 1.4, or 2.8	1 × 240, 1440, or 2640	616
No change in viability or growth	Human lymphoblast cell lines (Daudi and HSB_2)	2.45 (CW)	100–5000	25–1200	1 × 15	617
Decreased macrophage phagocytosis	Mouse macrophage	2.45 (CW)	500	15 J/min	1 × 30	618
Liberation of intracellular hydrolytic enzymes and increased death	Rabbit granulocytes	3 (CW)	10 or 50	[a]	1 × 15, 30, or 60	619
No change in growth, CFU, or various strains of exposed cultures under several growth conditions	E. coli	2.45 (CW)	(Far-field)	0.0075–75	1 × 240	489
No change in survival curves (measuring CFU) of exposed cultures	E. coli	2.45 (CW)	(Microwave oven)	~400	1 × 1	620
No effect on colony-forming ability	B. subtilis spores E. coli	2.6–4.0 (CW)	(Waveguide)	29	1 × 480	52

TABLE 16
Continued

Effects	Species	Frequency (GHz)	Exposure Conditions: Intensity (W/m^2)	Exposure Conditions: SAR (W/kg)	Duration (days × min)	Ref.
Temporary decrease in virulence (>6 hr) of bacteria for its host cells; recovery within 24 hr at 37°C	A. tumefaciens	10 (CW)	(Cavity)	~1	1 × 30, 60, 230	621
No alteration in viability, unstimulated DNA synthesis or PHA-stimulated DNA synthesis	Human lymphocytes infected with influenza virus	2.45 (CW) 2.45 (PW)	(Waveguide)	0.29 and 4 W/l	1 × 120	444
No change in viability, unstimulated or PHA-stimulated DNA and protein synthesis	Human lymphocytes	2.45 (PW)	(Waveguide)	0.29 and 4 W/l	1 × 120	623
No change in viability, unstimulated or PHA-stimulated DNA, RNA and total protein synthesis	Human lymphocytes	2.45 (CW)	(Waveguide)	12 W/l	1 × 120	624
No change in viability; decreased unstimulated RNA and total protein synthesis; delayed PHA-stimulated DNA, RNA and total protein synthesis	Human lymphocytes	2.45 (CW)	(Waveguide)	22.5 W/l	1 × 120	624

Note: PHA = phytohemagglutinin, Con A = concanavaline A, LPS = lipopolysaccharide, and CFU = colony-forming unit.

[a] Unable to calculate SAR.

Adapted and expanded from U S. Environmental Protein Agency, Biological Effects of Radiofrequency Radiation, Elder, J. A. and Cahill, D. F., Rep. No. EPA-600/8-83-026F, Research Triangle Park, N.C., 1984.

considerable redundancy and adaptability. Perturbations of the immune system may not have clinical significance. Immunological effects of RF are summarized in Tables 15 and 16.

XII. AUDITORY RESPONSE

"Microwave hearing" is sensed in the head as a clicking or buzzing due to pulsed MW at rather low power densities.[460,461] It appears that this phenomenon is one of thermoelastic expansion.[328,462–466] This is discussed in detail in Chapter 12.

XIII. OCULAR EFFECTS

During the past 30 years, numerous investigations in animals and several surveys among human populations have been devoted to assessing the relationship of MW exposure to the subsequent development of cataracts.[5] It is significant that of the many experiments on rabbits by several investigators using various techniques, a power density above 1 kW/m^2 for 1 hr or longer appears to be the lowest time-power threshold in the tested frequency range of 0.2 to 10 GHz. In other species of animals such as dogs and nonhuman primates, the threshold for experimental MW-induced cataractogenesis appears to be even higher. This threshold is a time-power threshold, that is, the higher the power density, the shorter is the time threshold and vice versa, down to a certain minimum power density. All of the reported effects of MW radiation on the lens can be explained on the basis of thermal injury.

A. THRESHOLD FOR OPACITY IN RABBITS

The most extensive investigations in this area were performed by Carpenter and van Ummersen[467] and Carpenter et al.[468] Later work by Guy et al.[469,470,471] and Kramar et al.[472,473] has shown that in rabbits exposed to 2.45 GHz, the threshold for cataract production is 1.5 kW/m^2 for 100 min. The data suggest that an intraocular temperature of at least 43°C must be obtained to induce cataracts, although exceptions may be found in the literature.[471] These investigators also found that single potentially cataractogenic exposures will not injure the eye under conditions of controlled general hypothermia, and exposure to 1 kW/m^2, 2 hr/day, for 4 to 9 days produced no cataracts as evidenced by periodic examinations for 6 months after exposure. Guy et al.[469] exposed rabbits to 0.918 GHz, 4.66 kW/m^2 for 15 min and 1.17 W/m^2 for 100 min with no evidence of cataract, and concluded the threshold for cataractogenesis is higher for this frequency in comparison to 2.45 GHz.

Appleton et al.[474] exposed rabbits to 3 GHz, 1 or 2 kW/m^2 for 15 to 30 min. Examination daily for 14 days, weekly for 1 month, and monthly for a year revealed no ocular changes. At power densities of 3, 4, or 5 kW/m^2 for 15 min, acute ocular changes involving, especially, the conjunctiva and iris occurred during exposure. In a comparable study by Hirsch et al.,[475] rabbits were repeatedly exposed to 3 GHz once daily for a month. Clinical examinations were carried out for 1 year afterward. No changes occurred at power densities under 3 kW/m^2.

B. BIOCHEMICAL CHANGES

Reports of decreased enzymatic activity[467,476] may quite likely be due to thermal inactivation with resultant alterations in metabolism. Decreases in ascorbic acid concentration in the lens have been cited as being the first biochemical indication of opacity formation;[477,478] progressive clouding of the lens is associated with decreases in ascorbic acid below 60 $\mu g/g$ of lens tissues. All of these effects could be fully reparable until the altered metabolism has produced a permament opacification of the lens. Latent periods and time-power thresholds would be in agreement with a mechanism of this nature.

C. THERMAL ASPECT OF MICROWAVE CATARACTOGENESIS

Guy et al,[469,470,471] Kramar et al.,[472,473] and Taflove and Brodwin[479] have computed the MW energy deposition and induced temperatures in the eyes of rabbits and a model of the human eye. They have indicated that a distinct hot spot exceeding 40.4°C probably occurs deep within the eye at a frequency of 1.5 GHz, when the power density would be cataractogenic (i.e. greater than 1.99 kW/m^2).

Paulsson[480] measured the absorption of 0.915-, 2.45-, and 9.0-GHz MW energy in a model of the human head. The absorbed power showed an essentially exponential decrease with the distance from the cornea at 9.0 GHz. At 2.45 and 0.915 GHz maximum absorbed power occurred within the eyeball.

D. PULSE-WAVE (PW) EXPOSURES

Kues and Monahan[481] investigated effects of pulsed MW on the eyes of monkeys in vivo. Exposures were at 2.45 GHz (10 microsec, 100 pps) at an average incident power of 100 W/m^2 (SAR = 2.65 W/kg) on three consecutive days for four hours each day. Damage to the corneal endothelium and an increase in iris vascular permeability was found persisting for more than 72 hours after exposure. This report suggests that PW exposures might have significantly lower thresholds for damage than CW exposures, especially high-peak-power exposures with low SAR resulting from a sufficiently slow repetition rate. Johnson et al.[482] also exposed monkeys to 1.25 GHz MW, 0.5 μsec-wide pulses, 16 pps at an average ocular SAR of 3.5 to 4.0 W/kg, 7 four-hr exposures spread over several weeks. At these high-peak, low repetition rate exposures retinal damage was found by electroretingraphy, suggesting cone photoreceptor damage. Creighton et al.[483] have also found evidence of an enhanced effect of pulsed over continuous MW at equivalent average power level by observing cataract formation in the rat lens *in vitro*.

E. CUMULATIVE EFFECT

The possibility of a cumulative effect on the lens from repeated "subthreshold" exposures of the rabbit eyes to MW was proposed. Such an effect presumably would result from clinically inapparent damage from any given exposure which would eventually manifest clinically after repeated exposures.

McAfee et al.[484] trained monkeys to face an RF source and then exposed them to 9.3-GHz pulsed RF at an average power density of 1.5 kW/m^2. Each of 12 monkeys was exposed for up to 20 min/day for 30 to 40 sessions over several months. A total of 75 monkeys, non-exposed and sham-exposed, served as controls. No cataracts or corneal lesions were seen in any of the 12 exposed animals up to 12 months after exposure.

Chou et al.,[485] exposed two groups of six rabbits to 2.45-GHz CW/PW (10-μsec pulses, PPS 100, peak power density 150 kW/m^2) at 150 W/m^2 for 2 hr/day over several months; a third group was sham irradiated. Periodic eye examinations for cataract formation yielded no statistically significant differences among the three groups.

Guy et al.[486] exposed four rabbits to 2.45-GHz RF at 100 W/m^2 (maximum whole-body SAR of 17 W/kg) 7 days/week, 23 hr/day, for 6 months. For controls, four other rabbits were sham exposed for the same durations. Periodic eye examinations with a slit-lamp microscope showed normal aging changes in the lenses and no significant differences between the exposed and control groups.

Although it is difficult to identify the time-power threshold precisely, varying the times and powers to produce such a curve, a general reading of the literature suggests that no one has been able to produce cataracts in animal models at power densities below 800 to 999 W/m^2 even by repetitive exposures. These values are well above those of the most widely used safety standards.[352,353]

In general the results from animal studies indicate that RF cataractogenesis is essentially a gross thermal effect that has a threshold power density at which the difference between the rates of heat generation by RF and heat removal is large enough to result in damage to the lens of the eye. The case of very high peak powers at low repetition rates may prove to be an exception. Thermoelastic expansion has been proposed as a mechanism in this case.[463]

F. EXTRAPOLATION TO THE HUMAN

After 30 years of studies of the effects of MWs on the ocular lens, primarily in the rabbit, the principal conclusions are

1. The acute thermal insult from high-intensity MW fields is cataractogenic in the rabbit if intraocular temperature is greater than 43°C.
2. The MW exposure threshold is between 1 and 1.5 kW/m^2 applied for about 60 to 100 min.
3. There does not appear to be a cumulative effect from MW exposure unless each single exposure is sufficient to produce some irreparable degree of injury to the lens.

That opacity of the ocular lens can be produced in rabbits by exposure to MW is well established. Extrapolating results from animal studies to humans is difficult, because the conditions, durations and intensities of exposure are usually quite different. Reports of adverse reactions to exposures are necessarily "uncontrolled" and exposure conditions are generally impossible to determine or reconstruct. It is also difficult to relate cause and effect, because lens imperfections do occur in otherwise healthy individuals, especially with increasing age. Numerous drugs, industrial chemicals, and certain metabolic diseases are associated with cataracts. Because of the many uncertainties, workmen's compensation and negligence cases are usually settled out of court with the details of the settlement withheld from the public. Lipman et al.[487] have reviewed the conditions under which MW might produce injury to the lens.

XIV. CRITIQUE

Elucidation of the biological effects of RF exposure requires study and analysis of the available literature. This literature has been accumulating, in a sustained way, since the late 1940s with the development and application of radar in World War II. The literature is not particularly "user-friendly" in that comparisons of studies must often rest on assumptions of uncertain validity. Since the 1940s there has been a steady improvement in measurement capabilities, dosimetric measurements, and standards of quality in the performance of investigations. This improvement has been so dramatic that it makes contemporary investigators mistrustful of the older literature, resting as it did on a less sophisticated technology and approach. The older literature also contained a major contribution from the closed societies of eastern Europe which was hobbled by barriers to free and open communication and which may have been "tainted" by political interference. It was commonly the case that reports from countries dominated by the former Soviet Union were inadequately documented and the results slanted to conform to expectations of authorities. This may have had the effect of endowing RF effects at low power with greater and perhaps more sinister efficacy in producing specific biological effects, especially on the nervous system, than the contemporary concensus would support.

The general improvement in the quality of research has produced areas of agreement on effects and potential hazards, but also left areas of disagreement. The issue of "window" effects produced by modulated RF on calcium exchange in the nervous system is still permeated with disagreement. See Chapter 12. The issue of an increase in intrinsic permeability of the blood-brain barrier at non-thermal levels of RF has been resolved to the satisfaction of a majority of students of the subject but some dissent still remains. General concensus is that there is no such increase at non-thermal levels and that brain temperature must rise to 41 to 43°C, rapidly producing tissue necrosis and morbidity, in order to see transfer of hydrophilic molecules into brain parenchyma. Studies of effects on the neuroendocrine system always pose interpretative problems because of the labile response of the system to stress. Even the most "gentled" animals may be responding to the stress of handling and this may potentiate an RF effect which otherwise may not be observed.

Even if the effects can be validated and the dose/response worked out quantitatively in a manner to pass the most stringent review, there remains the problem of what constitutes a hazard. One objective definition of injury is an irreversible change in biological function producing a functional deficit as observed at the organ or system level. With this definition, it is possible to define a hazard as a probability of injury on a statistical basis. It is important to distinguish between the intensities of RF energy at which injury may be sustained and those that produce nonharmful effects with no further pathophysiologic consequences. All effects are not necessarily hazards. Some effects may have beneficial application under appropriately controlled condition. For that reason therapeutic application of RF energy by use of diathermy technology to produce frank heating is excepted from safety standards prescribing permissible exposure limits. In this case risk/benefit analyses are applied.

It is important to determine whether an observed effect is irreparable, transient, or reversible, disappearing when the electromagnetic field is removed either immediately or after some interval of time. Of course, even reversible effects are unacceptable if they transiently impair the ability of the individual to function properly or to perform a required task.

Although there is a much greater appreciation in recent years of the need for rigorous experimental design, the challenges are still formidable. An investigator may think he is observing a low-level or nonthermal effect in one animal because the incident power is low, while in actuality the animal may be exposed to as much absorbed power in a specific region of the body due to resonance or focusing effects as another larger animal is with much higher incident power at certain frequencies. A detailed understanding of interspecies scaling factors is essential.

The review of the literature as a whole has its own problem. The process of comparing studies is usually not straightforward especially if the studies seem to contradict each other. Even an investigation which has as a specific goal the confirmation or non-confirmation of a previous study inevitably is different in some way. The significance of the differences are debated at scientific meetings and these debates are usually not reflected in the literature. If a follow-up study does not use exactly the same frequency, the question of the ability to extrapolate from one frequency to the other arises.

Pickard and Foster,[1] in 1987, proposed that it may be time to limit further RF biological effects research by formulating guidelines to restrict further work. They advanced this proposal because of the "inherent noisiness" of the field and the apparent inability to resolve controversies in a timely manner. In subsequent meeting forums many scientists took issue with the proposal, arguing that more (and more careful) research was needed, not less.

As mentioned in the introduction to this chapter the rate of RF research output has declined within the last five years. The U.S. Defense Department, the major underwriter of RF research, has decreased its support. Public concerns have directed more support to ELF research. In 1993 concerns that the RF output of cellular telephones might play a role in the initiation of tumors have, in turn, directed some support back to RF-effects research.

Although the public policy rationale undergoes periodic shifts, overall support for continuing research in electromagnetic effects will undoubtedly continue, motivated substantially by safety considerations and exploration of potential medical therapies.

ACKNOWLEDGMENTS

The author (EE) thanks Mrs. Doris Michaelson for her support of this revision of Dr. Michaelson's classic monograph, which first appeared in the 1986 edition of the Handbook, and also thanks Dr. S.-T. Lu for invaluable assistance. The work was supported by the Walter Reed Army Institute of Research. The opinions or assertions contained are private views of the author and are not to be construed as reflecting the official views of the Department of the Army or the Department of Defense.

REFERENCES

1. **Pickard, W. F. and Foster, K. R.,** The risks of risk research, *Nature,* 330, 531, 1987.
2. IEEE Standard for Safety Levels with Respect to Human Exposure to Radio Frequency Electromagnetic Fields, 3 kHz to 300 GHz, IEEE Standards Coordinating Committee 28, Institute of Electrical and Electronics Engineers, Inc., New York, NY, 1991.
3. **Petersen, R. C.,** Radiofrequency/microwave protection guides, *Health Phys.,* 61, 59, 1991.
4. **Sliney, D. H., Wolbarsht, M. L., and Muc, A. M.,** Differing radiofrequency standards in the microwave region–implications for future research, *Health Physics,* 49, 677, 1985.
5. Biological Effects of Radiofrequency Radiation, EPA-600/8-83-026F, Health Effects Research Laboratory, United States Environmental Protection Agency, Research Triangle Park, N.C., 1984, p 5–76.
   Hardy, J. D., Wolff, H. G., and Goodell, H., *Pain Sensations and Reactions,* Hafner Publishing Co., New York, NY, Chapter X.
6. **Girirajan, K. S., Young, L., and Prohofsky, E. W.,** Vibrational free energy, entropy, and temperature factors of DNA calculated by a helix lattice approach, *Biopolymers,* 28, 1841, 1989.
7. **Durney, C. H., Johnson, D. C., Barber, P. W., Massoudi, H., Iskander, M. F., Lords, J. L., Ryser, D. K., Allen, S. J., and Mitchell, J. C.,** *Radiofrequency Radiation Dosimetry Handbook,* 2nd ed., USAF Report SAM-TR-78-22, Brooks Air Force Base, TX., 1978.
8. **Gandhi, O. P.,** Conditions of strongest electromagnetic power deposition in man and animals, *IEEE Trans. Microwave Theory Tech.,* 23, 1021, 1975.
9. **Gandhi, O. P.,** State of knowledge for electromagnetic absorbed dose in man and animals, *Proc. IEEE,* 68, 24, 1980.
10. **Gandhi, O. P., Hunt, E. L., and D'Andrea, J. A.,** Deposition of electromagnetic energy in animals and in models of man with and without grounding and reflector effects. *Radio Sci.,* 12(6S), 39, 1977.
11. **Guy, A. W.,** Quantitation of induced electromagnetic field patterns and associated biologic effects, in *Biologic Effects and Health Hazards of Microwave Radiation,* Czerski, P. et al., Eds., Polish Medical Publishers, Warsaw, 1974, 203.
12. **Guy, A. W., Webb, M. D., and Sorensen, C. C.,** Determination of power absorption in man exposed to high frequency electromagnetic fields by thermographic measurements on scale models, *IEEE Trans. Biol. Med. Eng.,* 23, 361, 1976.
13. **Massoudi, H.,** Long Wavelength Analysis of Electromagnetic Power Absorption by Prolate Spheroidal and Ellipsoidal Models of Man, Ph.D thesis, University of Utah, Salt Lake City, 1976.
14. **Massoudi, H., Durney, D. H., and Johnson, C. C.,** Long-wavelength electromagnetic power absorption in ellipsoidal models of man and animals, *IEEE Trans. Microwave Theory Tech.,* 25, 47, 1977.
15. **Guy, A. W. and Chou, C. K.,** Effects of high-intensity microwave pulse exposure of rat brain, *Radio Sci,* 17, (5S) 169, 1982.
16. **Johnson, C. C.,** Recommendations for specifying EM wave irradiation conditions in bioeffects research, *J. Microwave Power,* 10, 249, 1975.
17. **Anne, A., Saito, M., Salati, O. M., and Schwan, H. P.,** Relative microwave absorption cross sections of biological significance, in *Proc. 4th Annu. Tri-Service Conf. Biol. Effects of Microwave Radiating Equipment; Biological Effects of Microwave Radiation,* Peyton, M. F., Ed., Plenum Press, New York, 1962, 153.
18. **Durney, C. H., Massoudi, H., Iskander, M. F.,** Radiofrequency Radiation Dosimetry Handbook, USAFSAM-TR-85-73, USAF School of Aerospace Medical Division (AFSC), Brooks Air Force Base, TX 78235-5301, December 1985.

19. **Gandhi, O. P.** Electromagnetic energy absorption in humans and animals, in *Biological Effects and Medical Applications of Electromagnetic Energy,* Gandhi, O. P., Ed., 1990, p 175.
20. NCRP, Radiofrequency Electromagnetic Fields: Properties, Quantities and Units, Biophysical Interactions and measurements, Rep No. 67, National Council on Radiation Protection and Measurements, Washington DC, 1981.
21. **Stuchly, M. A.,** Health Aspects of Radiofrequency and Microwave Radiation Exposure, Part 2, Department of National Health and Welfare, Ottawa, 1978.
22. **Heller, J. H.,** Cellular effects of microwave radiation, Biological Effects and Health Implications of Microwave Radiation, Symp. Proc., Cleary, S. F., Ed., Public Health Service BRH/DBE 70-21, U.S. Department of Health, Education, and Welfare, 1970, 116.
23. **Janes, D. D., Leach, W. M., Mills, W. A., Moore, R. T., and Shore, M. L.,** Effects of 2450 MHz microwaves on protein synthesis and on chromosomes in Chinese hamsters, *Nonioniz. Radiat.,* 1, 125, 1969.
24. **Yao, K. T. S.,** Microwave radiation-induced chromosomal aberrations in corneal epithelium of Chinese hamsters, *J. Heredity,* 69, 409, 1978.
25. **Chen, K. M., Samuel, A., and Hoopingavner, R.,** Chromosomal abberrations of living cells induced by microwave radiation, *Environ. Lett.,* 6, 37, 1974.
26. **Yao, K. T. S.,** Cytogenetic consequences of microwave irradiation on mammalian cells incubated *in vitro, J. Heredity,* 73, 133, 1982.
27. **Huang, A. T., Engle, M. E., Elder, J. A., Kinn, J. B., and Ward, T. R.,** The effect of microwave radiation (2450 MHz) on the morphology and chromosomes of lymphocytes, *Radio Sci.,* 12(6S), 173, 1977.
28. **Yao, K. T. S.,** Cytogenetic consequences of microwave incubation of mammalian cells in culture, *Genetics,* 83 (Suppl.), 584 (Abstr.), 1976.
29. **Manikowska, E., Luciani, J. M., Servantie, B., Czerski, P., Obrenovitch, J., and Stahl, A.,** Effects of 9.4 GHz microwave exposure on meiosis in mice, *Experientia,* 35, 388, 1979.
30. **Savage, J. R. K.,** Use and abuse of chromosomal aberrations as an indicator of genetic damage, *Int. J. Environ. Stud.,* 1, 233, 1977.
31. **Webb, S. J. and Booth, A. D.,** Absorption of microwaves by microorganisms, *Nature,* 222, 1199, 1969.
32. **Webb, S. J. and Booth, A. D.,** Microwave absorption by normal and tumor cells, *Science,* 174, 72, 1971.
33. **Stamm, M. E., Winters, W. D., Morton, D. L., and Warren, S. L.,** Microwave characteristics of human tumor cells, *Oncology,* 29, 294, 1974.
34. **Dardanoni, L., Torregrossa, V., Tamburello, C., Zanforlin, L., and Spalla, M.,** Biological effects of millimeter waves at spectral singularities, in *Electromagnetic Compatibility,* Wydawnictwo Politechniki, Wroclawskiej, Breslau, Poland, 1976, 308.
35. **Lee, R. A. and Webb, S. J.,** Possible detection of *in vivo* viruses by fine-structure millimeter microwave spectroscopy between 68 and 76 GHz, *IRCS Med. Sci.,* 5, 222, 1977.
36. **Devyatkov, N. D., et al.,** Highlights of the papers presented on millimeter wave biological effects, *Sov. Phys.-USPEKHI,* 4, 568, 1974.
37. **Stamm, M. E., Warren, S. L., Rand, R. W., et al.,** Microwave therapy experiments with B-16 murine melanoma, *IRCS Med. Sci.,* 3, 392, 1975.
38. **Berteaud, A. J., Dardalhon, M., Rebeyrotte, N., and Averbeck, D.,** Action d'un rayonnement electromagnetique a longueur d'onde milimetrique sur la croissance bacterienne, *C.R. Acad. Sci. Ser. D,* 281, 843, 1975.
39. **Grundler, W. and Keilmann, F.,** Sharp resonances in yeast prove nonthermal sensitivity to microwaves, *Phys. Rev. Lett.,* 51, 214, 1983.
40. **Grundler, W. and Keilmann, F.,** Nonthermal effects of millimeter microwaves on yeast growth, *Z. Naturforsch.,* 33C(1/2), 15, 1978.
41. **Gandhi, O. P., Hagmann, M. J., Hill, D. V., Partlow, L. M., and Bush, L.,** Millimeter-wave absorption spectra of biological samples, *Bioelectromagnetics,* 1, 285, 1980.
42. **Swicord, M. L., and Davis, C. C.,** Microwave absorption of DNA between 8 and 12 GHz, *Biopolymers,* 21, 2450, 1982.
43. **Rhee, K. W., Lee, C. F., Davis, C. C., Sagripanti, J. L., and Swicord, M. L.,** Further studies of the microwave absorption characteristics of different forms of DNA in solution, 10th Annual Meeting of the Bioelectromagnetics Society, Stamford, CT, June 19–23, 1988.
44. **Gabriel, C., Grant, E. H., Tata, R., Brown, P. R., Gestblom, B., Noreland, E.,** Microwave absorption in aqueous solutions of DNA, *Nature,* 328, 145, 1987.
45. **Sagripanti, J., and Swicord, M. L.,** DNA structural changes caused by microwave radiation, *Int. J. Radiat. Biol.,* 50, 47, 1986.
46. **Anderstam, B., Hamnerius, Y., Hussain, S., and Ehrenberg, L.,** Studies of possible genetic effects in bacteria of high frequency electromagnetic fields, *Hereditas,* 98, 11, 1983.
47. **Varma, M. M. and Traboulay, E. A.,** Evaluation of dominant lethal test and DNA studies in measuring mutagenicity caused by nonionizing radiation, Biological Effects of Electromagnetic Waves, Vol. 1, Publ.

(FDA) 77–8010, Johnson, C. C. and Shore, M. L., Eds., U.S. Department of Health Education, and Welfare, Rockville, MD, 1976, 386.

48. **Mittler, S.,** Failure of 2 and 20 meter radio waves to induce genetic damage in *Drosophila melanogaster, Environ. Res.,* 11, 326, 1976.
49. **Pay, T. L., Beyer, E. C., and Reichelder, C. F.,** Microwave effects on reproductive capacity and genetic transmission in *Drosophila melanogaster, J. Microwave Power,* 7, 75, 1972.
50. **Anderstam, B., Hamnerius, Y., Hussain, S., and Ehrenberg, L.,** Studies of possible genetic effects in bacteria of high frequency electromagnetic fields, *Hereditas,* 98, 11, 1983.
51. **Baranski, S., Debiec, H., Kwarecki, K., and Mezykowski, T.,** Influence of microwaves on genetical processes of *Aspergillus nidulans, J. Microwave Power,* 11, 146, 1976.
52. **Correlli, J. C., Gutmann, R. J., Kohazi, S., and Levy, J.,** Effects of 2.6–4.0 GHz microwave radiation on *E. coli B, J Microwave Power,* 12, 141, 1977.
53. **Blackman, C. F., Surles, M. C., and Benane, S. G.,** The effects of microwave exposure on bacteria mutation reduction, Vol. 1, Publ. (FDA) 77–8010, Symp. Biol. Eff. of E.M. Waves, U.S. Department of Health, Education and Welfare, Rockville, MD, 1976, 406.
54. **Dutta, S. K., Nelson, W. H., Blackman, C. F., and Brusick, D. J.,** Lack of microbiol genetic response to 2.45 GHz CW and 8.5 to 9.6 GHz pulsed microwaves, *J. Microwave Power,* 14, 275, 1979.
55. **Livingston, G. K., Johnson, C. C., and Dethlefsen, L. A.,** Comparative effects of water bath and microwave-induced hyperthermia on cell survival and sister chromatid exchange in Chinese hamster ovary cells, in Abstr. 1977 Int. Symp. on the Biological Effects of Electromagnetic Waves, Airlie, VA, 1977, 106.
56. **McRee, D. I., MacNichols, G., and Livingston, G. K.,** Incidence of sister chromatid exchange in bone marrow cells of the mouse following microwave exposure, *Radiat. Res.,* 85, 340, 1981.
57. **Saunders, R. D., Kowalczuk, C. I., Beechey, C. V., and Dunford, R.,** Studies of the induction of dominant lethals and translocations in male mice after chronic exposure to microwave radiation, *Int. J. Radiat. Biol.,* 53, 983, 1988.
58. **Meltz, M. L. and Walker, K. A.,** Genetic effects of microwave exposure on mammalian cells *in vitro,* in USAF Radiofrequency Radiation Bioeffects Research Program–A Review, Mitchell, J. C., Ed., Aeromed Rev. 4-18, Rep No. SAM-Tr-81-30, 1981, 201.
59. **Meltz, M. L. and Walker, K. A.,** Radiofrequency (microwave) radiation exposure of mammalian cells during UV-induced DNA repair synthesis, *Radiat. Res.* 110, 255, 1987.
60. **Brown, R. F., Marshall, S. V., and Hughes, C. W.,** Effects of RFR on excision-type DNA repair *in vivo,* in USAF Radiofrequency Radiation Bioeffects Research Program–A Review, Mitchell, J. C., Ed., Aeromed. Rev. 4-18, Rep. No SAM-TR-81-30, 1981, 184.
61. **Mickey, G. H.,** Electromagnetism and its effect on the organism, *N. Y. State J. Med.,* 63, 1935, 1963.
62. **Varma, M. M. and Traboulay, E. A.,** Evaluation of dominant lethal test and DNA studies in measuring mutagenicity caused by nonionizing radiation, in Biological Effects of Electromagnetic Waves, Vol. 1, Publ. (FDA) 77–8010, Johnson, C. C. and Shore, M. L. Eds., U.S. Department of Health, Education, and Welfare, Rockville, MD, 1976, 386.
63. **Saunders, R. D., Darby, S. C., and Kowalczuk, C. I.,** Dominant lethal studies in male mice after exposure to 2.45 GHz microwave radiation, *Mutation Res.,* 117, 345, 1983.
64. **Berman, E., Carter, H. B., and House, D.,** Tests of mutagenesis and reproduction in male rats exposed to 2450-MHz (CW) microwaves, *Bioelectromagnetics,* 1, 65, 1980.
65. **Baranski, S. and Czerski, P.,** *Biological Effects of Microwaves,* Dowden, Hutchinson & Ross, Stroudsburg, PA, 1976.
66. **Varma, M. M., Dage, E. L., and Joshi, S. R.,** Mutagenicity induced by nonionizing radiation in Swiss male mice, in Biological Effects of Electromagnetic Waves, Vol. 1, Johnson, C. C. and Shore, M. L., Eds. Publ. (FDA) 77–8010, U.S. Department of Health, Education, and Welfare, Rockville, MD, 1976, 397.
67. **Leach, W. M.,** On the induction of chromosomal aberrations by 2450 MHz microwave radiation. *J. Cell Biol.,* 70 (S), 387a (Abstr.), 1976.
68. **Liu, L. M., Nickless, F. G., and Cleary, S. F.,** Effects of microwave radiation on erythrocyte membranes, *Radio Sci.,* 14 (6S), 109, 1979.
69. **Janiak, M. and Szmigielski, S.,** Injury of cell membranes in normal and SV40-virus transformed fibroblasts exposed *in vitro* to microwave (2450 MHz) or water-bath hyperthermia (43°C), in Abstr. of 1977 Int. Symp. on the Biological Effects of Electromagnetic Waves, Airlie, VA, 1977.
70. **Ciaravino, V., Meltz, M. L., and Erwin, D. N.,** Absence of a synergistic effects between moderate-power radio-frequency electromagnetic radiation and Adriamycin on cell-cycle progression and sister-chromatid exchange, *Bioelectromagnetics,* 12, 289, 1991.
71. **Ciaravino, V., Meltz, M. L., and Erwin, D. N.,** Effects of radiofrequency radiation and simultaneous exposure with mitomycin C on the frequency of sister chromatid exchanges in Chinese Hamster Ovary cells, *Environmental Mutagenesis,* 9, 393, 1987.

72. **Lloyd, D. C., Saunders, R. D., Moquet, J. E., and Kowalczuk, C. E.,** Absence of chromosome damage in human lymphocytes exposed to microwave radiation with hyperthermia, *Bioelectromagnetics,* 7, 235, 1986.
73. **Hamnerius, Y., Rasmuson, A., and Rasmuson, B.,** Biological effects of high-frequency electromagnetic fields on *Salmonella Typhimurum* and *Drosophila melanogaster, Bioelectromagnetics,* 6, 405, 1985.
74. **Stodolnik-Baranska, W.,** The effects of microwaves on human lymphocyte cultures, in *Biologic Effects and Health Hazards of Microwave Radiation,* Czerski, P. et al., Eds., Polish Medical Publishers, Warsaw, 1974, 189.
75. **Dutta, S. K., Das, B., Ghosh, B., and Blackman, C. F.,** Dose dependence of acetylcholinesterase activity in neuroblastoma cells exposed to modulated radio-frequency electromagnetic radiation, *Bioelectromagnetics,* 13, 317, 1992.
76. **Saffer, J. D., Profenno, L. A.,** Microwave-specific heating affects gene expression, *Bioelectromagnetics* 13, 75, 1992.
77. **Saffer, J. D., Profenno, L. A.,** Microwave-specific heating affects gene expression, *Bioelectromagnetics,* 10, 75, 1992.
78. **Saffer, J. D., and Profenno, L. A.,** Sensitive model with which to detect athermal effects of non-ionizing electromagnetic radiation, *Bioelectromagnetics,* 10, 347, 1989.
79. **Parker, J. E., Kiel, L. J., and Winters, W. D.,** Effect of radiofrequency radiation on mRNA expression in cultured rodent cells, *Physiol. Chem. Phys.* NMR, 20, 129, 1988.
80. **D'Inzeo, G., Bernardi, P., Eusebi, F., Grassi, F., Tamburello, C., and Zani, B. M.,** Microwave effects on acetylcholine induced channels in cultured chick myotubes, *Bioelectromagnetics,* 9, 363, 1988.
81. **Rotkovska, D., Vacek, A., and Bartonickova, A.,** Effects of microwaves on the colony forming capacity of haemopoietic stem cells in mice, *Acta Oncologica,* 26, 233, 1987.
82. **Rotkovska, D., Bartonickova, A., and Kautska, J.,** Effects of microwaves on membranes of hematopoietic cells in their structural and functional organization, *Bioelectromagnetics,* 14, 79, 1993.
83. **Cleary, S. F., Liu, L., and Merchant, R. E.,** Glioma proliferation modulated 83.24 in vitro by isothermal radiofrequency radiation exposure, 83.24 *Radiat. Res.,* 121, 38, 1990.
84. **Brown, R. F. and Marshall, S. V.,** Differentiation of murine erythroleukemic cells during exposure to microwave radiation, *Radiat. Res.,* 108, 12, 1986.
84. **Liburdy, R. P., and Vanek, P. F. Jr.,** Microwaves and the cell membrane, *Radiat. Res.,* 102, 190, 1985.
85. **Liburdy, R. P., and Magin, R. L.,** Microwave-stimulated drug release from liposomes, *Radiat. Res.,* 103, 266, 1985.
86. **Phelan, A. M., Lange, D. G., Kues, H. A., and Lutty, G. A.,** Modification of membrane fluidity in melanin-containing cells by low-level microwave radiation, *Bioelectromagnetics,* 13, 131, 1992.
87. **Sandblom, J., and Theander, S.,** The effect of microwave radiation on the stability and formation of Gramicidin-A channels in lipid bilayer membranes, *Bioelectromagnetics,* 12, 9, 1991.
88. **Philippova, T. M., Novoselov, V. I., Bystrova, M. F., and Alekseev, S. I.,** Microwave effect on camphor binding to rat olfactory epithelium, *Bioelectromagnetics,* 9, 347, 1988.
89. **Liburdy, R. P., and Penn, A.,** Microwave bioeffects in the erythrocyte are temperature and pO_2 dependent: cation permeability and protein shedding occur at the membrane phase transition, *Bioelectromagnetics,* 5, 283, 1984.
90. **Allis, J. W., and Sinha-Robinson, B. L.,** Temperature-specific inhibition of human red cell Na+/K+ ATPase by 2,450-MHz microwave radiation, *Bioelectromagnetics,* 8, 203, 1987.
91. **Byus, C. V., Kartun, K., Pieper, S., and Adey, W. R.,** Increased ornithine decarboxylase activity in cultured cells exposed to low energy modulated microwave fields and phorbol ester tumor promoters, *Cancer Res.,* 48, 4222, 1988.
92. **Prausnitz, S., Susskind, C., and Vogelhut, P. O.,** Effects of chronic microwave irradiation on mice, *I.R.E. Trans. Biomed. Electron.,* 9, 104, 1962.
93. **Szmigielski, S., Szydinski, A., Pietraszek, A., and Bielec, M.,** Acceleration of cancer development in mice by long-term exposition to 2450 MHz microwave fields, in *URSI Int. Symp. Proc., Ondes Electromagnetiques et Biologie,* Berteaud, A. J., and Servantie, B., Eds., Paris, 1980.
94. **Szmigielski, S., Szudzinski, A., Pietraszek, A., Bielec, M., Jahiak, M., and Wrembel, J. K.,** Accelerated development of spontaneous and benzopyrene-induced skin cancer in mice exposed to 2450 MHz microwave radiation, *Bioelectromagnetics,* 3, 179, 1982.
95. **Preskorn, S. H., Edwards, W. D., and Justensen, D. R.,** Retarded tumor growth and augmented longevity in mice after fetal irradiation by 2450 MHz, *J. Surg, Oncol.,* 10, 483, 1978.
96. **Chou, C-K., Guy, A. W., Kung, L. L., John, R. B., Crowley, J. J., and Krupp, J. H.,** Long-term, low-level microwave irradiation of rats, *Bioelectromagnetics* 13, 469, 1992.
97. **Balcer-Kubiczek, E. K., and Harrison, G. H.,** Evidence for microwave carcinogenesis *in vitro, Carcinogenesis,* 6, 859, 1985.
98. **Balcer-Kubiczek, E. K. and Harrison, G. H.,** Neoplastic transformation of C3H/10T1/2 cells following

exposure to 120-Hz modulated 2.45-GHz microwaves and phorbol ester tumor promoter, *Radiat. Res.*, 126, 65, 1991.

99. **Overgaard, J.**, The design of clinical trials in hyperthermic oncology, in *Physics and Technology of Hyperthermia*, Field, S. B., and Franconi, C., Ed., Martinus Nijhoff Dordrecht, 598.
100. **Lehman, J. F.**, *Therapeutic Heat and Cold*. Rehabilitation Medicine Library, Baltimore, Williams & Wilkins, 1990.
101. **Dumansky, Y. D. and Rudichenko, V. F.**, Dependence of the functional activity of liver mitochondria with super-high frequency radiation, *Hyg. Sanit.*, 16, 1976.
102. **Ward, T. R., Allis, J. W., and Elder, J. A.**, Measure of enzymatic activity coincident with 2450 MHz microwave exposure, *J. Microwave Power*, 10, 315, 1975.
103. **Allis, J. W.**, Irradiation of bovine serum albumin with a crossed-beam exposure-detection system, *Ann. N.Y. Acad. Sci*, 247, 312, 1975.
104. **Belkhode, M. L., Johnson, D. L., and Muc, A. M.**, Thermal and athermal effects of microwave radiation on the activity of glucose-6-phosphate dehydrogenase in human blood, *Health Phys.*, 26, 45, 1974.
105. **Bini, M., Checcucci, A., Ignesti, A., Millanta, L., Rubino, N., Camici, S., Manao, G., and Tamponi, G.**, Analysis of the effects of microwave energy on enzymatic activity of lactate dehydrogenase (LDH), *J. Microwave Power*, 13, 96, 1978.
106. **Albert, E. N., McCullars, G., and Short, M.**, The effect of 2450 MHz microwave radiation on liver adenosine triphosphate (ATP), *J. Microwave Power*, 9, 205, 1974.
107. **Sanders, A. P., Schaefer, D. J., and Joines, W. T.**, Microwave effects on energy metabolism of rat brain, *Bioelectromagnetics*, 1, 171, 1980.
108. **Sanders, A. P., Joines, W. T., and Allis, J. W.**, Effects of continuous-wave, pulsed, and sinusoidal-amplitude-modulated microwaves on brain energy metabolism, *Bioelectromagnetics*, 6, 89, 1985.
109. **Dumansky, Y. D., Serdyuk, A., M., Litvinova, C. I., Tomashevskaya, L. A., and Popovich, V. M.**, Experimental research on the biological effects of 12-centimeter low-intensity waves, in *Health in Inhabited Localities*, 2nd ed., Kiev, 1972, 29.
110. **Malyshev, V. T. and Tkachenko, M. I.**, Activity of ferments on the mucous membrane of the small intestine under the influence of an SHF field, in *Physiology and Pathology of Digestion*, Kishenev, 1972, 186.
111. **Wangemann, R. T. and Cleary, S. F.**, The *in vivo* effects of 2.45 GHz microwave radiation on rabbit serum components, *Radiat. Environ. Biophys.*, 13, 89, 1976.
112. **Swiecicki, W. and Edelwejn, Z.**, The influence of 3 and 10 cm microwave irradiation in blood proteins in rabbits, *Med. Lotnicza*, 11, 54, 1963.
113. **Baranski, S.**, Histological and histochemical effects of microwave irradiation on the central nervous system of rabbits and guinea pigs, *Am. J. Phys. Med.*, 51, 182, 1972.
114. **Nikogosyan, S. V.**, Influence of UHF on the cholinesterase activity in the blood serum and organs in animals, in *The Biological Action of Ultrahigh Frequencies*, Letavet, A. A., and Gordon, A. V., Eds., JPRS 12471, 1962.
115. **Revutsky, E. L. and Edelman, F. M.**, Effects of centimeter and meter electromagnetic waves in the content of biologically active substances in human blood, *Philos. J. Ukr. Acad. Sci.*, 10, 379, 1964.
116. **Olcerst, R. B. and Rabinowitz, J. R.**, Studies on the interaction of microwave radiation with cholinesterase, *Radiat. Environ. Biophys.*, 15, 289, 1978.
117. **Mitchell, C. L., McRee, D. I., Peterson, N. J., Tilson H. A., Shandala, M. G., Rudnev, M. I., Varetskii, V. V., and Navakatikyan, M. I.**, Results of a United States and Soviet Union Joint Project on Nervous System Effects of Microwave Radiation, *Environmental Health Perspectives*, 81, 201, 1989.
118. **Ho, H. S. and Edwards, W. P.**, Oxygen-consumption rate of mice under differing dose rates of microwave radiation, *Radio Sci.*, 12 (6S), 131, 1977.
119. **Mirutenko, V. I.**, Investigating local thermal effect of electromagnetic (3 cm) waves on animals, *Fiziol. Ah. Akad. Nauk. UKR SSR*, 8, 382, 1962.
120. **Mirutenko, V. I.**, Effect of blood circulation on the distribution of heat, and the magnitude of the thermal effect during action of a SHF-UHF electromagnetic field on animals, *Fiziol. Zh. Akad. Nauk. UKR SSR*, 10, 641, 1964.
121. **Cook, H. F.**, The pain threshold for microwave and infra-red radiation. *J. Physiol.*, 118, 1, 1952.
122. **Herrick, J. F. and Krusen, F. H.**, Certain physiologic and pathologic effects of microwaves, *Electron. Eng.*, 72, 239, 1953.
123. **Krusen, F. H., Herrick, J. F., Leden, U., and Wakim, K. G.**, Microkymatotherapy: preliminary report of experimental studies of the heating effect of microwaves (radar) in living tissues, *Proc. Mayo Clin.*, 22, 209, 1947.
124. **Rae, J. W., Jr., Herrick, J. F., Wakim, K. G., and Krusen, F. H.**, A comparative study of temperatures produced by microwave and short wave diathermy, *Arch. Phys. Med.*, 30, 199, 1949.
125. **Semenov, A. I.**, The effect of UHF on the temperature of rabbit femoral tissues, *Byull. Eksper. Biol. Med.*, 60, 64, 1965.

126. **Engle, J. P., Herrick, J. F., Wakim, K. G., Grindlay, J. H., and Krusen, F. H.,** The effects of microwaves on bone and bone marrow and on adjacent tissues, *Arch. Phys. Med.,* 31, 453, 1949.
127. **Gersten, J. W., Wakim, K. G., Herrick, J. F., and Krusen, F. H.,** Effect of microwave diathermy on the peripheral circulation and on tissue temperature in man, *Arch. Phys. Med.,* 30, 7, 1949.
128. **Lehmann, J. F., Guy, A. W., Johnston, V. C., Brunner, G. D., and Bell, J. W.,** Comparison of relative heating patterns produced in tissues by exposure to microwave energy at frequencies of 2450 and 900 megacycles, *Arch. Phs. Med.,* 43, 69, 1962.
129. **Lehmann, J. F., McMillan, J. A., Brunner, G. D., and Johnston, V. C.,** Heating patterns produced in speciments by microwaves of the frequency of 2456 megacycles when applied with the "A," "B," and "C" directors, *Arch. Phys. Med.,* 43, 538, 1962.
130. **Gordon, Z. V.,** The problem of the biological action of UHF, *Tr. Gig. Tr. Prof. AMN SSR,* 1, 5, 1960.
131. **Gordon, Z. V. and Lobanova, Y. E.,** The temperature reaction of animals to exposure to SHF, in *The Biological Action of Super High-Frequency Fields,* Letavet, A. A. and Gordon, A. V., Eds., Academy of Medical Science, Moscow, 1960.
132. **Tyagin, N. W.,** Study of the thermal effect of SHF-UHF electromagnetic fields on various animals using the thermometric method, *Tr. Voyenno Med. Akad. Kirov, (U.S.S.R.),* 73, 9, 1957.
133. **Tyagin, N. V.,** Change in the blood of animals subjected to a SHF-UHF field, *Voyenno-Medit. Akad. Kirov Leningrad,* 73, 116, 1957.
134. **Lehmann, J. F., Brunner, G. D., McMillan, J. A., Silverman, D. R., and Johnston, V. C.,** Modification of heating patterns produced by microwaves at the frequencies of 2450 and 900 Mc by physiologic factors in the human, *Arch. Phys. Med. Rehabil.,* 45, 555, 1964.
135. **Lehmann, J. F., Johnston, V. C., McMillan, J. A., Silverman, D. R., Brunner, G. D., and Rathbun, L. A.,** Comparison of deep heating by microwaves at frequencies of 2450 and 900 megacycles, *Arch. Phys. Med.,* 46, 307, 1965.
136. **Lehmann, J. F., Silverman, D. R., Baum, B. A., Kirk, N. L., and Johnston, V. C.,** Temperature distributions in the human thigh, produced by infrared, hot pack and microwave application, *Arch. Phys. Med. Rehabil.,* 47, 291, 1966.
137. **Ely, T. S., Goldman, D., Hearon, J. Z., Williams, R. B., and Carpenter, H. M.,** Heating characteristics of laboratory animals exposed to ten centimeter microwaves. U.S. Naval Medical Research Institute (Res. Rep. Proj. NM 001-056.13.02), Bethesda, MD., IEEE Trans. Biomed. Eng., 11, 123, 1964.
138. **Gorodetskaya, S. F.,** The effect of centimeter radio waves on mouse fertility, *Fiziol. Zh.,* 9, 394, 1963.
139. **Imig, C. J., Thomson, J. D., and Hines, H. M.,** Testicular degeneration as a result of microwave irradiation, *Proc. Soc. Exp. Biol.,* 69, 382, 1948.
140. **Miro, L., Loubiere, R., and Pfister, A.,** Studies of visceral lesions observed in mice and rats exposed to UHF waves; a particular study of the effects of these waves on the reproduction of these animals, *Re. Med. Aeronaut. (Paris),* 4, 37, 1965.
141. **Bereznitskaya, A. N. and Kazbekov, I. M.,** Studies on the reproduction and testicular microstructure of mice exposed to microwaves in *Biological Effects of Radiofrequency Electromagnetic Fields,* Gordon, A. V., Ed., No. 4, Moscow 221–229 (JPRS63321), 1974.
142. **Saunders, R. D. and Kowalczuk, C. I.,** Effects of 2.45 GHz microwave radiation and heat on mouse spermatogenic epithelium, *Int. J. Radiat. Biol.,* 40(6), 623, 1981.
143. **Lebovitz, R. M., and Johnson, L.,** Acute, whole-body microwave exposure and testicular function of rats, *Bioelectromagnetics* 8, 37, 1987.
144. **Lebovitz, R. M., Johnson, L, and Samson, W. K.,** Effects of pulse modulated microwave radiation and conventional heating on sperm production, *J. Appl. Physiol.,* 62, 245, 1987.
145. **Cleary, S. F., Liu, L. M., Graham, R., and East, J.,** *In vitro* fertilization of mouse ova by spermatazoa exposed isothermally to radio-frequency radiation, *Bioelectromagnetics,* 10, 361, 1989.
146. **Carpenter, R. L. and Livstone, E. M.,** Evidence for nonthermal effects of microwave adiation: abnormal development of irradiated insect pupae, *IEEE Trans. Microwave Theory Tech.,* 19, 173, 1971.
147. **Van Ummersen, C. A.,** The effect of 2450 mc radiation on the development of the chick embryo, in *Biological Effects of Microwave Radiation,* Vol. 1, Peyton, M. F., Ed., Plenum Press, New York, 1961, 201.
148. **Van Ummersen, C. A.,** An Experimental Study of Development Abnormalities Induced in the Chick Embryo by Exposure to Radio Frequency Waves, Ph. D. thesis, Tufts University, Medford, Mass., 1963.
149. **Bereznitskaya, A. N.,** Research on the reproductive function in female mice under the impact of low intensity radio waves of different ranges, in Industrial Health and Biological Effects of Radiofrequency Electromagnetic Waves, Material of the 4th All-Union Symp., October 17 to 19, 1972, Moscow, 1972.
150. **Rugh, R., Ginns, E. I., Ho, H. S., and Leach, W. M.,** Are microwaves teratogenic? in *Biologic Effects and Health Hazards of Microwave Radiation,* Czerski, P. et al., Eds., Polish Medical Publishers, Warsaw, 1974, 98.
151. **Rugh, R., Ginns, E. I., Ho, H. S., and Leach, W. M.,** Responses of the mouse to microwaves radiation during estrous cycle and pregnancy, *Radiat. Res.,* 62, 225, 1975.

152. **Olsen, R. G.,** Constant dose microwave irradiation of insect pupae, *Radio Sci.,* 17(5S), 145, 1982.
153. **Lindauer, G. A., Liu, L. M., Skewes, G. W., and Rosenbaum, F. J.,** Further experiments seeking evidence of nonthermal biological effects of microwave radiation, *IEEE Trans. Microwave Theory Tech.,* 22, 790, 1974.
154. **Liu, L. M., Rosenbaum, F. J., and Pickard, W. F.,** The relation of teratogenesis in *Tenebrio molitor* to the incidence of low level microwaves, *IEEE Trans. Microwave Theory Tech.,* 23, 929, 1975.
155. **Olsen, R. G.,** Insect teratogenesis in a standing-wave irradiation system, *Radio Sci.,* 12(6S), 199, 1977.
156. **Pickard, W. F. and Olsen, R. G.,** Developmental effects of microwaves in *Tenebrio molitor:* experiments to detect possible influences of radiation frequency and of culturing protocols, in *Abstr. of the 1977 USNC/URSI Symp. on Biological Effects of Electromagnetic Waves,* National Academy of Sciences, Washington, DC, 1977, 66.
157. **McRee, D. I., Hamrick, P. E., Zinkl, J. E., Thaxton, P., and Parkhurst, C. R.,** Some effects of exposure of the Japanese quail embryo to 2.45 GHz microwave radiation, *Ann. N.Y. Acad. Sci.,* 247, 377, 1975.
158. **McRee, D. I. and Hamrick, P. E.,** Exposure of Japanese quail embryos to 2.45 GHz microwave radiation during development, *Radiat. Res.,* 71, 355, 1977.
159. **Spiers, D. E., and Baummer, S. C.,** Thermal and metabolic responsiveness of Japanese quail embryos following periodic exposure to 2,450 MHz microwaves, *Bioelectromagnetics,* 12, 225, 1991.
160. **Laskey, J., Dawes, D., and Howes, M.,** Progress report on 2450 MHz irradiation of pregnant rats and the effect on the fetus, in Radiation Bioeffects Summary Rep. PHS, Publ. BRH/DBE-70, U.S. Department of Health, Education and Welfare, Rockville, MD, 1970, 167.
161. **Chernovetz, M. E., Justesen, D. R., King, N. W., and Wagner, J. E.,** Teratology, survival, and reversal learning after fetal irradiation of mice by 2450 MHz microwave energy, *M. Microwave Power,* 10, 391, 1975.
162. **Chernovetz, M. E., Justesen, D. R., and Oke, A. F.,** A teratologic study of the rat: microwave and infrared radiations compared, *Radio Sci.,* 12(6S), 191, 1977.
163. **Berman, E., Kinn, J. B., and Carter, H. B.,** Observations of mouse fetuses after irradiation with 2.45 GHz microwaves, *Health Phys.,* 35, 791, 1978.
164. **Johnson, R. B., Mizumori, S., and Lovely, R. H.,** Adult behavioral deficit in rats exposed prenatally to 918-MHz microwaves, in Developmental Toxicology of Energy-Related Pollutants, DOE Symp., Ser. 47, Department of Energy, Washington, DC, 1977, 281.
165. **Nawrot, P. S., McRee, D. I., and Staples, R. E.,** Effects of 2.45 GHz CW microwave radiation on embryofetal development in mice, *Teratology,* 24, 303, 1981.
166. **Jensh, R. P., Weinburg, L, and Brent, R. L.,** Teratologic studies of prenatal exposure of rats to 915 MHz microwave radiation, *Radiat. Res.,* 92, 160, 1982.
167. **Rugh, R. and McManaway, M.,** Can electromagnetic waves cause congenital anomalies? in Int. IEEE/AP-S USN/URSI Symp., Amherst, MA, 1976, 143.
168. **Rugh, R. and McManaway, M.,** Anesthesia as an effective agent against the production of congenital anomalies in mouse fetuses exposed to electromaagnetic radiation, *J. Exp. Zool.,* 197, 363, 1976.
169. **Lin, J. C., Nelson, J. C., and Ekstrom, M. E.,** Effects of repeated exposure to 148 MHz radiowaves on growth and hematology of mice, *Radio Sci.,* 14, 173, 1979.
170. **Nelson, J. C., Lin, J. C., and Ekstrom, M. E.,** Teratogenic effects of RF radiation on mice, presented at Bioelectromagnetics Symp., Seattle, WA, June 18 to 22, 1979.
171. **Dietzel, F. and Kern, W.,** Abortion following ultra-shortwave hyperthermia animal experiments, *Arch. Gynakol.,* 209, 445, 1970.
172. **Dietzel, F., Kern, W., and Steckenmesser, R.,** Deformity and intrauterine death after short-wave therapy in early pregnancy in experimental animals, *Muench. Med. Wochenschr.,* 114, 228, 1972.
173. **Dietzel, F.,** Effects of nonionizing electro-magnetic radiation on the development and intrauterine implantation of the rat, *Ann. N.Y. Acad. Sci.,* 247, 367, 1975.
174. **Brown-Woodman, P. D. C., Hadley, J. A., Waterhouse, J, and Webster, W. S.,** Teratogenic effects of exposure to radiofrequency radiation (27.12 MHz) from a shortwave diathermy unit, *Industrial Health,* 26, 1, 1988.
175. **Michaelson, S. M., Guillet, R., Catallo, M. A., Small, J., Inamine, G., and Heggeness, F. W.,** Influence of 2450 MHz microwaves on rats exposed *in utero, J. Microwave Power,* 11, 165, 1976.
176. **Smialowicz, R. J., Kinn, J. B., and Elder, J. A.,** Perinatal exposure of rats to 2450 MHz (CW) microwave radiation: effects on lymphocytes, *Radio Sci.,* 14, 147, 1979.
177. **Jensh, R. P., Vogel, W. H., and Brent, R. L.,** Postnatal functional analysis of prenatal exposure of rats to 915 MHz microwave radiation, *J. Am. Coll. Toxicol,* 1, 73, 1982.
178. **Jensh, R. P., Weinberg, I., and Brent, R. L.,** An evaluation of the teratogenic potential or protracted exposure of pregnant rats to 2450 MHz microwave radiation, I. Morphologic analysis at term, *J. Toxicol. Environ. Health,* 11, 23, 1983.
179. **Jensh, R. P., Vogel, W. H., and Brent, R. L.,** An analysis of the teratogenic potential or protracted exposure of pregnant rats to 2450 MHz microwave radiation. II. Postnatal psychophysiologic analysis, *J. Toxicol. Environ. Health,* 11, 37, 1983.

180. **Jensh, R. P. and Ludlow, J.,** Behavioral teratology: application in low dose chronic microwave irradiation studies, in *Advances in the Study of Birth Defects,* Vol. 4, Persand, T. V. N., Ed., MTP Press, Lancaster, England, 1980, chap. 8.
181. **Jensh, R. P., Vogel, W. H., Ludlow, J., and McHugh, T.,** Studies concerning the effects of low dosage prenatal 6000 MHz microwave radiation on growth and development in the rat, *Teratology,* 19(2), 32A (Abstr.) 1979.
182. **Lary, J. M., Conover, D. L., Foley, E. D., and Hanser, P. L.,** Teratogenicity of 27.12 MHz radiofrequency radiation in rats, *Teratology,* 19, 36A, 1979.
183. **Lary, J. M., Conover, D. L., Foley, E. D., and Hanser, P. L.,** Teratogenicity of 27.12 MHz radiofrequency radiation in rats, *Bioelectromagnetics,* 2, 402, 1980.
184. **Conover, D. L., Lary, J. M., and Hanser, P. L.,** Thermal threshold for teratogenic response in rats irradiated at 17.12 MHz, *Bioelectromagnetics,* 1, 204, 1979.
185. **Lacy, K. K., Desesso, J. M., and Lary, J. M.,** A comparison of the teratogenic effects of radiofrequency radiation and hyperthermia: gross evaluation, *Teratology,* 21, 51A, 1980.
186. **Lacy, K. K., Desesso, J. M., Sadler, T. W., and Lary, J. M.,** A comparison of the teratogenic effects of radiofrequency radiation and hyperthermia: light microscopic evaluation, *Teratology,* 21, 52A, 1980.
187. **Conover, D., Lary, J. M., and Foley, E.,** Induction of teratogenic effects in rats by 27.12 MHz RF radiation, presented at the 1978 Symp. on Electromagnetic Fields in Biological Systems, Ottawa, June 17 to 30, 1978.
188. **Boak, R. A., Carpenter, C. M., and Warren, S. L.,** Studies on the physiological effects of fever temperaures II. The effect of repeated short wave (30 meter) fevers on growth and fertility of rabbits, *Exp. Med.,* 56, 725, 1932.
189. **Roux, C, Elefant, E., Gaboriaud, G., Jaullery, C., Gardette, J., Dupuis, R., and Lambert, D.,** Association of microwaves and ionizing radiation: potentiation of teratogenic effects in the rat, *Radiat. Res.,* 108, 317, 1986.
190. **Kaplan, J. N.,** Study of the lethal effects of microwaves in the developing squirrel monkey, Final Report for Contract No. 68-02-3210, U.S. Environmental Protection Agency, 1981.
191. **Kaplan, J. N., Polson, P., Rebert, C., Lunan, K., and Gage, M.,** Biological and behavioral effects of prenatal and postnatal exposure to 2450-MHz electromagnetic radiation in the squirrel monkey, *Radio Sci.,* 17(5S), 135, 1982.
192. **O'Connor, M. E.,** Mammalian teratogenesis and radiofrequency fields, *Proc. IEEE,* 68, 56, 1980.
193. **Hsu, C. Y.,** Influence of temperature on development of rat embryos, *Anat. Rec.,* 100, 79, 1948.
194. **Fernandez-Cano, L.,** Effect of increase or decrease of body temperature and hypoxia on pregnancy in the rat, *Fertil. Steril.,* 9, 455, 1958.
195. **Pleet, H., Graham, J. M., and Smith, D. W.,** Central nervous system and facial defects associated with maternal hyperthermia at four to 14 weeks gestation, *Pediatrics,* 67, 785, 1981.
196. **Edwards, M. J.,** Congenital malformations in the rat following induced hyperthermia during gestation, *Teratology,* 173, 1968.
197. **Edwards, M. J.,** Congenital defects in guinea pigs: fetal resorptions, abortions, and malformations following induced hyperthermia during early gestation, *Teratology,* 2, 313, 1969.
198. **Howarth, B., Jr.,** Embryonic survival in adrenalectomized rabbits following exposure to elevated ambient temperature and constant humidity, *J. Anim. Sci.,* 28, 80, 1969.
199. **Michaelson, S. M.,** Physiologic regulation in electromagnetic fields, *Bioelectromagnetics,* 3, 91, 1982.
200. **Way, W. I., Kritikos, H., and Schwan, H.,** Thermoregulatory physiologic responses in the human body exposed to microwave radiation, *Bioelectromagnetics,* 2, 341, 1981.
201. **Wilson, G. J.,** Experimental studies on congenital malformations, *J. Chron. Dis.,* 10, 111, 1959.
202. **Gruenwald, P.,** Mechanisms of abnormal development, *Arch. Pathol.,* 44, 398, 1947.
203. **Cameron, J. A.,** Termination of early pregnancy by artificial fever, *Proc. Soc. Exp. Biol. Med.,* 52, 76, 1943.
204. **Edwards, M. J.,** Congenital defects in guinea pigs following induced hyperthermia during gestation, *Arch. Pathol.,* 84, 42, 1967.
205. **Garrison, L. H.,** The effect of fever on the development of the rat incisor, *J. Dent. Res.,* 19, 215, 1940.
206. **Martson, L. V. and Voronina, V. M.,** Experimental study of the effect of a series of phosphororganic pesticides (Dipterex and Imidan) on embryogenesis, *Environ. Health Perspect.,* 13, 121, 1976.
207. U.S. Environmental Protection Agency, Biological Effects of Radiofrequency Radiation, Elder, J. A. and Cahill, D. F., Eds., Rep No. EPA-600/8-83-026F, Research Triangle Park, NC, 1984.
208. **Gildersleeve, R. P., Galvin, M. J., McRee, D. I., Thaxton, J. P., and Parkhurst, C. R.,** Reproduction of Japanese Quail after microwave irradiation (2.45 GHz CW) during embryogeny, *Bioelectromagnetics,* 8, 9, 1987.
209. **Gordon, Z. V., Lobanova, Y. A., and Tolgskaya, M. S.,** Some data on the effect of centimeter waves (experimental studies), *Gig. Sanit. (U.S.S.R.),* 12, 16, 1955.
210. **Novitskiy, Y. I., Gordon, Z. V., Presman, A. S., and Kholodov, Y. A.,** Radio Frequencies and Microwaves, Magnetic and Electrical Fields, National Aeronautics and Space Administration (NASA TT F-14.021), Washington, DC, 1971.

211. **Yakovleva, M. I., Shlyafer, I. P., and Tsvetkova, I. P.,** On the question of condition cardiac reflexes, the functional and morphological state of cortical neurons under the effect of superhigh-frequency electromagnetic fields, *Zh. Vyssh. Nervn. Deyat. (U.S.S.R.),* 18, 973, 1968.
212. **Kitsovskaya, I. A.** An investigation of the interrelationships between the main nervous processes in rats on exposure to SHF fields of various intensities, *Tr. Gig. Tr. Prof. AMN SSR,* 1, 75, 1960.
213. **Kitsovskaya, I. A.,** The effect of radiowaves of various ranges on the nervous system (sound stimulation method), in *On the Biological Effect of Radio-Frequency Electromagnetic Fields,* Moscow, 1968, 81.
214. **Lobanova, Y. A. and Gordon, Z. V.,** The study of olfactory sensitivity in persons exposed to SHF, *TR. Gig. Prof, AMN SSR,* 1, 52, 1960.
215. **Fukalova, P. P.,** The sensitivity of olfactory and optic analyzers in persons exposed to the effect of constantly-generated SW and USW, *Tr. Gig. Prog. AMN SSR (Moscow),* 2, 144, 1964.
216. **Goncharova, N. N., Karamyshev, V. B., and Maksimenko, N. V.,** Occupational hygiene problems in working with ultrashort-wave transmitters used in TV and radiobroadcasting, *Gig. Tr. Prof. Zabol.,* 10, 10, 1966.
217. **Tolgskaya, M. S. and Gordon, Z. V.,** Changes in the receptor and interoreceptor apparatuses under the influence of UHF, in *The Biological Action of Ultrahigh Frequencies,* Letavet, A. A. and Gordon, Z. V., Eds., Academy of Medical Science, Moscow, 1960, 104.
218. **Marha, K.,** Biological effects of rf electromagnetic waves, *Prac. Lek. (Prague,),* 15, 387, 1963.
219. **Tolgskaya, M. W. and Gordon, Z. V.,** *Pathological Effects of Radio Waves,* Meditsina Press, Moscow, 1971; transl. Consultants Bureau, New York, 1973.
220. **Thompson, W. D. and Bourgeois, A. E.,** Effects of Microwave Exposure on Behavior and Related Phenomena, Primate Behavior Lab., Aeromedical Research Lab. Report, (ARL-TR-65-20; AD 489245), Wright-Patterson AFB, OH, 1975.
221. **Yermakov, Y. V.,** On the mechanism of developing asthenovegetative disturbance under the chronic effect of a SHF-field, *Voyenno-Medit. Zh. (U.S.S.R.),* 3, 42, 1969.
222. **Gvozdikova, Z. M., Annan'yev, V. M., Zenina, I. N., and Zak, V. I.,** Sensitivity of the rabbit central nervous system to a continuous (nonpulsed) ultrahigh frequency electromagnetic field, *Byull. Eksp. Biol. Med. (Moscow),* 58, 63, 1964.
223. **D'Andrea, J. A., Thomas, A., and Hatcher, D. J.,** Rhesus monkey behavior during exposure to high-peak-power 5.62-GHz microwave pulses, *Bioelectromagnetics,* 15, 163, 1994.
224. **Petin, V. G., Dubovick, B. V., Rojkov, M. F., and Komarov, V. P.,** Lethal effects in laboratory animals and time intensity factor in exposure to electromagnetic fields, Proceedings of the 14th Annual International Conference of the IEEE Engineering in Medicine and Biology Society, Paris, Oct 29 to Nov 1, 1992, 289.
225. **Brown, D. O., Lu, S-T., and Elson, E. C.,** Characteristics of microwave evoked body movements in mice, *Bioelectromagnetics*, 15, 143, 1994.
226. **Wachtel, H., Adey, G. R., Chalker, R., Barnes, F.,** Temperature rise rate as a causal factor of rapid neural responses to microwave absorption, Fourth Annual Scientific Session, Bioelectromagnetics Society, Los Angeles, June 28 to July 2, 1982, 34.
227. **Akyel, Y. and Raslear, T. G.,** Further characterization of the microwave-evoked whole-body movement, Proceedings of the 14th Annual International Conference of the IEEE Engineering in Medicine and Biology Society, San Diego, October 1993.
228. **Baldwin, M. S., Bach, S. A., and Lewis, S. A.,** Effects of radiofrequency energy on primate cerebral activity, *Neurology,* 10, 178 1960.
229. **Kholodov, Y. A.,** Changes in the electrical activity of the rabbit cerebral cortex during exposure to a UHF-HF electromagnetic field. II. The direct action of the UHF-HF field on the central nervous system. *Byull. Eksp. Biol. Med. (Moscow),* 56, 42, 1963.
230. **Kholodov, Y. A.,** The influence of a VHF-HF electromagnetic field on the electrical activity of an isolated strip of cerebral cortex, *Byull. Eksp. Biol. Med. (Moscow),* 57, 98, 1964.
231. **Kholodov, Y. A.,** *The Effect of Electromagnetic and Magnetic Fields on the Central Nervous System,* NASA TT-F-465, Nauka Press, Moscow, 1966.
232. **Livanov, M. N., Tsypin, A. B., Grigoriev, Y. G., Kruschev, U. G., Stepanov, S. M., and Anen'yev, A. M.,** The effect of electromagnetic fields on the bioelectric activity of cerebral cortex in rabbits, *Byull. Eksp. Biol. Med.,* 49, 63, 1960.
233. **Baranski, S. and Edelwejn, Z.,** Electroencephalographic and morphological investigations on the influence of microwaves on the central nervous system, *Acta Physiol. Pol.,* 18, 423, 1967.
234. **Bawin, S. M., Gavalas-Medici, R. J., and Adey, W. R.,** Effects of modulated very high frequency fields on specific brain rhythms in cats, *Brain Res.,* 58, 365, 1973.
235. **Servantie, B., Servantie, A. M., and Etienne, J.,** Synchronization of cortical neurons by a pulsed microwave field as evidenced by spectral analysis of EEG from the white rat, *Ann. N.Y. Acad. Sci.,* 247, 82, 1975.
236. **Suvorov, N. B. and Kukhtina, G. V.,** Spatial dynamics of brain electrical activity during prolonged contact with physical factors, *Human Physiol.,* 10, 395, 1984.

237. **Chizhenkova, R. A.,** Slow potentials and spike unit activity of the cerebral cortex of rabbits exposed to microwaves, *Bioelectromagnetics,* 9, 337, 1988.
238. **Baranski, S. and Edelwejn, Z.,** Experimental morphologic and electroencephalographic studies of microwave effects on the nervous system, *Ann. N.Y. Acad. Sci.,* 247, 109, 1975.
239. **Johnson, C. C., and Guy, A. W.,** Non-ionizing electromagnetic wave effects in biological materials and systems, *Proc. IEEE,* 60, 692, 1972.
240. **Tyazhelov, V. V., Tigranian, R. E., and Khizhniak, E. P.,** New artifact-free electrodes for recording of biological potentials in strong electromagnetic fields, *Radio Sci.,* 12(6S), 121, 1977.
241. **Frey, A. H.,** A coaxial pathway for recording from the cat brain during illumination with UHF energy, *Physiol. Behav.,* 3, 363, 1968.
242. **Bawin, S. M. and Adey, W. R.,** Calcium binding in cerebral tissue, in Symp. on Biological Effects and Measurement of Radio Frequency/Microwaves, Hazzard, D. G., Ed., Publ. (FDA) 77–8026, U. S. Department of Health, Education and Welfare, Rockville, MD, 1977.
243. **Blackman, C. F., Elder, J. A., Weil, C. M., Benane, S. G., Eichinger, D. C., and House, D. E.,** Induction of calcium ion efflux from brain tissue by radio-frequency radiation: effects of modulation frequency and field strength, *Radio Sci.,* 14(6S), 93, 1979.
244. **Blackman, C. F., Benane, S. G., Elder, J. A., House, D. E., Lampe, J. A., and Faulk, J. M.,** Induction of calcium-ion efflux from brain tissue by radiofrequency radiation: effect of sample number and modulation frequency on the power-density window, *Bioelectromagnetics,* 1, 35, 1980.
245. **Tolgskaya, M. S.,** Morphological changes in animals exposed to 10 cm microwaves, *Vopr. Kurortol. Fizioter. Lech. Fiz. Kult.,* 1, 21, 1959.
246. **Tolgskaya, M. S. and Gordon, Z. V.,** Comparative morphological characterization of action of microwaves of various ranges, *Tr. Gig. Tr. Prof. AMN SSR,* 2, 80, 1964.
247. **Albert, E. N. and Sheriff, M.,** Morphologic changes in cerebellum of neonatal rats exposed to 2.45 GHz microwaves, *Electromagnetic Fields and Neurobehavioral Function,* O'Connor, M. E., and Lovely, R. H., Eds., Alan R. Liss, Inc., 1988, 135.
248. **Accinni, L., De Martino, C., and Mariutti, G.,** Effects of radiofrequency radiation on rabbit kidney: a morphological and immunological study, *Experimental and Molecular Pathology,* 49, 22, 1988.
249. **Bilokrynytsk'ky, V. S.,** Changes in the tigroid substance of neurons under the effect of radio waves, *Fiziol. Zh.,* 12, 70, 1966.
250. **Baranski, S., Czekalinski, L., Czerski, P., and Haduch, S.,** Experimental research on fatal effect of micrometric wave electromagnetic radiation, *Rev. Med. Aeronaut. (Paris),* 2, 108, 1966.
251. **Dolina, L. A.,** Morphological changes in the central nervous system due to the action of centimeter waves on the organism, *Arkh. Patol.,* 23, 51, 1961.
252. **Minecki, L. and Bilski, R.,** Histopathological changes in internal organs of mice exposed to the action of microwaves, *Med. Pr. (Poland),* 12, 337, 1961.
253. **Albert, E. N. and DeSantis, M.,** Do microwaves alter nervous system-structure? *Ann. N.Y. Acad. Sci.,* 247, 87, 1975.
254. **Albert, E. N.,** Light and electron microscopic observations on the blood brain barrier after microwave irradiation, in Symp. on Biological Effects and Measurements of Radiofrequency/Microwaves, Hazard, D. S., Ed., Publ. (FDA) 77–8026, U.S. Department of Health, Education and Welfare, Rockville, MD, 1977, 294.
255. **Albert, E. N.,** Ultrastructural pathology associated with microwave induced alterations in blood-brain barrier permeability, Proc. Biol. Eff. E. M. Waves, 19th Gen. Assembly, Int. Union Radio Sci., Helsinki, August 1978.
256. **Albert, E. N., Grau, L., and Kerns, J.,** Morphologic alterations in hamster blood-brain barrier after microwave irradiation, *J. Microwave Power,* 12, 43, 1977.
257. **Albert, E. N., Brainard, D. L., Randal, J. D., and Jannata, F. S.,** Neuropathological observations on microwave-irradiated hamsters, Proc. Biol. Eff. E. M. Waves, 19th Gen. Assembly, Int. Union Radio Sci., Helsinki, August 1978.
258. **Albert, E. N., Sherif, M. F., Papadopoulos, N. J., Slaby, F. J., and Monahan, J.,** Effect of nonionizing radiation on the purkinje cells of the rat cerebellum, *Bioelectromagnetics,* 2, 247, 1981.
259. **Albert, E. N., Sherif, M. F., and Papadopoulos, N. J.,** Effect of nonionizing radiation on the purkinje cells of the uvula in squirrel monkey cerebellum, *Bioelectromagnetics,* 2, 241, 1981.
260. **Sutton, C. H., Nunnaly, R. L., and Carroll, F. B.,** Protection of the microwave-irradiated brain with body-core hypothermia, *Cryobiology,* 10, 513, 1973.
261. **Albert, E. N.,** Reversibility of the blood brain barrier, *Radio Sci.,* 14(6S), 323, 1979.
262. **Albert, E. N. and Kerns, J. M.,** Reversible microwave effects on the blood-brain barrier, *Brain Res.,* 230, 153, 1981.
263. **Neubauer, C., Phelan, A. M., Kues, H., and Lange, D. G.,** Microwave irradiation of rats at 2.45 GHz activates pinocytotic-like uptake of tracer by capillary endothelial cells of cerebral cortex, *Bioelectromagnetics,* 11, 261, 1990.

264. **Oscar, K. J. and Hawkins, T. D.,** Microwave alteration of the blood-brain barrier system of rats, *Brain Res.,* 126, 281, 1977.
265. **Oldendorf, W. H.,** Measurement of brain uptake of radiolabeled substances using a tritiated waterinternal standard, *Brain Res.,* 24, 372, 1970.
266. **Oscar, K. J., Gruenau, S. P., Folker, M. T., and Rapoport, S. I.,** Local cerebral blood flow after microwave exposure, *Brain Res.,* 204, 220, 1981.
267. **Preston, E., Vavasour, E. J., and Assenheim, H. M.,** Permeability of the blood brain barrier to mannitol in the rat following 2450-MHz microwave irradiation, *Brain Res.,* 174, 109, 1979.
268. **Chang, B. K., Huang, A. T., Joines, W. T., and Kramer, R. S.,** The effect of microwave radiation (1.0 GHz) on the blood-brain barier in dogs, *Radio Sci.,* 17 (5S), 165, 1982.
269. **Williams, W. M., Hoss, W., Formanick, M., and Michaelson, S. M.,** Effects of 2450 MHz microwave energy on the blood-brain barrier to hydrophobic molecules. A. Effect on the permeability to sodium fluoride, *Brain Res. Rev.,* 7, 175, 1984.
270. **Williams, W. M., Del Cerro, M., and Michaelson, S. M.,** Effects of 2450 MHz microwave energy on the blood-brain barrier to hydrophobic molecules. B. Effect on the permeability to HRP, *Brain Res., Rev.,* 7, 171, 1984.
271. **Williams, W., M., Platner, J., and Michaelson, S. M.,** Effects of 2450 MHz microwave energy on the blood-brain barrier to hydrophobic molecules. C. Effect on the permeability to [^{14}C], *Brain Res. Rev.,* 7, 183, 1984.
272. **Williams, W. M., Lu, S.-T., Del Cerro, M., and Michaelson, S.,** Effects of 2450 MHz microwave energy on the blood-brain barrier to hydrophobic molecules. D. Brain temperature and blood-brain barrier permeability to hydrophilic tracers, *Brain Res. Rev.,* 7, 192, 1984.
273. **Williams, W. M., Lu, S.-T., Del Cerro, M., Hoss, W., and Michaelson, S. M.,** Effects of 2450 MHz microwave energy on the blood-brain barrier: an overview and critique of past and present research, *IEEE Trans. Microwave Theory Tech.,* 32, 808, 1984.
274. **Merritt, J. H., Chamness, A. F., and Allen, S. J.,** Studies on blood-brain barrier permeability after microwave-radiation, *Radiat. Environ. Biophys.,* 15, 367, 1978.
275. **Lin, J. C. and Lin, M. F.,** Studies on microwave and blood-brain barrier interaction, *Bioelectromagnetics,* 1, 313, 1980.
276. **Lin, J. C. and Lin, M. F.,** Microwave hyperthermia-induced blood-brain barrier alterations, *Radiat. Res.,* 89, 77, 1982.
277. **Neilly, J. P., and Lin, J. C.,** Interaction of ethanol and microwaves on the blood-brain barrier of rats, *Bioelectromagnetics,* 7, 405, 1986.
278. **Rozdilsky, B. and Olszewski, J.,** Permeability of cerebral blood vessels studies by radioactive iodinated bovine albumin, *Neurology,* 7, 270, 1957.
279. **Blasberg, R. G.,** Problems of quantifying effects of microwave irradiation on the blood-brain barrier, *Radio Sci.,* 14 (6S), 335, 1979.
280. **Rapoport, S. I., Ohno, K., Fredricks, W. R., and Pettigrew, K. D.,** A quantitative method for measuring altered cerebrovascular permeability, *Radio Sci.,* 14 (6S), 345, 1979.
281. **Baranski, S. and Edelwejn, Z.,** Studies on the combined effect of microwaves and some drugs on bioelectric activity of the rabbit CNS, *Acta Physiol. Pol.,* 19, 37, 1968.
282. **Edelwejn, Z.,** An attempt to assess the functional state of the cerebral synapses in rabbits exposed to chronic irradiation with microwaves, *Acta Physiol. Pol.,* 19, 897, 1968.
283. **Thomas, J. R. and Maitland, G.,** Microwave radiation and dextroamphetamine: evidence of combined effects on behavior of rats, *Radio Sci.,* 14 (6S), 253, 1979.
284. **Servantie, B., Bertharion, G., July, R., Servantie, A. M., Etienne, J., Dreyfus, P., and Escoubet,** Pharmacologic effects of a pulsed microwave field, in *Biologic Effects and Health Hazards of Microwave Radiation,* Czerski, P. et al, Eds., Polish Medical Publishers, Warsaw, 1974, 36.
285. **Lai, H., Horita, A., Chou, C.-K., and Guy, A. W.,** Low-level microwave irradiations affect central cholinergic activity in the rat, *Journal of Neurochemistry,* 48, 40, 1987.
286. **Lai, H., Horita, A., Chou, C.-K., and Guy, A. W.,** Effects of low-level microwave irradiation on hippocampal and frontal cortical choline uptake are classically conditionable, *Pharmacol. Biochem. Behav.,* 27, 635, 1987.
287. **Lai, H., Horita, G., and Guy, A. W.,** Acute low-level microwave exposure and central cholinergic activity: studies on irradiation parameters, *Bioelectromagnetics,* 9, 355, 1988.
288. **Lai, H., Carino, M. A., Horita, G., and Guy, A. W.,** Opioid receptor subtypes that mediate a microwave-induced decrease in central cholinergic activity in the rat, *Bioelectromagnetics,* 13, 237, 1992.
289. **Lai, H. Carino, M. A., Horita, G., and Guy, A. W.,** Low-level microwave irradiation and central cholinergic activity; a dose-response study, *Bioelectromagnetics,* 10, 203, 1989.
290. **Lai, H., Carino, M. A., Wen, Y. F., Horita, A, and Guy, A. W.,** Naltrexone pretreatment blocks microwave-induced changes in central cholinergic receptors, *Bioelectromagnetics,* 12, 27, 1991.

291. **Lai, H.,** Research on the neurological effects of nonionizing radiation at the University of Washington, *Bioelectromagnetics,* 13, 513, 1992.
292. **Lai, H., Carino, M. A., Horita, A., and Guy, A. W.,** Corticotropin-releasing factor antagonist blocks microwave-induced decreases in high-affinity choline uptake in the rat brain, *Brain Res. Bull.,* 25, 609, 1990.
293. **Bruce-Wolfe, V., and Justesen, D. R.,** Microwaves retard the anesthetic action of pentobarbital, *Abstr. Bioelectromagnetics Soc.,* 7, 47, 1985.
294. **Jauchem, J. R., Frei, M. R., and Heinmets,** Effects of psychotropic drugs on thermal responses to radiofrequency radiation, *Aviat. Space Environ. Med.,* 56, 1183, 1985.
295. **Michaelson, S. M., Thomson, R. A. E., and Howland, J. W.,** Physiological aspects of microwave irradiation of mammals, *Am. J. Physiol.,* 201, 351, 1961.
296. **Lobanova, Y. A.,** Investigations on the susceptibility of animals to microwave (MW) irradiation following treatment with pharmacologic agents. In Biologic Effects of Radiofrequency Electromagnetic Fields, Gordon, A. V., Ed. Joint Publications Reprints Service, Washington, DC, 201.
297. **Jauchem, J. R.,** Effects of drugs on thermal responses to microwaves, *Gen. Pharmacol.,* 16, 307, 1985.
298. **Ashani, Y., Henry, F. H., and Catravas, G. N.,** Combined effects of anticholinesterase drugs and low-level microwave radiation, *Radiat. Res.,* 84, 469, 1980.
299. **Quock, R. M. Kouchich, F. J., Ishii, T. K., and Lange, D. G.,** Microwave facilitation of Domperidone antagonism of Apomorphine-induced stereotypic climbing in mice, *Bioelectromagnetics,* 8, 45, 1987.
300. **Quock, R. M., Fujimoto, J. M., Ishii, T. K., and Lange, D. G.,** Microwave facilitation of methylatropine antagonism of central cholinomimetic drug effects, *Radiation Res.,* 105, 328, 1986.
301. **Lai, H., Horita, A., Chou, C.-K., Guy, A. W.,** Ethanol-induced hypothermia and ethanol consumption in the rat are affected by low-level microwave irradiation, *Bioelectromagnetics,* 5, 213, 1984.
302. **Neilly, J. P. and Lin, J. C.,** Interaction of ethanol and microwaves on the blood-brain barrier of rats, *Bioelectromagnetics,* 7, 405, 1986.
303. **Hjeresen, D. L., Francendese, A., and O'Donnell, J. M.,** Microwave attentuation of ethanol-induced hypothermia: ethanol tolerance, time course, exposure duration, and dose response studies, *Bioelectromagnetics,* 9, 63, 1988.
304. **Hjeresen, D. L., Francendese, A., and O'Donnell, J. M.,** Microwave attentuation of ethanol-induced interaction with noradrenergic neurotransmitter systems, *Health Phys.,* 56, 767, 1989.
305. **Modak, A. T., Stavinoha, W. B., Dean, U. P.,** Effect of short electromagnetic pulses on brain acetylcholine content and spontaneous motor activity in mice, *Bioelectromagnetics,* 2, 39, 1981.
306. **Gandhi, V. C. and Ross D. H.,** Alterations in alpha-adrenergic and muscarinic cholinergic receptor binding in rat brain following nonionizing radiation, *Radiation Research,* 109, 90, 1987.
307. **Inaba, R., Shishido, K., Okada, A., and Moroji, T.,** Effects of whole body microwave exposure on the rat brain contents of biogenic amines, *Eur. J. Appl. Physiol.,* 65, 124, 1992.
308. **Browning, M. D. and Haycock, J. W.,** Microwave radiation in the absence of hyperthermia has no detectable effect on synapsin I levels or phosphorylation, *Neurotoxicology and Teratology,* 10, 461, 1988.
309. **Gandhi, C. R., and Ross, D. H.,** Microwave induced stimulation of ^{32}Pi incorporation into phosphoinositides of rat brain synaptosomes, *Radiat. Environ. Biophys.,* 28, 223, 1989.
310. **Lai, H., Horita, A., Chou, C.-K., and Guy, A. W.,** A review of microwave irradiation and actions of psychoactive drugs, *IEEE Engineering in Medicine and Biology Magazine,* March 1987, 31.
311. **Chou, C.-K. and Guy, A. W.,** Effects of electromagnetic fields on isolated nerve and muscle preparations, *IEEE Trans. Microwave Theory Tech.,* 26, 141, 1978.
312. **Wachtel, H., Seaman, R., and Joines, W.,** Effects of low-intensity microwaves on isolated neurons, *Ann. N.Y. Acad. Sci.,* 247, 46, 1975.
313. **Seaman, R. L. and Wachtel, H.,** Slow and rapid responses to CW and pulsed microwave radiation by individual *Aplysia* pacemakers, *J. Microwave Power,* 13, 77, 1978.
314. **McRee, D. I. and Wachtel, H.,** Pulse microwave effects of nerve vitality, *Radiat. Res.,* 91, 212, 1982.
315. **Khramov, R. N., Sosunov, E. A., Koltun, S. V., Ilyasova, E. N., and Lednev, V. V.,** Millimeter-wave effects on electric activity of crayfish stretch receptors, *Bioelectromagnetics,* 12, 203, 1991.
316. **Bolshakov, M. A. and Alekseev, S. I.,** Bursting responses of Lymnea neurons to microwave radiation, *Bioelectromagnetics,* 13, 119, 1992.
317. **Pakhomov, A. G., Dubovick, B. V., Kolupayev, V. E., Degtyariov, I. G., and Pronkevich, A. N.,** Search for microwave-sensitive structures and functions in the nervous system, *Fourteenth Annual IEEE/EMBS,* Paris, 297, 1992.
318. **Pakhomov, A. G., Dubovick, B. V., Kolupayev, V. E., and Pronkevich, A. N.,** Microwave-induced changes in nerve conduction: effect of modulation, *Fifteenth Annual IEEE/EMBS,* San Diego, 293, 1993.
319. **Ginsburg, K. S., Lin, J. C., and O'Neill, W. D.,** Microwave effects on input resistance and action potential firing of snail neurons, *IEEE Transactions on Biomedical Engineering,* 39, 1011, 1992.
320. **Arber, S. L., and Lin. J. C.,** Microwave enhancement of membrane conductance in snail neurons: role of temperature, *Physiol. Chem. and Phys. and Medical NMR.,* 15, 160, 1983.

321. **Arber, S. L., and Lin, J. C.,** Microwave enhancement of membrane conductance: effects of EDTA, caffeine and tetracaine, *Physiol. Chem. and Phys. and Medical NMR,* 16, 469, 1984.
322. **Arber, S. L.,** Microwave enhancement of membrane conductance: calmodulin hypothesis, *Physiol, Chem. and Phys. and Med. NMR,* 17, 227, 1985.
323. **Arber, S. L., and Lin. J. C.,** Extracellular calcium and microwave enhancement of membrane conductance in snail neurons, *Radiat. Environ. Biophys,* 24, 149, 1985.
324. **Arber S. L., Neilly, J. P., Lin, J. C., and Kriho, V.,** The effect of 2450 MHz microwave radiation on the ultrastructure of snail neurons, *Phyiol. Chem. and Phys. and Med. NMR,* 18, 243, 1986.
325. **Sanza, J. N. and de Lorge, J.,** Fixed interval behavior of rats exposed to microwaves at low power densities, *Radio Sci.,* 12(6S), 273, 1977.
326. **deLorge, J.,** Operant Behaviofr and Colonic Temperature of Squirrel Monkeys (Saimiri sciureus) during Microwave Irradiation, NAMRL-1236 Naval Aerospace Medical Res. Lab., Pensacola, FL, 1977.
327. **D'Andrea, J. A., Gandhi, O. P., and Lords, J. L.,** Behavioral and thermal effects of microwave radiation at resonant and nonresonant wave lengths, *Radio Sci.,* 12(6S), 251, 1977.
328. **Lin, J. C.,** On microwave-induced hearing sensation, *IEEE Trans. Microwave Theory Tech.,* 25, 605, 1977.
329. **Gage, M.,** Behavior in rats after exposure to various power densities of 2450 MHz microwaves, *Neurobehav. Toxicol.,* 1, 137, 1979.
330. **Galloway, W. D.,** Microwave dose-response relationships on two behavioral tasks, *Ann. N.Y. Acad. Sci.,* 247, 410, 1975.
331. **Justesen, D. R. and King, N. W.,** Behavioral effects of low level microwave irradiation in the closed space situation, in Biological Effects and Health Implications of Microwave Radiation, Cleary, S. F., Ed., Symp. Proc., Public Health Service BRH/DBE 70-2, U.S. Department of Health, Education and Welfare, 1970, 154.
332. **Hunt, E. L., King, N. W., and Phillips, R. D.,** Behavioral effects of pulsed microwave adiation, *Ann. N.Y. Acad. Sci.,* 247, 440, 1975.
333. **Lobanova, Y. A.,** Survival and development of animals at various intensities and duration of SHF action, *Tr. Gig. Prof. AMN SSR,* 1, 61, 1960.
334. **Lebovits, R. M.,** Pulse modulated and continuous wave microwave radiation yield equivalent changes in operant behavior of rodents, *Physiol. Behav.,* 30, 891, 1983.
335. **Gage, M. I. and Guyer, W. M.,** Interaction of ambient temperature and controlled behavior in the rat, *Radio Sci.,* 17(5S), 179, 1982.
336. **deLorge, J. O.,** Effects of Microwaves on Animal Operant Behavior, Rep. No. NAMRL-1285, Naval Aerospace Medical Research Lab, Pensacola, FL, 1982.
337. **deLorge, J. O.,** Operant behavior and colonic temperature of Rhesus monkeys, Macaca mulatta, exposed to microwaves at frequencies above and near whole-body resonance, Rep. No. NAMRL-1289, Naval Aerospace Medical Research Lab., Pensacola, FL, 1983.
338. **Liddle, C. G., Putnam, J. P., Lewter, O. L., Lewis, J. Y., Bell, B., West, M. W., and Stead, A.,** Effect of 9.6 GHz pulsed microwaves on the Orb Web spinning ability of the Cross Spider, *Araneus diadematus, Bioelectromagnetics,* 7, 101, 1986.
339. **Lin, J. C., Guy, A. W., and Caldwell, L. T.,** Thermographic and behavioral studies of rats in the near field of 918-MHz radiations, *IEEE Trans. Microwave Theory Tech.,* 25, 833, 1977.
340. **Diachenko, J. A., and Milroy, W. C.,** The Effects of High Power Pulsed and Low Level CW Microwave Radiation on an Operant Behavior in Rats, Naval Surface Weapons Center, Dahlgren Laboratory, Dahlgren, VA, 1975.
341. **Thomas, J. R., Finch, E. D., Fulk, D. W., and Burch, L. S.,** Effects of low level microwave radiation on behavioral baselines, *Ann. N.Y. Acad. Sci.,* 247, 425, 1975.
342. **Roberti, B., Heebels, G. H., Hendricx, J. C. M., De Greef, A. H. A. M., and Wolthuis, O. L.,** Preliminary investigations of the effects of low-level microwave adiation on spontaneous motor activity in rats, *Ann. N.Y. Acad. Sci.,* 247, 417, 1975.
343. **Galloway, W. D. and Waxler, M.,** Interaction between microwave and neuroactive compounds, in Symp. on Biological Effects and Measurement of Radio Frequency/Microwaves, Hazzard, D., Ed., Publ. (FDA) 77-8026, U.S. Department of Health, Education and Welfare, Rockville, M, 1977, 62.
344. **D'Andrea, J. A., DeWitt, J. R., Gandhi, O. P., Stensaas, S., Lords, J. L., and Nielson, H. C.,** Behavioral and physiological effects of chronic 2,450-MHz microwave irradiation of the rat at 0.5 mW/cm^2, *Bioelectromagnetics,* 7, 45, 1986.
345. **D'Andrea, J. A., DeWitt, J. R., Emmerson, R. Y., Bailey, C., Stensaas, S., and Gandhi, O. P.** Intermittent exposure of rats to 2450 MHz microwaves at 2.5 mW/cm^2: behavioral and physiological effects, *Bioelectromagnetics,* 7, 315, 1986.
346. **DeWitt, J. R., D'Andrea, J. A., Emmerson, R. Y., and Gandhi, O. P.,** Behavioral effects of chronic exposure to 0.5 mW/cm^2 of 2,450-MHz microwaves, *Bioelectromagnetics,* 8, 149, 1987.

347. **Galvin, M. J., Tilson, H. A., Mitchell, C. L., Peterson, J., and McRee, D. I.,** Influence of pre- and postnatal exposure of rats to 2.45-GHz microwave radiation on neurobehavioral function, *Bioelectromagnetics,* 7, 57, 1986.
348. **Mitchell, C. L., McRee, D. I., Peterson, N. J., and Tilson, H. A.,** Some behavioral effects of short-term exposure of rats to 2.45-GHz microwave radiation, *Bioelectromagnetics,* 9, 259, 1988.
349. **Lebovitz, R. M.,** Pulse modulated and continuous wave microwave radiation yield equivalent changes in operant behavior of rodents, *Physiology & Behavior,* 30, 391, 1983.
350. **Akyel, Y., Hunt, E. L., Gambril, C., Vargas, Jr. C.,** Immediate post-exposure effects of high-peak-power microwave pulses on operant behavior of Wistar rats, *Bioelectromagnetics,* 12, 183, 1991.
351. **deLorge, J.,** Disruption of behavior in mammals of three different sizes exposed to microwaves: extrapolation to larger mammals, in 1978 Symp. on Electromagnetic Fields in Biological Systems, Stuchly, S. S., Ed., Ottawa, 1978, 215.
352. Safety Levels with Respect to Radio Frequency Electromagnetic Fields, 300 kHz to 100 GHz, *American National Standards Institute,* Report No. ANSI C95.1-1982, New, The Institute of Electrical and Electronic Engineers, Inc.
353. Biological Effects and Exposure Criteria for Radiofrequency Electromagnetic Fields, *National Council on Radiation Protection and Measurement,* Report No. 86, Bethesda, MD, National Council on Radiation and Measurements.
354. **Thomas, J. R., Schrot, J., and Banvard, R. A.,** Comparative effects of pulsed and continuous-wave 2.8-GHz microwaves on temporally defined behavior, *Bioelectromagnetics,* 3, 227, 1982.
355. **D'Andrea, J. A., Cobb, B. L., and de Lorge, J. O.,** Lack of behavioral effects in the Rhesus monkey: high peak microwave pulses at 1.3 GHz, *Bioelectromagnetics,* 10, 65, 1989.
356. **Hjeresen, D. L., Hoeberling, R. F., Kinross-Wright, J., Umbarger, K. O.,** Behavioral Effects of 1300-MHz High- Peak-power-Microwave Pulsed Irradiation, USAFSAM-TR-90-6, United States Air Force School of Aerospace Medicine, Brooks Air Force Base, TX, August 1990.
357. **Hjeresen, D. L., and Umbarger, K. O.,** Lack of behavioral effects of high-peak-power microwave pulses from an axially extracted virtual cathode oscillator, USAFSAM-TR-89-24, United States Air Force School of Aerospace Medicine, Brooks Air Force Base, TX, November 1989.
358. **Akyel, Y., Belt, M., Raslear, T. G., and Hammer, R. M.,** The effects of high-peak power pulsed microwaves on treadmill performance. In Blank, M. Ed., *Electricity and Magnetism in Biology and Medicine,* San Francisco, San Francisco Press, 1993, 668.
359. **Raslear, T. G., Akyel, Y., Bates, F., Belt, M., and Lu, S.-T.,** Temporal bisection in rats: The effects of high-peak power pulsed microwave irradiation, *Bioelectromagnetics,* 14, 459, 1993.
360. **Raslear, T., Akyel, Y., Serafini, R., Bates, F., and Belt, M.,** Memory consolidation in the rat following high-power microwave irradiation, Proceedings of the 13th Annual International Conference of the IEEE Engineering in Medicine & Biology Society, 13, 958, 1991.
361. **Raslear, T. G., Akyel, Y., Serafini, R., Bates, F., and Belt, M.,** Food demand and circadian rhythmicity following high-peak pulsed microwave radiation, *Proceedings of the Annual International Conference of the IEEE Engineering in Medicine and Biology Society,* 13, 962, 1991.
362. **Stolwijk, J. A. J.,** Responses to the thermal environment, *Fed. Proc., Fed. Am. Soc. Exp. Biol.,* 36, 1655, 1977.
363. **Stern, S., Margolin, L., Weiss, B., Lu, S.-T., and Michaelson, S. M.,** Microwaves: effect on thermoregulatory behavior in rats, *Science,* 206, 1198, 1979.
364. **Adair, E. R. and Adams, B. A.,** Adjustments in metabolic heat production by squirrel monkeys exposed to microwaves, *J. Appl. Physiol.,* 52, 1049, 1982.
365. **Lotz, W. G. and Michaelson, S. M.,** Temperature and corticosterone relationship in microwave exposed rats, *J. Appl. Physiol.,* 44, 476, 1963.
366. **Hamilton, C. L.,** Interactions of food intake and temperature regulation in the rat, *J. Comp. Physiol. Psychol.,* 56, 476, 1963.
367. **Bligh, J.,** Physiologic responses to heat, in *Fundamental and Applied Aspects of Nonionizing Radiation,* Michaelson, S. et al., Eds., Plenum Press, New York, 1975, 143.
368. **Adair, E. R., Adams, B. W., and Akel, G. M.,** Minimal changes in hypothalamic temperature accompany microwave-induced alteration of thermoregulatory behavior, *Bioelectromagnetics,* 5, 13, 1984.
369. **Corbit, J. D.,** Thermal motivation, In "Neural Control of Motivated Behavior." A report based on a NRP work session, *Neurosci. Res. Prog. Bull.,* 11, 4, 1973.
370. **Smialowicz, R. J., Riddle, M. M., Brugnolotti, P. L., Rogers, R. R., and Comptom, K. L.,** Detection of microwave heating in 5-hydroxytryptamine-induced hypothermic mice, *Radiat. Res.,* 88, 108, 1981.
371. **Adair, E. R.,** Microwaves and thermoregulation, in USAR Radiofrequency Radiation Bioeffects Research Program-A Review, Mitchell, J. C., Ed., Review 4-81, USAF School of Aerospace Medicine, San Antonio, TX, 1981.
372. **Adair, E. R. and Adams, B. W.,** Microwaves induce peripheral vasodilation in squirrel monkey, *Science* 207, 131, 1980.

373. **Lu, S. T., Lotz, W. G., and Michaelson, S. M.,** Advances in microwave-induced neuroendocrine effects: the concept of stress, *Proc. IEEE,* 68, 73, 1980.
374. **Petrov, I. R. and Syngayevskaya, V. A.,** Endocrine glands, in *Influence of Microwave Radiation on the Organism of Man and Animals,* Petrov, I. R., Ed., (NASA TT F0-708), Meditsina Press, Leningrad, 1970, 31.
375. **Demokidova, N. K.,** The effects of radiowaves on the growth of animals, in Biological Effects of Radiofrequency Electromagnetic Fields, Gordon, A. V., Ed., U.S. Joint Publications Research Service No. 63321, Arlington, VA, 1974, 237.
376. **Lenko, J., Dolatowski, A., Gruszecki, L., Klajman, S., and Januszkiewicz, L.,** Effect of 10-cm radar waves on the level of 17-ketosteroids and 17-hydroxycorticosteroids in the urine of rabbits, *Pregl. Lek.,* 22, 296, 1966.
377. **Mikolajczyk, H.,** Hormone reactions and changes in endocrine glands under influence of microwaves, *Med. Lotnicza,* 39, 39, 1972.
378. **Parker, L. N.,** Thyroid suppression and adrenomedullary activation by low-intensity microwave radiation, *Am. J. Physiol.,* 224, 1388, 1973.
379. **Lu. S.-T. Pettit. S., Lu, S.-J., and Michaelson, S. M.** Effects of microwaves on the adrenal cortex, *Radiat. Res.,* 107, 234, 1986.
380. **Milroy, W., C. and Michaelson, S. M.,** Thyroid pathophysiology of microwave radiation, *Aerosp. Med.,* 43, 1126, 1972.
381. **Baranski, S., Ostrowski, K., and Stodolnik, Baranska, W.,** Functional and morphological studies of the thyroid gland in animals exposed to microwave irradiation, *Acta Physiol. Pol.,* 23, 1029, 1972.
382. **Magin, R. L., Lu, S.-T., and Michaelson, S. M.,** Stimulation of dog thyroid by local application of high intensity microwaves, *Am. J. Physiol.* 233, E363, 1977.
383. **Magin, R. L., Lu, S.-T., and Michaelson, S. M.,** Microwave heating effect on the dog thyroid, *IEEE Trans. Biomed. Eng.,* 24, 522, 1977.
384. **Vetter, R. J.,** Neuroendocrine response to microwave irradiation, *Proc. Natl. Electron. Conf.,* 30, 237, 1975.
385. **Lu, S.-T., Lebda, N. A., Michaelson, S. M., Pettit, S., and Rivera, D.,** Thermal and endocrinological effects of protracted irradiation of rats by 2450 MHz microwaves, *Radio Sci.,* 12(6S), 145, 1977.
386. **Michaelson, S. M., Thomson, R. A. E., and Howland, J. W.,** Physiologic aspects of microwave irradiation of mammals, *Am. J. Physiol.,* 201, 351, 1961.
387. **McLees, B. D., and Finch, E. D.,** Analysis of the Physiologic Effects of Microwave Radiation, U.S. Naval Medical Research Institute, Proj. MF12 24.015-0001B, Rep. No. 3, Bethesda, MD, 1971.
388. **Michaelson, S. M.,** Biological effects of microwave exposure, Biological Effects and Health Implications of Microwave Radiation, Symp. Proc., Cleary, S. F., Ed., Public Health Service BRH/DBE 70-2, U.S. Department of Health, Education and Welfare, 1970, 35.
389. **Lu, S.-T., Lebda, N., Pettit, S., and Michaelson, S. M.,** Delineating acute neuroendocrine responses in microwave-exposed rats, *J. Appl. Physiol. Respir. Environ. Exercise Physiol,* 48(6), 927, 1980.
390. **Abhold, R. H., Ortner, M. J., Galvin, M. J., and McRee, D. I.,** Studies on acute *in vivo* exposure of rats to 2450-MHz microwave radiation. II. Effects on thyroid and adrenal axes hormones, *Radiat. Res., 88,* 448, 1981.
391. **Michaelson, S. M.,** Physiologic regulation in electromagnetic fields, *Bioelectromagnetics,* 3, 91, 1982.
392. **Sacchikova, M. N.,** Clinical manifestations of reactions to microwave irradiation in various occupational groups, in *Biological Effects and Health Hazards of Microwave Irradiation,* Warsaw, Poland, Polish Medical Publishers, 1973, 261.
393. **Presman, A. S. and Levitina, N. A.,** Nonthermal action of microwave on cardiac rhythm. Communiccation I. A study of the action of continuous microwaves, *Bull. Exp. Biol. Med.,* 53, 36, 1962.
394. **Basset, M. A. and Kenny, R. A.,** Cardiac response to whole-body heating, *Aviat. Space Environ. Med.,* 50, 387, 1979.
395. **Lu, S.-T., Brown, D. O., Johnson, C. E., Mathur, S. P., and Elson, E. C.,** Abnormal cardiovascular responses induced by localized high power microwave exposure, *IEEE Transactions on Biomedical Engineering,* 39, 484, 1992.
396. **Jauchem, J. R., Frei, M. R. and Heinmets, F.,** Heart rate changes to 5.6 GHz radiofrequency radiation: relation to average power density, *Proc. Soc. Exp. Biol. Med.,* 177, 383, 1984.
397. **Rasbury, J. A. and Seaman,** Transient cardiovascular changes in rats with microwave exposure of carotid sinus region, Proceedings of the 14th Annual IEEE EMBS Conference, Paris, Oct 1992, p 292.
398. **Cooper, T., Pinakatt, T., and Richardson, A. W.,** Effect of microwave induced hyperthermia on the cardiac output of the rat, *Physiologist,* 4, 21, 1961.
399. **Cooper, T., Jellinek, M., Pinakatt, T., Richardson, A. W., and Cooper, T.,** Effects of adrenalectomy, vagotomy and ganglionic blockade on the circulatory response to microwave hyperthermia, *Aerosp. Med.,* 33, 794, 1962.
400. **Pinakatt, T., Cooper, T., and Richardson, A. W.,** Control of ouabain on the circulatory response to microwave hyperthermia in the rat, *Aerosp. Med.,* 34, 497, 1963.

401. **Pinakatt, T., Richardson, A. W., and Cooper, T.,** The effect of digitoxin on the circulatory response of rats to microwave irradiation, *Arch. Int. Pharmacodyn, Ther.,* 156, 151, 1965.
402. **Goncharova, N. N., Karamyshev, V. B., and Maksimenko, N. V.,** Occupational hygiene problems in working with ultrashort-wave transmitters used in TV and radio broadcasting, *Gig. Tr. Prof. Zabol.,* 10, 10, 1966.
403. **Sadchikova, M. N. and Orlova, A. A.,** Clinical picture of the chronic effects of electromagnetic microwaves, *Ind. Hyg. Occup. Dis. (U.S.S.R.),* 2, 16, 1958.
404. **Michaelson, S. M., Thomson, R. A. E., Tamami, M. Y. E., Seth, H. S., and Howland, J. W.,** Hematologic effects of microwave exposure, *Aerosp. Med.,* 35, 824, 1964.
405. **Michaelson, S. M., Thomson, R. A. E., and Howland, J. W.,** Comparative studies on 1285 and 2800 mc/sec pulsed microwaves, *Aerosp. Med.,* 36, 1059, 1965.
406. **Michaelson, S. M., Thomson, R. A. E., and Howland, J. W.,** Biological Effects of Microwave Exposure, Griffiss Air Force Base, Rome Air Development Ctr. (ASTIA Doc. No. AD 824-242), Rome, NY, 1967.
407. **Kitsovskaya, I. A.,** The effect of centimeter waves of different intensities on the blood and hemopoietic organs of white rats, *Gig. Tr. Prof. Zabol.,* 8, 14, 1964.
408. **Baranski, S. and Czerski, P.,** Investigations of the behavior of corpuscular blood constituents in persons exposed to microwaves, *Lek. Woisk.,* 42, 903, 1966.
409. **Baranski, S.,** Effect of chronic microwave irradiation on the blood forming system of guinea pigs and rabbits, *Aerosp. Med.,* 42, 1196, 1971.
410. **Baranski, S.,** Effect of microwaves on the reactions of the white blood cell system, *Acta Physiol. Pol.,* 23, 685, 1972.
411. **Sacchitelli, F. and Sacchitelli, G.,** Protection of personnel exposed to radar microwaves, *Folia Med. (Naples),* 43, 1219, 1960.
412. **Vacek, D. R., A.,** Effect of high-frequency electromagnetic field upon haemopoietic stem cells in mice, *Folia Bio. (Praha),* 18, 292, 1978.
413. **Czerski, P., Paprocka-Slonka, E., Siekierzynski, M., and Stolarska, A.,** Influence of microwave radiation on the hematopoietic system, in *Biologic Effects and Health Hazards of Microwave Radiation,* Czerski, P. et al., Eds., Polish Medical Publishers, Warsaw, 1974, 67.
414. **Czerski, P., Paprocka-Slonka, E., and Stolarska, A.,** Microwave irradiation and the circadian rhythm of bone marrow cell mitosis, *J. Microwave Power,* 9, 31, 1974.
415. **Ivanov, A. I.,** Changes of phagocytic activity and mobility of neutrophils under the influence of microwave fields, in Summaries of Reports, Questions of the Biological Effect of a SHF-UHF Electromagnetic Field, Kirov Order of Lenin Military Medical Academy, Leningrad, 1972, 24.
416. **Hyde, A. S. and Friedman, J. J.,** Some effects of acute and chronic microwave irradiation of mice, in *Thermal Problems in Aerospace Medicine,* Hardy, J. D., Ed., Unwin, Surrey, England, 1968, 163.
417. **Miro, L., Loubiere, R., and Pfister, A.,** Effects of microwaves on the cell metabolism of the reticuloendothelial system, in *Biologic Effects and Health Hazards of Microwave Radiation,* Czerski, P. et al., Eds., Polish Medical Publishers, Warsaw, 1974, 89.
418. **Serdiuk, A. M.,** Biological effect of low-intensity ultrahigh frequency fields, *Vrach. Delo,* 11, 108, 1969.
419. **Spalding, J. F., Freyman, R. W., and Holland, L. M.,** Effects of 800 MHz electromagnetic radiation on body weight, activity, hematopoiesis and life span in mice, *Health Phys.,* 20, 421, 1971.
420. **Michaelson, S. M.,** Effects of exposure to microwaves: problems and perspectives. *Environ. Health Perspect.,* 8, 133, 1974.
421. **Smialowicz, R. J., Riddle, M. M., Brugnolotti, P. L., Sperazza, J. M., and Kinn, J. B.,** Evaluation of lymphocyte function in mice exposed to 2450 MHz (CW) microwaves, in *Proc. 1978 Symp. on Electromagnetic Fields in Biological Systems,* Stuchly, S. S. Ed., Ottawa, Can., 1979, 122.
422. **Liburdy, R. P.,** Effects of radio-frequency radiation on inflammation, *Radio Sci.,* 12(6S), 179, 1977.
423. **Wiktor-Jedrzejczak, W., Ahmed, A., Sell, K. W., Czerski, P., and Leach, W. M.,** Microwaves induce an increase in the frequency of complement receptor-bearing lymphoid spleen cells in mice, *J. Immunol.,* 118, 1499, 1977.
424. **Wiktor-Jedrzejczak, W., Ahmed, A., Czerski, P., Leach, W. M., and Sekk, K. W.,** Immunologic response of mice of 2450 MHz microwave radiation: overview of immunology and empirical studies of lymphoid spleen cells, *Radio Sci.,* 12(6S), 209, 1977.
425. **Smialowicz, R. J.,** The effect of microwaves (2450 MHz) on lymphocyte blast transformation *in vitro,* In Biological Effects of Electromagnetic Waves, Johnson, C. C. and Shore, M. L., Eds., Publ. (FDA) 77-8010, U.S. Department of Health, Education and Welfare, Rockville, MD, 1976, 472.
426. **Stodoolnik-Baranska, W.,** The effects of microwaves on human lymphocyte cultures, in *Biologic Effects and Health Hazards of Microwave Radiation,* Czerski, P. et al., Eds., Polish Medical Publishers, Warsaw, 1974, 189.
427. **Schlagel, C. J., Sulek, K., Ho, H. S., Leach, W. M., Ahmed, A., and Woody, J. N.,** Biological effects of

microwave exposure. II Studies on the mechanisms controlling susceptibility to microwave-induced increases in complement receptor positive spleen cells, *Bioelectromagnetics,* 1, 405, 1980.

428. **Smialowicz, R. J., Brugnolotti, P. L., and Riddle, M. M.,** Complement receptor positive spleen cells in microwave (2450 MHz) irradiated mice, *J. Microwave Power,* 16, 73, 1981.
429. **Yang, H. K., Cain, C. A., Lockwood, J., and Tompkins, W. A.,** Effects of microwave exposure on the hamster immune system. I. Natural killer cell activity, *Bioelectromagnetics,* 4, 123, 1983.
430. **Roberts, N. J., Lu, S.-T., and Michaelson, S. M.,** Human leukocyte functions and the U.S. safety standard for exposure to radiofrequency radiation, *Science,* 220, 318, 1983.
431. **Liburdy, R. P.,** Radiofrequency radiation alters the immune system: modulation of T- and B-lymphocyte levels and cell-mediated immunocompetence by hyperthermic radiation, *Radiat. Res.* 77, 34, 1979.
432. **Huang, A. T. and Mold, N. G.,** Immunologic and hematopoietic alterations by 2450-MHz electromagnetic radiation, *Bioelectromagnetics,* 1, 77, 1980.
433. **Wiktor-Jedrzejczak, W., Ahmed, A., Czerski, P., Leach, W. M., and Sell, K. W.,** Increase in the frequency of Fc receptor (FcR) bearing cells in the mouse spleen following a single exposure of mice to 2450 MHz microwaves, *Biomedicine,* 27, 250, 1977.
434. **Sulek, K., Schlagel, C. J., Wiktor-Jedrzecjzak, W., Ho, H. S., Leach, W. M., Ahmed, A., and Woody, J. N.,** Biologic effects of microwave exposure. I. Threshold conditions for the induction of the increase in complement positive (CR+) mouse spleen cells following exposure to 2450-MHz microwaves, *Radiat. Res.,* 83, 127, 1980.
435. **Liburdy, R. P.,** Radiofrequency radiation alters the immune system. II Modulation of *in vivo* lymphocyte circulation, *Radiat. Res.,* 83, 66, 1980.
436. **Hamrick, P. E., McRee, D. I., Thaxton, P., and Parkhurst, C. R.,** Humoral immunity of Japanese quail subjected to microwave radiation during embryogeny, *Health Phys.,* 33, 23, 1977.
437. **Szmigielski, S., Jeljaszewicz, J., and Wiranowska, M.,** Acute staphyloccal infections in rabbits irradiated with 3-GHz microwaves, *Ann. N.Y., Acad. Sci.,* 247, 305, 1975.
438. **Czerska, E. M., Elson, E. C., Davis, C. C., Swicord, M. L., and Czerski, P.,** Effects of continuous and pulsed 2450-MHz radiation of spontaneous lymphoblastoid transformation of human lymphocytes *in vitro, Bioelectromagnetics,* 13, 247, 1992.
439. **Cleary, S. F., Liu, L., and Merchant, R. E.,** *In vitro* lymphocyte proliferation induced by radiofrequency electromagnetic radiation under isothermal conditions, *Bioelectromagnetics,* 11, 47, 1990.
440. **Kiel, J. L., Wong, L. S., and Erwin, D. N.,** Metabolic effects of microwave radiation and convection heating on human mononuclear leukocytes, *Physiol. Chem. and Phys. and Med. NMR,* 18, 181, 1986.
441. **Huang, A. T., Engle, M. E., Elder, J. A., Kinn, J. B., and Ward, T. R.,** The effect of microwave radiation (2450 MHz) on the morphology and chromosomes of lymphocytes, *Radio Sci.,* 12(6S), 173, 1977.
442. **Zafra, C., Pena, J., de la Fuente, M.,** Effect of microwaves on the activity of murine macrophages *in vitro, Int. Archs. Allergy Appl. Immun.,* 85, 478, 1988.
443. **Czerski, P. and Siekierzynski, M.,** Analysis of occupational exposure to microwave radiation, in *Fundamental and Applied Aspects of Non-Ionizing Radiations,* Michaelson, S. M., Miller, M W., Magin, R., and Carstensen, E. L., Eds., Plenum Press, New York, 1975, 367.
444. **Roberts, N. J. Jr., Michaelson, S. M., and Lu, S.-T.,** Mitogen responsiveness after exposure of influenza virus-infected human mononuclear leukocytes to continuous or pulse-modulated radiofrequency radiation, *Radiat. Res.,* 110, 353, 1987.
445. **Gildersleeve, R. P., Thaxton, J. P., Parkhurst, C. R., Scott, T. R., Galvin, M. J., and McRee, D. I.,** Leukocyte numbers during the humoral and cell-mediated immune response of Japanese quail after microwave irradiation *in ovo, Comp. Biochem. Physiol.* 87A, 375, 1987.
446. **Hamrick, P. E.,** Thermal denaturation of DNA exposed 2450 MHz CW microwave radiation, *Radiat. Res.,* 56, 400, 1973.
447. **Budd. R. A. and Czerski, P.,** Modulation of mammalian immunity by electromagnetic radiation, *J. Microwave Power Electromagn. Energy,* 20, 217, 1985.
448. **Czerski, P.,** Microwave effects on the blood-forming system with particular reference to the lymphocyte, *Ann. N. Y. Acad. Sci.,* 247, 232, 1975.
449. **Gordon, Z. V.,** Biological Effect of Microwaves in Occupation Hygiene, Izd. Med., Leningrad, TT 70-50087, NASA TT F-633, 1970.
450. **Petrov, I. R., Ed.,** *Influence of Microwave Radiation on the Organism of Man and Animals,* (NASA TT F-708), Meditsina Press, Leningrad, 1979.
451. **Phillips, R. D., Hunt, E. L., Castro, R. D., and King, N. W.,** Thermoregulatory, metabolic and cardiovascular response of rats to microwaves, *J. Appl. Physiol.,* 38, 630, 1972.
452. **Liburdy, R. P.,** Effects of radiofrequency radiation on peripheral vascular permeability, (Abstr.), in Annu. Meet. Int. Union of Radio Science, Amherst, MA., 1976.
453. **Liburdy, R. P.,** Suppression of allograft rejection by whole-body microwave hyperthermia, *Fed. Am. Soc. Exp. Biol.,* Fed. Proc., (Abstr.), 37, 1281, 1978.

454. **Atkins, E. and Bodel, P.,** Fever, *N. Engl. J. Med.,* 286, 27, 1972.
455. **Bennett, I. L., Jr. and Nicastri, A.,** Fever as a mechanism of resistance, *Bacteriol. Rev.,* 24, 16, 1960.
456. **Kluger, M. J., Ringler, D. H., and Anver, M. R.,** Fever and survival, *Science,* 188, 166, 1975.
457. **Frenkel, J. K. and Caldwell, S. A.,** Specific immunity and nonspecific resistance to infection: listeria, protozoa, and viruses in mice and hamsters, *J. Infect. Dis.,* 131, 201, 1975.
458. **Mandell, G. L.,** Effect of temperature on phagocytosis by human polymorphonuclear neutrophils, *Infect. Immunity,* 12, 221, 1975.
459. **Roberts, N. J., Jr. and Steigbigel, R. T.,** Hyperthermia and human leukocyte functions: effects on response of lymphocytes to mitogen and antigen and bactericidal capacity of monocytes and neutrophils, *Infect. Immunity,* 18, 673, 1977.
460. **Frey, A. H.,** Auditory system response to RF energy, *Aerosp. Med.,* 32,, 1140, 1961.
461. **Frey, A. H.,** Human auditory system response to modulated electromagnetic energy, *J. Appl. Physiol.,* 17, 689, 1962.
462. **Chou, C.-K., Galambos, R., Guy, A. W., and Lovely, R. H.,** Cochlear microphonics generated by microwave pulses, *J. Microwave Power,* 17, 361, 1975.
463. **Chou, C.-K., Guy, A. W., and Galambos, R.,** Auditory perception of radiofrequency electromagnetic fields, *J. Acoust. Soc. Am.,* 71, 1321, 1982.
464. **Foster, K. R. and Finch, E. E.,** Microwave hearing; evidence for thermoacoustical auditory stimulation by pulsed microwaves, *Science,* 185, 256, 1974.
465. **Guy, A. W., Chou, C. K., Lin, J. C., and Christensen, D.,** Microwave induced acoustic effects in mammalian auditory systems and physical material, *Ann. N.Y. Acad. Sci.,* 247, 194, 1975.
466. **Lin, J. C.,** Microwave auditory effect—a comparison of some possible transduction mechanisms, *J. Microwave Power,* 11, 77, 1976.
467. **Carpenter, R. L. and van Ummersen, C. A.,** The action of microwave radiation on the eye, *J. Microwave Power,* 3, 3, 1968.
468. **Carpenter, R. L., Biddle, D. K., and van Ummersen, C. A.,** Biological effects of microwave adiation with particular reference to the eye, *Proc. Third Int. Conf. Med. Electron. (London),* 3, 401, 1960.
469. **Guy, A. W., Lin, J. C., Kramar, P. O., and Emery, A. F.,** Measurement of absorbed power patterns in the head and eyes of rabbits exposed to typical microwave sources, Proc. 1974 Conf. on Precision Electromagnetic Measurements, London, 1974, 255.
470. **Guy, A. W., Lin, J. C., Kramar, P. O., and Emergy, A. F.,** Quantitation of Microwave Radiation Effects on the Eyes of Rabbits at 2450 MHz and 918 MHz, Scientific Rep. No. 2, University of Washington, Seattle, January 1974.
471. **Guy, A. W., Lin, J. C., Kramar, P. O., and Emery, A. F.,** Effect of 2450 MHz radiation on the rabbit eye, *IEEE Trans. Microwave Theory Tech.,* 23, 492, 1976.
472. **Kramar, P. O., Emery, A. F., Guy, A. W., and Lin, J. C.,** The ocular effects of microwaves on hypothermic rabbits: a study of microwave cataractogenic mechanisms, *Ann. N.Y. Acad. Sci.,* 247, 155, 1975.
473. **Kramar, P. O., Guy, A. W., Emery, A. F., Lin, J. C., and Harris, C. A.,** Quantitation of Microwave Radiation Effects on the Eyes of Rabbits and Primates at 2450 MHz and 918, MHz, University of Washington, Bioelectromagnetics Research Laboratory Scientific Rep. No. 6, Seattle, WA, 1976.
474. **Appleton, B.,** Comment, *Ann. N.Y. Acad. Sci.,* 247, 133, 1975.
475. **Hirsch, S. E., Appleton, B., Fine, B. S., and Brown, P. V. K.,** Effects of repeated microwave irradiations to the albino rabbit eye, *Invest. Ophthalmol. Vis. Sci.,* 16, 315, 1977.
476. **Kinoshita, J. H., Merola, L. D., Dikmak, E. D., and Carpenter, R. L.,** Biochemical changes in microwave cataracts, *Doc. Ophthalmol.,* 20, 91, 1966.
477. **Merola, L. O. and Kinoshita, J. H.,** Changes in the ascorbic acid content in lenses of rabbit eyes exposed to microwave radiation, in *Biological Effects of Microwave Radiation, Vol. 1, Peyton, M. F., Ed., Plenum Press, New York, 1961, 285.*
478. **Weiter, J. J., Finch, E. D., Schultz, W., and Frattali, V.,** Ascorbic acid changes in cultured rabbit lenses after microwave irradiation, *Ann. N. Y. Acad. Sci.,* 247, 175, 1975.
479. **Taflove, A. and Brodwin, M. E.,** Computation of the electromagnetic fields and induced temperatures within a model of the microwave-irradiated human eye, *IEEE Trans. Microwave Theory Tech.,* 23, 88.
480. **Paulsson, L. E.,** Measurements of 0.915, 2.45, and 9.0 FHz absorption in the human eye, presented at the 6th European Microwave Conf., Rome, 1976.
481. **Kues, H. A. and Monahan, J. C.,** Microwave-induced changes to the primate eye, Johns Hopkins APL Technical Digest, 13, 244, 1992.
482. **Johnson, M. A., McLeod, D. S., Lutty, G. A., D'Anna, S. A., Perry C. R., Kues, H. A.,** Microwave induced retinal damage in the primate, *Investigative Ophthalmology and Visual Science,* 33, 1310, 1992.
483. **Creighton, M., Larsen, L. E., Steward-DeHaan, P. F., Jacobi, J. H., Sanwal, M., Baskerville, J. C., Bassen, H. E., Brown, D. O., and Trevithick, J. R.,** *In vitro* studies of microwave-induced cataract. Comparison of damage observed for continuous wave and pulsed microwaves, *Exp. Eye Res.,* 45, 357, 1987.

484. **McAfee, R. D., Longacre, A., Jr., Bishop, R. R., Elder, S. T., May, J. G., Holland, M. G., and Gordon, R.,** Absence of ocular pathology after repeated exposure of unanesthetized monkeys to 9.3 GHz microwaves, *J. Microwave Power,* 14, 41, 1979.
485. **Chou, C.-K., Guy, A. W., McDougall, J. A., and Han, L.-F.,** Effects of continuous and pulsed chronic microwave radiation on rabbits, *Radio Sci.,* 17(5S), 185, 1982.
486. **Guy, A. W., Kramar, P. W., Harris, C. A., and Chou, C.-K.,** Long-term 2450-MHz CW microwave irradiation of rabbits: methodology and evaluation of ocular and physiologic effects, *J. Microwave Power,* 15, 37, 1980.
487. **Lipman, R. M., Tripathi, B. J., and Tripathi, R. C.,** Cataracts induced by microwave and ionizing radiation, *Survey of Ophthalmology,* 33, 200, 1988.
488. **Dutta, S. K., Hossain, M. A., Ho, H. S., and Blackman, C. F., and Brusick, D. J.,** Effects of 8.6-GHz pulsed electromagnetic radiation on an *Escherichia coli* repair-deficient mutant, in *Electromagnetic Fields in Biological Systems,* Stuchly, S. S., Ed., IMPI, Edmonton, Alberta, Canada, 1979, 76.
489. **Blackman, C. F., Benane, S. G., Weil, C. M., and Ali, J. S.,** Effects of nonionizing electromagnetic radiation on single-cell biologic systems, *Ann. N.Y. Acad. Sci.,* 247, 352, 1975.
490. **Hamnerius, Y., Oloffson, H., Rasmuson, A., and Rasmuson, B.,** A negative test for mutagenic action of microwave radiation in *Drosophila melanogaster, Mutation Res.,* 68, 217, 1979.
491. **Anderstam, B., Hamnerius, Y., Hussain, S., and Ehrenberg, L.,** Studies of possible genetic effects in bacteria of high frequency electromagnetic fields, *Hereditas,* 98, 11, 1983.
492. **Mittler, S.,** Failure of chronic exposure to nonthermal FM radio waves to mutate *Drosophila, J. Heredity,* 68, 257, 1977.
493. **Smolyanskaya, A. A. and Vilenskaya, R. L.,** Effects of millimeter-band electromagnetic radiation on the functional activity of certain genetic elements of bacterial cells, *Trans. Usp Fiz. Nauk,* 110, 571, 1973.
494. **Mickey, G. H. and Koerting, L.,** Chromosome breakage in cultured Chinese hamster cells induced by radiofrequency treatment, *Environ. Mutagen Soc.,* 3, 25, 1970.
495. **Manikowska, E., Luciani, J. M., Servantie, B., Czerski, P., Obrenovitch, J., and Stahl, A.,** Effects of 9.4 GHz microwave exposure on meiosis in mice, *Experientia,* 35, 388, 1979.
496. **Blevins, R. D., Crenshaw, R. C., Houghland, A. E., and Clark, C. E.,** The effects of microwave radiation and heat on specific mutants of *Salmonella typhimurium* LT2, *Radiat. Res.,* 82, 511, 1980.
497. **McRee, D. I., MacNichols, G., and Livingston, G. K.,** Incidence of sister chromatid exchange in bone marrow cells of the mouse following microwave exposure, *Radiat. Res.,* 85, 304, 1981.
498. **Alam, M. T., Barthakur, N., Lambert, N. G., and Kasatiya, S. S.,** Cytological effects of microwave radiation in Chinese hamster cells *in vitro, Can. J. Genet. Cytol.,* 20, 23, 1978.
499. **McLees, B. D., Finch, E. D., and Albright, M. L.,** An examination of regenerating hepatic tissue subjected to radiofrequency irradiation, *J. Appl. Physiol.,* 32, 78, 1972.
500. **Dutta, S. K., Nelson, W. H., Blackman, C. F., and Brusick, D. J.,** Cellular effects in microbial tester strains caused by exposure to microwaves or elevated temperatures, *J. Environ. Pathol. Toxicol.,* 3, 195, 1980.
501. **Dardalhon, M., Averbeck, D., and Berteaud, A. J.,** Determination of thermal equivalent of millimeter microwaves in living cells, *J. Microwave Power,* 14, 307, 1979.
502. **Dardalhon, M., Averbeck, D., and Berteaud, A. J.,** Action des Ondes Centimetriques Seules ou Combinees avec les Rayons Ultra Violets sur les Cellules Eucaryotiques, in *URSI Int. Symp. Proc., Ondes Electromagnetiques et Biologie,* Berteaud, A. J. and Servantie, B., Eds., Paris, 1980, 17.
503. **Dardalhon, M., Averbeck, D., and Berteaud, A. J.,** Studies on possible genetic effects of microwaves in procaryotic and eucaryotic cells, *Radiat. and Environ. Biophys.* 20, 37, 1981.
504. **Allis, J. W. and Fromme, M. L.,** Activity of membrane-bound enzymes exposed to sinusoidally modulated 2450-MHz microwave radiation, *Radio Sci.,* 14 (6S), 85, 1979.
505. **Elder, J. A. and Ali, J. S.,** The effect of microwaves (2450 MHz) on isolated rat liver mitochondria, *Ann. N.Y. Acad Sci.,* 247, 251, 1975.
506. **Elder, J. A., Ali, J. S., Long, M. D., and Anderson, G. E.,** A coaxial air line microwave exposure system: respiratory activity of mitochondria irradiated at 2–4 GHz, in Biological Effects of Electromagnetic Waves, Publ. (FDA) 77–8010, Vol. 1, U.S. Department of Health, Education and Welfare, Rockville, MD, 1976, 352.
507. **Deichmann, W. B., Miale, J., and Landeen, K.,** Effect of microwave radiation on the hemopoietic system of the rat, *Toxicol. Appl. Phaarmacol.,* 6, 71, 1964.
508. **Paulsson, L.-E., Hamnerius, Y., and McLean, W. G.,** The effects of microwave radiation on microtubules and axonal transport, *Radiat. Res.,* 70, 212, 1977.
509. **Millar, D. B., Christopher, J. P., Hunter, J., and Yeandle, S. S.,** The effect of exposure of acetylcholinesterase to 2450-MHz microwave radiation, *Bioelectromagnetics,* 5, 165, 1984.
510. **Timm, R. R., Shilling, R. and Lange, D. G.,** Inhibition of ATPase activity in rat canicular membrane following chronic microwave irradiation, *Fed. Proc.,* 1985, 516 (abst.).
511. **Allis, J. W. and Fromme, M. L.,** Activity of membrane enzymes exposed to sinusoidally modulated 2450-MHz microwave radiation, *Radio Sci.* 14(6S), 85, 1979.

512. **Wang, X. H., Lou, Y. and Chiang, H.,** Effect of microwave exposure on acid phosphatase in WBC of living systems, *Eighth Ann. BEMS Meeting,* 1986, 76 (abst.).
513. **Chang, H. and Wang, M.,** Morphological change in mouse sperm following microwave exposure, *J. Bioelectricity,* 3, 367, 1984.
514. **Allis, J. W., Fromme, M. L., and Janes, D. E.,** Pseudosubstrate binding to ribonuclease during exposure to microwave radiation at 1.70 and 2.45 GHz, in Biological Effects of Electromagnetic Waves, Vol. 1, Johnson, C. C. and Shore, M. L., Eds., HEW Publ. (FDA) 77–8010, Department of Health, Education and Welfare, Rockville, MD, 1976, 366.
515. **Henderson, H. M., Hergenroeder, K., and Stuchly, S. S.,** Effect of 2450 MHz microwave radiation on horseradish peroxidase, *J. Microwave Power,* 10, 27, 1975.
516. **Lovely, R. H., Myers, D. E., and Guy, A. W.,** Irradiation of rats by 918-MHz microwaves at 2.5 mW/cm2: delineating the dose-response relationship, *Radio Sci.,* 12(6S), 139, 1977.
517. **Chamness, A. F., Scholes, H. R., Sexauer, S. W., and Frazer, J. W.,** Metal ion content of specific areas of the rat brain after 1600 MHz radiofrequency irradiation, *J. Microwave Power,* 11, 333, 1976.
518. **Ismailov, E. Sh.,** Effect of ultrahigh frequency electromagnetic radiation on electrophoretic mobility of erythrocytes (transl.), *Biophysics,*22, 510, 1978; *Biofizika,* 22, 493, 1977.
519. **Ismailov, E. Sh.,** Mechanism of effects of microwaves on erythrocyte permeability for potassium and sodium ions, *Biol. Nauki,* 3, 58, 1971; (English trans.: JPRS 72606, p. 38, January 12, 1979.)
520. **Hamrick, P. E. and Zinkl, J. G.,** Exposure of rabbit erythrocytes to microwave radiation, *Radiat. Res.,* 62, 164, 1975.
521. **Peterson, D. J., Partlow, L. M., and Gandhi, O. P.,** An investigation of the thermal and athermal effects of microwave irradiation on erythrocytes, *IEEE Trans. Biomed. Eng.,* 26, 428, 1979.
522. **Olcersst, R. B., Belman, S., Eisenbud, M., Mumford, W. W., and Rabinowitz, J. R.,** The increased passive efflux of sodium and rubidium from rabbit erythrocytes by microwave radiation, *Radiat. Res.,* 82, 244, 1980.
523. **Ho, H. S. and Edwards, W. P.,** The effect of environmental temperature and average dose rate of microwave radiation on the oxygen-consumption rate of mice, *Radiat. Environ. Biophys.,* 16, 325, 1979.
524. **Sanders, A. P., Schaefer, D. J., and Joines, W. T.,** Microwave effects on energy metabolism of rat brain, *Biolelectromagnetics,* 1, 171, 1980.
525. **Sanders, A. P., Joines, W. T., and Allis, J. W.,** Effect of continuous-wave, pulsed, and sinusoidal-amplitude-modulated microwaves on brain energy metabolism, *Bioelectromagnetics,* 6, 89, 1985.
526. **Phillips, R. D., Hunt, E. L., Castro, R. D., and King, N. W.,** Thermoregulatory metabolic and cardiovascular response of rats to microwaves, *J. Appl. Physiol.,* 38, 630, 1975.
527. **Adair, E. R. and Adams, B. W.,** Adjustments in metabolic heat production by squirrel monkeys exposed to microwaves, *J. Appl. Physiol.,* 52(4), 1049, 1982.
528. **Boggs, R. F., Sheppard, A. P., and Clark, A. J.,** Effects of 2450-MHz microwave radiation on human blood coagulation processes, *Health Phys.,* 22, 217, 1972.
529. **Cairnie, A. B., Hill, D. A., and Assenheim, H. M.,** Dosimetry for a study of effects of 2.45-GHz microwaves on mouse testes, *Bioelectromagnetics,* 1, 325, 1980.
530. **Muraca, G. J., Ferri, E. S., and Buchta, F. L.,** A study of the effects of microwave irradiation of the rat testes, in Biological Effects of Electromagnetic Waves, Vol. 1, Johnson, C. C. and Shore, M. L., Eds., DHEW Publ. (FDA) 77–8010, Department of Health, Education and Welfare, Rockville, MD, 1976, 484.
531. **Prausnitz, S. and Susskind, C.,** Effects of chronic microwave irradiation in mice, *IRE Trans. Biomed. Electron.,* 9, 194, 1962.
532. **Haidt, S. J. and McTighe, A. H.,** The effect of chronic, low-level microwave radiation on the testicles of mice, in 1973 IEEE-G-MIT Int. Microwave Symp., Maley, S. W., Ed., 1973, 324.
533. **Gunn, S. A., Gould, T. C., and Anderson, W. A. D.,** The effect of microwave radiation on morphology and function of rat testis, *Lab. Invest.,* 10(2), 301, 1961.
534. **Smialowicz, R. J., Ali, J. S., Berman, E., Bursian, S. J., Kinn, J. B., Liddle, C. G., Reiter, L. W., and Weil, C. M.,** Chronic exposure of rats to 100-MHz (CW) radiofrequency radiation: assessment of biological effects, *Radiat. Res.,* 86, 488, 1981.
535. **McAfee, R. D., Braus, R., Jr., and Fleming, J., Jr.,** The effect of 2450 MHz microwave irradiation on the growth of mice, *J. Microwaave Power,* 8, 111, 1973.
536. **Guillet, R. and Michaelson, S. M.,** The effect of repeated microwave exposure on neonatal rats, *Radio Sci.,* 12(6S), 125, 1977.
537. **Stavinoha, W. B., Medina, M. A., Frazer, J., Weintraub, S. T., Ross, D. H., Modak, A. L., and Jones, D. J.,** The effects 19 megacycle irradiation on mice and rat, in Biological Effects of Electromagnetic Waves, Vol 1, Johnson, C. C. and Shore, M. L., Eds., Publ. (FDA) 77–8010, Department of Health, Education and Welfare, Rockville, MD, 1975, 431.
538. **Hamrick, P. E. and McRee, D. I.,** Exposure of the Japanese quail embryo to 2.45-GHz microwave radiation during the second day of development, *J. Microwave Power,* 10, 211, 1975.

539. **Chernovetz, M. E., Justesen, D. R., and Levinson, D. M.,** Acceleration and deceleration of fetal growth of rats by 2450-MHz microwave radiation, in *Electromagnetic Fields, in Biological Systems,* Stuchly, S. S., Ed., IMPI, Edmonton, Alberta, Canada, 1979, 175.
540. **McRee, D. I. and Hamrick, P. E.,** Exposure of Japanese quail embryos to 2.45 GHz microwave radiation during development, *Radiat. Res.,* 71, 355, 1977.
541. **Berman, E., Ccarter, H. B., and House, D.,** Observations of rat fetuses after irradiation with 2450-MHz (CW) microwaves, *J. Microwave Power,* 16, 9, 1981.
542. **Johnson, R. B., Mizumori, S., and Lovely, R. H.,** Adult behavior deficit in rats exposed prenatally to 918-MHz microwaves, in Developmental Toxicology of Energy-Related Pollutants, Mahlum, D. C., Sikov, M. R., Hackett, P. L., and Andrew, F. D., Eds., DOE Symp. Ser. 47, Department of Energy, Washington, DC, 1978, 281.
543. **Shore, M. L., Felten, R. P., and Lamanna, A.,** The effect of repetitive prenatal low-level microwave exposure on development in the rat, in Symp. on Biological Effects and Measurement of Radio Frequency/ Microwaves, Hazzard, D. G., Ed., HEW Publication (FDA), 77–8026, Department of Health, Education and Welfare, Rockville, MD, 1977, 280.
544. **Goldstein, L. and Sisko, Z.,** A quantitative electroencephalographic study of the acute effects of X-band microwaves in rabbits, in *Biologic Effects and Health Hazards of Microwave Radiation,* Czerski, P. et al., Eds., Polish Medical Publishers, Warsaw, 1974, 128.
545. **Thomas, J. R., Burch, L. S., and Yeandle, S. S.,** Microwave radiation and chlordiazepoxide: synergistic effects on fixed-interval behavior, *Science,* 203, 1357, 1979.
546. **Thomas, J. R., Schrot, J., and Banvard, R. A.,** Behavioral effects of chlorpromazine and diazepam combined with low-level microwaves, *Neurobehav. Toxicol.,* 2, 131, 1980.
547. **Merritt, J. H., Hartzell, R. H., and Frazer, J. W.,** The effect of 1.6 GHz radiation on neurotransmitters in discrete areas of the rat brain, in Biological Effects of Electromagnetic Waves, Vol., 1, Johnson, C. C. and Shore, M. L. Eds., Hew Publ. (FDA) 77–8010, Department of Health, Education and Welfare, Rockville, MD, 1976, 290.
548. **Merritt, J. H., Chamness, A. F., Hartzell, R. H., and Allen, S. J.,** Orientation effects on microwave-induced hyperthermia and neurochemical correlates, *J. Microwave Power,* 12, 167, 1977.
549. **Zeman, G. H., Chaput, R. L., Glazer, Z. R., and Gershman, L. C.,** Gamma-aminobutyric acid metabolism in rats following microwave exposure, *J. Microwave Power,* 8, 213, 1973.
550. **Albert, E. N. and DeSantis, M.,** Histological observations on central nervous system, in Biological Effects of Electromagnetic Waves, Vol. 1, Johnson, C. C. and Shore, M. L., Eds., HEW Publ. (FDA) 77–8010, Department of Health, Education and Welfare, Rockville, MD, 1976, 299.
551. **Switzer, W. G. and Mitchell, D. S.,** Long-term effects of 2.45 GHz radiation on the ultrastructure of cerebral cortex and on hematologic profiles of rats, *Radio Sci.,* 12 (6S), 287, 1977.
552. **Frey, A. H. and Feld, S. R.,** Avoidance by rats of illumination with low power nonionizing electromagnetic energy, *J. Comp. Physiol. Psychol.,* 89, 183, 1975.
553. **Baranski, S.,** Histological and histochemical effect of microwave irradiation on the central nervous system of rabbits and guinea pigs, *Am. J. Phys. Med.,* 51, 182, 1972.
554. **Sutton, C. H. and Carol, F. B.,** Effects of microwave-induced hyperthermia on the blood-brain barrier of the rat, *Radio Sci.,* 14(6S), 329, 1979.
555. **Takashima, S., Onaral, B., and Schwan, H. P.,** Effects of modulated RF energy on the EEG of mammalian brains, *Radiat. Environ. Biophys.,* 16, 15, 1979.
556. **Bawin, S. M., Galavas-Medici, R. J., and Adey, W. R.,** Effects of modulated very high frequency fields on specific brain rhythms in cats, *Brain Res.,* 58, 365, 1973.
557. **Gordon, Z. V.,** Biological effect of microwaves in occupational hygiene, Israel Program for Scientific Translations, NASA TT F-633, TT 70–50087; NTIS Nj1-14632, 1970, Jerusalem, Israel.
558. **Austin, G. N. and Horvath, S. M.,** Production of convulsions in rats by high frequency electrical currents, *Am. J. Phys. Med.,* 33, 141, 1954.
559. **Snyder, S. H.,** The effect of microwave irradiation on the turnover rate of serotonin and norepinephrine and the effect of monoamine metabolizing enzymes, Final Report, Contract No. DADA 17–69–C–9144, (NTIS AD-729 161), U.S. Army Medical Research and Development Command, Washington, DC, 1971.
560. **Mitchell, D. S., Switzer, W. G., and Bronaugh, E. L.,** Hyperactivity and disruption of operant behavior in rats after multiple exposure to microwave radiation, *Radio Sci.,* 12(6S), 263, 1977.
561. **Moe, K. E., Lovely, R. H., Meyers, D. E., and Guy, A. W.,** Physiological and behavioral effects of chronic low level microwave radiation in rats, in Biological Effects of Electromagnetic Waves, Vol. 1, Johnson, C. C. and Shore, M. L., Eds., HEW Publ. (FDA) 77–8010, Department of Health, Education and Welfare, Rockville, MD, 1976, 248.
562. **D'Andrea, J. A., Gandhi, O. P., Lords, J. L., Durney, C. H., Johnson, C. C., and Astle, L.,** Physiological and behavioral effects of chronic exposure to 2450-MHz microwaves, *J. Microwave Power,* 14, 351, 1979.
563. **D'Andrea, J. A., Gandhi, O. P., Lords, J. L., Durney, C. H., Astle, L., Stensaas, L. J., and Schoenberg,**

A. A., Physiological and behavioral effects of prolonged exposure to 915-MHz microwaves, *J. Microwave Power,* 15, 123, 1980.

564. **Rudnev, M., Bokina, A., Eksler, N., and Navakatikyan, M.,** The use of evoked potential and behavioral measures in the assessment of environmental insult, in Multidisciplinary Perspectives, in Event-Related Brain Potential Research, Otto, d. A., Ed., EPA-600/9–77–043, U.S. Environmental Protection Agency, Research Triangle Park, NC, 1978, 444.
565. **Bermant, R. I., Reeves, D. L., Levinson, D. M., and Jusesen, D. R.,** Classical conditioning of microwave-induced hyperthermia in rats, *Radio Sci.,* 14(6S), 201, 1979.
566. **D'Andrea, J. A., Gandhi, O. P., and Kesner, R. P.,** Behavioral effects of resonant electromagnetic power absorption in rats, in Biological Effects of Electromagnetic Waves, Vol. 1, Johnson, C. C. and Shore, M. L., Eds., HEW Publ. (FDA) 77–8010, Department of Health, Educcation and Welfare, Rockville, MD, 1976, 257.
567. **deLorge, J. O.,** The effects of microwave radiation on behavior and temperature in rhesus monkeys, in Biological Effects of Electromagnetic Waves, Vol., 1, Johnson, C. C. and Shore, M. L., Eds., HEW Publ. (FDA) 77–8010, Department of Health, Education and Welfare, Rockville, MD, 1976, 158.
568. **deLorge, J.,** Disruption of behavior in mammals of three different sizes exposed to microwaves: extrapolation to larger mammals, in *Electromagnetic Fields in Biological Systems,* Stuchly, S. S., Ed., IMPI, Edmonton, Alberta, Canada, 1979, 215.
569. **deLorge, J.,** Operant behavior and rectal temperature of squirrel monkeys during 2.45-GHz microwave irradiation, *Radio Sci.,* 14(6S), 217, 1979.
570. **deLorge, J. and Ezell, C. S.,** Observing-responses of rats exposed to 1.28- and 5.62-GHz microwaves, *Bioelectromagnetics,* 1, 183, 1980.
571. **Scholl, D. M. and Allen, S. J.,** Skilled visual-motor performance by monkeys in a 1.2-GHz microwave field, *Radio Sci.,* 14(6S), 217, 1979.
572. **Thomas, J. R., Yeandle, S. S., and Burch, L. S.,** Modification of internal discriminative stimulus control of behavior by low levels of pulsed microwave radiation, in Biological Effects of Electromagnetic Waves, Vol. 1, Johnson, C. C. and Shore, M. L., Eds., HEW Publ. (FDA) 77–8010, Department of Health, Education and Welfare, Rockville, MD, 1976, 201.
573. **Schrot, J., Thomas, J. R., and Banvard, R. A.,** Modification of the repeated acquisition of response sequences in rats by low-level microwave exposure, *Bioelectromagnetics,* 1, 89, 1980.
574. **King, N. W., Justesen, D. R., and Clarke, R. L.,** Behavioral sensitivity to microwave irradiation, *Science,* 172, 398, 1971.
575. **Johnson, R. B., Meyers, D. E., Guy, A. W., Lovely, R. H., and Galambos, R.,** Discriminative control of appetitive behavior by pulsed microwave radiation in rats, in Biological Effects of Electromagnetic Waves, Vol. 1. Johnson, C. C. and Shore, M. L., Eds., HEW Publ. (FDA) 77–8010, Department of Health Education and Welfare, Rockville, MD, 1976, 238.
576. **Frey, A. H., Feld, S. R., and Frey, B.,** Neural function and behavior: defining the relationship, *Ann. N.Y. Acad. Sci.,* 247, 433, 1975.
577. **Hjeresen, D. L., Doctor, S. R., and Sheldon, R. L.,** Shuttle box side preference as mediated by pulsed microwave and conventional auditory cues, in *Electromagnetic Fields in Biological Systems,* Stuchly, S. S., Ed., IMPI, Edmonton, Alberta, Canada, 1979, 194.
578. **Monahan, J. C. and Ho, H. S.,** Microwave induced avoidance behavior in the mouse, in Biological Effects of Electromagnetic Waves, Vol. 1, Johnson, C. C. and Shore, M. L., Eds., HEW Publ. (FDA) 77–8010, Depaartment of Health, Education and Welfare, Rockville, MD, 1976, 174.
579. **Monahan, J. C. and Ho, H. S.,** The effect of ambient temperature on the reduction of microwave energy absorption by mice, *Radio Sci.,* 12(6S), 257, 1977.
580. **Gage, M. I., Berman, E., and Kinn, J. B.,** Videotape observation of rats and mice during an exposure to 2450-MHz microwave radiation, *Radio Sci.,* 14(6S), 227, 1979.
581. **Carroll, D. R., Levinson, D. M., Justesen, D. R., and Clarke, R. L.,** Failure of rats to escape from a potentially lethal microwave field, *Bioelectromagnetics,* 1, 101, 1980.
582. **Gage, M. I.,** Microwave irradiation and ambient temperature interact to alter rat behavior following overnight exposure, *J. Microwave Power,* 14, 389, 1979.
583. **Adair, E. R. and Adams, B. W.,** Microwaves modify thermoregulatory behavior in squirrel monkey, *Bioelectromagnetics,* 1, 1, 1980.
584. **Lu, S., Lebda, N., Pettit, S., and Michaelson, S. M.,** Microwave-induced temperature, corticosterone, and thyrotropin interrelationships, *J. Appl. Physiol.: Respir. Environ. Exercise Physiol.,* 50, 399, 1981.
585. **Mikolajczyk, H.,** Microwave-induced shifts of gonadotrophic activity in anterior pituitary gland of rats, in Biological Effects of Electromagnetic Waves, Vol. 1, Johnson, C. C. and Shore, M. L., Eds., Publ. (FDA) 77–8010, U.S. Department of Health, Education and Welfare, Rockville, MD, 1977, 377.
586. **Lotz, W. G. and Michaelson, S. M.,** Effects of hypophysectomy and dexamethasone on rat adrenal response to microwaves, *J. Appl. Physiol.: Respir. Environ. Exercise Physiol.,* 47, 1284, 1979.
587. **Birenbaum, L., Kaplan, I. T., Metlay, W., Rosenthal, S. W., and Zaret, M. M.,** Microwave and infra-

red effects on heart rate, respiration rate and subcutaneous temperature of the rabbit, *J. Microwave Power,* 10, 3, 1975.

588. **Kaplan, I. T., Metlay, W., Zaret, M. M., Birenbaum, L., and Rosenthal, S. W.,** Absence of heart-rate effects in rabbits during low-level microwave irradiation, *IEEE Trans, Microwave Theory Tech.,* 19, 168, 1971.
589. **Paff, G. H., Boucek, R. J., Nieman, R. E., and Deichmann, W. B.,** The embryonic heart subject to radar, *Anat. Rec.,* 147, 379, 1963.
590. **Hamrick, P. and McRee, D. I.,** The effect of 2450-MHz microwave irradiation on the heart rate of embryonic quail, *Health Phys.,* 38, 261, 1980.
591. **Liu, L. M., Rosenbaum. F. J., and Pickard, W. F.,** The insensitivity of frog heart rate to pulse modulated microwave energy, *J. Microwave Power,* 11, 225, 1976.
592. **Frey, A., H. and Seifert, E.,** Pulse modulated UHF illumination of the heart associated with change in heart rate, *Life Sci.,* 7, 505, 1968.
593. **Chou, C. K., Han, L. F., and Guy, A. W.,** Microwave radiation and heartbeat rate of rabbits, *J. Microwave Power,* 15, 87, 1980.
594. **Tinney, C. E., Lords, J. L., and Durney, C. H.,** Rate effects in isolated turtle hearts induced by microwave irradiation, *IEEE Trans. Microwave Theory Tech.,* 24, 18, 1976.
595. **Olsen, R. G., Lords, J. L. and Durney, C. H.,** Microwave-induced chronotropic effects of the isolated rat heart, *Ann. Biomed. Eng.,* 5, 395, 1977.
596. **Clapman, R. M. and Cain, C. A.,** Absence of heart rate effects in isolated frog heart irradiated with pulsed modulated microwave energy, *J. Microwave Power,* 10, 411, 1975.
597. **Deichmann, W. B., Bernal, E., Stephens, F., and Landeen, K.,** Effects on dogs of chronic exposure to microwave radiation, *J. Occup. Med.,* 5, 418, 1963.
598. **Djordjevic, Z. and Kolak, A.,** Changes in the peripheral blood of the rat exposed to microwave radiation (2400 MHz) in conditions of chronic exposure, *Aerosp. Med.,* 44, 1051, 1973.
599. **Djordjevic, Z., Lazarevic, N., and Djokovic, V.,** Studies on the hematologic effects of long-term, low dose microwave exposure, *Aviat. Space Environ. Med.,* 48, 516, 1977.
600. **McRee, D. I., Faith, R., McConnell, E. E., and Guy, A. W.,** Long-term 2450-MHz CW microwave irradiation of rabbits: evaluation of hematological and immunological effects, *J. Microwave Power,* 15, 45, 1980.
601. **Rotkovska, D. and Vacek, A.,** The effect of electromagnetic radiation on the hematopoietic stem cells of mice, *Ann. N.Y. Acad. Sci.,* 247, 243, 1975.
602. **Rotkovska, D. and Vacek, A.,** Modification of repair of X-irradiation damage of hemopoietic system of mice by microwaves, *J. Microwave Power,* 12, 119, 1977.
603. **Michaelson, S. M., Thomson, R. A. E., Odland, L. T., and Howland, J. W.,** The influence of microwaves on ionizing radiation exposure, *Aerosp. Med.,* 34, 111, 1963.
604. **Michaelson, S. M., Thomson, R. A. E., El Tamami, M. Y., Seth, H. W., and Howland, J. W.,** The hematologic effects of microwave exposure, *Aerosp. Med.,* 35, 824, 1964.
605. **Lappenbusch, W. L., Gillespie, L. J., Leach, W. M., and Anderson, G. E.,** Effect of 2450-MHz microwaves on the radiation response of X-irradiated Chinese hamsters, *Radiat. Res.,* 54, 294, 1973.
606. **Smialowicz, R. J., Weil, C. M., Kinn, J. B., and Elder, J. A.,** Exposure of rats to 425-MHz (CW) radiofrequency radiation: effects on lymphocytes, *J. Microwave Power,* 17, 211, 1982.
607. **Pazderova-Vejlupkova, J. and Josifkp. M.,** Changes in the blood count of growing rats irradiated with a microwave pulse field, *Arch. Environ. Health,* 34, 44, 1979.
608. **Shah, S. A. and Dickson,J. A.,** Effect of hyperthermia on the immune response of normal rabbits, *Cancer Res.,* 38, 3518, 1978.
609. **Szmigielski, S., Janiak, M., Hryniewicz, W., Jeljaszewicz, J., and Pulverer, G.,** Local microwave hyperthermia (43°C) and stimulation of the macrophage and T-lymphocyte systems in treatment of Guerin epitheliuma in rats, *Z. Krebsforsch.,* 91, 35, 1978.
610. **Marmor, J. B., Hahn, N., and Hahn, G. M.,** Tumor cure and cell survival after localized radiofrequency heating, *Cancer Res.,* 37, 879, 1977.
611. **Szmigielski, S., Pulverer, G., Hryniewicz, W., and Janiak, M.,** Inhibition of tumore growth in mice by microwave hyperthermia, streptolysin S and colcemide, *Radio Sci.,* 12(6S), 185, 1977.
612. **Roszkowski, W., Wrembel, J. K., Roszkowski, K., Janiak, M., and Szmigielski, S.,** The search for an influence of whole-body microwave hyperthermia on anti-tumor immunity, *J. Cancer Res. Clin. Oncol.,* 96, 311, 1980.
613. **Shah, S. A. and Dickson, J. A.,** Effect of hyperthermia on the immune response of normal rabbits, *Cancer Res.,* 38, 3518, 1978.
614. **Lin, J. C., Ottenbreit, M. J., Wang, S., Inoue, S., Bollinger, R. O., and Fracassa, M.,** Microwave effects on granulocyte and macrophage precursor cells of mice *in vitro, Radiat. Res.,* 80, 292, 1979.
615. **Stodolnik-Baranska, W.,** Lymphoblastoid transformation of lymphocytes *in vitro* after microwave irradiation, *Nature,* 214, 102, 1967.

616. **Hamrick, P. E. and Fox, S. S.**, Rat lymphocytes in cell culture exposed to 2450-MHz (CW) microwave radiation, *J. Microwave Power,* 12, 125, 1977.
617. **Lin, J. C. and Peterson, W. D.**, Cytological effects of 2450 MHz CW microwave radiation, *J. Bioeng.*, 1, 471, 1977.
618. **Mayers, C. P. and Habeshaw, J. A.**, Depression of phagocytosis: a nonthermal effect of microwave radiation as a potential hazard to health, *Int. J. Radiat. Biol.*, 24, 449, 1973.
619. **Szmigielski, S.**, Effect of 10-cm (3 GHz) electromagnetic radiation (microwaves) on granulocytes *in vitro, Ann. N.Y. Acad. Sci.*, 247, 275, 1975.
620. **Goldblith, S. A. and Wang, D. I. C.**, Effect of microwaves on *Escherichia coli* and *Bacillus subtilus, Appl. Microbiol.*, 15, 1371, 1967.
621. **Moore, H. A., Raymond, R., Fox, M., and Galsky, A. G.**, Low-intensity microwave radiation and the virulence of *Agrobacterium tumefaciens* strain B6, *Appl. Environ. Microbiol.*, 37, 127, 1979.
622. **Toler, J., Popovic, V. Bonasera, S., Popovic, P., Honeycutt,. C, and Sgoutas D.**, Long-term study of 435 MHz radio-frequency radiation on blood-borne end points in cannulated rats, Part II: methods, results, and summary, *J. Microwave Power EM Energy,* 23, 105, 1988.
623. **Roberts, N. J. Jr., Michaelson, S. M., and Lu, S.-T.**, Exposure of human mononuclear leukocytes to microwave energy pulse modulated at 16 or 60 Hz, *IEEE Transactions on Microwave Theory and Techniques,* MTT-32, 803, 1984.
624. **Roberts, N. J., Jr., Lu, S.-T., and Michaelson, S. M.**, Hyperthermia and human leukocyte functions: DNA, RNA, and total protein synthesis after exposure to $<41°$ or $>42.5°$ hyperthermia, *Cancer Research,* 45, 3076, 1985.

Chapter 12

MODULATED FIELDS AND "WINDOW" EFFECTS

Elliot Postow and Mays L. Swicord

CONTENTS

I. INTRODUCTION

As the body of scientific literature describing the effects of electromagnetic fields increases, so does the variety and complexity of reported effects. Because much bioelectromagnetics research has been motivated by the need to delineate hazardous conditions, experiments often were designed to determine a threshold, the unique value of an experimental variable above and below which the experimental outcomes are different (or begin to differ). Since it has been generally assumed that for all values below the threshold the experimental outcome will be the same, protocols were designed in which the variable under study differed significantly (often by as much as a factor of ten) from experimental series to experimental series. Experiments aimed at documenting a threshold are appropriate for the study of effects that depend on field strength, or on the rate or amount of energy absorbed. However, these experiments are not adequate to investigate most effects that depend upon specific modulation conditions. Of particular concern are effects that are reported to vary nonmonotonically over very small regions of the independent variable. As expected, experiments directed toward establishing a threshold did not uncover such effects. However, when frequency, field strength and/or modulation conditions were sequentially varied by small increments, some biological

0-8493-0641-8/96/$0.00+$.50

outcomes were observed at lower thresholds when exposure was to modulated fields, or occurred only when specific values of the independent variable were chosen. In the latter case, other nearby values above and below the response range produced different outcomes. The condition in which a specific result is found only when the independent variable is within certain well-characterized narrow ranges of values, is described as a "window" effect. Both frequency and field strength "windows" have been reported in the bioelectromagnetics literature and both types will be discussed below.

"Windows" are not new to biology. The existence of competing processes can result in a "window." An easily observed example of this phenomenon is the dependence of cell growth on temperature. As temperature increases from the minimum compatible with growth to about 35°C, growth rate increases monotonically, because the biochemical reactions required for growth proceed more rapidly. As the optimal temperature for a given type of cell is approached, the increase in growth with increase in temperature becomes smaller until finally, above the optimal temperature, increases in temperature produce lower growth rates. At higher temperatures, competing processes related to the thermal lability of critical chemical constituents of the cell become important. The competition of two processes can produce a single "window." However, the existence of multiple "windows" over a range of a single variable has also been reported. This cannot be explained as simple competition between two processes and a more complex theoretical explanation is required.

The existence of "window" effects, especially "multiple window" effects, is highly significant for several reasons. Most important among these are the implications for safety standards. Consider the case where a well-documented hazardous condition occurs within a defined range of exposure values, but is not observed at points outside the range. In this case, it would be difficult to define a "safe region," because it could always be argued that there may be another "window" at other exposure conditions within which the same hazardous situation would be produced. In this case, safety could be demonstrated only when the location of all "windows" could be predicted with a high degree of confidence, or when an exhaustive search for additional "windows" failed to discover any. The former requires a tested theory and the latter a tedious, perhaps impossible, series of experiments. In either case, it is imperative, certainly for standards-setting purposes, to perform experiments over a range of exposure conditions (e.g., frequency, modulation, dose, dose rate, peak power) to determine the existence or validate the absence of effects.

Medical applications provide another reason for studying "windows." Beneficial effects may be produced only by unique conditions of exposure. If these effects are to be of clinical use, the exposure conditions required to produce them must be carefully characterized.

Frequency and amplitude "windows" have been observed in various biological-effects experiments. These experiments, which will be reviewed below, include genetic, immune, hematologic, nervous system, and reproductive responses. "Window" responses have been observed over a wide range of frequencies from extremely low frequency (ELF) to millimeter waves. Most of the reported "window" effects have been observed following exposure to millimeter-wave radiation, ELF-modulated radiofrequency (RF) radiation, or ELF fields.

One of the first, and still important, applications of microwave radiation is radar, which typically requires that the radiation be pulse modulated. Therefore, many hazard analyses have considered pulsed radiation. Only a very limited, usually just one, set of pulse parameters is considered by an investigator. Therefore, experiments using pulsed irradiation are not considered to be evaluations of "window" effects. However, if the result of an experiment using pulse-modulated radiation is significantly different from one using continuous-wave exposure, a "window" effect may be present, and additional research is required. Comparisons of continuous wave (CW) and pulse-modulated irradiation will also be reviewed later in this chapter.

II. "WINDOW" EFFECTS IN THE MILLIMETER-WAVE REGION

The suggestion that radiation in the millimeter-wave region could have unique frequency-specific effects on biological systems was made in the mid-1970s.[1] A detailed review of millimeter-wave research has been provided by Motzkin.[2] Fröhlich[3] predicted that strong macromolecular oscillations in the 100- to 1,000-GHz frequency range are of biological significance. Early experimental support for Fröhlich's ideas came from Webb and Dodds[4] who showed that irradiation at 136 GHz slowed the growth of *E. coli*. This was followed by the same group's observation[5,6] that, depending on the frequency, the growth of microwave-irradiated *E. coli* was either enhanced or retarded. Within 2 or 3 GHz, the effect changed from a 50% increase to a 50% decrease. Berteaud et al.[7] also reported a marked decrease in the growth of bacteria exposed at 70.5 and 73 GHz. They examined several specific frequencies with replicate experiments at 73 GHz and showed both frequency dependence of growth and a high degree of variability among replicates under the same experimental conditions. In a follow-up attempt to generalize these results and investigate a possible mechanism, Berteaud's group used a diploid strain of yeast that is readily mutated (D_5 Zimmermann).[8] No evidence of altered survival, impaired function, or structural injury was observed in cells exposed to either 70.5- or 73-GHz radiation at power densities as high as 600 W/m^2, leading the authors to conclude that the previously reported effects of millimeter waves on bacteria are not mediated by direct action on cellular DNA. Blackman et al.[9] attempted to replicate the dramatic report of Webb and Dodds,[3] but found only a small enhancement of cell growth which was attributed to the slightly elevated temperature of the irradiated cells.

In other experiments, Webb and Booth[5] reported variations with frequency (in the 64- to 75-GHz frequency range) of the microwave-absorption properties of water and solutions containing DNA, RNA, and proteins. The fine structure of water is not expected to produce these effects; however, the ratio method employed raises questions. Generally, in the ratio method absorption losses of the sample holder are eliminated by first measuring the loss in the empty holder and comparing this with the loss observed in the filled sample holder. In Webb's experiment,[10] the sample holder was composed of two mica windows approximately 1 mm apart, located in a waveguide. Accurate absorption measurements using a ratio method can only be obtained when the sample thickness is much less than a quarter wavelength. Otherwise, reflections at both back and front surfaces of the sample will cause large standing waves within the system not accounted for by measurements with the empty sample holder, making accurate measurements impossible. The 1-mm sample was about a quarter wavelength for the frequencies used by Webb and Booth.[5] Gandhi et al.[11] reported experiments in this frequency range using a ratio-absorption technique and did not find variations of the type reported by Webb and Booth.[5] It should also be noted that the ratio method employed by both groups may not be sensitive enough to detect small perturbations in a highly absorbing medium such as water. Thus, contributions to absorption by a few large molecules suspended in highly absorbing water would go unnoticed.

The effects of microwaves on protein synthesis by mammalian cells (BHK-21/c13) were assessed at 0.1-GHz intervals over the ranges of 38 to 48 and 65 to 75 GHz. The exposure system allowed simultaneous examination of possible power density "windows," because the microwave power within the plane of the cell culture varied as the cosine, permitting simultaneous investigation of a wide range of power densities. Using the system described by Partlow et al.,[12] Bush et al.[13] did not detect acute (1 hr) effects of millimeter waves at power densities up to 3.5 kW/m^2. Webb[14] also reported that the induction of lambda prophage can be effected by millimeter-wave radiation (70.5 GHz) and that the frequency dependence shows a strong resonance. While an attempted replication produced negative results,[15] more recently, Lukashevsky and Beayaev[16] found a 30-fold increase in prophage induction caused by exposure to 70.45–70.75 GHz radiation with a resonance at 70.62 GHz. It appears that after 21 years Webb's results may have been replicated.

A series of papers from the former Soviet Union[1] reported very sharp resonance effects of millimeter-wave radiation at about 10 W/m^2. In many cases, the resonance was very sharp. The systems examined include: survivability of bacteria (41.96 GHz is most harmful);[17] stimulation of cell division in *Rhodotorula rubia* (maximum at 42 GHz);[18] damage of cell membrane, degeneration of protoplasm, and increase in cell size (46 GHz);[19] decrease in fertility of *Drosophila* (maximum at 43 GHz) and increased mutations in the second generation (maximum at 46 GHz);[20] and lower weight and retarded development of chicks born from irradiated eggs (maximum at 42 GHz).[20] These reports are especially terse in discussing experimental methods, making it difficult to rule out artifacts.

One of the most interesting reports[21] describes multiple resonances in the colicin synthesizing capability of a specific plasmid-containing strain of *E. coli.* At the resonant frequencies, the increase was as much as 300%. Repeated experiments (25) were conducted under a single set of experimental conditions resulting in statistically significant results ($p < .001$).

Smolyanskaya et al.[22] reviewed the earlier reports on colicinogenesis and suggested that the resonant effects result from the action of microwaves on rotational states of segments of macromolecules which alters the structure, conformation, and ultimately, the function of the macromolecules. Additional results were presented describing effects on the synthesis of an exogenous enzyme, but at slightly higher frequencies. These results indicate that slight changes in wavelength ($\pm$0.005 mm) may cause the observed inhibition of synthesis to disappear. It was also reported that a tenfold decrease in bacterial concentration resulted in a fourfold increase in the magnitude of the effect. Extensive attempts to replicate the work of Smolyanskaya and Vilenskaya[21] have not been successful.[22] However, the authors emphasize the frequency specificity of the effect and claim a capability of measuring wavelength to an accuracy of 0.01%. Whether or not additional frequency components, due to an unstable source, could suppress the observed response is not known.

The reports of Sevast'yanova and Vilenskaya[24,25] are also noteworthy because of the presence of several windows between 38 and 45 GHz, often separated by as little as 10 MHz, and the bimodal distribution of the data (see Figure 1). Depending on the frequency, millimeter-wave pretreatment either afforded bone marrow cells protection against the effects of x-rays or had no effect. In the latter case, the number of bone marrow cells was 70% greater than in the case without millimeter-wave pretreatment. Exposures at 100 to 750 W/m^2 were found to be equally effective, but exposures between 10 and 100 W/m^2 did not mitigate the effects of x-rays. These reports left an observer of resonance effects somewhat perplexed.[26] These data were later reinterpreted and are said to demonstrate two periodic processes offset by a period of 0.05 mm.[27] The paucity of data (a single data point for each node) does not support the two-processes hypothesis. Sevast'yanova et al.[28] also reported a resonance effect on the proliferation of bone marrow lymphocytes that had been incubated with the antineoplastic agent cyclophosphane prior to exposure to millimeter wave radiation. These resonances were also very sharp.

Belyaev and colleagues used a unique experimental method (anomalous viscosity time dependence (AVTD)) to investigate resonance effects of millimeter wave radiation[29–34] and, recently, frequency-dependent responses in the ELF range.[35] This method is based on the radial migration of large DNA-protein complexes, such as a genome complex, in the high gradient hydrodynamic field of a rotary viscosimeter. Radial migration of molecular complexes towards the rotating rotor causes anomalous changes of viscosity that can be registered by measuring the rotation period as a function of time. The assumption is that the AVTD depends strongly on the genome conformational state, which in turn depends on DNA molecular weight, micromedium, DNA-protein bonds, RNA, membranes, etc. If exposure to electromagnetic radiation results in an epigenetic response then this should result in conformational changes in the genome complex. The authors present experimental evidence and suggest that it is indicative of a direct interaction of the field with the genome. They suggest that field interactions with the membrane cannot explain the frequency dependence of the work reported by them and others. Recently, Belyaev, et al.[34] reported that millimeter wave radiation impeded the repair of x-ray damaged genomes of *E. coli.* Exposures within two frequency ranges, 41.25 to 41.50 GHz and 51.62 to 51.84 GHz, were effective.

Irradiation at 46.31 and 46.38 GHz (1.5 W/m^2) lowered penicillinase activity of *S. aureus,* while similar exposures at 46.22, 46.34, or 46.43 GHz had no detectable effect.[36] Multiple-resonance curves of the effect of x-band and millimeter-wave radiation on growth of the bacteria *B. mesentericus* and *P. fluorescens* were reported.[37] Evaluation of a RF device used to rewarm frozen blood led to the finding that when 46-GHz radiation was used for heating, the amount of free hemoglobin was minimal,[38] but cell injury varied more dramatically at frequencies greater than 46 GHz. Exposures were at 10 W/m^2.

Three additional review articles discuss Russian reports of resonance effects in biological systems irradiated with millimeter-wave radiation.[26,39,40] In general, resonant effects are characterized by: (1) sharp resonances (often multiple) where the width is 0.1 to 1% of the mean frequency, (2) insensitivity to power density, and (3) requirement for extended or multiple exposures (for microorganisms this may correspond to irradiation over several cell divisions).[39] Very different results, e.g., beneficial and hazardous effects or stimulation and inhibition, were reported to occur at the same frequency when different test systems were used.

Grundler and Keilmann[41] investigated and extended the work of Devyatkov.[18] However, they grew yeast in stirred cell suspensions, rather than on agar plates, to provide a more homogeneous thermal environment in the neighborhood of 32°C. At a given temperature, the growth rates of controls varied within ±3% of the mean value, while the growth rates of microwave-exposed samples varied by as much as 20%, depending on the frequency of the radiation. Multiple resonances were reported between 41.64 and 41.79 GHz with line widths of 10 MHz. (Experimental variability in frequency was estimated as 3%, and the absolute value of the frequency was known to 20 MHz.)

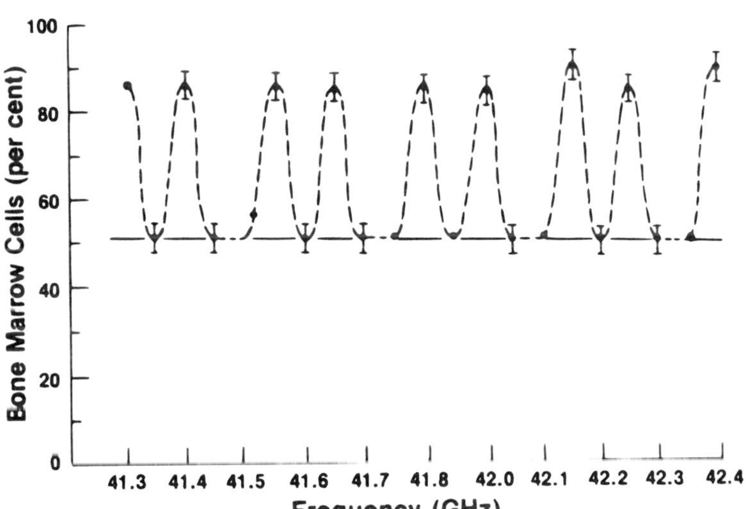

FIGURE 1 Relative number of bone marrow cells as a function of frequency of microwave radiation used for pretreatment. Cell counts in animals exposed to microwaves only represents 100. Animals exposed to x-rays (without microwave pretreatment) show a 50% reduction in the number of bone marrow cells. Pretreatment with millimeter waves either affords no protection or partial protection from the effect of x-rays, depending on the frequency of the millimeter wave radiation. (Figure redrawn from Sevast'yanova, L. A. and Vilenskaya, R. L., *Biol. Nauk.* 6, 48, 1974.)

Grundler et al.[42,43] improved on their dosimetric and temperature-monitoring techniques. Forward and reflected power were monitored and the reflected power minimized at each frequency with the aid of an x-y tuner. They presented data from four series of experiments, some of which were replicates at the same frequency, showing a very strong frequency dependence of growth rate. Grundler et al.[44] subjected their results to extensive statistical analysis which showed that the two studies[41–43] yield the same frequency dependence when a +11-MHz correction is applied to the data in the earlier study.[41]

Contrary to the results discussed above, Furia et al.[45] found that millimeter waves did not induce detectable changes in either the growth rate or the viability of yeast cells (*S. cerevisiae*) exposed at frequencies between 41.650 and 41.798 GHz for 4 h. They evaluated 15 frequencies in the range where the largest effects had been reported. In each experiment, the frequency was stabilized to ±50 Hz. Using DNA-repair-deficient mutants of bacteria (*E. coli*) and yeast (*S. cerevisiae*) Dardalhon et al.[46] found that irradiation at frequencies between 70 and 75 GHz neither decreased survival nor induced mutations. Belyaev[29] also found that millimeter wave radiation was not mutagenic for *E. coli.*

Raman spectroscopy is an excellent way of determining whether high frequency resonances exist. The presence of anti-Stokes lines of approximately the same intensity as the Stokes lines would substantiate the existence of coherently excited modes. Sharp, distinct Stokes Raman spectra of mammalian cells have been reported in the 15 to 300 cm^{-1} region (450 to 9000 GHz)[47] (the crucial anti-Stokes spectra have not been reported). However, the peaks are at different frequencies for normal and tumor cells in mammary tissue which is consistent with the report by Webb and Booth[10] of differences in the millimeter-wave absorption of normal and tumor cells. Webb and Stoneham[48] found narrow Stokes lines, in the region between 100 and 200 cm^{-1} (3000 to 6000 GHz), in a bacterial cell preparation, but only when the cells were actively dividing in synchrony. Webb et al.[49,50] then examined the anti-Stokes spectrum of synchronized, actively metabolizing *E. coli* and noted variations in the spectra with both time and frequency. Anti-Stokes intensities similar to the Stokes intensities indicate the existence of coherently excited modes.

Similar results of a preliminary study were reported by Drissler and Santo[51] who provided a systematic series of observations of variations in the Stokes-Raman spectra during irradiation. Furia and Gandhi[52] attempted to reproduce the earlier experiment of Webb[48] using *B. megaterium,* but didn't find Raman lines in the region of 20 to 300 cm^{-1} (600 to 9000 GHz). The work of Cooper and Amer[53] suggests that the time variations in the spectra observed by Webb et al.[49] were due to clumping of the synchronous cells which produced changes in the total light scattered (Mie scattering from clumped cells) and were not due to specific frequency components.

In general, Raman spectroscopy of cell suspensions is difficult. Several effects can produce spurious Raman lines as discussed by O'Sullivan and Santo.[54] At frequencies below 3000 GHz, the probability of experimental artifact in Raman spectroscopy is significant. Grating ghosts and artifactual scattering due to cell division are two possible causes of error.[55,56] Furthermore, when cell suspensions are examined, media that do not fluoresce must be used. Additional problems are encountered, because the cells settle producing time-dependent intensity fluctuations in the scattered light. Gandhi et al.[57] used nonfluorescing media and suspensions of cells in agarose. They did not detect Raman spectra in the region of 15 to 1500 cm^{-1} (450 to 4500 GHz) when examining baby hamster kidney cells (BHK-21/C13), chick fibroblast cells, or *E. coli.*

To avoid the appearance of false Raman peaks that were due to baseline variations, Layne et al.[58] used a polychrometer equipped with an optical multichannel analyzer. This system is not affected by baseline instability because it examines a broad spectral range at one time; it does not scan. Metabolically synchronous *E. coli* were observed from 100 to 2100 cm^{-1} (3000 to 62,000 GHz). Raman lines attributable to metabolic process or to the cells themselves were not found. However, the cells were more fluorescent during a limited phase in the division cycle, probably due to a variation of the redox state of some molecules within the bacteria or to a variation in the intracellular optical field. Layne et al., therefore, suggest that the Raman spectra of synchronous bacteria that were observed with scanning monochrometers, may be due to time-dependent variations in bacterial fluorescence.

Using still another experimental system, Kremer et al.[59] irradiated the salivary glands of the midge *Acricotopus lucidus* with millimeter waves at <50 W/m^2 and observed that the giant chromosome puffs that express genes for a secretory protein were reduced in size. While an exhaustive evaluation of the frequency dependence of this effect was not undertaken, swept-frequency irradiation (64.1 to 69.1 GHz) was observed to be less efficient than was irradiation at 67.2 or 68.2 GHz. Khramov et al.[60] reported decreases in the spontaneous firing rate of crayfish stretch receptors exposed to 34 to 78-GHz radiation at 2.5 kW/m^2. Although the frequency was scanned, frequency specificity was not found; at this dose, a thermal mechanism is probable.

The differential millimeter-wave absorption spectra of normal tissue slices and those from tumors were first reported by Webb and Booth.[10] Cells from a variety of tumors were reported to absorb more strongly at 69, 72, and 75 GHz and normal cells to absorb more strongly at 66, 68, and 70 GHz. It was suggested that this difference is the result of differences in the nucleic acid complement of tumor and normal cells. Stamm et al.[61] investigated effects of irradiation at higher frequencies (77 to 85 GHz) and reported that the absorption spectra of normal tissue and tumors were different and specific to the type of tissue. These experiments suffer from the use of thick samples, as discussed above. The use of millimeter-wave absorption to diagnose cancer is an attractive—yet unproven—application of microwave radiation.

As can be seen in Table 1, most reports of frequency-specific effects of millimeter-wave radiation demonstrate several closely spaced "frequency windows." Of the reports reviewed, the most common experiment is an evaluation of the growth of bacteria or yeast, but a wide variety of effects, including biochemical, genetic, and developmental alterations, have been found in these prokaryotic cells. Two results have potential medical application, i.e., protection against the effects of x-rays and differentiation between tumor and normal cells.

Very sharp resonances in the millimeter-wave region have been reported by Webb and Booth,[5] Berteaud,[7] Grundler and Keilmann,[41–44] Devyatkov,[18] Smolyanskaya and Vilenskaya,[21] etc. If these sharp resonances are due to effects on DNA, as is likely for the types of effects observed, they would suggest interaction with specific molecules or molecular systems which could result in the removal of a repressor molecule from the DNA, allowing transcription to take place. The sharpness of the line widths surprisingly suggests only mildly dampened absorption. This, in turn, indicates some form of isolation of the modes induced in the large molecule from the absorption modes of the surrounding media. Mode dampening would result in considerable line broadening and the absence of rapid variation of absorption with frequency.

It is difficult to draw overall conclusions concerning the numerous reported effects of millimeter-wave radiation because of: (1) inadequate description of experimental methods, (2) insufficient attention to millimeter-wave engineering, and (3) lack of replication. The work of Grundler et al.[44] stands as an example of excellent reporting of experimental details with careful attention paid to engineering design. However, even this work can be criticized because of possible spacial variation of energy distribution with frequency within the sample holder. Points close to the wave guide surface may not be sufficiently stirred and thus selected cells may remain longer in locations that have different exposure levels at different frequencies.

III. WINDOW EFFECTS OF ELF AND ELF-MODULATED FIELDS

The subject of frequency-specific and power-density-dependent windows has been reviewed by Adey[62–70], Adey and Bawin[71] and by Myers and Ross.[72] The topic is important for both theoretical and practical reasons and has, therefore, interested a number of scientists. However, there exist a variety of experimental results and a broad spectrum of opinion on this subject. Research has centered on four areas: electrophysiology, calcium movement, immunology, and hearing.

A. AUDITORY SYSTEM

During the late 1940s and 1950s, microwave workers were familiar with the anecdotal reports of "hearing radar," but not until 1961 was this phenomenon investigated in a serious manner[73] and not until 1974 was the physical explanation of the effect described.[74] Frey[73] demonstrated that a subject with otosclerosis which caused a 50-dB loss in air-conduction heard RF pulses normally, but that subjects with hearing loss above 5 kHz (bone-conduction dependent) could not hear microwaves. He extended this work examining various frequencies, pulse durations, and power densities determining thresholds in terms of peak power density. These thresholds vary with frequency, depending on penetration of the radiation into the head.[75,76] After considering several possible mechanisms, Frey[75] suggested that a direct interaction of RF radiation with neurons was the mechanism by which individuals heard microwaves. Sommer and Von Gierke[77] suggested that the mechanism was radiation pressure,

TABLE 1
Summary of Experimental Results in the Millimeter Region

Effect	Type of Frequency Window	Ref.
Bacterial growth	Single window	4, 17
Bacterial growth	Multiple windows	5, 60, 37
Bacterial growth	No window	9, 46
Stimulation of cell division	Single window	18
Decreased cell growth	Single window	7
Stimulation of cell growth	Multiple windows	41–44
Cell damage	Single window	19
Growth of mammalian cells	No window	13
Damage to red blood cells	Multiple windows	38
Bacterial synthesis of protein	Multiple windows	21
Bacterial synthesis of protein	No window	23
Decreased enzyme activity of bacteria	Multiple windows	36
DNA, RNA, protein absorption	Multiple windows	10
DNA, RNA, cell absorption	No window	55
Raman spectroscopy	Multiple windows	1, 40, 47, 50
Raman spectroscopy	No window	52, 57, 58
Genetic mutations in yeast	No window	8, 46
Growth of yeast	No window	45, 46
Induction of lambda prophage	Single window	14, 16
Increased mutation of *Drosophila*	Single window	20
Chromosomal effects in midge	Probable windows	59
Decreased fertility of *Drosophila*	Single window	20
Retarded development of chicks	Single window	20
Protection against x-ray damage	Multiple windows	24
Repair of x-ray damage	Multiple windows	34
Modification of cyclophosphane effect	Multiple windows	28
Differences between tumor and normal cells	Multiple windows	47, 61
Neuron firing in crayfish	No window	60

but this was disputed because of errors in computation.[78,79] Additional experiments[80,81] established that, for pulses shorter than 30 μs, the threshold is 0.4 J/m^2 per pulse regardless of pulse width or peak power. (This produces a peak-specific absorption of 16 mJ/kg, elevating the temperature in the head by 5×10^{-5} °C.) However, Frey and Messenger[82] reported that perceived loudness of 1.245-GHz pulses is peak-power dependent for 10- to 70-μs pulses.

Behavioral experiments used to determine if animals sense pulsed RF radiation often measure the movement of the animal from an environment that includes radiation to an otherwise identical habitat that is free of radiation. It has been reported that rats prefer the absence of pulsed microwaves[83,84] and that rats[85] and cats[79] will appropriately respond to pulsed microwaves when used as a conditioned stimulus. However, these experiments do not prove auditory mediation of the behavioral outcome. Food-deprived rats trained to respond to an acoustic signal hesitantly respond to microwave pulses at the same repetition rate as the acoustic signal.[86] This is the strongest behavioral evidence that rats are sensitive to modulation frequency and that they interpret these stimuli similar to the manner in which they would sound. A single 0.8–1.0-μs pulse of 1.25-GHz radiation presented 100 ms prior to a startle-inducing acoustic stimulus modifies the rat's startle response.[87] Hearing pulsed microwave radiation may be responsible for some reported effects on behavior, especially when small decrements in the performance of the animal are observed during or immediately following irradiation with pulsed microwaves.[88,89] In fact, Lai et al.[90] suggest that the reason some physiological effects in the CNS are elicited by pulsed microwaves, but not by CW radiation at the same dose rate, may be that the auditory system acts as the afferent system affected by the microwaves.

The response of an animal to auditory stimulation can be documented via electrophysiologic measurements including action potentials of the auditory cortex, thalamus, and auditory nerve as well as cochlear microphonics. Electrophysiologic measurements, therefore, can be used to determine whether pulsed microwaves produce an auditory sensation in laboratory animals. Furthermore, if measurements at all sites along the auditory pathway, for both microwave-induced and ordinary sound, give similar results, the microwave-induced auditory response must be initiated prior to the initial neurologic event. Therefore, several groups have attempted to measure electrophysiological responses to microwave pulses. Frey[91] measured microwave-induced neural responses of the auditory system of guinea pigs and cats, but could not detect the cochlear microphonic response. Using a more advanced electrode system, Guy et al.[80] were also unable to measure a microwave-induced cochlear microphonic. However, they showed that bilateral ablation of the cochlea eliminated previously observed microwave-induced potentials in the auditory system. Using a circularly polarized waveguide which more efficiently couples microwave energy into the guinea pig head, Chou et al.[92,93] detected a microwave-induced cochlear microphonic response at the round window. Microwave-induced cochlear microphonics were also recorded from cats.[94] For both cats and guinea pigs, the greater the length of the brain cavity the lower the frequency of the microwave-induced cochlear microphonic.[95] This relationship is consistent with the thermoelastic expansion model of the microwave hearing phenomenon (see below). The time between pressure waves corresponds to the time of propagation of acoustic waves in the medium. If thermoelastic waves are generated by absorption of a microwave pulse in the brain, then larger brain cavities would produce lower frequency cochlear microphonics. Extrapolation of data from guinea pigs[92,93] and cats[94] predicts that, in man, the microwave-induced cochlear microphonic will be between 7 and 10 kHz.[95] This is consistent with the inability of individuals to hear microwave pulses if they have neural hearing loss above 5 kHz. (It has been suggested[96] that the cochlear microphonics recorded during microwave irradiation are due to vibration artifacts and that the hearing of microwaves is mediated via direct neural stimulation by the radiation. The fact that the vibrations from piezoelectric crystals and laser pulses produce similar cochlear microphonics[93] strengthens the argument that the observed potential is due to bulk absorption in the brain.)

Other neurophysiologic studies of microwave hearing in cats include recording of extracellular potentials from: single eighth-nerve fibers[97,98] and brain stem using coaxial electrodes,[91] medial geniculate body and eighth nerve using glass microelectrodes,[80,99] inferior colleculus using glass microelectrodes, and cortex using high-resistance carbon electrodes.[79] (The use of metal electrodes during exposure to microwave radiation presents the possibility of field intensification at the tip of the electrode.[100])

Using microwave-transparent carbon-loaded Teflon electrodes to record auditory brain-stem-evoked responses from the skin of guinea pigs, Chou et al.[93] showed that microwave-induced brain-stem-evoked responses were similar to those produced by acoustic/pulses. Both Cain and Rissmann[81] and Lin et al.[101] reported similar results on the auditory brain-stem-evoked response in cats. Using the auditory brain-stem-evoked response to determine if guinea pigs hear microwave pulses, Chou and Guy[102] determined that: (1) for pulses shorter than 30 μs, microwave hearing depends on the energy per pulse; (2) for pulses between 50 and 500 μs, microwave hearing depends on peak power.

Wilson et al.[103] used autoradiographic techniques to show that the microwave hearing phenomenon does not involve the middle ear. This is consistent with the results of studies on humans[73] and auditory brain-stem-evoked experiments involving guinea pigs.[104] With the exception of a single report of the absence of a response,[91] all of the neurophysiologic data are consistent with a mechanical model of the microwave hearing phenomenon.

White[105] predicted that pulsed electromagnetic energy could produce pulses of pressure waves in absorbing solid or liquid material. Thermal expansion due to the absorption of electromagnetic energy can produce strains in the absorbing medium, leading to the production of a stress wave that propagates away from the heated region. Later analysis[106] showed that, for single long pulses, the induced stress wave is a function of peak power density; for shorter pulses, it is a function of the product of peak power density and pulse width (or energy per pulse). Foster and Finch[74] suggested that a thermoelastic wave of the type proposed by White[105] and Gournay[106] was the mechanism by which pulses of microwave radiation elicited auditory sensation. Foster and Finch[74] used a sensitive hydrophone to demonstrate that 2-to 27-μs pulses of 2.45-GHz radiation produced pressure waves in distilled water and the pressure wave changed as the temperature of the water cooled below 4°C. At 4°C the signal disappeared completely because the coefficient of thermal expansion of the water is zero at this temperature. This dependence on the coefficient of thermal expansion argues strongly for a microwave-induced thermoelastic mechanism of microwave hearing. Consistent with the earlier reported loudness-pulse width data, they found that, for short pulses, the intensity of the pressure wave recorded within a band width corresponding to audible sound depends on the energy per pulse. For longer pulses, the significant parameter is peak power.

The absorption of microwave radiation in a dielectric material can produce acoustic energy via three mechanisms: radiation pressure, electrostrictive force (elastic deformation of a dielectric material in an applied electromagnetic field), and thermoelastic stress.[80,107–109] Lin analyzed this problem by considering a one-dimensional model in which a plane wave impinges normally on a semi-infinite slab of homogenous dielectric material. The equation of motion of a particle in a medium responding to an applied pressure was solved by D'Almberts method for the three mechanisms stated above. In all materials, thermoelastic stress exceeds other types of stress by a very wide margin. It was found that "microwave-induced thermoelastic stress exceeds radiation pressure by almost three orders of magnitude; pressure generated by electrostriction is about ten times as high as that produced by the radiation pressure mechanism."[109] Numerical solutions were obtained for 10-μs pulses of 2.45-GHz radiation impinging on different dielectric materials and are provided in Table 2.

TABLE 2
Ratio of Microwave Energy[a] Deposited in Tissue[b] by Different Acoustic Mechanisms

Material	Electrostricture/ Radiation Pressure	Thermoelestic Stress/Radiation Pressure	Thermoelastic Stress/ Electrostricture
Brain	10.7	1301	122
Muscle	15.7	1290	82
Water	26.0	1225	41

[a] 10-μsec pulse of 2.45-GHz radiation.
[b] semi-infinite slab models of tissue

From Lin, J. C., Microwave Auditory Effects and Applications, Charles C Thomas, Springfield, Ill., 1978. With permission.

For a semi-infinite slab, regardless of the mechanism analyzed the peak stress (pressure) is usually at the surface being irradiated.[109]

Since the head is more closely approximated by a sphere than by a semi-infinite slab and calculations of microwave absorption in spheres have been made for homogenous and layered spheres,[100,110–113] Lin[114,115] then considered the thermoelastic propagation in a spherical model of the brain. A plane wave pulse of microwave radiation is considered to produce volume heating when absorbed in the sphere. The thermoelastic equation of motion, without shear stress, is then solved for the sound wave using boundary value techniques and Duhamel's principle. It is found that the frequency of the acoustic wave is independent of both the position within the sphere where the peak absorption of microwave energy occurs and the frequency of the microwave radiation. The acoustic range of frequencies depends only on the size of the sphere (varying inversely with the radius) and the propagation velocity of the acoustic wave in the spherical material. Two conditions were analyzed: (1) the surface of the sphere being rigidly constrained and (2) the surface of the sphere being free to move. Qualitatively similar solutions were obtained for both cases. However, the calculated peak acoustic pressure is higher for the case where the surface of the sphere is constrained. The constrained case gives better agreement with experimental results. Theoretical results also indicate that there is an optimal microwave pulse width for the conversion of microwave to acoustic energy which varies with sphere size. For a 7-cm radius sphere (model of the human head) the smallest optimal pulse width is 5 μs. A decade after publication of the theory,[114,115] Lin et al.,[116] using a small hydrophone inserted stereotaxically into the brain of a cat, measured directly the acoustic pressure wave produced by the absorption of pulsed microwave radiation (2.45 GHz, 2-μs pulse width, 15-kV peak power). Olsen[117] and Su and Lin[118] have suggested that biological materials can be imaged using an array of hydrophones to pick up microwave-induced acoustic waves.

Microwave hearing is a universally accepted low-level effect of microwave radiation. It enjoys this unassailable position because it has been experienced by many microwave workers, has been demonstrated and replicated in laboratory animals, and is completely predicted by classical physics. The theoretical explanation of microwave hearing begins with a thermally induced effect of microwave absorption, yet the hearing effect is modulation dependent and is produced only under specific modulation conditions.

B. OTHER EFFECTS ON THE BRAIN AND NERVOUS SYSTEM

The first reports of dependence of a biological effect of modulation frequency were by Bawin and colleagues,[119–121] who showed that weak VHF fields that are amplitude modulated at brain-wave frequencies influence spontaneous and conditioned electroencephalograph (EEG) patterns in the cat. This work has been criticized for the presence (during irradiation) of metallic electrodes which would produce an "antenna effect."[72] Glass or metal electrodes located in the RF field can cause field distortions and hence result in artifacts, due to the difference in electromagnetic properties of the electrode material and the solution bathing the nerve. While this artifact may be significant in investigating the mechanism of interaction, it is unlikely that an electrode artifact would favor specific modulation frequencies from among those in the ELF region that were considered.[119–121]

In a related experiment, Gavalas et al.[122] showed that a monkey's subjective estimate of the interval between events was shortened by as much as 0.4 s during exposure to a 7-Hz electric field at 7 V/m (peak-to-peak outside the head of the animal). Expanding on this finding, they later showed that, for exposure at 7 Hz, the threshold was lower and the magnitude of the effect larger than at 45, 60, or 75 Hz.[123]

Rabbits exposed to 1 to 10 MHz radiation (amplitude modulated at 15 Hz) for 2 hr/day over a period of several weeks showed changes in their EEG patterns (enhanced low frequency components and decreased higher frequency components). Modulation at lower (4 to 5 Hz) or higher (60 Hz) frequencies was not as effective.[123] This finding is reminiscent of an earlier report by Servantie et al.[88] that a microwave signal illuminating an animal introduced onto the EEG a component that was the same as the pulsing frequency of the irradiating signal. However, since only repetition rates of 500 and 600 Hz were investigated, these results cannot be used to support or reject the existence of a frequency "window."

The initial report of frequency-dependent windows of electrophysiological responses[122] was followed a few years later by the report[124] of similar frequency-selective responses of brain calcium. Alterations in the distribution of calcium are of great interest, because of the pivotal role played by calcium in the control of vital processes in the central nervous system (CNS) and other organ systems.

Cerebral hemispheres from 4- to 8-day-old chicks were incubated for 30 min in a physiological solution containing radioactive calcium and exposed to 147-MHz radiation, sinusoidally amplitude modulated (80 to 90% depth of modulation). Exposed brain sections exhibited increased concentrations of preloaded calcium (interpreted as increased calcium efflux) that were dependent on the modulation frequency of the 147-MHz radiation. The 9- to 20-Hz region was effective with maximal efflux being measured at 16 Hz; frequencies both higher and lower were ineffective.[124] Identical results were obtained following incubation with sodium cyanide, indicating that the cytochrome system is not required for this process.[124] From theoretical considerations, Tenforde[125] concluded that a mechanism other than thermal energy transfer was responsible for the observed alterations in calcium binding. The original experimental finding was corroborated and extended by Blackman et al.[126,127] who reported that the response occurred only within a narrow range of intensities (intensity "window") around 8.3 W/m^2. The observations were extended to 450-MHz carrier-wave frequency with similar results.[128] Again, 16-Hz sinusoidal modulation was effective in producing increased levels of calcium in the medium when exposures were at 1 or 10 W/m^2, but not at 0.5 or 50 W/m^2. Later, Blackman et al.[129] extended these experiments to 50 MHz where they found two intensity "windows" (one between 14.4 and 16.7 W/m^2 and the second around 36 W/m^2). In experiments using ELF-modulated RF fields marked variability was observed in sham specimens.[127,129] Expanding on their earlier work,[129] Blackman et al.[130] reported four new power density "windows" (for a total of six) in which the efflux of calcium from isolated chick brains is enhanced by exposure to 50-MHz radiation amplitude-modulated at 16 Hz. As power density increased, there appeared to be a progressive decrease in the interval between values of power density capable of producing an effect. This led them to suggest a model based on chaos theory.[130]

Both frequency and field-strength dependence of calcium efflux in excised chick forebrain[131] and in the cortex of anesthetized cats exposed to ELF fields have been reported.[132] The maximum effect was observed following exposure to 6- to 16-Hz fields at field strengths of 10 to 56 V/m in air (producing an electric gradient of the order of 10^{-5} V/m in tissue). Blackman et al.[133] also suggested the existence of a "window" at 16 Hz (by examining 1, 16, and 30 Hz) in the efflux of calcium from chick brain tissue. They then proceeded to demonstrate two power density windows for the effect at 3 to 8 and 40 to 50 V/m (field strengths between 1 and 70 V/m were investigated) (Figure 2). Blackman et al. found that the 16-Hz sinusoid produces the same type of effect, i.e., increased calcium efflux, as is reported when the 16-Hz signal is used to modulate an RF carrier. This is in contrast to the earlier report[131] that 16-Hz fields reduce calcium efflux. Blackman et al.[134] went on to demonstrate a number of "windows," both in frequency and field strength, in the ELF region (see Figure 3). Although very comprehensive, this study of ELF "windows" on calcium efflux could not evaluate all combinations of frequency and field strength. There may exist very broad windows between 60 and 105 Hz. On the other hand, it may be that at any frequency there exists a range of field strengths that affect calcium levels in the chick brain, and perhaps other animal tissues. The width and position of this "field-strength window" vary with frequency in an as yet undetermined manner. If this is the case, it would be inappropriate to distinguish "frequency windows" from "field strength windows." It has been suggested that the earth's magnetic field influences the location of calcium-efflux "windows." Evidence for this is discussed later in this chapter.

In order to provide additional information helpful in assessing the significance of the experimental results obtained using chick brain sections, studies were conducted using anesthetized cats. Cats immobilized under local anesthesia were surgically prepared in order to allow the introduction of radioactive calcium at the pial surface. Following surgery, the cats were exposed to fields similar to those used earlier (450-MHz fields, amplitude modulated at 16 Hz) except at 30 W/m^2 (specific absorption rate [SAR] = 0.29 W/kg). Efflux of preloaded calcium from the cerebral cortex was greater in exposed cats than in controls. Using a one-tailed t test, the difference was significant at $p < .032$. Different modulation frequencies or field strengths were not evaluated. Therefore, these results alone do not

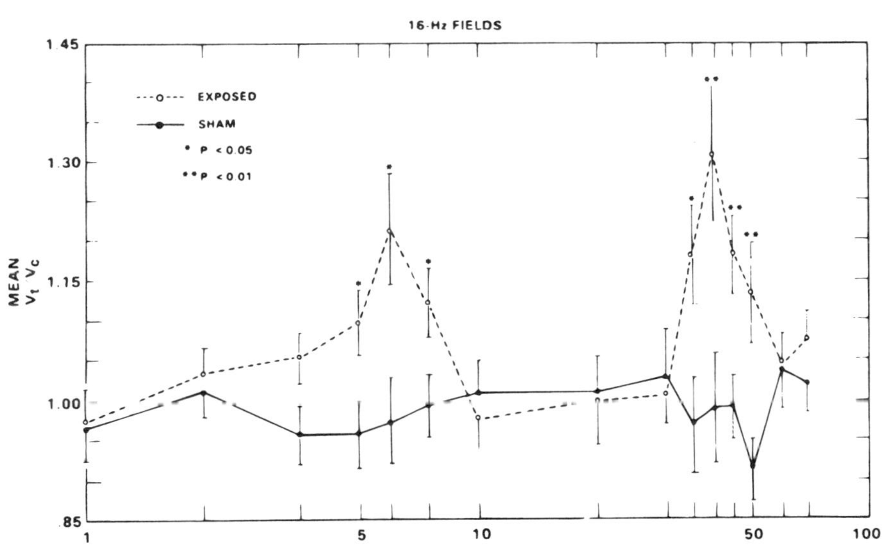

FIGURE 2 Mean relative quantity of preloaded calcium ions released from chick brain tissue. Treatment (ELF or sham exposure) is relative to water-bath treatment at 37°C. Electric field strength is peak-to-peak. Bars represent one standard error. The probability of sham and ELF-treatments being different is * for $p < 0.05$, ** for $p < 0.01$. (From Blackman et al. *Radiat. Res.*, 92, 510, 1982. With permission.)

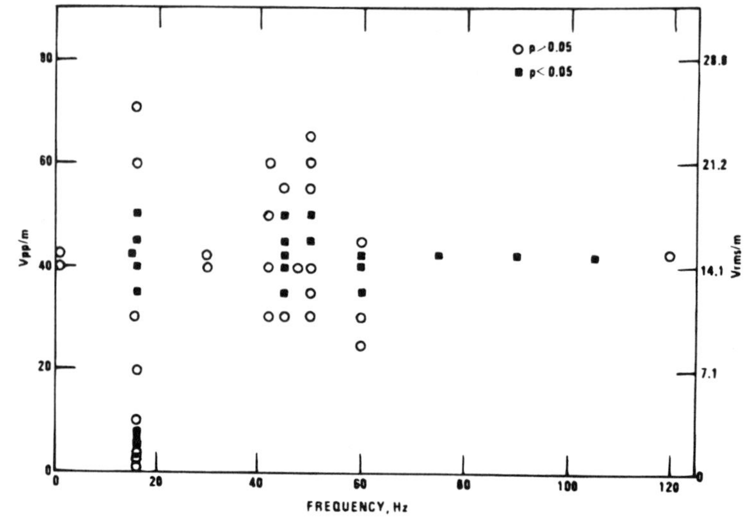

FIGURE 3 Experimental conditions for the release of preloaded calcium in chick brains is different from sham exposure. Field strength is in V/m peak-to-peak. Filled squares indicate conditions of frequency and field strength for which the efflux of calcium is significantly greater in ELF-treated brain tissue ($p < 0.05$). (Figure courtesy of C.F. Blackman.)

demonstrate a window in either frequency or field strength, but in the context of earlier work by this group it must be interpreted as supporting the existence of "windows."[132]

By considering the average electric field strength within a homogenous spherical sample, Joines and Blackman[135,136] were able to bring some order to the proliferation of intensity "windows." They reported that the intensity windows at 50, 147, and 450 MHz occurred at the same field strength within the specimen as calculated using their homogeneous spherical model of the exposed brain section. This analysis reduced the number of reported intensity windows from six to four. Even though the scaling technique was criticized because of the high degree of uncertainty with which internal field strength could be calculated,[137] it was used successfully to predict the presence of additional intensity "windows." Direct measurements of the electric field in chick brains exposed at 50, 147, and 450 MHz were compared with various theoretical predictions.[138] For a constant external field, the ratio of the measured internal field strength in samples exposed at 50 and 147 MHz was close to that predicted for a layered model.[136] Agreement with predictions for the higher frequency irradiation or using a simpler model[135] was not good. It must be pointed out that this approach provides no information on how many additional "windows" might exist or at what field strengths they could be expected.

Blackman et al.[139] later exposed chick brain tissue to electromagnetic fields (15 V/m, 59–69 nT) at 35 frequencies between 1 and 510 Hz. No pattern in the efflux of Ca^{++} as a function of frequency could be discerned. However, when *p*-values, a function that considers at each frequency both the difference between the means of the exposed and the sham groups and the variance of each group, were examined the data separated into three frequency-dependent groups. Frequencies between 15 and 315 Hz, except 165 Hz, fall into one group for which $p < .01$, a second group of intermediate *p*-values includes 60, 90, 180, and 405 Hz, and the third group includes the remaining frequencies.

Blackman et al.[140] reported that, depending on brain temperature, a 16-Hz electromagnetic field may increase, decrease, or have no effect on the release of Ca^{++} from chick brain tissue. Some of the early calcium efflux results should be re-analyzed in the light of this report. This work underscores the importance of controlling all variables and reporting their values.

Blackman et al.[141] reported that the brains of chicks that developed from eggs exposed to 60-Hz fields during incubation demonstrated a response to 50-Hz, but not to 60-Hz fields. In contrast, the brains of chicks exposed during incubation to 50-Hz fields were not affected by either 50- or 60-Hz fields. These results suggest that exposure during development may affect sensitivity to electromagnetic fields later in life.

Albert et al.[142] exposed chick brains to 147-MHz radiation amplitude modulated at 16 Hz at a power density of 7.5 W/m^2, but did not find any alteration in the flow of calcium in the tissue. Motivated by questions of safety, Merritt et al.[143] exposed the brains of anesthetized rats to microwave radiation (1, 2.06, or 2.45 GHz) modulated (16 20-ms pulses/s) at 5 to 100 W/m^2. Calcium levels were not altered by the microwave treatment. While these results may be important in a practical sense, because the exposure parameters are similar to those of operating systems, they do not help us to understand the phenomenon reported by others, because the experimental conditions are so different from those reported to produce "window" effects. Significant differences between the experiment performed by Merritt et al. and those performed by Blackman et al. are the value of the internal electric field strength, modulation conditions, and the carrier frequency. Blackman et al. state that the specific absorption rates (SARs) that produce the most significant responses are 1.3 mW/kg at 50 MHz and 1.4 mW/kg at 147 MHz. These absorption rates correspond to an average internal electric field of about 0.05 V/m. Merritt reported SARs of 0.29 to 2.9 W/kg at 1 GHz and 0.3 W/kg at 2.45 GHz. These SARs correspond to an average internal electric field strength of about 0.4 to 1.6 V/m, a factor of ten or more greater than the reported maximum effective level. Blackman's data indicate that changes of less than a factor of two in field strength can lead to different responses. Thus, it is possible that Merritt used an "ineffective" exposure level. It is also possible that no level of exposure will elicit such responses at carrier frequencies greater than 1 GHz. Thus, Merritt's results are consistent with those of Blackman et al, and Bawin et al. The reported response of the biological systems[131] is due to the modulation frequency and not the carrier frequency. Therefore, it seems reasonable to assume that the carrier wave is being detected or rectified by a nonlinear component such as the cell membrane. Since the molecular system has momentum, and at some high frequency will be unable to keep up with the rapidly time-varying field, it is reasonable to assume that such a rectification process would relax or cease to respond as the carrier frequency increases. Unfortunately, no experiments have been performed to date that establish the upper bound on the carrier frequency for this response.

In an effort to investigate the hypothesis that the RF-induced increase in calcium efflux in brain slices was the result of stimulated release from membrane binding sites, synaptosome preparations were studied.[144] The field strength was 43 V/m (in air) in all cases. While unmodulated, or 60-Hz amplitude-modulated, 450-MHz fields did not differ from sham irradiation, exposures in which the modulation was 16 Hz produced an increase in calcium efflux. Calcium release was measured as a function of time and was found to increase more rapidly following exposure to 16-Hz modulated fields than following injection of $CaCl_2$ (which releases calcium from the intracellular compartment). Therefore, the authors suggest that intracellular calcium is not involved in the RF effect on calcium movement. Membrane-bound calcium is the probable source of calcium released by electromagnetic radiation.

Dutta et al.[145] reported an RF-enhanced efflux of calcium ions from human neuroblastoma cells. Monolayer cultures were exposed to 915-MHz fields, with or without sinusoidal amplitude modulation (80%) at frequencies between 3 and 30 Hz. Ten exposures between 0.01 and 5.0 W/kg were used. Significant increase in calcium efflux was observed only at 0.05 and 1.0 W/kg. The increased efflux at 0.05 W/kg was dependent on the 16-Hz modulation but not at 1.0 W/kg. The same group later studied human neuroblastoma cells exposed to amplitude-modulated 147-MHz radiation.[146] Calcium efflux was enhanced by 13, 16, 57.5, or 60 Hz modulation, but not by 9, 18, 20, 55, 62.5, or 65 Hz. Ten dose rates between 0.0001 and 1 W/kg were studied in cells exposed to radiation modulated at 16 Hz, the modulation frequency that produced the greatest effect. Only two exposure conditions, 0.05 and 0.005 W/kg, produced statistically significant differences from unirradiated controls ($p < .003$). Using a rodent neuroblastoma cell line, Dutta et al.[147] showed that 147-MHz radiation amplitude-modulated at 16 Hz enhanced the digestion of acetylcholine. Of the six exposure conditions examined between 0.001 and 0.10 W/kg, only 0.02 and 0.05 W/kg produced statistically significant enhancements. Dutta et al.[148] also studied the activity of enolase, an enzyme involved in glycolysis, in human and rodent neuroblastoma cell lines that were exposed to 147 MHz radiation amplitude modulated at 16 Hz. Enolase activity was enhanced by a dose rate of 0.05 W/kg and depressed by 0.01 and 0.08 W/kg, while other dose rates, 0.001, 0.02, 0.07, 0.09, 0.10, and 0.50 W/kg, had no effect.

Several mechanisms through which RF or ELF fields may alter calcium levels in brain tissue have been proposed,[67,149–151] and are discussed later in this chapter. Most suggestions involve cooperative phenomena. However, following a more classical approach, Spiegel et al.[152] proposed an electromechanical model. They suggest that an electromagnetically induced force on the surface of the brain causes the brain to vibrate, dislodging calcium ions. Spiegel's model is consistent with the suggestion of Joines and Blackman[135] for extrapolating among carrier frequencies, because that model was based on establishing identical electric fields inside the brain tissue. Two experimental conditions that produced the same electric field would, in Spiegel's model, give the same electric-field-induced surface pressure.[152]

Recently, Blackman expanded on his finding that nerve outgrowth factor-stimulated neurite outgrowth in PC-12 cells can be inhibited by magnetic fields.[153,154] Blackman et al.[155] reported experimental verification of the prediction of the ion paramagnetic resonance (IPR) model regarding the position of a field-strength "window" in the applied AC magnetic field. The IPR model predicts specific responses depending on the flux densities of the component of the AC magnetic field parallel to the DC magnetic field and the frequency of the AC field. Resonance conditions are determined by the magnitude of the DC field and the frequency of the AC field in a manner discussed in the section on theoretical considerations. At resonance, the inhibition of neurite outgrowth varies in a complex manner depending on the ratio of the peak AC field strength to the DC field strength. At off-resonance conditions neurite outgrowth does not depend on the magnitude of the AC magnetic field.

Sanders et al.[156] examined energy metabolism in surgically uncovered brains of rats during and immediately after exposure to 200- or 591-MHz radiation. Radiation increased the fluorescence of reduced nicotinamide adenine dinucleotide (NADH) (during exposure) in a dose-dependent manner for exposures between 10 and 100 W/m^2. Exposures at 2.45 GHz were without effect. Post-exposure measurement of adenosine triphosphate (ATP) and creatine phosphate (CP) levels in the whole brain showed that ATP levels were decreased by 200- and 591-MHz radiation, but not by 2.45-GHz radiation, and that CP levels were affected only by 591-MHz radiation. SAR was comparable for irradiation at all three frequencies. The authors speculate that 200- and 591-MHz radiation inhibits specific, probably divalent-metal-ion-requiring, enzymes in the electron transport chain, but that 2.45-GHz radiation is not similarly effective. Although Sanders et al.[156] reported an unexpected frequency dependence of RF radiation on the alteration of brain bioenergetics, these results did not demonstrate the presence of a "window." In a follow-up experiment, rats were exposed to CW, or to sinusoidal amplitude-modulated (at 4–32 Hz) or square-wave pulse-modulated (250 or 500 pps) 591-MHz radiation at 0.2 W/m^2 for up to 5 min.[157] Brain NADH fluorescence, and ATP and CP levels were measured. All exposure conditions produced similar decreases in ATP and CP levels. A field-strength-dependent increase in NADH fluorescence was observed. This effect varied weakly with modulation frequency, being greatest between 16 and 24 Hz for amplitude-modulated exposures, and 500-pps for pulsed exposures. Sanders et al.[157] suggest that microwave inhibition of ATP in mitochondria could account for the observed increase in NADH and the decreased ATP and CP levels. In another study, membrane-bound cytochrome oxidase and ATPase were irradiated in vitro with 2.45-GHz radiation at 26 mW/g.[158] None of the modulation frequencies employed (16, 30, 90, or 120 Hz) altered the activity of these enzymes.

Rats exposed to 915-MHz microwaves in a TEM cell (field strength not stated) demonstrated a decreased blood brain barrier (BBB), especially to the passage of albumin. CW exposures were significantly less effective than were modulated exposures. All modulation frequencies investigated (8, 16, 50 and 200 Hz) produced similar results.[159,160]

Kuznetsov et al.[161] investigated the effect of amplitude modulated 800-MHz radiation at 1 W/m^2 on the concentration of GABA receptors in rat cortex. Of the modulation frequencies studied (0, 3, 5, 7, 16, and 30 Hz), only 16 Hz caused significant reduction in the concentration of receptors.

Exposure of rat olfactory epithelium to CW or pulse-modulated (1–100 Hz) 900-MHz radiation at SARs of 3 W/kg or greater decreased specific ^{3}H-camphor binding to membranes. However the effect did not vary with modulation frequency.[162] Mollusc neurons were exposed to CW or square-wave modulated (0.5–100 Hz) 900-MHz radiation. CW exposures of 0.5 to 4.0 W/kg did not change firing-rate patterns; however, modulated exposures in the same SAR range altered the firing patterns of two types of neurons.[163]

C. EFFECTS IN OTHER TISSUES AND ORGAN SYSTEMS

Because of its central role in a wide range of medical problems, the immune system has been actively studied. Therefore, it is understandable that a significant portion of the bioelectromagnetics literature is concerned with possible effects on the immune system. Several of these studies have considered "window" effects.

Huang et al.[164] reported that microwave irradiation (2.45 GHz, CW) of Chinese hamsters produced an increase in the transformation of circulating lymphocytes into blastic cells. The magnitude of this effect increased with increasing field strength between 50 and 300 W/m^2. However, the incidence of transformed cells observed following exposure at 450 W/m^2 was below that found for exposure at 50 W/m^2. The authors therefore suggest that this effect is another, albeit thermal, microwave "window" effect. Dose-response curves of the lymphocyte transformation index are commonly of the "inverted-U" shape.[165] Therefore, it is not necessary to postulate a unique type of microwave effect.

Reversible inhibition of allogenic cytotoxicity by murine T lymphocytes (CTLL-1) was reported in the presence of a 450-MHz field that was amplitude modulated at 60 Hz.[166] Modulation frequencies between 3 and 40 Hz (including 16 Hz) as well as 80 and 100 Hz produced inhibitions of cytotoxicity that were not as large as the 20% inhibition observed at 60-Hz modulation. Unmodulated carrier waves did not influence cytotoxicity. Pretreatment with the RF field was also effective when it was modulated at 60 Hz. Both the requirement for calcium in the initial stage of the cytotoxic reaction and reports that ELF-modulated 450-MHz fields affect calcium levels in brain tissue led Lyle et al.[166] to suggest the involvement of calcium in the observed effect. Adey's group extended their immunologic research to investigate the effect of amplitude-modulated RF fields (450 MHz) on protein kinase activity of human tonsil lymphocytes in culture.[167] While the activity of cyclic-AMP-dependent protein kinase was not affected by ELF-modulated fields, noncyclic-AMP-dependent protein kinase was markedly decreased for a brief period of time (returning to control values in 45 min even though RF exposure continued). This effect is described as being "windowed" in both modulation frequency (16 to 60 Hz) and duration of exposure.

Another calcium-dependent response of lymphocytes is the capping of antigen-antibody complexes in which the membrane-bound antigen-antibody complexes migrate to form a polar cap on the surface of the cell. Sultan et al.[168] investigated the effect of 147-MHz fields, amplitude-modulated at ELF frequencies on the capping phenomenon in mouse B lymphocytes. No effect was observed with RF fields modulated at 9, 16, or 60 Hz at a variety of power densities between 1 and 480 W/m^2. As expected, a temperature dependence of capping was observed. The capping of antigen-antibody complexes is dependent on intracellular calcium. These findings are, therefore, not inconsistent with those experimental results in which the efflux of calcium was altered by ELF-modulated RF fields, for in the latter experiments it is membrane-bound calcium that was affected.

Cultures of human lymphocytes were exposed at 10 W/m^2 to a 450-MHz field amplitude modulated at frequencies between 3 and 100 Hz for periods up to 60 minutes. Non-cAMP-dependent kinase activity was decreased by as much as half in cultures under 16-, 40-, and 60-Hz modulation conditions; CW and 3-, 6-, 80-, and 100-Hz modulation were without effect. No effects on cAMP-dependent kinase were found. An exposure duration "window" was also reported; exposures of 15 or 30 minutes produced the effect, but by 45 minutes control values were again observed.[169] Human HL-60 cells were exposed to ELF (15, 45, 60, 90, or 150 Hz) electromagnetic fields for 20 min. While all exposures increased the quantity of c-myc and H2B transcripts, the effect was greatest for exposure at 45 Hz. This led Wei et al.[170] to suggest that these results may be explained by a model based on some aspect of membrane-dependent signal transduction.

Veyret et al.[171] exposed mice for 5 days to low levels (15 mW/kg) of pulsed (1,000 1-μs pulses/s, 5-ns rise time) 9.4-GHz radiation both with and without concurrent amplitude modulation by a sine wave at discrete frequencies between 14 and 41 MHz. The mice were immunized prior to irradiation and their antibody responses were measured following the exposure. In the absence of amplitude modulation, microwaves did not significantly alter the immune response. However, a modulation-frequency-dependent augmentation or weakening of immune responses was observed. At two modulation frequencies (21 and 32 MHz) the number of plaque-forming cells, IgM levels, and IgG levels were all greater than in the sham-exposed mice.

In a preliminary study, Ning et al.[172] used flow cytometry to measure the effect of ELF fields on lymphocytes. Both 100-μT 50-Hz fields and pulsed electromagnetic fields, produced by an EBI generator (see Chapter 5), stimulated lymphocytes and Daudi lymphoma cells, but not HL-60 promyelogenous leukemia cells. The effect was greater for cells exposed to the pulsed electromagnetic field.

Schwartz et al.[173] exposed isolated frog hearts to a 240-MHz field either CW or sinusoidally modulated at 0.5 or 16 Hz and measured the movement of Ca^{++}. Neither CW radiation nor radiation that was modulated at 0.5 Hz, which is close to the beating frequency of the frog heart, altered the flow of calcium ions. However, 16-Hz modulated radiation increased calcium efflux at 0.15 and 0.30 mW/kg. Exposures at 0.24 and 0.36 mW/kg had no effect. In a similar experiment, but using atrial strips of frog heart and a 1-GHz carrier frequency, Schwartz et al.[174] did not detect a modulation-dependent effect on the movement of Ca^{++}. It was noted that in the first experiment a large number of atrial trabeculae were parallel to the magnetic field, however, in the second experiment they were largely perpendicular to the magnetic field. Schwartz et al.[174] suggest that this difference in experimental design may be responsible for the different results.

Yee et al.[175] exposed isolated rat hearts to 16-Hz amplitude-modulated 2.45-GHz microwaves that were pulsed (100 10-μs pulses/s). The SAR was 2–10 W/kg. Heart rate was not affected; this was attributed to careful thermal control of the hearts at 37.7°C. Caddemi et al.[176] exposed chick embryo hearts to CW or pulse-modulated (2–14 Hz) 2.45-GHz radiation at 4.2 W/kg. CW exposures caused a slight bradycardia, while modulated radiation induced stimulation or regularization of heartbeat in arrhythmias or fibrillation. Locking of cardiac rhythm to the modulation frequency was seen in several experiments.

Monolayers of embryonic chick calvarial cells were exposed to low frequency electric fields in a parallel plate capacitor for 30 min.[177] (The voltage gradient in the culture media was estimated to be about 10^{-5} V/m.) Exposure at 16 Hz was most effective in stimulating cell proliferation. Stimulation was greatest in cultures with the highest levels of alkaline phosphatase. Electric fields increased cell proliferation only when the cells were incubated in culture media in which bone cells had been exposed to the field, which is consistent with mediation by a mitogen.

Byus et al.[178] exposed Reuber H35 hepatoma cells to amplitude-modulated 450-MHz radiation at a peak intensity of 10 W/m^2. Ornithine decarboxylase activity (ODA) was increased up to 50% by 10-, 16-, and 20-Hz modulation, but was not affected by 6-, 60-, or 100-Hz modulation. Cells with microwave-enhanced ODA showed an altered response to the tumor promoter, 12-*o*-tetradecanoylphorbol-13-acetate (TPA). Balcer-Kubiczek and Harrison[179] exposed C3H cells to 2.45-GHz radiation, amplitude modulated at 120 Hz (other modulation conditions were not studied) at 0.1, 1.0 and 4.4 W/kg. While microwaves alone had no effect on cell survival or the induction of neoplastic transformation, in conjunction with TPA they produced a field-strength-dependent enhancement of transformation.

In another tissue, rat pancreas, increased efflux of Ca^{++} was demonstrated[180] following exposure to low levels (SAR estimated to be less than 0.075 W/kg) of 147-MHz radiation, amplitude modulated with a 16-Hz sine wave. The magnitude of the effect was similar to that observed in chick brains.[127,131] The efflux was not accompanied by the release of secretory proteins, as is common for pancreatic tissue. This was due to the fact that the small volume of medium in which the tissue was incubated did not contain sufficient nutrients to support metabolism.

Bawin et al.[124] examined chick skeletal muscle, but did not observe changes in calcium levels following exposure to modulated 147-MHz radiation identical to that which produced changes in chick brain tissue.

RF radiation dilates arterioles in frog webs. The observed effect is greatest when the radiation consists of a square wave, with each 0.05-ms burst being comprised of 500 individual pulses.[181] However, the result of selecting alternative pulse parameters was not discussed. This effect, which does not appear to be related to temperature, may be associated with decreased intracellular Ca^{++} in arteriolar smooth muscles.

Montaigne and Pickard[182] monitored the membrane potential of isolated algeal cells *Chara braunii* and *Nitella flexilis* while the cells were exposed to amplitude-modulated 147-MHz radiation at power densities between 2 and 1000 W/m^2. The importance of modulation frequencies from 4 to 64 Hz, at approximately 90% depth of modulation, was evaluated. Changes in the membrane potential occurring in phase or in phase quadrature with either the modulation signal or a signal at twice the modulation frequency (using a procedure described by Barsoum and Pickard[183]) were not found under any of the experimental conditions employed. The absence of reported effects of amplitude-modulated RF radiation on algeal cells *C. braunii*,[182] *C. covallina*,[184] and *N. flexilis*[182] may be due to the fact that these cells use calcium in a different manner than do mammalian cells.

Even though calcium plays an important role in many organ systems, one is struck by the fact that calcium plays a central role in almost all of the systems that have been reported to exhibit a "window" effect. The cyclotron resonance model (discussed below) is an attempt to explain the pivotal role of calcium ions in these effects.

IV. OTHER TYPES OF "WINDOW" EFFECTS

Pulsed electromagnetic fields have been used in clinical practice to accelerate or facilitate the healing of a bone nonunion as well as the growth of soft tissue. Clinical empirical observations have not led to an understanding of the basic mechanism whereby pulsed electromagnetic fields aid in such healing processes. However, several studies have examined the effect of various pulse characteristics in an attempt to maximize the biological response and to gain an understanding of the mechanism. These studies support the existence of a window effect, i.e., there are specific pulse characteristics that seem to promote healing of separated bones and others that do not. These studies are reviewed in chapter 5.

Pulse characteristics similar to those used by Cristel et al.[185] to promote bone healing have been used to investigate several cellular functions. One example is the work of Goodman et al.[186] in which two pulse forms that are in clinical use, a single pulse repeated at 72 Hz and a pulse train repeated at a rate of 15 Hz, were employed. Effects on RNA transcription patterns in salivary gland chromosomes of a dipteran, *Sciara coprophila,* were studied in three ways: (1) nascent RNA chains were identified by conventional autoradiography; (2) transcriptively active regions of chromosomes were identified using deoxyribonuclease I; and (3) the size and classes of synthesized RNAs were examined. It was shown that each type of pulse activates a different set of genes, resulting in a different RNA synthesis pattern.

Further research by the same group on salivary gland chromosomes[187–189] extended and confirmed these results. The transcriptional products from these cells have been altered following exposure to asymmetric pulsed 1.5-, 15-, or 72-Hz or sinusoidal 60- or 72-Hz magnetic fields. Magnetic field values varied from 0.8 μT to 8 mT. The response of the chromosomes varied with frequency, but the effect seemed to be an increase in synthetic activity.[190] In complementary studies,[189] two-dimensional gel electrophoresis was used to analyze changes in proteins. Additional studies[191–193] also showed variation in response depending on the frequencies used. Blank[194,195] collaborated with Goodman, Henderson and colleagues to study further change in the protein distribution patterns of cells exposed to various electromagnetic fields. They reported that the distribution patterns of proteins synthesized by the salivary gland cells of *Sciara* exposed to EM fields resemble those found following heat shock. Their conclusion is that EM fields act as a general stress on cells, but that this occurs for considerably smaller amounts of energy then is required for a heat shock response.

In addition to the pulsed wave forms mentioned above, CW exposures at various ELF frequencies were investigated. Maximum effect was observed at 45 Hz.[196,197] The amplitude of the magnetic field used in these experiments apparently varied with frequency with the level being about 7 mG at 45 Hz.[196] Dose response analysis is needed at each frequency to be more definitive. Further evidence of "windows" in amplitude and frequency were found on transcription in mammalian cells, specifically with chinese hamster ovary cells.[198] More extensive work was done with an acute promyelocytic leukemia cell line, HL-60.[196,199–202] HL-60 cells exposed to 200- to 380-μs pulses or sinusoidal fields were observed to have increased mRNA levels for c-myc, β-actin, c-src, β-tubulin, and histone H2B, all of which are normally expressed in HL-60 cells. Transcript increases were dependent on both field strength and exposure time. A "window" effect was observed with maximum transcript levels induced by a 20-min exposure at a magnetic flux density of 5.7 μT (rms), 60-Hz magnetic fields.[203] Transcript levels decreased to near control levels after a 20-min exposure at 570 μT. The authors pointed out that the observed effects for ELF exposure were not directly proportional to exposure duration or field strength. It is interesting to note that the response to histone, β-actin and c-myc were very similar in time and intensity.[99] Histone and β-actin are normally used as negative controls in this type of experiment and are not expected to change. The lack of internal controls (nonresponding transcript) raises questions concerning specificity of the technique to select specific transcripts, but not necessarily of the field response.

A close examination of the time/intensity response observed with, for example, c-myc[199,200] suggests an explanation for this "window effect." Litovitz[204] noted that the apparent rise and fall of the response with time was dependent on the amplitude of the applied magnetic field.

Although the data taken at various times was limited, it appears that the higher the field strength, the more rapid is the rise and subsequent fall of the response. For this analysis it is assumed that the first observation at 20 minutes was too late to observe the peak value of the response for the highest levels of exposure. The higher field strengths would cause a response that would peak before 20 minutes.

Litovitz[204] noted that this rise and fall can be mathematically modeled as an alteration in the rate constant(s) of two competing processes. For example, suppose that the amount of transcript for c-myc is in equilibrium in the cell; then the rate of creation of mRNA is equal to the rate of destruction or use. If the rate of creation can be increased, there would be a subsequent exponential rise in the amount of mRNA. This increase will continue until the supply of base units is depleted (or the rate of use of base units exceeds the rate of supply). At that point the rate of transcript creation will decrease until a new equilibrium value will be reached. This new equilibrium value may or may not be the same as the original equilibrium transcript level.

The fact that a higher field strength exposure results in an earlier peak messenger response may lead to the perception that a "window" exists. If the system is assayed at a single point in time well after the peak signal for the large exposure field, the system may have already declined to a value (i.e. transcript level) that is lower than the value produced over a longer time course by a weaker signal (see Figure 4). This would create an apparent "power window" which is but an artifact of the sampling time. Further testing of this model requires detailed time/dose response experiments with measurements at early times (a few minutes). Repeatable results may be difficult for such short times.

More recent work by Goodman, Henderson and colleagues[205] focused on the specific genes, such as the c-myc promotor region, which apparently are being affected by ELF fields. They also reported that exposure of simian virus-40-transformed human cells to magnetic fields results in increased levels of T-antigen mRNA and protein.[206]

Gene expression-related studies have been performed by other groups: Phillips[207–210] E. Goodman et al.[211], Czerska et al.[212], Ning et al.[172], and Greene et al.[213] None of these have performed studies as extensive as Goodman and Henderson. Although different positive results (complementary studies) were reported by all groups, none were successful in repeating the work of Goodman and Henderson on HL-60. Saffer and Thurston[214] attempted to replicate the work of Goodman and Henderson on c-myc expression in HL-60 cells and the work of Czerska et al. on Daudi cells. They failed to observe any changes at magnetic field conditions

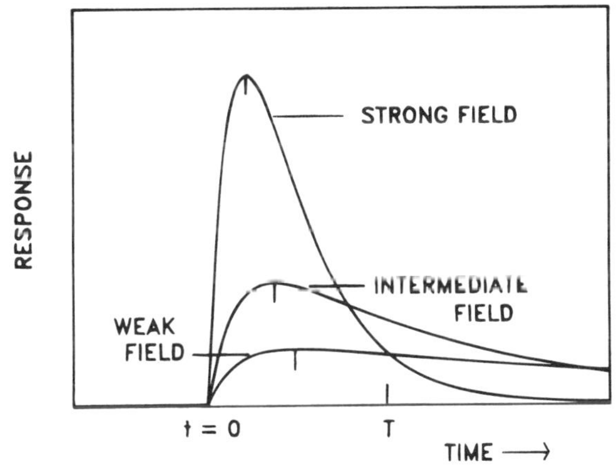

FIGURE 4 Schematic representation of a transient response by an hypothetical system to three different levels of exposure to an electromagnetic field turned on at time T = 0. (From Litovitz et al. *Bioelectromagnetics* 11, 298, 1990.

ranging from 8 μT to 1 mT and for concentrations ranging from 4×10^5 to 10^6 cells/ml. Also, Kropinski et al.[215] reported negative results for experiments on gene expression, but these studies differed significantly from the protocols of Goodman et al.

Delgado et al.[216] performed numerous experiments to show that weak magnetic fields at extremely low frequencies affect various biological systems, depending both on the amplitude of the field and on the frequency. For example; they suggest that embryological changes can be induced in developing embryos when exposed to low-level magnetic fields. They exposed chick embryos to 10-, 100-, or 1000-Hz fields at 0.12, 1.2, or 12 μT. The greatest response was found for exposure to 100-Hz fields at 1.2 μT, which was characterized as a "consistent and powerful inhibitory effect on embryogenesis." Again, a field-strength "window" is suggested.

Further investigations by this group,[217–218] indicated that pulse shape was an important factor. They found effects on embryogenesis that depended on the rise and fall times of the pulse. These variations in response with the rise and fall times suggest a biological response that is dependent on both the frequency and amplitude of the electric field induced within the cells. Further studies[219] suggest that the value and direction of the earth's magnetic field could also be significant factors. Juutilainen[220] reported similar findings suggesting frequency and amplitude dependence, but only a small number of eggs were used in each experiment, making it difficult to demonstrate statistical significance between results obtained under different exposure regimes.

The Office of Naval Research sponsored a concerted effort to establish whether these interesting findings could be replicated in several different laboratories.[221] Six independent experiments with a common protocol were performed in Canada, Spain, Sweden and the United States. The embryos were exposed to a unipolar, pulsed (100 50-μs pulses/s), magnetic field (1 μT peak) for 48 hours. Two laboratories reported highly significant increases in the number of abnormal embryos in the exposed groups (four-fold increase in one laboratory and two-fold increase in the other), one laboratory reported a marginally significant (9%) increase in abnormalities in the exposed group ($p = .08$), and the remaining three labs, including one that used a different strain of animal, showed no significant difference ($p =$ .6, .6, and .4). When data from all six labs were pooled, approximately 25% of exposed embryos were abnormal, while approximately 19% of controls were abnormal, a significant difference ($p < .001$).

It has been noted that the significance of the effect is related to the level of abnormals in the controls. Both Juutilainen[222] and Litovitz[223] reported that the background levels of abnormals significantly affected outcome. Litovitz noted that when abnormals were greater than 20% in the controls, the results were not statistically significant. There seemed to be an upper limit of about 25% abnormals, regardless of whether they exist in the background or are field induced. Both groups performed dose response experiments, and observed a threshold at about 1 μT. The fact that the six-laboratory study[221] operated at this 1-μT threshold level also may have contributed to variability in the results.

A "window" of magnetic field strengths that alters the growth rate of bacteria was reported by Aarholt et al.[224] During log-growth phase, bacteria were exposed to square waveform 16.66- or 50-Hz sinusoidal magnetic fields. At 50 Hz, decreased mean growth rates were found for exposures between 0.4 and 0.8 mT; exposures greater than 0.8 mT yielded a mean growth rate that was almost the same as for controls, and at higher field strengths growth was again depressed by the magnetic field. Exposure to a 16.66-Hz magnetic field at 0.8 mT or greater also caused a decrease in growth rate. Again the growth rate decreased or increased depending on the field strength. Moore[225] found that bacteria and yeast were either stimulated or inhibited by magnetic fields and that the effect depended on both field strength and perhaps frequency in the case of pulsed fields. However, the 30% increase in growth at

20 mT is well documented in at least four different strains of bacteria. At 10 mT the effect disappeared and at 30 mT it reversed, producing a 30% decrease in growth.

Aarholt et al.[22] also reported a power "window" in the effect of 50-Hz magnetic fields on the lac operon system. A sharp decrease in the rate of β-galactosidase synthesis was found at 0.27 mT, rapidly returning to control values as field strength increased. At 0.51 mT synthesis began to increase reaching a value twice that of controls in bacteria exposed at 0.54 mT. Control values of the enzyme were again found when exposure was increased to 0.56 mT. Thus, increased or decreased rates of β-galactosidase synthesis were effected by magnetic fields that differed only slightly in field strength.

It is tempting to compare results by simply calculating induced internal field strength and relating the results to the magnitude of the induced currents. However, this analysis would not provide an adequate explanation. For example, claims of different types of effects are made for small changes in low-frequency repetition rate with little change in average current density.

V. THEORETICAL CONSIDERATIONS

Mechanisms by which the observed "window effects" could be produced have been proposed.[150,152,226–268] These mechanisms have been reviewed by Taylor.[229] Some proposed mechanisms suggest interaction with complex membrane structures requiring cooperative phenomena and/or Bose-Einstein-type condensation of energy into a single mode. Others propose interaction with large molecular systems through vibrational modes or nonlinear interactions that result in local disturbances.

Fröhlich[3,226,230] explained that order or organization exists in many different forms in a system. Order is not confined to spatial order and order of motion can exist in thermal equilibrium. These "other" forms of order provide a basis for physical theories of frequency-selective responses (windows) that occur in a random 37°C thermal environment. One such system of order stems from membrane structure. A membrane, which is about 10^{-8} m thick, maintains a potential on the order of 10 to 100 mV. An oscillation of the membrane perpendicular to its surfaces will generate an oscillating electric dipole, the frequencies of which will be the natural resonances of the system which will be determined by the speed of sound (acoustic velocity) in the membrane. Fröhlich assumed the sound velocity to be about 10^3 m/s. The width of the membrane, 10^{-8} m, determines the wave length and will be equal to an odd multiple of half wave lengths (resonance for a violin string). Thus, the lowest natural vibrational mode will have a frequency on the order of 50 GHz. Fröhlich suggests that similar types of resonant structures will exist in large molecular systems such as protein and DNA. It is assumed that damping is not unduly high in these systems. Numerous oscillating polar systems of this type will interact with one another generating a series of normal modes of the system. Fröhlich suggests that the energy for the maintenance of these oscillations is supplied by metabolic processes. The model implies that very little energy is required to excite the various modes. Thus, coherent electric vibrations can exist in the biological system, provided that the energy supplied to ensembles of normal modes exceeds some critical value, so that a Bose condensation occurs, coherently exciting a single mode, presumably, the lowest. (Bose condensation, or Bose-Einstein condensation, refers to a system of indistinguishable particles capable of oscillating in many different modes; all particles will tend to occupy the lowest energy level, i.e., the mode with the lowest frequency.) Fröhlich offers the highly frequency-dependent biological effects of millimeter waves discussed in this chapter as supportive of not only the presence of these modes, but as evidence of the excitation of these modes by external sources.

The limit-cycle theory[231] was expanded by Kaiser,[234] who proposed that free internal oscillations can be entrained by an appropriate external field. Competition between slow external and fast internal oscillations leads to entrainment through a series of quasi-periodic states. This process is both frequency- and field-strength-dependent and is proposed to account for the "windows" that have been observed in the millimeter region of the spectrum.

More recently, Kaiser[233] extended his theoretical work relating Fröhlich's concepts of coherent excitations to general aspects of nonlinear systems. He suggested that relatively weak external fields can affect "chaotic motion" within biological systems. Eichwald and Kaiser[234] considered inositol-(1,4,5)triphosphate-induced calcium oscillations as a model. They suggest that AC and DC fields affect these oscillations and, in one calculation, predict a delay in the calcium response to an external stimulus. This is suggestive of the results reported by Byus et al.[167] in which levels of protein kinase were reduced in the presence of a modulated microwave field. A brief review of this work and calculations of how calcium oscillations will be affected by externally applied fields at several different frequencies is given by Grundler et al.[235] This model suggests ways in which a biological system may respond differently to external signals of various frequencies (frequency windows), but does not address amplitude dependence.

The reported genetic effects of ELF[186] or millimeter-wave[42] exposures that exhibit window or resonant frequency responses could be the result of direct interaction of the electromagnetic field with the membrane, inhibiting or enhancing membrane control function; or such effects may be due to the direct interaction of the applied field with the chromosome or DNA molecule. Some calculations indicate that the DNA molecule may show strong resonant absorption at microwave frequencies.[228] These calculations considered longitudinal and transverse vibrational modes (acoustic or optical) in dehydrated double-stranded DNA of varying lengths. The calculated resonant modes were length dependent, with the longitudinal acoustic vibrational modes being the most likely in the microwave frequency range. Van Zandt et al.[236] extended this work to consider critically dampened absorption of DNA in aqueous solutions.

The theoretical work of Lu et al.[237] on DNA absorption is not confined to the microwave region, but extends well into the IR where the theoretical predictions have been experimentally demonstrated.[238] Swicord et al.[239] reported considerably enhanced absorption in the x-band region from highly concentrated aqueous solutions containing DNA with a wide distribution of lengths. The same group[240] reported resonant absorption by DNA strands of a uniform length at frequencies that agree with theory.[228] The electric component of the microwave field causes the charged DNA molecule to vibrate at the microwave frequency. If this frequency corresponds to a resonant frequency of the DNA, a condition similar to a mechanical standing wave is established in the molecule. For linearized strands, fundamental and harmonic resonant frequencies are observed; they are related to odd multiples of a half wavelength of the excited sound wave. For circular DNA, resonant frequencies are related to multiples of the wavelength (λ) of the excited sound wave. Resonances are not observed below a few hundred megahertz. At lower frequencies a standing wave cannot be established because the molecule is smaller than $\lambda/2$.

Attempts by others[241,242] to replicate the reported resonant absorptions in DNA[240] were unsuccessful. However, the results indicated that DNA absorbed at least as strongly as highly absorbing water. Saif and colleagues[243] made absorption measurements in the frequency range of 1 MHz to 1 GHz. These measurements of conductivity fit a sum of three Debye relaxations at 7, 33, and 200 MHz, and are consistent with an assumption of absorption due to counter-ions condensed around DNA. The results suggest that the deposition of microwave energy per unit volume was as much as 100 times greater in the immediate vicinity of DNA as in the solution. These latter results are not necessarily supportive of a window effect, but could lead to variations in frequency response over a broad range of frequencies. The work of Sagripanti et al.[244] on plasmid damage due to microwave exposure is supportive of a strong molecular interaction.

The Prohofsky model[228] suggests a means of strongly coupling energy into a large molecule. Even though these modes may have short lifetimes, the continuous, coherent excitation of the mode could, in effect, maintain the molecule in an excited state. Whether, over a long period of time, this continual excitation and relaxation would result in disruption of function, has not been determined. Long-duration (100 ps or longer) excitation of a molecule implies isolation from the thermal environment. Prohofsky[245] has suggested that this model of absorption could explain the observations of Grundler et al.[44] The length of a nucleosome in the yeast chromosome is such that resonant absorption will occur at about 40 GHz. Small variations in frequency could result in rapid shifts into and out of resonance for the same or different nucleosomes.

A pulse-type solution for certain nonlinear partial differential equations is called "soliton" and has been used in solid-state physics, hydrodynamics, and elementary particle theory. Solitons are localized disturbances that are able to propagate through the nonlinear medium unattenuated. They are not wave trains, thus dispersion does not occur. The soliton (solitary wave) was first described by Scott-Russell[246] more than a century ago when he observed the unusual behavior of a wave generated by, separating from, and proceeding in front of a barge in a shallow canal. Several applications of soliton dynamics have been made to biological systems. The most notable example is the application to the α-helical protein first considered by Davydov[247] and then reviewed and expanded by Scott.[248]

The Davydov model relies on the nonlinearity of hydrogen bonding which potentially couples the basic structure of the α-helical protein with the three "spines." This excitation of amide-I (or CO stretching) vibrations, which under linear analysis would rapidly disperse, is coupled to longitudinal sound waves and is propagated as a soliton. However, the amide-I and nonlinear sound wave must be strongly coupled and the amide-I vibrations must exceed a threshold value in order to generate the local disturbance or soliton. Davydov points out that the probability of a soliton causing a photon emission, or vice versa, is very small. It seems reasonable to argue that excitation of the molecule by a photon would result in nonlocalized (e.g., vibrational) modes that would not directly cause a localized disturbance. However, one could postulate that coherent excitation of vibrational modes, as modeled by Prohofsky and Van Zandt,[228,236] could lead to large amplitude excitations even in a highly dampened system. Such large excitations could then possibly result in localized excitations, disturbances, or solitons. Furthermore, the Fröhlich and soliton models are complementary explanations of the storage and transport of energy in biological systems.[249]

Lawrence and Adey[151] suggested that solitons provide an explanation of the way signals are conducted through membranes. A chemical event, the arrival of a protein molecule at the surface of the receptor or channel site, creates a soliton in proteins associated with membrane channels and the soliton propagates along the protein and through the membrane. The soliton then supplies energy for the second chemical event inside the cell. It is suggested that this process may be altered by the presence of external fields, thus causing a bioelectromagnetic response.

Although most theories of interaction deal with induced or permanent dipole responses to electric fields, evidence has been presented suggesting that magnetic fields, through nuclear spin coupling, affect the dielectric properties of cells. Jafary-Asl et al.[250] reported resonant absorption peaks in the dielectric spectra of yeast cultures in solution at frequencies corresponding to proton NMR frequencies. A very sharp resonance (about 1-Hz bandwidth) was observed in the dielectric spectra a little above 2 kHz when a 50-μT magnetic field was applied. The resonant frequency increased proportionally with the applied field strength from 100 to 200 μT. They observed a pronounced drop in the dielectrophoretic yield (for a discussion of dielectrophoresis[251] see also chapter 2) of live yeast cells exposed to 2-kHz fields. This corresponds to the proton nuclear magnetic resonance frequency in the magnetic field of the earth. Kearley et. al[252] utilized vibrational spectroscopy to measure the frequency of proton oscillations within potential wells. This new technique has been used to establish the existence of double minimum potential wells. Hydrogen bonds form a two-dimensional network with protons being delocalized between two equivalent states. Preliminary work has led to the conclusion that proton transfer, which is a common process in biology, occurs naturally as a result of the hydrogen bond. Furthermore, it seems that even weak perturbations can bias the system causing the proton to localize in one of two equivalent wells. In general, it is difficult to explain how the state of the nucleus of the hydrogen atom can affect the macroscopic behavior of cells. However, this innovative work merits further development and analysis and could provide the basis for a theoretical explanation of a frequency "window."

It has been suggested that DC magnetic fields may interact with biological systems to produce resonance-type effects in the ELF frequency region.[253,254] Blackman et al.[253] reported that the field-induced change in calcium levels in samples from chick brain[124,131,133,134] is dependent on the magnitude of the local geomagnetic field. A 16-Hz signal produced changes in calcium efflux when the local geomagnetic field was 38 μT but became ineffective when the local geomagnetic field was reduced to 19 μT. Furthermore, the ineffective (at 38 μT) 30-Hz signal altered calcium efflux when the local geomagnetic field was changed to ±25.3 or ±76 μT. These results led Blackman et al. to postulate a resonance-like relationship for the alteration of calcium efflux requiring that the frequency of the impressed electromagnetic field be proportional to the product of the local geomagnetic field density and an index ($2n + 1$, where $n = 0, 1 \ldots$). The relation between the frequency of the impressed electromagnetic field and the local geomagnetic field required to affect calcium efflux was not suggested.

Grodsky[150] proposed a cooperative model to explain the release of bound calcium from excitable membranes that are exposed to ELF or ELF-modulated RF signals. This model is similar to the Ising model which was originally formulated for ferromagnetism, but which has also been applied to biological problems.[255] Grodsky envisions two two-dimensional parallel sheets; the first is the phospholipid sheet in the plasma membrane and the second is a lattice of "calcium sites." The phospholipid layer is a lattice of "dipole-like sites" where at each site an entity (pointing either up or down) combines with its nearest neighbor to control the occupation (binding) of the calcium sites in the parallel sheet. The "calcium site" sheet is more sparsely populated than the phospholipid-dipole sheet. Changing the state of a single dipole-like entity would thus not immediately affect the occupation of the calcium site, but could potentially lead to such an event. The energy required to change n neighboring dipole-like sites is less than n times the energy required to change one site.

In the Grodsky model[150] four states exist which can be depicted by giving a direction to both the phospholipid dipole-like sites and the calcium binding sites. Each (dipole and binding site) can point either up or down. Those directions, for example, could represent conditions of attraction or repulsion. The four states of the two sheets would then be: (1) both pointing up, (2) both pointing down, (3) pointing toward each other, and (4) pointing away from one another. Each configuration is characterized by an energy state. Shifts between these four bands, or energy states, thus produce resonant conditions. Grodsky suggests that the lowest resonant frequency state will possess the highest average amplitude, be the most likely to be excited, and will fall in the 1- to 30-Hz range. This cooperative model suggests four bands of frequencies which may or may not overlap. The lowest frequency of the lowest band is the most susceptible to resonance stimulation. Thus, small perturbations from an ELF signal may "resonate" with the structure resulting in a number of altered states and, consequently, the release of bound calcium. Similar cooperative models could be applied to other membrane properties or to interactions with any large macromolecular system or subsystem.

The theories described in this section indicate that frequency- or amplitude-specific effects are conceptually possible, although none has been shown to explain, in an unambiguous manner, the experimentally observed effects in biological systems. Liboff[256] suggested that the results of Blackman et al.[253] and subsequent studies by Thomas et al.[257] and McLeod et al.[258] are consistent with or obey ion cyclotron resonance (CR) conditions. These conditions are met when the angular frequency ω ($\omega = 2\pi f$) of the applied AC magnetic field meets the resonance conditions for a particular ion (e.g. Ca^{++}), i.e., $\omega = (q/m)B$, where B is the value of the local or applied DC magnetic field. The magnetically induced electric field presumably produces an acceleration, as in a cyclotron. Liboff noted that the resonance frequencies of certain biologically important ions, e.g., Na^+, K^+, Mg^{++}, and Ca^{++}, are in the ELF frequency range when B is the geomagnetic field. This led Liboff and his colleagues to investigate several calcium-dependent biological processes[259–263] including movement of the diatom *Amphora coffeaeformis.*[264–266] This diatom is reported to have a calcium concentration threshold that is affected by the application of AC and DC magnetic fields. Smith et al.[265] reported significant stimulation of movement in 0.5 and 0.25 mM Ca^{++} at 16 and 48 Hz AC; 32 or 64 Hz had no effect. Elimination of either the AC or DC field resulted in no effect on motility. Furthermore, no effect was observed when the AC and DC fields were perpendicular. Several investigators[267–269] could not replicate the diatom results, thus challenging the CR theory. The CR theory triggered a number of studies utilizing other biological endpoints. Some of these were designed to test the proposed theory and some just utilized the proposed CR conditions in the belief that this might enhance the possibility of positive results. The work of Grande et al.[270] and Smith[271] supported the proposed theory, however, several studies[272–279] failed to support the CR theory.

The application of CR to biological systems has been challenged on theoretical grounds,[280] and Liboff has discussed some limitations of the CR model.[281] Sandweiss[280] points out that at the resonant frequency of 16 Hz in a 35-μT magnetic field with an ion energy of 3.45 eV, the radius of the ion's orbit would be 48 m—a radius too large for biological systems. Furthermore, if resonance is damped for a sufficiently narrow frequency band, as was suggested by Liboff and McLeod,[282] then the mean collision time of the circulating ions cannot be much less than 0.023 s. However, Sandweiss estimated the actual mean free time between collisions to be on the order of 10^{-11} s. Thus, Sandweiss concludes that to accept the CR mechanism as an explanation of ELF EMF effects on ion-transport would call for radical changes in our understanding of matter. McLeod et al.[283] responded to these criticisms with a discussion of gated channels and helical paths through membranes.

The fact that some experimental data of Blackman, Liboff and others fit the relation $\omega = (q/m)$ B is of interest and perhaps important. However, this relation is not unique to CR. This fact, the limitations of the CR hypothesis and the lack of solid confirmatory experimental data led Lednev[284] to propose an alternative theory. This theory is based on earlier, but little known, work by Podgoretskii[285] who analyzed interactions of excited states in the presence of parallel DC and AC magnetic fields. Vibrational mode degeneracy of charged particles occurs in the presence of a magnetic field. The energy difference between the two degenerate states is $\Delta E = \omega(h/2\pi)$ where $\omega = (q/m)B$. This splitting of energy states, Zeeman splitting, is a phenomenon first studied in intense magnetic fields where the energy difference is in the optical range. Lednev proposed that these observations are applicable to ELF fields and that an AC field parallel to the DC field can affect the probability of transition from one of the degenerate states to another state. In this case the magnitude of the DC and AC fields are similar. A parallel AC field varies the energy difference between the two states which affects the probability of ion transition to the ground state. Ledenev does not consider absorption of energy from the field as necessary to drive the process but suggests that the field biases the system in one direction. Lednev discussed the example of Ca^{++} binding to protein complexes. The physical basis of this theory is not clear; what do the ground and excited states refer to? This could refer to the binding and unbinding of Ca^{++} to the protein complex or simple conformational changes which affect function. In either case there are limitations. The model assumes that damping or line broadening due to collisions does not occur. Lednev suggests that the location of an ion inside a firm protein matrix may eliminate collision broadening. This work has been criticized by Adair[286] whose main concerns are collision damping, already mentioned, and the improbability of radiative transitions. He states that "this effect can only be relevant through an electromagnetic transition; an object as heavy as ^{40}Ca is not likely to radiate strongly." This interpretation and criticism again points out a lack of agreement on the physical model being described. It seems that Lednev is addressing the question of biasing a transition that is normally taking place (conformational change that is changing from one functional state to another). Adair seems to address the question of the improbability of a transition. Blanchard and Blackman[287] examined the work of Lednev and found errors. (They use the term ion parametric resonance to avoid the criticism associated with the ion CR.) This re-examination observed that Lednev had omitted a factor of $2(-1)^n$ in his equations. It should be noted that Lednev partially corrected these problems in a subsequent paper[286] where he included the $(-1)^n$ term. According to Blanchard and Blackman,[287] these differences are very important to the expected resonance conditions. This is particularly true for selection of the appropriate amplitudes of the AC and DC fields. Incorrect selection will invalidate the resonance condition and "windows" of the biological response will not be observed. An important extension of Lednev's theory by Blanchard and Blackman was generalization of the results and consideration of other ions, such as Mg^{++}. In a companion paper, Blackman et al.[155] presented experimental confirmation of these predictions.

Weaver and Astumian[289–290] evaluated limits to the detection of electric fields by biological systems. Considering the limitations imposed by cell size and thermal noise, at first they predicted that the smallest electric field detectable by a cell is about 0.1 V/m. A more complex model that considers narrow banding (e.g., a cell receiving signals with a 10-Hz frequency band) and signal averaging (e.g., a signal that is present for 10^4 or 10^5 s) increases the maximum sensitivity of the cell to about 10^{-4} V/m. This model shows that low sensitivity is possible, but averaging times are required which are much longer than the time periods that have been reported for the appearance of observeable effects due to very weak fields.

Adair[291] performed an extensive classical analysis of the limits of detection of ELF electromagnetic fields by biological systems. He discusses coupling of the fields to tissue, thermal electrical noise, noise fields in tissue, noise fields in cell membranes, resonance conditions including cyclotron resonance, and nuclear magnetic resonance. He concludes that external weak ELF electromagnetic fields can not significantly affect biological processes at the cellular level. Weak fields are static magnetic fields smaller than the earth's field of 50 μT and time varying fields less than 4 μT which he compares to an equivalent response of an individual walking through the earth's field. (Of course, circulating currents cannot be produced by uniform motion through a constant magnetic field.) Thus, he discounts the effect of narrow banding, considered by Weaver and Astumian,[289] as well as resonance effects as being incompatible with cell characteristics and the requirement that the mean resonance energy must be greater than kT. Adair's approach leads to the conclusion that explanation of the effect of a weak ELF field on cells must lie outside the scope of conventional physics. It should be noted that Adair considered a very idealized model of the cell.

Litovitz and colleagues[292–296] presented theoretical arguments and experimental results suggesting that the cell may have special characteristics that allow it to receive signals that are below background noise levels. Signal averaging may take place over seconds,[292] time periods much shorter than suggested by Weaver and Astumian.[289] Litovitz and colleagues postulate the need for temporal and spacial coherence. The cell must be exposed to the coherent signal for a period long enough for recognition (perhaps averaging over a few seconds). Litovitz further speculates that spacial coherence is the key condition. The entire cell experiences the same externally applied signal at all locations and at the same time. This means that any or all receptors that are affected by this signal will be simultaneously affected. Time may be an important factor, not for averaging, but for collisions between affected receptor sites, or some other important membrane entity. This agrees with experimental data presented by Litovitz and colleagues indicating a transient increase in ornithine decarboxylase activity when preparations are exposed to 60-Hz ELF fields[292] or to ELF-modulated 915-MHz microwave fields.[293] No effects were observed when cells were exposed to CW microwave radiation. These effects and postulated theory may not suggest a window response, but rather an ELF-modulation-dependent effect.

VI. PULSE-MODULATION-DEPENDENT EFFECTS OF RF RADIATION

Even in the early years of research on the potential hazards of microwave radiation, several scientists considered the possibility that pulsed radiation could produce effects different from those elicited by continuous-wave radiation at the same average power. However, because of the limited availability of pulsed and CW equipment that operate at the same frequency, only a few investigators explicitly searched for such differences. Where comparisons were made, the pulsed equipment was most commonly a radar transmitter. Therefore, the pulse width generally ranged from 1 to 10 μs and the pulse repetition rate from 100 to 1500 pps. For comparison purposes, the average power, or later the SAR, was the same in both microwave-exposed groups.

Although the total amount of energy delivered to the absorber is the same for CW and pulsed irradiation at the same frequency and with the same time-averaged power, the time course of energy deposition and dissipation is different. Depending on the pulse parameters and the size, shape, thermal environment, etc. of the absorbing material, this difference may be significant and may alter physiological responses in a way that depends on the conditions of pulse modulation.

The earliest suggestion that CW and pulsed irradiation might elicit different effects on animals is an anecdotal report by Boysen[297] that pulsed radiation was absorbed more strongly than was CW radiation. Several early Russian experiments in which the effects of pulsed and continuous-wave irradiation were compared were reviewed by Gordon.[298] In one experiment, the mean survival time of mice was more than tenfold greater following pulsed VHF irradiation than following CW irradiation at the same average power density. However, when 3-GHz radiation was used, lethality was higher and more rapid with pulsed waves. Exposure to 3-GHz CW radiation at 100 W/m^2 produced higher core temperatures in rats compared with exposure to pulsed radiation at the same frequency and the same average power. Both pulsed and CW 3-GHz radiation produced similar degrees of hypotension in rats following prolonged exposure (26 weeks). On the whole, these early Russian studies do not suggest that pulse-modulated microwaves are more potent in their biological effect than is CW radiation.

Young adult New Zealand rabbits were exposed (2 hr/day for 3 months) to either CW or pulsed (100 10-μs pulses/s) 2.45-GHz radiation at 15 W/m^2 (1.6 W/kg).[299] A variety of clinical variables were monitored throughout the exposure period, and following exposure a complete pathological examination was completed. Because the sample size was small (six) and the data have large variances, differences between experimental groups are not significant. At the same average power, pulsed and CW 2.8-GHz radiation were shown to have the same effect on respiration rate, heart rate, and subcutaneous temperature of rabbits.[300] Similarly, in a study of the heart rate of embryonic quail, no differences between CW and pulsed radiation were reported.[301]

Frei et al.[302] exposed ketamine-anesthetized rats to CW or pulsed (500 0.5-μs pulses/s) 2.8-GHz radiation. They found that exposures at 300 W/m^2 or greater produced a significant increase in heart rate for pulsed irradiation, but not for CW irradiation. Using the same animal model, this group later reported that CW or pulsed (500 0.5-μs pulses/s) 2.8-GHz radiation,[303] or CW or pulsed (500 2-μs pulses/s) 9.3-GHz radiation[304] produced the same affect on heart rate, blood pressure, ECG and respiration rate. The reason for the difference is not clear, however, both experiments were conducted at dose rates where thermal effects dominate. Lu et al.[305] exposed the heads of ketamine-anesthetized rats to CW or pulsed (16 1μ-s pulses/s) 1.25-GHz radiation at very high levels, 6.4 W. Similar changes in pulse pressure and heart rate were observed for the two irradiation conditions.

Seaman and DeHaan[306] studied the inter-beat intervals of aggregated cardiac cells from chicken embryos during 190-s exposures to 2.45-GHz radiation. CW and two pulse modulation conditions were compared: 10,000 10.9-μs pulses/s and square-wave modulation at 16 Hz or 0.8–1.7 Hz (50% duty cycle). Exposures at 42 W/kg and higher decreased the inter-beat interval, which is consistent with a thermal mechanism. However, exposures at and below 12.2 W/kg increased the inter-beat interval for CW exposures, but decreased it for pulsed exposures. These effects were very small, but statistically significant.

Chronic exposure (3 hr/day for 3 months) of guinea pigs and rabbits to microwaves at 35 W/m^2 was associated with increased lymphocyte count in peripheral blood and in abnormalities in nuclear structure and mitosis in both erythroblastic and lymphoid cells.[307] In general, the effects of pulsed and CW radiation, at the same average power density, were the same. The author suggests the nonthermal nature of the effect, however, the results of this experiment do not address this hypothesis. Czerska et al.[308] investigated spontaneous lymphoblastoid transformation in normal human lymphocytes following exposure to either CW or pulsed (100–1,000 1-μs pulses/s) 2.45-GHz radiation. At thermogenic levels both CW and pulsed irradiation increased transformation with pulsed radiation showing a larger effect. At exposures not expected to increase temperature, only pulsed irradiation enhanced transformation.

Rabbits exposed to 2.95-GHz radiation showed a significant decrease in erythrocyte production and incorporation of iron into erythrocytes.[309–310] Animals exposed under pulse-modulated conditions (reported variously as 1200 1-μs pulses/s[309] and 400 1-s pulses/s[310]) showed a greater effect of the radiation. In fact, the effect was comparable to CW exposures of more than twice the duration.

Rabbit erythrocytes were exposed to CW or pulsed (parameters not provided) 8.42-GHz radiation at 65 W/kg. Both exposure conditions produced an increase in K^+ efflux, with pulsed microwaves showing a greater effect, but one that was not statistically significant.[311]

The effects of CW and pulsed (10-μs pulse width) 2.45-GHz radiation (at 250 and 100 W/m^2) on serum components in the rabbit were compared.[312] The mean colonic temperature was higher in rabbits exposed to CW radiation at 250 W/m^2, but the authors suggest that this is due to differences in exposure geometry leading to significant differences in SAR. Results of the analyses of serum chemistry are consistent with nonspecific thermal stress and do not differ for the two irradiation conditions.

McLees et al.[313] examined the effects of CW and pulsed 13.12-MHz radiation on the mitotic activity and chromosomal aberrations in regenerating rat liver and found no difference between treatment and control values for either exposure condition.

Several neurochemistry experiments comparing the effects of CW and pulsed microwave radiation were described by Baranski[314] and Baranski and Czerski.[310] Guinea pigs and rabbits were irradiated and the levels of acetylcholinesterase and succinic acid dehydrogenase were measured. Although exposure to CW and pulsed radiation often produced statistically different results, the data were highly variable. In some instances the effect was an increase or decrease depending on the choice of an experimental parameter. Because of the high variability in the data, caution must be exercised in using these results.

Lai et al.[315–316] compared the effect of CW and pulsed (500 2-μs pulses/s) 2.45-GHz radiation on sodium-dependent high affinity uptake of choline in the four regions of the brain containing the highest concentrations of cholinergic innervations—striatum, frontal cortex, hippocampus and hypothalamus. Two exposure systems, cylindrical waveguide and miniature anechoic chamber, were used to expose rats at 0.6 W/kg. Only in the hippocampus was the difference between CW and pulsed irradiation statistically significant; pulsed exposure reduced choline uptake while CW exposure had no effect. This was observed for both exposure systems. Both CW and pulsed radiation reduced choline uptake in the frontal cortex while results in the striatum and the hypothalamus depended on the exposure system. Lai et al.[90] suggest that the reason why some CNS effects are elicited by pulsed microwave radiation, but not by CW exposures at the same dose rate, may be that the auditory system acts as the afferent system affected by the microwaves. (Discussed in section III A of this chapter.)

Rabbits were exposed 2 h daily for 3 or 4 months to CW or pulsed (1,200 1-μs pulses/s) 2.95-GHz radiation at 50W/m^2. After irradiation, EEG recordings were obtained and histologic, histochemical, and metabolic studies were undertaken. Exposed animals demonstrated an increase in the high voltage waves in the EEG and decreased acetylcholinesterase levels throughout the brain, but these effects did not depend on the type of exposure. However, the uptake of phosphorus was reduced more by exposures to pulsed microwaves than to CW radiation.[317]

Greater quantities of iv-injected fluorescein were observed in the brains of rats exposed to pulse-modulated microwaves (1.2 GHz, 0.5-ms pulses, 1-kHz sawtooth wave shape) than in CW exposed rats.[318] Following up on these results, Oscar and Hawkins[319] found that specific modulation conditions (5 10-μs pulses/s) were more effective than CW radiation in compromising the blood-brain barrier. The interpretation of these results is complicated by a later study[320] in which it was demonstrated that low-power pulsed (500 2-μs pulses/s) microwave radiation markedly increases the local cerebral blood flow in rats. In this experiment, the effect was due to increased temperature of the brain. An attempted replication of the critical aspects of the Oscar and Hawkins[319] experiment has not led to confirmation of the difference between CW and pulsed irradiation.[321]

The threshold of microwave exposure required[322] to alter rabbit EEGs is about the same for CW or pulsed irradiation. While preliminary investigation of the effect of microwaves on the functioning of *Aplysia* neurons[323] did not demonstrate differences between CW and pulsed irradiation, further study[324] revealed that pulsed 1.5- and 2.45-GHz radiation induced rapid changes in firing rate more readily than did CW irradiation at the same specific absorption rate. This conclusion is based on a small amount of somewhat subjective data and is not amenable to statistical analysis. Isolated frog sciatic nerves exposed to pulsed (50 10-μs pulses/s) 2.45-GHz radiation[325] exhibited the same reduction in nerve viability as did cells exposed to CW radiation.[326] Synchronizing the microwave pulse with various segments of the firing cycle of the nerve was of no consequence. In all cases, the loss in vitality of the exposed nerves was highly significant $p < .01$) when compared to that of control cells. Seaman and Lebovitz[327] reported that neurons in the cat cochlear nucleus respond to 915-MHz pulses at per pulse thresholds of 11.1 mW/g and 0.6 μJ/g. This is similar to the threshold for eighth-nerve fibers, but is as much as an order of magnitude smaller than for many other potentials.

Kholodov[322] stated that, under identical experimental conditions, pulsed radiation was more effective than is CW radiation in altering behavior. This conclusion is based on a series of experiments, of which only one is discussed. In this experiment, a conditioned active avoidance reaction was developed in white mice following which they were exposed to microwave radiation (pulsed or CW) 2 hr/day for 8 days and then retested. In general, mice exposed to the CW or pulsed field required a greater number of trials to attain training criterion than did control mice. The difference in behavior between animals exposed to pulsed fields and those exposed to CW fields was most marked for exposure at the highest level, i.e., 10 W/m^2. Statistical analysis of the data was not provided.

Frey et al.[318] reported that rats avoided pulse-modulated (1000 0.5-ms pulses/s, sawtooth wave shape) 1.2-GHz radiation, but not CW radiation, even though the average power density of the pulsed exposure was one tenth that of the CW exposure. In a preliminary comparison of the effects of pulsed (770 1.3-μs pulses/s) and CW 3-GHz radiation, Roberti et al.[328] found no difference in the spontaneous motor activity of irradiated rats. Low levels of CW and pulsed S-band microwaves yield different effects on the behavior of trained rats.[329] However, the differences were not consistent. In addition, two different frequencies were used (2.45 GHz for CW irradiation and 2.88 GHz for pulsed irradiation), further complicating the interpretation of these results.

Both CW and pulsed (600 1-μs pulses/s) 1.3-GHz radiation, at 3.6–6.7 W/kg, affected operant behavior of rats to a similar degree.[330] Microwave-evoked body movements were studied in mice exposed either to pulsed (80 10-μs pulses/s) or gated CW 1.25-GHz radiation. Although the incidence of microwave-evoked body movements increases with SAR, it did not vary with pulse characteristics.[331] Lebovitz[332] investigated the effect of 1.3-GHz radiation, delivered either CW or pulsed (600 1-μs pulses/s), on rats performing a multi-component operant task. Under both thermal (5.8 or 6.7 mW/g) and non-thermal (3.6 mW/g) exposure conditions, microwave-induced changes in behavior were the same for CW and pulsed irradiations.

Thomas et al.[89] reported that alterations in temporal discrimination were more pronounced when rats were exposed to pulse-modulated radiation compared with instances of CW exposure at the same average power density. Radar-type pulses (500 2-μs pulses/s) of carrier frequency 2.8 GHz were used. Although the differences reported were small they are among the best documented. In interpreting the results of these experiments, and all other behavioral experiments in which animals are exposed to pulsed radiation, one must first consider the possibility that the animals were responding to the auditory sensation produced by the pulsed microwaves (discussed in Section III). This remains to be done.

The cataractogenic potential of pulse-modulated microwaves has been shown to be the same as that of CW microwaves.[333,334] However, irradiation of in vitro preparations of rat lenses with 915 MHz at 120 to 1200 W/kg pulsed irradiation (0.1 to 0.5-μs pulses) was more effective than was CW radiation in altering lens morphology. Capsular and epithelial defects were observed using scanning electron microscopy.[335,336] It was proposed that the increased potency of pulsed microwaves is due to the additional mechanical stress produced by thermoacoustic expansion. Extension of this research to longer (10-μs) pulses (918-MHz radiation) again demonstrated greater lens damage following pulsed radiation.[337] Weiter et al[338] found reduced ascorbic acid levels in rabbit lenses exposed to 2.45-GHz CW or 2.86-GHz (500 1-μs pulses/s) radiation. The effects were similar for both irradiation conditions.

Kues et al.[339] irradiated the eyes of anesthetized cynomolgus monkeys with CW or pulsed (100 10-μs pulses/s) 2.45-GHz microwaves. Lesions in the corneal endothelium were observed by specular microscopy and validated histologically. Pulsed microwaves at 100–150 W/m^2 (2.6–3.9 W/kg) had a greater effect than did CW microwaves at 200–300 W/m^2 (5.3–7.8 W/kg).

Although several reports indicate that pulsed radiation is more likely to produce biological effects than is CW radiation at the same average incident power density (see Table 3), a careful reading of the results may lead to a more qualified conclusion. Differences between the effects of CW and pulsed irradiation are small and sometimes conflicting. Statistical analysis of the results is often omitted and independent corroboration is lacking. However, thermoelastic expansion, which is responsible for effects in the auditory system, may mediate other effects of pulsed radiation. Behavioral changes and alterations in cellular architecture are among those most likely to result from thermoelastic expansion. Mechanisms that depend on a threshold value of field strength or power could be more sensitive to pulsed radiation than to CW radiation. It is expected that, for the biological response to be manifested, it will

TABLE 3
Comparison of the Effects of Continuous-wave (CW) and Pulsed Microwave (PW) Radiation at the Same Average Power Density

Effect	Maximum with PW/CW	Ref.
Survival of mice	PW/CW	298
Hypotension in rats	—	298
Heart and respiration rates in rabbits	—	300
Heart rates in rats	PW	302
Heart and respiration rates in rats	—	303, 304
Heart rate and pulse pressure in rats	—	305
Heart rate of embryonic quail	—	301
Cardiac cell activity in chickens	different	306
Erythrocyte production in rabbits	—	309
K^+ efflux from erythrocytes	—	311
Iron incorporation into rabbit erythrocytes	PW	309
Rabbit blood chemistry	—	312
Lymphoblastoid transformation	PW	308
Chromosomal aberrations and mitotic activity in rat liver (13 MHz)	—	313
Acetylcholinesterase and succinic dehydrogenase levels in rabbits and guinea pigs	—	314
Neurochemical changes in rabbits	PW	317
Viability of frog sciatic nerves	—	325
EEG in rabbits	—	322
Firing rates of *Aplysia* neurons	PW	300, 301
Evoked body movements in mice	—	331
Avoidance by rats	PW	318
Operant behavior in rats	—	330
Spontaneous motor activity of rats	—	328
Trained behavior in rats	PW	81, 329
Conditioned avoidance in mice	PW	322
Multi-component performance of rats	—	332
Blood-brain barrier alteration in rats	PW	324, 319, 321
Metabolism in rat brains	—	157
Choline uptake in rat brains	PW	315
Ascorbic acid levels in rabbit lens	—	338
Cataractogenesis in rabbits	—	333, 334
Rat lens in tissue culture	PW	335, 336
Damage to lens	PW	337
Corneal lesions in monkeys	PW	339

be necessary that some threshold pulse duration be exceeded. Furthermore, the threshold of pulse duration may depend on pulse intensity in a complex manner. Therefore, for some effects, lower thresholds may be found for exposures to pulsed radiation.

VII. CONCLUSIONS

Numerous biological responses to electric and magnetic fields that involve frequency or amplitude "windows" have been reported. The majority of these responses have been observed when exposure is to ELF, ELF-modulated RF, or millimeter-wave signals. In some cases, not enough data exist to be used as the basis of an explanation of the observed responses, or even to establish firmly the nature of the response.

There is no consistent repeatable pattern of millimeter-wave responses. Very few experiments have been repeated by a second investigator; in some cases where duplication has been attempted, the original results were not confirmed. However, these experiments require a high degree of precision in engineering aspects of the experimental design, and lack of such precision could be a reason for the lack of reproducibility.

The existence of highly specific, nonmonotonic responses to electromagnetic radiation is both unexpected and jarring to our sense of simplicity and order and has elicited a response of skepticism by some observers. Models proposed to explain these effects are but a start in the right direction; at present they are either incomplete or inconsistent. The extant body of experimental data is in need of independent replication. However, we hope that the dialogue that has been initiated by these models will be constructive and will lead to the formulation of critical tests.

Only a few reports make persuasive cases for greater sensitivity to pulsed radiation in the microwave region of the spectrum. However, thermoelastic expansion and other possible models argue for lower thresholds for pulsed than for CW radiation, at least in some areas.

REFERENCES

1. **Devyatkov, N. D.**, Ed., Scientific Session of the Division of General Physics and Astronomy, U.S.S.R. Academy of Sciences (17–18 January 1973), *Upekhi Fizicheskikh Nauk,* 110(3), 452, 1973; *Sov. Phys. Usp.,* 16, 568, 1974.
2. **Motzkin, S. M.,** Biological effects of millimeter-wave radiation, in *Biological Effects and Medical Applications of Electromagnetic Energy,* Gandhi, O. P., Ed., Prentice Hall, Englewood Cliffs, NJ, 1990, 373.
3. **Fröhlich, H.,** Long-range coherence and energy storage in biological systems, *Int. J. Quantum Chem.,* 2, 64, 1968.
4. **Webb, S. J. and Dodds, D. E.,** Inhibition of bacterial cell growth by 136-GHz microwaves, *Nature,* 218, 374, 1968.
5. **Webb, S. J. and Booth, A. D.,** Absorption of microwaves by microorganisms, *Nature,* 222, 1199, 1969.
6. **Webb, S. J.,** *Time and Motion in Metabolism and Genetics,* Charles C. Thomas, Springfield, IL., 1976.
7. **Berteaud, A. J., Dardalhon, M., Rebeyrotte, N., and Averbeck, D.,** Action d'un rayonnement electromagnetique a onde millemtrique sur le croissance bacterienne, *C. R. Acad. Sci. Paris,* 281D, 843, 1975.
8. **Dardalhon, M., Averbeck, D., and Berteaud, A. J.,** Determination of a thermal equivalent of millimeter microwaves in living cells, *J. Microwave Power,* 14, 307, 1979.
9. **Blackman, C. F., Beane, S. G., Weil, C. M., and Ali, J. S.,** Effect of nonionizing electromagnetic radiation on single-cell biological systems, *Ann. N.Y. Acad. Sci.,* 247, 352, 1975.
10. **Webb, S. J. and Booth, A. D.,** Microwave absorption by normal and tumor cells, *Science,* 174, 12, 1971.
11. **Gandhi, O. P., Hagmann, M. J., Hill, D. V., Parttow, L. M., and Bush, L.,** Millimeter-wave absorption spectra of biological samples, *Bioelectromagentics,* 1, 285, 1980.
12. **Partlow, L. M., Bush, L. C., Stensaas, L. J., Hill, D. W., Riazi, A., and Gandhi, O. P.,** Effects of millimeter-wave radiation on monolayer cell cultures. I. Design and validation of a novel exposure system, *Bioelectromagnetics,* 2, 123, 1981.
13. **Bush, L. C., Hill, D. W., Riazi, A., Stensaas, L. J., Partlow, L. M., and Gandhi, O. P.,** Effects of millimeter-wave radiation on monolayer cell cultures. III. A search for frequency-specific athermal biological effects on protein synthesis, *Bioelectromagenics,* 2, 151, 1981.
14. **Webb, S. J.,** Factors affecting the induction of lambda prophages by millimeter microwaves, *Phys. Lett.,* A73, 145, 1979.
15. **Kremer, F.,** personal communication, 1984.
16. **Lukashevsky, K. V. and Belayev, I. Ya.,** Switching of prophage genes in *Escherichia coli* by milimeter waves, *Mea. Sci. Res.* 18, 933, 1990.
17. **Kondrat'eva, V. F., Chistyakova, E. N., Shmakova, I. F., Ianova, N. B., and Treskunov, A. A.,** Effects of millimeter-band radiowaves on certain properties of bacteria, *Sov. Phys. Usp.,* 16, 572, 1974.
18. **Devyatkov, N. D.,** Influence of millimeter-band electromagentic radiation on biological objects, *Sov. Phys. Usp.,* 16, 568, 1974.
19. **Kesselev, R. I. and Zalyubovskaya, N. P.,** Millimeter-band electromagentic waves in the cell and certain structural elements of the cell, *Sov. Phys. Usp.,* 16, 576, 1974.

20. **Zalyubovskaya, N. P.,** Reactions of living organisms to exposure to millimeter-band electromagnetic waves, *Sov. Phys. Usp.*, 16, 574, 1974.
21. **Smolyanskaya, A. Z. and Vilenskaya, R. L.,** Effects of millimeter-band electromagnetic radiation on the functional activity of certain genetic elements of bacterial cells, *Sov. Phys. Usp.*, 16, 571, 1974.
22. **Smolyanskaya, A. Z.,** Action of millimeter electromagnetic waves on fat cells, in *Nonthermal Effects of Millimeter Radiation*, Devyatkov, N. D., Ed., U.S.S.R. Academy of Sciences, Moscow, 1981, chap. 7.
23. **Motzkin, S. M., Benes, L., Block, N., Israel, B., May, N., Kuriyel, J., Birenbaum, L., Rosenthal, S., and Han, Q.,** Effects of low-level millimeter waves on cellular and subcellular systems, in *Coherent Excitations in Biological Systems*, Fröhlich, H. and Kremer, F., Eds., Springer-Verlag, Berlin, 1983, 47.
24. **Sevast'yanova, L. A. and Vilenskaya, R. L.,** Mouse bone marrow reaction to altered UHF millimeter irradiation, parametric variation, *Biol. Nauki.*, 6, 48, 1974.
25. **Sevast'yanova, L. A. and Vilenskaya, R. L.,** A study of the effects of millitmeter-band microwaves on the bone marrow of mice, *Sov. Phys. Usp.*, 16, 570, 1974.
26. **Smolyanskaya, A. Z., Gel'vich, E. A., Golant, M. B., and Makhov, A. M.,** Resonance phenomena in the action of millimeter waves on biological objects, *Usp. Sovrem. Biol.*, 87(3), 381, 1979.
27. **Sevast'yanova, L. A.,** Specific action of millimeter radiowaves on biological systems, in *Nonthermal Effects of Millimeter Radiation*, Devyatkov, N. D., Ed., U.S.S.R. Academy of Sciences, Moscow, 1981, chap. 5.
28. **Sevast'yanova, L. A., Borodkina, A. G., Zubenkova, E. S., Golant, M. B., Rebrova, T. B., and Iskritskiy, V. L.,** Resonance effect of millimeter-range radiowaves on biological systems, in *Biological Effects of Millimeter-Wave Radiation*, ed N. D. Deryatkov, USSR Academy of Sciences, Moscow 1983 (English translation by the Foreign Broadcast Information Service 1985).
29. **Belyaev, I. Ya.,** Some biophysical aspects of the genetic effect of low-intensity millimeter waves, *Bioelectrochem. and Bioenerg.*, 27, 11, 1992.
30. **Belyaev, I. Ya., Shcheglov, V. S., and Alipov, Ye. D.,** Selection rules on helicity during discrete transitions of the genome conformational state in intact and x-rayed cells of *E. coli* in millimeter range of electromagnetic field, *Charge and Field Effects in Biosystems*, M. J. Allen, S. F. Cleary, A. E. Sowers, D. D. Shillady, Eds., Boston: Birkhauser, 1992, 115.
31. **Belyaev, I. Y., Alipov, Y. D., and Shcheglov, V. S.,** Chromosome DNA as a target of resonant interaction between *Escherichia coli* cells and low-intensity millimeter waves, *Electro. Magnetobiol.*, 11, 97, 1992.
32. **Belyaev, I. Y., Alipov, Y. D., Polunin, V. A., and Shcheglov, V. S.** Evidence for dependence of resonant frequency of millimeter wave interaction with *Escherichia coli* k12 cells on haploid genome length, *Electro. Magnetobiol.*, 12, 39, 1993.
33. **Belyaev, I. Ya. and Kravchenko, V. G.,** Resonance effect of low-intensity millimeter waves on the genome conformational state of rat thymocytes (MEETING ABSTRACT), European Bioelectromagnetics Assoc., 2nd Congress, 9–11 December, Bled, Slovenia, p. 82 1993.
34. **Belyaev, I. Y., Alipov, Y. D., Shcheglov, V. S., and Lystov, V. N.,** Resonance effect of microwaves on the genome conformational state of *E. coli* cells, *Z. Naturforsch.*, 47c, 621, 1992.
35. **Belyaev, I. Ya., Alipov, Ye. D., Matronchik, A. Yu., Radko, S. P., and Aizenberg, O. A.,** Possible physical mechanism of *E. coli* cell response to weak dc and ac magnetic fields (MEETING ABSTRACT), European Bioelectromagnetics Assoc., 2nd Congress, 9–11 December, Bled, Slovenia, p. 49 1993.
36. **Smolyanskaya, A. Z., Makhov, A. M., Gel'vich, E. A., and Golant, M. B.,** Influence of electromagnetic waves in the millimeter band on the induction of synthesis of penicillinase by *Staphlococcus aureus, Biol. Nauki.*, 5, 24, 1981.
37. **Ostapenkov, A. M., Matiso, V. A., Kaptreva, I. V., Belovolov, A. V., and Laurova, V. L.,** Investigation of the effects of low intensity UHF electromagnetic waves on *Bacillus mesentericus and Pseudomonas fluorescens, Biol. Nauki.*, 6, 47, 1976.
38. **Zalyubovskaya, N. P., Gordiyenko, O. I., and Kisselev, P. I.,** Action of electromagentic fields of superhigh frequency on erythrocytes preserved at low temperature, *Probl. Gemalol. Perelivaniya Krovi*, 20(4), 31, 1974.
39. **Devyatkov, N. D., Betskiy, O. V., Gel'vich, E. A., Golant, M. B., Mahkov, A. M., Rebrova, T. B., Sevast'yanova, L. A., and Smolyanskaya, A. Z.,** Effects on biological systems of electromagnetic oscillations in the millimeter range of wavelengths, *Radiobiologica*, 21(2), 163, 1981.
40. **Sevast'yanova, L. A.,** Distinctive features of biological effects of millimeter-range radiowaves and possible uses thereof in medicine, *Vestn. Akad. Med. Nauk. U.S.S.R.*, 2, 65, 1979.
41. **Grundler, W. and Keilmann, F.,** Nonthermal effects of millimeter microwaves on yeast growth, *Z. Naturforsch.*, 33c, 15, 1978; **Grundler, W., Keilmann, F., and Fröhlich, H.,** Resonant growth rate response of yeast cells irradiated by weak microwaves, *Phys. Lett.*, 62A, 463, 1977.
42. **Grundler, W., Keilmann, F., and Strube, D.,** Resonant-like dependence of yeast growth rate on microwave frequencies, *Br. J. Cancer.*, 45 (Suppl. 5), 206, 1982.
43. **Grundler, W. and Keilmann, F.,** Sharp resonances in yeast growth prove nonthermal sensitivity to microwaves, *Phys. Rev. Lett.*, 51, 1214, 1983.
44. **Grundler, W., Keilmann, F., Putterlik, V., Santo, L., Strube, D., and Zimmermann, I.,** Non-thermal

resonant effects of 42-GHz microwaves on the growth of yeast cultures. in Coherent *Excitations in Biological Systems,* Fröhlich, H. and Kremer, F., Eds., Springer-Verlag, Berlin, 1983, 21.
45. **Furia, L., Hill, D. W., and Gandhi, O. P.,** Effects of Millimeter-wave irradiation on growth of *Saccharomyces cerevisiae, IEEE Trans. Biomed Eng.,* 11, 993, 1986.
46. **Dardalhon, M., Averbeck, D., and Berteaud, A. J.,** Studies on possible genetic effects of microwaves on procaryotic and eucaryotic cells, *Rad. Environ. Biophy.,* 20, 37, 1981.
47. **Webb, S. J., Lee, R., and Stoneham, M.,** Possible viral involvement in human carcinomas, *Int. J. Quantum Chem.,* Suppl. 4, 277, 1977.
48. **Webb, S. J. and Stoneham, M.,** Resonance between 10^{11} and 10^{12} Hz in active bacterial cells as seen by laser Raman spectroscopy, *Phys. Lett.,* 60A, 261, 1977.
49. **Webb, S. J., Stoneham, M., and Fröhlich, H.,** Evidence for nonthermal excitation of energy levels in active biological systems, *Phys. Lett.,* 63A, 407, 1977.
50. **Webb, S. J.,** Coherent excitations in biological systems, *Phys. Rep.,* 60, 201, 1980.
51. **Drissler, F. and Santo, L.,** Coherent excitations and Raman effects, in *Coherent Excitations in Biological Systems,* Fröhlich, H. and Kremer, F., Eds., Springer-Verlag, Berlin, 1983, 6.
52. **Furia, L. and Gandhi, O. P.,** Absence of biologically related Raman lines in cultures of *Bacillus megasterium, Phys. Lett.,* 102A, 380, 1984.
53. **Cooper, M. S. and Amer, N. M.,** The absence of coherent vibrations in the Raman spectra of living cells, *Phys. Lett.,* 98A, 138, 1983.
54. **O'Sullivan, R. A. and Santo, J.,** Experimental aspects of Raman spectroscopy of microorganisms, *Can. J. Spectrosc.,* 26, 143, 1981.
55. **Illinger, K. H.,** Electromagnetic-field interaction with biological systems in the microwave and far-infrared region: physical basis, in *Biological Effects of Nonionizing Radiation,* Illinger, K. H., Ed., American Chemical Society, Washington, DC, 1981, 1.
56. **Illinger, K. H.,** Spectroscopic properties of in vivo biological systems: boson radiative equilibrium with steady state non-equilibrium molecular systems, *Bioelectromagentics,* 3, 9, 1982.
57. **Gandhi, O. P., Hill, D. W., Riazi, A., Wahid, P., Wang, C. H., and Iskander, M. F.,** *Biological Efffects of Millimeter Wave Irradiation,* Final report USAFSAM-TR-82-49, Brooks Air Force Base, TX, 1982.
58. **Layne, S. P., Bigio, S. J., Scott, A. C., and Lomdahl, P. S.,** Transient fluorescence in synchronously dividing *Escherichia coli, Proc. Natl. Acad. Sci. USA,* 82, 7599, 1985.
59. **Kremer, F., Koschnitzke, C., Santo, L., Quick, P., and Poglitsch, A.,** The non-thermal effect of millimeter wave radiation on the puffing of giant chromosomes, in *Coherent Excitations in Biological Systems,* Fröhlich, H. and Kremer, F., Eds., Springer-Verlag, Berlin, 1983, 10.
60. **Khramov, R. N., Sosunov, E. A., Koltun, S. V., Ilyasova, E. N. and Lednev, V. V.,** Millimeter-wave effects on electric activity of crayfish stretch receptors, *Bioelectromagnetics,* 12, 203, 1991.
61. **Stamm, M. E., Winters, W. D., Morton, D. J., and Warren, S. L.,** Microwave characteristics of human tumor cells, *Oncology,* 29, 294, 1974.
62. **Adey, W. R.,** Models of membranes of cerebral cells as substrates for information storage, *Biosystems,* 8, 163, 1976.
63. **Adey, W. R.,** Neurobiological effects of radiofrequency and microwave radiation, *Ann. N.Y. Acad. Med.,* 55, 1079, 1979.
64. **Adey, W. R.,** Long-range electromagnetic field interaction at brain cell surfaces, in *Magnecic Field Effects in Biological Systems,* Tenforde, T. S., Ed., Plenum Press, New York, 1979, 56.
65. **Adey, W. R.,** Frequency and power windowing in tissue interactions with weak electromagnetic fields, *Proc. IEEE,* 68, 119, 1980.
66. **Adey, W. R.,** Tissue interactions with nonionizing electromagnetic fields, *Physiol. Rev.,* 61, 435, 1981.
67. **Adey, W. R.,** Ionic nonequilibirum phenomena in tissue interactions with electromagnetic fields, in *Biological Effects of Nonionizing Radiation,* Illinger, K. H., Ed., American Chemical Society, Washington, DC, 1981, 271.
68. **Adey, W. R.,** Electromagnetic fields and the essence of living systems, in *Modern Radio Science 1990,* Bach, J., Ed., Oxford University Press, Oxford, UK, 1990, 1.
69. **Adey, W. R.,** Radiofrequency electromagnetic radiation, in *Electromagnetic Interactions with Biological Systems,* Lin, J. C., Ed. Plenum Publishing Co., NYC, 1989, 109.
70. **Adey, W. R.,** Cell membranes: The electromagnetic environment and cancer promotion, *Neurochem. Res.,* 13, 38, 1987.
71. **Adey, W. R. and Bawin, S. M.,** Binding and release of brain calcium by low level electromagnetic fields: a review, *Radio Sci.,* 17(5S), 149, 1982.
72. **Myers, R. D. and Ross, D. H.,** Radiation and brain calcium: a critique, *Neurosci. Neurobehav. Rev.,* 5, 503, 1981.
73. **Frey, A. H.,** Auditory system response to RF energy, *Aerosp. Med.,* 32, 1140, 1961.

74. **Foster, K. R. and Finch, E. F.,** Microwave hearing: evidence for thermoacoustic auditory stimulation by pulsed microwaves, *Science,* 185, 256, 1974.
75. **Frey, A. H.,** Human auditory system response to modulated electromagnetic energy, *J. Appl. Physiol.,* 17, 689, 1962.
76. **Frey, A. H.,** Some effects on human subjects of ultra-high-frequency radiation, *Am. J. Med. Electron.,* 2, 28, 1963.
77. **Sommer, R. C. and Von Gierke, H. E.,** Hearing sensations in electric fields, *Aerosp. Med.,* 35, 834, 1964.
78. **Ingalls, C. E.,** Sensation of hearing in electromagnetic fields, *N.Y. State J. Med.,* 67, 2992, 1967.
79. **Frey, A. H.,** Biological functions as influenced by low power modulated RF energy, *IEEE Trans. Microwave Theory Tech.,* 19, 153, 1971.
80. **Guy, A. W., Chou, C-K., Lin, J. C., and Christensen, D.,** Microwave-induced acoustic effects in mammalian auditory systems and physical materials, *Ann. N.Y. Acad. Sci.,* 247, 194, 1975.
81. **Cain, C. A. and Rissman, W. J.,** Mammalian auditory responses to 3.0 GHz microwave pulses, *IEEE Trans. Biomed. Eng.,* 25, 288, 1978.
82. **Frey, A. H. and Messenger, R.,** Human perception of illumination with pulsed ultra high frequency electromagnetic energy, *Science,* 181, 356, 1973.
83. **Frey, A. H. and Feld, S. R.,** Avoidance by rats of illumination with low power nonionizing electromagnetic energy, *J. Comp. Physiol. Psychol.,* 89, 183, 1975.
84. **Hjeresen, D. L., Doctor, S. K. and Shelton, R. L.,** Shuttle box side preference as mediated by pulsed microwaves and conventional auditory cues, in *Proc. Symp. on Electromagnetic Fields in Biological Systems,* International Microwave Power Institute, Edmonton, Alberta, Canada, 1979, 194.
85. **King, N. W., Justesen, D. R., and Clarke, R. L.,** Behavioral sensitivity to microwave radiation, *Science,* 172, 398, 1971.
86. **Johnson, A. B., Myers, D. E., Guy, A. W., Lovely, R. H., and Galambos, R.,** Discriminative Control in Appetitive Behavior by Pulsed Microwaves, HEW Publ. FDA77-8010, U.S. Department of Health, Education, and Welfare, Rockville, MD, 1976, 238.
87. **Seaman, R. L. and Beblo, D. A.,** Modification of acoustic startle by microwave pulses in the rat: a preliminary report, *Bioelectromagnetics,* 13, 323, 1992.
88. **Servantie, B., Servantie, A. M., and Etienne, J.,** Synchronization of cortical neurons by a pulsed microwave field as evidenced by spectral analysis of electrocorticograms from the white rat, *Ann. N.Y. Acad. Sci.,* 247, 82, 1975.
89. **Thomas, J. R., Schrot, J., and Banvard, R. A.,** Comparative effects of pulsed and continuous wave 2.8-GHz microwaves on temporally defined behavior, *Bioelectromagnetics,* 3, 227, 1982.
90. **Lai, H., Horita, A., Chou, C. K., and Guy, A. W.,** A review of microwave irradiation and actions of psychoactive drugs, *IEEE Trans. Eng. Med. Biol.,* Mar, 31, 1987.
91. **Frey, A. H.,** Brain-stem evoked responses associated with low-intensity pulsed VHF energy, *J. Appl. Physiol.,* 23, 984, 1967.
92. **Chou, C-K., Galambos, R., and Lovely, R. H.,** Cochlear, microphonics generated by microwave pulses, *J. Microwuve Power,* 10, 361, 1975.
93. **Chou, C-K., Guy, A. W., and Galambos, R.,** Microwave induced auditory responses: cochlear microphonics, in *Biologicul Effects of Electromagnetic Waves,* Vol. 1, Johnson, C. C. and Shore, M. I., Eds., Department of Health Education and Welfare, Rockville, MD, 1976, 89.
94. **Chou, C-K., Guy, A. W., and Galambos, R.,** Microwave-induced cochlear microphonics in cats, *J. Microwave Power,* 11, 171, 1976.
95. **Chou, C-K. and Guy, A. W.,** Auditory perception of radiofrequency electromaunetic fields, *J. Acoust. Soc. Am.,* 71, 1321, 1982.
96. **Frey, A. H. and Coren, E.,** Holographic assessment of a hypothesized microwave hearing mechanism, *Science,* 206, 232, 1979.
97. **Lebovitz, R. M. and Seaman, R. L.,** Single auditory unit responses to weak pulsed microwave radiation, *Brain. Res.,* 126, 370, 1976.
98. **Lebovitz, R. M. and Seaman, R. L.,** Microwave hearing: the response of single auditory neurons in the cat to pulsed microwave radiation, *Radio Sci.,* 12(6S), 229, 1977.
99. **Taylor, E. M. and Ashleman, B. T.,** Analysis of central nervous system involvement in the microwave auditory effect, *Brain Res.,* 14, 201, 1974.
100. **Johnson, C. C. and Guy, A. W.,** Nonionizing electromagnetic wave effects in biological materials and systems, *Proc. IEEE,* 60, 692, 1972.
101. **Lin, J. C., Meltzer, R. J., and Redding, F. K.,** Microwave evoked brain stem potentials in cats, *J. Microwave Power,* 14, 291, 1979.
102. **Chou, C-K. and Guy, A. W.,** Microwave-induced auditory responses in guinea pigs: relation of threshold and microwave-pulse duration, *Radio Sci.,* 14(6S), 193, 1979.

103. **Wilson, B. S., Zook, J. M., Joines, W. T., and Cassedy, J. H.,** Alterations in activity at auditory nuclei of the rat induced by exposure to microwave radiation: audiographic evidence using ^{14}C 2-deoxy-d-glucose, *Brain Res.,* 187, 291, 1980.
104. **Chou, C-K. and Galambos, R.,** Middle ear structures contribute little to auditory perception of microwaves, *J. Microwave Power,* 14, 321, 1979.
105. **White, R. M.,** Generation of elastic waves by surface heating, *J. Appl. Physics.,* 34, 3559, 1963.
106. **Gournay, L. S.,** Conversion of electromagnetic to acoustic energy by surface heating, *J. Acoust. Soc. Am.,* 40, 1322, 1966.
107. **Lin, J. C.,** Microwave auditory effect—a comparison of some possible transduction mechanisms, *J. Microwave Power,* 11, 77, 1976.
108. **Borth, D. E. and Cain, C. A.,** Theoretical analysis of acoustic signal generation in materials generated with microwave energy, *IEEE Trans. Microwuve Theory Tech.,* 25, 944, 1977.
109. **Lin, J. C.,** *Microwave Auditory Effects and Applications,* Charles C Thomas, Springfield, IL, 1978.
110. **Shapiro, A. R., Lutomirski, R. F., and Hura, H. T.,** Induced fields and heating within a cranial structure irradiated by an electromagnetic plane wave, *IEEE Trans. Microwave Theory Tech.,* 19, 187, 1971.
111. **Kritikos, H. and Schwan, H. P.,** Hot spots generated in conducting spheres by E. M. waves and biological implications, *IEEE Trans. Biomed. Eng.,* 19, 53, 1972.
112. **Kirtikos, H. and Schwan, H. P.,** The distribution of heating potential inside lossy spheres, *IEEE Trans. Biomed. Eng.,* 22, 451, 1973.
113. **Weil, C. M.,** Absorption characteristics of multilayered sphere models exposed to UHF/microwave radiation. *IEEE Trans. Biomed. Eng.,* 22, 468, 1975.
114. **Lin, J. C.,** Microwave hearing sensation, *IEEE Trans. Microwave Theory Tech.,* 25, 605, 1977.
115. **Lin, J. C.,** Further studies on the microwave auditory effect, *IEEE Trans. Microwave Theory Tech.,* 25, 938, 1977.
116. **Lin, J. C., Su, J.-L., and Wang, Y.,** Microwave-induced thermoelastic pressure wave propagation in the cat brain, *Bioelectromagnetics,* 9, 141, 1988.
117. **Olsen, R. G.,** Generation of Acoustic Images from the absorption of pulsed microwave energy, in *Acoustic Imaging Vol. 11,* Powers, J. P., Ed., Plenum, NYC, 1982, 53.
118. **Su, J-L. and Lin, J. C.,** Theremoelastic signatures of tissue phantom absorption and thermal expansion, *IEEE Trans. Biomed. Eng.,* 34, 179, 1987.
119. **Bawin, S. M.,** *Cat EEG and Behavior in Very High Frequency Electric Fields Amplitude-Modulated at Brain Wave Frequencies,* Ph.D. dissertation, University of California, Los Angeles, 1972.
120. **Bawin, S. M., Gavalas-Medici, R. J., and Adey, W. R.,** Effect of modulated very high frequency fields on specific brain rhythms in cats, *Brain Res.,* 58, 365, 1973.
121. **Bawin, S. M., Gavalas-Medici, R. J., and Adey, W. R.,** Reinforcement of transient brain biorhythms by amplitude-modulated VHF fields, in *Biological and Clinical Effects of Low Frequency Magnetic and Electric Fields,* Llaurado, J. G., Sances, A., Jr., and Battocletti, J. H., Eds.. Charles C Thomas, Springfield, IL., 1975, 172.
122. **Gavalas, R. J., Walter, D. O., Hammer, J., and Adey, W. R.,** Effect of low-level low-frequency electric fields on EEG and behavior in *Macaca nemestrina, Brain Res.,* 18, 491, 1970.
123. **Gavalas-Medici, R. J. and Day-Magdaleno, S. R.,** Extremely low frequency, weak electric fields affect schedule-controlled behavior of monkeys, *Nature,* 261, 256, 1976.
124. **Bawin, S. M., Kaczmarek, K. L., and Adey, W. R.,** Effects of modulated VHF lields on the central nervous system, *Ann. N.Y. Acad. Sci.,* 247, 74, 1975.
125. **Tenforde, T. S.,** Thermal aspects of electromagnetic field interactions with bound calcium at the nerve cell surface *J. Theor. Biol.,* 83, 517, 1980.
126. **Blackman, C. F., Elder, J. A., Weil, C. M., Benane, S. G., Eichinger, D. C., and House, D. E.,** Induction of calcium ion efflux from brain tissue by radiofrequency radiation: effects of modulation, frequency and field strength, *Radio Sci.,* 14(6S), 93, 1979.
127. **Blackman, C. F., Benane, S. C., Elder, J. A., House, D. E., Lampe, J. A., and Faulk, J. M.,** Induction of calcium ion efflux from brain tissue by radiofrequency radiation: effect of sample number and modulation frequency on the power density window, *Bioelectromagnetics,* 1, 35, 1980.
128. **Sheppard, A. R., Bawin, S. M., and Adey, W. R.,** Models of long-range order in cerebral macromolecules: effect of sub-ELF and of modulated VHF and UHF fields, *Radio Sci.,* 14(6S), 141. 1979.
129. **Blackman, C. F., Benane, S. G., Joines, W. T., Hollis, M. A., and House, D. E.,** Calcium ion efflux from brain tissue: power density versus internal field intensity dependencies at 50 MHz RF radiation, *Bioelectromagnetics,* 1, 211, 1980.
130. **Blackman, C. F., Kinney, L. S., House, D. E., and Joines, W. T.,** Multiple power-density windows and their possible origin, *Bioelectromagnetics,* 10, 115, 1989.
131. **Bawin, S. M. and Adey, W. R.,** Sensitivity of calcium binding in cerebral tissue to weak environmental electric fields oscillating at low frequencies, *Proc. Natl. Acad. Sci. U.S.A.,* 73, 1999, 1976.

132. **Adey, W. R., Bawin, S. M., and Lawrence, A. F.,** Effects of weak amplitude-modulated microwave fields on calcium efflux from awake cat cerebral cortex, *Bioelectromagnetics,* 3, 295, 1982.
133. **Blackman, C. F., Benane, S. G., Kinney, L. S., Joines, W. T., and House, D. E.,** Effect of ELF on calcium-ion efflux from brain tissue in vivo, *Radiat. Res.,* 92, 510, 1982.
134. **Blackman, C. F., Benane, S. G., House, D. E., and Joines, W. T.,** Effects of ELF (1–120 Hz) and modulated (50 Hz) RF fields on the efflux of calcium ions from brain tissue in vivo, *Bioelectromagnetics,* 6, 1, 1985.
135. **Joines, W. T. and Blackman, C. F.,** Power density, field intensity and carrier frequency as determinants of RF-energy-induced calcium ion efflux from brain tissue, *Bioelectromagnetics,* 1, 271, 1980.
136. **Joines, W. T. and Blackman, C. F.,** Equalizing the electric field intensity within chick brain immersed in buffer solution at different carrier frequencies, *Bioelectromagnetics,* 2, 411, 1981.
137. **Athey, T. W.,** Comparison of RF-induced calcium efflux from chick brain tissue at different frequencies: do the scaled power density windows align?, *Bioelectromagnetics,* 2, 407, 1981.
138. **Weil, C. M., Spiegel, R. S., and Joines, W. T.,** Internal field strength measurements in chick forebrains at 50, 147 and 450 MHz, *Bioelectromagnetics,* 5, 293, 1984.
139. **Blackman, C. F., Benane, S. G., Elliott, D. J., House, D. E., and Pollock, M. M.,** Influence of electromagnetic fields on the efflux of calcium ions from brain tissue in vitro: a three-model analysis consistent with the frequency response up to 510 Hz, *Bioelectromagnetics,* 9, 215, 1988.
140. **Blackman, C. F., Benane, S. G., and House, D. E.,** The influence of temperature during electric- and magnetic-field-induced alteration of calcium-ion release from in vitro brain tissue, *Bioelectromagnetics,* 12, 173, 1991.
141. **Blackman, C. F., Benane, S. G., House, D. E., Joines, W. T., and Speigel, R. J.,** Effect of ambient levels of power-line-frequency electric fields on a developing vertebrate, *Bioelectromagnetics,* 9, 129, 1988.
142. **Albert, E. N., Slaby, F., Roche, F., and Loftus, J.,** Effect of amplitude-modulated 147-MHz radiofrequency radiation on calcium ion efflux from avian brain tissue, *Rad. Res.,* 109, 19, 1987.
143. **Merritt, J. G., Shelton, W. S., and Chamnes, A. F.,** Attempts to alter $^{45}Ca^{2+}$ binding to brain tissue with pulse-modulated microwave energy, *Bioelectromagnetics,* 3, 475, 1982.
144. **Lin-Liu, S. and Adey, W. R.,** Low frequency amplitude-modulated microwave fields change calcium efflux rates from synaptosomes, *Bioelectromagnetics,* 3, 309, 1982.
145. **Dutta, S. K., Subramoniam, A., Ghosh, B., and Parshad, R.,** Microwave radiation-induced calcium ion efflux from human neuroblastoma cells in culture, *Bioelectromagnetics,* 5, 71, 1984.
146. **Dutta, S. K., Ghosh, B., and Blackman, C. F.,** Radiofrequency-induced calcium ion efflux enhancement from human and other neuroblastoma cells in culture, *Bioelectromagnetics,* 10, 197, 1989.
147. **Dutta, S. K., Das, K., Ghosh, B., and Blackman, C. F.,** Dose dependence of acetylcholinesterase activity in neuroblastoma cells exposed to modulated radiofrequency electromagnetic radiation, *Bioelectromagnetics,* 13, 317, 1992.
148. **Dutta, S. K., Watson, B., and Das, K. P.,** Intensity dependence of enolase activity by modulated radiofrequency radiation, *Bioelectrochem. and Bioenerg.,* 27, 179, 1992.
149. **Adey, W. R. and Bawin, S. M.,** Brain interactions with weak electric and magnetic fields, *Neurosci. Res. Prog. Bull.,* 15(1), 1977.
150. **Grodsky, I. T.,** Possible physical substrates for the interaction of electromagnetic fields with biologic membranes, *Ann. N.Y. Acad. Sci.,* 247, 117, 1975.
151. **Lawrence, A. F. and Adey, W. R.,** Non-linear wave mechanisms in interaction between excitable tissue and electromagnetic fields, *Neurol. Res.,* 4, 115, 1982.
152. **Spiegel, R. J., Joines, W. T. and Blackman, C. F.,** Calcium-induced efflux from isolated brain tissue: is it caused by an electromagnetically-induced pressure wave? (Abstr.), in *4th Annual Meeting Bioelectromagnetics Society,* 1982.
153. **Blackman, C. F., Benane, S. G., and House, D. E.,** Evidence for direct effect of magnetic fields on neurite outgrowth, *FASEB J.,* 7, 801, 1993.
154. **Blackman, C. F., Benane, S. G., House, D. E. and Pollock, M. M.,** Action of 50-Hz magnetic fields on neurite outgrowth in preochromocytoma cells, *Bioelectromagnetics,* 14, 273, 1993.
155. **Blackman, C. F., Blanchard, J. P., Benane, S. G., and House, D. E.,** Empirical test of an ion parametric resonance model for magnetic field interactions with PC-12 cells, *Bioelectromagnetics,* 15, 239, 1994.
156. **Sanders, A. P., Joines, W. T., and Allis, J. W.,** The differential effects of 200, 591 and 2450 MHz radiation on rat brain energy metabolism, *Bioelectromagnetics,* 5, 419, 1984.
157. **Sanders, A. P., Joines, W. T. and Allis, J. W.,** Effects of continuous wave, pulsed and sinusoidal-amplitude-modulated microwaves on brain energy metabolism, *Bioelectromagnetics,* 6, 89, 1985.
158. **Allis, J. W. and Fromme, M. L.,** Activity of membrane-bound enzymes exposed to sinusoidally modulated 2450-MHz microwave radiation, *Radio Sci.,* 14 (6S) 85, 1979.
159. **Persson, B. R. R., Salford, L. G., Brun, A., Eberhardt, J. L., and Malmgren, L.,** Increased permeability of the blood-brain barrier induced by magnetic and electric fields, *Ann. N.Y. Acad. Sci.,* 649, 356, 1992.

160. **Salford, L. G., Brun, A., Eberhardt, J., Malmgren, L., and Persson, B., R., R.,** Electromagnetic field-induced permiability of the blood-brain barrier shown by immunohistochemical methods, in *Interaction Mechanisms of Low-Level Electromagnetic Fields in Living Systems,* Ramel, C. and Norden, B., Eds. Oxford Science Publications, Oxford, 1992, 251.
161. **Kuznetsov, V. I., Iurinskaia, N. M., Kolomytkin, O., V., and Akoev, I. G.,** Action of microwaves with different modulation frequencies and exposure times on GABA receptor concentration in the cerebral cortex of rats, *Radiobiologica,* 31, 257, 1991.
162. **Philloppova, T. M., Novoselov, V. I., Bystrova, M. F., and Alekseev, S. I.,** Microwave effects on camphor binding to rat olfactory epithelium, *Bioelectromagnetics,* 9, 347, 1988.
163. **Bolshakov, M. A. and Alekseev, S. I.,** Bursting Responses of *Lymnea* neurons to microwave radiation, *Bioelectromagnetics,* 13, 119, 1992.
164. **Huang, A. T., Engle, M. E., Elder, J. A., Kinn, J. B., and Ward, T. B.,** The effect of microwave radiation (2450 MHz) on the morphology and chromosomes of lymphocytes, *Radio Sci.,* 12(6S), 173, 1977.
165. **Janosy, C. and Greaves, M. F.,** Lymphocyte activation: II. Stimulation of lymphocyte subpopulations by phytomitogens and heterotogous antilymphocyte sera, *Clin. Exp. Immun.,* 10, 525, 1972.
166. **Lyle, D. B., Schecter, P., Adey, W. R., and Lundak, R. L.,** Suppression of T-lymphocyte cytotoxicity following sinusoidally amplitude-modulated fields, *Bioelectromagnetics,* 4, 281, 1983.
167. **Byus, C. V., Lundak, R. L., Fletcher, R. M., and Adey, W. R.,** Alterations in protein kinase activity following exposure of cultured human lymphocytes to modulated microwave electromagnetic fields, *Bioelectromagnetics,* 5, 341, 1984.
168. **Sultan, M. F., Cain, C. A., and Tompkins, N. A. F.,** Immunological effect of amplitude-modulated radiofrequency radiation: B-lymphocyte capping, *Bioelectromagnetics,* 4, 157, 1983.
169. **Byus, C. V., Lundak, R. L., Fletcher, R. M. and Adey, W. R.,** Alterations in protein kinase activity following exposure of cultured human lymphocytes to modulated microwave fields, *Bioelectromagnetics,* 5, 341, 1984.
170. **Wei, L-X., Goodman, R., and Henderson, A.,** Changes in levels of c-myc and histone H2B following exposure of cells to low-frequency sinusoidal electromagnetic fields: evidence for a window effect, *Bioelectromagnetics,* 11, 269, 1990.
171. **Veyret, B., Bouthert, C., Deschaux, P., de Seze, R., Geffard, M., Joussot-Dubien, J., le Diraison, M., Moreau, J.-M., and Caristan, A.,** Antibody responses of mice exposed to low-power microwaves under combined, pulse-and-amplitude modulation, *Bioelectromagnetics,* 12, 47, 1991.
172. **Ning, J., Casamento, J., Czerska, E., Swicord, M., Al-Barazi, H., Davis, C., and Elson, E.,** Comparison of the effect of ELF on total RNA content in normal and transformed human cells, *Ann. N.Y. Acad. Sci.,* 649, 353, 1992.
173. **Schwartz, J-L., House, D. E., and Mealing, G. A. R.,** Exposure of frog hearts to CW or amplitude-modulated VHF fields:selective efflux of calcium ions at 16 Hz, *Bioelectromagnetics,* 11, 349, 1990.
174. **Schwartz, J-L. and Mealing, G. A. R.,** Calcium-ion movement and contractility in atrial strips of frog heart are not affected by low-frequency-modulated, 1 GHz electromagnetic radiation, *Bioelectromagnetics,* 14, 521, 1993.
175. **Yee, K-C., Chou, C-K., and Guy, A. W.,** Influence of microwaves on the beating rate of isolated rat hearts, *Bioelectromagnetics,* 9, 175, 1988.
176. **Caddemi, A., Tamburello, C. C., Zanforlin, L., and Torregrossa, M. V.,** Microwave effects on isolated chick hearts, *Bioelectromagnetics,* 7, 359, 1986.
177. **Fitzsimmons, R. J., Farley, J. R., Adey, W. R., and Baylink, D. J.,** Frequency dependence of increased cell proliferation, in vitro, in exposures to a low-amplitude, low-frequency electric field: evidence for dependence on increased mitogen activity released into culture medium, *J. Cell Physiol.,* 139, 685, 1989.
178. **Byus, C. V., Kartun, K., Pieper, P., and Adey, W. R.,** Increased ornithine decarboxylase activity in cultured cells exposed to low energy modulated microwave fields and phorbol ester promoters, *Cancer Res.* 48, 4222, 1988.
179. **Balcer-Kubiczek, E. K. and Harrison, G. H.,** Neoplastic transformation of C3H/10T(1/2) cells following exposure to 120-Hz modulated 2.45-GHz microwaves and phorbol ester tumor promoter, *Rad. Res.,* 126, 65, 1991.
180. **Albert, E., Blackman, C. F., and Slaby, F.,** Calcium-dependent secretory protein release and calcium efflux during RF irradiation of rat pancreatic tissue slices, in *Ondes Electromagnetiques et Biologie,* Berteaud, A. J. and Servantie, B., Eds., Paris, 1980, 325.
181. **Mitsushinko, M. and Okada, O.,** Non-thermal vasodilation by radiofrequency burst-type electromagnetic field radiation in the frog, *J. Physiol.,* 435, 257, 1991.
182. **Montaigne, K. and Pickard, W. F.,** The membrane potential of charcean cells exposed to amplitude-modulated low power 147-MHz radiation, *Bioelectromagnetics,* 5, 353, 1984.
183. **Barsoum, Y. A. and Pickard, W. F.,** The vacuolar potential of characean cells exposed to electromagnetic radiation in the range 200–8200 MHz, *Bioelectromagnetics,* 3, 393, 1982.

184. **Liu, L.-M., Garber, F., and Cleary, S. F.,** Investigation of the effects of continuous-wave, pulse- and amplitude-modulated microwaves on single excitable cells of *Chara corallina, Bioelectromagnetics,* 3, 203, 1982.
185. **Christel, P., Ceff, C., and Pilla, A. A.,** Modulation of rat radial osteotomy repair using electromagnetic current induction, in *Mechanisms of Growth and Control,* Becker, R. O., Ed., Charles C Thomas, Springfield, IL, 1981.
186. **Goodman, R., Bassett, C. A. L., and Henderson, A. S.,** Pulsing electromagnetic field induces cellular transcription, *Science,* 220, 1283, 1983.
187. **Goodman, R., Abbott, J., and Henderson, A. S.,** Transcriptional patterns in the x chromosome of *Sciara coprophila* following exposure to magnetic fields, *Bioelectromagnetics,* 8, 1, 1987.
188. **Goodman, R. and Henderson, A. S.,** Stimulation of RNA synthesis in the salivary gland cells of *Sciara coprophila* by an electromagnetic signal used for treatment of skeletal problems in horses, *J. Bioelectr.,* 6, 37, 1987.
189. **Goodman, R. and Henderson, A. S.,** Exposure of salivary gland cells to low-frequency electromagnetic fields alters polypeptide synthesis, *Proc. Natl. Acad. Sci. USA,* 85, 3928, 1988.
190. **Goodman, R., Weisbrot, D., Uluc, A., and Henderson, A.,** Transcription in Drosophila melanogaster salivary gland cells is altered following exposure to low-frequency electromagnetic fields: analysis of chromosome 3R, *Bioelectromagnetics,* 13, 111, 1992.
191. **Goodman, R. and Henderson, A. S.,** Sine waves enhance cellular transcription, *Bioelectromagnetics,* 7, 23, 1986.
192. **Goodman, R., Bassett, A. L., and Henderson, A. S.** Pulsing Electromagnetic Fields Induce Cellular Transcription, *Science,* 220, 1283, 1983.
193. **Goodman, R. and Henderson, A. S.,** Some biological effects of electromagnetic fields, *Bioelectrochem. and Bioenerg.,* 1985.
194. **Blank, M., Soo, L. Lin, H., Henderson, A. S., and Goodman, R.,** Changes in transcription in HL-60 cells following exposure to alternating currents from electric fields, *Bioelectrochem. and Bioenerg.,* 28, 301, 1992.
195. **Blank, M., Khorkova, O. and Goodman, R.,** Changes in polypeptide distribution stimulated by different levels of electromagnetic and thermal stress, *Bioelectrochem. and Bioenerg.,* 33, 109, 1994.
196. **Wei, L-X., Goodman, R., and Henderson, A.,** Changes in levels of c-myc and histone h2b following exposure of cells to low-frequency sinusoidal electromagnetic fields: evidence for a window effect, *Bioelectromagnetics,* 11, 269, 1990.
197. **Goodman, R., Blank, M., Lin, H., Dai, R., Khorkova, O., Soo, L., Weisbrot, D., and Henderson, A. S.,** Increased levels of hsp 70 transcripts induced when cells are exposed to low frequency electromagnetic fields, *Bioelectrochem. and Bioenerg.,* 33, 115, 1994.
198. **Goodman, R., Abbott, J., Krim, A., Henderson, A. S.,** Nucleic acid and protein synthesis in cultured chinese hamster ovary (cho) cells exposed to the pulsed electromagnetic fields, *J. Bioelectr.,* 4, 565, 1985.
199. **Goodman, R., Wei, L-X., Xu, J-C., and Henderson, A.,** Exposure of human cells to low-frequency electromagnetic fields results in quantitative changes in transcripts, *Biochim. Biophys. Acta.,* 1009, 216, 1989.
200. **Goodman, R., Bumann, J., Wei, L-X., and Shirley-Henderson, A.,** Exposure of human cells to electromagnetic fields: effect of time and field strength on transcript levels, *Electro. Magnetobiol.,* 11, 19, 1992.
201. **Goodman, R. and Henderson, A. S.,** Transcription and translation in cells exposed to extremely low frequency electromagnetic fields, *Bioelectrochem. and Bioenerg.,* 25, 335, 1991.
202. **Goodman, R., Chizmadzhev, Y. M., and Henderson, A. S.,** Electromagnetic fields and cells, *J. Cell. Biochem.* 51, 436, 1993.
203. **Goodman, R., Wei, L-X, Bumann, J., and Henderson, A. S.,** Exposure to electric and magnetic (EM) fields increases transcripts in HL-60 cells: does adaptation to EM fields occur? *Bioelectrochem. and Bioenerg.,* 29, 185, 1992.
204. **Litovitz, T. A., Montrose, C. J., Goodman, R., and Elson, E. C.,** Amplitude windows and transiently augmented transcription from exposure to electromagnetic fields, *Bioelectromagnetics,* 11, 297, 1990.
205. **Lin, H., Goodman, R., and Henderson, A. S.,** Specific region of the c myc promoter is responsive to electric and magnetic fields, *J. Cell. Biochem.* 54, 281, 1994.
206. **Gold, S., Goodman, R., and Henderson, A.,** Exposure of simian virus-40-transformed human cells to magnetic fields results in increased levels of T-antigen mRNA and protein, *Bioelectromagnetics* 15, 329, 1994.
207. **Phillips, J. L.,** Effect of 72 Hz pulsed magnetic field exposure on macromolecular synthesis in ccrf-cem cells, *Biochem. Biophys.,* 12, 1, 1991.
208. **Phillips, J. L.,** Effects of electromagnetic field exposure on gene transcription, *J. Cell. Biochem.,* 51, 381, 1993.
209. **Phillips, J. L., Haggren, W., Thomas, W. J., Jones, T. I., and Adey, W. R.,** Magnetic field-included changes in specific gene transcription, *Biochim. Biophys. Acta,* 1132, 140, 1992.
210. **Phillips, J. L., Haggren, W., Thomas, W. J., Jones, T. I., and Adey, W. R.,** Effect of 72 Hz pulsed magnetic field exposure on RAS p21 expression in ccrf-cem cells, *Cancer Biochem. Biophys.* 13, 187, 1993.

211. **Goodman, E. M., Greenebaum, B., and Marron, M. T.,** Altered protein synthesis in a cell-free system exposed to a sinusoidal magnetic field, *Biochim. Biophys. Acta.,* 1202, 107, 1993.
212. **Czerska, E., Casamento, J., Ning, J., Swicord, M., Al-Barazi, H., and Davis, C.,** Comparison of the effect of ELF on c-myc oncogene expression in normal and transformed human cells, *Ann. N.Y. Acad. Sci.,* 649, 340, 1992.
213. **Greene, J. J., Skowronski, W. J., Mullins, J. M., Nardone, R. M., Penafiel, M., and Meister, R.,** Delineation of electric and magnetic field effects of extremely low frequency electromagnetic radiation on transcription, *Biochem. Biophys. Res. Commun.,* 174, 742, 1991.
214. **Saffer, J. D. and Thurston, J. S.,** Alteration of gene expression by ELF fields, In Abstracts of the Sixteenth Annual Meeting of the Bioelectromagnetics Society, June 12–17, 1994.
215. **Kropinski, A. M., Morris, W. C., and Szewczuk, M. R.,** Sinusodial 60-Hz electromagnetic fields failed to induce changes in protein synthesis in Escerchia Coli, *Bioelectromagnetics* 15, 283, 1994.
216. **Delgado, J. M. R., Leal, J., Monteagudo, J. L., and Garcia, M. C.,** Embryological changes induced by weak extremely low frequency electromagnetic fields, *J. Anat.,* 143, 533, 1982.
217. **Ubeda, A., Leal, J., Trillo, M. A., Jimenez, M. A., and Delgado, J. M. R.,** Pulse shape of magnetic fields influence chick embryogenesis, *J. Anat.,* 137, 513, 1983.
218. **Ubeda, A.,** Authors' correction of data, *J. Anat.,* 140, 721, 1985.
219. **Leal, J., Shamsaifar, K., Trillo, M. A., Ubeda, A., Abraira, V., and Chacon, L.,** Embryonic development and weak changes of the geomagnetic field, *J. Bioelectr.,* 7, 141, 1989.
220. **Juutilainen, J. and Saali, K.,** Development of chick embryos in 1 Hz to 100 kHz magnetic fields, *Radiat. Environ. Biophys.,* 25, 135, 1986.
221. **Berman, E., Chacon, L., House, D., Koch, B. A., Koch, W. E., Leal, J., Lovtrup, S., Mantiply, E., Martin, A. H., Martucci, G. I., Mild, K. H., Monahan, J. C., Sandstrom, M., Shamsaifar, K., Tell, R., Trillo, M. A., Ubeda, A., and Wagner, P.,** Development of chicken embryos in a pulsed magnetic field, *Bioelectromagnetics,* 11, 169, 1990.
222. **Juutilainen, J.,** Effects of low-frequency magnetic fields on embryonic development and pregnancy, *Scand. J. Work. Environ. Health,* 17, 149, 1991.
223. **Litovitz, T. A., Montrose, C. J., and Litovitz, T. L.,** On a genetic predisposition to EMF-induced effects in chick embryos, Bioelectromagnetics Society, 13th Annual Meeting, 23–27 June, Salt Lake City, UT, p. 51 1991.
224. **Aarholt, E., Flinn, E. A., and Smith, C. W.,** Effects of low frequency magnetic fields on bacterial growth rate, *Phys. Med. Biol.,* 26, 613, 1981.
225. **Moore, R. L.,** Biological effects of magnetic fields: studies with microorganisms, *Can. J. Microbiol.,* 25, 1149, 1979.
226. **Fröhlich, H.,** The biological effects of microwaves and related questions, *Adv. Electron. Electron Phy.,* 53, 85, 1980.
227. **Davydov, A. S.,** Solitons in molecular systems, *Phys. Scr.,* 20, 387, 1979.
228. **Kohli, M., Mei, W. N., Prohofsky, E., and Van Zandt, L. L.,** Calculated microwave absorption of double helical B-conformation, poly(dg), poly(dc), *Biopolymers,* 20, 853, 1981.
229. **Taylor, L. S.,** The mechanisms of athermal microwave biological effects, *Bioelectromagnetics,* 2, 259, 1981.
230. **Fröhlich, H.,** Coherent electric vibrations in biological systems and the cancer problem, *IEEE Trans. Microwave Theory Tech.,* 26, 613, 1978.
231. **Scott, A.,** *Active and Nonlinear Wave Propagation in Electronics,* Wiley-Interscience, New York, 1970.
232. **Kaiser, F.,** Coherent oscillations in biological systems: interaction with extremely low frequency fields, *Radio Sci.,* 17(5S), 17, 1982.
233. **Kaiser, F.,** Biophysical models related to Fröhlich Excitations, *Nanobiology* 1, 149, 1992.
234. **Eichwald, C. and Kaiser, F.,** Model for receptor-controlled cytosolic calcium oscillations and for external influences on the signal pathway, *Biophys. J.* 65, 2054, 1993.
235. **Grundler, W., Kaiser, F., Keilmann, F., and Walleczek, J.,** Mechanisms of electromagnetic infractions with cellular systems, *Naturwissenschaften* 79, 551, 1992.
236. **Van Zandt, L. L., Kohli, M., and Prohofsky, E. W.,** Absorption of microwave radiation by DNA in aquo, *Biopolymers,* 21, 1465, 1982.
237. **Lu, K. C., Prohofsky, E., and Van Zandt, L. L.,** Vibrational modes of A-DNA, B-DNA, and A-RNA backbones; an application of a Green's function refinement procedure, *Biopolymers,* 16, 2491, 1977.
238. **Small, E. W., Peticolas, N., and Warner, L.,** Conformational dependence of the Raman scattering intensities from polynucleotides. III. Order-disorder changes in helical structures, *Biopolymers,* 10, 1377, 1971.
239. **Swicord, M. L., Edwards, G. S., Sagripanti, J. L., and Davis, C. C.,** Chain-length-dependent microwave absorption of DNA, *Biopolymers,* 22, 2513, 1983.
240. **Edwards, G. S., Davis, C. C., Saffer, J. P., and Swicord, M. L.,** Resonant microwave absorption of selected DNA molecules, *Phys. Rev. Lett.,* 53, 1284, 1984.

241. **Gabriel, C.; Grant, E. H., Tata, R., Brown, P. R., Gestblom, B., and Noreland, E.,** Dielectric behavior of aqueous solutions of plasmid DNA at microwave frequencies, *Biophys. J.,* 55, 29, 1989.
242. **Foster, K. R., Epstein, B. R., and Gealt, M. A.,** "Resonances" in the dielectric absorption of DNA?, *Biophys. J.,* 52, 421, 1987.
243. **Saif, B., Litovitz, T. A., Mohr, R. K., and Montrose,** On the origin of microwave absorption in DNA solutions, Bioelectromagnetics Society, 12th Annual Meeting, 10–14 June, San Antonio, TX, p. 41 1990.
244. **Sagripanti, J-L., Swicord, M. L., and Davis, C. C.,** Microwave effects on plasmid DNA, *Radiat. Res.,* 110, 219, 1987.
245. **Prohofsky, E.,** personal communication.
246. **Scott-Russell, J.,** Report on waves, *Proc. R. Soc. (Edinburgh),* 319, 1844.
247. **Davydov, A. S.,** *Biology and, Quantum Mechanics,* Pergamon Press, Elmsford, NY, 1982.
248. **Scott, A. C.,** Dynamics of Davydov solitons, *Phys. Rev. A,* 26, 578, 1982.
249. **Scott, A. C.,** Solitons in biological molecules, *Comments Mol. Cell. Biophys.,* 3, 15, 1985.
250. **Jafary-Asl, A. H., Solanki, S. N., Aarholt, E., and Smith, C. W.,** Dielectric measurements on live biological materials under magnetic resonance conditions, *J. Biol. Phys.,* 11, 13, 1982.
251. **Pohl, H. A.,** *Dielectrophoresis, The Behavior of Matter in Nonuniform Electric Fields,* Cambridge University Press, Cambridge, England, 1978.
252. **Kearley, G. J., Fillaux, F., Baron, M.-H., Bennington, S., and Tomkinson, J.,** A new look at proton transfer dynamics along the hydrogen bonds in amides and peptides, *Science,* 264, 1285, 1994.
253. **Blackman, C. F., Benane, S. G., Rabinowitz, J. R., House, D. E., and Joines, W. T.,** A role for the magnetic field in the radiation-induced efflux of calcium ions from brain tissue in vitro, *Bioelectromagnetics,* 6, 321, 1985.
254. **Liboff A. R.,** Cyclotron resonance mechanisms for electromagnetic energy transfer to cells, *Bioelectromagnetics Soc. 7th Annu. Meet., (Abstr.),* 1985.
255. **Thompson, C. J.,** *Mathematical Statistical Mechanics,* Princeton University Press, Princeton, NJ, 1972, chap 5 to 7.
256. **Liboff, A. R.,** Cyclotron resonance in membrane transport, in *Interactions Between Electromagnetic Fields and Cells,* A. Chiabrera, C. Nicolini, and H. P. Schwan, Eds., Plenum, New York City, 1985.
257. **Thomas, J. R., Schrot, J., and Liboff, A. R.,** Low-intensity magnetic fields alter operant behavior in rats, *Bioelectromagnetics,* 7, 349, 1986.
258. **McLeod, B. R. and Liboff, A. R.,** Dynamic characteristics of membrane ions in multi-field configurations of low-frequency electromagnetic radiation, *Bioelectromagnetics,* 7, 177, 1986.
259. **Smith, S. D., McLeod, B. R., and Liboff, A. R.,** Effects of resonant magnetic fields on chick femoral development *in vitro, J. Bioelectr.,* 10, 81, 1991.
260. **Liboff, A. R. and and Parkinson, W. C.,** Search for ion-cyclotron resonance in a Na^+-transport system, *Bioelectromagnetics,* 12, 77, 1991.
261. **Liboff, A. R., McLeod, B. R., and Smith, S. D.,** Rotating magnetic fields and ion cyclotron resonance, *J. Bioelectr.,* 8, 119, 1989.
262. **Liboff, A. R., Thomas, J. R., and Schrot, J.,** Intensity threshold for 60-Hz magnetically induced behavioral changes in rats, *Bioelectromagnetics,* 10, 111, 1989.
263. **Liboff, A. R., Rozek, R. J., Sherman, M. L., McLeod, B. R., and Smith, S. D.,** Ca^{++}-45 cyclotron resonance in human lymphocytes, *J. Bioelectr.,* 6, 13, 1987.
264. **McLeod, B. R., Smith, S. D., and Liboff, A. R.,** Calcium and potassium cyclotron resonance curves and harmonics in diatoms (*A. coffeaeformis*), *J. Bioelectr.,* 6, 153, 1987.
265. **Smith, S. D., McLeod, B. R., Liboff, A. R., and Cooksey, K.,** Calcium cyclotron resonance and diatom mobility, *Bioelectromagnetics,* 8, 215, 1987.
266. **McLeod, B. R., Smith, S. D., Cooksey, K. E., and Liboff, A. R.,** Ion cyclotron resonance frequencies enhance Ca^{++}-dependent motility in diatoms, *J, Bioelectr.,* 6, 1, 1987.
267. **Parkinson, W. C. and Sulik, G. L.,** Diatom response to extremely low-frequency magnetic fields, *Radiat. Res.,* 130, 319, 1992.
268. **Reese, J. A., Frazier, M. E., Morris, J. E., Buschbom, R. L., and Miller, D. L.,** Evaluation of changes in diatom mobility after exposure to 16-Hz electromagnetic fields, *Bioelectromagnetics,* 12, 21, 1991.
269. **Saalman, E., Galt, S., Hamnwnrius, Y., and Norden, B.,** Diaton motility: replication study in search of cyclotron resonance effects, in *Interaction Mechanisms of Low-Level Electromagnetic Fields in Living Systems,* Norden, B., Ramel, C., Eds., Oxford University Press, 1992, 280.
270. **Grande, D. A., Magee, F. P., Weinstein, A. M., and McLeod, B. R.,** The effect of low-energy combined AC and DC magnetic fields on articular cartilage metabolism, *Ann. N.Y. Acad. Sci.,* 635, 404, 1991.
271. **Smith, R. F.,** Lithium as a normal metabolite: some implications for cyclotron resonance of ions in magnetic fields, *Bioelectromagnetics,* 9, 387, 1988.

272. **Coulton, L. A. and Barker, A. T.,** Magnetic fields and intracellular calcium: Effects on lymphocytes exposed to conditions for cyclotron resonance, *Phys. Med. Biol.,* 38, 347, 1993.
273. **Durney, C. H., Kaminski, M., Anderson, A. A., Bruckner-Lea, C., Janata, J., and Rappaport, C.,** Investigation of AC-DC magnetic field effects in planar phospholipid bilayers, *Bioelectromagnetics,* 13, 19, 1992.
274. **Bell, G., Marino, A. A., Chesson, A., and Struve, F.,** Electrical states in the rabbit brain can be altered by light and electromagnetic fields, *Brain. Res.,* 570, 307, 1992.
275. **Prasad, A. V., Miller, M. W., Carstensen, E. L., Cox, Ch., Azadniv, M., and Brayman, A. A.,** Failure to reproduce increased calcium uptake in human lymphocytes at purported cyclotron resonance exposure conditions, *Radiat. Environ. Biophys.,* 30, 305, 1991.
276. **Bruckner-Lea, C., Durney, C. H., Janata, J., Rappaport, C., and Kaminski, M.,** Calcium binding to metallochromic dyes and calmodulin in the presence of combined, AC-DC magnetic fields, *Bioelectromagnetics,* 13, 147, 1992.
277. **Markov, M. S., Ryaby, J. T., Kaufman, J. J., and Pilla, A. A.,** Extremely weak AC and DC magnetic fields significantly affect myosin phosphorylation, *Charge and Field Effects in Biosystems,* M. J. Allen, S. F. Cleary, A. E. Sowers, and D. D. Shillady, Eds., Birkhauser, Boston, 1992, 225.
278. **Ngo, F. Q. H., Blue, J. W., and Roberts, W. K.,** The effects of a static magnetic field on DNA synthesis and survival of mammalian cells irradiated with fast neutrons, *Magn. Reson. Med.* 5, 307, 1987.
279. **Parkinson, W. C. and Hanks, C. T.,** Search for cyclotron resonance in cells in vitro, *Bioelectromagnetics,* 10, 129, 1989.
280. **Sandweiss, J.,** On the cyclotron resonance model of ion transport, *Bioelectromagnetics,* 11, 203, 1990.
281. **Liboff, A. R.,** The cyclotron resonance hypothesis: experimental evidence and theoretical constraints, in *Interaction Mechanisms of Low-Level Electromagnetic Fields in Living Systems,* Norden, B., Ramel, C., Eds., Oxford University Press, 1992, 280.
282. **Liboff, A. R. and McLeod, B. R.,** Kinetics of channelized membrane ions in magnetic fields, *Bioelectromagnetics,* 9, 39, 1988.
283. **McLeod, B. R., Liboff, A. R., and Smith, S. D.,** Electromagnetic gating in ion channels, *J. Theor. Biol.,* 158, 15, 1992.
284. **Lednev, V. V.,** Possible mechanism for the influence of weak magnetic fields on biological systems, *Bioelectromagnetics,* 12, 71, 1991.
285. **Podgoretskii, M. I. and Khrustalev, O. A.,** Interference in quantum transitions, *Sov. Phy. Uspekhi,* 6, 682, 1963.
286. **Adair, R. K.,** Criticism of Lednev's mechanism for the influence of weak magnetic fields on biological systems, *Bioelectromagnetics,* 13, 231, 1992.
287. **Blanchard, J. P. and Blackman, C. F.,** Clarification and application of an ion parametric resonance model for magnetic field interactions with biological systems, *Bioelectromagnetics,* 15, 217, 1994.
288. **Lednev, V. V.,** Parametric resonance model for the interaction of magnetic fields with biosystems: theoretical predictions and their experimental verification, First World Congress for Electricity and Magnetism in Biology and Medicine, 14–19 June, Lake Buena Vista, FL, 1993.
289. **Weaver, J. C. and Astumain R. D.,** The response of living cells to very weak electric fields: the thermal noise limit, *Science,* 247, 459, 1990.
290. **Weaver, J. C., and Astumian, R. D.,** Estimates for ELF effects: noise-based thresholds and the number of experimental conditions required for empirical searches, *Bioelectromagnetics Suppl.* 1, 119, 1992.
291. **Adair, R. K.,** Constraints on bilogical effects of weak extremely-low-frequency electromagnetic fields, *Phys. Rev. A,* 43, 1039, 1991.
292. **Litovitz, T. A., Krause, D., and Mullins, J. M,** Effect of coherence time of the applied magnetic field on ornithine decarboxylase activity, *Biochem. Biophys. Res. Commun.,* 178, 862, 1991.
293. **Litovitz, T. A., Krause, D., Penafiel, M., Elson, E. C., and Mullins, J. M.,** The role of coherence time in the effect of microwaves on ornithine decarboxylase activity, *Bioelectromagnetics,* 14, 395, 1993.
294. **Farrell, J. M., Barber, M., Doinov, P., Krause, D., and Litovitz, T. A.,** Superposition of a temporally incoherent magnetic field suppresses the change in ornithine decarboxylase activity in developing chick embryos induced by a 60-Hz sinusoidal field, *Electricity and Magnetism in Biology and Medicine,* M. Blank, Ed., San Francisco Press, Inc., 1993, 342.
295. **Mullins, J. M., Krause, D., and Litovitz, T. A.,** Simultaneous application of a spatially coherent noise field blocks response of cell cultures to a 60-Hz electromagnetic field, *Electricity and Magnetism in Biology and Medicine,* M. Blank, Ed., San Francisco Press, Inc., 1993, 345.
296. **Litovitz, T. A., Montrose, C. J., and Doinov, P.,** Spatial and temporal coherence affects the response of biological systems to electromagnetic fields, *Electricity and Magnetism in Biology and Medicine,* M. Blank, Ed., San Francisco Press, Inc., 1993, 339.
297. **Boysen, J. E.,** Hyperthermic and pathologic effect of EMR (350 Mc), *AMA Arch. Ind. Hyg.,* 38, 261, 1953.

298. **Gordon, Z. V.,** *Biological Effects of Microwaves in Occupational Hygiene,* Kaner, N., Translator, Keter Press, Jerusalem, 1970.
299. **Chou, C-K., Guy, A. W., McDougall, J. A., and Han, L-F.,** Effects of continuous and pulsed chronic microwave exposure on rabbits, *Radio Sci.,* 17(5S), 185, 1982.
300. **Birenhaum, L., Kaplan, I. T., Metaly, W., Rosenthal, S. W. and Zaret, M. M.,** Microwave and infrared effects on heart rate, respiration rate and sub-cutaneous temperature of the rabbit, *J. Microwave Power,* 10, 3, 1975.
301. **Hamrick, P. E. and McRee, D. I.,** The effect of 2450 MHz microwave irradiation on the heart rate of embryonic quail, *Health Phys.,* 38, 261, 1980.
302. **Frei, M. R., Jauchem, J. R., and Heinmets, F.,** Physiological effects of 2.8 GHz radio-frequency radiation: A comparison of pulsed with continuous wave radiation, *J. Microw. Pow. Elect. Ener.,* 23, 85, 1988.
303. **Frei, M. R., Jauchem, J. R., and Padilla, J. M.,** Thermal and physiological changes in rats exposed to CW and pulsed 2.8 GHz radiofrequency radiation in E and H orientations, *Int. J. Radiat. Biol.,* 6, 1033, 1989.
304. **Frei, M. R., Jauchem, J. R., and Heinmets, F.,** Thermoregulatory response of rats exposed to 9.3-GHz radiofrequency radiation, *Radiat. Environ. Biophy.,* 28, 67, 1989.
305. **Lu, S-T., Brown, D. O., Johnson, C. E., Mathur, S. P., and Elson, E. C.,** Abnormal cardiovascular responses induced by local high power microwave exposure, *IEEE Trans. Biomed. Eng.,* 39, 484, 1992.
306. **Seaman, R. L. and DeHaan, R. L.,** Inter-beat intervals of cardiac-cell aggregates during exposure to 2.45-GHz CW, pulsed, and square-wave modulated microwaves, *Bioelectromagnetics,* 14, 41, 1993.
307. **Baranski, S.,** Effect of chronic microwave irradiation on the blood forming system of guinea pigs and rabbits, *Aerosp. Med.,* 42, 1196, 1971.
308. **Czerska, E. M., Elson, E. C., Davis, C. C., Swicord, M. L. and Czerski, P.,** Effects of continuous and pulsed 2450-MHz radiation on spontaneous lymphoblastoid transformation of human lymphocytes *in vitro, Bioelectromagnetics,* 13, 247, 1992.
309. **Czerski, P., Paprochy-Stanka, E., Srekierzynski, M., and Stolarska, A.,** Influence of microwave radiation on the hematopoietic system, in *Biological Effects and Health Hazards of Microwave Radiation,* Czerski, P., Ostrowski, K., Shore, M. L., Silverman, C., Suess, M. J., and Waldeskog, B., Eds., Polish Medical Publishers, Warsaw, 1974, 67.
310. **Baranski, S. and Czerski, P.,** *Biological Effects of Microwaves,* Dowden, Hutchinson & Ross, Stroudsburg, PA, 1976.
311. **Cleary, S. F., Garber, F., and Liu, L. M.,** Effect of X-band microwave exposure on rabbit erythrocytes, *Bioelectromagnetics,* 3, 453, 1982.
312. **Wangemann, R. T. and Cleary, S. F.,** The *in vivo* effect of 2.45 GHz microwave radiation on rabbit serum components and sleeping times, *Radiat. Environ. Biophys.,* 13, 89, 1976.
313. **McLees, B. D., Finch, E. D., and Albright, M. L.,** An examination of regenerating hepatic tissue subjected to radiofrequency radiation, *J. Appl. Physiol.,* 32, 78, 1972.
314. **Baranski, S.,** Histological and histochemical effects of microwave irradiation on the central nervous system of rabbits and guinea pigs, *Am. J. Phys. Med.,* 54, 182, 1972.
315. **Lai, H., Horita, A., and Guy, A. W.,** Acute low-level microwave exposure and central cholinergic activity: studies of irradiation parameters, *Bioelectromagnetics,* 9, 355, 1988.
316. **Lai, H., Horita, A., Chou. C-K., and Guy, A. W.,** Low-level microwave irradiations affect central cholinergic activity in rats, *J. Neurochem.,* 48, 40, 1987.
317. **Baranski, S. and Edelwijn, Z.,** Experimental morphologic and electroencephalographic studies of microwave effects on the central nervous system, *Ann. N.Y. Acad. Sci.,* 247, 109, 1975.
318. **Frey, A. H., Feld, S. R., and Frey, B.,** Neuronal function and behavior: defining the relationship, *Ann. N.Y. Acad. Sci.,* 247, 433, 1975.
319. **Oscar, K. J. and Hawkins, T. D.,** Microwave alterations on the blood-brain-barrier of rats, *Brain Res.,* 126, 281, 1977.
320. **Oscar, K. J., Grueneau, S. P., Folker, M. T., and Rapoport, S. I.,** Local cerebral blood flow after microwave exposure, *Brain Res.,* 204, 220, 1981.
321. **Merritt, J. H., Chamnes, A. F., and Allen, S. J.,** Studies on blood-brain-barrier permeability after microwave irradiation, *Radiat. Environ. Biophys.,* 15, 361, 1978.
322. **Kholodov, Yu. A.,** *Reaction of the Nervous System to Electromagnetic Fields,* Moscow, 1975.
323. **Wachtel, H., Seaman, R., and Joines, W. T.,** Effects of low intensity microwave on isolated neurons, *Ann. N.Y. Acad. Sci.,* 247, 46, 1975.
324. **Seaman, R. and Wachtel, H.,** Slow and rapid response to CW and pulsed microwave radiation by individual *Aplysia* pacemakers, *J. Microwave Power,* 13, 77, 1978.
325. **McRee, D. I. and Wachtel, H.,** Pulse microwave effects on nerve vitality, *Radiat. Res.,* 91, 212, 1982.
326. **McRee, D. I. and Wachtel, H.,** The effect of microwave radiation on the vitality of isolated frog sciatic nerves, *Radiat. Res.,* 82, 536, 1980.

327. **Seaman, R. L. and Lebovitz, R. M.,** Thresholds of cat cochlear nucleus neurons to microwave pulses, *Bioelectromagnetics,* 10, 147, 1989.
328. **Roberti, B., Heebels, C. H., Hendricx, J. C. M., deGreef, A. H. A. M., and Wolthuis, O. L.,** Preliminary investigation of the effect of low-level microwave radiation on spontaneous motor activity in rats, *Ann. N.Y. Acad. Sci.,* 247, 417, 1975.
329. **Thomas, J. R., Finch, E. D., Folk, D. W., and Burch, L. S.,** Effects of low level microwave radiation on behavioral baselines, *Ann. N.Y. Acad. Sci.,* 247, 425, 1975.
330. **Lebovitz, R. M.,** Pulse-modulated and continuous-wave microwave radiation yield equivalent changes in operant behavior of rodents, *Physiol. Bhav.,* 30, 891, 1983.
331. **Brown, D. O., Lu, S-T., and Elson E. C.,** Characteristics of microwave evoked body movements in mice, *Bioelectromagnetics,* 15, 143, 1994.
332. **Lebovitz, R. M.,** Pulse modulated and continuous wave radiation yield equivalent changes in operant behavior of rodents, *Physiol. and Behavior,* 30, 891, 1983.
333. **Birenbaum, L., Kaplan, I. T., Metaly, W., Rosenthal, S. W., Schmidt, H., and Zaret, M. M.,** Effect of microwaves on the rabbit eye, *J. Microwave Power,* 4, 232, 1969.
334. **Birenbaum, L., Grosof, C. M., Rosenthal, S. W., and Zaret, M. M.,** Effect of microwaves on the eye, *IEEE Trans. Biomed. Eng.,* 16, 7, 1969.
335. **Stewart-DeHaan, P. J., Creighton, M. D., Larsen, L. E., Jacobi, J. H., Ross, W. M., and Trevithick, J. R.,** Microwave and temperature effects on marine occular lenses in vitro, *1980 IEEE MTT-S Int. Microwave Symp. Dig.,* 341, 1980.
336. **Stewart-DeHaan, P. J., Creighton, M. D., Larsen, L. E., Jacobi, J. H., Ross, J. H., Sanwall, W. M., Guo, T. C., Guo, W. W., and Trevithick, J. R.,** In vitro studies of microwave-induced cataracts: separation of field and heating effects, *J. Exp. Eye Res.,* 36, 15, 1983.
337. **Creighton, M. O., Larsen, L. E., Stewart-DeHaan, P. J., Jacobi, J. H., Sanwal, M., Baskerville, J. C., Bassen, H. E., Brown, D. O., and Trevithick J. R.,** In vitro studies of microwave-induced cataract. II. Comparison of damage observed for continuous wave and pulsed microwaves, *Exp. Eye Res.,* 45, 357, 1987.
338. **Weiter, J. J., Finch, E. D., Schultz, W, and Frattali, V.,** Ascorbic acid changes in cultured rabbit lenses after microwave irradiation, *Ann. N.Y. Acad. Sci.,* 247, 175, 1975.
339. **Kues, H. A., Hirst, L. W., Lutty, G. A., D'Anna, S., and Dunkelburger, G. R.,** Effects of 2.45-GHz microwaves on primate cornea, *Bioelectromagnetics,* 6, 177, 1985.

APPENDIX 1

IMPORTANT CONSTANTS AND FREQUENTLY USED UNITS OF MEASUREMENT

A. Constants

Boltzman constant	k	1.380662 (10^{-23}) J/K 0.861735 (10^{-4}) eV/K
Planck constant	h	6.626176 (10^{-34}) J/H
Elementary (electron) charge	e	1.6021892 (10^{-19}) C
Speed of light in vacuum	c	2.99792458 (10^{8}) m/sec
Magnetic permeability of vacuum	μ_0	$4\pi\ (10^{-7})$ H/m
Dielectric permittivity of vacuum	ϵ_0	$= \dfrac{1}{\mu_0 c^2} = 8.85418782\ (10^{-12})$ F/m $\approx \dfrac{1}{36\pi}\ 10^{-9}$ F/m
Avogadro's number	N_A	6.022045 (10^{23}) Molecules/g mol

B. Abbreviations for Prefixes

p	pico	$= 10^{-12}$
n	nano	$= 10^{-9}$
μ	micro	$= 10^{-6}$
m	milli	$= 10^{-3}$
c (as in cm)	centi	$= 10^{-2}$
k (as in kg)	kilo	$= 10^{3}$
M	mega	$= 10^{6}$
G	giga	$= 10^{9}$

C. Units of Measurement

Standard SI unit			Commonly used fractional or multiple units
Length	meter	m	$1\ \text{m} = 100\ \text{cm} = 1000\ \text{mm} = 10^{10}$ Angstrom $= 10^{10}$ Å
Area	square meter	m^2	$1\ \text{m}^2 = 10^4\ \text{cm}^2$
Mass	kilogram	kg	$1\ \text{kg} = 10^3\ \text{g}$
Energy	Joule	J	$1\ \text{J} = 10^7\ \text{erg} = 2.389\ (10^{-4})$ kcal 1 electron volt = 1 ev $\approx$ 1.6 (10^{-19}) J
Power = time rate of change of energy 1 W = 1 Joule per second = 1 J/s	Watt	W	

0-8493-0641-8/96/$0.00+$.50

Power per unit mass		1 W/kg	1 W/kg = 1 mW/g
Power per unit area		W/m^2	1 W/m^2 = 0.1 mW/cm^2
Frequency (symbol f)	Hertz	Hz	1 Hz = 1 cycle per second
Radian frequency $\omega = 2\pi f$	Radians per second	rad/sec	
Electric resistance	Ohm	Ω	
Electric conductance	Siemens	S	
Electric resistivity	Ohm-meter	Ωm	1 Ωm = 10^2 Ωcm
Electric conductivity (symbol g or σ)	Siemens per meter	S/m	1 S/m = 10^{-2} S/cm 0.1 S/m = 1 mS/cm Note: in older nomenclature 1 S = 1 Mho = 1 Ω and 1 Ω/m = 10^{-2} Ω/cm; 0.1 Ω/m = 1 mΩ/cm
Electric capacitance	Farad	F	
Relative dielectric permittivity			
$\epsilon_r = \epsilon' - j\epsilon''$	Dimensionless ratio		
$\epsilon'' = \dfrac{\sigma}{\omega\epsilon_o}$	Dimensionless ratio		
Dielectric permittivity of any material $\epsilon = \epsilon_o\epsilon_r$	Farad/meter	F/m	
Loss tangent $= \dfrac{\text{conduction current}}{\text{displacement current}}$			
$= \dfrac{\epsilon''}{\epsilon'} = \dfrac{\sigma}{\omega\epsilon_o\epsilon'}$	Dimensionless ratio		
Electric potential difference	Volt	V	
Electric current	Ampere	A	
Electric current density	Ampere/m^2	A/m^2	1 A/m^2 = 10^{-4} A/cm^2 = 0.1 mA/cm^2
Electric field intensity	Volt per meter	V/m	1 V/m = 10^{-2} V/cm = 10 mV/cm
Magnetic field intensity H	Ampere per meter	A/m	
Magnetic flux density or magnetic induction B = μH (where μ = magnetic permeability of material; in vacuum $\mu = \mu_o$)	Tesla	T	1 T = 10^4 Gauss = 10^4 G Note: if $\mu = \mu_0$ then H = 1 A/m corresponds to B = $4\pi(10^{-7})$ T = 4π mG 1 gamma = 10^{-5} G = 10^{-9} T = 1 nT; 1 milligamma = 1 pT
(For a more detailed discussion of magnetic units see Appendix to Chapter 3.)			

Electric dipole moment	Coulomb m	Cm	1 Debye = $3.33(10^{-30})$ Cm
Electric mobility (drift velocity of charge per unit of applied electric field)	m^2/Volt-second	m^2/Vs	
Coefficient of viscosity (shearing stress/rate of shearing strain)	Newton second/m^2	Ns/m^2	1 poise = 10^{-1} N sm^{-2}

APPENDIX 2

SAFETY STANDARDS

The standards, proposed standards, and recommendations summarized in the tables of this appendix represent the concensus of various committees and private or government agencies. Standards promulgated by different agencies and at different times are inconsistent in several instances and are therefore reproduced here only for reference and not because they are endorsed by the editors of this book. Standards are usually developed by committees after deliberations that can last several years and are generally revised only after additional periods of several years. In areas of active research they are therefore seldom based on the latest available information. For detailed considerations of safety provided by any particular electromagnetic environment the appropriate chapter of this book as well as the current scientific literature should be consulted.

A-Low Frequency Exposure Standards (0 Hz to 10 kHz)

TABLE 1
Contact currents

Standard	Population to which it applies	Frequency Hz	Exposure limit ma (rms)	Comments
U.K. Natl. Radiological Protection Board Draft Standard 1992	general population with no possibility of children being exposed	0.1 − 1000 10^3 − $(1.3)10^3$	1.0 $f^{0.7}$ f in kHz	These are "investigation levels" which are not limits on exposure. If investigation levels are exceeded "it is necessary to investigate compliance with the basic restrictions." For "basic restrictions" see current density, Table 2.
	where children could be exposed	0.1–370 370 − (70) 10^3	0.5 $f^{0.7}$ f in kHz	
European Prestandard ENV 50166-1 1995	workers	0.1–10000	3.5	"Contact currents arising from touching objects in alternating fields." "Reference levels" based on "controlling factors" listed in Table 2.
	general public	0.1–7500 7500–10,000	1.5 0.2f f in kHz	

0-8493-0641-8/96/$0.00+$.50

TABLE 2
Current densities

Standard	Population to which it applies	Frequency Hz	Exposure limit mA/m^2 (rms)	Comments
U.K. Natl. Radiological Protection Board Draft Standard 1992	general public (including children and workers)	1–10 Hz 10 Hz–1 kHz 1 kHz–100 kHz	100/f* 10 f/100	These are the "basic restrictions" on which contact current and electric field "investigation levels" are based.
European Prestandard ENV 50166-1 1995	workers	0.1–1 1–4 4–1000 1000–10000	40 40/f 10 f/100	Current density to be averaged over 100 mm^2 in the head and heart region (These values are the controlling factors for "reference levels" for electric and magnetic fields)**
	general public	0.1–1 1–4 4–1000 1000–10000	16 16/5 4 f/250	

* f in Hz.
** "Based on well-established short-term effects, which, depending on the frequency, include the stimulation of electrically-excitable cells in nerve and muscle tissue and heating."
Cardiac pacemakers and other active implantable devices and some metallic prosthetic implants may be affected below the given reference levels.

TABLE 3
Electric Fields

Standard	Population to which it applies	Frequency Hz	Field kV/m	Exposure duration t hours	Comments
U.K. Natl. Radiological Protection Board Draft Standard 1992	general public	< 24 24–600 600–600 × 10^3	25* 600/f	Not specified	"Investigation levels" (see tables 1, 2)
European Prestandard ENV 50166-1 1995	workers	0–0.1 0.1–50 50–150 150–1500 1500–10000	42 30 1500/f 1500/f 1	t** < 112/E t ≤ 80/E t ≤ 80/E	These are "reference levels" (see Table 2) E = electric field level
	general public (G. P.)	0–0.1 0.1–60 60–1500 1500–10000	14 10 600/f 0.4	Not specified	

TABLE 3—continued
Electric Fields

Standard	Population to which it applies	Frequency Hz	Field kV/m	Exposure duration t hours	Comments
ANSI/IEEE C95.1—1991 (USA)	"Controlled environment"	$3000–10^4$	0.61		An "uncontrolled environment" is one where individuals have no control over their exposure.
	"Uncontrolled environment"	$3000–10^4$	0.61		
American Conference of Governmental Industrial Hygienists	G. P.	0–30 kHz	1		"Threshold limits"
	G.P.	< 100 Hz	25		
		1000–4000	2500/f		
International Radiation Protection Association (IRPA): 1990	occupational	50/60	10	whole day	Basis: to limit body current
			30	short term	
	G.P.	50/60	5	whole day	
			10	few hrs/day	
World Health Organization (WHO)	G. P.	50/60	10	Intermittent exposure	
Florida (New Transm. lines 230 kV or smaller).	G. P.	60	8, 3		First figure for inside "Right Of Way" of Transmission or Distribution Line. Second figure at edge of ROW
500 kV Transm. line.		60	10, 2		
Lake Tarpon-Kathleen transm. line		60	8.94, 1.56		
Minnesota Environmental Quality Board, 1976	G. P.	60	8		
Montana Board of Natural Resources and Conservation, 1984	G. P.	60	1		For transmission lines 69 kV and above at edge of ROW
New Jersey Commission on Radiation Protection, 1981	G. P.	60	3		At edge of ROW
New York Public Service Commission, 1978	G. P.	60	1.6		At edge of ROW

*f in Hz.
**Total time that can be spent in any 8-hour period above any field level subject to max. field in column 4

TABLE 4
Magnetic flux density

Standard	Population to which it applies	Frequency Hz	Exposure limit micro Tesla	Exposure duration hrs	Comments
U.K. Natl. Radiological Protection Board Draft Standard 1992	general public	0–1	2×10^5	24	Average over 24 hrs.
					Max value
	(G. P.)		2×10^6		Max value, limbs only
			5×10^6		---------------------
		0.4–1000	$8 \times 10^4/f$*		"Investigation levels" (see Table 1)
		$1000 - 535 \times 10^3$	80		
European Prestandard ENV 50166-1 1995	workers	0–0.1	2×10^6 (5×10^6)		"Reference levels" (see Table 2)
		0.1–0.23	1.4×10^6 (3.5×10^6)		
		0.23–1	$3.2 \times 10^5/f$		3.5×10^6 for limbs
		1–4	$3.2 \times 10^5/f$		0.1–0.36 Hz
		4–1500	$8 \times 10^4/f$		$1.25 \times 10^6/f$ for limbs
		1500–10000	53 (83)		0.36–1500 Hz
	G. P.	0–0.1	40000(10^5)		"Reference levels"
		0.1–1.15	28000		71000 for limbs
		1.15–1500	32000/f		0.1–7.1 Hz
		1500–10000	21 (330)		5(10^5)/f for limbs
					7.1–1500 Hz
Sweden SWEDAC (Swedish procurement agency) Dec. 31, 1990	G. P.	5–2000	0.25		Applies to Video Display Terminals purchased by the state
ANSI/IEEE C95.1-1991 (USA)	"controlled environments" and G. P.	3000–10^4	200		A "controlled environment" is a location where exposed persons are aware of the potential for exposure.
American Conference of Govermental Industrial Hygienists	G. P.	< 1	6×10^4	Daily TWA	whole body TWA = time weighted average
		0	6×10^5	Daily TWA	Extremity
		0	20×10^5		Ceiling value
		1–30000	60000/f		Ceiling
	Pacemaker	< 6	1000		Ceiling
	Weakers	6–30000	6000/f		
International Radiation Protection Association (IRPA) 1990	Occupational	50/60	500	whole day	Basis: To limit induced body current
			5000	short term	
			25000	Limbs	
	G. P.	50/60	100	whole day	
			1000	few hrs/day	

TABLE 4—continued
Magnetic flux density

Standard	Population to which it applies	Frequency Hz	Exposure limit micro Tesla	Exposure duration hrs	Comments
California City of Irvine, May 15, 1989	G. P.	60	0.4		
Florida Department of Environmental Regulation, 1989, amended Jan 7, 1993	G. P. (At edge of transmission line right of way)	60	15		transmission lines below 230 kV
			20		500 kV transm. lines
			25		Double circuit tr. line
			3.5*		Lake Tarpon-Kathleen line
			22.9**		(*below and above, **500 MW load)
New York Public Service Commission-1990	G. P.	60	20		at edge of right of way (ROW) of transmission lines
Tennessee City Commission of Brentwood April 1991	G. P.	60	0.4		at ROW of 120 kV and larger transmission lines

B-RF and Microwave Exposure Standards

TABLE 5

5A-IEEE[a] Maximum Permissible Exposure for Controlled Environments[b] Electromagnetic Fields[c]

Frequency Range (MHz)	Electric Field Strength (E) (V/m)	Magnetic Field Strength (H) (A/m)	Power Density (S) E-Field; H-Field (W/m^2)	Avering Time $\|E\|^2$, $\|H\|^2$, or S (minutes)
0.003–0.1	614	163	(1,000; 10,000,000)†	6
0.1–3.0	614	16.3/f	(1,000; $100{,}000/f^2$)†	6
3–30	1842/f	16.3/f	($9{,}000/f^2$; $100{,}000/f^2$)†	6
30–100	61.4	16.3/f	(10.0; $100{,}000/f^2$)†	6
100–300	61.4	0.163	10.0	6
300–3,000			f/30	6
3,000–15,000			100	6
15,000–300,000			100	$616{,}000/f^{1.2}$

5B-Induced and Contact Radiofrequency Currents[d]

Frequency Range	Maximum Current (mA)		
	Through both feet	Through each foot	Contact
0.003–0.1 MHz	2,000f	1,000f	1,000f
0.1–100 MHz	200	100	100

TABLE 6

6A-IEEE[a] Maximum Permissible Exposure for Uncontrolled Environments[b] Electromagnetic Fields[c]

Frequency Range (MHz)	Electric Field Strength (E) (V/m)	Magnetic Field Strength (H) (A/m)	Power Density (S) E-Field; H-Field (W/m^2)	Averaging Time $\|E\|^2$, S or $\|H\|^2$ (minutes)	
0.003–0.1	614	163	(1,000; 10,000,000)†	6	6
0.1–1.34	614	16.3/f	(1,000; 100,000/f^2)†	6	6
1.34–3.0	823.8/f	16.3/f	(1,800/f^2; 100,000/f^2)†	f^2/0.3	6
3–30	823.8/f	16.3/f	(1,800/f^2; 100,000/f^2)†	30	6
30–100	27.5	158/$f^{1.668}$	(2.0; 9,400,000/$f^{3.336}$)†	30	0.0636/$f^{1.337}$
100–300	27.5	0.0729	2.0	30	30
300–3,000			f/150	30	
3,000–15,000			f/150	90,000/f	
15,000–300,000			100	616,000/$f^{1.22}$	

6B-Induced and Contact Radiofrequency Currents[d]

Frequency Range	Maximum Current (mA) Through both feet	Through each foot	Contact
0.003–0.1 MHz	900f	450f	450f
0.1–100 MHz	90	45	45

TABLE 7

IEEE[a] Relaxation for Partial Body Exposures

	Frequency (MHz)	Peak Value of Mean Squared Field Strength‡	Equivalent Power Density W/m^2
Controlled Environment	0.1–30	$<20\|E\|^2$ or $<20\|H\|^2$	
	30–600		<20
	600–96,000		$<20(167f)^{0.25}$
	96,000–300,000		400
Uncontrolled Environment	0.1–30	$<20\|E\|^2$ or $<20\|H\|^2$	
	30–600		40
	600–96,00		667f
	96,0000–300,000		200

f = frequency in MHz

[a]Institute of Electrical and Electronic Engineers Standard (C95.1-1991) successor to the ANSI (American National Standards Institute Standard

[b]Occupational exposures or exposures that are the incidental result of passage through an area.

[c]Values are spatial averages over the vertical cross-section of the human body.

[d]Values given may not protect against the startle response or against burns from transient discharges when contactinng an energized object.

†Values of plane-wave equivalent power density are provided for comparison.

‡ Values of mean squared field strength are spatially averaged values from the above tables.

— For simultaneous exposure at several frequencies for which there are different values of the maximum permissable exposure (MPE), the sum of the fractions of the MPE values should not exceed unity. Exposure to pulsed electromagnetic fields in the range of 0.1 to 300,000 MHz may not exceed a peak value of 100 kV/m.

TABLE 8

8A-IRPA[a] Occupational Exposure Limits for RF Fields

Frequency (MHz)	Unperturbed RMS Field Strength		Equivalent Plane Wave Power Density (W/m^2)
	Electric (V/m)	Magnetic (A/m)	
0.1–1	614	1.6/f	
1–10	614/f	1.6/f	
10–400	61	0.16	10
400–2,000	$3f^{0.5}$	$0.008f^{0.5}$	f/40
2,000–300,000	137	0.36	50

8B-IRPA[a] General Public Exposure Limits for RF Fields

Frequency (MHz)	Unperturbed RMS Field Strength		Equivalent Plane Wave Power Density (W/m^2)
	Electric (V/m)	Magnetic (A/m)	
0.1–1	87	$0.23/f^{0.5}$	
1–10	$87/f^{0.5}$	$0.23/f^{0.5}$	
10–400	27.5	0.073	2
400–2,000	$1.375f^{0.5}$	$0.037f^{0.5}$	f/200
2,000–300,000	61	0.16	10

[a]International Radiation Protection Association.
— For simultaneous exposure at several frequencies for which there are different values of the maximum permissible exposure (MPE), the sum of the fractions of the MPE values should not exceed unity.
— Maximum values are for six-minute averages.
— Body current shall not exceed 200 mA. However, if the current at the point of contact exceeds 50 mA there is a risk of burns.
— For pulsed fields, electric and magnetic field strengths, as averaged over the pulse width, should not exceed 32 times the values given above (1000 times for power density).
— For near field exposures above 10 MHz, values above may be exceeded so long as $5(E^2/120\pi)/6 + (120\pi H^2) \leq S$ where S is the plane wave power density.
— f in MHz

TABLE 9
Canadian Ministry of National Health and Welfare

9A-Occupational Exposure Limits

Frequency (MHz)	Unperturbed RMS Field Strength		Equivalent Plane Wave Power Density (W/m^2)
	Electric (V/m)	Magnetic (A/m)	
0.01–1	600	4.9	
1–10	600/f	4.9/f	
10–30	60	4.9f	
30–300	60	0.163	10
300–1,500	$3.46f^{0.5}$	$0.0093f^{0.5}$	f/30
1,500–300,000	140	0.36	50

9B-General Public Exposure Limits

Frequency (MHz)	Unperturbed RMS Field Strength		Equivalent Plane Wave Power Density (W/m^2)
	Electric (V/m)	Magnetic (A/m)	
0.01–1	280	2.19	
1–10	280/f	2.19/f	
10–30	28	2.19/f	
30–300	28	0.073	2
300–1,500	$1.616f^{0.5}$	$0.004f^{0.5}$	f/150
1,500–300,000	62	0.16	10

9C-Contact Current Limits

Frequency (MHz)	RMS Current	
	Workers	Public
0.01–0.1	400f	150f
0.1–30	40	15

— For simultaneous exposure at several frequencies for which there are different values of the maximum permissible exposure (MPE), the sum of the fractions of the MPE values should not exceed unity.
— Values are averages over a 0.1-hour period. However, time-averaged values need be calculated from multiple measurements only if the field changes significantly (more than 20%) within the 0.1-hr period.
— f in MHz

TABLE 10
European Prestandard (January 1995)

10A-Exposure of workers to continuous radiation for periods exceeding 6 minutes

Frequency (MHz)	Electric Field Strength (V/m)	Magnetic Field Strength (A/m)	Power Density (W/m^2)
0.01–0.038	1000	42	
0.038–0.61	1000	1.6/f	
0.61–10	614/f	1.6/f	

TABLE 10—continued
European Prestandard (January 1995)

10A-Exposure of workers to continuous radiation for periods exceeding 6 minutes

Frequency (MHz)	Electric Field Strength (V/m)	Magnetic Field Strength (A/m)	Power Density (W/m^2)
10–400	61.4	0.16	10
400–2000	$3.07f^{0.5}$	$0.00814f^{0.5}$	f/40
2000–150,000	137	0.364	50
150,000–300,000	$0.354f^{0.5}$	$0.00094f^{0.5}$	0.0003334f

— For frequencies less than 0.61 MHz, exposures are assumend to be to either electric or magnetic fields. If exposure is to both electric and magnetic, fields the exposure should be consideed to be a simultaneous exposure to two sources.
— Parts of the body may be exposed to higher field strengths if the specific absorption rate (SAR) does not exceed 20 W/g for any 10 g of tissue in the hands, wrists, ankles, or feet; or 10 W/g for any other tissue of the body.
— For higher frequencies (greater than 0.614 MHz for electric fields or 0.038 MHz for magnetic fields) and exposures less than 6 minutes the reference levels may be relaxed.
— f in MHz

10B-Exposures for workers exposed to pulsed radiation

Frequency (MHz)	Peak Electric Field Strength (V/m)	Peak Magnetic Field Strength (A/m)	Peak Power Density (W/m^2)
0.01–0.23	4760	200	
0.23–3.73	4760	46/f	
3.73–10	17,750/f	46/f	
10–400	1775	4.6	8160
400–2000	$88.8f^{0.5}$	$0.23f^{0.5}$	20.4f
2000–150,000	3970	10.3	40,890
150,000–300,000	$10.3f^{0.5}$	$0.0266f^{0.5}$	0.274f

f in MHz

10C-Exposure of the general public to continuous wave radiation for periods exceeding 6 minutes

Frequency (MHz)	Electric Field Strength (V/m)	Magnetic Field Strength (A/m)	Power Density (W/m^2)
0.01–0.042	400	16.8	
0.042–0.68	400	0.7/f	
0.68–10	275/f	0.7/f	
10–400	27.5	0.07	2
400–2000	$1.37f^{0.5}$	$0.00364f^{0.5}$	f/200
2000–150,000	61.4	0.163	10
150,000–300,000	$0.158f^{0.5}$	$0.000421f^{0.5}$	0.0000667f

— For frequencies less than 0.68 MHz, exposures are assumend to be to either electric or magnetic fields. If exposure is to both electric and magnetic fields, the exposure should be consideed to be a simultaneous exposure to two sources.
— Parts of the body may be exposed to higher field strengths if the specific absorption rate (SAR) does not exceed 20 W/g for any 10 g of tissue in the hands, wrists, ankles, or feet; or 10 W/g for any other tissue of the body.
— For higher frequencies (greater than 0.614 MHz for electric fields or 0.038 MHz for magnetic fields) and exposures less than 6 minutes the reference levels may be relaxed.
— For pulses of duration $\leq$ 30 μs derived value of energy density of 0.02 J/m^2 is specified.
— f in MHz

10D-Exposures for the general public exposed to pulsed radiation

Frequency (MHz)	Peak Electric Field Strength (V/m)	Peak Magnetic Field Strength (A/m)	Peak Power Density (W/m^2)
0.01–0.25	1936	80	
0.25–4.16	1936	20/f	
4.16–10	7940/f	20/f	
10–400	794	2	1588
400–2000	$39.7f^{0.5}$	$0.1f^{0.5}$	3.97f
2000–150,000	1775	4.17	7934
150,000–300,000	$45.8f^{0.5}$	$0.0115f^{0.5}$	0.053f

f in MHz

TABLE 11
Recommendation of the National Radiation Protection Board (United Kingdom, 1993)

Frequency (MHz)	Electric Field Strength (V/m)	Magnetic Field Strength (A/m)	Power Density (W/m^2)
0.001–0.535		64	
0.535–12		$18/f^2$	
12–200	50	0.13	6.6
200–400	0.25f	0.00066f	$0.000165f^2$
400–800	100	0.26	26
800–1,550	0.125f	0.00033f	$0.000041f^2$
1,550–300,000	194	0.52	100

f in MHz

TABLE 12
German Free-Access Exposure Limits (1990)

Frequency (MHz)	Electric Field Strength (V/m)	Magnetic Field Strength (A/m)	Power Density (W/m^2)
0.10–0.14	300	16	
0.14–0.92	300	2.19/f	
0.92–10	275/f	2.19/f	
10–30	27.5	2.19/f	
30–400	27.5	0.07	2
400–2000	$1.37f^{0.5}$	$0.00364f^{0.5}$	f/200
2000–30,000	61.4	0.16	10

f in MHz

TABLE 13
Austrian Exposure Limits for the General Public

Frequency (MHz)	Electric Field Strength (V/m)	Magnetic Field Strength (A/m)
0.01–0.03	1500	350
0.03–2	1500	$7.057/f^{1.133}$
2–30	$3405/f^{1.1825}$	$7.057/f^{1.133}$
30–3000	61	0.16
3000–12,000	$0.57f^{0.5836}$	$0.001139f^{0.5929}$
12,000–3,000,000	137	0.36

f in MHz

TABLE 14
Sweedish National Board of Occupational Health and Safety Exposure Limits (1987)

	Continuous Wave Fields (averaged over 6 minutes)		Pulsed Fields (averaged over 1 second)	
Frequency (MHz)	Electric Field Strength (V/m)	Magnetic Field Strength (A/m)	Electric Field Strength (V/m)	Magnetic Field Strength (A/m)
3–30	140	0.40	300	0.8
30–300	60	0.16	300	0.8
300–300,000	60		300	

REFERENCES FOR SUMMARY OF EXPOSURE STANDARDS

Tables 1 to 4 are based on

(1) CENELEC—European Committee for Electrotechnical Standardization (Central Secreteriat: rue de Strassart 35, B-1050 Brussels, Belgium)
Document ENV 50166-1 (European Prestandard)
Human exposure to electromagnetic fields, Low frequency (0 Hz to 10 kHz).

(2) U.S. Department of Transportation, Federal Railroad Administration, Final Report DOT/FRA/ORD-93/27 August 1993: Safety of High Speed Guided Transportation Systems. Available as PB94-117819 from the U.S. National Technical Information Agency, Springfield, VA 22161.

(3) Barnabas Kunsch: The new European Pre-Standard ENV 50166 "Human exposure to electromagnetic fields". Austrian research Center Seibersdorf, A-2444 Seibersdorf, Austria, March, 1995.

INDEX

0-8493-0641-8/95/$0.00+$.50